Wetlands, Streams, and Other Waters

Wetlands, Streams, and Other Waters

Regulation • Conservation Mitigation Planning

Paul D. Cylinder
Kenneth M. Bogdan
April I. Zohn
Joel B. Butterworth

Wetlands, Streams, and Other Waters

Regulation • Conservation
Mitigation Planning

Paul D. Cylinder
Kenneth M. Bogdan
April I. Zohn
Joel B. Butterworth

May 2004

Solano Press Books
Post Office Box 773
Point Arena, California 95468
www.solano.com

tel (800) 931-9373
fax (707) 884-4109
email spbooks@solano.com

Cover design by Christy Anderson
and Solano Press Books

Book design by Solano Press Books

ISBN-923956-76-X

Notice

Before you rely on the information in this book, be sure you are aware that some changes in the statutes or case law may have gone into effect since the date of publication. The book, moreover, provides general information about the law. Readers should consult their own attorneys before relying on the representations found herein.

Chapters at a Glance

Contents

Contents

Contents

Contents

Short Articles continued

Appendices

Tables

Figures

Preface

Growth and development pressure throughout the United States have collided with efforts to protect our wetlands, streams, and other waters. Agencies, corporations, and individuals wishing to develop or protect these valuable and dwindling resources can become lost in the maze of federal, state, and local laws and regulations. We have attempted to unravel the intricacies of the regulatory programs that control activities in streams, wetlands, and other waters; and offer advice, drawn from extensive professional experience, on how most effectively and efficiently to navigate the regulatory process. We have written this book for:

- Land use planners
- Project managers for agencies preparing environmental compliance documents
- Developers
- Landowners
- Regulatory agency personnel
- Elected officials
- Environmental consultants
- Members of environmental organizations
- Lawyers
- Water suppliers
- Growers and others involved in agricultural production
- Mine operators
- Foresters
- Ranchers
- Environmentally concerned citizens

Readers will find this book and its extensive appendices to be a valuable and reliable reference on regulation and the environmental permitting process for wetlands, streams, and other waters. The federal regulatory process is presented in detail; state regulatory programs are summarized, and additional information is provided for states with the most extensive regulatory programs. Section 404 of the Clean Water Act and Section 10 of the Rivers and Harbors Act are examined in detail, and other federal laws related to the regulation of wetlands and other waters are introduced. Readers will find information and advice on:

- Ecology of wetlands and other waters
- Identification of wetlands and other waters
- Federal and state regulatory programs for wetlands and other waters
- Best approaches to the permit process and permitting agencies
- Mitigation planning and implementation
- Regional wetland conservation planning

We hope that the advice in this practical guide will prove helpful in pursuing projects and programs involving wetlands, streams, and other waters throughout the United States.

Paul D. Cylinder
Kenneth Bogdan
April Zohn
Joel Butterworth

March 2004

About the Authors

Paul D. Cylinder, Ph.D., is a Principal with Jones & Stokes. Dr. Cylinder uses his training and experience in ecology and environmental regulations to assist clients in complying with Section 404 of the Clean Water Act, the federal and state Endangered Species Acts, the National Environmental Policy Act (NEPA), and various other federal and state environmental regulations that address natural resources. Dr. Cylinder has conducted ecological studies, biological resources impact assessments, and endangered species surveys; facilitated wetland delineation and permitting; developed habitat conservation plans, natural resources management plans, habitat restoration plans, biological resource mitigation plans, endangered species recovery plans; and prepared biological and physical resource databases and maps. Dr. Cylinder received his Ph.D. in botany from the University of California, Berkeley, and his B.A. in biological sciences from the University of Chicago.

Kenneth Bogdan, J.D., is Environmental Counsel with Jones & Stokes. He specializes in compliance issues related to environmental laws and regulations, including Section 404 of the Clean Water Act, federal and state Endangered Species Acts, NEPA, and the California Environmental Quality Act (CEQA). Mr. Bogdan regularly teaches and speaks at workshops and conferences on wetlands, endangered species, NEPA, and CEQA. He received his J.D. from King Hall, University of California, Davis, and his B.S. in environmental management from Rutgers University.

April Zohn is a Regulatory Compliance Specialist with Jones & Stokes, where she focuses on wetland regulatory issues involving Sections 401 and 404 of the Clean Water Act, the Coastal Zone Management Act, state coastal regulations, and federal and state endangered species regulations. Ms. Zohn specializes in developing permitting strategies and implementation plans for sensitive water resource and coastal zone management projects, and conducts training courses on the federal wetland permitting program and implementation of the Coastal Zone Management Act. Ms. Zohn received her B.S. in marine science from the United States Coast Guard Academy, New London, Connecticut.

Joel Butterworth is a Senior Wetland Scientist with Jones & Stokes, with particular expertise in soil science and wetland ecology. Mr. Butterworth has prepared conceptual and detailed wetland and riparian habitat mitigation plans, evaluated the suitability of candidate habitat mitigation sites, conducted wetland delineations, and prepared erosion and sediment control plans. Mr. Butterworth received his M.S. in geography from Oregon State University and his B.A. in geography from the University of California, Santa Barbara.

Acknowledgments

We would like to thank several individuals without whom this book could not have been produced in the clear, useful style that was achieved. Larry Goral provided proofreading, editing, and rewriting assistance to improve the style and clarity of all chapters. The contributions of Alan Barnard and Christy Anderson, who coordinated graphics production, bring the pages of this document to life by displaying complex ideas in understandable figures, charts, and photographs. We greatly appreciate the diligence of Brooke Fraschetti in assembling the appendices that present laws and regulations in their original text, as well as other information that makes this handbook useful and user friendly. We thank Tom Adams for preparing the case summaries presented in appendix K. We would also like to thank Mark Matthies, Michele Waltz, Susan Lee, Hunt Durey, Karen Shaffer, and Ginger Fodge for contributing ideas and reviewing draft text.

CHAPTER ONE

Introduction

Wetlands, streams, and other waters are highly productive and complex ecosystems. Once considered of little or no use, wetlands are now center stage as citizens and politicians alike acknowledge their great importance and extraordinary rate of loss. Protection of wetlands is a national challenge. An estimated 53 percent have been lost in the contiguous 48 states over the last 200 years (Dahl 1990), and the country is currently losing nearly 60,000 acres per year (U.S. Fish and Wildlife Service 2000). This continuing decline has prompted federal, state, and local governments to regulate activities that threaten these special natural resources.

An estimated 53 percent of wetlands have been lost in the contiguous 48 states over the last 200 years, and the current rate of loss is nearly 60,000 acres per year.

This book is a practical guide and desktop reference for anyone seeking to understand how wetlands, streams, and other waters are regulated in the United States. Federal wetland laws are described in detail, with particular attention to Section 404 of the Clean Water Act (CWA), the predominant protection law for wetlands and other waters. Extensive appendices present the text of federal laws, key regulations, and regulatory guidance. In addition, because most states have some form of wetland regulation, a comprehensive overview of relevant state laws and programs is provided.

CWA = Clean Water Act

What Are Wetlands?

Wetlands are areas of land that are wet either permanently or seasonally and support specially adapted vegetation. To regulate activities in wetlands, federal and state agencies have developed specific definitions and methods for identifying their boundaries. Identification methods vary among agencies, focussing on hydrologic, soil, and vegetative parameters. To be regulated as a wetland under federal law, a site must have specific indicators of conditions for each of these three categories. Changes in identification methods have been controversial because they have resulted in changes in the extent of areas considered subject to jurisdiction. Chapter 2, Ecology of Wetlands

Wetlands are areas of land that are wet either permanently or seasonally that support specially adapted vegetation.

and Other Waters, explains the basic ecological concepts underlying the distinguishing characteristics of wetlands. Not all wetlands are easily recognizable; for example, those occurring in seasonally dry or desert areas can be particularly difficult to identify.

The definitions federal agencies use to identify wetlands for regulatory purposes are presented in chapter 3, Jurisdictional Limits of Wetlands and Other Waters. The identification of other regulated water bodies, such as rivers, streams, lakes, and bays, is also described.

What Is the Value of Wetlands?

Many wetland functions are valuable, although not all are obvious.

Wetlands affect our lives and livelihoods in many ways. While valuable, not every wetland function is obvious. For example, wetlands provide flood protection by slowing flows and storing water. They serve as the recharge site for groundwater that is a source of the public water supply. Toxics and other pollutants passing through wetlands are transformed and removed, and water quality is improved. They protect stream banks and shorelines from erosion, and are essential to food production because they provide food, spawning, and nursery areas for many commercial fish and shellfish.

Wetlands offer recreational opportunities, open space, and aesthetic possibilities. Boating, swimming, fishing, hunting, hiking, photography, bird and other wildlife observation, and scientific study are activities that take place in wetlands or depend on them to enrich human existence.

How Are Wetlands and Other Waters Regulated?

USACE = United States Army Corps of Engineers

The U.S. Army Corps of Engineers (USACE), through the authority of Section 404 of the CWA, is the federal agency most involved in wetland regulation. Wetlands are only one type of water body that is regulated under federal law. USACE regulates many other waters, including streams, lakes, ponds, bays, and portions of the oceans that meet specific criteria (*see* chapter 3, Jurisdictional Limits of Wetlands and Other Waters). Federal regulation of wetlands and other waters under Section 404 is described in chapter 4, Federal Regulation of Wetlands and Other Waters, and the permitting process is described in chapter 5, Section 404 Permitting Process. Summaries of important cases highlighting judicial interpretations of Section 404 clarify the circumstances under which the federal government can regulate privately owned wetlands (*see* appendix K, Case Law Summaries).

Some states also regulate activities in wetlands and other waters.

Some states also regulate activities in wetlands and other waters. In some cases, state agencies may regulate wetlands and other aquatic resources where the federal government does not exert jurisdiction. In parts of some states, such as the coastal zone, wetlands are regulated more strictly. State laws that regulate activities in wetlands around the nation are summarized in chapter 6, State Wetland Laws.

Can Wetland Losses Be Mitigated?

The mitigation of impacts on wetlands can be a complicated affair. It requires good scientific information, careful planning, close coordination of all concerned parties, and effective mitigation design and implementation. Creating, restoring, or enhancing wetlands on project sites can be successful if mitigation is properly planned and implemented; alternatively, mitigation banking–the use of preapproved and established sites where wetland habitat compensates for wetland impacts elsewhere–can be used in the mitigation of wetland impacts or incorporated into regional wetland conservation plans. Chapter 7, Mitigation Planning, describes the wetland mitigation process, describes key issues, offers important recommendations, and presents a framework for effective and efficient mitigation planning and implementation.

The mitigation of impacts on wetlands requires good scientific information, careful planning, close coordination of all concerned parties, and effective mitigation design and implementation.

The Future of Wetland Regulation

The future of wetland regulation involves breaking away from project-specific wetland planning and impact mitigation. The regional or watershed approach to wetland conservation planning is rapidly gaining favor. Early identification of resources and designation of important wetland conservation areas provides a context for assessing the relative impacts of individual projects within a larger planning area. Regional wetland conservation planning can help project proponents situate their projects more effectively and can aid regulatory agencies in streamlining the permitting process. As of this writing, regional wetland conservation plans are in development across the United States. Chapter 8, Regional Wetland Conservation Planning, discusses important planning concepts and key elements for successful development of regional wetland conservation plans.

Great egrets feeding in seasonal freshwater marsh

Chapter 9, Epilogue, summarizes some of the present and future policy challenges surrounding wetland regulation and conservation planning that face federal, state, and local agencies; landowners; project proponents; and environmentally concerned citizens.

CHAPTER TWO

Ecology of Wetlands and Other Waters

Wetlands, streams, lakes, ponds, and other waters are highly variable physical and biological systems that support a wide range of ecological functions and typically play a key role in the larger ecosystems in which they occur. Once abundant, wetland habitats in the United States have shrunk to less than half their historical extent. Rivers and streams have been filled, diverted, and channelized, and lakes and bays have been converted to dry land throughout the nation. This chapter explains the distinctive features of wetlands and other waters and describes some ways in which different types of wetlands and other waters are classified. This chapter also presents examples of human activities that can alter or destroy wetlands.

Wetlands, streams, lakes, ponds, and other waters are highly variable physical and biological systems that support a wide range of ecological functions and typically play a key role in the larger ecosystems in which they occur.

Ecology of Wetlands

Wetlands are characterized by distinctive physical, chemical, and biological features, including hydrology, soils, and vegetation types that typify these specialized habitats.

Wetlands are characterized by distinctive physical, chemical, and biological features, including hydrology, soils, and vegetation types that typify these specialized habitats.

Wetland Hydrology

Wetland ecosystems often develop in a transitional zone between upland and deepwater habitats (Figure 2-1). Sites that support wetlands are frequently flooded or ponded or have permanently or seasonally saturated soils. Unlike the well-drained soils of upland habitats that may become saturated for short intervals, wetland soils are poorly drained and remain waterlogged for long periods. Water levels in wetlands, characterized by daily, seasonal, or yearly fluctuations, are generally more shallow than in deepwater habitats such as ponds, lakes, and bays.

Knowing how these wet conditions develop is key to understanding wetlands. The water in wetlands derives from direct precipitation, overland or near surface flow, shallow groundwater, or some combination of these

Figure 2-1
Wetlands Cross Section

Wetlands are transitional habitats between uplands and deep water habitats.

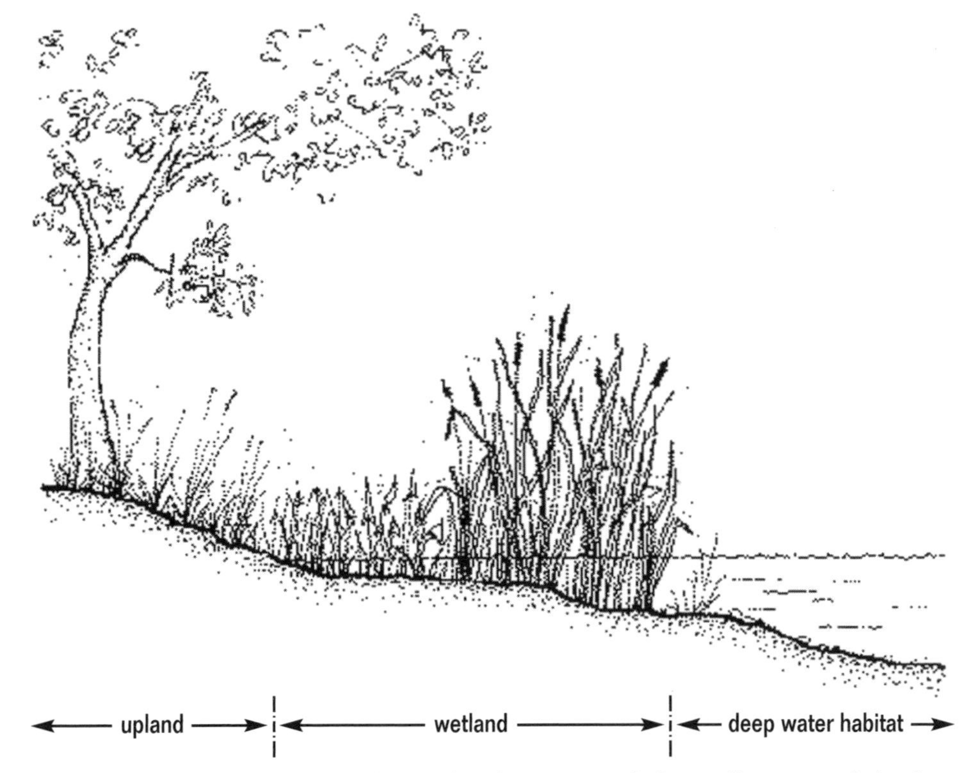

processes. Ponding of water in wetlands may result from direct precipitation or from groundwater rising above the surface, while flooding may be due to headwater or backwater flooding of rivers and streams or to tidal action. Soil saturation may result from a high water table or from the aftermath of ponding and flooding. Water is lost from wetlands through evaporation, plant transpiration, surface or subsurface flow, percolation into the groundwater, or tidal action. The processes by which water reaches and leaves a site are collectively referred to as *wetland hydrology*.

Wetland Soils

Soils that are saturated for long periods undergo chemical and physical changes that set them apart from well-drained upland soils.

Soils that are saturated for long periods undergo chemical and physical changes that set them apart from well-drained upland soils. The most immediate effect of soil saturation is a rapid loss of oxygen. Soils typically have many pores through which oxygen diffuses with relative rapidity, keeping the soil well oxygenated; plant roots and soil microorganisms consume this oxygen.

When soils become saturated, the pores fill with water. Because oxygen moves through water very slowly, plant roots and soil microorganisms quickly consume the available oxygen. The resulting oxygen-deficient state is referred to as an *anaerobic* condition. Anaerobic conditions prevail in wetland soils, and soils with water-induced anaerobic conditions are referred to as *hydric* soils.

The anaerobic conditions of hydric soils significantly reduce microbial activity, thereby reducing the rate of decomposition of dead plant material. Many wetland soils, such as peats and mucks, have a very high organic matter content as a result of the dead plant material that has accumulated over centuries, or even millennia.

In the absence of oxygen, other soil changes take place. Nutrients important to plants and microorganisms, such as nitrogen, iron, manganese, and sulfur, are converted to unusable forms. The altered metals and chemicals become more mobile, more capable of being leached from the soil, and sometimes reach toxic levels (Mitsch and Gosselink 1993). These chemical changes also result in physical changes, most noticeably in soil color. Wetland mineral soils are typically dark gray or black in contrast to the bright red, brown, and yellow mineral soils of upland habitats. The dark-colored soil of some wetlands (especially seasonal wetlands) are mottled in the upper part with patches of bright-colored material where iron and manganese ions become concentrated and oxidized (Vepraskas 1992).

In the absence of oxygen, nutrients important to plants and microorganisms, such as nitrogen, iron, manganese, and sulfur, are converted to unusable forms.

Wetland Plants

Plants that grow in wetlands have adapted to the anaerobic conditions of saturated soils. To survive, these plants have evolved adaptations to supply their roots with oxygen. Many wetland plants, for example, have shallow root systems. In saturated soils, more oxygen is present near the soil surface, where it can enter the soil from the atmosphere; plants with shallow root systems can take advantage of this thin surface layer of oxygenated soil (Figure 2-2). Other wetland species have hollow stems through which oxygen can be transported from the shoots to the roots. Wetland plants also have developed unique physiological mechanisms to cope with anaerobic soil conditions, such as temporarily shutting down their metabolism, developing alternative chemical pathways, and storing chemical intermediates for later use (Mitsch and Gosselink 1993).

Figure 2-2
Wetland Plant Adaptations

A shallow root system is a common adaptation seen in wetland plants. Shallow roots take advantage of the thin oxygenated surface layer of soil.

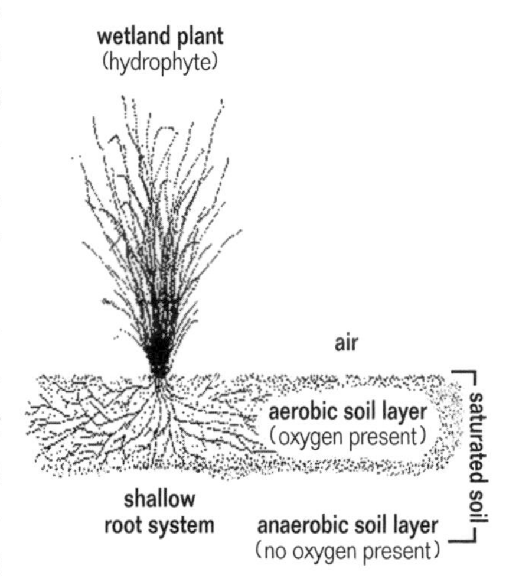

Plants that have adapted to wetland conditions are known as *hydrophytes* (meaning *water plants*). Some hydrophytes, called *obligates*, specifically require wetland conditions for survival and reproduction. Other hydrophytes are tolerant of wetland conditions but can also survive in nonwetland upland habitats. Because they can live under a variety of conditions, these plants are known as *facultative* species. Upland plants will not survive in soils that are frequently saturated or inundated for a long period.

The U.S. Fish and Wildlife Service (USFWS) has developed a classification system for hydrophytic plants (Reed 1988) (Table 2-1). Under this system, plant species are separated into *indicator status categories* on the basis of each species' probability of occurring in wetland rather than upland habitat.

USFWS = **United States Fish and Wildlife Service**

The Variety of Wetlands in the United States

A wide variety of wetland habitats occur in the United States. Different wetland types are usually defined by the types of plants and animals they support. These biotic communities can vary considerably, depending on the site's hydrologic regime, substrate, water source, and water quality.

Wetlands habitats can vary considerably, depending on the site's hydrologic regime, substrate, water source, and water quality.

Table 2-1. Wetland Plant Indicator Categories

Classification	Symbol	Definition
Obligate Wetland	OBL	Almost always occurs in wetlands (estimated probability, less than 99 percent) under natural wetlands conditions
Facultative Wetland	FACW	Usually occurs in wetlands (estimated probability 67 to 99 percent), but occasionally found in nonwetlands
Facultative	FAC	Equally likely to occur in wetlands or nonwetlands (estimated probability, 34 to 66 percent)
Facultative Upland	FACU	Usually found in nonwetlands (estimated probability 67 to 99 percent), but occasionally found in wetlands (estimated probability, 1 to 33 percent)
Obligate Upland	UPL	Occur almost always (estimated probability, less than 99 percent) in nonwetlands under natural conditions

Reed 1988

Water may reach a wetland from many sources, including direct rainfall, surface runoff, rising groundwater, percolation, tidal flooding, overbank flooding, and backwater flooding. The water may be fresh, brackish, saline, or hypersaline. It may also be high or low in nutrients and acidic, neutral, or alkaline.

Wetlands can have different hydrologic regimes, or patterns of occurrence of water, varying in frequency, duration, depth, scouring action, and seasonal timing.

Different wetland types have different hydrologic regimes, or patterns of occurrence of water. These regimes may vary in frequency, duration, depth, scouring action, and seasonal timing. For example, tidal salt marshes are typically inundated twice daily, but riparian forests may be inundated only following major runoff events. Mangrove swamps in south Florida are flooded or saturated year-round, while vernal pools in the Central Valley of California may hold water for only several weeks each spring. Prairie potholes of the upper Midwest may be dry for several years or support freshwater marsh and rich waterfowl habitat for several years, depending on cycles of wet periods and droughts. Wet alpine meadows usually receive just enough water to saturate the soil, while tidal freshwater marshes may be inundated to depths of six feet.

Bald cypress swamps are found in permanently and semi-permanently flooded bottomlands along rivers in the Southeast.

The substrates upon which wetlands develop include cobbles, gravels, sands, loams, dense clays, organic material, and combinations of these. Substrates may vary in thickness from several inches to tens of feet, and they can vary greatly in nutrient content, alkalinity, acidity, and chemical composition.

Water source, water quality, hydrologic regime, and substrate properties are not independent factors; rather, each factor interacts with all the others. Different combinations result in different types of wetlands. Four examples are discussed below.

Cypress swamps are freshwater, tree-dominated wetlands of the southeastern United States. They are associated with poorly drained areas in floodplains, backwaters, and depressions of the coastal plan and Mississippi River Valley. Cypress swamps are dominated by bald cypress, pond cypress, water tupelo, and black gum. These habitats have ponded water for most or all of the year, typically with fluctuating levels. Water is generally standing or slow flowing; scour is minimal but nutrient input can be quite high. The soil is usually sandy or clay.

Riparian forests of the western United States occur along riverbanks and are dominated by trees such as cottonwoods and willows. Their source of water is usually from overbank or backwater flooding, and the water is fresh with neutral to alkaline pH. Seasonal flooding takes place in late winter and early spring and can last for a few weeks to a month or more (Figure 2-3). The habitat may flood annually or less frequently, perhaps 50 of every 100 years, and strong scouring can occur. The summertime water table generally lies within 20 feet of the surface. The substrate is typically deep loams to gravelly sands, highly enriched with nutrients carried in with floodwater.

Riparian forest

Tidal salt marshes, dominated by low, perennial plants such as cordgrasses, glassworts, blackgrass, needlerush, and salt grass, are another type of wetland habitat. The primary source of water is tidal flooding from the ocean; the water is saltiest during summer, with neutral to alkaline pH. The hydrologic regime is characterized by flooding twice a day; this tidal flooding varies in duration from one to several hours depending on elevation (Figure 2-3). Monthly variations in the level of highest and lowest tides are associated with lunar influences. The soil, typically of fine loams to clay, is saturated at all times, and tidal scouring can be strong.

Vernal pools, dominated by small annual plants such as meadowfoam, popcornflower, and goldfields, are wetland habitats endemic to certain lowlands of California and southern Oregon. Vernal pools are natural depressions filled by winter rains. Warm spring temperatures bring colorful wildflower displays and slow evaporation of the pools. By summer the pools are completely dry and the flowers gone. Most plant and invertebrate animal species of the vernal pools wait out the dry summer and fall as dormant seeds, cysts, and eggs. Vernal pool soils are shallow and are underlain by an impervious layer of dense clay, hardpan, or bedrock that causes rainwater to be held in the depression.

Figure 2-3
Typical Annual Hydrologic Cycles for Tidal Salt Marsh and Riparian Forest

Schematic only; not from actual data

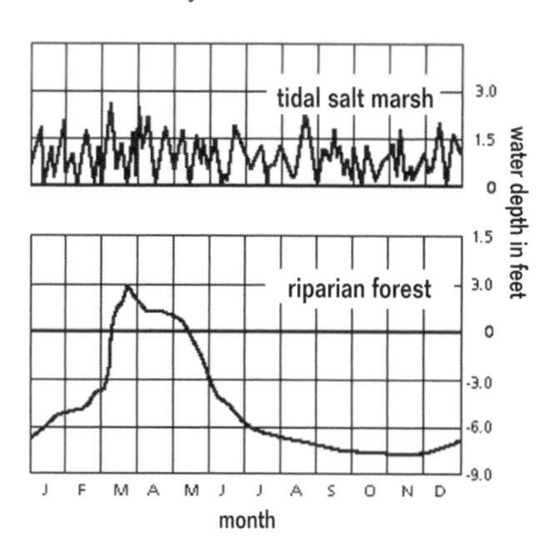

Ecology of Other Waters

While wetlands are vegetated, other waters are typically unvegetated areas supporting flowing, flood, ponded, or tidal water. Other waters include rivers, streams, lakes, ponds, bays, and oceans. This section provides a general overview of stream, lake, and pond ecology.

Stream Ecology

Streams include a wide range of linear features that convey water—everything from large rivers to small creeks, intermittent drainages, and desert arroyos. Most streams are fed by runoff and tend to be highly dynamic. Flow may vary dramatically through the year and between years, depending on precipitation and runoff conditions. Extreme variation is found in desert washes in the southwestern part of the country. Streams may be dry for

Streams include a wide range of linear features that convey water–everything from large rivers to small creeks, intermittent drainages, and desert arroyos.

several years, but a sudden storm may transform them into a raging torrent. By contrast, spring-fed creeks tend to have a relatively constant flow throughout the year. Most rivers and creeks periodically flood their banks and the adjacent floodplain, depositing sediment and nutrients to the overflowed area. Frequently flooded lands tend to support wetlands, and biological productivity in floodplains tends to be relatively high. Floodplains may support extensive forests, such as the bottomland hardwood forests of the Southeast, or may support narrow stringers of riparian vegetation such as those in the arid West.

The precipitation rate, size of the watershed, and substrate infiltration rate determine the amount of runoff that reaches streams within a watershed.

The ecology of streams is highly dependent on the conditions of the watershed (the basin that drains into the stream). The precipitation rate, size of the watershed, and substrate infiltration rate determine the amount of runoff that reaches streams within a watershed. Runoff from the watershed generally provides most of the organic material carried into the stream. In upstream river reaches, the food chain mostly starts with dead organic material that enters the stream as leaves, wood, and other organic material that has fallen directly into the river or is transported into the river by surface runoff. Streams in forested watersheds and tree-lined streams in general have a higher biological productivity than rivers that run through more sparsely vegetated areas. Many of the insect and fish species that live in the upper watershed eat dead organic material. In downstream reaches of larger rivers, production of aquatic plants and plankton (suspended unicellular organisms) gains in importance, and plant-eating fish become more important than in upstream reaches. Nutrients are generally transported downstream, and the flow of natural rivers becomes more nutrient-rich as it moves downstream. Eventually, the nutrients end up in highly productive estuaries along the bay or ocean shores. The reverse of this nutrient pattern can be caused by fish species such as salmon that spend most of their lives in the ocean, then migrate up rivers to spawn and die. In Alaska, for example, salmon carcasses contribute a substantial amount of nutrients to coastal stream habitats.

Pickleweed or glasswort is an important vegetative component of coastal salt marshes.

Because of their linear nature, streams have a much greater interface with upland habitat than do lakes or ponds. Along this interface, streams erode their banks and deposit sediment downstream, frequently causing a stream to meander or migrate across the landscape. Stream banks erode at the outside of a meander, while sediment is deposited in point bars at the inside of the bend. A pool, the deepest part of a stream's reach, is usually found in the outside bend, while the shallowest areas are found along the straight portion between meanders, where fast-moving riffles result, characterized by a surface of choppy or undulating water. Riffles are oxygen-rich and support high algae and invertebrate production, providing food for fish. Deeper, lower-velocity pools provide cover and holding areas for fish.

Oxygen-rich riffles support high algae and invertebrate production, providing food for fish. Deeper, lower-velocity pools provide cover and holding areas for fish.

Streams are often divided into high-gradient and low-gradient categories. High-gradient streams of more mountainous regions tend to have higher flow velocities, less pronounced meanders, narrow floodplains, and deep V-shaped valleys. By contrast, low-gradient streams usually have lower-flow velocities, wide meanders, broad floodplains, and move across wide, flat valleys. Low-gradient streams in desert areas often form braided patterns with numerous channels.

A wide variety of wetlands are found along the water-land interface of a stream, including swamps, marshes, and riparian forests. Often important for fish breeding or nurseries, these habitats can be especially rich in wildlife, and the organisms living there are a major nutrient source for aquatic life. In urbanized or agricultural landscapes these wetland habitats may form important corridors of movement between otherwise isolated forested areas. Stream bank vegetation that often shades the stream and lowers the water temperature may be essential during the summer months for the survival of coldwater fish such as trout and salmon.

Coastal salt marsh

As a result of artificial control measures, few rivers are allowed to inundate their historical floodplains. Major dams and reservoirs capture flood peaks, and flood control structures such as levees and floodwalls protect lands adjacent to rivers from inundation. Although these measures have greatly reduced property damage and loss of life, flood control structures have severed the ecological connection between rivers and their historical floodplain, thereby reducing the productivity of the land and, especially in the West, reducing the availability of soil water. Because floodplains temporarily store floodwaters, the potential for downstream flooding is reduced. Consequently, the construction of flood control levees in upstream areas may lead to increased flooding in downstream areas. The propensity for flooding may be affected by a watershed's land surface conditions, especially in smaller watersheds. For example, rainfall mostly infiltrates the soils of naturally vegetated watersheds, while urbanization reduces infiltration and increases the proportion of rainfall that rapidly enters streams as surface runoff, causing higher and more sudden flood peaks.

Although artificial control measures have greatly reduced property damage and loss of life, flood control structures have severed the ecological connection between rivers and their historical floodplain.

Lake and Pond Ecology

Natural lakes and ponds occur in basins that hold permanent or seasonal standing water. The ecological conditions in lakes and ponds are generally determined by size, depth, geographic location, climate, salt content, and input of nutrients. Lakes are often classified by their biological productivity. Highly productive lakes (usually shallow, with warm water dominated by algae) are referred to as *eutrophic* ("good foods") lakes, and low-productivity lakes (usually deep with cold, clear water) are called *oligotrophic* ("few foods") lakes (Odum 1971). Desert salt and alkali lakes occur in undrained basins of desert regions where evaporation exceeds precipitation. Although they may only support a few species, desert salt lakes may be highly productive, producing huge amounts of brine shrimp and fly larvae (Whittaker 1975). Lakes in arid regions may expand and contract substantially over time, and some of these lakes may dry completely on an annual or supra-annual cycle.

Vernal pools in the Central Valley of California. Pools in the foreground are disturbed by construction vehicles.

Glacial ponds, such as this one in Maine, are common in the Northeast and upper Midwest.

Water depth and clarity determine a lake's light penetration, and light penetration determines whether submerged plants will grow on the lake bottom. A *pond* is typically defined as a shallow waterbody in which light penetrates to the bottom and the bottom is vegetated, in contrast to a *lake* where light usually does not reach the bottom and bottom vegetation is absent except at the lake's edge. In many deeper lakes, water temperature stratifies in warm months, as surface waters become warm, and colder, more dense water remains at the bottom. In cold months, the surface water cools, and the dense water sinks, creating vertical circulation (Odum 1971).

The United States can be divided into major lake regions based on geography and climate:

- Mountain region of the Pacific Northwest, coastal Alaska, the Rockies, and California
- Glacial terrain region of the northern Midwest and Northeast
- Riverine region of the Mississippi River watershed and the Southeast
- Desert region of the Southwest
- Subarctic and Arctic regions of central and northern Alaska, respectively (Rockwell 1998)

Environmental conditions in the interior of a lake or pond primarily change vertically. Temperatures become more constant at greater depths, and the amount of light decreases with depth.

Lakes and ponds with shallow, fluctuating shorelines may support large marshes, swamps, or riparian forests. Lakes and ponds with steeper shores exhibit rapidly changing environmental conditions as water depth increases from the shoreline to the interior of the water body. Environmental conditions in the interior of a lake or pond primarily change vertically. Temperatures become more constant at greater depths, and the amount of light decreases with depth. The size and surface of a lake or pond can also greatly affect its ecological condition. Larger watersheds or watersheds with urban or agricultural development, tend to provide more nutrients to a lake than smaller or undeveloped watersheds. A lake or pond with higher nutrient input tends to produce more suspended algae and may be less clear than a lake that is nutrient poor.

An artificial impoundment results in an pond or lake that is artificial. A large impoundment behind a dam used for water supply is typically referred to as a *reservoir*. Rapid and frequent fluctuation in water levels and the steep shoreline of a reservoir often results in little or no vegetation along the impoundment's edge. An artificial impoundment that is allowed to fluctuate naturally or is managed to more stable or slowly fluctuating water levels may support more natural lake and pond ecological conditions.

Classification of Wetlands and Other Waters

Wetlands are seasonal or perennial depending on the duration of saturation or inundation.

Wetlands can be classified by a variety of criteria and methods. They are seasonal or perennial (permanent) depending on the duration of saturation or inundation. Riparian forest, vernal pool, and seasonal freshwater marsh are examples of seasonal wetlands, while mangrove swamp, cypress swamp, tidal salt marsh, bog, and permanent freshwater marsh are examples of perennial wetlands.

Dominant vegetation type is another criterion that can be used to classify wetlands. Cypress swamp and mangrove swamp are examples of woody wetland habitat, while cattail marsh, tidal salt marsh, and bog are examples of herbaceous wetland habitat.

A typical classification of wetlands in the United States based on general vegetative appearance and hydrology is provided in Mitsch and Gosselink (1993); this system is summarized in Table 2-2. USFWS uses a more complex system for mapping wetlands and other waters for the National Wetlands Inventory (NWI). USFWS uses the Cowardin system (Cowardin *et al.* 1979) (Table 2-3), which classifies both wetland and deepwater habitats (other waters). The Cowardin system is based hierarchically on the large-scale ecosystem, hydrology, vegetative cover, and substrate.

HGM = **Hydrogeomorphic**
NWI = **National Wetlands Inventory**

An approach to classifying wetlands called the hydrogeomorphic (HGM) classification uses physical factors rather than vegetative characteristics to classify wetlands (Brinson 1993, Smith *et al.* 1995) (Table 2-4). Three fundamental factors are used in HGM.

- Geomorphic setting (landscape position of wetlands)
- Hydrology (water source)
- Hydrodynamics (flow and fluctuation)

The HGM classification groups wetlands by similar functions and was intentionally created to be independent of the distribution and associations of species. While not an assessment method, the HGM classification system was intended to support the development of methods for assessing wetland functions. The HGM approach is also used to classify streams and other waters and to assess the functions of these waters.

The HGM classification groups wetlands by similar functions and was intentionally created to be independent of the distribution and associations of species.

Functions and Values of Wetlands and Other Waters

Functions are the environmental conditions and processes that occur in wetlands and other waters. Examples of functions of wetlands and other waters are:

- Floodflow storage and conveyance
- Groundwater recharge and discharge
- Wave and erosion attenuation
- Sediment capture
- Pollutant filtering
- Fish and shellfish production
- Waterfowl and shorebird habitat
- Endangered species habitat
- Biogeochemical cycling

Many types of forested wetlands occur across the United States and are especially abundant in the Southeast.

Table 2-2. General Classification of North American Wetlands

Coastal Wetland Ecosystems	Inland Wetland Ecosystems
Tidal salt marshes	Freshwater marshes
Tidal freshwater marshes	Northern peatlands
Mangrove wetlands	Southern deepwater swamps
	Riparian wetlands

Adapted from Mitsch and Gosselink 1993

Table 2-3. Classification of Wetlands and Deep Water Habitats Used by the U.S. Fish and Wildlife Service

System	Subsystem	Class
Marine	Subtidal	Rock bottom Unconsolidated bottom Aquatic bed Reef Open water
	Intertidal	Aquatic bed Reef Rocky shore Unconsolidated shore Estuarine
Estuarine	Subtidal	Rock bottom Unconsolidated bottom Aquatic bed Reef Open water
	Intertidal	Aquatic bed Reef Streambed Rocky shore Unconsolidated shore Emergent Scrub-shrub Forested
Riverine	Tidal	Rock Unconsolidated bottom Streambed Aquatic bed Rocky shore Unconsolidated shore Emergent Open water
	Lower perennial	Rock Unconsolidated bottom Streambed Aquatic bed Rocky shore Unconsolidated shore Emergent Open water

Table 2-3. Classification of Wetlands and Deep Water Habitats Used by the U.S. Fish and Wildlife Service continued		
System	**Subsystem**	**Class**
Riverine continued	Upper perennial	Rock Unconsolidated bottom Streambed Aquatic bed Rocky shore Unconsolidated shore Emergent Open water
	Intermittent	Rock Unconsolidated bottom Streambed Aquatic bed Rocky shore Unconsolidated shore Emergent Open water
Lacustrine	Unknown perennial	Rock Unconsolidated bottom Streambed Aquatic bed Rocky shore Unconsolidated shore Emergent Open water
	Limnetic	Rock bottom Unconsolidated shore Aquatic bed Open water
	Littoral	Rock bottom Unconsolidated bottom Aquatic bed Rocky shore Unconsolidated shore Emergent Open water
Palustrine		Rock bottom Unconsolidated bottom Aquatic bed Rocky shore Unconsolidated shore Moss-lichen Emergent Scrub-shrub Forested Open water

Cowardin et al. *1979*

NOTE. *Classes are further divided into subclasses based on substrate or plant and animal (e.g., corals, mussels, barnacles) cover. Modifiers, which describe hydrologic regimes, water chemistry, soils, and disturbances, also may be added to the classification of sites. See also* www.state.ma.us/mgis/nwi_clas.txt.

Cattail marsh. Cattails can be found in wetlands throughout the United States.

Table 2-4. Hydrogeomorphic Classification of Wetlands

Hydrogeomorphic Class geomorphic setting	Water Source dominant	Hydrodynamics dominant	Examples of Regional Subclass: Eastern USA	Examples of Regional Subclass: Western USA and Alaska
Riverine	Overbank flow from channel	Unidirectional, horizontal	Bottomland hardwood forests	Riparian forested wetlands
Depressional	Return flow from groundwater and interflow	Vertical	Prairie pothole marshes	California vernal pools
Slope	Return flow from groundwater	Unidirectional, horizontal	Fens	Avalanche chutes
Mineral soil flats	Precipitation	Vertical	Wet pine flatwoods	Large playas
Organic soil flats	Precipitation	Vertical	Peat bogs; portions of Everglades	Peat bogs
Estuarine fringe	Overbank flow from estuary	Bidirectional, horizontal	Chesapeake Bay marshes	San Francisco Bay marshes
Lacustrine fringe	Overbank flow from lake	Bidirectional, horizontal	Great Lakes marshes	Flathead Lake marshes

Smith et al. *1995*

Wetland functions may be measured among wetlands of the same type to determine the relative function of each wetland. Such an analysis is called a functional assessment.

Wetland functions may be measured among wetlands of the same type to determine the relative function of each wetland. Such an analysis is called a *functional assessment*. The various functions of a wetland can be scored (quantitatively or qualitatively) and, in aggregate, used to measure the wetland's ecosystem integrity. Wetlands with high integrity are those areas that exhibit the full range of physical, chemical, and biological attributes and processes (functions) that characterize the specific wetland type in the same region prior to artificial alteration (Smith 2000). Wetlands with high integrity can be used as reference wetlands; the functions of these reference wetlands can be used as standards to assess other wetland sites that may or may not have the same level of integrity. Measures of wetland functions are important tools in wetland impact analyses, mitigation planning, and monitoring (*see* chapter 7, Mitigation Planning).

Functions of wetlands and other waters are assessed independently of any value such functions might provide to people, while wetland values assess precisely those benefits that do accrue to people.

Functions of wetlands and other waters are assessed independently of any value such functions might provide to people, while *values* assess precisely those benefits that do accrue to people. Examples of values of wetlands and other waters are:

- Recreation (e.g., fishing, swimming, nature watching, hunting)
- Water supply
- Water quality improvement
- Flood protection
- Food production (e.g., fisheries, shellfish beds, agriculture)
- Timber production
- Education and research use
- Open space and visual aesthetics

Context is an important determining factor in evaluating the values of a specific wetland function. For example, a wetland that functions to store floodflows upstream of a populated area would have value for flood protection, while a wetland with similar functions that is downstream of the populated area would not have similar value.

Loss of Wetlands

Human activities have greatly reduced the historical extent of wetlands in the United States. For most of the past two centuries, wetlands were viewed as impediments to agricultural and urban expansion. The "reclamation" of wetland habitat to usable upland habitat was considered beneficial. Dahl (1990) estimates that over the last 200 years, 53 percent of wetlands in the 48 contiguous states have been lost (Figure 2-4). Of the many factors contributing to this loss, the most important have been conversion to agricultural land, flood control projects, water diversions, and urban development. Of an estimated 221 million acres of wetlands in the contiguous 48 states in the 1780s, only an estimated 104 million acres remained in the 1980s (Dahl 1990). Alaska, with the greatest amount of wetlands (approximately 170 million acres) of any state, has lost less than one percent of its historical total (Dahl 1990). Except for Alaska, Hawaii, and New Hampshire, no state has lost less than 20 percent of its original wetlands (Dahl 1990). Florida, Louisiana, Minnesota, and Texas have the greatest extent of remaining wetlands in the 48 contiguous states. Arkansas, California, Connecticut, Illinois, Iowa, Kentucky, Maryland, Missouri, and Ohio have each lost more than 70 percent of their historical wetlands (Dahl 1990).

For most of the past two centuries, wetlands were viewed as impediments to agricultural and urban expansion. The reclamation of wetland habitat to usable upland habitat was considered beneficial.

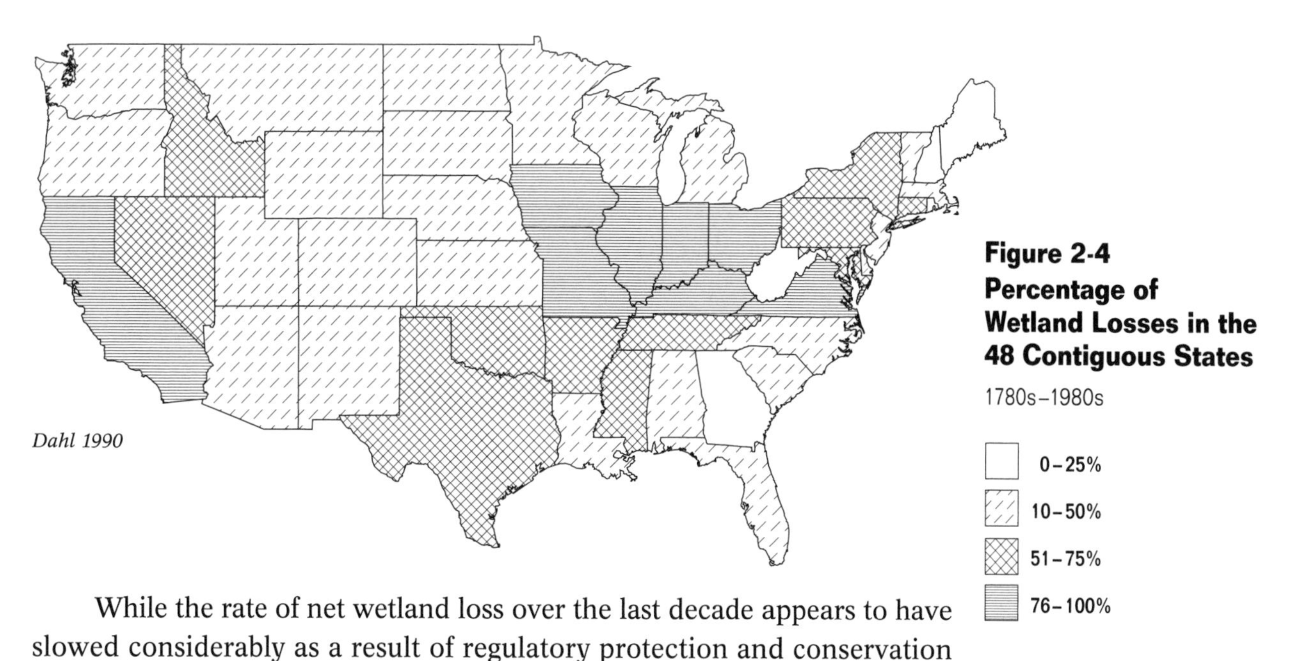

Dahl 1990

Figure 2-4
Percentage of Wetland Losses in the 48 Contiguous States
1780s–1980s

While the rate of net wetland loss over the last decade appears to have slowed considerably as a result of regulatory protection and conservation incentive programs, wetlands were still declining at an estimated 58,500 acres per year between 1986 and 1997 (U.S. Fish and Wildlife Service 2000).

This rate of loss compares favorably, though, to the estimated average annual loss of 458,000 acres from the mid-1950s to the mid-1970s, and the 290,000-acre average annual loss between the mid-1970s and the mid-1980s (U.S. Department of Interior 1994).

Losses among different types of wetlands vary, with forested and freshwater wetlands continuing to decline disproportionately (U.S. Fish and Wildlife Service 2000). Moreover, while the areal extent of wetland loss may be declining, the reduction in function and ecosystem integrity of remaining wetlands continues. This deterioration of function and integrity cannot be assessed by measuring only the areal extent of wetlands (U.S. Fish and Wildlife Service 2000).

While the areal extent of wetland loss may be declining, the reduction in function and ecosystem integrity of remaining wetlands continues.

Activities Affecting Wetlands

While the general causes of wetland losses have been addressed (*see* Loss of Wetlands above), numerous specific activities can destroy or substantially alter the hydrology, soil, vegetation, or wildlife of wetlands. These activities include:

- Pumping water or excavating ditches, which can cause wetlands to drain
- Placing fill material, which can severely disrupt or eliminate wetland hydrology by raising bottom elevations
- Excavating so that the resulting water level is too deep to support characteristic wetland vegetation and hydrology
- Construction and management of dams, diversions, and levees, which can alter or destroy the wetland by changing the frequency, timing, or duration of inundation
- Deep plowing or ripping through a claypan or hardpan, causing a wetland supported by a perched water table to drain
- Mowing, plowing, burning, or otherwise removing plants and vegetation, thereby degrading or destroying the wetland's wildlife habitat function
- Grazing, which can remove excessive amounts of vegetation and, consequently, degrade a wetland's function as wildlife habitat

Activities in locations away from wetlands may also destroy or alter hydrology, soils, vegetation, or wildlife. These indirect impacts, that typically manifest more slowly, include:

Activities in locations away from wetlands may also destroy or alter hydrology, soils, vegetation, or wildlife.

- Sediments deposited in the wetland from upslope erosion, potentially raising bottom elevations and affecting hydrology
- Erosion of wetland substrate caused by changes in hydrology (e.g., increased input of concentrated runoff) that can modify bottom contours
- Flooding resulting from dam impoundment or increased streamflows, potentially drowning vegetation
- Reductions in effective watershed size caused by construction of stormwater drainage or other systems, decreasing the amount of water that flows to the wetland

- Increases in impervious surfaces in the watershed resulting from urban development, causing increased peak flows and reducing sediment input
- Reductions in effective floodplain size caused by construction of flood control channels, levees, or dams, decreasing the amount of overbank flows to the wetland
- Shading from structures such as bridges, potentially causing loss of wildlife habitat and vegetative cover
- Introduction of nonnative plant and animal species that can outcompete or consume native species
- Contamination by pesticides, herbicides, fertilizers, heavy metals, oils or other chemicals in runoff from mining sites, agricultural land, urban development, industrial waste, and oil drilling sites–contamination that can poison plants and animals, make the soil infertile, cause overgrowth of plants, destroy the invertebrate food base, and result in bioaccumulation of toxic materials in the food chain

CHAPTER THREE

Jurisdictional Limits of Wetlands and Other Waters

Many federal and state laws regulate activities in wetlands, streams, and other waters. Implementing regulations and guidelines for these laws explain the regulatory processes, starting with the determination of jurisdictional boundaries. This chapter addresses the means by which various regulatory agencies define jurisdictional boundaries, with an emphasis on regulation under Section 404 of the Clean Water Act (CWA). This chapter also presents a practical approach for integrating the identification of aquatic resources regulated by multiple federal and state agencies. Such an approach is recommended to ensure compliance with all applicable federal and state regulations; moreover, integration is both cost and time effective, offering many benefits to project proponents.

CWA = Clean Water Act
USACE = United States Army Corps of Engineers

Full discussions of federal regulatory authority and the Section 404 permit process, as well as summaries of various state regulatory authorities over wetlands and waters appear in chapter 4, Federal Regulation of Wetlands and Other Waters, chapter 5, Section 404 Permitting Process, and chapter 6, State Wetland Laws.

Intertidal coastal waters provide valuable foraging habitat for many shorebirds.

Clean Water Act Section 404: Waters of the United States

Section 404 of the Clean Water Act regulates activities that result in the discharge of dredged or fill material into *waters of the United States*. This broad category of water bodies encompasses both wetland and nonwetland aquatic habitats, such as streams, rivers, lakes, ponds, bays, and oceans. These nonwetland waters are collectively referred to as *other waters*. Whether Congress has the authority to delegate its legislative role to USACE in granting the ability to define "waters of the United States" was addressed in *U.S. v. Mills*, 36 F. 3d 1052 (11th Cir. 1994). The court held that, despite criminal charges that could possibly result from violations

Under Section 404 of the Clean Water Act, "waters of the United States" also encompasses nonwetland aquatic habitats–such as streams, rivers, lakes, ponds, bays, and oceans–that are collectively referred to as "other waters."

Saturated soils in freshwater marsh habitat. Water quickly filled this unlined soil pit, indicating that the water table is at the surface and the soil is saturated to the surface.

CFR = Code of Federal Regulations
EPA = United States Environmental Protection Agency

Soil pit in riparian habitat for a wetland delineation

of the CWA, Congress had not unconstitutionally delegated it legislative authority to USACE.

Waters of the United States are defined by USACE regulations (33 CFR 328.3(a), parts 1–8) as:

(1) All waters that are currently used, or were used in the past, or may be susceptible to use in interstate or foreign commerce, including all waters that are subject to the ebb and flow of the tide;

(2) All interstate waters including interstate wetlands;

(3) All other waters such as intrastate lakes, rivers, streams (including intermittent streams), mudflats, sandflats, wetlands, sloughs, prairie potholes, wet meadows, playa lakes, or natural ponds, the use, degradation or destruction of which could affect interstate or foreign commerce including any such waters:

  (i) That are or could be used by interstate or foreign travelers for recreational or other purposes; or

  (ii) From which fish or shellfish are or could be taken and sold in interstate or foreign commerce; or

  (iii) That are used or could be used for industrial purpose by industries in interstate commerce;

(4) All impoundments of waters otherwise defined as waters of the United States under the definition;

(5) Tributaries of waters identified in paragraphs (1) through (4) of this section;

(6) The territorial seas;

(7) Wetlands adjacent to waters (other than waters that are themselves wetlands) identified in paragraphs (1) through (6) of this section. Waste treatment systems, including treatment ponds or lagoons designed to meet the requirements of CWA (other than cooling ponds as defined in 40 CFR 123.11(m) which also meet the criteria of this definition) are not waters of the United States.

(8) Waters of the United States do not include prior converted cropland. Notwithstanding the determination of an area's status as prior converted cropland by any other agency, for the purposes of the Clean Water Act, the final authority regarding Clean Water Act jurisdiction remains with EPA.

33 CFR 328.3(a); 40 CFR 230.3(s)

As waters of the United States, wetlands are regulated under the CWA (*see* subsections 3 and 7 above). Under federal regulations, wetlands are identified as:

> [T]hose areas that are inundated or saturated by surface or ground water at a frequency and duration sufficient to support, and that under normal circumstances do support, a prevalence of vegetation typically adapted

for life in saturated soil conditions. Wetlands generally include swamps, marshes, bogs, and similar areas. 33 CFR 328.3(b); 40 CFR 230.3(t)

Using the definition of waters of the United States in the Code of Federal Regulations, essentially all natural bodies of water are included under the definition of waters of the United States (however, see discussion of link to interstate commerce, below):

- Oceans
- Bays
- Estuaries
- Tidal wetlands
- Lakes
- Ponds
- Desert playas
- Seasonal ponds
- Rivers
- Perennial streams
- Intermittent streams
- Desert washes and arroyos
- Ephemeral swales
- Perennial wetlands
- Seasonal wetlands
- Vernal pools and prairie potholes

Many artificial or disturbed water bodies can also fall under this definition, including:

- Reservoirs
- Farm or stock ponds fed by direct rainfall or impoundment of a stream (not by pumped water)
- Artificial wetlands that receive water without artificial controls (such as pumps, valves, or gates)
- Farmed wetlands

However, some bodies of water can be excluded from Section 404 regulation:

- Irrigation ditches (not considered tributaries of waters of the U.S.)
- Drainage ditches excavated in uplands
- Temporary sediment basins on construction sites
- Reflecting pools
- Wastewater systems, including treatment ponds and lagoons
- Ponds and wetlands created as part of an ongoing mining operation (unless created as mitigation for past impacts)
- Isolated ponds and wetlands that do not have a nexus to interstate commerce (as discussed below)

In general, federal regulatory authority over actions within a particular state is derived from the interstate commerce clause of the U.S. Constitution. Unless specifically

Figure 3-1
Waters of the United States
Section 404 of the Clean Water Act

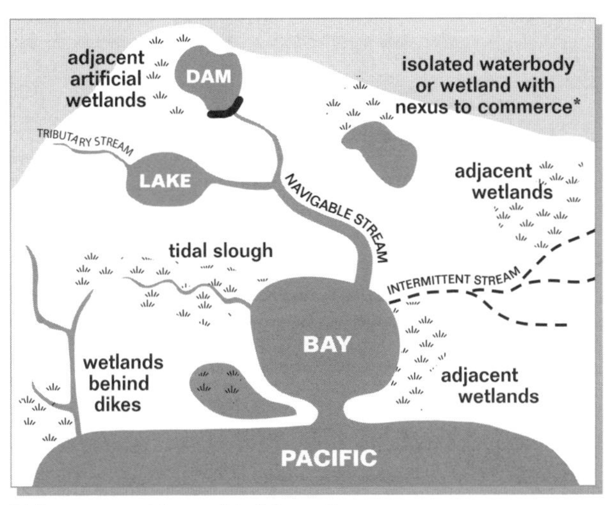

* May or may not be regulated depending on the final interpretation of the SWANCC case

Farmed wetlands continue to support wetland hydrology and soils, but natural vegetation is replaced by crops.

APA	=	Administrative Procedures Act
CZMA	=	Coastal Zone Management Act
CZMP	=	Coastal zone management program
NEPA	=	National Environmental Policy Act
NPDES	=	National Pollutant Discharge Elimination System
NRCS	=	Natural Resources Conservation Service
NRDC	=	National Resources Defense Council
SWANCC	=	Solid Waste Agency of Northern Cook County
USFWS	=	United States Fish and Wildlife Service

Figure 3-2
Data Form—Routine Wetland Determination

1987 USACE Wetlands Delineation Manual

DATA FORM
ROUTINE WETLAND DETERMINATION
(1987 COE Wetlands Delineation Manual)

Project/Site: ______________________ Date: __________
Applicant/Owner: ______________________ County: __________
Investigator: ______________________ State: __________

Do Normal Circumstances exist on the site?	Yes	No	Community ID: ________
Is the site significantly disturbed (Atypical Situation)?	Yes	No	Transect ID: ________
Is the area a potential Problem Area? (If needed, explain on reverse.)	Yes	No	Plot ID: ________

VEGETATION

Dominant Plant Species	Stratum	Indicator	Dominant Plant Species	Stratum	Indicator
1. ________	____	____	9. ________	____	____
2. ________	____	____	10. ________	____	____
3. ________	____	____	11. ________	____	____
4. ________	____	____	12. ________	____	____
5. ________	____	____	13. ________	____	____
6. ________	____	____	14. ________	____	____
7. ________	____	____	15. ________	____	____
8. ________	____	____	16. ________	____	____

Percent of Dominant Species that are OBL, FACW or FAC (excluding FAC–)

HYDROLOGY

_ Recorded Data (Describe in Remarks):
- ___ Stream, Lake, or Tide Gauge
- ___ Aerial Photographs
- ___ Other

_ No Recorded Data Available

Field Observations:

Depth of Surface Water: ______ (in.)

Depth to Free Water in Pit: ______ (in.)

Depth to Saturated Soil: ______ (in.)

Wetland Hydrology Indicators

Primary Indicators:
- ___ Inundated
- ___ Saturated in Upper 12 Inches
- ___ Water Marks
- ___ Drift Lines
- ___ Sediment Deposits
- ___ Drainage Patterns in Wetlands

Secondary Indicators (2 or more required):
- ___ Oxidized Root Channels in Upper 12 Inches
- ___ Water-Stained Leaves
- ___ Local Soil Survey Data
- ___ FAC-Neutral Test
- ___ Other (Explain in Remarks)

Remarks:

SOILS

Map Unit Name
(Series and Phase): ______________________________

Drainage Class: ____________
Field Observations
Confirm Mapped Type? Yes No

Taxonomy (Subgroup): ______________________________

Profile Description

Depth (inches)	Horizon	Matrix Color (Munsell Moist)	Mottle Colors (Munsell Moist)	Mottle Abundance/Contrast	Texture, Concretions Structure etc.
______	______	__________	__________	__________	__________
______	______	__________	__________	__________	__________
______	______	__________	__________	__________	__________
______	______	__________	__________	__________	__________
______	______	__________	__________	__________	__________
______	______	__________	__________	__________	__________

Hydric Soil Indicators:

___ Histosol
___ Histic Epipedon
___ Sulfidic Odor
___ Aquic Moisture Regime
___ Reducing Conditions
___ Gleyed or Low-Chroma Colors

___ Concretions
___ High Organic Content in Surface Layer in Sandy Soils
___ Organic Streaking in Sandy Soils
___ Listed on Local Hydric Soils List
___ Listed on National Hydric Soils List
___ Other (Explain in Remarks)

Remarks:

WETLAND DETERMINATION

Hydrophytic Vegetation Present? Yes No (Circle)
Wetland Hydrology Present? Yes No
Hydric Soils Present? Yes No

(Circle)
Is this Sampling Point Within a Wetland? Yes No

Remarks:

Approved by HQUSACE 3/92

Development of Definition of "Waters of the United States" Through the Courts

Jurisdiction of USACE under the Rivers and Harbors Act of 1899 is limited to "navigable waters of the United States." Navigable waters are those waters subject to the ebb and flow of the tide and/or those that have been or could be used to transport interstate or foreign commerce. Therefore, when Congress first enacted the CWA, USACE's interpretation was to relate Section 404 directly to their authority under the Rivers and Harbors Act. The initial Section 404 regulations issued by USACE after 1972, interpreted the CWA's directive to issue permits for discharges in waters that were navigable or fit under the definition of "navigable waters of the United States" as it was defined by the Rivers and Harbors Act.

In 1975, the Natural Resources Defense Council (NRDC) challenged USACE's limitation of Section 404 jurisdiction in federal court. Agreeing with the NRDC that Congress intended the jurisdiction of the CWA to extend further than the traditional notion of navigable waters of the United States, the district court ordered USACE to revise its regulations to reflect the full regulatory mandate of the CWA. In an attempt to regulate the discharge of pollutants on small, nonnavigable tributaries, the court stated that Congress, through the CWA, had asserted overall federal jurisdiction over the "waters of the United States." Although the language of the CWA appeared to equate it with "navigable waters of the United States," the court held that the mandate of the CWA was to greatly expand the federal jurisdiction to the maximum extent permissible under the commerce

reserved in the U.S. Constitution, the federal government has delegated all rights to govern state issues to the sovereign states and retained authority over only those issues that relate to interstate commerce, because one sovereign state should not be regulating issues that affect another sovereign state. It is with this authority that Congress passed the CWA (as well as other environmental regulations, such as the Clean Air Act and Endangered Species Act).

Section 404's language provides that USACE shall issue permits "for the discharge of dredged or fill material into *navigable waters* at specified disposal sites." Section 404(a). The CWA defines "navigable waters" as "the *waters of the United States,* including the territorial seas." 33 USC 1362(7). USACE issued regulations to implement Section 404 and includes the definition of waters of the United States provided above.

Historically, federal control over water (including USACE's jurisdiction under the Rivers and Harbors Act) was restricted to navigable waters (those that might carry foreign or interstate commerce) and was not focused on wetlands and other water resources. However, by the broad definition outlined in its regulations, USACE extended its Section 404 authority over "other waters." USACE regulations also identify that waters of the United States could include any waters that "the use, degradation or destruction of which could affect interstate commerce." USACE lists examples to include waters "that are or could be used by interstate or foreign travelers for recreational or other purposes; or from which fish or shellfish are or could be taken and sold in interstate or foreign commerce; or which are used or could be used for industrial purpose by industries in interstate commerce" (33 CFR 328.3(a)(3)). The court has interpreted the CWA's Section 404 jurisdiction to extend to desert gullies or "arroyos" based on occasional flows providing a surface connection with navigable waters. *Quivira Min. Co. v. U.S. Environmental Protection Agency,* 765 F. 2d 126 (10th Cir. 1985).

Another example can be seen in a case concerning Sections 301 and 402 of the CWA (NPDES permit) and the regulation of aquatic herbicides the court extended the jurisdiction of the CWA to irrigation canals situated in such a way to be tributary to waters of the United States. *Headwaters, Inc. v. Talent Irr. Dist.,* 243 F. 3d 526 (9th Cir. 2001). 33 CFR 328.3(a)(3).

Because Congress did not use the word "wetlands" when the CWA was enacted, to understand how USACE jurisdiction applies, it is necessary to understand how the term "navigable waters" (the actual language of Section 404(a) of the CWA) is related to "waters of the United States" (as used in regulatory and judicial interpretations of the CWA), and how it is related to interstate and foreign commerce.

According to USACE's current regulations implementing the CWA, wetlands are an integral part of the definition of waters of the United States (33 CFR 328.3). The CWA's jurisdiction is directly related to whether the area in question falls within the definition of a wetland and whether "the use, degradation or destruction" of it "could affect interstate commerce." However, as described below, the issue of whether wetlands need to be linked to a navigable water of the United States, in order to be considered "waters of the United States" may not yet be fully resolved.

Relation to Interstate Commerce. It is clear on its face that, except for one, all the subsections of 33 CFR 328.3(a) defining jurisdictional waters relate to the federal government's interest in regulating interstate commerce. The USACE regulations state that "other waters" that are totally intrastate (including wetlands) are only considered jurisdictional waters when their use, degradation, or destruction could affect interstate or foreign commerce. Therefore, wetlands and "other waters" that are purely *intra*state (that is located entirely within a state's boundaries and *not* connected to *navigable* waters of the United States) may be included within the definition of waters of the Unites States as long as the intrastate waters are tied to interstate commerce.

Solid Waste Agency of Northern Cook County v. U.S. Army Corps of Engineers. In early 2001, the U.S. Supreme Court issued a landmark ruling regarding USACE's regulation of isolated intrastate waters in *Solid Waste Agency of Northern Cook County v. U.S. Army Corps of Engineers (SWANCC).* Between the years of 1985 and 2001, USACE generally extended its jurisdiction over wetlands beyond "adjacent wetlands" and regulated the discharge of dredged or fill material into any intrastate wetlands and isolated waters, whether or not they had a link to navigable waters of the United States, in the spirit of *NRDC v. Callaway* and *Riverside Bayview Homes* (*see* sidebars). These isolated, intrastate wetlands were considered waters of the United States because the use, degradation, or destruction of these wetlands could affect interstate or foreign commerce, as stated in USACE regulations (33 CFR 328.3(a)(3)).

In response to *Riverside Bayview,* USACE in 1986 adopted new regulations to clarify the scope of its jurisdiction under Section 404. As noted above, the CWA jurisdiction defines the term "navigable waters" to include the "waters of the United States, including the territorial seas." 33 USC 404, 502(7). In its 1986 regulations, USACE defined the "waters of the United States" as including any waters "the use, degradation, or destruction of which could affect interstate or foreign commerce" (33 CFR 328.3(a)(3)). When it published its 1986 regulations in the Federal Register, USACE included the preamble language, known as the "Migratory Bird Rule," that *intra*state waters had a sufficient link to *inter*state commerce, and were therefore considered waters of the United States, if the waters are or would be used:

clause (*National Resources Defense Council v. Callaway,* 392 F. Supp. 685 [D. D.C. 1975]). In 1977, USACE regulations were revised to be consistent with this decision.

The 1977 USACE regulations were the first to acknowledge that the Section 404 jurisdictional authority covered not only truly navigable waters, but also tributaries to navigable waters and wetlands adjacent to navigable waters, and isolated wetlands and other waters with some connection to interstate commerce. USACE's assertion of jurisdiction over isolated waters generated a considerable degree of controversy and, later in the year, the House of Representatives passed a bill that would have restricted USACE's Section 404 jurisdiction to traditional navigable waters and adjacent wetlands and eliminated jurisdiction over nonadjacent isolated wetlands. After lengthy debate, the Senate rejected this legislation.

Another court decision in the 1970s interpreted the extent of the CWA's authority and affected USACE jurisdiction over the placement of dredged or fill material into waters of the United States. In *United States v. Holland,* 373 F. Supp. 665 (M.D. Fla. 1974), decided prior to *NRDC v. Callaway,* the court concluded that the CWA extended federal jurisdiction to all waters that might affect commerce, without regard to the traditional notions of navigability. The court held that tidelands are considered "waters of the United States" and should be regulated under the CWA (including Section 404 as regulated by USACE), even though the discharge activity in question was beyond the mean high water mark and therefore beyond traditional navigable waters. ■

Figure 3-3
Scope of USACE Section 10 and Section 404 Regulatory Jurisdiction

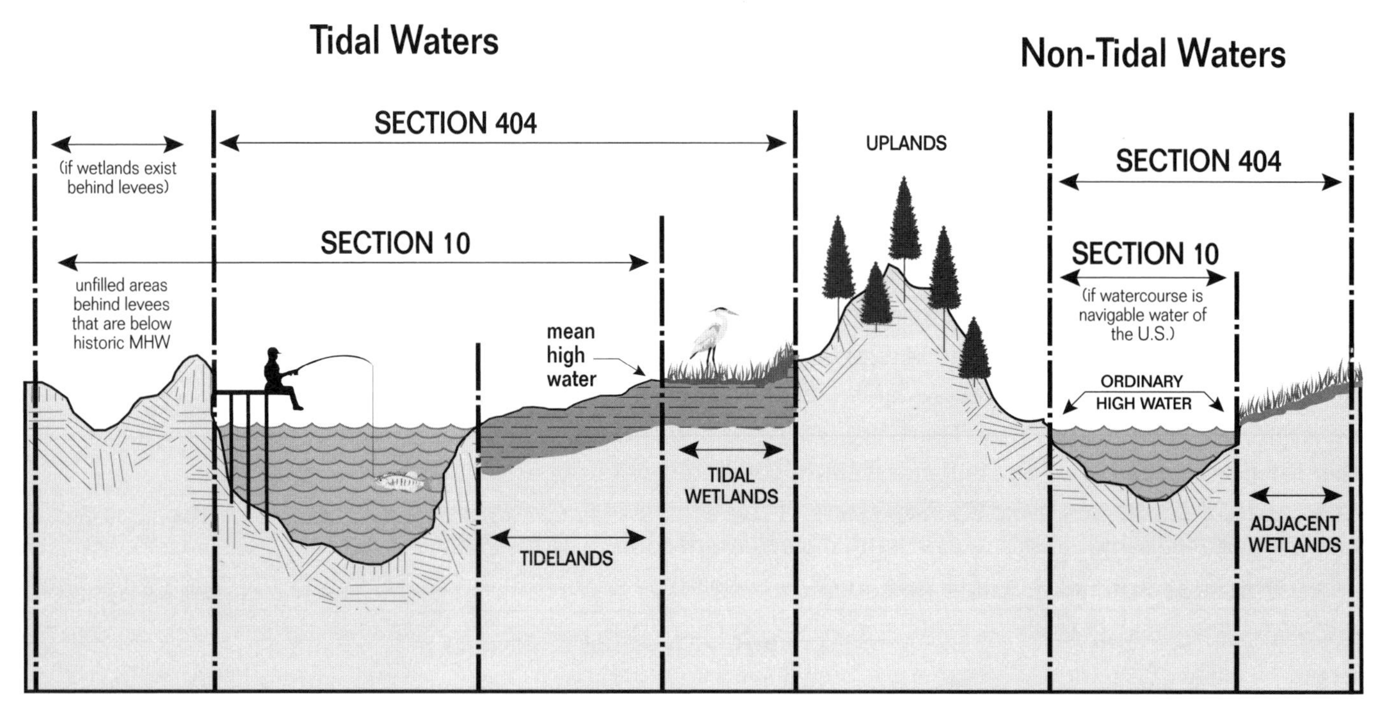

- As habitat by birds protected by Migratory Bird Treaties
- As habitat by other migratory birds that cross state lines
- As habitat for endangered species
- To irrigate crops sold in interstate commerce

The Migratory Bird Rule would have allowed USACE to assert jurisdiction over isolated wetlands that goes beyond traditional notions of commerce and extends to areas that merely have the potential to be used as habitat for migratory birds.

The rule, as generally applied, would allow USACE to assert jurisdiction over isolated wetlands that goes beyond traditional notions of "commerce" and even extends to areas that merely have *the potential* to be used as habitat for migratory birds (the rationale being that migratory birds are considered interstate and foreign commerce because of links to recreation, other industries, and international treaties signed under the Migratory Bird Treaty Act, and therefore if an activity is in a water that could be habitat to a migratory bird the activity could affect interstate or foreign commerce). In effect, this rule allowed USACE to regulate any intrastate wetland or other waters that met USACE's delineation criteria (because these areas are presumably always habitat for migratory birds).

Open water habitat provides important loafing and foraging areas for many species of birds.

In January 2001, however, the U.S. Supreme Court narrowed the rule regarding the link between interstate commerce and wetlands. The *SWANCC* decision involved a challenge by the Solid Waste Agency of North Cook County (the agency) to USACE's exercise of jurisdiction over a number of nonnavigable, isolated, intrastate ponds located on abandoned mining land that the agency was attempting to develop as a solid waste disposal site. The abandoned sand and gravel pit included remnant excavation trenches that had returned over time to largely natural conditions through the growth of considerable vegetation and the development of a number of scattered permanent and seasonal ponds. USACE eventually asserted jurisdiction over these isolated ponds based upon the Migratory Bird Rule and, accordingly, the agency requested a Section 404 permit from USACE to fill some of them. USACE denied the permit on several grounds.

After being denied the permit, the agency filed suit in federal court challenging USACE's exercise of jurisdiction over the ponds as beyond the authority delegated by Congress under Section 404. After a favorable ruling in the trial court but unfavorable ruling in the Seventh Circuit, the agency appealed to the U.S. Supreme Court.

The U.S. Supreme Court on January 19, 2001, by a narrow 5–4 majority, reversed the Seventh Circuit. In what appears by its holding to be a narrow ruling, the U.S. Supreme Court held that USACE's jurisdiction under Section 404 of the CWA does not extend to nonnavigable, isolated, intrastate waters based solely on the fact that these waters are used as habitat by migratory birds. The language of the Court's opinion is, however, somewhat broad and casts a large amount of uncertainty as to the constitutionality of portions of USACE's authority to regulate certain isolated wetlands and other water resources (as well as casting doubt as to whether other federal environmental regulatory agencies have authority over purely intrastate resources).

As for the congressional intent to authorize USACE to extend Section 404 jurisdiction to wetlands that are not navigable waters, the U.S. Supreme Court had already spoken in *Riverside Bayview Homes* (*see* adjacent sidebar). However, the Court in *SWANCC* stated that "Congress' separate definitional use of the phrase 'waters of the United States' [does not] constitute a basis for reading the term 'navigable

Court Analysis of Jurisdictional Waters

Traditionally, courts have upheld USACE jurisdictional determinations where the regulated waters or wetlands could be linked directly to interstate commerce. USACE (at least prior to 2001) did not consistently consider the link of wetlands to interstate or foreign commerce, let alone navigable waters, when making a jurisdictional determination. In *Avoyelles Sportsmen's League, Inc. v. Marsh,* 715 F. 2d 897 (5th Cir. 1983), the court upheld USACE's assertion that because the forested land in question (on which land clearing and other discharge activities were planned) was subject to the average annual flood in a river basin, it is considered waters of the United States and regulated under Section 404 of the CWA. The court held that vegetation "typically adapted for life in saturated soils" is not limited to species surviving their entire life cycle in saturated soils. Therefore, even though it did not resemble typical wetland traits, the forested land could still be proven to be waters of the United States.

A 1985 U.S. Supreme Court decision gave a more definitive answer to the outer limits of USACE's jurisdictional boundaries related to interstate commerce. In *United States v. Riverside Bayview Homes,* 474 U.S. 121 (1985), the Supreme Court upheld USACE's interpretation that waters of the United States include adjacent wetlands (33 CFR 328.3(a)(7)). In the underlying case to the Supreme Court decision, USACE considered low-lying, marshy land next to a navigable water to be an adjacent wetland and therefore waters of the United States. USACE's jurisdictional determination for the area in question was based on the vegetation, saturated soils, and hydrologic connection to adjacent navigable waters of the marshy land, even though the land was not subject to flooding by adjacent navigable waters.

The Supreme Court held that USACE's regulations clearly state that "saturation by either surface or ground water is sufficient to bring an area within the category of wetlands, provided that the saturation is sufficient to and does support wetland vegetation" and that this regulation is consistent with the Congressional intent of the CWA, demonstrating the "evident breadth of concern for protection of water quality and aquatic ecosystems." The Court also found meaningful the 1977 Senate's rejection of the House of Representatives' bill to narrow USACE jurisdiction to truly navigable waters, finding that Congress implicitly acquiesced in USACE's broad reading of Section 404 as reflected in the 1977 regulations.

Following the Supreme Court's 1985 approval of USACE's broad assertion of federal jurisdiction in *Riverside Bayview,* lower courts, in an overwhelming majority of cases, approved the extension of USACE's CWA jurisdiction over isolated wetlands and water bodies, if a functional link between the wetlands and interstate commerce exists (that is, if evidence shows that the activity affecting the wetland would affect interstate commerce). ■

waters' out of the statute. We said in *Riverside Bayview Homes* that the word 'navigable' in the statute was of 'limited effect' and went on to hold that Section 404(a) extended to nonnavigable wetlands adjacent to open waters. But it is one thing to give a word limited effect and quite another to give it *no effect whatever* [emphasis added]. The term 'navigable' has at least the import of showing us what Congress had in mind as its authority for enacting the CWA: its traditional jurisdiction over waters that were or had been navigable in fact or which could reasonably be so made." Therefore, the Supreme Court in *SWANCC* distinguished *Riverside Bayview Homes* by emphasizing that it was the "significant nexus" between the wetlands at issue in the earlier case and other navigable waters.

Although the U.S. Supreme Court does not expressly state it in *SWANCC,* it appears that in cases where the intrastate, isolated waters have no significant nexus to navigable waters, Section 404 of the CWA does not apply. Where such a nexus does exist, then the waters in question are appropriately subject to federal regulation. However, several questions remain regarding the definition of "significant nexus to navigable waters." As of this writing, there has been no official guidance from EPA or USACE regarding interpretation of the *SWANCC* decision. However, on January 19, 2001, ten days after the Court issued its decision in *SWANCC,* the Chief Counsel for USACE and the General Counsel of the EPA issued a joint memorandum identifying the aspects of the regulatory definition of "waters of the United States" affected by *SWANCC* (*see* sidebar).

The *SWANCC* ruling has resulted in greater attention being paid to the determination of whether wetlands are isolated or are adjacent to waters of the United States. Adjacent wetlands are identified as jurisdictional under subsection 7 in the definition of waters of the United States. *Adjacent* is defined to mean:

> [B]ordering, contiguous, or neighboring. Wetlands separated from other waters of the United States by man-made dikes or barriers, natural river berms, beach dunes and the like are "adjacent wetlands."
>
> 33 CFR 328.3(c)

EPA and USACE Interpretation of SWANCC Ruling. On January 10, 2003, EPA and USACE issued an Advance Notice of Proposed Rulemaking on the Clean Water Act Regulatory Definition of "Waters of the United States" (EPA/USACE Notice) stating that:

> In view of SWANCC, neither agency will assert CWA jurisdiction over isolated waters that are both intrastate and non-navigable, where the sole basis available for asserting

The Migratory Bird / Interstate Commerce Case History

In 1989 the Fourth Circuit was asked to rule whether the Section 404 jurisdiction was sufficiently linked to interstate commerce over isolated waters, based on the *possibility* of use by migratory birds. USACE was challenged on the application of the "migratory bird rule" because the rule was only contained in the preamble to the regulations and not part of the actual regulation subject to the notice and comment requirements of the Administrative Procedures Act.

In *Tabb Lakes v. United States,* the court held that the USACE application of the migratory bird rule was in fact "unofficial rulemaking," because it was substantive rather than interpretive, and was therefore required to go through APA's notice and comment procedures prior to enforcement. In reaction to this holding, USACE and EPA issued a memorandum stating that they believed the court incorrectly decided this case and would only suspend application of the migratory bird rule to determine jurisdiction over isolated waters in USACE districts within the boundaries of the fourth circuit (Virginia, Maryland, West Virginia, North Carolina, South Carolina).

In 1990, the Ninth Circuit, on the other hand, upheld USACE's jurisdiction over isolated waters based on the actual or potential use by migratory birds (*Leslie Salt Co. v. United States,* 896 F. 2d 354 (9th Cir. 1990)). In 1993, the Seventh Circuit reviewed the application of the migratory bird rule to specific facts, holding that the wetlands in question did not meet the criterion for interstate commerce. The court had originally ruled that the wetlands were not jurisdictional because they were isolated from navigable waters; however, the court vacated that ruling and instead applied the migratory bird rule and held that these isolated waters were not suitable habitat for migratory birds and therefore not jurisdictional waters under Section 404. *Hoffman Homes Inc. v. EPA,* 999 F. 2d 256 (7th Cir. 1993).

In a case that was somewhat of a predecessor to *SWANCC* (although not directly mentioned in the

CWA jurisdiction rests on any of the factors listed in the "Migratory Bird Rule." In addition, in view of the uncertainties after SWANCC concerning jurisdiction over isolated waters that are both intrastate and non-navigable based on other grounds listed in 33 C.F.R. § 328.3(a)(3)(i)–(iii), field staff should seek formal project-specific Headquarters approval prior to asserting jurisdiction over such waters, including permitting and enforcement actions.

On December 16, 2003, the headquarters of EPA and USACE issued a joint press release stating that they would not be issuing a new rule on federal regulatory jurisdiction over isolated wetlands as had been stated in the January EPA/USACE Notice. While a new rule is no longer expected, the guidance provided in the EPA/USACE Notice is still instructive. The EPA/USACE Notice reiterates that traditionally navigable waters remain within the definition of waters of the U.S. and therefore are jurisdictional under Section 404 and that isolated, intrastate, navigable waters are also jurisdictional:

> In accord with the analysis in SWANCC, waters that fall within the definition of traditional navigable waters remain jurisdictional under the CWA. Thus, isolated, intrastate waters that are capable of supporting navigation by watercraft remain subject to CWA jurisdiction after SWANCC if they are traditional navigable waters, i.e., if they meet any of the tests for being navigable-in-fact.

The EPA/USACE Notice also states that wetlands adjacent to navigable waters are jurisdictional under the CWA. The EPA/USACE Notice indicates that CWA jurisdiction will generally remain over tributaries to navigable waters and wetlands adjacent to such tributaries. The EPA/USACE Notice directs field staff to seek formal, project-specific headquarters approval prior to asserting jurisdiction over isolated, non-navigable, intrastate waters based on links to commerce identified in current regulations (i.e., those waters that are used or could be used by interstate or foreign travelers for recreation or other purposes; for fish or shellfish harvest for sale in interstate or foreign commerce; for industrial purposes by industries in interstate or foreign commerce). The December 16 press release stated that EPA and USACE "have decided to preserve the federal government's authority to protect our wetlands" and reiterated the Administration's commitment to the goal of "no net loss" of wetlands in the United States. Following the press release, EPA's assistant administrator indicated that USACE Districts are still required to check with USACE Headquarters on issues of jurisdiction over isolated wetlands.

U.S. Supreme Court decision), the Fourth Circuit Court in 1997 was again asked to rule on whether it was an appropriate extension of Section 404 coverage to waters that are neither connected closely to interstate commerce nor navigable waters, per the directive under 33 CFR 328.3(a)(3).

In *United States v. Wilson*, 133 F. 3d 251 (4th Cir. 1997), the Fourth Circuit held that USACE did not have jurisdiction over wetlands that lacked direct or indirect surface connection to interstate waters, navigable waters or interstate commerce. In that case, USACE's regulations attempted to extend jurisdiction over intrastate, nonnavigable waters solely on the basis that the use, degradation, or destruction of such waters *could affect* interstate commerce. The Fourth Circuit concluded that the regulation presented "serious constitutional difficulties under the Commerce Clause" because USACE did not require that the regulated *activity* have a substantial effect on interstate commerce, or that the *intrastate waters* have a nexus with navigable waters. The court in *Wilson* also concluded that USACE improperly asserted jurisdiction over isolated wetlands that were located up to ten miles from a navigable water or tributary because these wetlands did not have a "direct or indirect surface connection" with interstate waters.

In reaction to the *Wilson* decision, USACE issued guidance to its district offices stating that its Section 404 jurisdiction should only be asserted where it can establish an "actual link between the water body and interstate or foreign commerce"; and where individually or combined, the use, degradation or destruction of isolated waters with such a link would have a substantial effect on interstate or foreign commerce. Guidance for USACE and EPA Field Offices Regarding Clean Water Act Section 404 Jurisdiction Over Isolated Waters in Light of *United States v. James J. Wilson* (guidance withdrawn after January 19, 2001 *SWANCC* decision). ■

USACE and EPA Guidance on "Waters of the United States" Affected by SWANCC

The Joint Memorandum issued by counsel for EPA and USACE states that, although SWANCC discussed federal CWA jurisdiction in broad terms, its holding was limited to invalidating USACE application of 33 CFR 328.3(a)(3), as clarified and applied to the ponds at issue pursuant to the Migratory Bird Rule. Therefore, SWANCC does not affect the scope of 33 CFR 328.3(a)(3) itself.

The Joint Memorandum advises field staff to no longer rely on the use of waters or wetlands by migratory birds as the sole basis for the assertion of jurisdiction under the CWA. Because SWANCC is limited to "nonnavigable, isolated, intrastate" waters, field staff are advised under the Joint Memorandum to exercise CWA jurisdiction over waters falling outside this category to the full extent of their authority under the CWA and regulations and consistent with court opinions. Because the Court did not overturn Riverside Bayview, traditionally navigable waters, interstate waters, their tributaries, and wetlands adjacent to each are still considered by USACE and EPA under the Joint Memorandum to be waters of the United States.

The Joint Memorandum identifies 33 CFR section 328.3(a)(3) as the only subsection of the regulatory definition of "waters of the United States" affected, or potentially affected, by SWANCC.

The Joint Memorandum advises that, because of the *SWANCC* decision, waters covered by subsection (a)(3) that could affect interstate commerce solely by virtue of their use as habitat by migratory birds are no longer considered "waters of the United States." With respect to other nonnavigable, isolated, intrastate waters, the Joint Memorandum suggests that other types of interstate commerce

Determining Jurisdictional Limits of Wetlands and Other Waters

Delineations of waters of the United States, including wetlands, are typically conducted by an applicant or the applicant's consultant and submitted to USACE for verification. Delineations for the purposes of the CWA must include, at a minimum, a map indicating the boundary of all waters of the United States and a report documenting the data collected and the conclusions reached regarding jurisdiction.

Because wetlands are typically the regulatory focus for determining USACE jurisdiction, all other forms of waters of the United States are often referred to collectively as *other waters*. The mapping of waters of the United States into two categories–wetlands and other waters–is a widely used convention in preparing USACE jurisdictional delineations. The mapping effort typically identifies wetlands, other special aquatic sites, other waters, navigable waters, and historically navigable waters as separate units. Many USACE field offices have published delineation guidance that is available at their offices or online.

The Wetland Delineation Method Used by USACE. In 1987, the USACE Waterways Experiment Station in Vicksburg, Mississippi, published the *Corps of Engineers Wetlands Delineation Manual* (1987 Manual) to be used by staff to identify and delineate wetland boundaries for the purpose of Section 404 regulation (U.S. Army Corps of Engineers 1987). The techniques described in the 1987 Manual constitute USACE's officially recognized method for delineating wetlands, and are based on the regulatory definition of wetlands as set forth in 33 CFR 328.3[b] and 40 CFR 230.3[t].

To determine whether a site is a wetland, the 1987 Manual uses a three-parameter test. The three parameters are vegetation, soils, and hydrology. If a site supports positive indicators of hydrophytic vegetation, hydric soils, and wetland hydrology, USACE is likely to consider the wetland to be within its jurisdiction under the authority of Section 404. Except in disturbed or abnormal circumstances, positive indicators for all three of these parameters must be present for a site to qualify as a jurisdictional wetland.

An example of a disturbed situation would be a site where hydrophytic vegetation is not present because the vegetation has been artificially removed from the wetland. An example of abnormal circumstances would be a situation in which wetland hydrology that would normally (i.e., during years of normal rainfall) be present is lacking as a result of prolonged drought.

The boundary between wetland and nonwetland habitats is defined as the location where positive indicators of one of the three parameters are no longer present. At its upper elevational boundary, a wetland gives way to upland habitat. A lower elevational boundary, where it exists, demarcates the wetland from deepwater habitat or other waters, in which wetland vegetation cannot survive. *See* chapter 2, Ecology of Wetlands and Other Waters, for a discussion of the ecology and characteristics of wetlands.

Because of the heightened attention paid to questions of adjacency as a result of the *SWANCC* ruling, documentation may also be required regarding the historic and present hydrologic connection of a wetland to tidal or nontidal waters. In preparation of a delineation report, information such as frequency of surface water connection to tidal or nontidal waters should be noted, as well as groundwater connection and biological relationship to other waters.

Boundaries of Waters of the United States in the Absence of Wetlands. Although some waters of the United States have associated wetlands adjacent to them, many do' not. In the absence of adjacent wetlands, USACE applies the following definitions for determining the jurisdictional boundaries of territorial seas, tidal waters, and nontidal waters:

(a) **Territorial Seas.** The limit of jurisdiction in the territorial seas is measured from the baseline (the line on the shore reached by ordinary low tides) in a seaward direction a distance of three nautical miles.

(b) **Tidal Waters of the United States.** The landward limit of jurisdiction in tidal waters:

(1) Extends to the high tide line (encompasses spring high tides and other high tides that occur with periodic frequency), or

(2) When adjacent nontidal waters of the United States are present, the jurisdiction extends to the limits identified in paragraph (c) below.

(c) **Nontidal Waters of the United States.** The limits of jurisdiction in nontidal waters:

(1) In the absence of adjacent wetlands, the jurisdiction extends to the ordinary high water mark (*see* definition below), or

(2) When adjacent wetlands are present, the jurisdiction extends beyond the ordinary high water mark to the limit of the adjacent wetlands.

(3) When the water of the United States consists only of wetlands the jurisdiction extends to the limit of the wetland.

33 CFR 328.4

connections may still be relied upon to assert CWA jurisdiction subject to review on a case-by-case basis. The memorandum suggests further that two factors should be considered to determine whether subsection (a)(3) may support CWA jurisdiction:

(1) With respect to waters that are isolated, intrastate, and nonnavigable—jurisdiction may be possible if their use, degradation, or destruction could affect other "waters of the United States," thus establishing a significant nexus between the water in question and other "waters of the United States";

(2) With respect to waters that, although isolated and intrastate, are navigable—jurisdiction may also be possible if their use, degradation, or destruction could affect interstate or foreign commerce (examples of ways the use, degradation or destruction of a water could affect such commerce are provided at 33 CFR 328.3(a)(3)(i)–(iii)).

Impoundments of (a)(3) waters, tributaries of (a)(3) waters, and wetlands adjacent to (a)(3) waters are to be analyzed in accordance with the above, and are to be considered "waters of the United States" if they impound, are tributary to, are adjacent to, or are themselves "waters of the United States."

It should be noted that there are many detractors of this Joint Memorandum. Some argue that the conclusion that Section 404 jurisdiction over intrastate, nonnavigable, isolated waters that have some Commerce Clause connection, other than migratory bird use, is still valid after the SWANCC decision; it is unsupported by any reasonable reading of the decision. The detractors argue that although the Supreme Court did not strike down 33 CFR 328.3(a)(iii) itself, it is clear from the opinion that nonnavigable waters are not subject to regulation unless they fall within the protections offered by Riverside Bayview. ■

Tidal areas include coastal areas, river mouths, deltas, estuaries, and bays. Nontidal waters of the United States include rivers, streams, lakes, and ponds. USACE jurisdiction in nontidal waters is measured to the ordinary high water mark, which is defined in the federal regulations to mean:

> [T]hat line on the shore established by the fluctuations of water and indicated by physical characteristics such as clear, natural line impressed on the bank, shelving, changes in the character of soil, destruction of terrestrial vegetation, the presence of litter and debris, or other appropriate means that consider the characteristics of the surrounding areas.
>
> 33 CFR 328.3(e)

Special Aquatic Sites

Particular kinds of waters of the United States, called *special aquatic sites,* receive special attention from USACE and EPA under CWA Section 404. *See* chapter 5, Section 404 Permitting Process, for a discussion of the guidelines pertaining to special aquatic sites. These waters are geographic areas, large and small, that possess distinctive characteristics of productivity, habitat, wildlife protection, or other important ecological values (40 CFR 230.3(q)(1)). These areas are generally recognized as significantly influencing or positively contributing to the overall environmental health or vitality of the entire ecosystem of a region. Special aquatic sites adversely affected by a project are subject to greater scrutiny than other waters in the determination of appropriate mitigation measures. The six types of special aquatic sites are:

- **Wetlands.** *See* the definition in Clean Water Act Section 404: Waters of the United States above.
- **Sanctuaries and refuges.** Sites designated by federal, state, or local jurisdictions that are managed for the purpose of fish and wildlife resources
- **Mudflats.** Periodically inundated and exposed unvegetated areas such as tidal coastal areas and the edges of inland lakes, ponds, and rivers
- **Vegetated shallows.** Permanently inundated sites with rooted, submerged vegetation
- **Coral reefs.** Calcium- or silica-based deposits produced by invertebrates that establish reefs in warm ocean waters (coral reefs are found in Florida and Hawaii)
- **Riffle and pool complexes.** High-quality fish and wildlife habitat on steep gradient portions of rivers or streams where a rapid current flowing over a coarse substrate results in turbulence (riffles), and a slower moving current in deeper areas results in smooth flow (pools).

Because these areas are afforded a higher level of protection, delineations of waters of the United States should have all special aquatic sites indicated on the maps and discussed in the delineation report.

Normal Circumstances and Prior Converted Cropland

Under USACE regulations, the definition of jurisdictional waters states that an area must contain the wetlands criteria (soil, vegetation, hydrology) under "normal circumstances" (33 CFR 328.3(b)).

Therefore, a question sometimes arises regarding the determination of the Section 404 jurisdiction for wetland areas that have been under some form of manipulation to the extent that the wetlands parameters are no longer evident. USACE issued further guidance to assist project proponents to understand that land use practices (e.g., vegetation removal or water pumping) that might temporarily remove one of the parameters for determining a wetland, will not remove it from consideration as waters of the United States. However, USACE would not assert jurisdiction over those areas that under normal circumstances would not exhibit the three parameters under the criteria for determining wetlands, even if these areas may have exhibited the three parameters in the past. *See* August 27, 1986 RGL, 86-9; April 10, 1990 Memorandum to all Division and District Counsel from Lance Wood, Assistant Chief Counsel.

It is important to point out that the regulatory definition of wetlands does not include areas that were transformed into dry land by 1975. These lands are prior converted wetlands and since the CWA does not apply retroactively, they fall outside of Section 404's jurisdiction as long as they

Rivers and Harbors Act, Section 10: Navigable Waters

Pursuant to Section 10 of the Rivers and Harbors Act of 1899, USACE regulates the construction of structures in, over, or under, excavation of material from, or deposition of material into *navigable waters*. *See* chapter 4, Federal Regulation of Wetlands and Other Waters, for a discussion of regulations under Section 10 of the Rivers and Harbors Act. Navigable waters of the United States are defined in the federal regulations as:

> [T]hose waters that are subject to the ebb and flow of the tide and/or are presently used, or have been used in the past, or may be susceptible for use to transport interstate or foreign commerce. A determination of navigability, once made, applies laterally over the entire surface of the water body, and is not extinguished by later actions or events which impede or destroy navigable capacity.
>
> 33 CFR 329.4

In tidal areas, the limit of navigable waters is the mean high tide line; in nontidal areas it is the ordinary high water mark. Navigable waters typically have the same boundaries as, or lie within the boundaries of, waters of the United States. Larger streams, rivers, lakes, bays, and oceans are navigable waters that may comprise all or a part of waters of the United States. Wetlands are typically not part of navigable waters and are therefore not regulated under the Rivers and Harbors Act. In some cases–such as where coastal salt marsh or tidal freshwater marsh lies below the mean high tide line–wetlands may exist within navigable waters.

In addition to the navigable waters described above, historically navigable waters are also subject to federal regulation under Section 10 of the Rivers and Harbors Act. Historically navigable waters are those areas that are no longer navigable as a result of artificial modifications, such as levees, dikes, and dams. To identify sites that were historically below mean high tide or the ordinary high water mark, and hence historically navigable, USACE uses records such as geological survey maps or other topographic maps. Each USACE district typically maintains a list of the specific navigable waters within its jurisdiction.

Food Security Act of 1985

The Food Security Act of 1985 contains a provision for penalizing agricultural producers who plant commodities on wetlands that have been drained, filled, or otherwise altered. *See* chapter 4, Federal Regulation of Wetlands and Other Waters, for a discussion of the Food Security Act. The Natural Resources Conservation Service (NRCS) makes the determination of whether a farm tract contains wetlands (*see* sidebar). NRCS defines a wetland as:

> [A]n area that has a predominance of hydric soils and that is inundated or saturated by surface or groundwater at a frequency and duration

were converted to dry land by 1975 and wetlands have not since that time been reestablished. *Golden Gate Audubon Soc., Inc. v. USACE,* 769 F. Supp. 1306 (N.D. Cal. 1992).

This issue also arises where there is a proposal to discharge dredged or fill material into waters of the United States where farming has occurred (*see* discussion below for Section 404(f) exemptions for farming activities), and the proposed activity is not exempt under Section 404(f) (because it is considered a change in use). The issue is then whether that wetland area still maintains the qualities of a "water of the United States." The focus is typically on the precise meaning of "normal circumstances" as it applies to wetlands that have been disturbed by farming activities.

In August 1993, USACE issued a final regulation clarifying that "prior converted cropland" is not subject to Section 404 jurisdiction (58 FR 45008, amending 33 CFR 228.3(a)(8)). Prior converted cropland is generally defined—consistent with the Natural Resources Conservation Service "Swampbuster" program—as wetlands that before December 23, 1985, were cropped and manipulated to remove excess water such that inundation lasts no more than 14 consecutive days during the growing season. Essentially, prior converted croplands have been effectively drained or filled and no longer exhibit wetland characteristics. By contrast, farmed wetlands might revert to full wetland function with cessation of farming activities. USACE Regulatory Guidance Letter 90–7. *See also* appendix H. ■

sufficient to support, and under normal circumstances does support, a prevalence of hydrophytic vegetation typically adapted for life in saturated soil conditions, except lands in Alaska identified as having a high potential for agricultural development and a predominance of permafrost soils.

Food Security Act Manual 1994

NRCS will make a determination of and/or delineate wetlands using the Food Security Act Manual on a farm tract for an agricultural producer whose intent is to keep the land in agricultural use. This delineation is valid for purposes of the Clean Water Act, should the agricultural activity require USACE authorization.

Coastal Zone Management Act of 1972

The Coastal Zone Management Act (CZMA) is designed to preserve, protect, develop, and–where possible–restore and enhance coastal zone resources, which could include wetlands. *See* chapter 4, Federal Regulation of Wetlands and Other Waters, for a discussion of CZMA. Of the 35 states with coastal resources, 33 have federally approved coastal zone management programs (CZMPs) with provisions for the protection of wetlands. (At this writing, the remaining two states, Illinois and Indiana, do not have federally approved CZMPs.) Some states have adopted the USACE definition of wetlands, while others have their own definitions to better meet the conservation requirements laid out in their CZMPs.

U.S. Fish and Wildlife Service

USFWS does not have direct responsibility for regulating activities in wetlands unless the wetland is related to habitat of a species listed as threatened or endangered under the federal Endangered Species Act. However, for purposes of advising federal agencies of a proposed action's effect on wetland resources as required by such laws as the National Environmental Policy Act (NEPA) and Fish and Wildlife Coordination Act, USFWS has developed a wetland definition that it uses when classifying habitat types:

> Wetlands are lands transitional between terrestrial and aquatic systems where the water table is usually at or near the surface or the land is covered by shallow water. Wetlands must have one or more of the following attributes:
>
> - At least periodically, the land supports predominantly hydrophytes
> - The substrate is predominantly undrained hydric soil, or
> - The substrate is non-soil and is saturated with water or covered by shallow water at some time during the growing season of each year
>
> Classification of Wetlands and Deepwater Habitats of the United States, Cowardin *et al.* 1979

NRCS Role in Wetland Delineation on Agricultural Land

In an effort to combine USACE CWA wetland mapping with NRCS Food Security Act wetland mapping, the 1995 U.S. Department of Agriculture (USDA) Farm Bill gave NRCS the responsibility of mapping wetlands on agricultural land and USACE the responsibility of mapping wetlands on nonagricultural land, regardless of the intended use of the land.

This new role for NRCS led to the development of a Memorandum of Agreement between NRCS, USACE, EPA, and USFWS requiring each state to develop interagency mapping conventions that would reconcile the differences between the 1987 Manual, the Food Security Act Manual, and the standard mapping practices of each agency. This process led to confusing and cumbersome mapping requirements resulting from the differences between the regulatory intent of the CWA and that of the Food Security Act. Developers seeking Section 404 permits to fill wetlands while converting agricultural land to a nonagricultural use now needed to bring NRCS into an already complex regulatory process.

Consequently, USDA published the 2000 Farm Bill, which indicated that NRCS would be responsible for making wetland determinations on agricultural land only when the intended land use would remain in agriculture. Any intended change from agriculture to another use triggers USACE responsibility for the wetland determination. ■

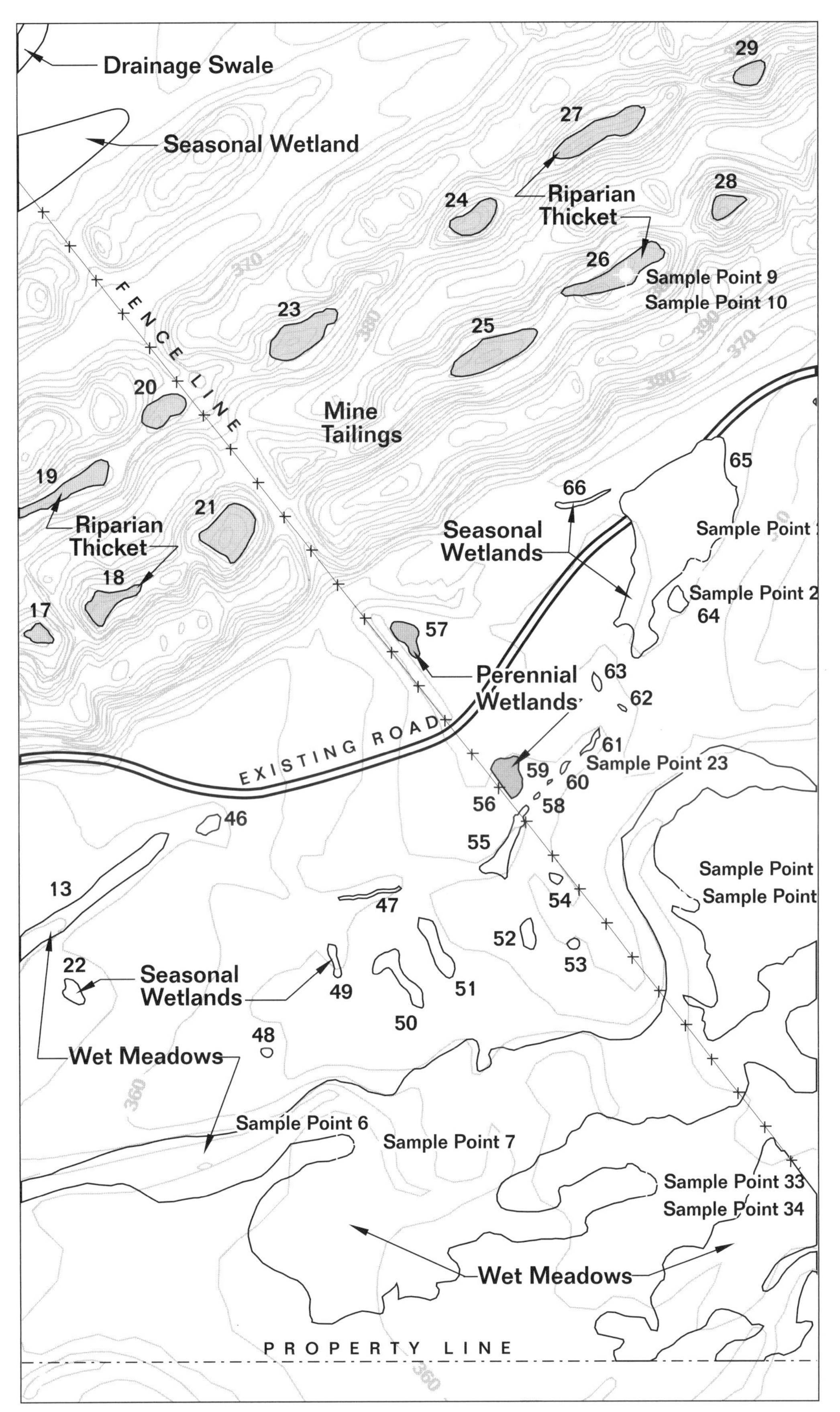

Figure 3-4
Example of a Wetland Delineation Map

Note that topography and cultural features are indicated, wetlands are identified by type and individually numbered, and delineation sample points are shown.

Tidal mudflats are waters of the United States. They are not considered wetlands under USACE's definition because mudflats are not vegetated. Mudflats are special aquatic sites designated in EPA's Section 404(b)(1) Guidelines.

State Regulation of Waters and Wetlands

Chapter 6, State Wetland Laws, presents summaries of state programs that regulate wetlands and other waters. Coastal protection programs and state game and fish regulations are common vehicles for state protection of wetlands and other waters. USACE web sites often have links to other regulatory agencies with jurisdiction in the associated region. When proposing to conduct activities that may affect a wetland in a state with one or more wetland protection programs, it is important to investigate the requirements of each regulatory agency with authority over the proposed activity in the wetland. Each agency is likely to have criteria to determine the extent of its jurisdiction. Jurisdictional boundaries under state regulation may encompass an area greater than USACE jurisdiction under Section 404.

Integrated Method for Identifying Wetland Resources

Rooted floating vegetation, called "vegetation shallows," are special aquatic sites under EPA's Section 404(b)(1) Guidelines.

Identification of wetlands at project sites should employ an integrated approach to address all potential regulatory boundaries and to identify other regulated water bodies and wetland-associated habitats. Delineating federal and state jurisdictional boundaries in a single effort is the most expeditious and cost-effective approach. Although a comprehensive field investigation should identify areas regulated by all agencies with authority over the water body or wetland, separate reports and mapping efforts are usually required by each agency with a regulatory interest.

Using an integrated approach to identifying federal and state jurisdictional boundaries will streamline the analysis of environmental effects, permit processing, and mitigation planning. Local agencies may have wetland policies or ordinances to be considered, and site maps should incorporate appropriate boundaries. Where jurisdictions overlap, a single impact analysis and mitigation plan may be used to resolve the needs of all the regulatory agencies involved. *See* chapter 7 for a discussion of mitigation planning. Mapping all jurisdictional boundaries simultaneously allows project planners to see the extent of regulatory constraints and opportunities for use of a given site.

Mapping all jurisdictional boundaries simultaneously allows project planners to see the extent of regulatory constraints and opportunities for use of a given site.

Although USACE recognizes wetlands as a single jurisdictional entity, many types of wetlands exist (*see* chapter 2, Ecology of Wetlands and Other Waters). Maps should identify each wetland by habitat type. The identification of the type and extent of wetland habitats that could be affected by a project provides valuable information for site planning, permit processing, and mitigation planning.

CHAPTER FOUR

Federal Regulation of Wetlands and Other Waters

The U.S. Army Corps of Engineers is the federal agency with the most involvement in regulating public and private land use activities in wetlands (Figure 4-1). Although USACE's authority focuses on *regulating activities* in waters of the United States (jurisdictional waters), the common perception is that USACE's authority focuses on *protecting wetlands.* However, USACE's authority extends to streams and many other waters besides wetlands (*see* chapter 3, Jurisdictional Limits of Wetlands and Other Waters). In addition, other federal agencies are involved in regulating activities that occur in wetlands and other waters, and in some instances these federal agencies may have more of a role than USACE in the protection of wetlands and other waters.

Farmed wetlands are regulated by NRCS under the Food Security Act and by USACE under Section 404 of the Clean Water Act.

The federal government, through implementation of several federal laws, regulations, and executive orders, regulates activities in jurisdictional waters because of the importance of these waters to water quality, navigation, flood control, and biological function as habitat for specific endangered, threatened, or other sensitive species; they are also regulated as coastal and agricultural resources. In connection with these various functions and uses, the federal government currently attempts to address the protection and use of jurisdictional waters through a number of detailed programs and statutes.

USACE = United States Army Corps of Engineers

This chapter reviews the various federal laws and federal agencies that regulate activities that may directly or indirectly affect wetlands, streams, and other waters of the United States (Table 4-1). As stated in earlier chapters, the terms *wetlands, jurisdictional waters, navigable waters,* and *waters of the United States* have definitions based on a number of criteria relating to hydrology, vegetation, and soils. In the context of federal regulations, these terms are interchangeable in some cases, depending on the applicable law. In other cases, however, they may have distinct and separate meanings and may be regulated differently, depending on how they are

The terms wetlands, jurisdictional waters, navigable waters, *and* waters of the United States *have definitions based on a number of criteria relating to hydrology, vegetation, and soils.*

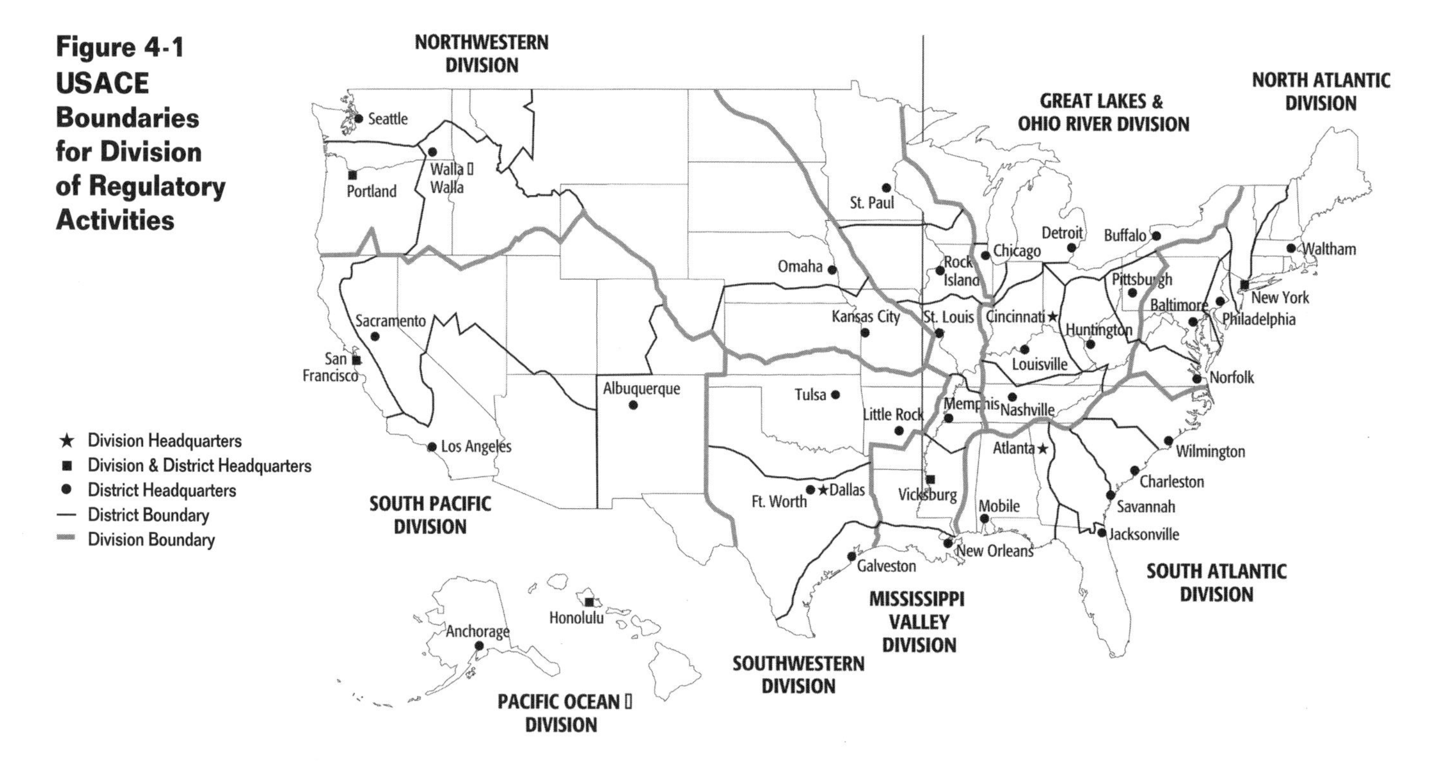

Figure 4-1 USACE Boundaries for Division of Regulatory Activities

defined and how they fit within the purview of federal agency enabling statutes. This chapter also describes the various terms, related regulatory language, and the way that implementation of the laws and regulations differentiates between the various concepts.

CWA = Clean Water Act
NEPA = National Environmental Policy Act
NPDES = National Pollutant Discharge Elimination System

Section 404 of the Clean Water Act (CWA) (33 USC 1251–1387) is the primary statute regulating activities in jurisdictional waters. The primary emphasis of this chapter is on Section 404 of the CWA (*see* appendix A for the full text of Section 404) and the regulation of the "discharge of dredged or fill material into waters of the United States." This chapter also reviews other components of the CWA–specifically Section 401 and the requirement for state water quality certification and Sections 301 and 402 and the issuance of permits under the National Pollutant Discharge Elimination System (NPDES) program. Additionally, this chapter addresses the relevant components of the Rivers and Harbors Act of 1899, the National Environmental Policy Act of 1969 (NEPA), the Fish and Wildlife Coordination Act, the Endangered Species Act, the Coastal Zone Management Act, and the Food Security Act.

History of Federal Wetlands Regulation

Historically, the attitude toward wetlands, at that time more commonly referred to as "swamps," was that they were nuisance areas to be eliminated if possible.

The federal government's concern for wetlands and other waters began in the nineteenth century with attempts to diminish rather than protect them. Historically, the attitude toward wetlands, at that time more commonly referred to as "swamps," was that they were nuisance areas to be eliminated if possible. Through the U.S. Department of Agriculture, the federal

Table 4-1. Federal Agencies That Regulate Activities in Wetlands and Other Waters

Agency	Regulation	Authority
United States Army Corps of Engineers (USACE)	Clean Water Act Section 404	Regulate the placement of dredged or fill material into waters of the United States
	Rivers and Harbors Act of 1899, Section 10	Regulate work in navigable waters of the United States
U.S. Environmental Protection Agency (EPA)	Clean Water Act	Enforcement of regulations, may veto Section 404 permit
	NEPA	Commenting authority
U.S. Fish and Wildlife Service (USFWS)	Fish and Wildlife Coordination Act	Reviews/comments on federal actions that affect surface waters, including wetlands and other waters; includes Section 404 permit applications
	Endangered Species Act	USACE must consult with USFWS if listed species may be affected
	NEPA	Commenting authority
National Marine Fisheries Service (NOAA Fisheries)	Fish and Wildlife Coordination Act	Reviews/comments on federal actions that affect coastal waters, including Section 404 permit applications
	Endangered Species Act	USACE must consult with NOAA Fisheries if listed anadramous fish or marine species may be affected
	NEPA	Commenting authority
Natural Resources Conservation Service (NRCS)	Food Security Act of 1985	Penalizes agricultural producers who plant agricultural crops in wetlands that have been altered

Swamp Lands Act of 1849, 1850, and 1860.

At first glance, it appears incongruous that USACE, a subdivision of the United States military, would ultimately develop the most comprehensive federal regulatory program to assist, at least in appearance, in protecting the ecological value of wetlands and other waters. Historically, USACE's regulatory focus was not on issues affecting jurisdictional waters but on more structural issues related to navigation of waters. It was in the 1960s that USACE first promulgated regulations that could be interpreted as a regulatory mechanism to protect the ecological value of jurisdictional waters. Since the mid-1970s, USACE's authority over certain activities located in jurisdictional waters has developed into the present permit system.

It was in the 1960s that USACE first promulgated regulations that could be interpreted as a regulatory mechanism to protect the ecological value of jurisdictional waters.

USACE is a division of the United States Army that was originally created by Congress to erect and maintain frontier forts and other military defense facilities (Act of March 16, 1802). As it continues today, the mission of USACE's *military program* is to provide engineering, construction, and environmental management services for the Army, Air Force, other assigned U.S. Government agencies, and foreign governments.

USACE, though an arm of the United States military, also has a civil program addressing water infrastructure, environmental management and restoration, and response to natural and anthropogenic disasters.

USACE, though an arm of the United States military, also has a *civil program* with a mission addressing three broad areas: water infrastructure, environmental management and restoration, and response to natural and anthropogenic disasters. USACE's activities in the civil works program focus on navigation, flood risk reduction, environmental management (ecosystem restoration, environmental stewardship, radioactive site cleanup), wetlands and waterways permitting (including the Section 404 regulatory program), real estate (for military and civil works activities of the Army and Air Force and for other federal agencies as requested), recreation, emergency response, research and development, hydroelectric power, shore protection, dam safety, and water supply.

Intermittent stream in spring

The portion of USACE's civil program that regulates activities in jurisdictional waters originated in part with the Act of May 10, 1824, in which Congress authorized USACE to undertake river and harbor improvements promoting navigation (with focus on both naval activities and interstate travel). In 1899, Congress enacted the Rivers and Harbors Act (33 USC 401–413), which further defined USACE's authority related to navigation. The Rivers and Harbors Act established a program (still in existence today) authorizing USACE to issue permits for activities in "navigable waters of the United States." USACE's function under the Rivers and Harbors Act is to protect commerce (through protecting navigability) in these navigable streams and waterways (*see* appendix B for portions of this Act). This authority includes the requirement of obtaining a permit from USACE for the discharge of "any refuse matter of any kind" into any navigable water of the United States (Rivers and Harbors Act, 33 USC 406 section 13).

The extension of USACE jurisdiction to protect jurisdictional waters as an environmental resource began in the 1960s with new interpretations of the Rivers and Harbors Act.

The extension of USACE jurisdiction to protect jurisdictional waters as an environmental resource began in the 1960s with new interpretations of the Rivers and Harbors Act. During the environmental movement in that decade, USACE moved towards using its permit authority under the Rivers and Harbors Act to deny permit applications based on the public interest in fish and wildlife resources. Specifically, USACE began to utilize Section 13 of the Rivers and Harbors Act to prosecute dischargers of pollution (such as industrial waste) for unpermitted discharges into navigable waters of the United States.

FWCA = Fish and Wildlife Coordination Act of 1958

During this same period, the combination of the Rivers and Harbors Act and the Fish and Wildlife Coordination Act of 1958 (FWCA) (16 USC 661, described in detail later in this chapter) raised further issues regarding USACE's regulatory authority over navigable waters as an environmental resource. The FWCA was enacted to ensure that fish and wildlife conservation receives equal consideration and is coordinated with other features of federal agency water resource development programs. Therefore, whenever a federal agency proposes to, or authorizes projects that may, alter or modify any body of water for any purpose, the federal agency must first consult and coordinate its actions and projects with the U.S. Fish and Wildlife Service (USFWS), National Oceanic and Atmospheric Administration (NOAA) National Marine Fisheries Service (NOAA Fisheries), and the affected state game and fish agency.

MOA = Memorandum of Agreement
NOAA = National Oceanic and Atmospheric Administration
USFWS = United States Fish and Wildlife Service

Pursuant to the FWCA, USACE adopted regulations in 1967 for issuance of permits under the Rivers and Harbors Act to include a public interest/fish and wildlife conservation review standard. Also in 1967, the Secretaries of the Army and Interior entered into a Memorandum of Agreement (July 13, 1967) (reprinted in 33 Fed. Reg. 18,672–73 (1968)) in which the Secretary of the Army agreed to consider the Secretary of the Interior's (i.e., USFWS's) views on the merits of proposed permit activities (the "1967 MOA"). Over the next few years, USACE continued to expand its role in regulating waters by relying on the 1967 MOA as a basis for its authority to regulate proposed activities based on ecological effects. The court upheld this expanded public interest review standard under the Rivers and Harbors Act in *Zabel v. Tabb,* 430 F. 2d 199 [5th Cir. 1970]; *cert. denied,* 401 U.S. 910 (1971). In this case, USACE denied a Rivers and Harbors Act permit to fill 11 acres of submerged land in Florida because of fish and wildlife concerns. The federal court upheld the challenge of USACE's denial of the permit, stating that USACE "was entitled, if not required, to consider ecological factors in its permitting process" because of the FWCA's directive.

Additionally, NEPA (described in detail later in this chapter) expanded the scope of USACE's environmental review of proposed Rivers and Harbors Act permit actions. When deliberating the framework for the CWA, Congress noted the conflicting questions raised by the Rivers and Harbors Act, FWCA, and NEPA, regarding the scope of USACE's authority to regulate certain waters as an environmental resource. After this period of uncertainty about the proper scope of USACE's authority to review a project's environmental effects, Congress passed the CWA in 1972. Over time, this act, specifically Section 404 of the CWA, has resolved these questions about USACE jurisdiction over jurisdictional waters and has proven to be the leading regulatory authority of the federal government to oversee activities in wetlands, streams, and other waters.

NEPA expanded the scope of USACE's environmental review of proposed Rivers and Harbors Act permit actions.

Section 404 of the Clean Water Act

The CWA is primarily focused on water quality issues, the large majority of which is unrelated to USACE's authority to regulate the discharge of dredged or fill material under Section 404. The CWA generally authorizes the U.S. Environmental Protection Agency (EPA) to regulate water quality through the regulation of discharges of pollution. For example, as discussed later in this chapter, Sections 301 and 402 of the CWA prohibit the discharge

EPA = United States Environmental Protection Agency

NPDES = National Pollutant Discharge Elimination System
RCRA = Resource Conservation and Recovery Act

of all pollution into waters of the United States unless permitted under the National Pollutant Discharge Elimination System (NPDES), either by EPA or through an EPA-approved state program. 33 USC 1311, 1342. The Congressional intent of the CWA makes it clear that USACE does not have the authority to prosecute dischargers of pollution under the Rivers and Harbors Act (or CWA) and that only the EPA is authorized to prosecute dischargers of pollution.

The authority of EPA under the CWA to regulate all "pollution" discharge activities affecting the "waters of the United States" is somewhat modified by Section 404 (as will be described below, federal courts have noted that Section 404 regulates only "clean discharge activities" and not pollution discharges). Although Section 404 identifies USACE as the principal authority to regulate discharges of dredged or fill material into waters of the United States (33 USC 1344), Congress has authorized EPA to have a specific oversight role over USACE's authority. The following section presents the framework of Section 404 and, along with the discussion in chapter 3, the extension of USACE's CWA jurisdiction to wetlands as well as other waters of the United States. The USACE permit process under Section 404 is discussed in chapter 5.

Although Section 404 identifies USACE as the principal authority to regulate discharges of dredged or fill material into waters of the United States, Congress has authorized EPA to have a specific oversight role over USACE's authority.

Urban encroachment and channelization of a stream

Overview of Section 404

Section 404 of the CWA (*see* appendix A for the full text) requires that private, state, and federal entities (other than USACE-implemented military and civil programs) obtain a permit from USACE before discharging dredged or fill materials into waters of the United States, which may include wetlands. *See* chapter 3 for definitions of "waters of the United States" and "wetlands."

Although Section 404 of the CWA gives USACE direct authority over proposed discharges of dredged or fill material into waters of the United States, EPA plays a significant role in developing regulations for the USACE Section 404 program and in reviewing the issuance of permits by USACE. Beyond the 404(b)(1) process and 404(c) veto authority (described in chapter 5), EPA has promulgated regulations governing implementation of Section 404 that parallel USACE regulations.

Discharge of Dredged or Fill Material

Section 404 of the CWA prohibits the discharge of dredged or fill material into the waters of the United States without a permit from USACE.

Section 404 of the CWA prohibits the discharge of dredged or fill material into the waters of the United States without a permit from USACE. The following section presents the definition of "discharge of dredged or fill material" and examples of what is considered a "discharge" activity subject to regulation under Section 404 (Figure 4-2).

Clean Water Act Definitions. In general, the CWA prohibits the discharge of any pollutant without a permit. "Discharge" of a pollutant is defined as the

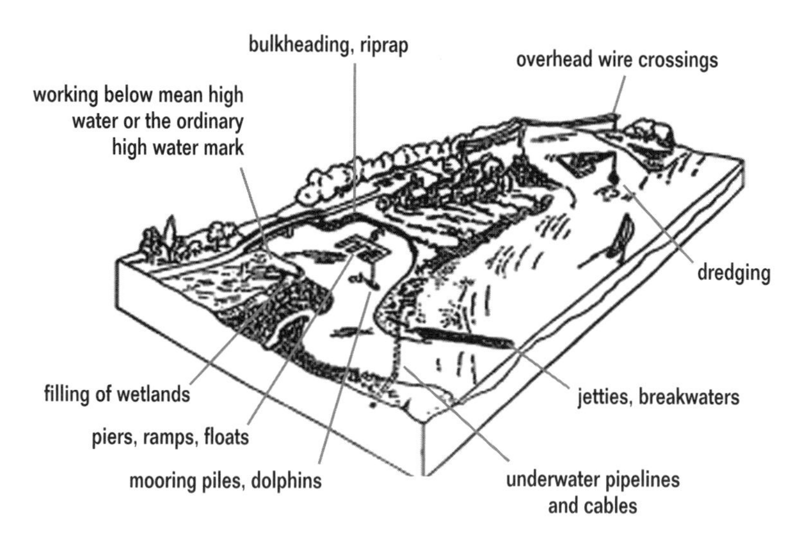

Figure 4-2
Activities Requiring a USACE Permit Under Section 404 of the Clean Water Act and Section 10 of the River and Harbors Act

addition of any pollutant to waters of the United States from any point source (33 USC 1362). Section 301 of the CWA prohibits all discharges of pollutants from a point source; Section 402 authorizes EPA (and state approved programs) to issue discharge permits.

Implicit in the CWA definition of "pollutant" is the inclusion of dredged and fill material regulated by Section 404 (33 USC 1362). *See,* for example, *U.S. v. Pozsgai,* 999 F. 2d 719 (3d Cir. 1993), that held dredged spoil, rock, and sand are pollutants pursuant to the CWA. Section 404 regulates the discharge of dredged or fill material into waters of the United States. The discharge of "dredged material" typically means adding into waters of the United States materials that were removed from waters of the United States. The discharge of "fill material" typically means adding into waters of the United States materials (such as concrete, dirt, rock, pilings, or side cast material) that are for the purpose or have the effect of either replacing an aquatic area with dry land or raising the elevation of an aquatic area. USACE has interpreted the term "discharge" broadly to include not only the direct placement of materials into wetlands, but also the secondary effects from what is typically considered a dredging or removal activity that may involve only very small discharges of dredged or fill material.

Examples of Regulated Activities. Typical activities regulated as a discharge under Section 404 include the use of heavy

Siting Landfills in Jurisdictional Waters May Not Be Regulated Under Section 404

The placement of garbage or landfill may not be considered dredged or fill material under Section 404. In 1998, the Ninth Circuit was asked to review USACE's authority over the siting of a landfill for solid waste disposal in jurisdictional waters in Washington state (*Resource Investments, Inc. v. U.S. Army Corps of Engineers,* 151 F. 3d 1162 (9th Cir. 1998)). The court reviewed authority vested in USACE under Section 404 to regulate discharges of dredged or fill material, as well as the authority vested in EPA under the CWA and Resource Conservation and Recovery Act (RCRA) to regulate the discharge of other wastes such as those involved in landfills. The court reviewed the definitions of dredged and fill material under CWA in relation to the definition of regulated fill under RCRA. The court interpreted Section 404 as basically regulating the discharge of *clean fill material* and that RCRA regulated this type of landfill material. The court noted also that EPA had issued separate regulations for wetlands protection under RCRA, similar to those under the CWA. Therefore, the court held that EPA (or the state with an EPA-approved program) was the sole entity to regulate the location of a landfill in jurisdictional waters. Therefore, USACE does not have regulatory authority over the placement of material into landfills.

It should be noted that a regulatory clarification issued after this case by USACE stated that fill activities associated with the landfill, such as placement of liners and other infrastructure in a landfill prior to accepting landfill material, are considered a discharge regulated under Section 404 (67 FR 31129–31143). Also, other facilities outside the landfill (e.g., access roads, operations buildings) may trigger the need for a section 404 permit if located in jurisdictional waters. As discussed separately, the excavation of an area to create a pit for the location of the landfill may not be regulated under Section 404 if the material excavated is removed and placed in an upland location and no fill activities occur beyond incidental fallback. ■

History of the Excavation Rule

The interpretation of the term "discharge" as it relates to Section 404 has a long history. Prior to 1993, the USACE regulation of discharge activities was typically limited to those actions where there was a clear placement of fill material. Certain excavation activities that occurred in jurisdictional waters that may have altered the soil and hydrology of the waters were not consistently regulated. In August 1993, as part of the settlement agreement to a case brought by an environmental group, USACE clarified its position regarding "discharges" (*North Carolina Wildlife Federation v. Tulloch* [E.D. N.C. 1992, Civil No. C90-713-CIV-5-BO. 58 Federal Register 45008, August 25, 1993; 33 CFR 323.2(d)(1)]). The August 1993 regulation, known as the "excavation rule" or "Tulloch rule," clarified that incidental redeposition of material (such as soil and rock) associated with mechanized land clearing, ditching, channelization, and other excavation activities that destroy or degrade wetlands or other waters of the United States *are considered discharges* of dredged or fill material and are therefore subject to Section 404 regulation. Excluded from regulation are *de minimis* discharges (considered incidental soil movement) that have only minimal environmental effects.

However, the burden was placed on the proposed discharger to prove that the effects of the discharge will be minimal. Certain activities remained as nondischarge activities, including certain types of land clearing (where the activity involved only the cutting or removal of vegetation above the ground, where the root system is not substantially disturbed or where no mechanized pushing, dragging, or similar activities that redeposit excavated materials are involved; a discharge by a wheeled or tracked vehicle while cutting vegetation off ➡

construction equipment such as bulldozers to place rock, sediment, organic debris, and other fill material into jurisdictional waters. Other activities such as covering, leveling, or grading sites (and erosion from construction sites) where jurisdictional waters occur are also considered discharge activities. *The Avoyelles Sportsmen's League, Inc. v. Marsh,* 715 F. 2d 897 (5th Cir. 1986), held that backhoes and bulldozers used to clear forest land fall within the CWA's definition of a "point source." As such, depositing pollutants in association with earthmoving activities in waters of the United States, requires a Section 404 permit.

Tractors conducting deep ripping have been held to be a point source that results in the discharge of pollutants when conducted in wetlands. *Borden Ranch Partnership v. USACE,* 261 F. 3d 810 (9th Cir. 2001). In *Borden* the court reasoned that deep ripping removes soil from the wetlands and transforms it into "dredged spoil." The court also rejected the application of the farming exemption because deep ripping radically alters the hydrological regime. *Borden* was denied *writ of certiorari* by the United States Supreme Court; therefore, the Ninth Circuit decision remains the controlling decision on this issue.

In addition to the common types of discharges, the following are some examples of activities that have been upheld as constituting a "discharge of pollutants" (and therefore a "discharge of dredged or fill material"):

- Redeposit of spoil dredged by boat propellers
- Land leveling either with earth and vegetative debris or by mechanized activities
- Removal of vegetation
- Nonestablished farming activities
- Deep ripping

Excavation and Incidental Fallback. It has long been questioned whether dredging activities fit under the definition of "discharge." Such excavation activities in jurisdictional waters have certainly had a deleterious effect on the function and value of these waters, and yet they have not been consistently regulated under Section 404. Dredging is not an activity that is explicitly regulated by the CWA. USACE's current policy, based by certain court directives (*see* sidebar), states that the regulated activity of a "discharge" includes excavation where the amount of dredged material redeposited in the jurisdictional waters is more than "incidental fallback." Incidental fallback is defined as the redeposit of small volumes of dredged material that is incidental to the excavation activity when such material falls back to substantially the same place as the initial removal. Therefore, unless project information shows that the activity will result in only incidental fallback, the use of mechanized earth-moving equipment to conduct landclearing, ditching, channelization, in-stream mining, or other excavation activities within jurisdictional waters is considered a discharge of dredged material requiring a permit from USACE under Section 404.

Draining. Draining activities within jurisdictional waters are often deleterious to the function and value, and yet are typically not regulated by USACE under Section 404. If the draining activity can be performed in a way where there is no discharge of dredged or fill material, it is not considered a discharge activity and therefore not regulated under Section 404. However, USACE will typically examine these drainage activities closely to determine whether there is some associated fill (e.g., pipeline or some kind of soil movement to create drainage slopes). It should also be noted that these drained areas may still qualify as jurisdictional waters, even though the hydrology no longer exists, because it is not considered "normal circumstances" (*see* April 10, 1990 Memorandum to All Division and District Counsel from Lance Wood, Assistant Chief Counsel in appendix G). In *Save Our Community v. U.S. Environmental Protection Agency,* 971 F. 2d 1155 (5th Cir. 1992), the court held that drainage of a wetland without evidence of a discharge of a pollutant is not within the jurisdiction of Section 404.

Pilings. The placement of pilings is considered a discharge where it would have the effect of a discharge (e.g., where pilings are so closely spaced to increase sedimentation rates or where they would reduce the reach or impair the flow of circulation of waters of the United States). The placement of pilings is not considered a discharge where it would not have the effect of the placement of fill (e.g., linear projects such as bridges, elevated walkways, and powerline structures). However, these activities may still require a permit under the Rivers and Harbors Act. 33 CFR 323.3(c).

Exemptions. Certain discharges of dredged or fill material into jurisdictional waters are specifically exempt from Section 404 permit requirements under congressional directive. In these cases, no permit is needed and USACE will not condition the action prior to the discharge activity. While some mistakenly consider the Section 404(e) general permit process, including Nationwide permits (*see* chapter 5), an exemption to Section 404, these are actually Section 404 permits issued for certain types of activities, and have conditions that require compliance (including in some instances notification to USACE) prior to being eligible for the permit and proceeding with the discharge activity.

above ground may, however, be sufficient to be considered a discharge activity regulated by Section 404 (33 CFR 323.2[d][2]).

As part of the *Tulloch* settlement agreement, USACE also clarified that the placement of pilings to construct structures in waters of the United States was subject to Section 404 regulation when that placement has the effect of a discharge of fill material (58 FR 45008, August 25, 1993; 33 CFR 323.3(c)). Examples include projects where the placement of pilings are so close that sedimentation rates increase or the bottom of the water body is effectively replaced. Applying this standard, linear projects constructed on pilings—such as bridges, piers, and power line towers—generally will not have the physical effect of fill material and will not be regulated under Section 404. This determination s, however, made by USACE and EPA, on a case-by-case basis.

The USACE current policy was shaped by litigation on the Tulloch rule. In 1998, the D.C. Circuit was asked to review the excavation rule as to whether Congress intended Section 404 to apply to certain types of dredging activities. In *National Mining Congress v. Army Corps of Engineers*, 145 F. 3d 1399 (D.C. Cir. 1998), the court concluded that the CWA regulates only the addition of pollutants, not the removal from jurisdictional waters. Therefore, the court held that Section 404 cannot be interpreted as regulating activities where material is removed from waters of the United States and only a small portion of it happens to fall back into the jurisdictional waters. The USACE current rule regarding these excavation activities stems from the court's note that Section 404 could apply to dredging activities where the redeposit of dredged material falls back in an area different from where it was dredged or when the dredged material is sidecast onto jurisdictional waters. ■

Court Interpretations of Section 404(f) Exemptions

Normal silviculture, farming, and ranching activities have been narrowly construed by the courts to allow exemptions from the Section 404 permitting process only when they are part of "ongoing, normal operations." Activities cannot be related to preparing the property for new uses separate from ongoing silviculture or agriculture operations (*Borden Ranch Partnership v. U.S. Army Corps of Engineers*, 261 F. 3d 810 (9th Cir. 2001), *cert. granted*, 122 S. Ct. 2355 (2002)). Therefore, the exemption does not apply to converting wetlands to new farmland or grazed land to farmland but does cover conventional rotation of fields, which may leave an area lying fallow for a season. ■

RGL = **Regulatory guidance letter**

Exemptions Under Section 404(f). The only outright exemptions to the USACE permit requirements contained in Section 404 for discharges are listed in Section 404(f)(1) (*see also* May 3, 1990 Memorandum for the Field, "Clean Water Act Section 404 Regulatory Program and Agricultural Activities," RGL No. 86-3, 87-7, 87-9 in appendix G). The following activities are exempted from regulation under Section 404:

- Normal farming, silviculture (timber or forestry), and ranching and may include plowing, minor draining, and harvesting. This exemption applies only to "normal" types of activities that are considered "continuing as of December 23, 1985." Therefore, if land supporting jurisdictional waters is not currently in use for agricultural, silvicultural, or ranching purposes, the conversion to agriculture, silviculture, or ranching is not exempt because it is not considered a continuing use.
- Constructing and maintaining stock or farm ponds and irrigation ditches, or maintaining (but not construction of) drainage ditches
- Constructing or maintaining farm, forest, or mining roads; or temporary roads for moving mining equipment
- Maintaining or reconstructing structures that are currently serviceable (this includes emergency repair of recently damaged, currently serviceable structures)
- Constructing temporary sedimentation basins for construction
- Activities regulated by an approved best management practices program authorized by Section 208(b)(4) of the CWA

A "recapture" of regulatory authority over these exempt activities is provided in Section 404(f)(2). The recapture provision states that the exemptions under Section 404(f)(1) do not apply if the activity would violate toxic effluent standards or if the activity has the purpose of bringing an area within jurisdictional waters into a use to "which it was not previously subject" and would constitute a new use impairing the flow, circulation, or reach of waters. Although USACE initially determines whether the activities fall under Section 404(f), EPA has ultimate administrative authority over the interpretation. The change in use from normal agricultural activities to other uses has been frequently used by USACE to invoke the recapture clause for the normal farming activities exemption. Therefore, when there is a proposal to change a current agricultural use in jurisdictional waters to a different use, such as residential development, that change in use, and the discharge of dredged or fill material associated with it, is not exempt from the requirements of Section 404.

Exemptions Under Section 404(r). Section 404(r) of the CWA exempts a narrow class of federal activities from Section 404 permit requirements. The discharge activities under this exemption must be part of the specific, congressionally authorized, federal construction project. The exemption must be explicitly identified in the legislation. Furthermore, before authorization and funding, the federal agency must submit to Congress an environmental impact statement (EIS), prepared pursuant to the National Environmental Policy Act, that includes consideration of the EPA's Section 404(b)(1) Guidelines.

EIS = **Environmental impact statement**

Other Federal Regulatory Authority Protecting Wetlands

Federal laws other than the Rivers and Harbors Act of 1899 and Section 404 of the CWA also specifically or indirectly regulate activities occurring in jurisdictional waters.

Chapter 5 describes those laws with which USACE is required to comply when issuing a permit under Section 404. However, federal laws, other than the Rivers and Harbors Act of 1899 and Section 404 of the CWA, specifically or indirectly regulate activities occurring in jurisdictional waters. A project proponent should prepare a strategy (*see* chapter 5) for compliance with all environmental regulations that may be necessary for activities occurring in wetlands, streams, and other waters before submitting a permit application to USACE. If properly addressed, these other federal laws should not prevent project implementation. However, compliance may make finalizing the project description, including mitigation requirements, more difficult and may delay the date the project can begin. In addition, other federal agencies may have different interpretations of methods of compliance with those laws that also are triggered by Section 404 (i.e., NEPA, Section 7 of ESA, Section 106 of NHPA, CZMA, and the Wild and Scenic Rivers Act). For example, even if USACE does not require an EIS for NEPA compliance for issuing the Section 404 permit, an EIS may be required by another federal agency's compliance with NEPA for the same action.

The cement lining of flood-control channels improves flow for flood control purposes, but provides no substrate on which wetland plants can root. In addition, the cement side wall deprives riparian habitats of water.

Rivers and Harbors Act of 1899

The Rivers and Harbors Act generally authorizes USACE to protect commerce in navigable streams and waterways, with three sections addressing permit authority:

- Section 9 requires a permit from USACE for the construction of a dike or dam in navigable waters of the United States.
- Section 10 requires a permit from USACE for any obstruction to "the navigable capacity of any waters of the United States" (including the placement of piers or the activities of dredging, filling, and other construction activities).
- Section 13 requires a permit from USACE for the discharge of "any refuse matter of any kind" into any navigable water of the United States.

USACE jurisdiction under the Rivers and Harbors Act of 1899 is limited to those activities affecting the navigable waters of the United States.

USACE jurisdiction under the Rivers and Harbors Act of 1899 (33 USC 401–413) is limited to those activities affecting the navigable waters of the United States (Figure 3-3). According to USACE 's current regulations implementing the Rivers and Harbors Act, navigable waters of the United States are defined as those waters subject to the ebb and flow of the tide shoreward to the mean high water mark and/or those that are presently used, have been used in the past, or may be susceptible to use to transport interstate or foreign commerce (33 CFR 329.4). *See* chapter 3 for a detailed discussion of the boundaries of USACE's jurisdiction under the Rivers and Harbors Act.

To a great extent the regulatory authority of USACE under the Rivers and Harbors Act has been superseded by the CWA. The jurisdiction of USACE under the CWA overlaps and extends beyond the geographic scope of its jurisdiction under the Rivers and Harbors Act; therefore, when discussing USACE regulatory authority over wetlands, the Rivers and Harbors Act of 1899 will generally be of little consequence, separate from Section 404 jurisdiction. In fact, an applicant uses one form (the Department of the Army Permit) to apply for authorization under both laws.

When discussing USACE regulatory authority over wetlands, the Rivers and Harbors Act of 1899 will generally be of little consequence, separate from Section 404 jurisdiction.

USACE's jurisdiction under the Rivers and Harbors Act differs from Section 404 in the following ways:

Fill activity in riparian scrub habitat

- **Areas regulated.** "Navigable waters" as regulated under the Rivers and Harbors Act is typically considered a subset of "waters of the United States" as regulated by the Clean Water Act. However, navigable waters under the Rivers and Harbors Act is defined to include those areas that "have been used in the past, or may be susceptible to use to transport interstate or foreign commerce." *See* 43 USC 333(e). Therefore, where the reach of historically navigable waters has been reduced by human activity, e.g., levees or dikes, the historically navigable areas may be considered jurisdictional under the Rivers and Harbors Act, even though not considered jurisdictional waters under the Clean Water Act.
- **Activities regulated.** Section 10 of the Rivers and Harbors Act regulates the construction, excavation, or deposition of materials in, over, or under navigable waters, or any work that would affect the course, location, condition, or capacity of those waters. This includes structures (e.g., piers, wharfs, breakwaters, bulkheads, jetties, weirs, transmission lines) and work such as dredging or disposal of dredged material, or excavation, filling, or other modifications to the navigable waters of the United States. However, Section 10 also regulates activities not considered a "discharge of dredged or fill material" and therefore not regulated by Section 404 of the CWA. These Section 10-only regulated activities in navigable waters include dredging activities (even where there is only incidental fallback), directional drilling underneath the waterway (even where there is no fill activities to enter or exit the drilling area), and overhead line crossings (even where there is no fill activities to place structures to hold up the overhead lines.)

Section 10-only regulated activities in navigable waters include dredging activities, directional drilling underneath the waterway, and overhead line crossings.

- **No exemptions.** The Section 404(f) exemptions (*see* above) that apply to certain discharge activities in waters of the United States for purposes of regulation under Section 404 of the CWA do not apply under Section 10 of the Rivers and Harbors Act if those activities occur in navigable waters (as opposed to other waters considered jurisdictional waters under the CWA). Therefore, if certain normal agricultural activities would include the discharge of fill material into navigable waters, a Section 10 permit would be required, even though no Section 404 permit is required.

- **No EPA oversight.** The USACE permit authority under the Rivers and Harbors Act is not subject to EPA oversight or any other restrictions specific to the CWA (e.g., compliance with the Section 404(b)(1) Guidelines). In addition, EPA has no veto authority over permits issued under Section 10. Therefore, in those cases, such as with certain exemptions under the CWA, the Rivers and Harbors Act alone will apply to activities occurring in navigable waters.

Clean Water Act Sections 301 and 402

The CWA, which provided a sweeping revision of earlier water pollution control laws, calls for restoring and maintaining the chemical, physical, and biological integrity of the nation's waters. Its principal goal is to address the problems of water pollution through the National Pollutant Discharge Elimination System. Section 301 prohibits the discharge of any pollutant without a permit, and Section 402 sets up the permit program administered by the EPA (or a state program approved by EPA) (*see* appendix A for text of CWA Sections 301 and 402). Although a project may not involve the deposit of dredged or fill material into waters of the United States, the Sections 301 and 402 provisions of the CWA may still apply and regulate aspects of a project–such as wastewater discharges during construction–that could provide some protection for wetlands. By generally regulating water pollution, the CWA provisions of Sections 301 and 402, beyond Section 404, have the effect of regulating activities that may affect jurisdictional waters.

The CWA, which provided a sweeping revision of earlier water pollution control laws, calls for restoring and maintaining the chemical, physical, and biological integrity of the nation's waters.

Clean Water Act Section 303

Section 303(d) of the Clean Water Act requires that states make a list of waters that are not attaining standards after the technology-based limits are put in place (also known as "impaired waters"). For waters on this list (and where the EPA administrator deems they are appropriate), Section 303 of the Clean Water Act requires states to develop total maximum daily loads (TMDLs). EPA is required to review and approve the list of impaired waters and each TMDL. If EPA cannot approve the list or a TMDL, states are required to establish one.

TMDL = Total maximum daily load

A TMDL must account for all sources of the pollutants that caused the water to be listed. Federal regulations require that the TMDL, at a

Providing controlled public access to protected wetlands and other waters is a key component to effective conservation.

minimum, account for contributions from point sources (federally regulated discharges) and contributions from nonpoint sources. Contrary to the acronym, the limitations contained in a TMDL may be other than "daily load" limits. There also can be multiple TMDLs on a particular water body, or there can be one TMDL that addresses numerous pollutants. The basis for grouping is whether or not there can be a common analytical approach to the assessment or a common management response to the impairment. Technical issues and the number of combinable pollutants affect the exact number of TMDLs that will be necessary to address the state's water quality problems. Again by regulating water quality within a particular stream, the provisions of Section 303 have the effect of regulating certain activities that may affect wetlands, streams, and other waters.

Clean Water Act Section 401

Section 401 of the CWA requires that all applicants for a federal license or permit that may result in any discharge into jurisdictional waters provide the federal agency with a certification from the state in which the discharge originates stating that the discharge will comply with the state's water quality plan (where the plan has been approved by EPA) (*see* appendix A for text of CWA Section 401). Where there is no state-approved water quality plan, EPA issues the certification. The federal agency is prohibited from issuing the license or permit until the certification has been obtained.

The majority of Section 401 certifications occur with Section 404 permits. However, all federal agencies are subject to this requirement where the action will involve discharges into jurisdictional waters. By generally regulating water pollution and compliance with state water quality standards, the CWA provisions of Section 401, beyond Section 404, have the effect of regulating activities that may affect jurisdictional waters (*see* discussion of Section 401 compliance in chapter 5, Section 404 Permitting Process).

By generally regulating water pollution and compliance with state water quality standards, the CWA provisions of Section 401, beyond Section 404, have the effect of regulating activities that may affect jurisdictional waters.

National Environmental Policy Act

NEPA and its implementing regulations direct all federal agencies to assess the effects of their actions on the "human environment." Action is broadly defined to include whenever a federal agency undertakes, funds (in whole or in part), or issues a permit grant, entitlement, or otherwise approves a new or continuing activity, including nonfederal projects. NEPA's directive is considered to be the basic national charter for environmental protection. NEPA analysis includes assessing the effects of a proposed project on a particular resource, mitigation measures available to reduce that effect, and alternatives to meeting the action's purpose and need. Federal agencies are also required to consider the cumulative effects of the planned action and other reasonably foreseeable projects.

NEPA analysis includes assessing the effects of a proposed project on a particular resource, mitigation measures available to reduce that effect, and alternatives to meeting the action's purpose and need.

NEPA compliance is set up in a phased process. First, the federal agency determines that the action is subject to NEPA and that there are no agency defined categorical exclusions that apply. Second, the agency determines if the action would have a significant effect on quality of the human environment through preparation of an environmental assessment (EA). The third step is the preparation of either a finding of no significant impact (FONSI) when the action (or modified action with mitigation incorporated) would not cause significant effects, or the preparation of an environmental impact statement (EIS). While NEPA does not specifically require their protection, certain jurisdictional waters are mentioned as a resource to be evaluated when determining whether a project will have a significant effect on the human environment. The Council on Environmental Quality's regulations to implement NEPA specifically state that an impact's significance is based on its "intensity," which may include the action's effect to the "unique characteristics of the geographical area, such as... wetlands, wild and scenic rivers, or ecologically critical areas" (40 CFR 1508.27).

EA = Environmental assessment
EIS = Environmental impact statement
FONSI = Finding of no significant impact

During the NEPA process, where USACE's authority is triggered under Section 404, the other federal agencies typically defer to USACE regarding the impact analysis and mitigation requirements for impacts to jurisdictional waters. In instances where a USACE permit is not required for activities adversely affecting wetlands–that is, no discharge of dredged or fill material is involved but there are impacts due to dredging or indirect activities–NEPA will require the federal agency to assess the project's effect on sensitive resources, such as certain jurisdictional waters. NEPA does not provide for uniform wetland protection, but does require the federal agency to evaluate a project's effects on wetlands, including its effects on wildlife and plant species of concern located in wetlands–and, where appropriate, recommend mitigation to reduce the effects of these impacts. When reviewing NEPA documents, USFWS and EPA typically focus on wetland impacts, indicating that mitigation should be provided when they are affected adversely. Also, the NEPA process has its own public interest review criteria that allow the public to review and comment on the method of analysis and the conclusions made in the environmental documentation, including assessment of impacts on wetlands.

During the NEPA process, where USACE's authority is triggered under Section 404, the other federal agencies typically defer to USACE regarding the impact analysis and mitigation requirements for impacts to jurisdictional waters.

Endangered Species Act

The purpose of the federal Endangered Species Act is "to provide a means whereby the ecosystems upon which endangered species and threatened species depend may be conserved." (16 USC 1531.) The Act establishes an official listing process for plants and animals considered to be in danger of extinction, establishes an official listing process for critical habitat necessary for the survival and recovery of those listed species, requires development of specific plans of action for the recovery of listed species, restricts activities perceived to harm or kill listed species or adversely affect habitat, and requires federal agencies to ensure that their actions do not jeopardize the continued existence of listed species. 16 USC 1532, 1536.

Jurisdictional waters, although a sensitive resource, are not in and of themselves subject to listing under the Endangered Species Act, but can be designated as critical habitat.

Jurisdictional waters, although a sensitive resource, are not in and of themselves subject to listing under the Endangered Species Act, but can be designated as critical habitat (e.g., many rivers and streams along the Pacific Coast are designated critical habitat for salmon and steelhead). The designation of critical habitat requires federal agencies to assess whether a proposed action would adversely modify or destroy critical habitat. In addition, because wetlands and other jurisdictional waters can provide habitat for species of threatened or endangered plants and wildlife, these areas may be protected through regulation of activities considered to "take" the species (where the habitat modification may lead to death or injury to the listed species). The Endangered Species Act may, therefore, prohibit projects from adversely affecting wetlands and other jurisdictional waters considered habitat for listed species, without consultation with or permission from USFWS or NOAA Fisheries.

Found mainly in the Northeast and Northwest, bogs are characterized by permanently ponded, acidic peat soils.

Fish and Wildlife Coordination Act

The Fish and Wildlife Coordination Act requires that all federal agencies consult with USFWS, NOAA Fisheries, and the state's wildlife agency for activities that affect, control, or modify waters of any stream or other bodies of water. Under the authority of the Fish and Wildlife Coordination Act, USFWS and NOAA Fisheries review applications for permits issued under Section 404, as well as other federal actions seen to modify waters, and prepare a coordination act report to document the coordination of issues between the federal agency and resource agencies.

Considerable overlap exists between the environmental review requirements contained in NEPA, the Endangered Species Act, and the Fish and Wildlife Coordination Act.

Considerable overlap exists between the environmental review requirements contained in NEPA, the Endangered Species Act, and the Fish and Wildlife Coordination Act. Through the Fish and Wildlife Coordination Act, USFWS and NOAA Fisheries have an expanded responsibility for project review that includes concerns about general plant and wildlife species that may not be addressed by NEPA and the Endangered Species Act. In particular, this expanded responsibility may include a project's secondary effects on wetlands and other jurisdictional waters.

Food Security Act of 1985

The Food Security Act of 1985 contains a provision for penalizing agricultural producers who plant commodities on wetlands that have been drained, filled, or otherwise altered (*see* chapter 3 for definition of wetland under the Food Security Act). Commonly known as "Swampbuster," this provision, which is administered by the U.S. Department of Agriculture through the National Resources Conservation Service (NRCS), removes the producer's eligibility for all government price and income support programs for violation of the Act. Rather than having their federal subsidies completely withheld, farmers who inadvertently drain wetlands may be fined $750–$10,000, and the landowner, to qualify to pay the fine, must restore the converted wetland.

NRCS = National Resources Conservation Service

As amended by the 1990 Farm Act, the Food Security Act of 1985 also authorizes a voluntary program for farmers to reduce water pollution from agricultural practices. Those reducing pesticide, fertilizer, and other pollutant drainage may get federal cost-sharing assistance. The Act created a Conservation Reserve Program that pays landowners to take highly erodible land out of crop production and to plant vegetation that controls soil erosion and helps wildlife, including vegetation typical of certain types of wetlands. The Act also created the Wetlands Reserve Program, which seeks out farmed and converted wetlands critical for migratory bird and other wildlife habitat for wetland conservation easements. The areas within the wetland conservation easement plan, which must be approved by both NRCS and USFWS, carry specific use prohibitions and allowances intended to further restore and protect the wetlands.

As amended by the 1990 Farm Act, the Food Security Act of 1985 also authorizes a voluntary program for farmers to reduce water pollution from agricultural practices.

Coastal Zone Management Act

The Coastal Zone Management Act of 1972 (CZMA) creates a broad program of land use management based on control by each coastal state. These state coastal management programs focus on land uses within the coastal zone and sensitive resources that co-exist with the different land uses. Typically states use permit programs established through approved coastal management programs to restrict certain uses within the coastal zone. The CZMA also requires all applicants for federal permits and federal agency project sponsors to obtain proof of certification from the coastal state that the proposed action is consistent with the state's approved coastal program (*see* chapter 6 for a list of state agencies that administer state coastal programs).

The Coastal Zone Management Act of 1972 creates a broad program of land use management based on control by each coastal state.

CZMA = Coastal Zone Management Act

After determining whether a project falls within the designated coastal zone, the project proponent should design it to be compatible with the state coastal act and the local coastal resources management plan. If a permit is required by a state coastal agency, it may have special conditions attached, particularly if the project could affect an environmentally sensitive habitat, such as a wetland. In addition, for certain federal actions within the coastal zone, project applicants must demonstrate that their project is consistent with the state's approved coastal management program before a federal agency can issue a permit or award funding.

Wild and Scenic Rivers Act

Any federal activity that could affect the characteristics of a designated wild and scenic river must coordinate with the federal agency responsible for that particular river to determine the effect on the river.

The National Wild and Scenic Rivers Act establishes a system to designate wild and scenic rivers within federally managed lands. Any federal activity that could affect the characteristics of a designated wild and scenic river must coordinate with the federal agency responsible for that particular river (U.S. Bureau of Land Management, National Park Service, USFWS, or U.S. Forest Service) to determine the effect on the river.

The federal agency responsible for management of the river will issue a letter of consent if the proposed action would not have a direct or indirect effect on the resources for which the river was designated under the Wild and Scenic Rivers Act. If the consent is denied, the federal agency may recommend alternatives that would not affect the river. In this way, the Wild and Scenic Rivers Act adds an additional method of regulating activities within jurisdictional waters.

Executive Orders

Executive Order 11990 requires federal agencies to follow avoidance/mitigation/preservation procedures with public input, before proposing new construction in wetlands.

Executive Order 11990: Protection of Wetlands (1977). This executive order provides an overall wetlands policy for all agencies managing federal lands, sponsoring federal projects, or providing federal funds to state or local projects. Executive Order 11990 requires federal agencies to follow avoidance/mitigation/preservation procedures with public input, before proposing new construction in wetlands. When federal lands are proposed for lease or sale to nonfederal parties, Executive Order 11990 requires that restrictions be placed in the lease or conveyance to protect and enhance the wetlands on the property. Although it does not apply to federal discretionary authority for nonfederal projects (other than funding) on nonfederal land, Executive Order 11990 has the effect of restricting the sale of federal lands with wetlands. Also, its restrictions apply to wetlands on property within military installations proposed for closure and on land taken over through the federal bail-out program for failed mortgage companies.

Executive Order 11988 requires that all federal agencies take action to reduce the risk of flood loss, to restore and preserve the natural and beneficial values served by flood plains, and to minimize the impact of floods on human safety, health, and welfare.

Executive Order 11988: Flood Plain Management (1977). This executive order is a flood hazard policy for federal agencies. Executive Order 11988 requires that all federal agencies take action to reduce the risk of flood loss, to restore and preserve the natural and beneficial values served by flood plains, and to minimize the impact of floods on human safety, health, and welfare. An agency's action must reflect consideration of alternatives to avoid adverse impacts in flood plains. Where location in a flood plain is unavoidable, the federal agency must modify the action to minimize the effects of a project. Because many wetlands are located in flood plains, Executive Order 11988 has the secondary effect of protecting wetlands.

CHAPTER FIVE

Section 404 Permitting Process

This chapter describes the various types of Section 404 permits, summarizes the required process for each, recommends a variety of approaches, and suggests avenues for challenging permit decisions. Section 404 of the Clean Water Act (CWA) prohibits the unauthorized discharge of dredged or fill material into wetlands and other waters of the United States. The Section 404 permit requirement applies to all private, state, and federal entities other than USACE. These permits are generally referred to as *404*, *wetlands*, or *fill* permits. However, the official title of a USACE-issued permit under Section 404 of the CWA is a *Department of the Army* or *DA* permit. Compliance with Section 10 of the Rivers and Harbors Act of 1899 may be combined with any of the permits described below.

CWA = Clean Water Act
DA = Department of the Army
EPA = Environmental Protection Agency
IP = Individual permit
LOP = Letter of permission
NWP = Nationwide permit
PGP = Programmatic general permit
RGP = Regional general permit
USACE = United States Army Corps of Engineers

In most states, USACE's permit process under the CWA currently applies to all activities that are regulated under Section 404. This process can change, however, where a state assumes the authority for administering a portion or all of the Section 404 program. *See* chapter 6, State Wetland Laws, for a discussion of state wetland regulations. Under Section 404(g), qualified states may apply to EPA to assume administration of the Section 404 program, except for navigable waters.

USACE issues two broad categories of permits under Section 404 to authorize the discharge of dredged or fill material into waters of the United States, including wetlands. Standard permits include individual permits and, where applicable, letters of permission (LOPs). General permits include nationwide permits (NWPs), regional general permits (RGPs), and programmatic general permits (PGPs).

Flood control channel. Channelized waterways typically support little or no wetland habitat because the banks are too steep and maintenance is often conducted to remove vegetation.

A standard permit for a specific activity may be issued only after an individual application is submitted and the formal review process is complete (Figure 5-1). Standard permits are issued either as individual permits (sometimes referred to as IPs) or as LOPs. A particular district of USACE

Figure 5-1
Section 404
Individual Permit
Process Flow Chart

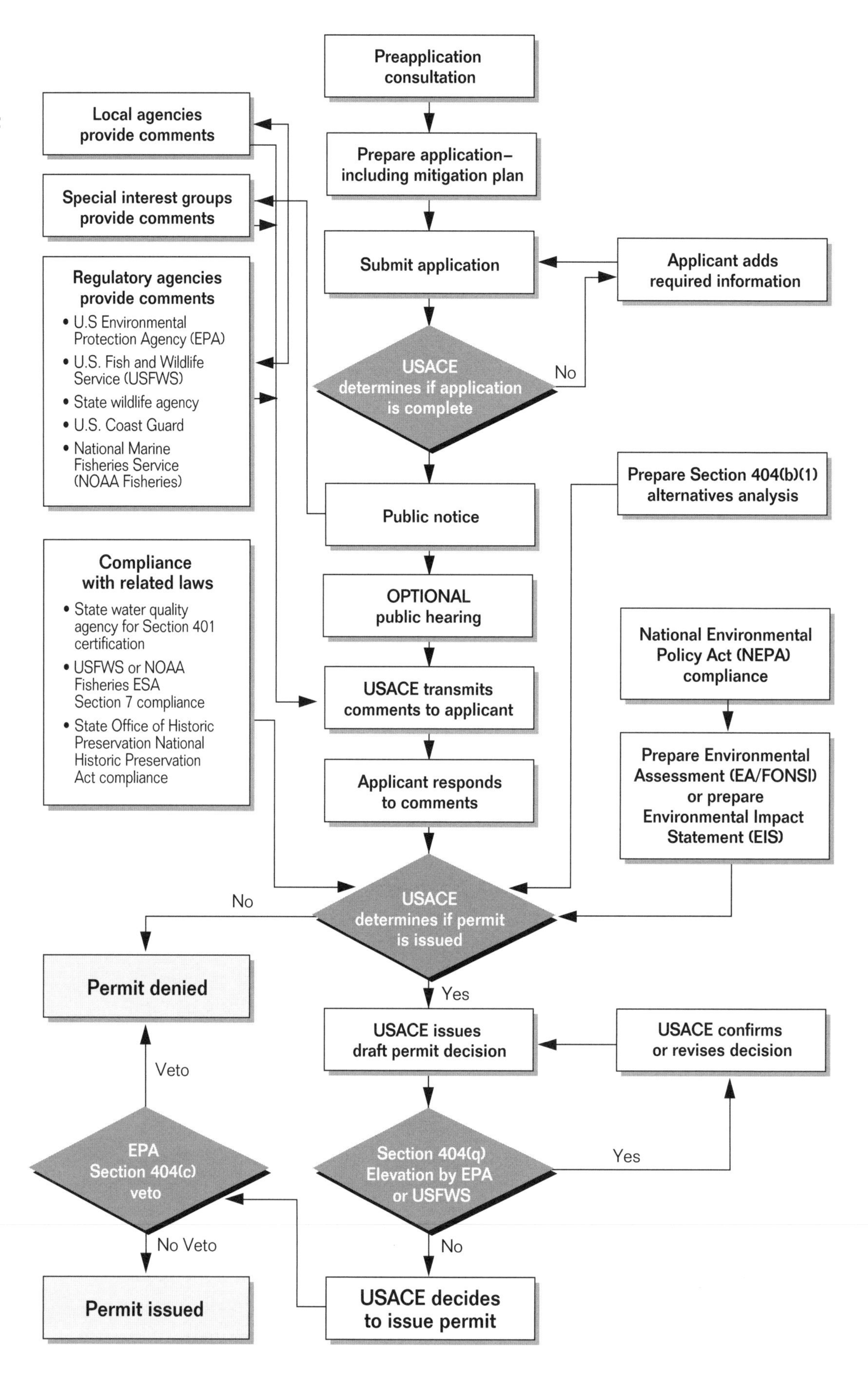

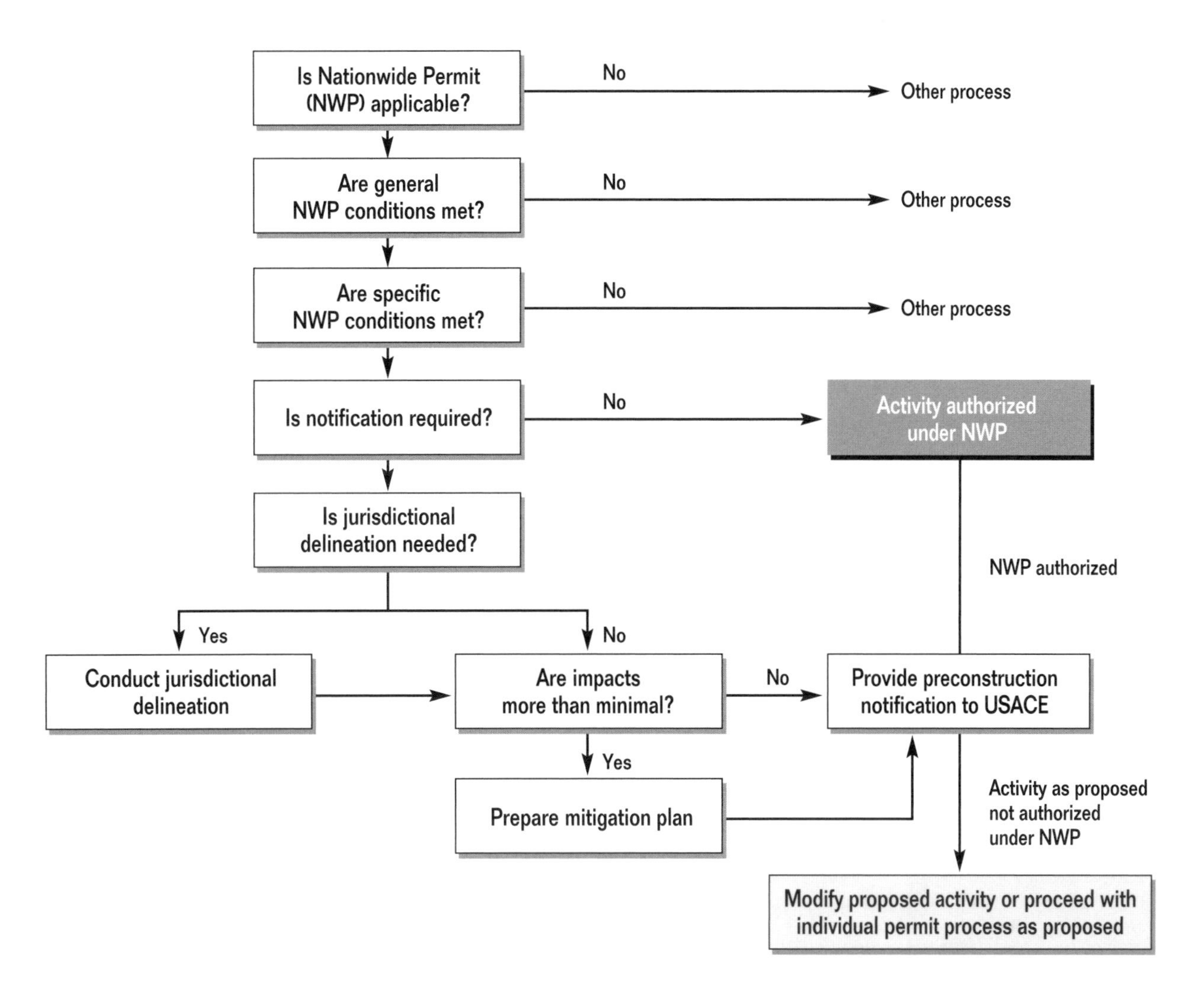

Figure 5-2
Section 404 Nationwide Permit Process Flow Chart

may adopt LOPs to facilitate a more streamlined process and are available for certain types of projects that have extensive coordination and support from resource agencies (*see* below).

Section 404(e) provides USACE with the ability to greatly reduce processing time by issuing "prior-authorized" general permits for specified categories of activities on a state, regional, or national basis. A general permit, which may have notification requirements, is issued prior to a request to authorize a specific activity. It is intended to authorize activities that USACE has determined would have minimal adverse impacts on aquatic resources, individually or cumulatively, and that would require minimal time, if any, for USACE review. NWPs are one category of general permit with a specific process determined by USACE (Figure 5-2).

As discussed below, activities authorized under either a standard or general permit require compliance with USACE's Section 404 regulations, EPA's Section 404(b)(1) Guidelines, NEPA, the federal Endangered Species Act, Section 106 of the National Historic Preservation Act, Section 401 of the CWA, and the Coastal Zone Management Act. Actions authorized under a general permit, however, do not require separate NEPA review and 404(b)(1)

NEPA = National Environmental Policy Act

compliance, because USACE prepares a NEPA document and conducts a 404(b)(1) Guidelines compliance review when the general permit is issued.

Although USACE's permit procedures are specified under both USACE and EPA Section 404 regulations (33 CFR 323, 40 CFR 230) (appendices C and D), the interpretations of various permit process compliance issues are sometimes subjective and dependent on the particular USACE district and USACE permit manager. USACE is divided into 38 districts, each administered by a district engineer (Figure 4-1). The district with permit authority over an action is determined by the location of the activity in relation to district boundaries. If an activity is on the boundary between two districts or occurs over a large area falling within two districts' boundaries, both districts should be notified to determine which will have authority (typically one district will assume lead permit authority over the project).

The USACE permit process under Section 404 of the CWA applies to all discharges of dredged or fill material into waters of the United States.

As will be discussed throughout this chapter, the USACE permit process under Section 404 of the CWA applies to all discharges of dredged or fill material into waters of the United States (*see* chapter 4 for exemptions). As mentioned, discharges for USACE-specific projects follow a different procedure that is outlined in this chapter. Although USACE does not literally issue permits to itself (that is, the regulatory branch does not issue permits for USACE Civil Works projects), USACE must comply with the same laws that apply to all other applicants for Section 404 permits.

Debris piles in a floodplain that previously supported riparian forest, with remnant riparian forest in the background

Jurisdictional Determination

A Section 404 permit is required only if the proposed activity would involve the discharge of dredged or fill material into waters of the United States. If a proposed activity will affect wetlands or other waters, the first step in the process is to determine the boundaries of jurisdictional waters of the United States, including wetlands, on the project site (*see* chapter 3, Jurisdictional Limits of Wetlands and Other Waters). The project proponent should obtain a verified jurisdictional determination from USACE before filing the application for a 404 permit. Once the amount of fill and the location and acreage of wetlands or other waters affected are known, the project proponent may be able to revise the project design such that the project can qualify for a general permit; design modification can even obviate a Section 404 permit altogether (by eliminating all fill of jurisdictional waters).

USACE is ultimately responsible for the accuracy of the jurisdictional wetland delineation. However, because of USACE staffing and funding limitations, the prospective applicant (or a consultant) normally prepares the delineation report documenting the Section 404 jurisdictional boundaries. The report, which typically includes a narrative portion, data forms, map, and site photographs, is then submitted to the appropriate USACE regulatory office (*see* appendix H) with a cover letter requesting USACE to verify that the enclosed documentation accurately depicts the boundaries of jurisdictional waters of the United States on the project site. As part of this verification process, the USACE regulatory staff often visits the project site to confirm the accuracy of the delineation. After revisions, if any, requested by USACE are incorporated into the delineation, USACE sends the project proponent a verification letter confirming the jurisdictional boundaries on the project site. This verification letter is valid for five years. The applicant may appeal the jurisdictional boundary determination (*see* administrative appeals later in this chapter).

Effects of Discharge Activities

Any activity that involves placement (or discharge) of dredged or fill material, unless considered "incidental fallback" during dredging activities (discussed in chapter 4), into wetlands or other waters of the United States requires Section 404 authorization from USACE before that dredged or fill material is placed into waters. Actions that have only minor direct, indirect, and cumulative environmental effects may be covered by a general permit (e.g., a nationwide permit) that has already been issued by USACE and that satisfies the proposed activity's Section 404 compliance requirement.

The EPA Section 404(b)(1) Guidelines (appendix D) and the 1990 Memorandum of Agreement (1990 MOA) between the Department of the Army and EPA on Mitigation Required under the 404(b)(1) Guidelines (appendix G) require that a project proponent, when entering the market for a suitable project site, assess alternative available sites to determine on which site the project could be completed with the least amount of adverse impacts on wetlands and other waters of the United States. In practice, however, the analysis of alternative sites is usually conducted after a project site has been chosen and rarely at the time of market entry, putting the project proponent at risk of having the permit application denied.

The Section 404(b)(1) Guidelines and the 1990 MOA require that projects should avoid or minimize adverse effects on jurisdictional waters of the United States. According to the MOA, the proper sequence of mitigation priority in project design is to:

- First, avoid adverse effects on jurisdictional waters of the United States
- Second, if avoiding adverse effects is not practicable, minimize effects on jurisdictional waters of the United States to the extent practicable, and
- Third, compensate for those impacts on jurisdictional waters of the United States that are unavoidable

If the discharge activities cannot avoid the jurisdictional areas, the project proponent should, according to the Section 404(b)(1) Guidelines and the 1990 MOA, strive to minimize the amount of acreage affected within the jurisdictional boundaries. In practice, performing this sequencing with regard to avoiding and minimizing impacts to waters of the United States could allow the project to qualify for authorization under a general permit, such as an

Indirect Impacts on Wetlands and USACE Authority

An irrigation district sought to build a dam and reservoir on a tributary to the South Platte River in Colorado. USACE determined that the deposit of fill material associated with construction of the dam did not meet the conditions required for a nationwide permit because the increased use of water facilitated by the reservoir would deplete the streamflow and adversely affect the critical habitat of the whooping crane, an endangered species. While the fill itself would not affect the whooping crane, the dam facility USACE would have to permit as necessary for dam operations would affect the species.

The circuit court upheld USACE's denial of the use of an NWP, finding that the CWA and USACE regulations "authorize the Corps to consider downstream effects of changes in water quantity in determining whether a proposed discharge qualified for a[n] NWP." The court stated that the fact that the reduction in water does not result from direct federal action does not lessen USACE's duty. USACE was "required to consider all effects, direct and indirect, of the discharge for which authorization was sought." *Riverside Irrigation District v. Andrews*, 758 F. 2d 508 (10th Cir. 1985). ■

MOA = Memorandum of agreement

NWP, with a reduced permitting process (*see* discussion of general permits below). Also, reduction in the amount of jurisdictional acreage affected will reduce the amount of mitigation that USACE requires.

Preapplication Meeting

NOAA = National Oceanographic and Atmospheric Administration
USFWS = United States Fish and Wildlife Service

Although not required, the regulations provide for a preapplication meeting with staff at USACE, EPA, and USFWS (and, as appropriate, NOAA Fisheries), state fish and wildlife agency, other relevant state resource agencies, and local and regional agencies with authority over land use at the project location). This meeting is especially recommended for larger projects to allow resource agencies to review a proposed project at a relatively early stage. At this meeting the project proponent may receive suggestions for modifications to the project design or mitigation features that, if incorporated into the final project design, might help expedite the permit process. Local USACE regulatory staff may be contacted at any time to answer questions about USACE jurisdiction over a project site or proposed activity (*see* appendix H for addresses and telephone numbers of USACE offices).

Before requesting a preapplication meeting, a project proponent should have an accurate wetland delineation and be fairly certain that a permit will be required for the proposed project.

Before requesting a preapplication meeting, a project proponent should have an accurate (preferably verified) wetland delineation and be fairly certain that a permit will be required for the proposed project. It is suggested that the proponent provide a summary of the project to USACE, EPA, USFWS, and other appropriate resource agencies before the meeting so that they can be familiar with basic information about the project before the meeting. The project proponent or authorized agent is usually asked to attend the preapplication meeting to describe the project (but not provide a "sales pitch" for it), to answer questions, and to hear any concerns or recommendations the agencies might have, including issues regarding alternatives and mitigation requirements. If possible, the proponent should try to address as many of these issues as possible before submitting the permit application to USACE.

Standard Permits

After preparing a wetland delineation, reviewing the regulations, and consulting with USACE, the project proponent can determine whether the proposed activity will require a standard permit, a general permit, or no permit at all (Figures 5-1 and 5-2). USACE may, under its own independent review, determine that the proposal requires an individual permit, even if the proposal, on its face, appears to meet the criteria for a general permit.

Individual Permits

If a standard permit is required, a separate decision must be made regarding whether the project would qualify for a Letter of Permission, if required, or if it must go through the Individual Permit process. IPs require submission of an IP application (appendix J, Department of the Army Application) and compliance with USACE's formal review process. This process provides opportunities for public notice and comment, requires preparation of an alternatives analysis as required by EPA Section 404(b)(1) Guidelines and NEPA, and requires compliance with NEPA's environmental review process. In general, USACE's decision to issue an IP is based on an evaluation of probable impacts of the proposed activity, analyzed according to the Section 404(b)(1) Guidelines, and the effect the proposed activity will have on the public interest.

IPs may be issued before or, in certain limited circumstances, after a discharge occurs. Permits issued after an unauthorized discharge has occurred, called "after-the-fact" permits, allow some or all of the fill to remain in place provided certain conditions are followed. If USACE determines through the IP process that an after-the-fact permit should not be issued, it will require that the unpermitted fill be removed. Performing fill activities prior to application to USACE can lead to both civil and criminal penalties (*see* discussion on enforcement below). Therefore, it is recommended that a project proponent never perform a fill activity in reliance of the expectation that USACE will issue an after-the-fact permit.

Timing of Overall Permit Process. After USACE determines that the submitted permit application is complete (*see* permit application requirements below), the minimum time necessary for USACE to process an IP is 60 days. Although USACE states that its staff strives to process most IPs within 120 days, six to twelve months is common in some districts. Processing a complicated IP (one involving a complex project, public controversy, impacts on threatened or endangered species, or significant environmental impacts) typically takes more than a year (or, in some instances, more than two years).

Permit Application Requirements. The application for a Department of the Army permit to discharge dredged or fill material into waters of the United States must include:

- Name, address, and telephone number of the project applicant and its authorized agent
- Project name, name of affected water body, location of project, and directions to project site
- Project description (all reasonably related activities should be included in the same permit application)
- Purpose of and need for the proposed project
- Purpose of the activities involving the discharge of dredged or fill material, type and quantity of material to be discharged, and the surface area, in acres, of wetlands or other waters filled
- Drawings, sketches, or plans sufficient for public notice (usually all submittals must be 8.5 by 11 inches with a one-inch border; detailed engineering drawings are typically not required)
- Names and addresses of owners of all adjoining properties

Recommended Content of an Individual Permit Application

- Cover letter summarizing materials in the application
- Completed application (ENG Form 4345), with additional sheets to complete longer descriptions for some information blocks (appendix J)
- Project maps and graphics (black and white, 8.5 by 11 inches) that USACE can use in the public notice
- Mailing list for USACE to use in distributing the public notice
- Wetland delineation report and map and a copy of USACE delineation verification letter
- Section 404(b)(1) alternatives analysis report
- Mitigation and monitoring plan
- Biological assessment to comply with Endangered Species Act (if necessary)
- Cultural resources inventory report to comply with National Historic Preservation Act (if necessary)
- Federal consistency determination to comply with Coastal Management Act (if necessary)

- Locations and dimensions of adjacent structures
- Authorizations required by other federal, interstate, state, or local agencies for the work (including dates of all approvals or denials already received), and
- Additional specific information for activities involving the construction of structures for certain improvements, such as compliance with dam safety criteria

JARPA = Joint aquatic resource permits application

Requirements for an IP application are specified by regulation (33 CFR 325), but the application form itself may vary among districts (some USACE districts have developed joint applications, sometimes called joint aquatic resource permits applications (JARPAs), with state or local agencies). Therefore, it is best to obtain the application form from either the local USACE office or from the local USACE district web site.

Although not required prior to USACE's issuance of a public notice for the proposed project, USACE may ask an applicant to submit additional information during the permit review process. This information typically includes a compensatory mitigation plan and a report demonstrating compliance with the Section 404 (b)(1) Guidelines (*see* below). Including these documents with the permit application package will contribute to a more informative public notice, which may result in the receipt of more substantive comments on the proposal. Mitigation plans may be conceptual or detailed at this stage in the permit review process; typically a detailed mitigation plan will be required prior to issuance of the IP (*see* chapter 7 for more detail on mitigation planning).

As of November 1, 2003, fees for USACE permits are $10 for noncommercial projects and $100 for commercial or industrial projects. Federal, state, and local government agencies pay no fees. Fees are collected at the time the IP is issued, not at the time of application.

Within 15 days of receiving a permit application, the district must either notify the applicant of any additional information necessary to make the application complete, or prepare and distribute a public notice.

Public Notice/Hearings. USACE regulations require that, within 15 days of receiving a permit application, the district must either notify the applicant of any additional information necessary to make the application complete, or prepare and distribute a public notice. As discussed in the Permit Application Requirements section above, a complete application includes enough information to describe the project and its impacts, but may not include all information needed to reach a permit decision, such as the mitigation plan or 404(b)(1) alternatives analysis.

Although each USACE district uses a slightly different format for its public notices, the notices all contain the basic information required for the IP application. Typically, the public notice contains sufficient project information to elicit comments on the issuance of a permit. For large or complex projects where other supporting documents may facilitate public comment, the public notice may refer interested parties to the location (e.g., district office, a public library, or web page) where such documents may be reviewed. The public notice also includes contact information for the

USACE district's project manager and/or the project proponent or agent, should someone wish to request further details.

The public notice is sent to the district's standard public notice mailing list (including EPA, USFWS, NOAA Fisheries, other interested federal and state agencies, and Indian tribes), those listed on the application as adjoining property owners, and any others requesting notice. Many USACE districts also publish their public notices on their web sites (*see* appendix H, USACE Districts Contact Information). After issuing a public notice, USACE must allow a minimum of 15 days, but usually allows 30 days, for written comments to be submitted. If the circumstances justify extension, USACE can grant an additional 30 days for comments. Extensions to the comment period are most often granted when reviewing agencies make the request and offer justification.

After issuing a public notice, USACE must allow a minimum of 15 days, but usually allows 30 days, for written comments to be submitted. If the circumstances justify extension, USACE can grant an additional 30 days for comments.

When responding to a public notice, written comments should always identify the subject activity by the project title and application number provided in the public notice. In order to assist USACE in identifying issues and potential solutions, comments should be as specific to the project as possible.

Desert ephemeral drainages or washes usually do not support wetlands, but are often determined to be waters of the United States by USACE.

Written comments may include a request for a public hearing. Although the regulations do not say when to hold a public hearing, and the receipt of a request for a hearing does not necessarily mandate that one be held, USACE will usually grant requests for hearings when sufficient demonstration of general public interest is given. The propensity to hold a hearing varies among district offices. Typically, USACE holds a public hearing when it believes that written comments have not elicited the full extent of issues. At least 30 days prior to the hearing, USACE issues a public notice of the date, time, and place, along with any additional information available pertaining to the permit application. USACE may also decide to hold a public hearing upon receipt of an application for a project expected to be controversial or to generate a large public response. In such cases, information regarding the time and location of the public hearing will be included in the initial public notice.

Public hearings are not considered formal legal proceedings. Although a presiding officer and a stenographer are typically present, these hearings are not considered judicial proceedings requiring the swearing-in of witnesses. USACE's presiding officer may ask questions of the project applicant or those submitting comments; however, cross-examination by adversarial parties is not permitted. Where possible, USACE public hearings can be combined with other required public hearings, such as those required for compliance with NEPA or state environmental laws. Statements made at the public hearing are transcribed and become part of the administrative record.

USACE must consider all comments submitted in response to the public notice and at the public hearing. USACE transmits the comment letters and comments submitted at the public hearing to the project applicant, who is then given an opportunity to respond.

USACE must consider all comments submitted in response to the public notice and at the public hearing.

Substantive Standards for USACE's Decision. The substantive requirements for the decision by USACE to issue a permit under Section 404 of the CWA are found in EPA's Section 404(b)(1) Guidelines and USACE's public interest review regulations. Issuance or denial of the permit is governed by these standards; moreover, compliance with NEPA, Section 7 of the Endangered Species Act, Section 401 of the Clean Water Act, Section 106 of the National Historic Preservation Act, the Coastal Zone Management Act, and the Wild and Scenic Rivers Act (discussed later in this chapter) may also affect the decision to issue the permit and the specific conditions listed in the permit.

The Section 404 regulations authorize USACE to add conditions to a permit to mitigate impacts on wetlands and other waters. In addition, the NEPA process requires that USACE, when issuing a permit, evaluate and disclose the effects of the permit decision and the alternatives on the environment. This process, and compliance with several other laws described below, may also involve the imposition of other conditions that, when applied to the project, would ensure that environmental impacts are mitigated. USACE typically relies on the opinions of USFWS, NOAA Fisheries, the state fish and wildlife agency, and the state water quality agency to assess the adequacy of the proposed mitigation.

EA	= Environmental assessment
EIS	= Environmental impact statement
LEDPA	= Least environmentally damaging practicable alternative
ROD	= Record of decision

USACE's analysis under EPA guidelines is usually performed simultaneously with the public review process, with an applicant typically supplying much of the information necessary for USACE to document its findings. The district must prepare a statement of findings or, when an environmental impact statement (EIS) is prepared under NEPA, a record of decision (ROD) on whether to issue or deny the permit. The findings or ROD must include a statement of facts, an analysis indicating conformity with the Section 404(b)(1) Guidelines, an environmental assessment (EA) or EIS (as required by NEPA), and the views of the district about how the project will affect the public interest.

EPA's Section 404(b)(1) Guidelines. Section 404(b) of the CWA directs EPA to issue regulations to guide USACE's issuance of permits under Section 404 (appendix D). Compliance with the Section 404(b)(1) Guidelines is qualitative and involves exercise of USACE's judgment in applying the Guidelines to a particular permit application. In general, the Guidelines state that USACE cannot issue a permit if there is a practicable alternative (also known as the *least environmentally damaging practicable alternative*, or LEDPA) to the proposed discharge that would have less adverse impact on the aquatic ecosystem (so long as that alternative does not have other significant adverse environmental consequences). If a project cannot comply with the Section 404(b)(1) Guidelines, USACE must deny the permit. In addition, even if USACE finds that a project complies with the Guidelines and issues a permit, EPA may independently veto that permit decision under Section 404(c) of the CWA, because of noncompliance with the Section 404(b)(1) Guidelines (*see* below).

If a project cannot comply with the Section 404(b)(1) Guidelines, USACE must deny the permit.

The Guidelines also afford particular protection to special aquatic sites, which include wetlands (*see* chapter 4, Jurisdictional Limits of Wetlands and Other Waters). The Guidelines state that when a proposed activity requiring a discharge into a special aquatic site is not water dependent, it is presumed that other practicable alternatives are available that do not involve the discharge into special aquatic sites, unless it can be clearly demonstrated otherwise (40 CFR 230.10(a)). Water dependency is based on evidence that the proposed action could not meet its basic project purpose (as justified by the "purpose and need" of the action) unless located in a special aquatic site; this would include activities directly associated with and reliant on waters of the United States. Examples of water-dependent activities may include marinas and logging activities where the logs are transported by waterway.

Water dependency is based on evidence that the proposed action could not meet its basic project purpose unless located in a special aquatic site.

The applicant must show that no other practicable alternatives exist that would have fewer adverse effects on the aquatic ecosystem. Depending on the complexity of the project and the number of alternatives considered, a separate 404(b)(1) alternatives analysis document is often prepared and submitted with or subsequent to the permit application. Conducting an alternatives analysis subsequent to permit application can be problematic. In *Bersani v. U.S. Environmental Protection Agency,* 850 F. 2d 36 (2d Cir. 1988), the court held that available alternatives were to be measured at the time the applicant entered the market. In this particular case, property that was available at the time the applicant entered the market but was subsequently purchased by a competitor was determined to be a practical (available) alternative despite no longer being on the market.

The project purpose is an important component of the alternatives analysis. The project purpose should be specific, but should not be so narrowly defined as to eliminate any other alternative location or design. Typically, the applicant must extensively justify the need for and purpose of the proposed project, as well as the selection of the project location as compared to offsite and onsite alternatives. While generally focusing on the project applicant's statement of purpose, USACE will typically exercise independent judgment in defining the purpose and need for the project from both the applicant's and the public's perspective. An example of this can be found in *Shoreline Association v. Marsh,* 555 F. Supp. 169 (D. Md. 1983). The court upheld the USACE's determination that an application describing a project as water dependent was actually primarily a townhouse project with boat storage facilities and launch being incidental. This resulted in the identification of upland alternatives with fewer impacts on wetlands.

The project purpose should be specific, but should not be so narrowly defined as to eliminate any other alternative location or design.

The practicability of alternatives is a function of financial, technical, and logistical factors, including availability of the site to the project proponent (which might be measured from the time the applicant entered the market to pursue site locations). A property not presently owned by the applicant that could be reasonably obtained, utilized, expanded, or managed in order to fulfill the basic purpose of the proposed action may be considered a

Wildlife viewing in riparian forest

practicable alternative by USACE and EPA. *See*, for example, *Hough v. Marsh*, 557 F. Supp. 74 (D. Mass. 1982) (holding letter from a realtor is not sufficient evidence to support a finding of no alternatives).

If cost is a factor in determining practicable alternatives, the applicant may be required to provide justification of financial commitments and expenditures. The consideration focuses on whether the projected cost for a particular alternative is substantially greater than the costs normally associated with the particular type of project. An illustrative case that dealt with practical alternatives is *James City County, Va. v. U.S. Environmental Protection Agency*, 955 F. 2d 254 (4th Cir. 1992). In *James City* the county was attempting to construct a drinking water project to provide for the demands of its growing population. USACE had determined that no practicable environmentally superior alternatives existed; however, the EPA disagree and exercised its veto authority. The EPA concluded that viable alternatives existed, and on appeal the court overturned the EPA's determination based on the alternatives involving increased costs being prohibited or not supported by applicable authorities, being experimental, and that it would not provide a sufficient quantity of water. Cost analyses often include proprietary information that a project proponent may not want included in the USACE public record. Although a request may be made to USACE to keep this information out of the public file, it may be subject to disclosure under the Freedom of Information Act (FOIA). Therefore, applicants who do not wish to disclose information regarding financial commitments and investment-backed interests are encouraged to use other reasons for determining the practicability of an alternative where possible.

FOIA = Freedom of Information Act

The following are guidelines from court decisions with regard to practicable alternatives.

- A letter from a realtor claiming that the site was the only one suitable for a project to build two houses and a tennis court is not a sufficient basis to conclude that no alternatives were available (*Hough v. Marsh*, 557 F. Supp. 74 (D. Mass. 1982)).
- The fact that other sites would offer a less attractive marketing package to purchasers was sufficient to conclude that feasible alternatives were not available (in *National Audubon Society v. Hartz Mountain Development*, 14 Env H. L. Rep. (Envtl. L. Inst.) 20724 (D. N.J. 1983)).
- Developer's upland property was determined to be the least environmentally damaging alternative to building a housing development in wetlands area. USACE rejected the claim of water dependency because the townhouse segment of the development would have included a boat storage facility (*Shoreline Associates v. Marsh*, 555 F. Supp. 169 (D. Md. 1983)).
- USACE guidelines impose an obligation on USACE to take the permit applicant's objectives into account when considering whether practicable alternatives exist for the proposed project (*Louisiana Wildlife Federation, Inc. v. York*, 761 F. 2d 1044 (5th Cir. 1985)).

- The fact that the project (a log storage and export facility) was water dependent allowed the preclusion of alternatives because they were too costly or logistically infeasible (*Friends of the Earth v. Hintz,* 800 F. 2d 822 (9th Cir. 1986)).
- The Federal Power Act does not preempt the application of the CWA with respect to assessing alternatives to the licensing of a hydroelectric power project. The court stated that Section 404 transmits a "crisp and unwavering message: all significant discharges, whether or not exempt from the permit requirement, must be subjected to Section 404(b)(1) scrutiny." *Monongahela Power Company v. Marsh,* 809 F. 2d 41 (D.C. Cir. 1987); *cert. denied,* 484 U.S. 816 (1987).
- The significant additional cost of an alternative may not by itself eliminate it from consideration as a practicable alternative (*Friends of the Earth v. Hall,* 693 F. Supp. 904 (W.D. Wash. 1988)).
- Availability of an alternative is not determined at the time of application, but is determined at least by the time the applicant begins the site selection process, also referred to as *market entry.* *See* sidebar, page 73.
- The proposed purpose of supplying two separate water districts from single water source was too narrowly defined and unnecessarily precluded alternatives (*Simmons v. United States Army Corps of Engineers,* 120 F. 3d 664 (7th Cir. 1997)).

Beyond proving that the proposed project is the least environmentally damaging practicable alternative, the Section 404(b)(1) Guidelines further restrict permitting of discharge activities that violate water quality or toxic effluent standards, jeopardize the continued existence of a species listed under the federal Endangered Species Act, or violate marine sanctuary protections (40 CFR 230.10(b)). The Section 404(b)(1) Guidelines also require certain findings regarding the immediate physical impacts caused by the fill activity. The Guidelines prohibit discharges that will "cause or contribute to significant degradation of the waters of the United States." USACE findings should include information about the effect of the fill on the water bottom, water flow and circulation, turbidity, the aquatic

Recommended Steps for Structuring an Alternatives Analysis in Compliance with Section 404(b)(1) Guidelines

1 Conduct preliminary planning

- Develop a basic purpose and need for action.
- Document this need with information from agency planning or other studies if relevant.
- Develop a range of available sites and possibly nonstructural alternatives that could meet purpose and need.

2 Make preliminary jurisdictional determination

- Determine the constraints of available sites regarding impacts to waters of the United States, paying particular attention to affecting special aquatic sites.
- Determine other environmental constraints of the project site (e.g., endangered species and cultural resource effects).

3 Develop screening criteria for the practicability of alternatives

- Define the screening criteria that will be used to evaluate the practicability of an alternative.
- Document reasons for cost, availability, technological, logistical, and other feasibility constraints.

4 Conduct "staged" screening process according to Section 404(b)(1) Guidelines (integrated with NEPA)

- Determine broad range of alternatives (onsite, offsite, and nonstructural) that could meet purpose and need.
- Apply screening criteria previously developed.
- At first stage of screening process, eliminate those alternatives that clearly do not meet the purpose and need or other screening criteria.
- At second stage of screening process, document detailed evaluation of the alternatives ability to meet the screening criteria (possibly in the draft EIS for NEPA compliance).
- At third stage of screening process, evaluate the practicable alternatives and environmental effects, and determine the least environmentally damaging practicable alternative (possibly in the final EIS for NEPA compliance).

5 Conduct preapplication consultation with USACE (along with EPA, USFWS, NOAA Fisheries, and state Fish and Game agency)

- Present overview of screening process and reasons for selecting the proposed action (as the LEDPA).

6 Submit permit application, alternatives analysis, and other supporting documents to show compliance with Section 404(b)(1) Guidelines

ecosystem and organisms, contamination of the water, and downstream resources (40 CFR 230.10(c)). USACE is required to deny a permit if the findings show that the proposed discharge, prior to mitigation, would result in "significant degradation." Indirect effects of projects are also to be evaluated to determine whether they will result in adverse impacts. These indirect effects are not limited to those associated with water quality but also can extend to impacts caused by decreases in water quantity resulting from a dam project requiring a USACE permit. *See,* for example, *Riverside Irr. Dist. v. Andrews,* 758 F. 2d 508 (10th Cir. 1985) (holding consideration of indirect effects of permit (dam) on critical habitat could be considered in assessing applicability of general verses individual Section 404 permit).

The Section 404(b)(1) Guidelines state that "no discharge shall be permitted unless appropriate and practicable steps have been taken to minimize potential adverse impacts of the discharge on the aquatic ecosystem."

The project must also be implemented in a way that minimizes adverse impacts. The Section 404(b)(1) Guidelines state that "no discharge shall be permitted unless appropriate and practicable steps have been taken to minimize potential adverse impacts of the discharge on the aquatic ecosystem." The Guidelines set forth standards to minimize various kinds of impacts. These rules apply to the mitigation plan, which should include not only methods to mitigate effects of the discharge, but also methods to compensate for impacts on wetlands for which impacts could not be avoided or minimized. 40 CFR 230.10(d). *See* appendix E, Regulatory Guidance Letters and MOUs/MOAs–USACE Regulatory Guidance Letters on guidance on flexibility of the 404(b)(1) Guidelines and mitigation banking.

MOU = Memorandum of understanding

Public Interest Review. USACE's decision to issue a permit is based on an evaluation of the probable environmental impacts of the proposed project and its intended use based on the public interest. USACE's determination of effect on the public interest is based on the benefits reasonably expected to accrue from the project weighed against the reasonably foreseeable detriments of the proposed project. Factors considered in the public interest review include conservation, economics, aesthetics, environmental quality, historic values, fish and wildlife values, flood control, land use, navigation, recreation, water supply and quality, energy needs, safety, food production, and the general public and private need and welfare (33 CFR 320.4). If the Section 404(b)(1) Guidelines are met, USACE will grant a permit unless issuing the permit is determined to be contrary to the public interest.

The public interest review includes a presumption against discharge into wetlands considered to perform functions important to the public interest.

The public interest review includes a presumption against discharge into wetlands considered to perform functions important to the public interest. Examples include wetlands in sanctuaries or refuges, wetlands that serve as valuable flood and stormwater storage areas, or wetlands that are significant in shielding other areas from wave action. A permit allowing a discharge into such wetlands cannot be granted unless the district engineer determines that the benefits of the discharge outweigh the damage to the wetland resource.

Section 404(q) Elevation Process. Section 404(q) of the CWA authorizes USACE to enter into an agreement with appropriate federal agencies to "minimize, to the maximum extent practicable, duplication, needless paperwork, and delays in the issuance of permits" under Section 404. USACE has signed MOAs with EPA, the U.S. Department of Interior (for USFWS review), and the U.S. Department of Commerce (for NOAA Fisheries review) to establish policies and procedures to implement Section 404(q). These MOAs define the process for EPA, USFWS, and NOAA Fisheries to use to elevate disputes with local USACE districts over both specific permit application decisions and general policy matters. The elevation process is intended to ensure higher level agency review in an expedited process.

The MOAs establish a threshold for the elevation process for those permit decisions involving discharges that will have "a substantial and unacceptable impact on aquatic resources of national importance (ARNIs)." The agencies must state during the public comment period, in response to the public notice, that they believe the project may result in a substantial impact on ARNIs. Absent this statement, the agencies have effectively waived their right to request 404(q) review. However, if USACE has received notification from at least one of the agencies, according to the MOAs, USACE must transmit a draft permit decision with proposed special conditions to address the agencies' concern and a draft EA to that agency for review. The agency or agencies must then indicate if they wish to continue the elevation process. The higher-level agency offices review local agency disputes and decide whether the district's decision should reach USACE's Assistant Secretary of the Army for Civil Works in Washington, D.C. USACE, however, retains the ultimate authority to issue the permit and can reject the arguments of other agencies.

ARNI = **Aquatic resource of national importance**

Wetlands provide habitat for a wide variety of wildlife.

This elevation process may not eliminate delays in every case. Although the MOAs link the permit elevation process to potential impacts on ARNIs, the MOAs and other federal regulations do not specify what constitutes an aquatic resource of national importance. Consequently, the elevation review process remains uncertain and could subject IP applications to longer delays while the agencies determine on a case-by-case basis whether the aquatic resource to be affected by the permitted activity is considered an ARNI.

Letter of Permission

The LOP is another type of standard permit process available under Section 404. Although it involves an abbreviated procedure for permit processing, it requires coordination with the federal and state fish and wildlife agencies and a public interest evaluation, but does not include publication of a public notice on a project-by-project basis.

In order to be able to issue LOPs under Section 404 of the CWA, a USACE district must first develop a list of categories of activities proposed to be authorized under LOP procedures *(note:* not all USACE districts have developed the list of activities eligible for the LOP process). This list must be developed in coordination with state and federal fish and wildlife agencies, EPA, the state water quality agency and, if applicable, the state coastal

In order to be able to issue LOPs under Section 404 of the CWA, a USACE district must first develop a list of categories of activities proposed to be authorized under LOP procedures.

zone management agency. A public notice must be published–this notice describes the proposed list of activities to be included in the LOP processing procedure, requests comments on the list, and offers the opportunity for a public hearing on the matter (similar to the process described for IP procedures). Projects proposed under an LOP must still comply with the Section 404(b)(1) Guidelines, NEPA, Section 7 of the Endangered Species Act, Section 106 of the National Historic Preservation Act, and the Coastal Zone Management Act, and must obtain Section 401 water quality certification prior to permit issuance. For NEPA compliance, proposed projects under an LOP are covered by a specific USACE categorical exclusion.

In order to be eligible for an LOP, the project typically must be proposed for the benefit of certain environmental resources.

In order to be eligible for an LOP, the project typically must be proposed for the benefit of certain environmental resources. The project should include extensive preapplication coordination with USFWS, NOAA Fisheries, and other relevant federal and state resource agencies. Habitat restoration or other environmental enhancement actions that benefit wetlands or endangered and threatened species are normally the type of projects eligible for a letter of permission.

EPA Veto

Section 404(c) of the CWA grants EPA the authority to "veto" a permit issued by USACE if EPA determines that the discharge would have unacceptable adverse impacts on water supplies or fishing, wildlife, or recreation areas.

Section 404(c) of the CWA grants EPA the authority to "veto" a permit issued by USACE if EPA determines that the discharge would have unacceptable adverse impacts on water supplies or fishing, wildlife, or recreation areas. This permit review authority is separate from EPA's involvement in USACE's proposed permit decision under the Section 404(q) MOA (*see* above). Whether or not EPA has voiced concern over ARNIs in the Section 404(q) process, after USACE issues a permit under Section 404, EPA may decide to invoke its veto authority under Section 404(c). EPA must give public notice and opportunity for public hearing prior to its veto decision. The final determination on USACE permits issued under Section 404 are made by EPA's administrator. However, EPA has no authority to review USACE's decision to issue a permit pursuant to the Rivers and Harbors Act of 1899.

EPA has vetoed permits issued under Section 404 only on rare occasions. These include:

- A recreational facility at a North Miami Landfill (46 FR 10203) (1981)
- A warehouse and storage yard in Mobile Bay, Alabama (49 FR 29142) (1984)
- An impoundment to manage water levels for waterfowl in Jehosee Island, South Carolina (50 FR 20291) (1985)
- Flood control and reclamation of land in southern Louisiana (50 FR 47268) (1985)
- Agricultural conversion in Dade County, Florida (53 FR 30093) (1988)
- A warehouse and office facility in New Jersey (*see Russo Development Corporation v. Thomas,* 735 F. Supp. 631 (D. N.J. 1989))
- Big River reservoir in Rhode Island (56 FR 10666) (1990)

- A recreational lake in Georgia (*see City of Alma v. United States*, 744 F. Supp. 1546 (S.D. Ga. 1990))
- A dam for a water supply project in Two Forks, Colorado (56 FR 76) (1991)

In most of these instances, EPA used Section 404(c) to enforce its interpretation of the Section 404(b)(1) Guidelines. However, as discussed in the case of *James City County, Virginia* (*see* sidebar), EPA may base its veto decision solely on the fact that the project would cause adverse environmental effects (after the court rejected EPA's use of violation of the Section 404(b)(1) Guidelines). Although its 404(c) veto authority is rarely invoked, EPA's threat to veto USACE's permit has led to EPA sometimes playing a major role in the permit process before USACE has even issued a permit (such as in cases where EPA uses the Section 404(c) threat to cause an applicant to withdraw a permit application).

General Permits

Section 404(e) of the CWA authorizes USACE to issue general permits on a regional, programmatic, or nationwide basis. General permits are designed to apply to categories of discharge activities that are similar in nature and will cause only minimal adverse environmental effects. By definition, general permits are permits that have already been issued; projects that meet the conditions of the general permit may proceed under that authorization. However, it is important to note that many general permits may require notification to USACE prior to initiating fill activities into waters of the United States (although, like IPs, verification may, in limited circumstances, be provided after the fact). All permits authorized under Section 404(e) must be reviewed by USACE every five years, at which time they may be reissued, modified, or revoked.

Regional General Permits

RGPs are a type of general permit issued by a USACE district or division for a specific geographic region, such as a watershed, city, county, or state. The activities covered under an RGP are for similar projects within a particular region that typically have minimal direct, indirect, and cumulative impacts, and are not covered by any other type of general permit.

The process for creating an RGP follows the basic IP procedures, including a public notice and comment period.

EPA Veto Actions and What Is a Practicable Alternative

In *Bersani v. United States Environmental Protection Agency*, 850 F. 2d 36 (2d Cir. 1988); *cert. denied*, 489 U.S. 1089 (1989), commonly referred to as the *Sweeden's Swamp* case (or Attleboro Mall case), a developer proposed the construction of a shopping mall in the Attleboro, Massachusetts area. After surveying several sites (including a site with only one acre of wetlands), the developer purchased the 49-acre red maple Sweeden's Swamp, determined by the EPA, USFWS, and USACE to have high wildlife habitat values. The developer's permit application to USACE for filling 32 acres of the swamp included measures to mitigate for the loss of wetlands by converting nine acres of uplands into marsh and creating another 36 acres of wetlands at a gravel pit offsite.

The EPA objected to the issuance of a permit under Section 404. USACE first indicated that the permit would be denied because another site (with only one acre of wetlands and previously surveyed by the developer) was available. When a competing developer purchased the alternative site, USACE reversed its decision and issued the permit claiming that, because the alternative site was lost and the suggested replacement mitigation was acceptable, no other site could be environmentally preferable.

Pursuant to its authority under Section 404(c), EPA vetoed the issuance of the permit because, even if successful, the proposed mitigation would not eliminate the adverse effects on wildlife. Given the experimental state of wetland creation, EPA would not allow mitigation measures based on artificially created wetlands to substitute for analysis of available alternative sites.

Because the loss could be avoided by using an available alternative site, the EPA considered the wetlands loss "unacceptable." The other site was still available to the applicant, according to the EPA, because the determination of availability for the Section 404(b)(1) alternatives analysis is not limited to the time of permit application but includes the developer's entire site-selection process, beginning with the project proponent's entry into the market. This has become known as the "market entry test."

The veto followed a court's rejection of a challenge by the developer that EPA did not have independent authority to disagree with USACE's findings. The applicant tried unsuccessfully to persuade a court in a district outside the Massachusetts circuit to stop the initiation of the EPA 404(c) proceedings (*Newport Galleria Group v. Deland,* 618 F. Supp. 1179 (D. D.C. 1985)). The developer contested EPA's veto on the grounds that the EPA had taken too long to make a permit decision, but the court rejected that argument in *Bersani v. Deland,* 640 F. Supp. 716 (D. Mass. 1986).

Finally, the developer challenged the reasonableness of the agency's decision. The court affirmed the EPA's independent authority to disagree with USACE and to apply its own independent interpretation of 'practicable alternatives'. The court declared the EPA's finding (the existence of a practicable, less damaging alternative for the mall) to be reasonable and specifically upheld the 'market entry' test for the availability of practicable alternatives. *Bersani v. United States Environmental Protection Agency,* 850 F. 2d 36 (2d Cir. 1988); *cert. denied,* 489 U.S. 1089 (1989).

In the case of *James City County, Va. v. United States Environmental Protection Agency,* 155 F. 2d 254 (4th Cir. 1992), USACE issued a permit to allow the placement of fill to construct a dam for a proposed reservoir in Virginia. USACE had determined that no environmentally preferable, practicable alternatives were identified to meet the project purpose of supplying the local community with water. The EPA vetoed the permit under Section 404(c), claiming that the presumption that the project would have an unacceptable adverse effect because less environmentally damaging, practicable alternatives were available was not rebutted. The EPA determined that USACE did not have sufficient information to conclude that no alternatives were available and that an alternative, such as a yet-to-be designed regional water source, was practicable.

The county proposing the reservoir challenged the EPA veto decision. Agreeing with the county that EPA ➡

Creation of an RGP may also include extensive coordination with local, state, or federal agencies. The RGP may require that USACE be notified prior to initiating fill activities and may be limited to activities within certain impact thresholds. Like all general permits, RGPs are valid for five years. The use of RGPs in regional wetland conservation planning is discussed in chapter 8, Regional Wetland Conservation Planning.

Programmatic General Permits

A PGP is a type of general permit that is based on an existing state, local, or other federal agency program. The purpose of a PGP is to reduce duplication of effort where strong wetland protection programs exist. PGPs may, like RGPs, be conditioned to require that USACE be notified prior to initiating fill activities, and may only authorize activities within certain impact thresholds. The use of PGPs in regional wetland conservation planning is discussed in chapter 8, Regional Wetland Conservation Planning.

Nationwide Permits

NWPs are general permits issued by USACE on a national level. Currently forty-three NWPs cover activities such as placement of navigational aids, fish and wildlife harvesting devices, outfall structures, linear transportation crossings, bank stabilization activities, and stream and wetland restoration projects (Table 5-1). USACE issued new NWPs on January 15, 2002, that went into effect on March 18, 2002 (for the full text, *see* appendix E, Nationwide Permits). These NWPs will expire on March 18, 2007. NWPs are most widely used for utility line activities, minor road crossings, and minor discharges associated with residential or commercial development.

Commonly used throughout the country, NWPs provide a streamlined approach to compliance with Section 404 for certain development activities. Because project authorization for an NWP is typically much less costly and time-consuming than for an IP (Figure 5-2), it is to the applicant's advantage to review a proposed activity carefully to determine if an NWP applies or if the project could be modified to enable authorization under an NWP.

Regulations authorizing NWPs give USACE the authority to modify, suspend, or revoke NWPs for specific activities or within specific geographic regions, as well as the authority for districts or divisions to add other conditions, called *regional conditions,* to the general conditions (33 CFR 330.1(d)). The list of regional conditions can be obtained by contacting local USACE offices or from district web sites.

Use of Multiple NWPs (and Restriction on Use with IP). Within the limits described in General Condition 15, two or more NWPs can be combined, or *stacked,* to authorize a single complete project. However, an NWP and an IP cannot be used for separate components of the same project. The two permit types can be used together only if the activity authorized under the NWP has independent utility and can function or meet its purpose independently of the larger project.

NWP Conditions. All NWPs are subject to general conditions (Table 5-2) that include compliance with several other federal laws triggered by a federal agency action (*see* below), as well as additional conditions specifically applicable to that particular NWP (*see* appendix C). If a project cannot comply with one or more conditions of an NWP, the project proponent must apply for an IP. It is important to carefully read and understand both the specific and general conditions of an NWP. However, because USACE has discretion as to their applicability, a project may comply with all conditions of an NWP and still be required to go through the IP process.

Some NWPs require that USACE always be notified prior to commencing fill activities (*see* Preconstruction Notification Process below), while other NWPs only require notification when impacts on waters of the United States exceed certain thresholds. Some NWPs do not require USACE notification at all, so long as all the conditions of the NWP are met.

Preconstruction Notification Process (PCN). The mechanism for notification in the NWP program is called a preconstruction notification and is described in General Condition 13 of the NWPs. Under the PCN process, for certain NWPs, the project proponent must notify USACE of the intent to perform fill activities that would be authorized under the particular NWP, and must follow the process described below prior to beginning the fill activity (Table 5-2). To avoid potential delays and reduce legal uncertainty about Section 404 authority over a project, many project proponents notify USACE of *any* NWP activity, whether or not a General Condition 13 is triggered and formal PCN is required. Typically, USACE will provide the applicant with written confirmation that the NWP is appropriate. However, it is suggested that the project proponent contact the local USACE district office before submitting the notification to be sure that the proponent can expect a timely

improperly presumed that available alternatives existed without proof to the contrary, the district court stated that this presumption applies only when the project is not water-dependent. The court found that the reservoir, to fulfill its basic purpose of impounding a stream, must be located in wetlands and was therefore water-dependent. The court, declaring that the presumption that practicable alternatives exist should not be applied, found instead that alternatives cited by EPA were not practicable even if its presumption was proper. Because EPA did not demonstrate the existence of available alternatives, the court held that EPA did not meet its statutory duty of showing that the discharge would have an unacceptable adverse impact.

The court sent the issue back to EPA to determine whether environmental considerations alone (and not the availability of practicable alternatives) would warrant a veto of the USACE permit. EPA reissued its veto decision on the basis of unacceptable adverse effects to the environment. Reviewing the district court's ruling that the EPA veto was not justified, the circuit court stated that the "veto based solely on environmental harms was proper," because EPA was only required to assess issues of water quality and not to incorporate the county's need for water when evaluating the project's unacceptable adverse effects.

The circuit court also reversed the district court's decision not to defer to EPA's determination that the environmental effects of the project were severe enough to warrant a permit veto. While the district court held that EPA did not accurately account for any mitigation implemented as part of the project, the circuit court stated that the record "supports EPA's conclusion that the project [even with incorporation of mitigation measures] nevertheless would cause significant harm." In upholding EPA's veto decision, the circuit court stated that "EPA based its veto decision on several factors, including harm to existing fish and wildlife species, damage to the ecosystem, destruction of wetlands, and inadequate mitigation. Its findings are supported by the administrative record, are not arbitrary and capricious, and, for that matter, are supported by substantial evidence" *Ibid.* ■

Table 5-1. Nationwide Permits*—Effective March 18, 2002; expire March 19, 2007						
Permit	Permit Title	Activities Authorized	Special Conditions	PCN Required	Delin. Required	Laws That Apply
1	**Aids to Navigation**	Placement of aids to navigation and regulatory markers.	Must be approved by and installed in accordance with USCG requirements.	No	No	Section 10
2	**Structures in Artificial Canals**	Structures constructed in artificial canals within principally residential developments.	The connection of the canal to a navigable water has been previously authorized.	No	No	Section 10
3	**Maintenance**	Repair, rehabilitation, or replacement of any previously authorized, currently serviceable, structure, or fill; excavation activities to remove accumulated sediment in the vicinity of an existing structure, and placement of new or additional riprap to protect the structure; discharges into waters of the United States needed to restore upland areas.	Cannot use structure or fill differently from original use. Removal of sediment from vicinity of structure must be minimum needed to restore waterway in vicinity of structure to dimensions when structure was built and cannot exceed 200 feet in any direction from structure. Permit cannot be used in conjunction with NWPs 18 or 19 to restore damaged upland areas and cannot be used to reclaim historic lands lost over an extended period of time to normal erosion.	Sometimes (many qualifiers)	No	Section 10 Section 404
4	**Fish and Wildlife Harvesting, Enhancement, and Attraction Devices and Activities**	Fish and wildlife harvesting devices and activities, and small fish attraction devices.	Does not authorize artificial reefs or impoundments designed to hold motile species such as lobster or the use of covered oyster trays or clam racks.	No	No	Section 10 Section 404
5	**Scientific Measurement Devices**	Devices to measure and record scientific data, and similar structures.	Discharge for small weirs and flumes constructed primarily to record water quantity and velocity; must be limited to 25 cubic yards	Sometimes (if discharge is 10 to 25 cubic yards)	No	Section 10 Section 404
6	**Survey Activities**	Survey activities including core sampling, seismic exploratory operations, exploratory bore holes, soil survey, sampling, and historic resources surveys.	Does not authorize activities associated with the recovery of historic resources, discharge of excavated materials from test wells for oil and gas exploration, fill placed for roads and other similar activities, or any permanent structures.	No	No	Section 10 Section 404
7	**Outfall Structures and Maintenance**	Construction of outfall structures and associated intake structures and maintenance excavation associated with outfall and intake structures.	Amount excavated or dredged must be minimum necessary; excavated or dredged material must be retained at upland site; proper erosion and sediment controls must be used. Construction of intakes only authorized if directly associated with outfall structure.	Yes	Yes w/ PCN	Section 10 Section 404
8	**Oil and Gas Structures**	Structures for the exploration, production, and transportation of oil, gas, and minerals within certain areas on the outer continental shelf.	Structures cannot be placed in designated shipping safety fairways or traffic separation schemes.	No	No	Section 10
9	**Structures in Fleeting and Anchorage Areas**	Structures, buoys, floats and other devices to facilitate moorage of vessels within USCG-designated anchorage or fleeting areas.	None.	No	No	Section 10
10	**Mooring Buoys**	Noncommercial, single-boat mooring buoys.	None.	No	No	Section 10
11	**Temporary Recreational Structures**	Temporary buoys, markers, small floating docks, and similar structures for recreational use.	Must be removed within 30 days after use has been discontinued. At USACE reservoirs, reservoir manager must approve each structure individually.	No	No	Section 10

Table 5-1. Nationwide Permits*—Effective March 18, 2002; expire March 19, 2007 continued						
Permit	**Permit Title**	**Activities Authorized**	**Special Conditions**	**PCN Required**	**Delin. Required**	**Laws That Apply**
12	**Utility Line Activities**	Activities required for construction, maintenance and repair of utility lines and associated facilities such as substations; foundations for overhead utility line towers, poles, and anchors; and access roads for the construction and maintenance of utility lines	For utility lines, preconstruction contours must be restored and exposed slopes and stream banks stabilized; wetlands cannot be drained. For substation facilities and access roads in nontidal waters, excluding nontidal wetlands adjacent to tidal waters, activity cannot result in greater than 0.5 acre loss of nontidal foundations for utility line towers, poles, and anchors authorized for all waters, provided foundations are minimum size necessary and separate footings are used where feasible.	Sometimes (many qualifiers)	Yes w/ PCN	Section 10 Section 404
13	**Bank Stabilization**	Bank stabilization activities necessary for erosion prevention	Activity cannot exceed 500 feet in length or an average of one cubic yard per running foot placed below the OHWM without notifying USACE. No material can be placed in a wetland or placed in a manner that impairs surface water flow into or out of a wetland.	Sometimes (if >500 feet or >one cubic yard average per foot)	No	Section 10 Section 404
14	**Linear Transportation Projects**	Activities required for the construction, expansion, modification, or improvement of linear transportation crossings, such as highways, railways, trails, airport runways, and taxiways	In nontidal waters, cannot cause the loss of greater than 0.5 acre of waters. In tidal waters, cannot cause the loss of greater than 0.33 acre of waters. Compensatory mitigation proposal required for discharges in a special aquatic site or discharges that cause more than 0.1 acre loss of waters. Width of fill limited to the minimum necessary for crossing. Cannot be used to authorize nonlinear features associated with transportation projects.	Sometimes (if >0.1 acre; if discharge in special aquatic site)	Yes w/ PCN	Section 10 Section 404
15	**USCG Approved Bridges**	Discharge incidental to the construction of bridges across navigable waters	Must be authorized by USCG as part of the bridge permit. Causeways and approach fills not authorized.	No	No	Section 404
16	**Return Water from Upland Contained Disposal Areas**	Return water from upland, contained dredged material disposal area	None. Quality of return water is controlled by the state through Section 401 process.	No	No	Section 404
17	**Hydropower Projects**	Discharges associated with hydropower projects	Must be (1) small projects licensed by FERC at existing reservoirs with generating capacity no more than 5000 kW, or (2) FERC-exempt projects.	Yes	No	Section 404
18	**Minor Discharges**	Minor discharges into all waters	Volume of discharge and area excavated cannot exceed 25 cubic yards below the OHWM. May not cause the loss of more than 0.1 acre of a special aquatic site.	Sometimes (if >10 cubic yards below OHWM; if discharge in special aquatic site)	Yes w/ PCN	Section 10 Section 404
19	**Minor Dredging**	Minor dredging activities	May not dredge more than 25 cubic yards below the OHWM. Restrictions on areas that can be dredged (e.g., dredging not allowed in anadromous fish-spawning areas)	No	No	Section 10 Section 404
20	**Oil Spill Cleanup**	Containment and cleanup of oil and hazardous substances	Must be done in accordance with national and state contingency plans.	No	No	Section 10 Section 404
21	**Surface Coal Mining Activities**	Discharges associated with surface coal mining and reclamation operations	Coal-mining activities must be authorized by DOI, or by states with approved programs.	Yes	Yes w/ PCN	Section 10 Section 404

Permit	Permit Title	Activities Authorized	Special Conditions	PCN Required	Delin. Required	Laws That Apply
Table 5-1. Nationwide Permits*—Effective March 18, 2002; expire March 19, 2007 continued						
22	**Removal of Vessels**	Temporary structures or minor discharges required for removal of wrecked, abandoned, or disabled vessels, or man-made obstructions to navigation.	Does not authorize removal of vessels listed or determined eligible for listing on the National Register of Historic Places.	No	No	Section 10 Section 404
23	**Approved Categorical Exclusions**	Activities undertaken, assisted, authorized, regulated, funded, or financed, in whole or in part, by another federal agency or department.	USACE must approve the federal agency's categorical exclusion prior to use.	No	No	Section 10 Section 404
24	**State-Administered Section 404 Programs**	Any activity permitted by a state administering its own Section 404 permit program is permitted under Section 10 of the Rivers and Harbors Act of 1899.	Does not authorize activities that do not involve a Section 404 state permit.	No	No	Section 10
25	**Structural Discharges**	Discharges of material into tightly sealed forms or cells for use as structural members for standard pile supported structures or for general navigation.	Does not authorize filled structural members that would support buildings, building pads, homes, house pads, parking areas, storage areas and other such structures.	No	No	Section 404
26	**(Reserved)**					
27	**Stream and Wetland Restoration Activities**	Activities associated with the restoration of former waters, enhancement of degraded tidal and nontidal wetlands and riparian areas, the creation of tidal and nontidal wetlands and riparian areas, and the restoration and enhancement of nontidal streams and nontidal open water areas.	Activity must be conducted on: (1) nonfederal public lands and private lands in accordance with a binding wetland enhancement agreement; (2) reclaimed surface coal mine lands in accordance with an OSM or applicable state agency permit; or (3) any other public, private, or tribal lands. Only native species should be planted onsite. Does not authorize the conversion of a stream or wetland to another aquatic use. Compensatory mitigation not required provided the authorized work results in a net increase in aquatic resource functions and values in the project area.	Sometimes (if activities occur on any public or private land that are not described by (1) or (2))	No	Section 10 Section 404
28	**Modifications of Existing Marinas**	Reconfiguration of existing docking facilities within an authorized marina area.	Does not authorize dredging, additional slips, dock spaces, or expansion of any kind within waters.	No	No	Section 10
29	**Single-family Housing**	Discharges into nontidal waters for the construction or expansion of a single-family home and attendant features.	Cannot cause the loss of more than 0.25 acre of nontidal waters. If part of a subdivision, may not exceed an aggregate total loss of waters of 0.25 acre for the entire subdivision.	Yes	Yes (formal wetland delineation for parcels >0.25 acre)	Section 10 Section 404
30	**Moist Soil Management for Wildlife**	Discharges and maintenance activities for moist soil management for wildlife.	Performed on nontidal federally owned or managed, state-owned or managed, or local government agency owned or managed property.	No	No	Section 404
31	**Maintenance of Existing Flood Control Facilities**	Discharge resulting from activities associated with the maintenance of existing flood control facilities.	For facilities previously authorized or exempted by USACE, or facilities constructed by USACE and transferred to a nonfederal sponsor. Activities limited to those conducted within previously approved "maintenance baseline." Mitigation required one time only and determined when maintenance baseline established.	Yes w/ PCN	Yes	Section 10 Section 404
32	**Completed Enforcement Actions**	Any structure, work or discharge remaining in place, or undertaken for mitigation, restoration, or environmental benefit for compliance with a nonjudicial settlement agreement with USACE or a federal court decision.	Unauthorized activity affected no more than five acres of nontidal or one acre of tidal wetlands. Settlement agreement provides aquatic resources of equal or greater value than those values lost.	No	No	Section 10 Section 404

Permit	Permit Title	Activities Authorized	Special Conditions	PCN Required	Delin. Required	Laws That Apply
Table 5-1. Nationwide Permits*—Effective March 18, 2002; expire March 19, 2007 continued						
33	**Temporary Construction, Access, and Dewatering**	Temporary structures, work, and discharges necessary for construction activities or access fills or dewatering of construction sites.	Associated primary activity must be authorized by USACE or USCG, or not subject to USACE or USCG regulations. Affected areas must be restored to pre-project conditions. Notification must include restoration plan of reasonable measures.	Yes	No	Section 10 Section 404
34	**Cranberry Production Activities**	Discharges associated with expansion, enhancement or modification activities at existing cranberry production operations.	Cumulative total acreage of disturbance per cranberry production operation may not exceed ten acres of waters. May not result in net loss of wetland acreage.	Yes	Yes w/ PCN	Section 404
35	**Maintenance Dredging of Existing Basins**	Excavation and removal of accumulated sediment for maintenance of existing marina basins, access channels to marinas or boat slips, and boat slips to previously authorized depths or controlling depths for ingress/egress, whichever is less.	Dredged material must be disposed of at an upland site, and proper siltation controls must be used.	No	No	Section 10
36	**Boat Ramps**	Activities required for the construction of boat ramps.	Discharge in waters may not exceed 50 cubic yards. Boat ramp may not exceed 20 feet in width. No material can be placed in special aquatic sites, including wetlands.	No	No	Section 10 Section 404
37	**Emergency Watershed Protection and Rehabilitation**	Work done by or funded by NRCS, USFS, or DOI under authorized programs.	Must be for NRCS emergency watershed protection, USFS emergency burn area rehabilitation, or DOI emergency wildfire burn area stabilization and rehabilitation.	Yes	No	Section 10 Section 404
38	**Cleanup of Hazardous and Toxic Waste**	Activities required to effect the containment, stabilization, or removal of hazardous or toxic waste materials performed, ordered, or sponsored by a government agency with established legal or regulatory authority.	Does not authorize the establishment of new disposal sites or the expansion of existing sites used for the disposal of hazardous or toxic waste.	Yes	Yes w/ PCN	Section 10 Section 404
39	**Residential, Commercial, and Institutional Developments**	Discharges into nontidal waters, excluding nontidal wetlands adjacent to tidal waters, for the construction or expansion of residential, commercial and institutional building foundation and pads and attendant features.	May not cause the loss of more than 0.5 acre of nontidal waters. Cannot cause the loss of more than 300 linear feet of a streambed, unless for an intermittent stream and USACE waives the 300-foot limit in writing. No linear-foot limit for ephemeral waters. Compensatory mitigation required. May be required to establish vegetated buffers next to open waters. For residential subdivisions, the aggregate total loss of waters cannot exceed 0.5 acre.	Sometimes (if loss of any open waters below OHWM; if loss is >0.1 acre or is >300 ft of intermittent stream)	Yes w/ PCN	Section 10 Section 404
40	**Agricultural Activities**	Discharges into nontidal waters, excluding nontidal wetlands adjacent to tidal waters, for improving agricultural production and the construction of building pads for farm buildings.	Discharge may not result in the loss of more than 0.5 acre of nontidal wetlands. Conditions for discharges dependent on whether or not the permittee is a USDA participant. Compensatory mitigation plan required. Does not authorize relocation of more than 300 linear feet of existing serviceable drainage ditches, unless USACE waives the 300-foot limit in writing. No linear-foot limit for ephemeral waters.	Sometimes (if loss is >0.1 acre; if >300 ft of impacts to existing ditches in intermittent streams; if construction of farm building pads impacts wetlands in agricultural production prior to Dec. 23, 1985)	Yes w/ PCN (unless wetlands were in agricultural production prior to Dec. 23, 1985)	Section 404

Permit	Permit Title	Activities Authorized	Special Conditions	PCN Required	Delin. Required	Laws That Apply
41	**Reshaping Existing Drainage Ditches**	Discharges into nontidal waters, excluding nontidal wetlands adjacent to tidal waters, to modify the cross-sectional configuration of currently serviceable drainage ditches constructed in waters.	Cannot increase drainage capacity beyond the original design capacity. Compensatory mitigation not required.	Sometimes (if modified by >500 feet)	Yes w/ PCN	Section 404
42	**Recreational Facilities**	Discharges into nontidal waters, excluding nontidal wetlands adjacent to tidal waters, for the construction or expansion of certain recreational facilities that are integrated into the natural landscape.	May not cause the loss of more than 0.5 acre of nontidal waters. Discharge may not cause the loss of more than 300 linear feet of a streambed, unless the stream is an intermittent stream and USACE waives the 300-foot limit in writing. No linear-foot limit for ephemeral waters. Compensatory mitigation required.	Sometimes (if loss is >0.1 acre; if discharge is >300 ft of impact to intermittent stream)	Yes w/ PCN	Section 404
43	**Storm Water Management Facilities**	Discharges into nontidal waters, excluding nontidal wetlands adjacent to tidal waters, for the construction and maintenance of stormwater management facilities.	Discharge for the construction of new facilities cannot cause the loss of more than 0.5 acre of water. The discharge may not cause the loss of more than 300 linear feet of a streambed, unless the stream is an intermittent stream and USACE waives the 300-foot limit in writing. No linear-foot limit for ephemeral waters. Notification must include a maintenance plan, wetland delineation, and compensatory mitigation proposal.	Sometimes (if loss is >0.1 acre; if loss >300 ft of intermittent stream)	Yes w/ PCN	Section 404
44	**Mining Activities**	Discharges associated with aggregate mining, hard rock mining, and associated support activities.	Mined area, plus the acreage loss of waters resulting from support activities, cannot exceed 0.5 acre. Compensatory mitigation required. Beneficiation and mineral processing for hard rock/mineral mining may not occur within 200 feet of the OHWM. Except for aggregate mining in lower perennial streams, no aggregate mining can occur within streambeds where the average annual flow is more than one cfs or in waters within 100 feet of the OHWM of headwater stream segments where the average annual flow is more than one cfs.	Yes	No	Section 10 Section 404

TABLE 5-1 ACRONYMS

DOI	=	Department of Interior
FERC	=	Federal Energy Regulatory Commission
NRCS	=	Natural Resources Conservation Service
OHWM	=	Ordinary high water mark
PCN	=	Preconstruction notification
USCG	=	United States Coast Guard
USFS	=	United States Forest Service

Source: 67 Federal Register 10, Tuesday, January 15, 2002. Department of the Army, USACE. Issuance of Nationwide Permits; Notice.

NOTE. *This table is a summary of NWP requirements. Please refer to appendix E (Nationwide Permits Issued January 15, 2002) for the specific regulatory requirement.*

response. Because of staff workloads, some USACE offices may not welcome notifications that are not required; accordingly, issuing response letters to NWPs where General Condition 13 is not triggered may have a relatively low priority.

The NWP process for activities requiring a PCN is summarized below (*see also* Figure 5-2). The fill activity cannot begin prior to completion of this process.

- Applicant submits PCN to USACE describing the activity and including documentation of compliance with general, regional, and NWP-specific conditions.

- Within 30 days from receipt of the PCN, USACE must determine whether the notification is complete.
- USACE may request additional information necessary to make the PCN complete; however, USACE can make only one request for additional information.
- If the applicant does not submit all the necessary information requested, USACE will notify them that the PCN is still incomplete and that the review process will not begin until the missing information has been submitted.
- Forty-five days from receipt of a complete PCN *(not from when USACE notifies the applicant that the PCN is complete),* USACE must either:
  - Notify the applicant in writing that the project may proceed under the NWP, subject to any special conditions USACE may have added to the authorization; or
  - Notify the applicant that the project cannot be authorized by an NWP and that an individual permit application must be submitted.
- If 45 days have passed after the applicant has submitted a complete PCN to USACE (including documentation of compliance with all general and specific conditions), and the applicant has not received written notice from USACE, the applicant may begin the fill activity.

If the project complies with all conditions of a particular NWP and USACE does not respond to the PCN within 45 days, USACE may then only modify, suspend, or revoke the NWP authorization in accordance with procedures outlined in 33 CFR 330.5(d)(2). However, it is recommended that the applicant attempt to notify USACE that no notice has been received and that the project will proceed. Moreover, other resource agencies may not have similar "deemed approved" processes and may require the applicant to provide USACE approval for other regulatory processes. Therefore, it is prudent to coordinate with USACE to obtain formal notice that a particular NWP is appropriate for Section 404 compliance.

If the project complies with all conditions of a particular NWP and USACE does not respond to the PCN within 45 days, USACE may then only modify, suspend, or revoke the NWP authorization.

It is important to note that obtaining compliance with other general conditions of an NWP may take more than 45 days; compliance with the Endangered Species Act or the National Historic Preservation Act are common examples. In these situations, USACE should notify the project proponent within the 45-day review period whether the project may proceed under the NWP authorization once these pending processes have been completed.

For projects that require a PCN and that would result in the loss of more than 0.5 acre of waters of the United States, USACE must immediately provide a copy of the PCN to the offices of EPA, USFWS, NOAA Fisheries, if appropriate, the state water quality agency, and the State Historic Preservation Officer (SHPO). The project proponent may assist USACE in this process by sending a copy of the PCN directly to these agencies. The agencies have 10 calendar days from the transmission date

SHPO = State Historic Preservation Officer

Table 5-2. Nationwide Permit General Conditions—effective March 18, 2002

Number	Title	Requirement
1	Navigation	Activity must not cause more than a minimal adverse effect on navigation.
2	Proper Maintenance	Any authorized structure or fill must be properly maintained
3	Soil Erosion and Sediment Controls	Appropriate soil erosion and sediment controls must be used during construction. Exposed soil, as well as work below the ordinary high water mark, must be permanently stabilized at the earliest practicable date. Permittees encouraged to complete work within waters of the United States during periods of low flow or no flow.
4	Aquatic Life Movements	Activity may not substantially disrupt the necessary life-cycle movements of those species of aquatic life indigenous to the waterbody. Culverts placed in streams must be installed to maintain low flow conditions.
5	Equipment	Heavy equipment working in wetlands must be placed on mats, or other measures must be taken to minimize soil disturbance.
6	Regional and Case-by-Case Conditions	Activity must comply with any regional conditions added by the Division Engineer, and any case-by- case conditions added by USACE, state or federal agency enforcing Section 401 of the CWA, and state agency requiring a consistency determination under the Coastal Zone Management Act.
7	Wild and Scenic Rivers	An activity cannot occur in a National Wild and Scenic River System unless the federal agency responsible for management of the river determines in writing that the proposed activity will not adversely affect the wild and scenic river designation. This applies to rivers officially designated by Congress as "study rivers" as well.
8	Tribal Rights	Activity or its operation cannot impair tribal rights, including reserved water rights and treaty fishing/hunting rights.
9	Water Quality	If a project has the potential to affect water quality, the applicant must obtain water quality certification from the appropriate state agency. For certain NWP's, if the state 401 certification program does not require or approve water quality management measures, the permittee must provide water quality management measures to USACE that will ensure minimal degradation of water quality.
10	Coastal Zone Management	In certain states, an individual state coastal zone management consistency concurrence must be obtained or waived.
11	Endangered Species	Activities cannot jeopardize the continued existence of a threatened or endangered species listed under the federal Endangered Species Act (ESA), or destroy or adversely modify the critical habitat of such species. Applicants must notify USACE if a project could affect a listed species or designated critical habitat.
12	Historic Properties	A prospective permittee must notify USACE if an authorized activity may affect any historic property listed, determined to be eligible, or which the prospective permittee has reason to believe, may be eligible for listing on the National Register of Historic Places. The activity can not begin until the permittee is notified by USACE that the requirements of the National Historic Preservation Act have been satisfied and that the activity is authorized.
13	Notification	When required by the terms of the NWP, a preconstruction notification (PCN) must be submitted to the District Engineer for review. Specifies contents of the PCN, including general applicant and project information (e.g., applicant name, project location and description), and may be required to include other additional information, such as a jurisdictional delineation, compensatory mitigation proposal, or restoration plan. Also discusses processing steps and timelines, notification forms, types of District Engineer decisions, agency coordination procedures, and wetland delineation requirements.
14	Compliance Certification	Every permittee who has received NWP verification from USACE must submit a signed certification stating that the work has been completed in accordance with the authorization and that the mitigation was completed in accordance with the permit conditions.

Table 5-2. Nationwide Permit General Conditions—effective March 18, 2002 continued

Number	Title	Requirement
15	Use of Multiple Nationwide Permits	Use of multiple NWPs is authorized provided the acreage loss of waters of the United States does not exceed the acreage limit of the most restrictive NWP.
16	Water Supply Intakes	No activity may occur in the proximity of a public water supply intake, except where the activity is for the repair of the intake structures or adjacent bank stabilization.
17	Shellfish Beds	No activity may occur in areas of concentrated shellfish populations, unless the activity is directly related to a shellfish harvesting activity authorized by NWP 4.
18	Suitable Material	Activities may not use "unsuitable" material (e.g., trash, debris, car bodies) and material used for construction or discharge must be free from toxic pollutants in toxic amounts.
19	Mitigation	Projects must be designed to avoid and minimize adverse effects to waters of the United States, and mitigation will be required to the extent necessary to ensure effects are minimal. Mitigation at a one-to-one ratio, typically in the form of restoration, will be required for all wetland impacts requiring a PCN unless the District Engineer determines in writing that another form of mitigation would be more environmentally appropriate and provides a project-specific waiver of this requirement. On a case-by-case basis, USACE may authorize use of vegetated buffers, preservation, mitigation banks or in-lieu fees in addition to, or as a replacement for, other required mitigation.
20	Spawning Areas	Activities in spawning areas during spawning seasons must be avoided to the maximum extent practicable, and activities that would result in the physical destruction of an important spawning area are not authorized.
21	Management of Water Flows	To the maximum extent practicable, activities that could affect waterflows must be designed to maintain preconstruction downstream flow conditions. In addition, the activity must not permanently restrict or impede the passage of normal or expected high flows.
22	Adverse Effects from Impoundments	If an activity creates an impoundment of water, adverse effects to the aquatic system due to the acceleration of the passage of water, and/or the restricting of the flow, shall be minimized to the maximum extent practicable.
23	Waterfowl Breeding Areas	Activities occurring in migratory waterfowl breeding areas must be avoided to the maximum extent practicable.
24	Removal of Temporary Fills	Any temporary fills must be removed in their entirety and the affected areas returned to preexisting elevations.
25	Designated Critical Resource Waters	Discharges of dredge or fill material into designated critical resource waters is prohibited under some NWPs, and requires notification under others. Designated critical resource waters include NOAA-designated marine sanctuaries, National Estuarine Research Reserves, National Wild and Scenic Rivers, critical habitat for federally listed threatened and endangered species, coral reefs, state natural heritage sites, and outstanding national resource waters or other waters officially designated by a state as having particular environmental or ecological significance.
26	Fills Within 100-year Floodplains	Restricts the use of certain NWPs if the authorized activity would result in (1) a discharge into waters of the United States within the mapped 100-year floodplain, below headwaters, or (2) a discharge into waters of the United States within the FEMA or locally mapped floodway, above headwaters. The 100-year floodplains will be identified through the existing FEMA Flood Insurance Rate Maps or FEMA-approved local floodplain maps.
27	Construction Period	Allows USACE to establish a project completion date beyond the expiration of an NWP if the activity has been verified. USACE may also extend this completion date upon request.

Ten-Step Regulatory Integration Process

1 Conduct preliminary constraints analysis to identify related environmental requirements

- Develop stable project purpose and need
- Conduct preliminary site evaluation
- Develop preliminary list of regulatory and permit requirements

2 Consult with regulatory agencies to:

- Confirm their jurisdiction over the proposed action
- Learn the specific steps in their review process
- Determine the scope of any necessary technical studies; survey protocol; mitigation guidelines
- Agree on integrated processing and review schedule

3 Prepare a comprehensive environmental compliance strategy that:

- Explains the major steps in each agency's review process
- Identifies parallel steps and common technical study requirements
- Contains a master schedule for integrated environmental review
- Identifies critical path issues
- Identifies responsible individuals within all agencies (or consulting firm) staff

4 Draft and sign a memorandum of understanding defining:

- Roles and responsibilities of signatories (including regulatory agencies)
- Timing of document preparation and review
- Conflict resolution mechanisms

5 Conduct necessary reviews and technical studies collaboratively

- Ensure studies conducted follow the unique requirements of each agency involved
- Incorporate protocol surveys, modeling techniques, impact assessment methodology, and mitigation standards into NEPA where appropriate
- Consider agency regulatory standard for NEPA significance thresholds
- Involve public and other stakeholder input as appropriate

6 Consolidate results of completed technical studies into draft NEPA document (if applicable)

- Consistent project objectives and alternatives
- Provide copies of the studies to lead agency
- Lead agency evaluates results for consistency and attempts to resolve any inconsistencies
- If necessary, lead agency should disclose draft preliminary results and explain ongoing consultation process

7 Conduct coordinated public and interagency review

- Conduct reviews in accordance with each lead agency's NEPA requirements
- Responsible and cooperating agencies may formally or informally resolve conflicts with the lead agency
- Feedback from general public, landowners, and other stakeholders should be incorporated early in review process
- Feedback may be solicited via press releases, newsletters, announcements, presentations, public workshops, etc.

8 Incorporate the results of any additional studies into the final NEPA document (if applicable)

- Resolve any lingering disagreements
- Document any expert disagreements and state reasoning for selection approach
- Prepare supplement and conduct new public review, if necessary

9 Certify and adopt the consolidated NEPA document (if applicable)

- Complete administrative record with all correspondence and comments on final document
- Incorporate appropriate conditions in project approval

10 Ensure that cooperating agencies and individual federal agencies use the NEPA document in their regulatory decisions (if applicable)

- Agencies make separate decisions depending on their legal and regulatory requirements
- Agencies should use the same NEPA document to make their decisions

to notify USACE (via telephone or facsimile transmission) that they intend to provide substantive, site-specific comments on the proposal. If so notified, USACE must wait an additional 15 calendar days before making a decision on the PCN.

PCN Requirements. The basic information required in a PCN includes: name, address, and telephone number of the project proponent; location of the proposed project; and a description, including purpose, direct and indirect adverse environmental effects the project may have, and any other USACE permits intended to be used for the project. Specific

NWPs have additional PCN requirements, such as preparation of a wetland delineation or a mitigation plan. Documentation of compliance with all general and specific conditions should be included in the PCN.

Ponds and lakes enhance recreational experiences.

USACE Review of Actions Authorized by NWPs. When reviewing a specific project authorized by an NWP, a USACE district can add special requirements to the relevant NWP authorization to provide additional environmental protection, such as the requirement for a mitigation and monitoring plan. USACE can also use discretionary authority and deny the use of an NWP if the project is determined to have more than a minimal adverse impact on the aquatic ecosystem, even if all NWP conditions are met. In such a circumstance, the project applicant would be instructed to apply for an IP to authorize the project as proposed, or may be advised to make modifications to the project in order to qualify for the NWP.

Federal Laws Related to Issuance of a USACE Permit

USACE permit decisions for both standard and general permits under Section 404 of the Clean Water Act require compliance with other federal laws.

As a federal agency action, USACE permit decisions for both standard and general permits under Section 404 of the CWA require compliance with other federal laws that are discussed in this section. Other sections of the CWA and the Rivers and Harbors Act, that either contain specific provisions for wetland protection or provide for wetland protection as a secondary purpose, are discussed in chapter 4, Federal Regulation of Wetlands and Other Waters.

Prior to submitting a permit application to USACE, applicants should prepare a strategy for compliance with all environmental regulations necessary for USACE to process a Section 404 permit, as well as for all other regulations necessary for wetland activities. The applicant's goal should be to integrate all permit conditions into a single project description and process to avoid duplicative, successive environmental reviews. In most cases, the environmental laws with which USACE must comply will not impede USACE from eventually issuing a permit. However, these federal laws may complicate analysis of alternatives, environmental baseline, project impacts, and mitigation for project effects, and will almost certainly delay the permit review process. Furthermore, if compliance with these other laws involves especially sensitive resources, USACE may not allow the project to proceed under an NWP or, in the case of an IP, may require the applicant to assist USACE in preparing an EIS, rather than the more typically required EA and FONSI for NEPA compliance.

FONSI = **Finding of no significant impact**

Environmental Permitting Strategy

Permit decisions of USACE under Section 404 of the CWA, as a federal discretionary action, require compliance with other federal laws. In addition, other federal laws contain specific provisions for, as a secondary purpose, protection of wetlands. Prior to submitting a permit application to USACE, applicants should prepare a strategy for compliance with all environmental regulations necessary for USACE to process a permit, as well as for all other regulations necessary for activities in wetlands and other waters.

The applicant's goal should be to integrate all permit conditions into a single project description and environmental permitting process to avoid duplicative, successive environmental reviews. In most cases, the environmental laws with which USACE must comply will not impede USACE from eventually issuing a permit. However, these federal laws may make mitigation for project effects more difficult and may delay the permit review process. Furthermore, if compliance with these other laws involves especially sensitive resources (e.g., endangered species), USACE may not allow the project to proceed under a streamlined permit process.

To avoid successive environmental reviews, the applicant should prepare a project description integrating all regulatory requirements with USACE's permit conditions. If high value wetlands are present, the project proponent should also investigate wetlands preservation programs applying to the project site to determine whether the proposed activities would be allowable or whether entering the site into these programs may be more advantageous than implementing the project. ■

National Environmental Policy Act

Whenever any federal agency proposes to undertake an action, including granting a permit, NEPA requires the agency to assess the effects of its action on the quality of the human environment (42 USC 4332; 40 CFR 1501). As described in detail in chapter 4, Federal Regulation of Wetlands and Other Waters, the process typically entails the preparation an EA to examine the potential environmental consequences of the proposed federal permit decision. Based on the EA, the federal agency either issues a finding of no significant impact (FONSI) where it determines that the action will have a significant effect on the human environment or determines that an EIS must be prepared.

NEPA Compliance for Standard Permits. If the action is proposed by a federal agency or if the private, local, or state project requires other federal agency discretionary review, USACE may not be the agency that takes the lead role in preparing the NEPA document. When USACE requires a standard permit for the proposed action, USACE's strategy for compliance with NEPA typically entails determining if actions by other federal agencies are necessary and whether NEPA compliance will be addressed by that federal agency. In practice, USACE does not usually assume the role of lead agency for permit actions unless no other federal agencies are involved.

When it is the lead agency, USACE typically seeks to avoid preparing an EIS, where legally appropriate, by preparing an EA that supports a FONSI. To facilitate USACE's NEPA review, a project proponent should first complete all the environmental studies required by other laws, and then furnish USACE with all relevant information regarding the environmental effects of the project. However, if the project will require preparation of an EIS, the project proponent may want to combine USACE's environmental review with other agencies' environmental review requirements.

In practice, most USACE standard permit decisions (LOPs are categorically excluded under USACE's NEPA regulations) are based on an EA and FONSI and not an EIS. Typically, an EIS is prepared for large projects with major environmental issues. Some projects may have major environmental issues at the beginning of the permit process but still avoid the NEPA requirement for preparation of an EIS due to the project proponent's incorporation of measures to avoid or substantially reduce significant effects on the environment. For example, if, during preparation of other environmental analyses (such as those required for compliance with a state environmental protection law), the project design is modified to reduce environmental impacts and measures to mitigate environmental impacts are incorporated into the project, the need to prepare an EIS might be obviated before beginning the USACE permit application process.

The scope of USACE's NEPA analysis is typically limited to those portions of the project within the control of USACE through the permit decision (i.e., where there is limited federal control or a "small federal handle"); that is, NEPA review is limited to those portions of the project within the jurisdictional boundaries of USACE, and not upland portions of the project unless those portions are directly related to USACE's discretionary authority. However, USACE, through the NEPA process, may address the impacts of upland portions of the permitted activity when addressing indirect and cumulative effects of the proposed project (33 CFR 325). In a case involving USACE issuance of a permit for the discharge of fill material in 11 acres of wetlands for a golf course in Nevada, the Ninth Circuit federal court upheld USACE's restriction of its NEPA analysis to only the environmental effects of the golf course and not to the associated resort hotel. The court found that the golf course and resort were not "joined," which would have required USACE to issue a permit for the golf course incorporating the resort hotel into the "major federal action." *Sylvester v. U.S. Army Corps of Engineers,* 871 F. 2d 817 (9th Cir. 1989). Another example can be seen in a case involving a large-scale mixed development, where the permit process had been divided into three separate applications corresponding to the phases of the overall project. USACE was sued for limiting its analysis to only those activities taking place within its jurisdiction as opposed to the whole project. The court upheld USACE's decision based on USACE regulations and evidence in the record that the phases were not connected actions because they each had independent utility. *Wetlands Action Network v. USACE,* 222 F. 3d 1105 (9th Cir. 2000).

The scope of USACE's NEPA analysis is typically limited to those portions of the project within the control of USACE through the permit decision.

NEPA Compliance for General Permits. NEPA review is performed as part of the administrative rule-making process at the time general permits, including NWPs, are issued; by definition, actions authorized under general permits must not have more than minimal adverse effects on the human environment (if they did, an IP would be required). USACE's NEPA compliance, which can be reviewed at the time the general permit is issued, is in all likelihood an EA supporting a FONSI. No further NEPA review is conducted by USACE on a project-specific basis for activities authorized under general permits.

NEPA review is performed as part of the administrative rule-making process at the time general permits, including NWPs, are issued.

Endangered Species Act

Section 7 of the federal Endangered Species Act requires that all federal agencies ensure that their actions do not jeopardize the continued existence of species listed as threatened or endangered or adversely modify or destroy the species' designated critical habitat. 16 USC 1536. The federal agency is required to fulfill this requirement through consultation with USFWS and/or NOAA Fisheries. Therefore, the federal Endangered

Species Act becomes an issue for activities disturbing wetlands and other waters only when the property contains a federally listed species or designated critical habitat that may be affected by a permit decision. In that event, USACE must initiate consultation with USFWS and/or NOAA Fisheries pursuant to Section 7 of the federal Endangered Species Act (16 USC 1536; 50 CFR Part 402).

Although compliance is solely the responsibility of USACE, all applicants for standard permits are required to assist USACE in completion of the ESA Section 7 process. USACE is also required to comply with Section 7 when issuing general permits. Accordingly, USACE requires project-specific compliance with Section 7 as a general condition for specific activities that are proposed to proceed under any NWP (*see* General Condition 11 for NWPs in appendix E).

USACE is required to notify USFWS and/or NOAA Fisheries of the proposed action to determine if any listed species or critical habitat could be affected by the proposed action. If there is a potential to affect a listed species or critical habitat, USACE must provide to USFWS/NOAA Fisheries all available information regarding the potential effect of the permit action on the species or habitat. This procedure may require USACE to prepare a biological assessment (BA) of the effect of the permit action (but not necessarily the effect of the entire project) on the listed species and critical habitat. USACE only prepares a BA where it is determined under NEPA that the proposed action significantly affects the quality of the human environment, triggering the preparation of an EIS. However, where no BA is deemed necessary, USACE must prepare a similar document (sometimes referred to as a biological evaluation) to determine whether the action could adversely affect the listed species. If USACE concludes in the BA (or similar document) that the proposed action could adversely affect the listed species or critical habitat, it must request formal consultation with USFWS/NOAA Fisheries. If formal consultation is required, USFWS and/or NOAA Fisheries will issue a biological opinion (BO) stating whether the permit action is likely to jeopardize the continued existence of the listed species or adversely modify or destroy critical habitat.

BA = Biological assessment
BO = Biological opinion
RPA = Reasonable and prudent alternative
RPM = Reasonable and prudent measure

The BO may include reasonable and prudent measures (RPMs) intended to minimize the effects on the listed species; these RPMs in turn must become special conditions to the Section 404 permit. However, if the BO concludes that the continued existence of a listed species would be jeopardized by the USACE action (issuance of a Section 404 permit), reasonable and prudent alternatives (RPAs) must be specified in the BO. USACE then must not issue the permit and may direct the applicant to resubmit an application proposing one of the RPAs. If there are no RPAs, USACE is required, under the Section 404(b)(1) Guidelines, to deny the permit.

A recommended approach for compliance with the federal Endangered Species Act is to assist USACE in the initial phases of consultation with

USFWS/NOAA Fisheries. The project proponent should request that USFWS/NOAA Fisheries provide a list of species and critical habitat protected under the federal Endangered Species Act that may occur in the project area. If required, appropriate surveys should be conducted early in the project planning phase to allow for sufficient time for consultation between USACE and USFWS/NOAA Fisheries. When possible, the project proponent should design a project to avoid any potential effect on listed species and critical habitat.

Section 106 of the National Historic Preservation Act

Section 106 of the National Historic Preservation Act requires a federal agency to review all actions that may affect a property listed or eligible for listing in the National Register of Historic Places. USACE has regulations to guide the review of permit decisions and specific procedures for initial review of permit applications, public notice, site investigations, eligibility determinations, determination of "area of potential effect" (APE), assessment of effects, consultation with the SHPO, the Tribal Historic Preservation Officer (THPO) if needed, and, in limited circumstances, the Advisory Council on Historic Preservation (ACHP), and finally decision making. The permit decision must reflect consideration of the effects of the project's fill activities on historic properties and should incorporate conditions to avoid or reduce those effects.

ACHP = Advisory Council on Historic Preservation
APE = Area of potential effect
NHPA = National Historic Preservation Act
THPO = Tribal Historic Preservation Officer

Although compliance is solely the responsibility of USACE, all applicants for standard permits are required to assist USACE in compliance with Section 106 of the NHPA. USACE is also required to comply with Section 106 when issuing general permits. In addition, USACE requires project-specific compliance with Section 106 of the NHPA as a general condition for specific activities requesting authorization under a particular NWP (*see* appendix E, General Condition 12 for NWPs).

Coastal streams, such as this one in Rhode Island, often support highly productive estuaries.

The project proponent should check the National Register of Historic Places for resources occurring in the project area (APE), and should survey the APE for those properties eligible for future listing. This process should begin as early as possible to avoid delays when coordinating between USACE's cultural resource staff, SHPO, THPO, and ACHP. When feasible, the project should be designed to avoid any potential affect on a property listed or eligible for listing.

Section 401 of the Clean Water Act

Section 401 of the CWA and the Section 404(b)(1) Guidelines require that the discharge of dredged or fill material into waters of the United

States not violate state water quality standards; Section 401 requires proof of that through a certification issued by the state or EPA. Although compliance is solely the responsibility of USACE, all applicants for standard permits are required to assist USACE in compliance with Section 401 of the CWA. USACE is also required to comply with Section 401 when issuing general permits. For certain general permits, the state may issue certification at the time of issuance and separate Section 401 certification for each project may not be required. However, USACE requires project-specific compliance with Section 401 of the CWA as a general condition for specific activities requesting authorization under a particular NWP where the state has not issued certification (*see* General Condition 9 for NWPs in appendix E).

Neither IPs nor general permits will be issued under Section 404 until the state has been notified and the applicant has obtained a certification of state water quality standards.

Neither IPs nor general permits will be issued under Section 404 until the state has been notified and the applicant has obtained a certification of state water quality standards (often called *401 certification* or *Water Quality Certification)*. The process for obtaining certification under Section 401 varies from state to state, and may vary with the type of permit requested. Several states have not issued approved water quality standards; in those states, EPA issues certification of compliance with Section 401 of the CWA. In many states, USACE notifies the state water quality agency when it receives an application for a Section 404 permit; in other states, the project proponent must apply separately for 401 certification. *See* chapter 6, State Wetland Laws, for a list of state agencies responsible for implementing the Section 401 program.

In many states, some or all NWPs are certified under Section 401 of the CWA, meaning permit applicants do not need to apply for 401 certification on a project-by-project basis. Check with the appropriate USACE District Regulatory Branch for information regarding 401 certification and any special conditions that may apply.

Coastal Zone Management Act

Section 404 requires USACE to obtain proof of certification that the proposed project is consistent with the state coastal management programs implemented under the CZMA of 1972.

Section 404 requires USACE to obtain proof of certification that the proposed project is consistent with the state coastal management programs implemented under the CZMA of 1972. The CZMA creates a broad program based on land development controls within coastal zones, incorporating state involvement through the development of programs for comprehensive state management of resources in these designated areas. Typically, states promulgate a law, as directed by the CZMA, enabling a designated regulatory agency (e.g., a coastal commission, local planning departments) to develop a state management program focused on resources within the coastal zone. *See* chapter 6, State Wetland Laws, for a summary of state coastal regulations and state agencies responsible for implementing approved coastal zone management programs. For projects in the coastal zone, Section 404 permit applicants must certify to USACE that the proposed activity complies with (and will be conducted in a manner consistent with) the CZMA and any of the state or local regulatory requirements governing coastal resources. Absent such certification, USACE cannot issue a Section 404 permit.

Although USACE has sole responsibility for compliance, all applicants for standard permits are required to assist USACE in determining consistency under the CZMA. USACE is also required to comply with the CZMA when issuing general permits (*see* General Condition 10 for NWPs in appendix E).

After determining whether the site falls within the designated coastal zone, the project proponent should design the project to be compatible with the state and local coastal programs. The project proponent should obtain necessary approvals under the CZMA prior to applying to USACE for a permit keeping in mind that, in many cases, the boundaries of coastal resources (e.g., wetlands) may be different than those defined by USACE. Project proponents should ensure that they are familiar with natural resource boundaries in the coastal zone when designing projects.

Wild and Scenic Rivers Act

The National Wild and Scenic Rivers Act (WSRA) establishes a system to designate wild and scenic rivers within federally managed lands. Any federal activity that could affect the characteristics of a designated wild and scenic river must coordinate with the federal agency responsible for the particular river (either the U.S. Bureau of Land Management, National Park Service, USFWS, or U.S. Forest Service) to determine the effect on the characteristics of the river's designation.

WSRA = National Wild and Scenic Rivers Act

Although compliance is solely the responsibility of USACE, all applicants for standard permits are required to assist USACE in compliance with the WSRA. USACE is also required to comply with the WSRA when issuing general permits. Accordingly, USACE requires project-specific compliance with the WSRA as a general condition for specific activities requesting authorization under a particular general permit (*see* General Conditions 7 and 25 for NWPs in appendix E).

The responsible federal agency will issue a letter of consent if the proposed action would not have a direct or indirect effect on the resources for which the river was designated under the WSRA. If the consent is denied, the federal agency may recommend alternatives that would not affect the river.

Administrative Appeals

The USACE regulations establish procedures for the administrative appeal of jurisdictional determinations, permit applications denied with prejudice, or declined individual permits (*see* 33 CFR Part 331, 65 FR 16486–16585). Permit decisions made by a division engineer or higher authority may be appealed to a USACE official at least one level higher than the decisionmaker. This higher Army official makes the decision on the merits of the appeal, and may appoint a qualified individual to act as a review officer within the district for all appeals.

Permit decisions made by a division engineer or higher authority may be appealed to a USACE official at least one level higher than the decisionmaker.

Requests for appeal may be submitted within 60 days after permit denial or declined individual permit. Examples of reasons for appeals include, but are not limited to: a procedural error, an incorrect application of law, regulation, or officially-promulgated policy, omission of material fact, incorrect application of the Section 404(b)(1) Guidelines, or use of incorrect data. The reviewing officer has 30 days to determine if the request for appeal is complete. After determining that it is complete, the reviewing officer has 30 days to determine whether a site investigation is necessary and another 30 days to complete the investigation. The reviewing officer has 60 days from submittal of a complete appeal to call an appeal conference.

Dense freshwater marsh vegetation provides cover for wildlife.

After a request for appeal has been submitted by the applicant, the reviewing officer conducts an independent review of the administrative record to address the reasons for the appeal cited by the applicant and to verify that the record provides an adequate and reasonable basis supporting the district engineer's decision, that facts or analysis essential to the district engineer's decision have not been omitted from the administrative record, and that all relevant requirements of law, regulations, and officially-promulgated USACE policy guidance have been satisfied. The reviewing officer may seek input from any employee of USACE or of another federal or state agency, or from any recognized expert, so long as that person had not been previously involved in the action under review. The division engineer will make a final decision on the merits of the appeal within 90 days of the receipt of the complete request for appeal.

No permit applicant may file a legal action against USACE in the federal courts based on a permit denial or declined individual permit until after a final USACE decision has been made and the appellant has exhausted all applicable administrative remedies under the administrative appeals process. The appellant is considered to have exhausted all administrative remedies when a final USACE decision is made in accordance with the administrative appeals process.

Review, Enforcement, and Violations

This section discusses the avenues for challenging USACE decisions under Section 404 of the CWA. This section also summarizes the enforcement functions of USACE and EPA and discusses the liabilities and criminal and civil penalties that may be involved with Section 404 violations.

Judicial Review

Federal courts have jurisdiction to review many USACE and EPA administrative actions under the CWA. However, the CWA does not provide for judicial review of certain decisions affecting wetlands, such as the method for determining the extent of jurisdictional wetlands on a proposed site. These decisions must be challenged through the administrative appeals process and other federal laws, such as the Administrative Procedures Act (*see* the case of *Tabb Lakes*) or federal questions of constitutional authority (*see Hoffman Homes*).

USACE's decision to grant or deny a permit under Section 404 of the CWA is subject to review by the federal courts. An applicant can challenge the permit denial, but cannot challenge USACE's permit conditions except by refusing to accept the entire permit. Other interested parties may challenge USACE's decision to issue a permit, as long as they meet federal requirements for standing. USACE and EPA regulations are subject to judicial review under the CWA as to whether they are consistent with the Act, and EPA's decision to veto a permit under Section 404(c) is also subject to judicial review.

USACE's decision to grant or deny a permit under Section 404 of the CWA is subject to review by the federal courts.

The courts have stated that they lack jurisdiction to review USACE's rejection of an applicant's request to proceed under an NWP. According to the courts, USACE's response is not considered to be an order or license. Even if it were an order, under the Administrative Procedures Act, it is not considered a final action with binding legal effect. *Avella v. United States Army Corps of Engineers*, 916 F. 2d 721(11th Cir. 1990); *Mulberry Hills Development Corporation v. U.S.*, 32 Env't. Rep. Cas. (BNA) 1195 (D.C. Md. 1989).

Under the "arbitrary and capricious" standard, to review a decision by USACE or EPA, federal courts give deference to the administrative agency–provided that the administrative record is complete and that the agency had a rational basis for its decision. The court must defer to the administrative agency even if the court does not agree with the agency decision.

The remedy available to a party challenging USACE and EPA decision errors is typically limited to the court entering a declaration of errors by USACE or EPA, reversal of the agency's decision, and remand of the decision back to the agency for revision in light of the court's findings. The court is not permitted to substitute itself and undertake the agency action.

The court is not permitted to substitute itself and undertake the agency action.

Because the CWA contains no statute of limitations for filing challenges to USACE and EPA actions under Section 404 or NEPA, the general five-year statute of limitations for civil actions against the government of the United States typically applies.

Enforcement

Section 404 of the CWA can be enforced through administrative orders and penalties, civil judicial enforcement by the government or interested citizens, and criminal prosecutions. EPA and USACE share administrative enforcement authority under Section 404. Under a 1989 MOA between USACE and EPA (*see* appendix F), USACE will conduct the initial investigations for enforcement cases. USACE or EPA may issue administrative orders requiring compliance with permit conditions, and may issue cease-and-desist orders against unpermitted discharges of dredged or fill material. In a case challenging the EPA's independent enforcement authority, the court held permitless discharges to be within the enforcement authority of the EPA. *Hobbs v. U.S.,* 947 F. 2d 941 (4th Cir. 1991). *See also Buttrey v. U.S.,* 690 F. 2d 1186 (5th Cir. 1982) (upholding USACE's authority under the Commerce Clause to regulate the discharge of fill material into navigable waters of the United States).

Courts will not review USACE or EPA compliance orders under the CWA until the federal government brings an enforcement proceeding. *Southern Pines Associates v. United States,* 31 ERC 2020 (4th Cir. 1990); *Hoffman Group, Inc. v. United States Environmental Protection Agency,* 902 F. 2d 567 (7th Cir. 1990); *Mulberry Hills Development Corporation; McGown v. United States,* 747 F. Supp. 539 (E.D. Mo. 1990).

Liability

CWA obligations apply to a *person,* which is defined as "an individual, corporation, partnership, association, state, municipality, commission, or political subdivision of a state, or any interstate body," for administrative and civil enforcement (33 USC 1362). For criminal enforcement, the term *person* also includes "any responsible corporate officer" (33 USC 1319).

In general, the test for liability under the CWA extends to anyone who performed, exercised control over, or had responsibility for the unpermitted discharge activity. *United States v. Board of Trustees of Florida Keys Community College,* 531 F. Supp. 267 (S.D. Fla. 1981). This includes the owner of the property, contractors who perform the physical work, and even design engineers. Although subsequent landowners are not liable for the penalties for an unpermitted discharge activity that took place on the land prior to ownership, illegal discharges are considered to be "continuing violations" and may still require removal or mitigation.

Civil and Criminal Actions

Under Section 309 of the CWA, both USACE and EPA have independent enforcement authority. Although administrative orders are not independently enforceable against a violator, USACE will refer the violation of the administrative order to the United States Department of Justice for either civil or criminal judicial enforcement. For cases determined to be appropriate, USACE and EPA may recommend criminal or civil actions to obtain penalties for violations, compliance with the orders and directives issued, and other relief as appropriate. The federal government is not required to elect between pursuing an administrative remedy or filing a judicial enforcement action. *See, for example, Deltona Corp. v. Alexander,* 682 F. 2d 888 (11th Cir. 1982) (holding that despite prior issuance of Rivers and Harbors Act Section 10 permits for same project, USACE's denial of

A Landowner Goes to Jail for Filling a Wetland

The United States brought action against a landowner in Pennsylvania under the CWA, alleging that fill had been discharged into wetlands without a permit pursuant to Section 404. The wetlands were adjacent to an unnamed tributary to the Pennsylvania Canal that flows into the Delaware River. The landowner had cleared old tires and other refuse and then filled and leveled part of the property to build a garage. Between April 1987 and November 1987, USACE repeatedly warned the landowner, both orally and by issuance of a cease-and-desist letter, not to continue filling the site until a permit was obtained. Each time the USACE biologist visited the site, however, he noted that substantial additional filling had taken place.

In December 1987, USACE issued an administrative directive instructing the defendant landowner to cease and desist the filling activity and to remove the fill within 45 days, or to cease and desist filling and apply for an "after the fact" permit within ten days. The landowner did not apply for a permit and continued to fill the site through 1988 (EPA actually videotaped the landowner filling wetlands after the cease-and-desist order was issued). In August 1988, the federal district court issued a temporary restraining order. After the landowner continued to fill the site, using a bulldozer to level the fill, the court issued a preliminary restraining order and a contempt of court order against the defendant landowner.

Also implicated in the unpermitted discharges of fill material at the site were a demolition, excavation, and hauling company and a demolition and metal and concrete recycling company that saved over

Section 404 permits was appropriate and cannot be estopped when exercising sovereign power for the benefit of the public).

USACE and EPA assess administrative penalties for violations of Section 404 under two classes. Class I violations are for less serious unpermitted activities and, under EPA authority, carry a maximum of $10,000 per violation, with a total maximum of $25,000. Class II violations are for more serious unpermitted activities and, under EPA authority, carry a maximum of $10,000 per day for each day during which the violation continues, with a total maximum of $125,000. The penalty authority of USACE does not distinguish between classes. Although USACE has less formal penalty proceedings for Class I violations, both classes are subject to penalties of up to $25,000 per day. Both class violations under USACE and EPA authority allow for notice and opportunity for hearing.

Under the federal court system, through prosecution by the United States Department of Justice, the typical remedy is an injunction to require a party either to stop further unpermitted discharge activities or to remove the unpermitted fill and restore the damaged area. Courts may award civil penalties of up to $25,000 per day per violation. The CWA does not provide a statute of limitations for civil actions. However, courts typically apply a five-year statute of limitations, which begins when the government *learns of the violation,* not on the date when the violation occurred. *United States v. Banks,* 115 F. 3d 916 (11th Cir. 1997) dealt with the statute of limitations for enforcement of violations of the CWA. USACE filed suit requesting injunctive relief at least eight years after Banks began developing his property. Banks asserted that equitable relief was barred by the fact that the statute of limitations in 28 USC 2462 for civil penalties had expired. The court held that the statute of limitations argument was not relevant in light of the federal government's sovereign capacity to protect the public interest.

Criminal enforcement for Section 404 violations are rarely pursued and usually apply to those unpermitted discharges that involve significant environmental harm, abusive conduct, continued illegal conduct after notice, and other serious knowing and willful violations. Negligent violations carry misdemeanor sanctions, including penalties of $2,500 to $25,000 per day and imprisonment of up to one year. Knowing violations carry felony sanctions, including penalties of $5,000 to $50,000 per day and imprisonment of up to three years. A knowing violation that places another person in danger of death or injury carries a maximum fine of $250,000 and 15 years imprisonment, and corporations can be fined up to $1,000,000. The requirement of criminal intent for violations of the CWA was discussed in *United States v. James J. Wilson,* 133 F. 3d 251 (4th Cir. 1997). In *Wilson* the defendants were charged with knowingly discharging fill material and

$142,800 and $30,262, respectively, by disposing of the fill at the landowner's site rather than at a permitted landfill.

Finding the defendants strictly liable for the unpermitted discharges, the court issued a permanent injunction against further filling of the wetlands site without first obtaining a permit under Section 404. The court also issued a USACE-submitted restoration order for the site, to be implemented by the defendants. *United States v. Pozsgai,* 31 ERC 1230 (E.D. Pa. 1990).

In the related criminal prosecution, the defendant landowner was convicted of 40 counts of illegal discharge of pollutants under the CWA and was sentenced to three years in jail, fined $202,000, and ordered to restore the wetlands in which the violations occurred. The Supreme Court refused to review the conviction and sentence. *United States v. Pozsgai,* 897 F. 2d 524 (3d Cir. 1990); *cert. denied,* 498 U.S. 812 (1990).

The landowner also challenged the constitutional reach of the CWA's imposition of civil sanctions for filling wetlands. The defendant argued that the wetland's only connection to interstate commerce was a nearby canal used to transport coal in the nineteenth century. The Supreme Court again refused to hear his appeal and allowed the third circuit ruling that in this case the CWA's jurisdiction was properly applied. *Pozsgai v. United States,* 999 F. 2d 719 (3d Cir. 1993); *cert. denied,* 510 U.S. 1110 (1994).

As of the Supreme Court's latest denial of the appeal, the landowner had served approximately two years of his three-year sentence and was paying off his civil penalty in installments. He is still seeking an after-the-fact permit. ■

Landowners Claim Corps' Permit Denial Is Taking of Private Property

A landowner in a Florida case, who had purchased thousands of acres of mangrove wetlands in coastal Florida in 1964, received permits from USACE under Section 10 of the Rivers and Harbors Act of 1899 to develop two of its five tracts (prior to enactment of the Clean Water Act). In 1976, the landowner applied for permits from USACE under Section 404 to fill the remaining tracts, approximately one-third of the landowner's total planned lots. USACE denied the permits, and the landowner claimed a regulatory taking.

The court of claims rejected the landowner's argument. The court ruled that the landowner failed to show that the denial left the landowner with "no economically viable use of its land." Because the denial was for only one-third of the proposed development, and even on that one-third some areas could be developed without a permit, the court rejected the argument that the property was denied its most profitable use. The court also rejected the allegation that the earlier Section 10 permits created a reasonable expectation of subsequent approvals. *Deltona Corp. v. United States*, 657 F. 2d 1184 (Ct. Cl. 1981); *cert. denied*, 455 U.S. 1017 (1982).

In another Florida case, a large-scale miner of limestone in 1972 purchased a tract of 1,560 acres for the sole purpose of mining limestone. USACE denied a permit under Section 404 that would have enabled the landowner to mine 98 acres of the tract. Rather than appeal the denial, the landowner claimed a regulatory taking.

Although the claims court agreed with the landowner that a taking existed and awarded more than $1 million (*Florida Rock Industries, Inc. v. United States*, 8 Cl. Ct. 160 (1985)), the Federal Circuit Court of Appeals overturned the decision because the denial of the permit only precluded mining and did not foreclose other economically viable uses, such as sale of the property. The court ruled that a valid defense to this type of takings claim was that the property's fair market value had increased (even if the reason for the increase was that speculators would be willing to buy the property in the hope that future regulatory policies might change). The

excavated dirt into wetlands in violation of the CWA. The court reversed the lower court's convictions on the grounds that it had failed to require proof of *mens rea* (mental state" as to intent to commit an act) with respect to each element of an offense defined by the CWA. The court held that *mens rea* requires the government to prove the defendant's knowledge of facts meeting each essential element of the substantive offense, but need not prove that the defendant knows his or her conduct to be illegal.

Suits by Citizens

Under Section 505 of the CWA, any person having an interest that is or may be adversely affected can bring a civil action against any person, including the United States, for violation of the CWA, including Section 404, or against the EPA or USACE for failure to perform nondiscretionary duties under the CWA. Citizens must first notify the violator and the state and federal governments of intent to sue. This must be done at least 60 days prior to filing suit and cannot proceed if either the federal or state government is prosecuting the violation.

Citizens may intervene in a case where the federal or state government has brought proceedings to prosecute a violation of Section 404. If a citizen proceeds with a suit, a copy of the complaint must be furnished to the government, and the government must be given 45 days to review a consent decree. The federal government may use its authority to intervene in the citizen suit or to present *amicus* (advice to the court) briefs to address matters affecting the government's authority. Plaintiffs who prevail in citizen suits may be awarded their costs, including attorney's fees.

The Takings Issue

The Fifth Amendment to the U.S. Constitution prohibits the government from *taking* private property without just compensation. Courts have recognized that governmental regulation of land use may sometimes be so restrictive that, by eliminating the economic use of the property, the action constitutes a taking. As a defense to imposition of permit requirements under Section 404 of the CWA, landowners may assert that a regulatory taking has occurred; however, this assertion has not been demonstrated to be a valid defense to violations of Section 404 of the CWA or to permit conditions imposed by USACE. Takings claims are more appropriate for project applicants who have been denied a permit. Although much judicial activity has involved alleged takings under many government activities, including Section 404, the takings defense is not likely to constitute a substantial restraint on implementation of Section 404.

Typically, if a link exists between the federal government's assertion of jurisdiction and the regulated activity, and if some economically viable use of the land remains, a taking is not considered to have occurred. Therefore, government requirements for a permit or strict permit conditions must be directly related to the reasons for regulating the activity and requiring a permit in the first place.

This issue is exemplified by a U.S. Supreme Court decision in which the court rejected the government's attempt to require public access to a privately owned marina-style residential community in Hawaii. After securing appropriate local approvals, the developer constructed a private marina on a pond adjacent to navigable waters and excavated a channel to connect the pond to the navigable waters. USACE claimed that the development changed the character of the land from private property to property that must be opened to the public; but the court held that USACE exceeded the bounds of ordinary regulation, and therefore such a requirement may be imposed only by paying compensation under the Fifth Amendment. *Kaiser Aetna v. United States,* 444 U.S. 164 (1979).

The forum to decide takings claims for permit denial is in the U.S. Court of Federal Claims (formerly the Court of Claims). Takings claims seeking monetary compensation in excess of $10,000 are initially heard in the claims court (as opposed to other CWA issues, which are brought in federal district courts), and appeals from the claims court are heard in the Federal Circuit Court of Appeals. The claims court will hear takings claims only after final permit denial–after denial has been appealed through the regulatory process. *See,* for example, *U.S. v. Tull,* 769 F. 2d 182 (4th Cir. 1985) (holding that the CWA's definition of wetlands was sufficiently definite to give fair notice of what the CWA prohibited or requires).

Although recent court decisions have favored landowners who claim a regulatory taking, landowners still must prove that property values diminish when a land use regulation that is imposed cannot be justified by the legal necessity to protect important natural resources. *1902 Atlantic*

case was remanded back to the claims court to determine the taking questions in light of fair market values. *Florida Rock Industries, Inc. v. United States,* 791 F. 2d 893 (Fed. Cir. 1986); *cert. denied,* 479 U.S. 1053 (1987).

The claims court, in deciding the question of the fair market value of the property after application of the regulation, accepted the landowner's valuation that the highest and best use of the property was only "future recreational/water management" (open space). The court rejected the government's argument (and ignored the direction of the circuit court) that the land had a much higher investment value because speculators may not fully comprehend the regulatory restrictions placed on the property. Comparing the value of the property before regulation (when mining was a permitted use) to its value after regulation (when open space was the best use), the court stated that the value of the property was reduced by 95 percent as a result of USACE's permit denial under Section 404. The court found that USACE's denial constituted a taking and awarded the plaintiff more than one million dollars. *Florida Rock Industries, Inc. v. United States,* 21 Cl. Ct. 161 (1990); *see also* a companion case, *Loveladies Harbor, Inc. v. United States,* 21 Cl. Ct. 153 (1990), in which the claims court awarded more than $2.6 million to a developer who was denied a Section 404 permit to develop 11.5 acres of wetlands.

For the second time, the Federal Circuit Court of Appeals reversed the claims court's calculation of the valuation of the property before and after imposition of the regulation. The circuit court directed the claims court to determine the fair market value of the land based on the speculative market in the project area (west of Miami). If the court finds that land has decreased in value, the claims court is required to find whether that reduction should be considered a "noncompensable diminution" in value or a taking of some portion of value. Within this finding, the claims court is required to assess the compensating benefits from the regulatory environment to the property and others similarly situated, whether the benefits are widely shared by society while the costs are borne by few, and whether alternative activities permitted are economically realistic. *Florida Rock Industries, Inc. v. United States,* 18 F. 3d 1560 (Fed. Cir., 1994); *cert. denied,* 513 U.S. 1109 (1995). ■

Ltd. v. Hudson, 574 F. Supp. 1381 (E.D. Va. 1983) held that denial of a USACE permit for the fill of an 11-acre tidally influenced borrow pit for industrial development amounted to a regulatory taking. In *Loveladies Harbor Inc. v. U.S.,* 28 F. 3d 1171 (Fed. Cir. 1994), a regulatory takings claim was upheld involving the denial of a Section 404 permit resulting in property values changing from $2.6 million to $12,500. *Florida Rock Industries, Inc. v. U.S. Court of Appeals for the Federal Court,* 18 F. 3d 1560 (Fed. Cir. 1994), involved the denial of a Section 404 permit to mine limestone and the resulting effect on economic use and value of the property. The appellate court overturned the lower court's determination that comparable sales values be discounted based on the notion that the purchasers were not knowledgeable of the regulatory scheme. The appellate court held that where a market provides a well-substantiated value a court may not substitute its own judgment as to what is a wise investment.

CHAPTER SIX

State Wetland Laws

Although some states rely solely on federal laws and regulations for protection of wetland resources, many have adopted an additional layer of oversight to protect habitats unique or particularly sensitive to their region. These state regulations are often more stringent than federal regulations, overlapping or even, at times, appearing to conflict with the federal standards. These uncertainties make it difficult for wetland permit applicants and environmentally concerned citizens to successfully navigate the complex laws governing wetland resource protection.

Although some states rely solely on federal laws and regulations for protection of wetland resources, many have adopted an additional layer of oversight to protect habitats unique or particularly sensitive to their region.

Table 6-1 presents an overview of each of the 50 states' wetland protection programs, including state agencies responsible for federal Clean Water Act Section 401 certification, wetland conservation oversight, and coastal resource protection. In addition, some of the more complex state wetland protection programs (i.e., those of California, Florida, Massachusetts, Michigan, New Jersey, and Washington) are discussed at a greater level of detail to illustrate the consideration and effort some states have invested in their resource management programs. The summaries presented in the table and in the discussions below do not represent an exhaustive treatment of state programs; the relevant state or local agencies should be contacted directly for additional information.

Forested wetland and cattail marsh

California

Several agencies in California enforce regulations that offer protection to wetlands. The California Department of Fish and Game (DFG) has jurisdictional authority over resources, including wetlands associated with rivers, streams, and lakes under the California Fish and Game Code, sections 1600–1616, sometimes referred to as the "streambed alteration agreement program." Under this statute, the DFG has the authority to regulate work that will substantially divert or obstruct the natural flow of, or substantially

DFG = California Department of Fish and Game

California Department of Fish and Game May Regulate Where USACE Does Not

The California Department of Fish and Game has jurisdiction over streamside habitats that, under the federal definition, may not qualify as wetlands. Depending on the particular situation and the type of fish or wildlife at risk, the California Department of Fish and Game can extend its lateral jurisdiction over a stream as far as the edges of the surrounding floodplain, an area that may be much larger than that regulated by USACE.

A common example of the California Department of Fish and Game's broader jurisdiction occurs around California's riparian forests. These habitats often extend outside the ordinary high water mark and frequently do not exhibit all three wetland indicators necessary for regulation under the Section 404 of the Clean Water Act (wetland hydrology, hydrophytic vegetation, and hydric soils). In such cases, the California Department of Fish and Game usually considers the outer edge of the riparian vegetation to be the limit of its jurisdictional authority; this limit is commonly outside USACE's jurisdiction. ■

CZMA = **Coastal Zone Management Act**
USACE = **United States Army Corps of Engineers**

change or use any material from the bed, channel, or bank of, any river, stream, or lake. Prior to construction, a project proponent must submit a notification of streambed alteration to the regional DFG office responsible for the project area. In addition to the formal application materials and fee, the proponent must include a copy of the appropriate environmental impact analysis document required for compliance with the California Environmental Quality Act. After reviewing the application materials, the DFG enters into a streambed alteration agreement with the applicant and imposes conditions on the agreement to ensure that no net loss of wetland values or acreage results from project construction. The DFG does not consider this streambed or lakebed alteration agreement a permit, but rather an agreement between the DFG and the project proponent.

The California Coastal Commission has jurisdiction over wetlands in the coastal zone under both state and federal legislation (California Coastal Act of 1976 (Public Resources Code Section 30000 et seq.) and CZMA, respectively. Generally, California's coastal zone extends seaward 3 miles and inland 1,000 yards; however, it can extend inland as much as 5 miles depending on the sensitivity of the resources in the surrounding area. A coastal development permit must be prepared for any proposed development that would modify land or water use in the coastal zone. Most of the California Coastal Commission's management responsibilities for coastal resources have been delegated to local governments with approved local coastal programs; consequently, the California Coastal Commission only reviews coastal development permit applications for projects that are likely to affect tidelands, submerged lands, or public trust lands; projects in areas without an approved local coastal program; or federal projects requiring a consistency determination under the CZMA.

The San Francisco Bay Conservation and Development Commission has jurisdiction over coastal activities occurring within the San Francisco Bay Area and Suisun Marsh. The commission's federal authority also stems from the CZMA, but its state authority is derived from the McAteer-Petris Act and the Suisun Marsh Preservation Act. Specifically, the San Francisco Bay Conservation and Development Commission's jurisdiction extends to certain waterways that flow into the bay; salt ponds and managed wetlands around the bay; a 100-foot "shoreline band" that surrounds the bay; and all areas that lie below the 10-foot contour line in Suisun Marsh.

The California State Lands Commission manages submerged lands, tidelands, and swamp and overflowed lands owned by the State of California. Under the doctrine of public trust, the state may preclude uses on submerged lands and tidelands that are inconsistent with the purposes of the public trust, which include commerce, navigation, fisheries, and recreation. If project proponents wish to use submerged lands and tidelands owned by the state, it may be necessary to obtain a land use lease from the State Lands Commission before proceeding with the project.

Table 6-1. Summary of State Wetland Protection and Regulatory Programs[1]

State	Responsible USACE District[2]/ Dist. Iden. Code	Section 401 Water Quality Certification Program[3]	State Wetlands Program[4]	Coastal Resources Program	Joint State/Fed. Permit App.[5]	Application available online?
ALABAMA	Mobile/sam Nashville/orn	**Alabama Dept. of Envir. Mgmt. (ADEM) Field Operations Div.** **www.adem.state.al.us** **TEL: 334/271-7714** Project applicants receive 401 certification through a joint 404/401 permitting process between USACE and ADEM. For individual permits, USACE issues a joint public notice and individual certification with best management practices. NWPs are automatically certified compliant with 401 when approved. Applications are sent to the ADEM Permit Coordination and Development Center (PCDC).	No state wetland regulations.	**ADECA, Coastal Programs Office** **www.adeca.state.al.us** **TEL: 334/432-6533** Alabama's coastal zone management program is administered by the Alabama Dept. of Economics and Community Affairs (ADECA) and ADEM. ADECA provides planning guidance for the overall program while ADEM issues all required permits. The coastal zone in Alabama is defined by the 10-foot contour line. Federal agencies must seek a federal consistency determination from ADEM for development projects in the coastal zone, and projects with no federal involvement must obtain a Coastal Use Permit. Currently, ADEM is requiring applicants to obtain a coastal use permit for wetlands "deregulated" by USACE as a result of SWANCC.	NO	YES
ALASKA	Alaska/poa	**Alaska Dept. of Env. Conser. (DEC) Div. of Air and Water Quality** **www.state.ak.us/local/akpages ENV.CONSERV** **TEL: 907/451-2101** DEC certifies all dredge and fill permits to ensure that discharges to navigable waters or wetlands resulting from federal permit activities will not violate state water-quality standards.	**Alaska Dept. of Fish and Game (DFG) Habitat Division** **www.state.ak.us/adfg** **TEL: 907/465-6160** DFG requires the following types of permits. It should be noted that these two permits are not "wetland" permits, but rather "habitat" permits that could be required in wetland areas. 1. Special Areas Permit—for work in a refuge, critical habitat area, or sanctuaries; and 2. Title 16—Fish Habitat Permit—for work that will alter the natural flow of a stream that supports anadromous fish. Alaska Dept. of Natural Resources (DNR) Div. of Mining, Land and Water www.dnr.state.ak.us/mlw/index.htm TEL: 907/269-8525 DNR requires a permit for any work in state-owned lands, including placement of dredge and fill into wetlands.	**Div. of Gov. Coordination** **www.gov.state.ak.us/dgc** **TEL: 907/465-3562** The Division of Governmental Coordination is responsible for completing consistency reviews to determine if projects in the coastal zone meet state standards and coastal district policies. There are 35 coastal districts in Alaska that have programs requiring other state and federal agencies to complete a consistency review process. An online "Coastal Project Questionnaire" is available to assist project applicants. All state permits are routed through the Division of Governmental Coordination	NO	YES
ARIZONA	Los Angeles/spl	**Arizona Dept. of Envir. Quality (DEQ)** **www.adeq.state.az.us** **TEL: 602/207-4525** USACE must release a public notice on a Section 404 action before a Section 401 application can be submitted.	No state wetland regulations.	Not applicable	NO	YES
ARKANSAS	Little Rock/swl Memphis/mvm Vicksburg/mvk	**Arkansas Dept. of Envir. Quality State Permits Branch, Water Div.** **www.adeq.state.ar.uswater/ branchpermits.htm#401** **TEL: 501/682-0645** Section 401 certification is based on compliance with Arkansas' Pollution Control and Ecology Commission regulations, which established water-quality standards for surface waters in Arkansas.	No state wetland regulations. The Arkansas Soil and Conservation Service administers a wetland and riparian zone tax credit program that gives income tax credit for developing or restoring wetlands/riparian zones on private land. www.accessarkansas.org/aswcc/Page18.html Arkansas Game and Fish Commission acts as a wetland restoration agency, without permit jurisdiction. For additional information on inter-agency wetland efforts, call 800/364-4263. www.mawpt.org/default.asp	Not applicable	NO	YES

Table 6-1. Summary of State Wetland Protection and Regulatory Programs[1] continued

State	Responsible USACE District[2]/ Dist. Iden. Code	Section 401 Water Quality Certification Program[3]	State Wetlands Program[4]	Coastal Resources Program	Joint State/Fed. Permit App.[5]	Application available online?
CALIFORNIA	Los Angeles/spl Sacramento/spk San Francisco/spn	**State Water Resources Control Board (SWRCB)** **www.swrcb.ca.gov** **TEL: 916/341-5254** The SWRCB has delegated authority for 401 certification to its nine Regional Water Quality Control Boards. *See* discussion in text.	**Calif. Dept. of Fish and Game (DFG)** **www.dfg.ca.gov** **TEL: 916/653-7664** DFG has authority over wetland resources associated with rivers, streams, and lakes. Through its Lake and Streambed Alteration Program, DFG regulates work that will substantially divert or obstruct the natural flow of, or substantially change or use any material from the bed, channel, or bank of, any river, stream, or lake. Applicants enter into an agreement with DFG to ensure no net loss of wetland values or acreages. In addition, the SWRCB has advised that any discharges to "waters of the state," which include isolated waters and wetlands, is regulated by the California Porter-Cologne Water Quality Act and requires a permit from one of the 9 Regional Water Quality Control Boards.	**Calif. Coastal Commission (CCC)** **www.coastal.ca.gov** **TEL: 415/904-5200** **San Francisco Bay Conservation and Development Commission (BCDC) (for coastal resources within San Francisco Bay and Suisun Marsh)** **www.bcdc.ca.gov** **TEL: 415/352-3600** Development activities that modify land or water use in the coastal zone require a permit from either the CCC, BCDC, or local government responsible for a defined area through an approved local coastal program (LCP). Federal applicants must also submit requests for consistency determinations to either the CCC or BCDC for approval.	NO	YES
COLORADO	Albuquerque/spa Omaha/now Sacramento/spk	**Colorado Dept. of Public Health and Envir. Water Quality Control Div.** **www.cdphe.state.co.us** **TEL: 303/692-3500**	No state wetland regulations. **Colorado Div. of Wildlife** **www.wildlife.state.co.us/habitat/wetlands/wetlandsindex.asp** **email: alexchappell@state.co.us** The Colorado Division of Wildlife heads a "wetland protection program" that "protects" wetlands through voluntary/incentive based programs. Participants include federal, state, and local agencies, as well as private citizens and stakeholders. They have no permit authority.	Not applicable	NO	YES
CONNECTICUT	New England/nae	**Connecticut Dept. of Envir. Protection (DEP)** **Bureau of Water Mgmt.** **www.dep.state.ct.us/wtr** 401 certification for all tidal waters, including tidal wetlands, is administered by the Office of Long Island Sound Programs (OLISP)—860/424-3034. 401 certification for all other waters is administered by Inland Water Resources Division—860/424-3019.	**DEP** **Bureau of Water Mgmt.** **Inland Water Resources Division** **www.dep.state.ct.us/wtr** **TEL: 860/424-3019** DEP requires an Inland Wetlands and Watercourses Permit for activities undertaken by state agencies and affecting inland wetlands and watercourses. Such activities include fill, dredge, and clearing activities.	DEP, through OLISP, has two permit programs for coastal projects: 1) for Structures, Dredging and Fill; and 2) for Tidal Wetlands. DEP also requires Water Diversion and Stream Encroachment Line Permits. *See* online guide. www.dep.state.ct.us/pao/userguid.htm TEL: 860/424-3003	NO	YES
DELAWARE	Philadelphia/nap	**Delaware Dept. of Natural Res. (DNR)** **www.dnrec.state.de.us** **TEL: 302/739-4691** DNR requires individual 401 certification for all USACE individual permit applications and certain NWPs. Other NWPs have been automatically approved or waived.	No state wetland regulations.	**DNR** **Div. of Soil & Water Conservation** **Coastal Zone Evaluation** **www.dnrec.state.de.us** **TEL: 302/739-3451** Coastal wetlands are managed by DNR, Coastal Zone Evaluations division.	NO	NO
FLORIDA	Jacksonville/saj	**Florida Dept. of Envir. Protection (DEP) and Water Mgmt. Divisions** **www.dep.state.fl.uswater/wetlands/index.htm** 401 certifications are issued through the wetland permitting process. *See* State Wetlands Program discussion at right. *See* discussion in text.	**Florida DEP** **Water Mgmt. Divisions** **www.dep.state.fl.us/water/wetlands/index.htm** The state regulates wetlands through two different permitting programs, dependent on the project's location in the state. Applicants apply to DEP through a joint permitting process, which meets the requirements of the federal Section 404 program, the state coastal program, and the state/regional ERP/WRPP programs. State approval of the permit application constitutes federal 401 certification, unless	**DEP** **Div. of Envir. Resource Permitting, Bureau of Beaches and Coastal Systems** **www.dep.state.fl.us/beaches** **TEL: 850/488-1262** Coastal resources are protected under the Beach and Shore Preservation Act, which is implemented by the Bureau of Beaches and Coastal Systems. There are four coastal regulatory programs:	YES	YES

Table 6-1. Summary of State Wetland Protection and Regulatory Programs[1] continued

State	Responsible USACE District[2]/ Dist. Iden. Code	Section 401 Water Quality Certification Program[3]	State Wetlands Program[4]	Coastal Resources Program	Joint State/Fed. Permit App.[5]	Application available online?
FLORIDA continued			certification is specifically waived in the permit conditions. Local governments also have supplemental processes that commonly exceed state requirements. Additionally, USACE has issued a State Programmatic General Permit for certain minor activities (*see* sidebar in text). **Envir. Resource Permit Program (ERP)** – Represents merging of former WRPP and management and storage of surface waters permitting program (MSSW). – Applies statewide except within the panhandle. *See* WRPP. – Permits for dredge/fill of tidal/freshwater wetlands and alterations of the landscape, including the creation or alteration of wetlands and other surface waters, alterations of uplands that affect flooding, and all stormwater management activities. – Implemented cooperatively at the state level by DEP and at the regional level by four of the state's five water mgmt districts. Jurisdiction outlined in operation agreements. Local governments may also enact additional wetland regulations. – Grandfathered, exempt and general permit activities. Grandfathered activities follow old system. *See* WRPP. **Wetland Resource Permitting Program (WRPP)** – Original program for Wetland permitting in Florida. – Applies in panhandle (Northwest Florida Water Management District). – Permits for dredge/fill and construction of structures in the landward extent of wetlands and other surface waters. – Implemented cooperatively at state level by DEP and at the regional level by five state water mgmt districts. Primary permitting process for Northwest Florida Water Mgmt. District—other districts use when activities are grandfathered. Jurisdiction outlined in operation agreements. Local governments may also enact additional wetland regulations. – Exempt and general permit activities.	1. Coastal Construction Permit Program, which regulates construction projects seaward of the mean high water line 2. Joint Coastal Permit Program, which covers the requirements of both the coastal construction permit and ERP programs 3. Coastal Construction Control Line Permit Program, which regulates projects that occur in the portion of beach/dune system that is subject to significant fluctuations caused by wind/wave forces; and 4. Coastal Zone Protection Program, which extends special siting and building code standards within the coastal zone (administered by Florida Department of Community Affairs (DCA))		
GEORGIA	Jacksonville/saj Savannah/sas	**Georgia Dept. of Natural Res. (DNR)** **Water Protection Branch** **Non-Point Source Program** **www.dnr.state.ga.us/dnr** **TEL: 404/675-6240**	No state wetland regulations.	**DNR** **Coastal Resources Div.** **www.dnr.state.ga.us/dnr/coastal** **TEL: 912/264-7218** Coastal Marshland Protection Act: authorizes DNR, Coastal Resources Division, to manage certain activities and structures in tidal wetlands and require Marshland Protection Committee Permits for other activities. The law identifies 14 tidal marshland plants that are used to delineate tidal wetlands and uplands. Shore Protection Act: authorizes DNR, Coastal Resources Division, to protect the state's shoreline, limiting activities in shore areas and requiring permits for certain activities and structures on the beach.	NO	YES

Table 6-1. Summary of State Wetland Protection and Regulatory Programs[1] continued

State	Responsible USACE District[2]/ Dist. Iden. Code	Section 401 Water Quality Certification Program[3]	State Wetlands Program[4]	Coastal Resources Program	Joint State/Fed. Permit App.[5]	Application available online?
HAWAII	Honolulu/poh	**Hawaii Dept. of Health (DOH) Envir. Mgmt. Div. Clean Water Branch www.hawaii.gov/doh/eh/cwb/forms TEL: 808/586-4309** DOH provides blanket certification for certain NWPs.	No state wetland regulations. **Hawaii Dept. of Land and Natural Resources (DLNR) www.state.hi.us/dlnr/cwrm TEL: 808/587-0214** DLNR, Commission on Water Resources, processes and enforces stream channel alteration permits, stream division work alteration permits, and amendments to interim instream flow standards. Each of these programs could indirectly affect wetland resources.	**Dept. of Business, Economic Development and Tourism (DBEDT) www.state.hi.us/dbedt/czm TEL: 808/587-2809** Hawaii's coastal zone management program is administered by the DBEDT Office of Planning. DBEDT has delegated to some counties and cities the authority to make federal consistency determinations.	NO	YES
IDAHO	Walla Walla/nww	**Idaho Dept. of Health and Welfare Div. of Envir. Quality Surface Water www2.state.id.usdeq/water/water1.htm** USACE issues joint public notice for 401 certifications associated with Individual Permits	No state wetland regulations NOTE: State web site with inclusive list of all state/federal permits. www.oneplan.org	Not applicable	NO	YES
ILLINOIS	Chicago/lrc Louisville/lrl Memphis/mvm Rock Island/mvr St. Louis/mvs	**Illinois Envir. Protec. Agency (EPA) Bureau of Water www.epa.state.il.us/water TEL: 217/782-9039** Applicants prepare a joint application for EPA, Department of Natural Resources (DNR), and USACE, although each agency retains jurisdiction over its permitting process. 401 certification should be requested in the cover letter to EPA.	**Illinois Dept. of Natural Resources (DNR) Office of Water Resources http://dnr.state.il.us/waterresources TEL: 217/782/3863** DNR regulates work in streams, lakes (including Lake Michigan), and rivers to protect public access, floodways, and endangered species. Although it has no specific wetland policy, it reviews the joint application for impacts on wetlands relative to the above topic areas.	**DNR Lake Michigan Programs Section TEL: 312/793-3123** The Lake Michigan Programs Section manages the state's interest in Lake Michigan. Illinois does not have a federally approved coastal zone management program.	YES	YES
INDIANA	Detroit/lre Louisville/lrl	**Indiana Dept. of Envir. Mgmt. (DEM) Office of Water Quality www.in.gov/idem/water/planbr/401/401home.html TEL: 317/233-2481** DEM requires individual certifications for all USACE Individual Permits and many NWPs.	No state wetland regulations. DEM and USACE are working on a regional general permit that will replace the majority of the new NWPs. Although there are no formal state wetland regulations, the Indiana Division of Fish and Wildlife promotes an Indiana Wetland Conservation Plan. Also, the Indiana Department of Natural Resources (DNR) requires permits for work that alters a shoreline or lake or that occurs in a floodplain.	DNR is currently developing a coastal program for the state of Indiana. TEL: 317/233-0132	NO	YES
IOWA	Rock Island/mvr	**Iowa Dept. of Natural Resources (DNR) Water Resource Section www.state.ia.us/government/dnr/organiza/epd TEL: 515/281-6615** Joint application for 401 certification and USACE 404 permit. If USACE requires an Individual Permit, the state reviews the 401 certification request. If USACE requires an NWP, no 401 certification is issued by the state.	No state wetland regulations.	Not applicable	NO	YES

Table 6-1. Summary of State Wetland Protection and Regulatory Programs[1] continued

State	Responsible USACE District[2]/ Dist. Iden. Code	Section 401 Water Quality Certification Program[3]	State Wetlands Program[4]	Coastal Resources Program	Joint State/Fed. Permit App.[5]	Application available online?
KANSAS	Kansas City/nwk Tulsa/swt	**Kansas Dept. of Health and the Envir. Bur. of Water, Non-Point Source Section** **www.kdhe.state.ks.us** **TEL: 785/296-4195** Joint application for 401 certification and USACE 404 permit. If USACE requires an Individual Permit, the state reviews the 401 certification request. The state and USACE have also applied regional conditions for many NWPs, all of which are found online at the Kansas City USACE website. The state also requires that applicants prepare a water quality protection plan (WQPP) if they propose to work within 0.5 mile of special aquatic life use waters (which include wetlands). The WQPP need not be approved, but will remain on file with the state to ensure that, in the event of a public complaint, the applicant is meeting all measures outlined in the WQPP/ state water-quality standards.	No state wetland regulations. Although there are no specific regulations in Kansas to protect wetlands, the Kansas Department of Wildlife and Parks (DWP) issues permits to protect threatened and endangered species on DWP-owned lands. DWP could protect wetlands if they provide habitat for these species.	Not applicable	NO	YES
KENTUCKY	Huntington/orh Louisville/lrl Memphis/mvm Nashville/orn	**Kentucky Natural Resources and Envir. Protection Cabinet** **Dept. of Envir. Protection (DEP)** **Div. of Water** **www.nr.state.ky.us** **TEL: 502/564-2150** For USACE Individual Permits, the public notice serves as the 401 application. For USACE NWPs, the project applicant must submit a request for certification to DEP. 401 certifications are good for one year.	No state wetland regulations. Although there are no additional state permit requirements, wetland mitigation plans and alternatives analysis submitted to USACE through the Section 404 program are reviewed by DEP.	Not applicable	NO	YES
LOUISIANA	New Orleans/mvn Vicksburg/mvk	**Louisiana Dept. of Envir. Quality (DEQ)** **Office of Envir. Services** **www.deq.state.la.us** **TEL: 225/765-0664**	No state wetland regulations. *See* notes for Coastal program.	**Louisiana Dept. of Natural Resources (DNR)** **Coastal Mgmt. Div.** **www.savelawetlands.org** **TEL: 800/267-4019** Coastal wetlands are protected under the Louisiana Coastal Resources Program, which requires coastal use permits (CUPs) for impacts on any coastal resources in the Louisiana coastal zone. DNR also completes federal consistency determinations.	NO	YES
MAINE	New England/nae	**Maine Dept. of Envir. Protection (DEP)** **Bureau of Land and Water Quality (BLWQ)** **Div. of Land Resource Regulation** **www.state.me.us/dep/blwq** **(Regional oversight—*see* website for points of contact)** **Maine Land Use Reg. Comm. (LURC)** **www.state.me.us/doc/ lurc/lurchome.htm** **TEL: 207/287-2631**	**DEP** **BLWQ Div. of Land Resource Regulation** **www.state.me.us/dep/blwq** Wetlands are protected under the Natural Resources Protection Act (NRPA). A permit is required if an activity is in, on, over, adjacent to, or operated in a manner that material or soil can be washed into a protected natural resource (which includes coastal/freshwater wetlands). There are four types of permits:	*See* State Wetlands discussion at left for information on coastal resources program.	YES	YES

Table 6-1. Summary of State Wetland Protection and Regulatory Programs[1] continued

State	Responsible USACE District[2]/ Dist. Iden. Code	Section 401 Water Quality Certification Program[3]	State Wetlands Program[4]	Coastal Resources Program	Joint State/Fed. Permit App.[5]	Application available online?
MAINE continued		401 certification is incorporated into the processes for other state permits. Responsibility is divided between BLWQ and LURC, depending on which agency is certifying the specific project.	1. Individual NRPA Permit: for activities with significant impacts, including those in coastal wetlands. Applications are filed with DEP and forwarded to USACE on joint permit application. Some activities require preapplication meeting. 2. Permit by Rule (PBR): for activities predetermined to have no significant impact. A notification form is sent to DEP; USACE has guidance stating that most PBRs qualify for Programmatic General Permits (PGPs), which are non-notifying. If a PBR does not qualify for a PGP, specific information must be sent to USACE independent of notification to DEP. Also, PBR application may require a "request for approval of timing of activity" from Department of Inland Fisheries and Wildlife, Department of Marine Resources, or Department of Atlantic Salmon Commission. 3. General Permit: for specific activities (i.e., in cranberry bogs in freshwater wetlands) with certain limitations. 4. Tier Review: used if only freshwater wetland (FWW) impacts will result. Level of review is dependent on level of impact (tiered system). Tier 1: activities in FWW that meet certain criteria and are less than 15,000 square feet. Tier 2: activities in FWW that meet certain criteria and are between 15,000 square feet and one acre. Level of review for Tier 2 activities is greater than for Tier 1. Shoreland Zoning Law: designed to protect water resources, including wetlands, from pollution, degradation, and flooding. Requires that municipalities adopt Shoreland Zoning Maps and Ordinances outlining what activities can occur in certain areas. Shorelands are defined as areas within 250 ft of the high water line/edge of coastal/freshwater wetlands, rivers, salt water bodies, and within 75 ft of the OHWM in streams. Administered through municipalities. DEP, BLWQ, Shoreland Zoning Unit offers regional assistance. In addition, in areas with no local government to establish land use controls (i.e., unorganized areas), the Maine Land Use Regulation Commission (LURC) requires that applicants obtain an LURC permit.	Also, Coastal Dredging Permit—Joint application between DEP/USACE specific to coastal waters.		
MARYLAND	Baltimore/nab Pittsburgh/lrp	**Maryland Dept. of Envir. (MDE)** **Wetlands and Waterways Program** **www.mde.state.md.us/wetlands** **Regulatory Services Coordination (RSC) Office** **TEL: 410/631-8087** Section 401 certification is incorporated into the wetlands joint application process.	**MDE** **Wetlands and Waterways Program** **www.mde.state.md.us/wetlands** **RSC Office** **TEL: 410/631-8087** Divisions within the Wetlands and Waterways Program— 1. Nontidal Wetlands and Waterways 2. Tidal Wetlands	**Maryland Dept. of Natural Resources (DNR)** **Coastal Zone Mgmt Division** **www.dnr.state.md.us/bay/czm/aboutczm.html** **TEL: 410/260-8730** Meet federal consistency requirements and provide guidance/oversight for projects in the coastal zone.	YES	YES

Table 6-1. Summary of State Wetland Protection and Regulatory Programs[1] continued

State	Responsible USACE District[2]/ Dist. Iden. Code	Section 401 Water Quality Certification Program[3]	State Wetlands Program[4]	Coastal Resources Program	Joint State/Fed. Permit App.[5]	Application available online?
MARYLAND continued			Activities that occur within wetlands and waters of the state must meet the requirements of Maryland's State Programmatic General Permit. This permit is administered by USACE and MDE through a joint application process. The joint application is used for tidal/nontidal activities that alter wetlands, including a 25-ft buffer around nontidal wetlands and 100-ft buffer around tidal wetlands/ wetlands of special state concern. Joint applications are processed through the RSC Office and distributed to all federal/ state agencies as appropriate.	Other relevant programs: Chesapeake Bay Critical Area Protection Program - Critical area defined as any area within 1,000 ft of the Bay or its tidal waters—requires 100 ft buffer from MHW - Critical area governed by local Critical Area Programs—Chesapeake Bay Critical Area Commission and DNR review/ comment on development proposals; local jurisdictions enforce using zoning regulations - Also has specific provisions for nontidal wetlands - For information by district, see: www.dnr.state.md.us/ criticalarea/faq.html		
MASSA-CHUSETTS	New England/nae New York/nan	**Massachusetts Dept. of Envir. Protection (DEP)** **Bureau of Resource Protection** **Div. of Wetlands** **www.state.ma.us/dep** **(telephone by regional office—*see* web page)** **DEP** **Bureau of Resource Protection** **Waterways Program** **www.state.ma.us/dep** **TEL: 617/292-5695** 401 certification for fill and excavation projects in waters and wetlands is issued by the Division of Wetlands; certification for dredging and dredge material disposal is issued by the Waterways Program. However, most projects that are approved by the local community conservation commission under the Wetlands Protection Act (WPA) do not need additional 401 review.	**DEP** **Bureau of Resource Protection** **Wetlands Protection Program** **www.state.ma.us/dep** **TEL: 617/292-5695** The WPA governs work in or within 100 feet of wetlands. It is administered locally by voluntary Community Conservation Commissions (Commissions) that are appointed by the city council. DEP oversees the WPA on the state level. DEP develops regulations, gives technical training to commissioners, and hears appeals. The WPA permit process is summarized below. 1. Applicant files 'Request for Determination of Applicability' with appropriate Commission (optional). 2. Applicant files 'Notice of Intent' with Commission if work will alter a resource area. 3. Commission reviews application, completes site visit, and presents at public hearing. 4. Commission issues 'Order of Conditions' to approve/deny project. 5. Appeals go to DEP. Other Wetland Programs: - Rivers Protection Act: regulates work within 200 ft of a riverfront area. Activities are not authorized if there is a practicable alternative or if there will be a significant adverse impact. Administered by Commissions/DEP - More than 100 Massachusetts communities have local wetland protection bylaws in addition to state/federal regulations. Check with Commissions. - NOTE: DEP Wetland Conservancy Program maps all state wetlands. There are permanent restrictions on certain wetlands.	The Massachusetts Office of Coastal Zone Management reviews federal activities in the coastal zone and acts as a liason for other agencies to consult on coastal zone issues. It is not a permitting authority. www.state.ma.us/czm TEL: 617/626-1200	NO	YES

Table 6-1. Summary of State Wetland Protection and Regulatory Programs[1] continued

State	Responsible USACE District[2]/ Dist. Iden. Code	Section 401 Water Quality Certification Program[3]	State Wetlands Program[4]	Coastal Resources Program	Joint State/Fed. Permit App.[5]	Application available online?
MICHIGAN	Detroit/lre Louisville/lrl	**Michigan Dept. of Env. Quality (DEQ)** **Surface Water Quality Div.** **www.deq.state.mi.us/swq** **TEL: 517/373-1949** 401 certification is generally completed through DEQ processing of other permits, but if there is only a federal action, DEQ will issue a 401 certification independent of any other permit process.	**DEQ** **Land and Water Mgmt. Div.** **www.deq.state.mi.us/lwm** **TEL: 517/373-1170** Wetlands are regulated primarily under Part 303 of the Natural Resources and Environmental Protection Act. A wetland construction permit is required if a project will dredge, drain, fill, or construct a structure in a wetland. Michigan is also the only state that has been given the authority by EPA to administer the federal 404 program, with certain exceptions for Section 10 waters and more complex projects/ impacts. All permit applications are filed with the Permit Consolidation Unit (PCU)	DEQ administers by issuing permits, as described in State Wetlands discussion at left.	YES	YES
MINNESOTA	St. Paul/mvp	**Minnesota Dept. of Natural Resources (DNR)** **Minnesota Pollution Control Agency** **www.pca.state.mn.us/ netscape4.html** **TEL: 651/296-6300** Minnesota's 401 program is administered by DNR, while local governments enforce the Minnesota Wetland Conservation Act. Applicants submit a joint permit to the appropriate local government, DNR, and USACE; this permit meets the requirements of the Minnesota Wetland Conservation Act, CWA Section 401, CWA Section 404, and DNR permit requirements for work in public waters.	Local Government Units (LGU); Minnesota Wetland Conservation Act *See* Section 401 discussion at left.	Minnesota's coastal zone management program is not approved. For additional information, contact DNR. www.dnr.state.mn.us/waters/ lakesuperior/index.html TEL: 218/834-6625	YES	YES
MISSISSIPPI	Memphis/mvm Mobile/sam Nashville/orn Vicksburg/mvk	**Mississippi Dept. of Envir. Quality (DEQ)** **Office of Pollution Control** **Envir. Permit Div.** **www.deq.state.ms.us/newweb/ homepages.nsf** **TEL: 601/961-5073**	**DEQ** **Office of Pollution Control** **Envir. Permit Div.** **www.deq.state.ms.us/newweb/ homepages.nsf** **TEL: 601/961-5073** Applicants submit a joint application to USACE, Mississippi Department of Wildlife, Fisheries, and Parks; Bureau of Marine Resources; and DEQ. A joint public notice is also issued through this process.	**Mississippi Dept. of Marine Resources (DMR)** **Coastal Ecology Permit Office** **www.dmr.state.ms.us** **TEL: 228/374-5000** DMR requires applicants to obtain a Coastal Zone Wetland Permit for activities in the coastal zone. This request is also submitted through the joint application process. DMR completes federal consistency determinations also.	YES	YES
MISSOURI	Kansas City/nwk Little Rock/swl Memphis/mvm Rock Island/mvr St. Louis/mvs Tulsa/spk	**Missouri Dept. of Natural Resources (DNR)** **Water Pollution Control Program** **www.dnr.state.mo.us/deq/ wpcp/homewpcp.htm** **TEL: 573/751-1300** DNR requires 401 certification any time an applicant proposes construction, earth movement, or placement/disposal of fill material in a wetland or other water body.	No state wetland regulations **DNR** **Water Resources Program** **TEL: 573/751-2867** **www.dnr.state.mo.us/dgls/wrp/wetlands.htm** DNR administers the state wetlands conservation plan through the Water Resources Program, which "encourages" the protection and restoration of wetlands. DNR also reviews projects that affect wetlands, although it does not have regulatory authority.	Not applicable	NO	YES

State	Responsible USACE District[2]/ Dist. Iden. Code	Section 401 Water Quality Certification Program[3]	State Wetlands Program[4]	Coastal Resources Program	Joint State/Fed. Permit App.[5]	Application available online?
MONTANA	Omaha/nwo	**Montana Dept. of Nat. Resources and Conservation (DNRC)** **Water Resources Div.** **www.dnrc.state.mt.us/permit.html** **TEL: 406/444-6601** DNRC issues 401 certifications under a "3A authorization," or short-term exemption from Montana's surface water-quality standards.	No state wetland regulations. Wetlands are protected indirectly through enforcement of state laws designed to manage other types of state resources (e.g., floodplains, streambeds). Applicants file a joint Application for Proposed Work in Montana's Streams, Wetlands, Floodplains and other Water Bodies. The permit application covers the following requirements: 1. Natural Streambed and Land Preservation Act Permits (310 permits). Issued by the local conservation districts for activities that modify the bed/bank of a stream. 2. Floodplain Permit. Issued by the County Floodplain Administrator for activities that occur in floodplains. 3. Stream Protection Act Permits (SP 124). Issued by the Montana Department of Fish, Wildlife and Parks for projects that affect the natural shape of a stream. 4. Section 404/10. Issued by USACE for impacts on waters of the United States. 5. 3A Authorization. Issued by DEQ. *See* 401 certification. 6. Navigable Rivers Land Use License/ Easement. Issued by DNRC for impacts on navigable waters.	Not applicable	NO	YES
NEBRASKA	Omaha/now Sacramento/spk	**Nebraska Dept. of Env. Quality (DEQ)** **Surface Water Section** **www.deq.state.ne.us** **TEL: 420/471-2186**	No state wetland regulations. Currently, DEQ "advises" applicants to mitigate impacts on wetlands outside USACE jurisdiction on the assumption that failure to do so could be a violation of the state wetland/water-quality standards.	Not applicable	NO	YES
NEVADA	Sacramento/spk	**Nevada Div. of Env. Protection (DEP)** **Bureau of Water Quality Planning** **www.ndep.state.nv.us/bwqp** **TEL: 775/687-4670**	No state wetland regulations. DEQ reviews projects with impacts on wetlands through the USACE Section 404 process and through Section 402 applications.	Not applicable	NO	YES
NEW HAMPSHIRE	New England/nae	**New Hampshire Dept. of Envir. Services (DES)** **Water Div.** **www.des.state.nh.us/wmb/ section_401.htm** **TEL: 603/271-3503**	**DES** **Water Division** **Wetlands Bureau** **www.des.state.nh.us/wetlands** **TEL: 603/271-2147** RSA 482-A (NH statute) requires applicants to obtain a permit if they propose to dredge, fill, or construct a structure in a wetland or other water of the state. There are two permit categories: 1) Standard Dredge and Fill Application and 2) Minimum Impact Expedited Application. RSA 482-A also permits activities in coastal resources in the state. For federal compliance, USACE has approved a New Hampshire Programmatic General Permit that integrates some minor Section 10/Section 404 permits into other New Hampshire permitting actions, but requires separate general permits for all other actions. NOTE: New Hampshire Conservation Commissions act on behalf of local municipalities to protect local community natural resources. They can request to review dredge/fill permit applications filed with the wetlands bureau and are the only municipal bodies with the authority to intervene in those permit actions. www.nhacc.org/nhacc.htm	**NH Office of State Planning** **Coastal Programs Office** **http://webster.state.nh.us/coastal** **TEL: 603/271-2155** The Coastal Programs Office implements the Coastal Zone Management Program and coordinates with DES, Water Division, for permitting in coastal areas. Also completes federal consistency reviews.	NO	YES

State	Responsible USACE District[2]/ Dist. Iden. Code	Section 401 Water Quality Certification Program[3]	State Wetlands Program[4]	Coastal Resources Program	Joint State/Fed. Permit App.[5]	Application available online?
Table 6-1. Summary of State Wetland Protection and Regulatory Programs[1] continued						
NEW JERSEY	New York/nan Philadelphia/nap	**New Jersey Dept. of Envir. Protec. (DEP)** **Div. of Water Quality** **www.state.nj.us/dep/dwq** TEL: **609/292-7977**	**DEP** **Land Use Mgmt. & Compliance Division** **Land Use Regulatory Program (LUR)** **www.state.nj.us/dep/landuse** Wetlands are protected under the New Jersey Wetland Protection Act. Applicants must obtain a Letter of Interpretation (LOI) from DEP to officially verify the presence, absence, and required transition area or buffer of freshwater wetlands before applying for a Freshwater Wetlands Permit. Most permits issued are "general" permits for minor activities; more complex projects may require an "individual" permit and extensive alternatives analysis.	DEP requires a Coastal Area Facility Review Act (CAFRA) Permit, Waterfront Development Permit, and/or a Coastal Wetlands Permit for activities within CAFRA boundaries, development activities in coastal wetlands, development activities in tidally flowed waterways anywhere in the state, or development activities in areas adjacent to tidally flowed waterways outside CAFRA's jurisdiction. All permits are issued by LUR, but require review by the County Board and County Environmental Commission. *See* text. The state coastal zone management program is administered by DEP, Office of Coastal Planning and Program Coordination. TEL: 609/777-3241 Additional protection for coastal wetlands is provided by the Tidelands Act (requires permit for development in tidally flowed lands) and the Flood Hazard Area Control Act (requires permit for construction activities in the floodplain).	NO	YES
NEW MEXICO	Albuquerque	**New Mexico Envir. Dept.** **Surface Water Quality Bureau (SWQB)** **Nonpoint Source Section** **www.nmenv.state.nm.us/swqb/swqb.html** TEL: **505/827-1041** Applicants must obtain 401 certification prior to work in stream channels, lakes, or wetlands. A joint application is submitted to the SWQB and USACE. Approval is granted independently by both agencies. Currently, SWQB issues blanket certifications for some NWPs in intermittent and ephemeral drainages. Individual certification is required for all work in perennial streams and all USACE Individual Permit applications.	No state wetland regulations.	Not applicable	YES	YES
NEW YORK	Buffalo/lrb New York/nan Philadelphia/nap Pittsburgh/lrp	**New York Dept. of Envir. Conservation (DEC)** **Div. of Envir. Permits** **www.dec.state.ny.us** TEL: **518/457-7424** 401 certification is issued as a Protection of Waters Permit. Required any time an applicant proposes to disturb the bed/bank of a stream or excavate or place fill in a navigable water adjacent to a wetland.	**DEC** **Div. of Envir. Permits** **www.dec.state.ny.us** TEL: **518/457-7424** **Adirondack Park Agency (APA)** TEL: **518/891-4050** Environmental Permits are issued under the Uniform Procedures Act (UPA), NY State Environmental Conservation Law, 1977. The following permits are submitted to DEC on a joint federal/state application, which is in turn forwarded to USACE: 1. Freshwater Wetland Permit. DEC maps all protected wetlands (outside Adirondack Park) and classes them according to their functions and values. Permits are required for activities that adversely affect wetlands larger than 12.4 acres, including a 100-ft buffer, or smaller wetlands that have unusual local importance.	**New York Dept. of State** **Div. of Coastal Resources (DCR)** **www.dos.state.ny.us/cstl/cstlwww.html** TEL: **518/474-6000** DCR completes federal consistency determinations for projects in the coastal zone. *See* State Wetlands discussion at left for additional information on coastal permitting requirements.	YES	YES

Table 6-1. Summary of State Wetland Protection and Regulatory Programs[1] continued

State	Responsible USACE District[2]/ Dist. Iden. Code	Section 401 Water Quality Certification Program[3]	State Wetlands Program[4]	Coastal Resources Program	Joint State/Fed. Permit App.[5]	Application available online?
NEW YORK continued			2. Tidal Wetlands Permit. Tidal wetlands occur anywhere tidal inundation occurs on a daily, monthly, or intermittent basis. Maps of all tidal wetlands are maintained by DEC regional offices and county clerk offices. DEC regulates any activity that adversely affects any tidal wetland, including a 300-ft buffer. Adirondack Park: freshwater wetlands less than one acre or adjacent to open water in Adirondack Park are regulated by APA.			
NORTH CAROLINA	Huntington/orh Norfolk/nao Wilmington/saw	**North Carolina Dept. of Envir. Health and Natural Resources (DENR)** **Div. of Water Quality** **Wetlands/401 Certification Unit** **http://h2o.enr.state.nc.us** **TEL: 919/733-9646** DENR issues an individual 401 certification if an individual Section 404 permit is required. A general certification is required by DNR for NWPs, general permits, and regional permits.	**DENR** **Div. of Water Quality** **Wetlands/401 Certification Unit** **http://h2o.enr.state.nc.us** **TEL: 919/733-9646** Wetlands are regulated primarily through the 401 process. In response to the SWANCC case, isolated wetlands and isolated surface waters no longer regulated by USACE are now regulated by DENR, Division of Water Quality.	**DENR** **Div. of Coastal Mgmt.** **http://dcm2.enr.state.nc.us** **TEL: 888/472-6278** Applicants must obtain a Coastal Area Management Act (CAMA) permit from DENR, Division of Coastal Management, if the development affects an "Area of Environmental Concern" and occurs within one of 20 coastal counties. Areas of Environmental Concern are categorized in four groups: 1. Estuaries and ocean system 2. Ocean hazard system 3. Public water supplies 4. Natural and cultural resource areas CAMA permits are categorized as major, general, or minor. DCM coordinates with 14 other state and federal agencies on all major permits.	NO	YES
NORTH DAKOTA	Omaha/nwo	**North Dakota Dept. of Health** **Envir. Health Section** **Div. of Water Quality** **www.health.state.nd.us/ndhd/environ/wq** **TEL: 701/328-5210** North Dakota does not actually have a 401 certification application. USACE forwards the Division of Water Quality a copy of its public notice, to which the division of Water Quality responds.	No state wetland regulations.	Not applicable	NO	NO
OHIO	Buffalo/lrb Huntington/orh Louisville/lrl Pittsburgh/lrp	**Ohio EPA** **Div. of Surface Waters** **www.epa.state.oh.us** **TEL: 614/644-2001**	**Ohio EPA** **Div. of Surface Waters** **www.epa.state.oh.us** **TEL: 614/644-2001** In response to the SWANCC case, Ohio passed Ohio House Bill 231 for additional protection of isolated wetlands. As a result, applicants proposing development in an isolated wetland must obtain a permit from Ohio EPA. All other wetlands, with the exception of those in the coastal zone, are only regulated by USACE. Ohio is currently reviewing another proposal to regulate all other waters/wetlands in the state.	**Dept. of Natural Resources (DNR)** **Real Estate and Land Mgmt. Div.** **Coastal Mgmt. Program** **Coastal Services Center** **www.dnr.state.oh.us/coastal** **TEL: 419/626-7980** Within Ohio's coastal zone, applicants use the Coastal Permit and Lease Application form to apply for coastal erosion area permits, shore structures permits, and submerged lands permits. These consolidated forms are submitted to DNR.	NO	YES

Table 6-1. Summary of State Wetland Protection and Regulatory Programs[1] continued

State	Responsible USACE District[2]/ Dist. Iden. Code	Section 401 Water Quality Certification Program[3]	State Wetlands Program[4]	Coastal Resources Program	Joint State/Fed. Permit App.[5]	Application available online?
OKLAHOMA	Tulsa/swt	**Oklahoma Dept. of Envir. Quality** **www.deq.state.ok.us** **TEL: 405/702-8194**	No state wetland regulations. **Oklahoma Conservation Commission** **TEL: 405/521-2384** **www.okcc.state.ok.us** The Oklahoma Conservation Commission (OCC) and Oklahoma's 88 Conservation Districts are responsible for conservation of renewable natural resources. In 1996, the OCC developed a wetland management strategy for the state, called the Oklahoma Comprehensive Wetland Conservation Plan. The plan promotes private and public cooperation in managing wetlands through education, technical assistance, and financial incentives. The OCC does not have permitting authority.	Not applicable	NO	NO
OREGON	Portland/nwp	**Oregon Dept. of Envir. Quality (DEQ)** **www.state.or.us** **TEL: 503/229-5371** *See* wetlands discussion at right.	**Oregon Div. of State Lands (DSL)** **http://statelands.dsl.state.or.us/doyouneed.htm** **TEL: 503/378-3805** An applicant proposing to dredge, fill, or otherwise alter a water of the state must obtain a "removal-fill" permit from DSL. DSL must authorize a project under the program if it would result in greater than 50 cy of fill material being placed in a wetland. A joint DSL/USACE permit application with USACE/DSL is used and forwarded to DEQ for 401 certification. An online permit handbook is available at DEQ's website. Several cities and counties in Oregon also provide protection for wetland resources.	**Dept. of Land Conservation and Development (DLCD)** **Ocean and Coastal Program** **http://www.lcd.state.or.us/coast** **TEL: 503/373-0050** OCP staff administers the statewide plan that guides development of local land use plans governing activities in Oregon's coastal zone. OCP staff review local and state permits and review federal consistency determinations. City/county planning departments are responsible for local permitting requirements. Other state agency programs, such as the Oregon Beach Bill (administered by Oregon Parks and Recreation) and the Removal/Fill Law (administered by DSL) are incorporated into the CZMP and provide additional wetland protections. The DLCP web site provides a list of other state agencies whose programs could indirectly/directly result in the protection of coastal wetlands.	YES YES	YES YES
PENNSYLVANIA	Baltimore/nab Pittsburgh/lrp Philadelphia	**Penn. Dept. of Envir. Protection (DEP)** **Bureau of Water Supply and Wastewater Mgmt.** **www.dep.state.pa.us/dep/deputate/watermgt/watermgt.htm** **TEL: 717/787-5017** 401 certification is integrated into other required permits and approvals from DEP. Certification requires compliance with Pennsylvania's Clean Streams Law. If no other permit is triggered, the Bureau of Water Supply and Wastewater Management will issue 401 certification on an individual basis.	**DEP** **Bureau of Watershed Mgmt.** **www.dep.state.pa.us/dep/deputate/watermgt/watermgt.htm** **TEL: 717/787-5267** Activities in wetlands and waters within the Commonwealth of Pennsylvania must comply with the Pennsylvania State Programmatic General Permit (PASPGP).	**DEP, Coastal Zone Mgmt. Program** **TEL: 717/789-2529** DEP completes consistency reviews and enforces local zoning laws through implementation of the state coastal zone management program.	YES	YES
RHODE ISLAND	New England/nae	**Rhode Island Dept. of Envir. Mgmt. (DEM)** **Bureau of Natural Resources** **Office of Water Resources** **www.state.ri.us/dem** **TEL: 401/277-6820**	**DEM** **Bureau of Natural Resources** **Office of Water Resources** **www.state.ri.us/dem** **TEL: 401/222-3961** DEM regulates all freshwater wetlands, including 50-ft buffers. DEM also regulates	**Coastal Resources Mgmt. Council (CRMC)** **TEL: 401/783-3370** **DEM Technical Assistance** **TEL: 401/222-6822** Coastal "Freshwater" Wetlands: DEM and CRMC have created a	YES	YES

Table 6-1. Summary of State Wetland Protection and Regulatory Programs[1] continued

State	Responsible USACE District[2]/ Dist. Iden. Code	Section 401 Water Quality Certification Program[3]	State Wetlands Program[4]	Coastal Resources Program	Joint State/Fed. Permit App.[5]	Application available online?
RHODE ISLAND continued		401 certification request processed with DEM freshwater wetland permit/USACE application (i.e., no application required). If there is no federal nexus, or if the project occurs in the coastal zone, an applicant must apply for a 401 certification directly from DEM.	100-ft corridors adjacent to either side of rivers/streams less than 10 ft wide, and 200-ft corridors adjacent to either side of rivers/streams more than 10 ft wide. Applicants must: 1. File a Request to Determine Presence of Wetlands; 2. File a Request to Verify the Delineated Edge of Wetlands; 3. File a Request for Preliminary Determination (where DEM determines if the project is outside its jurisdiction, is exempt, constitutes an "insignificant alteration, or is a "significant alteration.") 4. If the project constitutes a "significant alteration," file an Application to Alter Fresh Water Marsh. This requires a public review/approval. Joint application process: USACE has issued a programmatic general permit (PGP) for review of proposals in coastal and inland waters and wetlands; with some exceptions, any permit issued by DEM under the PGP serves as the federal permit as well.	line that generally follows certain state/local roads. If a project falls on the landward side of the line, a permit must be obtained from DEM (*see* state wetlands program). If a project falls shoreward of the line, CRMC will coordinate and oversee state and local agencies to ensure coastal zone management issues are incorporated into decisionmaking.		
SOUTH CAROLINA	Charleston/sac	**South Carolina Dept. of Health and Envir. Control (DHEC)** **Office of Envir. Quality Control** **Bureau of Water** **www.scdhec.net/water** **TEL: 803/898-4249**	No state wetland regulations.	**DHEC** **Office of Ocean and Coastal Resource Mgmt.** **www.scdhec.net/ocrm** **TEL: 843/744-5838** Wetlands in the state's coastal zone "critical area" are protected under the state's coastal zone management program. For wetland impacts, applicants apply to USACE for 404/401/coastal wetland permits. USACE coordinates with the state and issues a joint public notice. An online guide to wetland permitting can be found at http://scdhec.net/eqc/water/pubs/401guide	YES	YES
SOUTH DAKOTA	Omaha/nwo	**South Dakota Dept. of Envir. and Natural Resources** **Surface Water Quality Program** **www.state.sd.us/denr/ENVIRO/index.htm** **TEL: 605/773-3351**	No state wetland regulations. Management of wetlands is primarily delegated to NRCS.	Not applicable	NO	YES
TENNESSEE	Memphis/mvm Nashville/orn	**Tennessee Dept. of Envir. and Conservation (DEC)** **Div. of Water Pollution Control** **www.state.tn.us/environment/wpc** **TEL: 615/532-0625** If a federal permit is required (i.e., USACE NWP), the 401 process is initiated through the federal public notice process. If a federal permit is not required, Tennessee requires applicants to obtain an "Aquatic Resource Alteration Permit (ARAP)/Section 401 Certification." *See* state wetlands program at right.	**DEC** **Div. of Water Pollution Control** **www.state.tn.us/environment/wpc** **TEL: 615/532-0625** An ARAP is required if there is any physical alternation to a stream/wetland, including water withdrawal. ARAPs are divided into Individual and General Permits based on the type/extent of work. Additional permitting information can be obtained at the Environmental Assistance Center (888/891-TDEC), or online. www.state.tn.us/environment/permits/index.html	Not applicable	NO	YES

Table 6-1. Summary of State Wetland Protection and Regulatory Programs[1] continued

State	Responsible USACE District[2]/ Dist. Iden. Code	Section 401 Water Quality Certification Program[3]	State Wetlands Program[4]	Coastal Resources Program	Joint State/Fed. Permit App.[5]	Application available online?
TEXAS	Albuquerque/spa Fort Worth/swf Galveston/swg Tulsa/swt	**Texas Nat. Resource Conservation Commission (TNRCC)** **Attn: 401 Coordinator** **www.tnrcc.state.tx.us** **TEL: 512/239-1000** TNRCC uses the 401 process to "preserve wetland resources and the functions they provide." For smaller projects (i.e., less than 3 acres/1500 ft), an applicant can incorporate BMPs into the project description for automatic approval. All other requests require individual review and an alternatives analysis. In general, TNRCC does not allow discharges in wetlands if practicable alternatives with less adverse impacts exist. Joint public notice issued for Sections 401/404.	No state wetland regulations.	**Texas General Land Office (GLO)** **Coastal Div.** **www.glo.state.tx.us/coastal/cmp** GLO administers the Texas coastal zone management program.	NO	YES
UTAH	Sacramento/spk	**Utah Dept. of Envir. Quality (DEQ)** **Div. of Water Quality** **www.eq.state.ut.us** **TEL: 801/538-6329** 401 certifications are completed through review of the public notice issued by USACE. No actual permit application is submitted to DEQ.	No state wetland regulations.	Not applicable	NO	NO
VERMONT	New England/nae New York/nan	**Vermont Agency of Nat. Resources** **Dept. of Envir. Conservation (DEC)** **Water Quality Div., Wetlands Office** **www.anr.state.vt.us/dec/waterq** **TEL: 802/241-3770** Applicants should submit a letter to DEC requesting 401 certification (there is no permit application form).	**Vermont Agency of Nat. Resources DEC** **Water Quality Div.** **Wetlands Office** **www.anr.state.vt.us/dec/waterq** **TEL: 802/241-3770** Vermont Wetlands Rules establish a three-tier wetland classification system: – Class 1 and 2 wetlands, including their buffers, are protected under the wetland rules and identified on the Vermont Significant Wetland Inventory (VSWI) maps. Activities in Class 1 or 2 wetlands require a Conditional Use Determination (CUD) if the proposed use is not allowed under the wetland rules. CUDs are only approved if a project will not have an undue adverse effect on functions of significant wetlands. CUDs are submitted to the town clerk or regional planning commission for approval. – Class 3 wetlands do not provide significant functions and values; activities in Class 3 wetlands do not require a CUD. Local Protection: "Act 250" is a local act to protect lakes, streams, and wetlands specifically associated with major subdivisions and developments – DEC, Water Quality Division reviews Act 250 applications and makes recommendations on approval to the District Environmental Commissions (local government) and Environmental Board (local commission responsible for major subdivisions and developments in Vermont).	*See* State Wetlands discussion at left.	NO	YES

Table 6-1. Summary of State Wetland Protection and Regulatory Programs[1] continued

State	Responsible USACE District[2]/ Dist. Iden. Code	Section 401 Water Quality Certification Program[3]	State Wetlands Program[4]	Coastal Resources Program	Joint State/Fed. Permit App.[5]	Application available online?
VERMONT continued			Additional Permits: – Stream Alteration Permit: for construction in river/stream or within banks; DEC, Water Quality Division—by watershed. – Shoreland Encroachment Permit: for encroachment in public water; DEC, Water Quality Division —802/241-3791			
VIRGINIA	Huntington/orh Norfolk/nao	**Virginia Dept. of Envir. Quality (DEQ) Water Div.** **www.deq.state.va.us/ permits/water.html** **TEL: 804/698-4109** DEQ requires applicants to obtain a Virginia Water Protection (VWP) Permit to meet 401 requirements. Specifically, they must contact the Virginia Marine Resources Commission (VMRC) to file a joint permit application, which meets federal, local, and state requirements. VMRC will copy DEQ, the local wetland board, and USACE, all of which will respond separately.	**DEQ** **www.deq.state.va.us/permits/water.html** **TEL: 804/698-4109** Applicants must obtain a VMRC Permit, or "Permit for Construction in Waters in the Commonwealth and in Wetlands" to build, dump, or trespass upon or over, encroach upon, take or use material from beds of bays, oceans, rivers, streams, or creeks. This process is identical to that described for the VWP permit.	**Virginia Marine Resources Commission (VMRC)** **www.state.va.us/mrc** **TEL: 757/247-2200** VMRC manages/regulates tidal wetlands and coastal primary sand dunes in coordination with Virginia local wetland boards. Also: Chesapeake Bay Program for resources specific to Bay.	YES	YES
WASHINGTON	Portland/nwp Seattle/nws	**State Dept. of Ecology (Ecology) Envir. Review Section** **www.ecy.wa.gov/programs.html** **TEL: 360/407-6912** 401 certification is applied by Ecology in three forms for the NWP program: 1) approved; 2) denied without prejudice; and 3) partially denied without prejudice. Regional conditions apply to all NWPs, and most impacts greater than 0.25 acre require individual 401certification and USACE notification. EPA issues 401 certification for federal lands with exclusive jurisdiction within the state and most Native American Indian Tribal land. The 401 certification process begins with the receipt of public notice from USACE, which includes a request for 401 certification.	**Local Governments (oversight by Ecology, Shorelands and Wetlands Mgmt., TEL: 360/407-7272)** Wetlands are regulated through the Shoreline Management Act (SMA), which applies to all marine waters, submerged tidelands, lakes larger than 20 acres, streams, wetlands, marshes, swamps, and bogs (all with a 200-ft buffer). These resources are regulated through local shoreline master programs (regulatory/ planning documents) that are written by local governments with policy guidance from Ecology. Local governments issue Shoreline Substantial Development Permits if the proponent's application is in conformance with the local shoreline master program. Applicants complete a Joint Aquatic Resource Permits Application (JARPA) form for both federal and state permits. The JARPA covers water quality certification and Section 404, Section 10, and SMA permit requirements. Approvals are still issued independently by the different agencies. Wetlands are also protected under the state Water Pollution Control Act (WPCA), which prohibits discharges of pollutants to waters of the state, including wetlands. The WPCA contains the state water-quality standards applied during 401 certification, but in the absence of a federal nexus (e.g., Section 404 application), WPCA standards may be applied through another state water quality permitting process (e.g., wastewater discharge permit) to protect resource. Cities and counties also regulate development in wetlands through implementation of the Growth Management Act (GMA), which requires that each city/county adopt regulations and land use plans to protect critical areas, including wetlands.	Washington has an approved coastal zone management program that governs coastal resources in the 15 counties with saltwater shorelines. Because the program is based on the SMA, the only additional regulatory authority it provides is through the federal consistency provisions of the federal CZMA.	YES	YES

Table 6-1. Summary of State Wetland Protection and Regulatory Programs[1] continued

State	Responsible USACE District[2]/ Dist. Iden. Code	Section 401 Water Quality Certification Program[3]	State Wetlands Program[4]	Coastal Resources Program	Joint State/Fed. Permit App.[5]	Application available online?
WEST VIRGINIA	Huntington/orh Pittsburgh/lrp	**West Virginia Dept. of Envir. Protection (DEP) Div. of Water Resources http://129.71.240.41/webapp/_dep/ owrwebsite/homepage.cfm TEL: 304/759-0583**	No state wetland regulations.	Not applicable	NO	YES
WISCONSIN	St. Paul/mvp	**Wisconsin Dept. of Natural Resources (DNR) Div. of Water, Fisheries Mgmt., and Habitat Protection www.dnr.state.wi.us/org/ water/fhp/wetlands TEL: 608/267-7498** DNR issues 401 certifications through a joint wetland/401 permit process. An application for a Wetland Water Quality Certification should be submitted to the state to address 401/wetland impacts at the state level. The proposed project must also meet local zoning ordinances prior to implementation.	**DNR Div. of Water, Fisheries Mgmt. and Habitat Protection www.dnr.state.wi.us/org/water/fhp/wetlands TEL: 608/267-7498** Wisconsin wetlands are regulated at both the state and local levels. DNR helps determine wetland and OHWM boundaries and provides oversight for local zoning implementation. Cities and counties regulate general development and activities in wetlands adjacent to navigable waters through zoning restrictions for "shoreland and wetland" areas.	**Wisconsin Dept. of Administration (DOA) Coastal Mgmt. Program www.doa.state.wi.us/dhir/boir/coastal TEL: 608/267-3687** Wisconsin's coastal zone management program is administered by six state agencies and a coastal management council. DOA provides consistency oversight and works with proponents/other agencies to ensure that coastal policies are considered in the design process.	NO	YES
WYOMING	Omaha/nwo	**Wyoming Dept. of Envir. Quality Water Quality Div. Watershed Planning Program http://deq.state.wy.us TEL: 307/777-7081**	No state wetland regulations. Wyoming is currently pursuing legislation to protect isolated wetlands no longer regulated by USACE as a result of SWANCC.	Not applicable	NO	NO

NOTES

1. The information presented in this table is based on telephone conversations with state representatives and internet research.
2. USACE divisions follow watershed boundaries, not state boundaries, so states are often divided among different districts. *See* www.usace.army.mil/where.html#Divisions for more information on USACE district boundaries. District-specific web links are noted in the table as follows: Mobile/sam = www.sam.usace.army.mil
3. Refers to the state agency responsible for implementation of Section 401 of the Clean Water Act. If a state agency is not listed, authority for Section 401 in that state remains with EPA.
4. Refers to any state agency with a regulatory responsibility for protection of wetland resources. If several state agencies are listed, jurisdictions are either overlapping or divided by area or resource type. If no agency is listed, wetland resources are only protected by federal statute.
5. Indicates if a project applicant applies for a wetland permit through a joint federal/state application, which would meet both state wetland and federal Clean Water Act application requirements.

CWA = **Clean Water Act**
EPA = **Environmental Protection Agency**
ERP = **Environmental Resource Permit**
MGL = **Massachusetts General Law**
MSSW = **Management and Storage of Surface Waters Program**
SPGP = **State programmatic general permit**
WRPP = **Wetland Resource Permitting Program**

Florida

Wetland resources in Florida have been regulated by the Florida Department of Environmental Protection (previously known as the Department of Environmental Regulation) since the 1970s. Under current Florida law (Florida Environmental Reorganization Act of 1993, Chapter 93-213, Laws of Florida), the comprehensive Environmental Resource Permit (ERP) program governs alterations of landscape in the state (except within the limits of the Northwest Florida Water Management District), including dredge or fill activities in tidal or freshwater wetlands, creation or alteration of wetland habitats or surface water systems, and alterations of upland habitat that could affect flooding or stormwater management activities. This program is implemented cooperatively at the state level by the Florida Department of Environmental Regulation and at the regional level by four of the state's five water management districts. Operating agreements between the Florida Department of Environmental Protection and the water management districts specify the permit and review authorities for which each entity is responsible. In addition, many local governments have adopted separate local wetland regulations that augment, but do not replace, the ERP program.

The Florida Department of Environmental Protection's Bureau of Beach and Coastal Systems regulates activities along the coastlines of the Atlantic Ocean and the Gulf of Mexico under the authority of the Beach and Shore Preservation Act (Chapter 161, Florida Statutes) and the federal CZMA. Specifically, project applicants must obtain permits from the Bureau of Beach and Coastal Systems for construction activities that extend seaward of an established coastal construction control line as well as for activities waterward of mean high water (Florida Administrative Code, Chapters 62B-41, 62B-26, 62B-33, 62B-49, and 62B-36). As a result, wetlands that fall within Florida's coastal zone could be regulated under two separate state programs. Federal agencies must also, pursuant to the CZMA, seek a consistency determination from the Bureau of Beach and Coastal Systems for any projects that they propose, permit, or fund within Florida's coastal zone.

To streamline the permitting process, project applicants submit one joint wetland resource and ERP application to the Florida Department of Environmental Protection, which in turn acts as the "lead agency" for dissemination of information. The joint permit application meets the federal requirements of USACE, the state and regional ERP requirements, and coastal permitting requirements. Although the actual federal and state permitting processes remain separate, issuance of a state permit constitutes water quality certification under Section 401 of the CWA, unless such certification is specifically waived in the permit. Additional information on any of these programs or permitting processes can be found at www.dep.state.fl.us/water/wetlands/index.htm.

The state of Florida also requires authorization for any construction on or use of submerged lands owned by the state. Copies of the joint wetland resource and ERP applications are forwarded to the Board of Trustees of the Internal Improvement Trust Fund, which is charged with

Florida Merged Two Older Programs to Establish Its ERP Program

The ERP program represents a merger of two older permitting programs: the Wetland Resource Permitting Program (WRPP) and the Management and Storage of Surface Waters Program (MSSW). The WRPP regulated dredge and fill activities in all waters of the state that were connected, either by natural or artificial watercourses, to a "named" water of the state, where "named" waters included rivers, lakes, streams, the Gulf of Mexico, and the Atlantic Ocean. This program did not regulate activities in hydrologically isolated wetlands, and was implemented exclusively by the Florida Department of Environmental Protection.

The MSSW program regulated alterations to the landscape that affected surface water flows in uplands as well as wetlands. The MSSW program was implemented independently and exclusively by the five water management districts in Florida. In 1994, these permitting processes were consolidated into the ERP program, with the exception of those projects located within the jurisdiction of the Northwest Florida Water Management District, where the WRPP and MSSW permitting programs for agriculture, silviculture, and dam safety remain in effect. ■

Florida's Wetland Evaluation and Delineation Program

Florida does not use USACE's wetland delineation methodology to determine the landward extent of wetland habitats. Although the state's written definition of a wetland is almost identical to the one adopted by USACE, Florida requires that only two of the three parameters defined in its regulations (i.e., wetland vegetation, hydric soils, and wetland hydrology) be present for an area to qualify as a wetland. In addition, Florida has refined the federal list of wetland plants to develop a more accurate vegetative index for statewide wetland plant identification. This refined list can result in discrepancies between state and federal wetland boundary determinations. For example, Florida's vegetative index identifies slash pine (*Pinus elliotti*) and gallberry (*Ilex glabra*) as upland species, whereas USACE calls them facultative species.

Because these two plant species are extremely common in Florida, it is not unusual for a project proponent to have to work independently with the Florida Department of Environmental Protection and USACE to determine two different landward wetland limits. Additional information on wetland delineations in Florida can be obtained from the Florida Department of Environmental Protection through the Wetland Evaluation and Delineation Section (www.dep.state.fl.us/water/wetlands/delineation/index.htm). ■

Florida's State Programmatic General Permit

In 1997, the USACE Jacksonville District and the Florida Department of Environmental Protection, in conjunction with NOAA Fisheries, USFWS, EPA, and the Florida Game and Freshwater Fish Commission, developed an expanded State Programmatic General Permit (SPGP) to govern minor work in Florida's USACE jurisdictional waters. The new SPGP, called SPGP III-R1, expanded the coverage authorized under two previous SPGPs to include:

- Shoreline stabilization projects (including riprap, seawalls, and other shoreline stabilization projects)
- Boat ramp and boat launch area projects and their associated structures
- All dock, pier, marina, and associated facility projects
- Maintenance dredging projects for canals and channels
- Selected projects covered by Department of Environmental Protection exemptions, and
- Selected projects covered by Department of Environmental Protection General Permits

Project applicants file a joint federal and state application with Florida's Department of Environmental Protection. This joint application is screened to determine whether or not a project is a candidate for review under the SPGP III-R1. If the project is covered by SPGP III-R1, it is assigned to one of three categories—green, yellow, or red—based on the extent of its compliance with the SPGP conditions.

Projects that qualify for the "green" category are reviewed only by the Department of Environmental Protection. The appropriate state authorization is forwarded to the project applicant along with a notification stating that USACE will not require additional permitting under Section 404 of the Clean Water Act. Projects in the "yellow" category are reviewed cooperatively by USFWS, EPA, NOAA Fisheries, and USACE. These agencies draft a Combined Federal Position paper summarizing the federal position on the project and designating the project as either a "green" or "red" project under the SPGP. Projects in the "red" category do not qualify for the SPGP and are evaluated independently by USACE through the standard Section 404 permitting process. ■

determining how the public's interests may best be served. The largest projects are reviewed by the board, while staff of the Florida Department of Environmental Protection and the water management districts have been delegated the authority to take action on most authorizations.

Massachusetts

Both coastal and inland wetland resources in Massachusetts are protected under the Wetland Protection Act (Massachusetts General Law {MGL} Chapter 131, Sec. 40). Massachusetts has adopted an unusual regulatory structure in which volunteer Community Conservation Commissions, appointed by each municipality's executive board, administer the provisions of the Wetland Protection Act within their respective jurisdictional areas. With technical support from the Massachusetts Department of Environmental Protection, each Commission applies regulations developed by the Department of Environmental Protection to ensure that proposed projects do not adversely jeopardize wetland resources or the public interests they serve.

Project applicants proposing to remove, dredge, fill, or alter a wetland resource (including a 100-foot radius around it) must submit a Notice of Intent to the appropriate commission prior to beginning work. Prior to submitting the Notice of Intent, the applicant may file a Request for Determination of Applicability to determine if the Wetland Protection Act applies or to confirm the boundaries of a resource area. Once the Notice of Intent has been submitted, the Commission will review the application to ensure that the type and extent of work proposed meets the performance standards outlined in the Wetland Protection Act. After a site visit has been completed, the Commission will present the Notice of Intent at a public hearing, where it will subsequently issue an Order of Conditions or deny the project. Project applicants may appeal commission decisions to the Department of Environmental Protection.

Three other acts that are enforced through the Wetland Protection Act process provide additional protection for particularly sensitive wetlands. Both the Inland Wetlands Restrictions Act (MGL Chapter 131, Section 40A) and the Coastal Wetlands Restriction Act (MGL Chapter 130, Section 105) place permanent restriction orders on selected inland and coastal wetlands in more than 50 communities throughout the state. These restriction orders prohibit certain activities in selected wetlands, and guide the commissions in what types of work can be authorized under a Wetland Protection Act permit. The Massachusetts Rivers Protection Act (MGL Chapter 131, Section 40) provides protection for the 200-foot area that parallels the high water line of every naturally flowing perennial river or stream in the state. These *Riverfront Areas* are treated as wetland resources and managed by commissions in accordance with the processes set forth in the Wetland Protection Act.

In addition to the Wetland Protection Act, Wetlands Restriction Act, and Rivers Protection Act, more than 100 Massachusetts communities enforce additional local wetland protection bylaws.

Michigan

The Land and Water Management Division of Michigan's Department of Environmental Quality is responsible for managing the state's sensitive natural resources, including wetlands. In accordance with Part 303, Wetland Protection, of the Natural Resources and Environmental Protection Act (NREPA) (1994 PA 451), the Land and Water Management Division requires that a project applicant obtain a wetland construction permit before dredging, draining, filling, or constructing a structure in a regulated wetland. Regulated wetlands are wetlands that are contiguous with the Great Lakes, inland lakes, ponds, rivers, or streams; wetlands that are located within 500 feet of an inland lake, pond, river, or stream; and any wetland that is hydrologically isolated, larger than five acres, and located in a county with a population of more than 100,000 people.

In general, permit applications are submitted to the Permits Consolidation Unit of the Land and Water Management Division and forwarded to the appropriate district office for an onsite inspection and decision. An applicant may qualify for either a general or a *public notice* permit, depending on the impacts of the project. General permits are issued for categories of activities, as defined in Part 303 of the NREPA, that "...will cause only minimal adverse environmental effects when performed separately, and will have only minimal cumulative adverse effects on the environment." Section 30312(1) of 1994 PA 451. Public notice permits are issued when the criteria specified in each categorical general permit cannot be met, or when it is determined that public review of the project is prudent and necessary.

Michigan is unique in that EPA, under CWA Section 404(g), has given it the authority to administer the Section 404 permit program. With certain exceptions (*see* sidebar), approval for projects affecting state and federal jurisdictional wetlands has been delegated to the Land and Water Management Division. As a result, although the wetland construction permit application is a "joint" permit application, review and approval authority for most projects that impact wetlands remains with the state.

The Land and Water Management Division also offers indirect protection to wetlands through four other parts of the NREPA. Specifically,

Massachusetts Wetland Conservancy Program for Wetlands Mapping

To document the extent and conditions of the state's wetlands, the Massachusetts Department of Environmental Protection is currently using aerial photography and photograph interpretation to map all state wetlands larger than one quarter of an acre. These maps, produced under the auspices of the Wetlands Conservancy Program, will improve coordination among regulatory programs and will serve as a resource for community members and project applicants seeking additional information on wetland habitats within their community. Although these maps are not designed to replace project-specific wetland delineations, they will be useful in creating local wetland inventories and various planning tools. Copies of completed maps can be obtained from the Community Conservation Commission responsible for a given area, or from the Department of Environmental Protection. ■

NREPA = **Natural Resources and Environmental Protection Act**

Michigan Administers Section 404 Program, But USACE Retains Some Jurisdiction

Under a 1984 Memorandum of Agreement between the State of Michigan Department of Natural Resources (now referred to as the Department of Environmental Quality) and EPA, the State of Michigan was given authority over certain projects regulated under Section 404 of the Clean Water Act. However, certain complex projects with significant impacts must still be independently reviewed by the USACE Detroit District. Such projects include:

- Projects that would result in a "major" discharge of dredged or fill material; major discharge is defined as fill involving more than 10,000 cubic yards of material, new construction of breakwaters or seawalls that exceed 1,000 feet in total length, culvert enclosures of more than 100 feet with more than 200 cubic yards of fill, or channelization of more than 500 feet of river;
- Projects that would result in discharge of a toxic pollutant, including a hazardous substance, toxic substance, or hazardous waste;
- Projects that would result in a discharge of dredged or fill material into critical areas established under state and federal law, including fish and wildlife sanctuaries, national and historical monuments, wilderness areas and preserves, national and state parks, components of the National Wild and Scenic River Act system, designated critical habitat of threatened or endangered species, sites identified or proposed under the National Historic Preservation Act, or other sites identified by the EPA;
- Projects that would result in a discharge of dredged or fill material into a water of a state other than Michigan; and
- Projects that would result in significant adverse effects on a water unique to a particular geographic region; would significantly reduce the commercial or recreational value of the area; or would affect species that are federally listed or proposed for listing as endangered or threatened. Such a determination would be made by EPA in consultation with USACE and USFWS.

Part 301, *Inland Lakes and Streams,* regulates construction activities in inland waters of the state; Part 31, *Floodplain Hazard Management,* requires review of proposals to occupy, fill, or grade lands within the state's floodplains; Part 353, *Sand Dunes Protection and Management,* regulates development within designated critical dune areas along the Great Lakes shoreline; and Part 323, *Shorelands Protection and Management,* in combination with the Coastal Management Program, provides for designation and proper management of environmental areas, high risk erosion areas, and flood risk areas along the Great Lakes shoreline. Local governments can also regulate wetlands by ordinance, but cannot require permits for activities exempted from regulation under Part 303.

New Jersey

The New Jersey Department of Environmental Protection, Land Use Management and Compliance Division is responsible for the oversight of environmentally sensitive lands, including freshwater wetlands, coastal areas, and floodplains. Through its Land Use Regulation Program unit, the New Jersey Department of Environmental Protection issues permits under six New Jersey statutes: the Freshwater Wetlands Protection Act (New Jersey Statutes Annotated (NJSA) 13:9B); the Flood Hazard Area Control Act (stream encroachment) (NJSA 58:16A), the Coastal Area Facility Review Act (CAFRA) (NJSA 13:19); the Waterfront Development Act (NJSA 12:5-3); the Wetlands Act of 1970 (NJSA 13:9A); and the Tidelands Act (NJSA 12:3-1).

Project applicants proposing development in freshwater wetlands must obtain a Freshwater Wetland Permit from the Land Use Regulation unit prior to construction. This process begins by obtaining a Letter of Interpretation from the New Jersey Department of Environmental Protection that verifies the presence, absence, or boundaries of freshwater wetlands on a given site and determines the size of any required transition area or buffer.

Freshwater Wetland Maps, available at any municipal or county clerk's office, through any county or public library with a geographic information system (GIS) computer system, or at the New Jersey Department of Environmental Protection's Maps and Publications Office, can be consulted prior to application for a Letter of Interpretation.

The Freshwater Wetland Maps provide guidance on where these resources are found in New Jersey, but an official Letter of Interpretation must be obtained prior to applying for a Freshwater Wetland Permit.

If the Letter of Interpretation documents freshwater wetlands on the site and the proposed project would impact these wetlands, the project applicant must apply for a Freshwater Wetland Permit. Minor activities typically fall within the purview of a state General Permit (e.g., repair of an existing structure, stream bank stabilization, etc.), while more complicated projects may require a state Individual Permit and extensive alternatives analysis. Moreover, activities that occur within 150 feet of a wetland or in the transition area defined in the Letter of Interpretation must be addressed either through the Freshwater Wetland Permit Process or independently through the New Jersey Department of Environmental Protection's Transition Area Waiver process.

Development activities in the coastal waters of New Jersey are regulated by the New Jersey Department of Environmental Protection through implementation of CAFRA, the Waterfront Development Law, and the Wetland Act of 1970. CAFRA regulates development in southern New Jersey through issuance of CAFRA permits, which are based on zoning standards in predetermined CAFRA zones. CAFRA's area is generally defined as the area south of Cheesequake Creek and along the Delaware Bay south of Kilcohook National Wildlife Refuge. A map of the CAFRA boundary line can be found at www.state.nj.us/dep/landuse/coast/caframap/caframap.html.

CAFRA is complemented by the Waterfront Development Law, which regulates development in tidal waterways anywhere in the state and in areas adjacent to tidal waterways outside CAFRA's jurisdiction. Project applicants proposing development within any of these waterways must obtain a Waterfront Development Permit.

Finally, the Wetland Act of 1970 further requires the New Jersey Department of Environmental Protection to regulate development in coastal wetlands. Applicants proposing to excavate, dredge, fill, or place a structure in a coastal wetland must obtain a Coastal Wetland Permit from the New Jersey Department of Environmental Protection.

All of these permits are processed by the Land Use Regulation unit. Moreover, they require county review by both the County Planning Board and County Environmental Commission.

In addition, if a project involves construction in New Jersey's floodplain, the applicant must, under the Flood Hazard Control Area Act, obtain a Stream Encroachment Permit from the New Jersey Department of Environmental Protection. Floodplain boundary determinations can be made by the New Jersey Department of Environmental Protection's Flood Plain Management Office, or can be generally determined by review of federal flood insurance maps or state floodplain maps, both of which can be obtained at county municipal offices.

Michigan's Wetland Assessment Program

In 1998, Michigan's Department of Environmental Quality implemented the Wetland Assessment Program to assist private citizens in identifying, locating, and verifying the location of wetlands and uplands on their property. Although not required prior to submitting an application for a wetland construction permit, a completed wetland assessment can help expedite the time required for project review and determinations by the Department of Environmental Quality.

Additional information on this program is available from the Department of Environmental Quality, Land and Water Management Division, Inland Lakes and Wetlands Unit, 517/241-8485 (www.deq.state.mi.us/lwm).

The Wetland Act of 1970 further requires the New Jersey Department of Environmental Protection to regulate development in coastal wetlands.

CAFRA = Coastal Area Facility Review Act
GIS = Geographic information system
NJSA = New Jersey Statutes Annotated

Under the Tideland Act, an applicant must also get permission from the Bureau of Tidelands Management to use "riparian lands", or lands that are currently, or have been previously, inundated by the tide of a natural waterway. Permission for such use comes from the Bureau of Tidelands Management in the form of a tidelands license, lease, or grant.

Washington

JARPA = Joint Aquatic Resources Permit Application
RCW = Revised Code of Washington
SMA = Shoreline Management Act

Washington's Department of Ecology (Ecology), commonly in collaboration with local governments, protects state wetlands through several state environmental laws. The Shoreline Management Act (SMA) (RCW 90.58) established a cooperative program between local and state governments in which cities and counties develop and administer local shoreline master programs, with policy guidance, technical assistance, and oversight provided by the state. Shoreline master programs are based on state guidelines but are tailored by each city and county to meet the specific needs of the community. These programs act as both planning and regulatory documents and are generally designed to give preference to activities that protect water quality and the natural shoreline, are water-related or water-dependent, and preserve and enhance public access.

The Shoreline Management Act governs all marine waters, submerged tidelands, lakes larger than 20 acres, and all streams with a mean annual flow greater than 20 cubic feet per second.

The SMA, as reflected in each of the local master programs, governs all marine waters, submerged tidelands, lakes larger than 20 acres, and all streams with a mean annual flow greater than 20 cubic feet per second. Biological wetlands and some or all of the 100-year floodplain associated with such lakes, streams, and marine waters are also protected, as is a 200-foot wide shoreline area landward from the water's edge. Project applicants proposing development in any of these regulated areas must obtain a Substantial Development Permit from the local government with jurisdiction over the resource. In some circumstances, an applicant may qualify for a Conditional Use or Variance Permit or an exemption, but, in all cases, must still be able to demonstrate that the project has met all substantive policies and regulations of the local master program. Most of Ecology's involvement in the permitting process is focused on technical assistance to both the local government and project applicant prior to a local decision; however, Ecology does review all permits to determine if the local action (i.e., approval or disapproval) is consistent with the local master program and the policies of the SMA.

Mangrove swamps are found in brackish water at the southern tip of Florida.

Substantial Development Permit applications are submitted on a Joint Aquatic Resources Permit Application (JARPA) form, which combines many state and federal permitting applications. Specific to wetlands, the JARPA form can be used to apply for Shoreline Management Permits, Water Quality Certifications, and USACE Section 404 and Section 10 permits. Although the authority for approval of these different permits is retained by the issuing agency, the JARPA form consolidates state and federal permitting processes and facilitates a more concise project review.

The state Water Pollution Control Act (WPCA) (RCW 90.48) is another statute Ecology may use to protect wetlands and other waters. The WPCA "prohibits discharges of polluting matter to waters of the state," where waters of the state include streams, lakes, rivers, ponds, inland waters, salt waters, waters courses, and other surface and underground waters. The Surface Water Quality Standards (WAC 173-201A) are the primary implementation regulations of the Act that apply to wetland resources. In addition, the anti-degradation policy contained in the standards (WAC 173-201A-070) ensures that the existing beneficial uses of all wetlands are protected and that projects that would result in long-term harm to the environment are not permitted.

The WPCA is typically enforced through issuance of a state water quality certification pursuant to Section 401 of the federal CWA. However, if a proposed activity degrades a wetland but falls outside the jurisdiction of a federal agency (e.g., USACE), Ecology may use another state water quality permitting process to implement the act (e.g., wastewater discharge permit, short-term water quality modification).

The Growth Management Act (GMA) (RCW 36.70A) requires all cities and counties in the state of Washington that meet a specified threshold population size or growth rate to adopt regulations that protect critical areas, including wetlands. With guidance from the state Office of Community Development, local governments identify critical areas within their jurisdiction and adopt development regulations and land use implementation plans for protection of those critical areas. Most cities and counties require applicants to obtain permits for activities in or near critical areas, although the permit processes vary from jurisdiction to jurisdiction.

In addition to SMA, WPCA, and GMA, the states' coastal resources are protected by the federal CZMA and Washington's approved CZMP. The state CZMP is based primarily on the SMA, but applies only within the 15 counties with saltwater shoreline. Under the CZMA, Ecology has the authority to review federal actions for consistency with the CZMP, an authority which in many cases is the only avenue by which the state can influence federal projects.

Other state laws that could incidentally regulate wetlands include the State Environmental Policy Act (SEPA) (RCW 43.21C), which requires full disclosure of the potential environmental effects of any proposed action; and the State Hydraulic Code (RCW 75.20.100–160), which requires a Hydraulic Project Approval (HPA) permit from the Washington Department of Fish and Wildlife (WDFW) for all work that occurs below the ordinary high water mark of state waters, including portions of wetlands. The JARPA form can be used to apply to WDFA for the HPA permit. The HPA is used to ensure that fish habitat is protected from projects that occur in state waters, primarily streams and rivers.

Update: Washington's Shoreline Management Guidelines

On November 29, 2000, the Department of Ecology adopted new rules, referred to as the Shoreline Master Program Guidelines (Guidelines), in an attempt to better guide local government on how to achieve the level of protection the SMA required. These guidelines represent a five-year effort by state and federal agencies to review and update the twenty-year old state rule. They included new limitations on the amount of development allowed adjacent to streams, lakes, and marine waters, as well as requirements for replacement of natural vegetation along shorelines to facilitate erosion control and habitat enhancement activities.

On August 27, 2001, the Shoreline Hearings Board invalidated the new Shoreline Master Program Guidelines on the basis of potential conflicts with the federal Endangered Species Act and the Washington Administrative Procedures Act. Currently, state officials, environmentalists, and business interests are negotiating an agreement on adoption of the new Guidelines. In the interim, the original guidelines can be found at Chapter 173-16 of the Washington Administrative Code. ■

CZMP = Coastal Zone Management Program
GMA = Growth Management Act
HPA = Hydraulic Project Approval
SEPA = State Environmental Policy Act
WDFW = Washington Department of Fish and Wildlife
WPCA = Water Pollution Control Act

CHAPTER SEVEN

Mitigation Planning

This chapter describes the process of mitigation planning that is used to compensate for the impacts of a project on waters of the United States. The process is presented chronologically, beginning with how existing wetlands may affect the preliminary design of a proposed project, and culminating with the completion of a mitigation effort when permit obligations have been met. Although compensatory mitigation is the primary emphasis of this chapter, subjects such as impact quantification, the process of mitigation through impact avoidance and minimization, and mitigation through mitigation banks and payment of in-lieu fees are also discussed.

The focus of this chapter is on mitigation planning procedures and requirements promulgated by USACE and EPA. Because many states also have laws that regulate activities in wetlands and other aquatic areas, the reader should contact state and local regulatory authorities to determine the mitigation requirements for those jurisdictions (*see* chapter 6, State Wetland Laws). The authors recommend that mitigation carried out to satisfy USACE/EPA requirements be integrated with mitigation executed to satisfy state and local requirements.

EPA = Environmental Protection Agency
USACE = United States Army Corps of Engineers

Mitigation Through Impact Avoidance and Minimization

As described in chapter 4, Federal Regulation of Wetlands and Other Waters, and chapter 5, Section 404 Permitting Process, the Section 404(b)(1) Guidelines and the USACE/EPA 1990 Memorandum of Agreement (MOA) require a permit applicant to justify project-related impacts on waters of the United States and to implement mitigation to avoid, minimize, and compensate (in that order) for impacts on waters of the United States. Avoidance of impacts is the primary form of mitigation and minimization of impacts is the

MOA = Memorandum of agreement

secondary form of mitigation that USACE and EPA require under their 1990 MOA (*see* appendix F). These forms of mitigation should be implemented before compensatory mitigation is considered. Impact avoidance and minimization must be done to the extent practicable while meeting the basic project purpose and may entail reconfiguration of the basic project design. The applicant may work with a wetland consultant, using the delineation map, information on the wetlands' functions and values, and the consultant's experience with similar habitats to develop an effective impact avoidance and minimization strategy.

Bridge over seasonal wetland. Bridge crossings remove wetland habitat directly by placement of fill for pilings and indirectly by shading habitat under the bridge.

The goal of the mitigation planning process in the context of avoidance and minimization is to devise a project design that avoids or minimizes impacts on wetlands overall (particularly the highest value wetlands), wetlands that provide habitat for sensitive species, and wetlands in which compensatory mitigation (discussed below) may not be feasible. Often, the decision to affect a particular wetland or a given amount of wetlands is based upon the feasibility and cost of mitigating those impacts. Therefore, a developer may attempt to strike a balance between the revenue that would be generated by developing land that contains wetlands and the cost of mitigating those impacts. Regardless of the relationship between potential return and mitigation costs, USACE and EPA expect project proponents to avoid wetlands to the maximum extent practicable to select the least environmentally damaging practicable alternative for project location and design.

In addition to avoidance and minimization, the Section 404(b)(1) Guidelines require a developer to consider an alternative project site if the proposed development would result in unacceptable impacts on waters of the United States (*see* discussion in chapter 5, Section 404 Permitting Process). Moreover, those impacts that the project proponent cannot address through minimization and avoidance may be addressed through compensatory mitigation, described below.

Quantification of Wetland Impacts

Before mitigation planning can begin, the type and location of existing wetland habitats on a project site must be identified.

Before mitigation planning can begin, the type and location of existing wetland habitats on a project site must be identified. As described in chapter 3, Jurisdictional Limits of Wetlands and Other Waters, wetlands are identified and mapped during a wetland delineation. This information, sometimes accompanied by descriptions of the functions and values of each of the habitats that occur on the site (*see* chapter 2, Wetland Ecology), may be used in the preliminary design phase of a project.

Once the project layout has been determined following the avoidance and minimization process, the delineation map is overlaid with the project development map to calculate the acreage of waters of the United States that would be adversely affected, directly and indirectly, by the project. However, because many projects entail impacts on more than one wetland or other aquatic resource habitat type, it is usually not sufficient to simply quantify impacts on such areas in terms of "wetlands" or "waters of the United States." Accordingly, it is recommended that delineation maps differentiate between the various habitat types that occur on the project site, and that the impact quantification be conducted in the same manner. This information facilitates the mitigation planning process.

In quantifying impacts on aquatic areas, permanent impacts must be distinguished from temporary impacts. Temporary impacts are short-term disturbances such as vegetation removal and placement of temporary fill to accommodate movement of construction equipment. Unlike permanent impacts, temporary impacts involve aquatic resources being returned to their approximate preconstruction condition after construction activities are complete. Accordingly, temporary impacts do not result in a loss of acreage, but do result in a temporal loss of function and value. Depending on the type and duration of temporary impact and the amount of time required for the temporarily affected habitat to return to its original condition, USACE may or may not require compensatory mitigation. In any event, USACE will require assurances that measures will be implemented to revegetate or restore the affected area once construction is complete.

Temporary impacts are short-term disturbances such as vegetation removal and placement of temporary fill to accommodate movement of construction equipment.

Although habitat acreage has traditionally been used as the standard measure for quantifying impacts and mitigation requirements, USACE has recently promoted the concept of using a debit and credit system. USACE Regulatory Guidance Letter 02-2 (*see* appendix G, USACE Regulatory Guidance Letters/Memoranda to the Field) defines a debit and credit as follows:

> **Debit.** A unit of measure (e.g., functional capacity units in HGM) representing the loss of aquatic function at a project site; the measure of function is typically indexed to the number of acres [of aquatic resources] that are impacted by issuance of the permit.
>
> **Credit.** A unit of measure (e.g., functional capacity units in HGM) representing the gain of aquatic function at a compensatory mitigation site; the measure of function is typically indexed to the number of acres of resource restored, established, enhanced, or protected as compensatory mitigation.

HGM = Hydrogeomorphic

USACE Regulatory Guidance Letter 02-2 recommends that the evaluation of wetland impacts be conducted in a manner such that an identified debit can be offset by a credit. The method that is used to quantify debits should be the same as that used to quantify credits. Determination of debits and credits is made using USACE district-approved methods, such as the HGM approach or acre-for-acre mitigation-to-impact ratios.

In the absence of more definitive information about an aquatic site's functions, Regulatory Guidance Letter 02-2 allows a one-to-one acreage replacement to be a reasonable surrogate to ensure no net loss of functions.

The Guidance Letter also specifies that USACE districts require compensatory mitigation for streams to replace stream functions. However, where a functional assessment for a stream is not available, the Guidance Letter indicates that the mitigation provide one-for-one replacement of the linear footage of the stream.

Mitigation Implementation Timing Relative to Impacts

Most compensatory mitigation is implemented before or concurrently with project impacts on aquatic areas, providing for at least some establishment of the mitigation habitat and its functional benefit. This approach allows the functions and values lost at the impact area to be concurrently replaced at the mitigation site. However, in cases where financial assurances are provided and where the likelihood of mitigation success is high, USACE may allow the impacts to occur in advance of the mitigation implementation. In such cases, as specified by Regulatory Guidance Letter 02-2, USACE requires that the following requirements be met before any impacts on aquatic areas occur under an issued permit:

- The mitigation plan has been approved
- The permittee has obtained ownership of or otherwise has demonstrated authority to implement mitigation at the mitigation site
- A permanent source of water to support the mitigation has been identified
- The appropriate financial assurances to construct and maintain the mitigation site have been established

Collecting *Scirpus* plugs (underground stems) from freshwater marsh for transplanting to a marsh restoration site

Additionally, Regulatory Guidance Letter 02-2 specifies that initial physical (e.g., grading) and biological (e.g., planting) improvements usually be completed by the first full growing season following the permitted impacts on aquatic areas. In the event that this requirement cannot be met, USACE may require additional compensatory mitigation or other measures that compensate for the temporal loss in acreage, function, and value. These may include mitigation at a higher mitigation-to-impact ratio and increased financial assurances.

Circumstances in which USACE may allow impacts to occur in advance of mitigation may include, for example, where a simple restoration project is required, on certain federally-aided highway projects, and for emergencies and enforcement actions.

Conversely, for mitigation projects that are implemented and documented to be successful in advance of the authorized impacts, the mitigation-to-impact ratio may be reduced because temporal habitat loss and the risk of mitigation failure will have been eliminated.

Compensatory Mitigation

All impacts on aquatic areas subject to the jurisdiction of USACE that cannot be avoided or minimized must be offset through compensatory mitigation.

The goal of compensatory mitigation is to prevent a net loss of wetland or other aquatic resource acreage, function, and value. Although payment of in-lieu fees and mitigation at mitigation banks (discussed later in this chapter) also involve compensatory mitigation, this section is devoted to compensatory mitigation efforts implemented directly by the project proponent, rather than mitigation implemented through these other means.

Regulatory Guidance Letter 02-2 sets forth USACE's overall guidelines on compensatory mitigation. This Guidance Letter directs USACE districts to use watershed and ecosystem approaches when determining mitigation requirements, and to consider the resource needs of the watersheds in which the impacts occur, as well as any impacts on neighboring watersheds.

RGL 02-2 directs USACE districts to consider the resource needs of the watersheds in which the impacts occur, as well as any impacts on neighboring watersheds.

Types of Compensatory Mitigation

Several types of compensatory habitat mitigation are available to offset impacts on waters of the United States. Regulatory Guidance Letter 02-2 defines these as follows:

Establishment (creation). The manipulation of the physical, chemical, or biological characteristics present to develop a wetland on an upland or deepwater site, where a wetland did not previously exist. Establishment results in a gain of wetland acreage.

Established freshwater marsh (wetland) surrounding an open water area (other waters) constructed in an upland grassland area

Restoration. The manipulation of the physical, chemical, or biological characteristics of a site with the goal of returning natural or historic functions to a former or degraded wetland. For the purpose of tracking net gains in wetland acreage, restoration is divided into:

Re-establishment. The manipulation of the physical, chemical, or biological characteristics of a site with the goal of returning natural or historic functions to a former wetland. Re-establishment results in rebuilding a former wetland and results in a gain of wetland acreage.

Rehabilitation. The manipulation of the physical, chemical, or biological characteristics of a site with the goal of repairing natural or historic functions of a degraded wetland. Rehabilitation results in a gain in wetland function but does *not* result in a gain of wetland acreage.

Enhancement. The manipulation of the physical, chemical, or biological characteristics of a wetland (undisturbed or degraded) site to heighten, intensify, or improve specific function(s) or to change the growth stage or composition of the vegetation present. Enhancement is undertaken for specified purposes such as water quality improvement, flood water retention, and wildlife habitat. Enhancement results in a change in wetland function(s) and can lead to a decline in other wetland functions, but does not result in a gain of wetland acres. This term includes activities commonly associated with enhancement, management, manipulation, and directed alteration.

Enhancement results in a change in wetland function(s) and can lead to a decline in other wetland functions, but does not result in a gain of wetland acres.

Protection/maintenance (preservation). The removal of a threat to, or preventing the decline of, wetland conditions by an action in or near a wetland. This term includes the purchase of land or easements, repairing water-level control structures or fences, or structural protection such as repairing a barrier island. This term also includes activities commonly associated with the term preservation. Preservation does not result in a gain of wetland acreage. Except in extraordinary circumstances, habitat protection may serve as mitigation only when executed in conjunction with establishment, restoration, or enhancement activities.

Except in extraordinary circumstances, habitat protection may serve as mitigation only when executed in conjunction with establishment, restoration, or enhancement activities.

General Approach to Compensatory Mitigation

Compensatory mitigation, whether in the form of establishment, restoration, or enhancement, may be executed using a variety of techniques. These techniques may be divided into three broad categories:

- Restoring and managing wetland hydrology
- Eliminating or controlling chemical or other contaminants that are adversely affecting wetlands
- Restoring and managing native biota

Commission on Geosciences, Environment and Resources 1992

Techniques include:

- Grading in naturally upland areas to create the appropriate hydrologic conditions
- Removal of fill material from former wetlands
- Reestablishing streamflow
- Removing ditches and drain tiles in agricultural fields
- Breaching levees to restore tidal action
- Removing invasive nonnative species
- Controlling contaminant loading in inflow waters

Habitat enhancement may entail removal of an invasive species (for example, the giant reed in this photograph) from an area that meets the criteria for a wetland but provides limited habitat function and value.

In certain limited circumstances, USACE may provide mitigation credit for inclusion of nonwetland riparian and upland areas into the mitigation project. The upland area must enhance the function of the aquatic resources and increase the overall ecological function of the mitigation site or the function of aquatic resources in the watershed.

Irrespective of the technique employed to implement mitigation, USACE in recent years has encouraged permit applicants to adopt a watershed or ecological approach when developing compensatory mitigation plans. Further, Regulatory Guidance Letter 02-2 recommends that permit applicants include a mix of wetland, open water, and adjacent upland areas (including buffers) in their mitigation proposals, thereby offering a greater variety of functions. Watershed- or ecologically-based compensatory mitigation plans tend to be more successful and provide greater function than programs with a more narrow focus. In particular, in Appendix B of the Guidance Letter, USACE encourages project applicants to consider the following when developing a compensatory program:

Watershed- or ecologically-based compensatory mitigation plans tend to be more successful and provide greater function than programs with a narrower focus.

- **Consider the hydrogeomorphic and ecological landscape and climate.** Where practicable, locate the mitigation site in a setting of comparable landscape position and hydrogeomorphic class. Duplicate the features of reference wetlands or enhance connectivity with natural upland landscape elements.

  Natural disturbances, such as floods, droughts, muskrats, geese, and storms, should be accommodated in mitigation designs. The design should aim to restore a series of natural processes at a mitigation site to ensure that resilience to the disturbances will have been achieved.
- **Adopt a dynamic landscape perspective.** Consider both current and future watershed hydrology and wetland location. Account for future changes in land use. Select sites that are, and will continue to be, resistant to disturbances from the surrounding landscape–by preserving large buffers and conectivity to other wetlands, for example. Build on existing wetland and upland systems. If possible, locate the mitigation site to take advantage of refuges, buffers, green spaces, and other preserved elements of the landscape.
- **Whenever possible, choose wetland restoration over creation.** Select sites where wetlands previously existed or where nearby wetlands still exist. Restoration has been observed to be more feasible and sustainable than creation of wetlands. In restored sites, the proper substrate may be present, seed sources may be onsite or nearby, and the appropriate hydrological conditions may exist or be more easily established.

Whenever possible, choose wetland restoration over creation. In restored sites, the proper substrate may be present, seed sources may be onsite or nearby, and the appropriate hydrological conditions may exist or be more easily established.

- **Restore or develop naturally variable hydrological conditions.** Promote naturally variable hydrology, with emphasis on enabling fluctuations in water flow and level, duration, and frequency of change, representative of comparable wetlands in the same landscape setting. Preferably, natural hydrology should be allowed to reestablish itself rather than to have active engineering devices mimic a natural hydroperiod. When restoration is not an option, favor the use of passive devices with a higher likelihood of sustaining the desired hydroperiod over the long term. Avoid designing a system dependent on water-control structures or their artificial infrastructure that must be maintained in perpetuity in order for wetland hydrology to be present. In situations where direct (in-kind) replacement is desired, candidate mitigation sites should have the same basic hydrologic attributes as the impacted site.
- **Avoid over-engineered structures in the wetland design.** Design the system for minimal maintenance. Set initial conditions and let the system develop. The system of plants, animals, microbes, substrate, and water flows should be developed for self-maintenance and self-design. Whenever possible, avoid manipulating wetland processes through the use of an approach requiring continual maintenance. Avoid hydraulic control structures and other engineered structures that are vulnerable to chronic failure and require maintenance and replacement. When feasible, use natural recruitment sources for establishment of more resilient vegetation. Some systems, especially estuarine wetlands,

Avoid manipulating wetland processes through the use of an approach that requires continual maintenance.

Salvaging topsoil from a seasonal wetland that has been permited for removal. Topsoil salvage and reapplication to created wetlands increases the likelihood of mitigation success by providing appropriate substrate, soil microbes, invertebrates, and seeds of wetland plants.

are rapidly colonized, and natural recruitment is often equivalent or superior to plantings. Take advantage of native seed banks and use soil and plant material salvage whenever possible.

- **Pay particular attention to appropriate planting elevation, depth, soil type, and seasonal timing.** When planting is necessary, select appropriate genotypes. Genetic differences within species can affect the outcome of wetland restoration.
- **Provide appropriately heterogeneous topography.** The need to promote specific hydroperiods to support specific wetland plants and animals means that appropriate elevations and topographic variations must be present in restoration and creation sites. Slight differences in topography (e.g., micro- and meso-scale variations and presence and absence of drainage connections) can alter the timing, frequency, amplitude, and duration of inundation. Plan for elevations appropriate to plant and animal communities that are reflected in adjacent or nearby natural systems.
- **Pay attention to subsurface conditions, including soil and sediment geochemistry and physics, groundwater quantity and quality, and infaunal communities.** Inspect and characterize the soils in some detail to determine their texture and permeability. Characterize the general chemical conditions and variability of soils, surface water, groundwater, and tides. At a minimum, these should include chemical attributes that control critical geochemical or biological processes, such as pH, redox, nutrients (nitrogen and phosphorus species), organic matter content, and suspended matter.
- **Consider complications associated with creation or restoration in seriously degraded or disturbed sites.** A seriously degraded wetland, surrounded by an intensively developed landscape, may achieve its maximal function only as an impaired system that requires active management to support natural processes and native species. It should be recognized, however, that mitigation may optimize the functional performance of some degraded sites; considerations of this kind should be included if the goal of mitigation is water- or sediment-quality improvement, promotion of rare or endangered species, or other objectives best served by siting a wetland in a disturbed landscape.

Develop a monitoring plan that identifies potential problems and provides direction for corrective actions early in the process.

- **Conduct early monitoring as part of adaptive management.** Develop a thorough monitoring plan that identifies potential problems and provides direction for corrective actions early in the process. The monitoring of wetland structure, processes, and function from the onset of restoration or creation can detect potential problems. Process monitoring (e.g., water-level fluctuations, sediment accretion and erosion, plant flowering, and bird nesting) is particularly important because it may identify the source of a problem and how it can be remedied. Monitoring and control of nonindigenous species should be part of any effective adaptive management program. Assessment of wetland performance must be integrated with adaptive management.

In-Kind Versus Out-of-Kind Mitigation. The term *in-kind* refers to mitigation which results in habitats that provide the same characteristics and functions as the habitats the project will affect. In-kind mitigation is usually preferred over *out-of-kind* mitigation, where the compensation habitat is functionally different than that of the habitat affected. In some cases, only out-of-kind mitigation is possible because establishing, restoring, or enhancing certain types of habitat, such as an herbaceous wetland supported by a groundwater seep, is not feasible.

The term in-kind *refers to mitigation which results in habitats that provide the same characteristics and functions as the habitats the project will affect.*

If out-of-kind mitigation is considered, the ecological values provided should be equal to or greater than those of the affected wetland. Some out-of-kind mitigation may be unacceptable; for example, nontidal compensation wetlands usually should not be the mitigation method for impacts on tidal habitats. During the permit review process, USACE decides on a case-by-case basis whether to require or allow out-of-kind mitigation.

If out-of-kind mitigation is considered, the ecological values provided should be equal to or greater than those of the affected wetland.

Onsite Versus Offsite Mitigation. Until recently, USACE typically required that compensatory mitigation be implemented at the site on which the wetland impacts occur. However, USACE has increasingly embraced implementing compensatory mitigation for the development project offsite. In a 2001 National Research Council report on mitigation implemented under the Section 404 program, onsite compensatory mitigation was sometimes found not to be appropriate because of hydrologic alterations caused by site development (Committee on Mitigation Wetland Losses 1992). Additionally, onsite mitigation sites tend to be relatively small and fragmented, and subject to more human disturbance than larger, planned wetland conservation areas. In contrast to mitigation of wetlands implemented onsite, those located offsite may have higher ecological integrity, be more connected to existing, larger wetlands, and be located where land acquisition costs are lower. Finally, mitigation implemented onsite may be constrained by site conditions to the point that the attempted mitigation exceeds the project site's capacity to support it.

Mitigation of wetlands implemented offsite may have higher ecological integrity, be more connected to existing, larger wetlands, and be located where land acquisition costs are lower.

In general, USACE may favor offsite mitigation when practicable opportunities for onsite mitigation are few, or when offsite mitigation is otherwise environmentally preferable to onsite mitigation. However, when all other factors are equal, USACE usually prefers onsite to offsite mitigation.

Mitigation Goals

The goals of a mitigation project are developed in coordination with USACE and other regulatory agencies, as applicable. Mitigation goals are typically directed toward replacing the functions, values, and acreage of the habitats affected; however, regional aquatic resource requirements may also be a factor in the development of mitigation goals. The goals will usually state the minimum acceptable ratio of mitigation to impact acres, or credits to debits. As specified in the USACE and EPA MOA on mitigation (*see* appendix F, USACE and EPA Memoranda of Agreement):

In practice, ratios commonly range from 1.5:1 to 2.5:1, but may range beyond 3:1 for mitigation habitats that are slow to develop biological habitat functions and for which prior establishment and restoration efforts have been only marginally successful.

Although Regulatory Guidance Letter 02-2 supports the national policy of "no overall net loss" of wetlands, permits may be granted for individual projects that do not meet this goal. Rather, the Guidance Letter directs USACE districts to strive to meet this goal on a cumulative basis, and to achieve it programmatically.

NOAA = National Oceanic and Atmospheric Administration
NRCS = National Resources Conservation Service
USFWS = United States Fish and Wildlife Service

USACE may decide to coordinate with EPA, USFWS, NOAA Fisheries, and NRCS to solicit comments on the goals, technical adequacy, and overall appropriateness of the mitigation proposal. Additionally, tribal, state, and local regulatory authorities may participate in the goal-forming process and conceptual mitigation design. The opportunity for these entities to participate in reviewing a mitigation proposal is dictated by the type of USACE authorization being considered and the mitigation requirements associated with that authorization. For example, most mitigation proposals prepared in support of nationwide permit authorizations require review only by USACE. However, in all cases, the decision to grant nationwide permit authorization for a project rests with USACE, although USACE may consider other agencies' comments on the proposal.

Compensatory Mitigation Plans

Compensatory mitigation plans provide a detailed description of the physical, biological, and jurisdictional characteristics of the mitigation site and how the habitats will be constructed and maintained.

Compensatory mitigation plans provide a detailed description of the physical, biological, and jurisdictional characteristics of the mitigation site and how the mitigation habitats will be constructed and maintained. Such plans become part of the Section 404 permit as a special condition. USACE guidelines allow mitigation proposals that are submitted with permit applications and preconstruction notifications to be conceptual or detailed, depending on the amount of mitigation that is needed and on the reliability of the parties involved in ensuring follow-through on the mitigation program. Mitigation plan development begins with an assessment of the functions and values of the aquatic resources to be removed by the project, continues through evaluation and selection of a mitigation site, and concludes with preparation of the plan document.

Both USACE headquarters and some USACE districts have published guidelines on the content of compensatory mitigation plans. Guidelines from USACE headquarters level are found in Regulatory Guidance Letter 02-2, which identifies the following minimum components of a mitigation plan, whether conceptual or detailed:

- Baseline site information
- Goals and objectives of mitigation
- Considerations for site selection
- Mitigation work plan

- Performance standards
- Responsible party information
- Site protection measures
- Contingency plan
- Monitoring and long-term management plan
- Financial assurances

The mitigation work plan element should include details of mitigation construction, such as construction methods and schedule, water sources, plant materials to be used, site maintenance, and other relevant components of the particular plan.

Guidelines published by individual USACE districts are more comprehensive and detailed than the basic information identified by USACE headquarters. For example, guidelines published by the Louisville district (*see* sidebar) for projects in Kentucky require that the mitigation plan provide detailed information on mitigation site soil and hydrologic conditions, planting specifications, expected functions and values of the mitigation habitats based on the HGM approach, and other information.

In addition to mitigation plan guidelines and monitoring requirements for wetlands in general, certain other districts have published mitigation guidelines for specific aquatic resources, such as riparian (San Francisco District), freshwater wetland (Seattle District), and vernal pool (Sacramento District) habitats.

Evaluation and Selection of a Mitigation Site

A fundamental aspect of developing a mitigation plan is the selection of a site to implement the compensatory mitigation. Site selection is based on numerous factors, perhaps the most important being geographic location relative to the development project. Regulatory Guidance Letter 02-2 specifies that the mitigation generally should be within the same area (e.g., watershed, county) as the development site, such that the mitigation can be reasonably expected to provide appropriate compensation for impacts on the aquatic resource. Although mitigation implemented in nearby watersheds may be appropriate, the rationale for proposing this should be presented in the mitigation plan. The farther the proposed mitigation is from the impact site, the greater the scrutiny the proposed mitigation may receive. USACE may even require higher mitigation-to-impact ratios (credit-to-debit ratios) as the distance of the mitigation site increases from the impact site.

Regulatory Guidance Letter 02-2 lists important factors to consider when evaluating the suitability of candidate mitigation sites, including aquatic resource needs of the affected watershed,

Louisville District Wetland Compensatory Mitigation and Monitoring Plan Guidelines for Kentucky*

TABLE OF CONTENTS

CHAPTER 1 Development Site Description
- Introduction
- Location
- Identification of Responsible Parties
- Site Characterization

CHAPTER 2 Proposed Compensatory Mitigation Site Description
- Location
- Proposed Wetland Classification
- Functions and Values
- Soils
- Proposed Vegetation
- Hydrology

CHAPTER 3 Success Criteria and Performance Standards
- Construction Schedule
- Soils Parameters
- Vegetation
- Hydrology
- Water Quality
- Functions and Values of Compensatory Mitigation Site

CHAPTER 4 Monitoring
- Parameters
- Sampling Frequency
- Monitoring Reports

CHAPTER 5 Contingency Plan
- Reporting Protocol
- Alternative Locations for Contingency Mitigation

CHAPTER 6 Permanent Protection Measures

APPENDIX A
Maps, Plans, and Drawings

APPENDIX B
Functions and Values Checklist

APPENDIX C
Wetland Plant List

* Abbreviated. In its original form, this outline is considerably more detailed than what is presented here. Users should consult the full text (available at http://155.80.93.250/orf/info/wetguide.htm) before preparing a mitigation plan for the Louisville District.

the cost of implementation, and logistics. The Guidance Letter also indicates that mitigation sites with the potential to attract waterfowl and other bird species that could pose a threat to aircraft should not be located near airports.

The following steps are commonly taken when selecting sites to be used for mitigation.

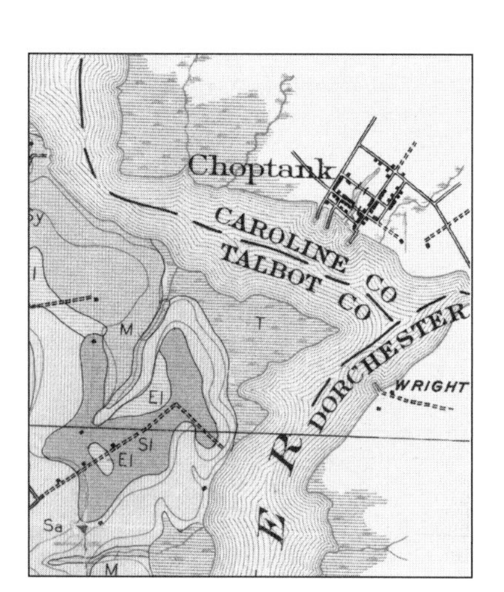

Figure 7-1
Soil Survey Map

Historic soil survey and topographic maps may provide clues to the former type and extent of wetlands that may have once existed at a site.

Literature Review. Knowledge of a candidate mitigation site's history may be important in understanding the present conditions at the site, and may assist in identifying the mitigation (i.e., habitat establishment, restoration, and enhancement) and the habitat type most appropriate for the site. Accordingly, in addition to field investigations of present conditions at a candidate mitigation site, information on historic conditions and management at the site should be gathered. Useful historic and contemporary materials include:

- Topographic maps
- Soil survey maps (*see* Figure 7-1)
- Aerial photographs
- Ground-level photographs
- Stream gage and rainfall records
- Groundwater well records
- Vegetation maps
- Wetland delineation maps
- Geotechnical reports

Aerial photographs, both historic and contemporary, are particularly useful in assessing past and present site conditions. The photographs may provide a basis for assessing such conditions as vegetation composition and variability over time, changes in the location of a stream channel, extent of flooding, past agricultural use, and other attributes.

In addition to the information listed above, discussions with landowners, land managers, and other individuals with long-term familiarity with the site can provide useful information on historic site conditions, land use, and management practices.

Field and Other Investigations. Once review of existing site information is completed, a field investigation is conducted to document and map baseline conditions at the candidate mitigation site. The field investigation is best performed by a multidisciplinary team, particularly one with expertise in plant ecology, wildlife biology, hydrology, and soil science. Members from other disciplines, such as fisheries biologists, water quality specialists, landscape architects, civil engineers, and others may be involved at this stage and during the design process. The investigation may entail compiling information on the site attributes listed below.

Existing wetlands and other vegetation communities. If a wetland delineation does not exist for the site, one may need to be conducted to determine if wetlands are present and whether it may be appropriate to restore or enhance the wetlands as part of the mitigation effort. A vegetation map of the entire candidate mitigation site is useful for integrating the mitigation habitat with existing habitats, including upland habitats, which may be as sensitive or as desirable as aquatic habitats.

- **Topography.** Because successful establishment or restoration of wetland hydrology is usually dictated by specific topographic conditions, a detailed topographic map will provide a basis for determining the amount of earthwork that may be required. Such a map is useful in the site evaluation stage, but because of the cost involved, it may not be prepared until after the site has been formally selected for mitigation purposes.

Although useful in the site evaluation stage, a detailed topographic map may not be prepared until after the site has been formally selected for mitigation purposes.

- **Hydrology.** Because hydrology is the driving force behind wetlands, observations of existing minimum and maximum water levels, existing extent of floodwater inundation, the presence of base flows, water quality, and groundwater level fluctuations may be important, depending on the wetland habitat type. This information may need to be augmented by projections of peak flow rates, runoff volumes, and floodplain extent. A water balance (water budget) may need to be calculated to confirm that the mitigation habitat has the proper hydroperiod to sustain the vegetation. At sites on which the mitigation entails expanding an existing habitat, a water balance may need to be calculated to determine if sufficient water would be present to sustain both the existing and the mitigation habitats.

Detailed soil analyses from backhoe pits are often necessary to confirm site suitability for wetland restoration.

- **Soils.** The accuracy of existing soil survey maps may need to be verified to confirm that land-disturbing activities such as excavation, filling, and erosion occurring after the survey was made have not resulted in soil conditions different than those described in the soil survey. In some cases, a soil map more detailed than that of the soil survey may need to be prepared for the mitigation effort. The soils at a site will partially dictate the type of mitigation habitat that may be appropriate and may reveal variations in groundwater levels. For some wetland habitats, it may be necessary to measure soil chemical and physical characteristics, such as permeability rates.
- **Wildlife.** Wildlife using the candidate mitigation site, particularly sensitive species protected by federal or state endangered species regulations, may partially dictate or constrain the type of mitigation that can be implemented. Surveys may be necessary to document the presence or absence of such species and their habitat.
- **Existing land use and other constraints.** Land use constraints such as ownership, easements, planning and zoning designations, cultural resources, water rights, geologically unstable areas, and the presence of hazardous materials may need to be identified. Setbacks may be required from such land uses and conditions.

Land use constraints such as ownership, easements, planning and zoning designations, cultural resources, water rights, geologically unstable areas, and the presence of hazardous materials may need to be identified.

- **Land use history.** Past uses or management such as cropping, grazing, burning, irrigation, excavation, filling, and herbicide use can all affect site conditions. Knowledge of site history is important in understanding existing site conditions and for planning mitigation efforts.

Opportunities and Constraints Evaluation. The information review and field investigation provide a basis to evaluate opportunities and constraints associated with achieving the goals of the mitigation plan. A map showing sensitive habitats and other site features that may constrain the mitigation should be prepared. When the candidate mitigation site would not meet some or all of the mitigation goals, additional sites or an alternative site may need to be identified. A meeting between the project proponent, the mitigation consultant, USACE, and other regulatory agencies may be necessary to determine the appropriate course of action.

When the candidate mitigation site would not meet some or all of the mitigation goals, additional sites or an alternative site may need to be identified.

Mitigation Site Design

Mitigation/Design Alternatives. Depending on the size and complexity of the selected mitigation site, it is sometimes appropriate to develop several mitigation alternatives. Based on the site evaluation, the alternatives should address identified constraints in the context of the mitigation goals. Alternatives may be presented to USACE and other relevant regulatory agencies for comment; this can be done, for example, at a preapplication meeting. The preferred alternative is the one that best addresses mitigation goals, is most sensitive to site conditions, and minimizes conflicts.

The preferred alternative is the one that best addresses mitigation goals, is most sensitive to site conditions, and minimizes conflicts.

Conceptual Design. The preferred design alternative can be refined into a conceptual mitigation design that may take the form of a graphic (Figure 7-2) supported by a schematic cross-section. The conceptual mitigation design generally includes such elements as the location of the mitigation site relative to the development project, "footprint" of the mitigation habitat(s), hydrologic inputs, public access considerations, and other such factors that will provide USACE and any other regulatory agencies with an accurate depiction of the proposed mitigation. To avoid misunderstandings and possible major revisions to the mitigation program at a more detailed, later stage of design, the conceptual plan should be presented to all relevant regulatory agencies for review and comment.

Detailed Design. The detailed mitigation design is different than the conceptual design in that it provides sufficient detail to enable a contractor or other party to construct the mitigation without significant direction from the designer. Because it often entails earthwork, the mitigation is usually best presented on a topographic base map of the site, which allows precise changes in grade to be shown. Because appropriate wetland hydrology for a given habitat type is often controlled by subtle changes in grade, a topographic map with a one-foot contour interval may be needed. In addition to the location and extent of each mitigation habitat, the detailed design may show public access trails, roads, and upland buffer zones.

Unlike the conceptual design, a detailed mitigation design provides sufficient detail to enable a contractor or other party to construct the mitigation without significant direction from the designer.

The planting plan is frequently developed separately from the grading plan element of the design. The planting plan shows the types and spacing

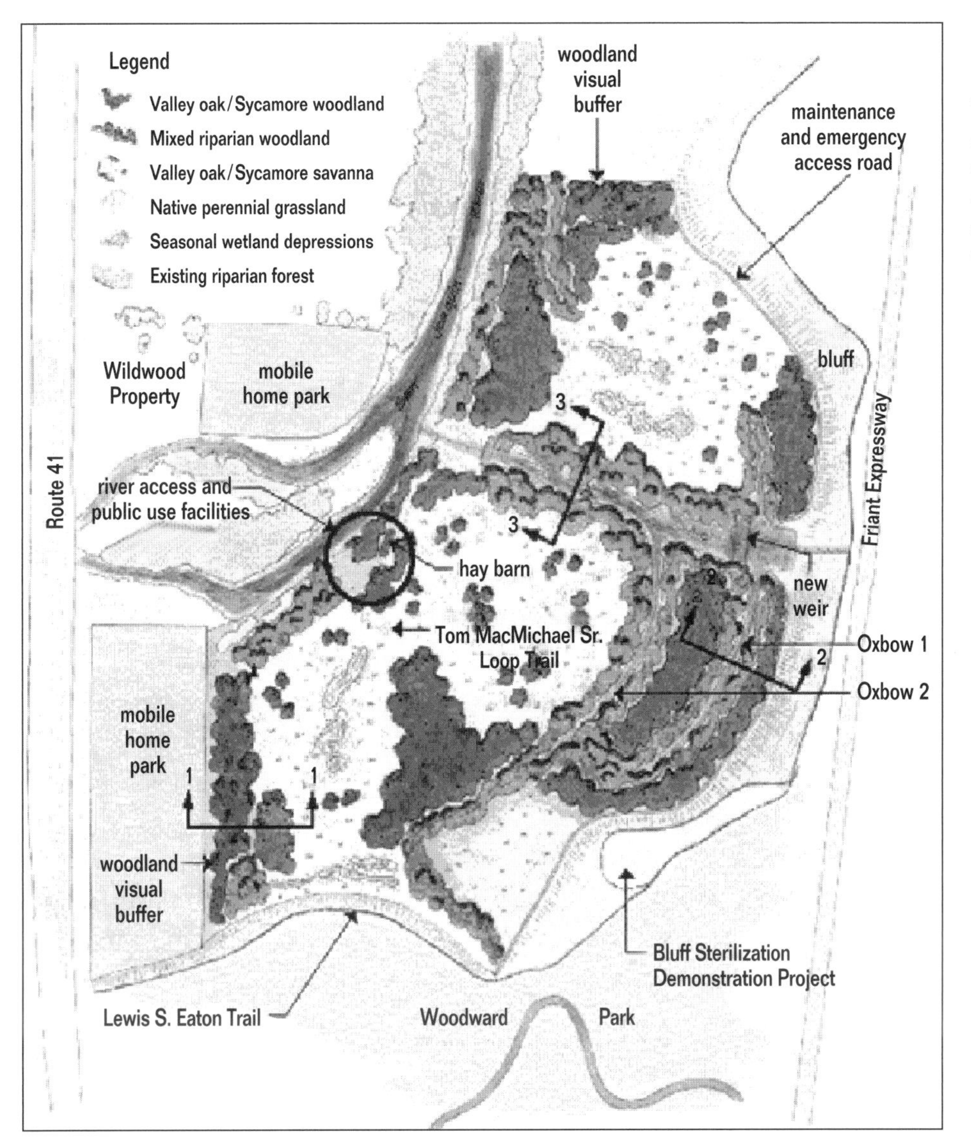

Figure 7-2
Mitigation Design

The conceptual mitigation design provides an overview of the planned habitats, adjacent land uses, buffer areas, public access, and other site conditions.

of plantings, areas to be seeded, and irrigation systems, if appropriate. USACE districts usually require use of only native species. Further, the use of seed and plant material derived from local sources, which is typically best adapted to site conditions, is increasingly standard practice.

For some designs, only plan sheets are required to construct the mitigation. However, for other projects, particularly those requiring competitive bids from contractors, a bid specifications document may also be required.

Establishing vegetation to restore a brackish tidal marsh may begin by transplanting plugs of emergent species from a nearby marsh.

The detailed design may include direction on such elements as construction schedule, grading, soil compaction, topsoil stripping, species and sources of seed and plant material to be used, locations of plantings, methods of planting, weed control, fencing, and maintenance measures. Because many mitigation sites are located near urban areas, designers may need to consult with local mosquito or vector control districts to incorporate appropriate strategies for management into the design.

Many designs inappropriately emphasize the planting and seeding aspects of the mitigation, and do not properly consider hydrology, the driving force behind wetlands. If the design does not allow for the creation or restoration of wetland hydrology, either the target wetland habitat type will not be established, or only upland habitats will result at the site. It is usually beneficial to identify the hydrologic conditions that support wetlands to be removed by the project and to emulate these conditions at the mitigation site to the extent practicable.

Mitigation planners should incorporate more wetland habitat acreage into their design than the project permit requires.

It is recommended that mitigation planners incorporate into their design more wetland habitat acreage than is required by the project permit. This provides some "insurance" in the event that part of the mitigation site does not meet the performance criteria. Moreover, establishing, restoring, or enhancing excess habitat acreage is usually less expensive to implement at the time of the overall mitigation construction than as a separate design and implementation effort later. The extent of the excess acreage is dictated by the degree of uncertainty associated with a particular habitat, the capacity of the site to support the excess acreage, and cost considerations.

Before mitigation construction, the detailed design should be discussed with the regulatory agencies to gain their approval.

Mitigation Construction

Habitat restoration, establishment, and enhancement efforts are usually constructed by general contractors capable of earthwork, irrigation system and plant material installation, and seeding.

Habitat restoration, establishment, and enhancement efforts are usually constructed by general contractors capable of earthwork, irrigation system and plant material installation, and seeding. Some contractors specialize in habitat construction; these are typically more sensitive to the nuances of grade levels, plant material handling requirements, and other factors that are critical to a successful habitat construction operation. Mitigation site maintenance work (described below) is typically conducted by the mitigation construction contractor.

Mitigation Performance Monitoring and Reporting

Performance standards are observable or measurable attributes that can be used to determine if a compensatory mitigation project meets its objectives.

Defining Performance Standards. As defined by the National Research Council (Committee on Mitigating Wetland Losses 2001), performance standards are "observable or measurable attributes that can be used to determine if a compensatory mitigation project meets its objectives." Performance standards are also known as success criteria, success standards, and release criteria.

Performance standards for compensatory mitigation wetlands may be based on (1) the characteristics of the wetlands to be affected, (2) the characteristics of a specific, naturally occurring habitat near the mitigation habitat, or (3) the characteristics of a typical habitat type in the project site vicinity. Such standards may be developed for a specific mitigation effort by an applicant's consultant, or they may be based on the guidelines published by some USACE districts for preparing mitigation plans. In any case, the performance standards are defined prior to permit issuance and are either specified as permit conditions or are detailed in the mitigation plan.

Performance standards are based on a variety of physical or biological conditions that are specific to the target habitat, such as those listed below.

Naturally establishing cottonwoods on the banks of an artificial oxbow

- Percent survivorship of plantings
- Height and stem diameter of plantings
- Percent cover by hydrophytic species
- Stem density of target vegetation
- Species richness (i.e., the number of different species that are present)
- Percent vegetative cover (especially of native species)
- Maximum percent cover by invasive nonnative species
- Depth and duration of flooding or soil saturation
- Benthic invertebrate populations
- Wildlife use (birds, mammals, amphibians, reptiles, fish)
- Water quality characteristics

Often tied to many of the standards above is the areal extent (e.g., acreage) of the mitigation site that must meet a given performance standard.

The mitigation wetland's performance standards are typically based on the functions of the wetlands removed by the project; the standards may also be based on existing wetlands on or near the mitigation site that have characteristics similar to the mitigation wetlands and that can be used during mitigation performance monitoring as *control* or *reference* wetlands. Reference wetlands, which may be more pristine than the affected habitats, serve as a point of comparison for interpreting the performance monitoring results of the mitigation wetland over time. For example, hydrology data may be simultaneously collected at the mitigation and reference wetlands, such that the effects that variable climate years have on the mitigation wetland may be compared with effects at the reference wetland. These observations allow the biological monitor to evaluate changes in the mitigation wetlands in the context of environmental variability (for example, to assess the effects of below average rainfall in the region). The reference wetlands may provide the monitor with a basis to conclude that there may not be a problem with mitigation wetland design, but rather that a failure to meet performance standards is a reflection of precipitation conditions.

Reference wetlands, which may be more pristine than the affected habitats, serve as a point of comparison for interpreting the performance monitoring results of the mitigation wetland over time.

In cases where the functions of wetlands removed by the project are lower than that of the mitigation wetlands (as reflected in the performance standards), reference wetlands of high function are necessary to measure the mitigation wetland's performance. For example, project mitigation measures may call for replacement of low-function wetlands with high-function wetlands of smaller size but equivalent total function.

Project mitigation measures may call for replacement of low-function wetlands with high-function wetlands of smaller size but equivalent total function.

Monitoring Methods. Monitoring methods should be described in the mitigation plan. The monitoring program defines the schedule of monitoring, intensity of data collection, methodology of data analysis, and specifications for

Created wetland with water control structure

reporting the results of performance monitoring. The methodology for data collection and analysis should be scientifically and statistically valid, but should also be cost effective. Most importantly, the methodology should be tailored to the specific mitigation habitat and designed to identify the mitigation habitat's progress toward achieving the performance standards.

Aerial and ground-level photographs may be taken of the mitigation site at prescribed intervals to help document and evaluate the progress of the mitigation effort. Although they may be expensive, aerial photographs may be cost effective in quantifying the acreage of the mitigation that meets the performance standards, especially for large mitigation sites. Ground-level photographs should be taken from fixed stations to provide a visual, chronological record of changes in hydrologic and vegetation conditions.

Schedule and Timing of Performance Monitoring. Most performance monitoring is conducted over a five- to ten-year period following mitigation implementation, although a longer period may be required for certain slow-developing habitats such as forested wetlands. The timing of monitoring depends on the type of wetland and the attributes to be monitored. Certain wildlife species, for example, may be present only for short periods during the year. For some habitats, particularly those that develop slowly, it may be appropriate and cost effective to monitor every other year.

Remediation Measures. Performance monitoring may identify the need to rectify problems at the mitigation site; for example, weed control measures may be necessary to control recruitment of undesired species, or grading may be necessary to modify hydrologic conditions. Generally, if mitigation is properly planned, the degree and frequency of necessary remediation measures will be minimal.

When significant remediation measures are required, the monitoring period may have to be extended. Permit conditions typically require additional monitoring time in such instances. If progress is observed but performance standards are not met after the designated time period, a viable remediation measure could be a longer monitoring period to give the wetland habitat more time to become established.

In some cases, no amount of remediation measures are sufficient to achieve the mitigation goals and performance standards for the required mitigation acreage. In this event, the required amount of mitigation should be implemented at an alternate mitigation site, which should be identified in the original mitigation plan. Part or all of the mitigation acreage may need to be implemented at the alternate site.

USACE usually requires that the results of the performance monitoring be documented in annual–and occasionally more frequent–reports.

Monitoring Reports. USACE usually requires that the results of the performance monitoring be documented in annual–and occasionally more frequent–reports. At a minimum, such reports describe the methods used to conduct the monitoring, the progress of the mitigation effort with respect to the performance standards, and corrective actions taken or needed. For mitigation efforts in which the monitoring period is considerably longer than five years, monitoring reports may sometimes be submitted less than annually, subject to USACE approval.

Maintenance and Protection of Mitigation Site

USACE generally requires the project proponent to identify the party responsible for a mitigation site's construction, monitoring, and short- and long-term maintenance and management. Financial assurances provided by the proponent are commensurate with the degree of resource impact and the amount and type of mitigation required. The financial assurances may be in the form of performance bonds, trusts, escrow accounts, letters of credit, or other instruments. The assurances may be phased out or reduced as the mitigation site increasingly meets the performance standards.

Usually, the project proponent is responsible for maintenance and performance monitoring for the first five to ten years after mitigation construction. The type of maintenance activities that are required depend on the mitigation habitat type and may include water management, replacement planting, weed and pest management, sediment and trash removal, and mosquito management.

The project proponent is usually responsible for maintenance and performance monitoring for the first five to ten years after mitigation construction.

Long-term ownership, habitat protection, and management of the site can be assigned to the permit holder, public agencies, or nonprofit organizations. Deed restrictions, conservation easements, or other instruments must often be placed on the mitigation site to prevent changes in use should ownership change. Often the project proponent will establish a trust fund for long-term operation and management to ensure that the mitigation site will be preserved as wildlife habitat.

Common Causes of Mitigation Failure

The art and science of wetland habitat establishment, restoration, and enhancement are still in their infancy and continue to evolve. Designers and contractors have learned many lessons from mitigation projects completed over the last 25 years or so–perhaps most from their failures and mistakes. Some common reasons for partial or complete failure include the following:

- Infeasible mitigation goals and performance standards
- Lack of data on soils or hydrology on which to base the project design
- Inadequate planning and design detail
- Lack of involvement of the mitigation designer during construction
- Lack of maintenance during the first few years following implementation

Preparing willow cuttings for planting in a riparian habitat restoration site

Many of these design and implementation errors are manifested at the mitigation site by the presence of upland plant species, excessive plant mortality, low plant vigor, and invasion of weeds. Other mitigation projects suffer from partial failure as a result of "forcing" an amount of habitat establishment that is beyond the site's capacity to support it.

Confirmation of Mitigation Success

Because compensatory mitigation wetlands should ideally be self-sustaining, many USACE districts require that the mitigation site meet the specified performance standards for a minimum of two consecutive years without human intervention, such as weed removal, irrigation, and replanting, before it is considered successful.

Some USACE districts include a special condition in the Section 404 permit requiring the applicant to provide financial assurance that mitigation will be implemented and that it will meet final performance standards.

Some USACE districts include a special condition in the Section 404 permit requiring the applicant to provide financial assurance that mitigation will be implemented and that it will meet final performance standards. Such assurances may be in the form of letters of credit, performance bonds, escrow accounts, casualty insurance, or other approved instruments. The assurance may be phased out or reduced as the mitigation project increasingly meets the performance standards or otherwise matures. Those mitigation efforts that are more experimental or have a historically lower rate of success are more likely to require a relatively higher level of financial assurances than those efforts in which the likelihood of success is more certain (Committee on Mitigating Wetland Losses 2001).

Mitigation Banks

Mitigation banks are sites in which wetlands and other aquatic areas have been restored, established, enhanced, or preserved expressly to provide compensatory mitigation in advance of authorized impacts on similar wetlands or other waters.

An alternative to compensatory mitigation implemented by the permittee is the use of a mitigation bank (*see* appendix G, USACE Regulatory Guidance Letters/ Memoranda to the Field, for guidance on mitigation banks). Mitigation banks are sites in which wetlands and other aquatic areas have been restored, established, enhanced, or (in exceptional cases) preserved expressly to provide compensatory mitigation in advance of authorized impacts on similar wetlands or other waters. The objective of a mitigation bank is to provide for the replacement of the chemical, physical, and biological functions of wetlands and other aquatic areas that are lost as a result of authorized impacts. Using a method approved by USACE and other cooperating agencies, the newly established wetland functions at a mitigation bank are quantified as credits; these credits are available for use by the bank sponsor or other parties to mitigate aquatic resource impacts (i.e., debits). Credits may be authorized to compensate for unavoidable impacts under Sections 404 and 10. Under Sections 404 and 10 and the Swampbuster programs, use of mitigation credits may be authorized by USACE and NRCS either when onsite compensation is not practicable or when the use of a mitigation bank is environmentally preferable.

Mitigation banks offer permit applicants greater flexibility than mitigation implemented at individual project sites in complying with mitigation requirements, and may provide ecological advantages as well as advantages to the applicant:

- The integrity of aquatic ecosystems may be better maintained by consolidating mitigation efforts into a single, larger parcel.
- Mitigation banks may combine financial resources and planning and scientific expertise not available or practicable with development of project-specific mitigation proposals. This combining of resources can increase the potential for the successful establishment and long-term management of wetlands at mitigation sites.
- Permit processing time can be reduced, and compliance with compensatory mitigation requirements may be more cost-effective than mitigation efforts at project sites.
- Mitigation habitats at mitigation banks are typically implemented and functioning in advance of project impacts, thereby reducing temporal losses of wetland functions and values and minimizing the uncertainty of mitigation success. USACE may accept a lower mitigation-to-impact ratio than it would require for project-specific mitigation.
- The project proponent is relieved of the obligation to monitor and maintain the mitigation site; long-term management is the responsibility of the mitigation bank operator.

Soil test pit to determine suitability for wetland creation

To establish a bank, prospective bank sponsors submit a prospectus to USACE or, when applicable, to NRCS, to initiate the process of planning and review by the appropriate regulatory and resource agencies. The information provided in the prospectus serves as the basis for establishing the mitigation banking instrument, which serves to document agency concurrence on the goals and administration of the bank. The instrument describes the physical and legal characteristics of the bank and how it will be operated. The signatory agencies collectively comprise the mitigation bank review team, that typically includes USACE, USFWS, and NOAA Fisheries; state and local agencies may also be signatories. The primary role of the mitigation bank review team is to facilitate the establishment of mitigation banks through the development of banking instruments. USACE is responsible for authorizing use of a particular bank for specific projects and for determining the number and availability of credits required to compensate for a project's impacts.

To establish a bank, prospective bank sponsors submit a prospectus to USACE or, when applicable, to NRCS, to initiate the planning and review process by the appropriate regulatory and resource agencies.

Each mitigation bank has a service area, within which it provides appropriate compensation for impacts on wetlands and other aquatic areas. This service area is defined in the banking instrument. The definition of the service area is based on hydrologic and biological criteria. As articulated in the USACE/EPA MOA on mitigation under Section 404 (1990), USACE and EPA usually prefer mitigation to be implemented onsite. However, this guidance does not preclude the use of a mitigation bank when there is no practicable opportunity for onsite mitigation, or when the use of a bank would be environmentally preferable to onsite mitigation. 60 FR 228, November 28, 1995.

Mitigation Through Payment of In-Lieu Fees and Cash Donations

An in-lieu fee arrangement allows a project proponent to shift legal responsibility for implementing the compensatory mitigation required for a project to another, USACE-approved party.

Another alternative for mitigation is payment of in-lieu fees. An in-lieu fee arrangement allows a project proponent to shift legal responsibility for implementing the compensatory mitigation required for a project to another, USACE-approved party (a sponsor). The sponsor is typically a natural resource management entity such as a park or open space district. With this approach to mitigation, the project proponent provides funds to the sponsor, who administers an in-lieu fee financial account and holds the funds until they are adequate for the mitigation to be constructed. The fees cover implementation of either specific or general wetland or other aquatic resource development projects. In-lieu fee mitigation may be used to compensate for impacts on waters of the United States that are authorized under both Section 404 general and individual permits, provided that certain criteria are met (65 FR 216, November 7, 2000). *See* appendix G for in-lieu fee guidance. In-lieu fee arrangements are different from mitigation banks, as in-lieu fees typically do not provide compensatory mitigation in advance of project impacts.

Some USACE districts allow project proponents to make cash donations on an ad hoc *basis to a natural resource management entity to satisfy compensatory mitigation requirements.*

Finally, some USACE districts allow project proponents to make cash donations on an *ad hoc* basis to a natural resource management entity to satisfy compensatory mitigation requirements. The cash donations are technically not considered in-lieu fee arrangements, as there is no formal agreement between USACE and the party receiving the donation. However, as in mitigation banking and in-lieu fee arrangements, the cash donations involve a shift of legal responsibility for the mitigation to the recipient of the funds.

Mitigation Effectiveness

It is usually much more cost effective to mitigate wetland impacts through minimization and avoidance than through compensatory mitigation.

Because reliable compensatory wetland mitigation techniques are still being developed, and because implementing wetland mitigation projects are often expensive, it is usually much more cost effective to mitigate wetland impacts through minimization and avoidance than through compensatory mitigation.

To date, compensatory mitigation projects involving habitat restoration and especially establishment (creation) have had variable degrees of success. Some habitats, such as freshwater emergent marshes, are more easily restored or established than other habitats, such as fens and bogs. Mitigation wetlands have generally been identified as having lower function and value, plant and animal species diversity, vegetative biomass, and soil fertility and organic carbon levels than natural wetlands of the same habitat type (Committee on Mitigating Wetland Losses 2001). Accordingly, USACE and other regulatory agencies may require high mitigation-to-impact ratios for habitats known to be difficult to restore or establish or to be slow to provide functional replacement.

Wetland mitigation construction constitutes only part of the total cost of a mitigation program, which also includes site evaluation, design, monitoring, maintenance, and sometimes land acquisition. Proper evaluation of a mitigation site to determine the feasibility of implementing mitigation, realistic mitigation goals, oversight of the construction operation by the designer, and proper maintenance all help to ensure that the mitigation project will be a success.

Wetland mitigation construction constitutes only part of the total cost of a mitigation program, which also includes site evaluation, design, monitoring, maintenance, and sometimes land acquisition.

Mitigation banks, payment of in-lieu fees, and cash donations all offer flexibility to project proponents, may be more cost effective than compensatory mitigation implemented on the project site, and commonly provide more effective and ecologically sound mitigation than onsite mitigation.

CHAPTER EIGHT

Regional Wetland Conservation Planning

The 1990s saw the beginning of a new emphasis on regional planning and cooperative efforts for protection of wetlands and other waters as an alternative to project-by-project permit battles. Regional wetland conservation planning is a proactive approach to addressing the conservation of wetlands and other waters in concert with economic growth and development over a large geographic area.

Regional wetland conservation planning is a proactive approach to addressing the conservation of wetlands and other waters in concert with economic growth and development over a large geographic area.

This form of proactive planning is in marked contrast with the project-specific permitting typified by the use of project-specific individual and nationwide general permits as described in chapter 4, Federal Regulation of Wetlands and Other Waters, and chapter 5, Section 404 Permitting Process. Regional planning and cooperative efforts for wetland protection offer several advantages over the traditional regulatory approach:

- The highest-value wetlands and other waters can be identified and protected.
- Landowners have more opportunity to integrate aquatic resources planning early in their development plans.
- Concerns of landowners, environmental groups, and other stakeholders can be addressed through a voluntary, collaborative process.
- Regional programs can accomplish more cost-effective and ecologically-effective wetland conservation than the project-by-project approach.
- Protection of wetlands and other waters can be more readily integrated with other comprehensive natural resource protection efforts, such as watershed planning, habitat conservation planning, and water quality protection.

Freshwater marsh in winter. Recreational use can adversely affect wildlife, but is also important to the development of public appreciation for wetlands.

EPA and USACE have regulatory and nonregulatory programs to assist local, state, and federal agencies in wetland conservation planning. A key principle of current federal wetland policy is to encourage nonregulatory programs such as advance planning, research, and public/private cooperative

EPA = Environmental Protection Agency
USACE = United States Army Corps of Engineers

Wet montane meadow dominated by corn lily

efforts. Federal agencies are providing increased technical assistance and promoting voluntary cooperative wetland planning and restoration efforts. Continued funding has been provided for the U.S. Department of Agriculture's Wetlands Reserve Program, administered by NRCS, which helps farmers restore wetlands on their properties.

This chapter presents an overview of tools and approaches for developing regional wetland conservation plans. The following topics are discussed:

- Section 404 tools for regional wetland conservation planning
- Key components of the regional wetland conservation planning process
- Elements of a regional wetland conservation plan
- Combined compliance with the Endangered Species Act and Section 404
- Advantages and disadvantages of regional plans

Section 404 Tools for Regional Wetland Conservation Planning

ADID = Advanced identification
CWA = Clean Water Act
NRCS = Natural Resources Conservation Service

Federal wetland policy promotes and supports advance planning for wetland resources. Advance planning for wetland protection can assume a variety of forms, but a regional, watershed-based approach offers the greatest opportunity for protecting the highest value wetlands, as well as for integrating wetland protection with water quality protection and flood control. Regulations and policies promulgated under Section 404 of the Clean Water Act (CWA) provide mechanisms for partnerships of local and state agencies with EPA and USACE. This section describes the following regulatory tools for regional wetland and watershed conservation planning:

- Advanced Identification
- Regional General Permit
- Programmatic General Permit
- Special Area Management Plan
- "Regional Individual Permit"

Advanced Identification

Advanced identification is a method of identifying the suitability of wetland sites for the future disposal of dredged or fill material.

Advanced identification (ADID) is a method, in accordance with EPA's Section 404(b)(1) Guidelines, of identifying the suitability of wetland sites for the future disposal of dredged or fill material. 40 CFR 230.80. In consultation with the state and in coordination with USACE, EPA identifies two types of sites under the ADID process:

- Possible future disposal sites, including existing disposal sites and nonsensitive areas; and
- Areas generally unsuitable for disposal site specifications (e.g., sites unsuitable for placement of dredged or fill material).

40 CFR 230.80

Advanced classification of sites in these categories provides information that can be used to support local land use planning and to facilitate the process for individual or general Section 404 permits. The identification of an area as possible future disposal sites does not constitute a permit for the discharge of dredged or fill material into wetlands, but does indicate to potential developers that issuance of a permit is likely. Conversely, the designation of a site as generally unsuitable for disposal suggests that issuance of a permit is unlikely or that substantial conditions will likely accompany any permit that is issued.

Information provided in an ADID allows EPA and USACE to focus their regulatory efforts on reducing wetland losses where resource values or scarcity are of greatest concern and, consequently, most likely to come into conflict with development pressures.

The ADID:

- Enables more effective land use planning
- Increases public awareness of the importance and value of aquatic ecosystems
- Provides the regulated community with an indication of the likelihood of permit issuance

Sullivan and Richardson 1993

Verde River Advanced Identification

Region 9 of EPA completed an advanced identification for the Verde River and its tributaries located northeast of Phoenix, Arizona. A detailed assessment of the functions and values of wetland resources of the Verde River was prepared. The Verde River was selected for an ADID based on three key factors:

- High wetland and riparian functions and values
- High probability of wetland loss or degradation without proper planning and management
- The opportunity to participate and work cooperatively in other comprehensive planning efforts

Sullivan and Richardson 1993

Following consultation with the state, local agencies may request that EPA initiate an ADID in their area. An ADID can be made more effective as a wetland conservation and development planning tool if it is followed by USACE issuance of a general permit (*see* discussion below) that streamlines regulatory requirements on nonsensitive sites. One disadvantage is that an ADID may require a great deal of time and cost to complete, involving considerable coordination among various agencies and interest groups.

Regional General Permits

PGP = Programmatic general permit
RGP = Regional general permit

Regional general permits (RGPs), which are issued by USACE district or division engineers, authorize a class of activities within a geographic region that are similar in nature and have minimal individual or cumulative environmental effects. 33 CFR 325.5(c)(1). RGPs serve to streamline the permitting process within the designated boundaries, avoiding the more complex and extended process of issuing individual permits on a project-by-project basis. The project proponents must meet special and general conditions included in the RGP in order for the project to qualify for authorization under the RGP. Local agencies can work in partnership with USACE to develop appropriate RGP conditions within all or a portion of the local jurisdiction.

For regional wetland planning purposes, conditions in the RGP can be developed to coincide with the goals and provisions of local wetland policies or a regional wetland conservation plan.

For regional wetland planning purposes, conditions in the RGP can be developed to coincide with the goals and provisions of local wetland policies or a regional wetland conservation plan. Once issued by USACE, RGPs are standing permits that all individuals can use if all stated conditions are met. In contrast to programmatic general permits (PGPs) (described

As with all general permits, RGPs are issued for five-year periods, subject to expiration, modification, or renewal by USACE.

below), permitting control under an RGP is not delegated to a local entity but remains the responsibility of USACE. As with all general permits, RGPs are issued for five-year periods, subject to expiration, modification, or renewal by USACE.

Programmatic General Permits

Cedar swamps, dominated by Atlantic White Cedar, are found along the coastal plain of the Atlantic and Gulf Coasts.

Programmatic general permits are issued by a USACE district or division engineer where a local, state, or other federal program provides protections for waters and wetlands that achieve the objectives of the Section 404 permit program. 33 CFR 325.5(c)(3). The local, state, or other federal agency assumes portions of USACE responsibility, as defined in the PGP, within its jurisdiction. PGPs offer a streamlined regulatory procedure under Section 404, because the agency holding the permit becomes the permitting authority and USACE is no longer involved in approval of individual permit use. As with all general permits, PGPs are reviewed by USACE every five years; at that point they may lapse or be reauthorized with or without modification.

Local governments may develop ordinances that are at least as stringent as Section 404 requirements and apply to USACE for a PGP. Wetland permitting could then be conducted by the local agency in conjunction with their standard land use authorizations. If the wetland ordinance is developed in coordination with a regional wetland conservation plan, then the local program would provide greater certainty with regard to wetland protection and development approvals.

Special Area Management Plans

CZMA = **Coastal Zone Management Act**
SAMP = **Special area management plan**

The development of special area management plans (SAMPs) was authorized under amendments to the 1980 CZMA. A SAMP is defined as:

> [A] comprehensive plan providing for natural resource protection and reasonable coastal-dependent economic growth containing a detailed and comprehensive statement of policies, standards, and criteria to guide public and private uses of lands and waters; and mechanisms for timely implementation in specific geographic areas within the coastal zone.

Although the SAMP process was originally intended to be applied to the coastal zone, USACE guidance issued in 1986 (and extended in 1992 through 1997; though expired, its tenets are still applied today) for the use of SAMPs stated that the process of collaborative interagency planning within a geographic area of special sensitivity is just as applicable in noncoastal areas. USACE Regulatory Guidance Letter 86-10. A successfully developed SAMP can:

- Reduce the problems associated with the traditional case-by-case review of wetland impacts and mitigation and individual permit applications
- Provide some predictability to the development process
- Address individual and cumulative impacts on wetlands in the context of broad ecosystem needs

One disadvantage is that SAMPs can require a great deal of time and effort to develop, involving substantial coordination among various agencies and interest groups. According to USACE guidance, the advantages of a SAMP may outweigh the disadvantages if the following conditions are present before a SAMP is proposed:

- The area is environmentally sensitive and under strong developmental pressure
- A sponsoring local agency ensures that the plan fully reflects local needs and interests
- Full public involvement is encouraged in the planning and development process
- All parties express a willingness at the outset to conclude the SAMP process with a definitive regulatory product

USACE Regulatory Guidance Letter 86-10

According to USACE guidance, the ideal SAMP concludes with two products:

- Appropriate local/state approvals and a USACE general permit or abbreviated processing procedure for activities in specifically defined situations, and
- A local/state restriction and/or an EPA Section 404(c) restriction (preferably both) for undesirable activities

USACE Regulatory Guidance Letter 86-10

Under Section 404(c), EPA may veto USACE issuance of a Section 404 permit (*see* chapter 4, Federal Regulation of Wetlands and Other Waters). A Section 404(c) restriction is an EPA action to prohibit discharges into designated waters of the United States prior to any permit application. Although USACE may still be requested to issue individual permits for activities that do not fall into either category above, individual permits should represent a small number of the total permit actions within the area covered by the SAMP.

Anchorage Special Area Management Plan

The Anchorage, Alaska SAMP is a good example of one that is considered successful. Under this SAMP, which was enacted in 1982, wetlands were classified into four categories.

- **Preservation.** Sites where no development is allowed except in special cases
- **Conservation.** Sites where some development is allowed
- **Developable.** Sites where development is not hindered by the presence of wetlands
- **Special study.** Sites with wetlands requiring additional study before they can be classified into any of the first three categories

Under this SAMP, USACE issued general permits for sites that were designated as developable, while those designated as preservation or conservation have been permitted only on a project-by-project basis. Since enactment of the Anchorage SAMP, most development has occurred in sites that were designated as developable. ■

Salvesen 1990

Regional Individual Permit

In instances where a single, large landowner (e.g., a military installation, state park, or private timber company) or land use controlling entity (e.g., a city or county) wishes to develop a regional wetland conservation plan, an individual Section 404 permit can be issued for the region. This "regional individual permit" would incorporate the elements of a regional wetland conservation plan with the requirements for obtaining an individual Section 404 permit. There are no federal regulations or policies for such a regional individual permit process, but it is not precluded and can provide many benefits to resources and landowners.

An RIP incorporates the elements of a regional wetland conservation plan with the requirements for obtaining an individual Section 404 permit.

A regional individual permit can be developed where a single permit applicant can provide sufficient information on the location of jurisdictional

APP = **Abbreviated permit process**

Comparison of ADIDs, RGPs, PGPs, SAMPs, and Regional Individual Permits

Wetland designations under ADIDs have no regulatory standing, but they do provide land developers and local agencies with valuable information for land purchase and planning decisions. Although the practice is not encouraged, permit applications can be filed and permits approved for placement of fill in wetland sites designated by an ADID as unsuitable for placement of fill material. Conversely, fill permits may be denied for wetlands designated as nonsensitive under an ADID.

In contrast, the approval process for SAMPs concludes with USACE and EPA regulatory decisions designating sites suitable for the abbreviated permit process, as well as sites where discharges into wetlands are restricted. SAMPs provide much of the same wetland locational information as do ADIDs, but with regulatory restrictions attached. SAMPs may designate wetland sites as legally restricted from fill activities (Section 404[c] restrictions). In addition, the SAMP specifically designates wetland sites where general or individual permits developed under the SAMP are applicable.

RGPs and PGPs may be issued without an ADID or SAMP process. Avoiding these formal processes may save in plan and permit development time and cost. However, completion of an ADID prior to the formulation of a general permit provides valuable information for establishing local agency responsibilities and the scope and conditions of the general permit. By providing specific data on wetland extent, location, and function within a region, an ADID also lends credibility to local wetland regulations that are based upon its findings, activities approved under the RGP or PGP, and the special and general conditions of the RGP or PGP.

While RGPs are permits with fixed conditions, PGPs allow local, state, and other federal agencies to assume some of USACE's authority within a particular region. This authority enables the controlling agency to regulate wetlands with greater flexibility and to maintain consistency with regionally specific approaches and priorities.

Regional individual permits provide the highest level of certainty for wetlands conservation and economic growth and development because of the highly specific resource and land use information they must contain. Regional individual permits are usually more costly to develop than other approaches because of the cost of developing the required information and of reaching the land use decisions necessary for permit approval. The increased cost to develop conservation plans for regional individual permits may be outweighed by the reduced costs of individual project approvals during plan implementation.

Table 8-1 lists examples of regional wetlands conservation planning efforts from around the United States. ■

wetlands and other waters; impacts on wetlands and other waters resulting from proposed activities; and mitigation measures to avoid, minimize, and compensate for those impacts. The differences between a regional individual permit application and a project-level permit application are that landscape-scale approaches to resource evaluation and conservation can be used, and fill activities and mitigation measures would take place over an extended time period. A regional individual permit must include an analysis of alternatives pursuant to Section 404(b)(1) Guidelines. Regional individual permits allow the flexibility for USACE to set the permit duration at the time of issuance; presumably, USACE could issue long-term permits allowing for impacts and mitigation measures to be implemented under a regional conservation plan over many years. Permit duration would not be restricted to renewals every five years, as are general permits.

Rather than issue a regional individual permit, USACE may develop an abbreviated permit process (APP) that allows the authorization of projects within the region under separate individual permits but does not require additional and separate Section 404(b)(1) alternatives analyses for each project. The alternatives analysis is conducted on the overall regional plan, and each future project need only show consistency with the regional plan.

A regional individual permit provides a high level of certainty that wetlands and other aquatic resources will be conserved. The specific location and function of resources to be removed and resources to be protected and restored are identified at the time of permit issuance. A regional individual permit provides a high level of certainty to land owners, resource users, and development interests because all uses are identified and authorized at the time of permit issuance, and authorization can be provided for a long time period (as long as permit conditions are met).

Table 8–1. Examples of Regional Wetlands Conservation Plans

Plan	Location
Mill Creek SAMP	King County, Washington
City of Superior SAMP	Douglas County, Wisconsin
Vincent Farms SAMP	Baltimore County, Maryland
Middle River Neck Peninsula SAMP	Baltimore County, Maryland
Hackensack Meadowlands SAMP	Bergen and Hudson Counties, New Jersey
Anchorage Wetland Plan (SAMP)	Anchorage, Alaska
Southwest Florida EIS	Lee and Collier Counties, Florida
Newhall Ranch–Santa Clara River	Valencia, California
Beale Air Force Base Habitat Conservation and Management Plan	Yuba County, California
Santa Rosa Plain Vernal Pools Ecosystem Conservation Plan	Santa Rosa, California
San Diego Creek Watershed SAMP	Orange County, California
San Juan Creek Watershed SAMP	Orange County, California
Otay River Watershed SAMP	San Diego County, California

Recommendations for Successful Regional Wetland Conservation Planning

Successful regional wetland conservation planning requires attention to a number of key components in the plan development process. Five key components of the planning process are discussed in this section.

- Physical and biological science
- Land and water use planning
- Regulatory compliance
- Economic analysis
- Public involvement and consensus

A successful planning process is one that results in a completed and approved regional wetland conservation plan and a Section 404 permit. Each of the planning components listed above and described below must receive ample attention from plan preparers if the planning process is to succeed.

A successful planning process is one that results in a completed and approved regional wetland conservation plan and a Section 404 permit.

Physical and Biological Science

Regional plans should use the best available physical and biological science to identify resources, establish goals and objectives, analyze impacts, and develop conservation measures. To support the planning process, the scientific disciplines of hydrology, hydrodynamics, geomorphology, sediment and water chemistry, and wetlands ecology must be combined with the applied sciences of ecosystem restoration and watershed management.

Lake Champlain on the Vermont and New York border is on the Atlantic Flyway, with an estimated 20,000 to 40,000 ducks and geese using its waters each year.

Principles of conservation biology should underpin the design and connectivity of managed wetland conservation areas. Application of these principles is essential to support the conservation of individual species, species diversity, and overall ecosystem functions.

The functional assessment of wetlands and other waters is a critical tool in wetland conservation planning. Chemical, physical, and biological attributes are quantitatively or qualitatively assessed at each aquatic resource site in a planning area. These attributes may include water source, quality, flow, and fluctuation; geographic, geologic, and soil conditions; vegetative cover and composition; species diversity; and human use and disturbance.

Assessment methods can be tailored to meet the needs of the particular regional wetland conservation plan. Reference wetlands and other waters that exhibit high-quality functions and values can serve to establish the scale by which other wetlands and waters of the same type are measured. Numerical models may be developed to score the relative functionality of wetlands and other waters.

Land and Water Use Planning

At the most fundamental level, regional conservation plans are land use plans and, where major streams and rivers are involved, are water use and water management plans. Regional conservation plans are never driven by resource conservation alone. Good land use planning and water management planning must parallel wetland conservation planning to develop a successful regional conservation plan. Quantitative or qualitative assessments of wetland values can be incorporated into the planning process along with functional assessments of wetlands (*see* Physical and Biological Science above).

Good land use planning and water management planning must parallel wetland conservation planning to develop a successful regional conservation plan.

Local government land use authority (e.g., general plans, master plans, comprehensive land use plans) is the typical means by which land use planning is combined with wetland conservation planning. Impact fees, open space zoning, transfer of development densities, and mitigation banks are just some of the planning tools available to local governments.

Most state agencies that own large tracts of natural lands have planning processes that can be dovetailed with wetland conservation plans. On federal lands, federal planning processes such as U.S. Forest Service land and resource management plans, U.S. Bureau of Land Management land use plans, and Department of Defense integrated natural resources management plans can include a focus on wetland conservation.

Federal and state agencies with responsibility for water supply and flood control play an important role in wetland and watershed conservation planning because operation and maintenance of river system control structures such as dams, weirs, and levees play an important role in wetland and watershed conservation planning.

Regulatory Compliance

Coastal salt marsh

Regional wetland conservation planning must involve a regulatory compliance strategy to be successful. Specifically, plan developers should develop a strategy to address compliance with Section 404. Some alternative approaches to Section 404 compliance are described above in Section 404 Tools for Regional Wetlands Conservation Planning.

In developing a regulatory compliance strategy, two broad concepts are important to keep in mind: "regulatory streamlining" and "one-stop shopping." Regulatory streamlining involves the simplification of the permit process, usually involving reducing regulatory overlap and processing complexity. Typically, a local agency assumes more control and USACE relinquishes control within the plan area (e.g., a PGP).

One-stop shopping involves combining two or more regulatory approvals into a single approval process. For example, the regional conservation plan might necessitate not only a Section 404 permit and the associated Section 401 certification, but also compliance with the federal Endangered Species Act (*see* sidebar, page 164), state endangered species and wetland protection laws, and local government requirements. Watershed plans under Section 303 of the CWA total maximum daily load (TMDL) development and approval process can also be combined with regional wetland planning and Section 404 permitting (*see* chapter 4).

TMDL = **Total maximum daily load**

Economic Analysis

Even if it is developed using the best scientific expertise and land use planning principles, a regional conservation plan is of little use if it is not economically feasible. A key component of conservation planning is determining the cost of plan implementation and how the plan will be funded. What is the economic impact of the plan on the local economy? Who will pay for plan implementation? How will costs be spread among those who benefit? These questions must be addressed early in the planning process, or a successful plan will never emerge.

Public Involvement and Consensus

Regional wetland conservation planning is a public process that must have public support for the plan is to be successful. Involvement of the public should begin early in the process, continue through the phases of planning and implementation, and include outreach and education efforts that publicize the planning process to facilitate a high level of inclusion. Plans based on consensus that bring all interested parties (stakeholders) to the table are most likely to succeed. Stakeholders typically include landowners, development interests, environmental groups, agricultural interests, resource user interests (e.g., timber, fisheries), recreational interests, and federal, state, and local regulatory and resource agencies.

Plans based on consensus that bring all interested parties (stakeholders) to the table are most likely to succeed.

Elements of a Regional Wetland Conservation Plan

HCP = Habitat conservation plan

In the previous section key components of the regional wetlands planning process were identified. This section recommends the essential elements of the plan document itself. The elements that must be included in a regional wetland conservation plan are similar to those required in a regional habitat conservation plan (HCP) under Section 10 of the Endangered Species Act. A wetland conservation plan should address the following:

Montane meadow, riparian scrub, and aspen riparian forest in the Sierra Nevada

- Geographic scope
- Purpose and need and resource goals and objectives
- Existing conditions
- Covered activities
- Permit duration
- Impact analysis
- Conservation strategy and mitigation measures
- Adaptive management plan
- Monitoring program
- Implementation costs and funding mechanisms
- Alternatives analysis

Geographic Scope

The planning area boundary is typically based on watershed, political, and land ownership boundaries, or some combination.

Determining the geographic scope and defining the planning area boundaries for a regional conservation plan are obviously important early decisions. The planning area boundary is typically based on watershed boundaries, political boundaries, land ownership boundaries, or some combination. Watershed boundaries are the most ecologically reasonable choice, while political and ownership boundaries (e.g., city and county jurisdictions) may lead to the simplest regulatory strategies. The complexity of the planning process increases with the geographic scope because of the greater diversity of wetlands and waters and the greater number of landowners and other stakeholders involved.

Statement of Purpose and Need and Resource Goals and Objectives

NEPA = National Environmental Policy Act

The statement of purpose and need for a regional wetland conservation plan should be clearly stated early in the planning process; moreover, such a statement is required for Section 404 and NEPA compliance. One example of a general project purpose for a regional conservation plan could be:

> [T]o provide a comprehensive plan for protecting, enhancing, and restoring aquatic resources while providing for reasonable future economic growth and development in accordance with local land use plans.

Resource goals and objectives provide more specific descriptions of expected outcomes in the context of the overall project purpose. At the onset of the planning process, the plan preparers should identify specific, measurable goals and objectives for aquatic resources. Goals and objectives for regional wetland plans are often based on the expected extent of the aquatic resources (i.e., acreage) and sometimes on functional assessment scoring. An overall goal of no net loss of wetlands/waters functions and values is required to meet USACE and EPA standards.

This channelized stream is surrounded by urban development. Even if wetland vegetation reestablishes itself in the channel, there is no buffer area to keep the development from encroaching on the stream.

Existing Conditions

Good regional planning processes are based on good data. The most valuable data for planning are those that identify and describe existing aquatic resources. Waters of the United States may be delineated using standard on-ground approaches (e.g., the routine method) as well as remote sensing methods involving aerial photography or satellite imagery. On-ground delineations may be prohibitively expensive for regional plans that cover very large areas.

In addition to the delineation of waters, a functional assessment may be conducted for each wetland and water body. The functional assessment may use a quantitative or a qualitative model to score the relative function for each wetland and water body against reference wetlands and waters for individual and overall functional parameters.

Covered Activities

In order to be in compliance with Section 404, the regional plan must identify the activities that will be covered by the plan. Covered activities should include all actions that could result in impacts on wetlands or waters for which the permit will be applicable. Plans may present covered activities in either or both of two ways: (1) a detailed list and specific descriptions of activities, and (2) a general description of activities likely to take place in the planning area. Detailed activity lists offer more specificity regarding impacts and mitigation measures, whereas general descriptions of activities allow plans to address future growth and development that cannot be specifically identified at the time of plan or permit approval.

Covered activities should include all actions that could result in impacts on wetlands or waters for which the permit will be applicable.

Permit Duration

RGPs and PGPs are limited to five-year durations, after which they must be either renewed or terminated. The duration of individual permits can be established in the permit based on the length of time necessary to complete covered activities (e.g., build-out of a city or county) and the response time of the resources affected by the plan (e.g., recovery time of a wetland habitat following disturbance or the time necessary to achieve wetland restoration success).

Impact Analysis

An impact analysis must be conducted to assess known or potential effects that activities covered under the regional wetland conservation plan could

The impact analysis should address all mechanisms of impacts on aquatic resources and not be limited to the impacts of placement of dredged or fill material.

have on wetlands and other waters. This impact analysis should address all mechanisms of impacts on aquatic resources; it should not be limited to the impacts of placement of dredged or fill material (*see* Activities Affecting Wetlands in chapter 2). Moreover, indirect and cumulative impacts on wetlands and other waters must also be assessed.

Freshwater seep

A common standard must be established for determining the magnitude and severity of impacts. In the simplest approach, impacts are measured on the basis of the extent of different aquatic resources removed or degraded (e.g., acreage of loss of different types of wetlands). More sophisticated approaches use a functional assessment method and numerical models to assess and quantify the loss of functions of specific wetlands and other aquatic resources (*see* Functions and Values of Other Waters in chapter 2). It is important that standard units of measure be used for describing existing conditions, setting resource goals and objectives, assessing impacts, developing conservation measures, and developing the monitoring program. This "common currency" allows for a straightforward analysis in determining the type and amount of impact, the type and amount of mitigation necessary to compensate for impacts, and the success in achieving goals and objectives.

Conservation Strategy and Mitigation Measures

The conservation strategy is the heart of a regional wetland conservation plan. Depending on the size and scope of the plan, the strategy may include a broad-based set of measures and policies or specific mitigation measures. The conservation strategy must incorporate an approach that meets the sequencing requirements of USACE/EPA policy (to first avoid, second minimize, and third compensate for impacts on wetlands and other special aquatic sites).

There are as many approaches to conservation strategies as there are regional wetland conservation plans. Plans may include landscape/watershed-level measures as well as wetland/water body-specific measures. The conservation strategy may include mitigation measures on a hierarchy of scale: region, watershed, sub-watershed, stream reach, wetland. Plans may specifically identify on a map the boundaries of conservation areas to be established ("map-based plan") or may describe a process, without identifying specific locations, by which conservation areas will be assembled during plan implementation ("process-based plan"). Map-based and process-based approaches may be combined in the same plan.

Conservation strategies typically include measures to preserve wetlands and other waters, to enhance and restore wetlands and other waters, and to preserve and restore ecosystem processes.

Conservation strategies typically include measures to preserve wetlands and other waters, to enhance and restore wetlands and other waters, and to preserve and restore ecosystem processes. The strategy may specify mitigation ratios (amount of restoration or preservation required for each unit of wetlands/waters affected) based either on acreage of affected wetlands or on a functional assessment scoring system. Mitigation banking is a tool that can be incorporated into a regional conservation strategy (*see* discussion of mitigation banks in chapter 7, Mitigation Planning). Where a regional general permit will be established, the conservation strategy should include statements that correspond to the general and specific conditions in the RGP.

Wetlands provide important wildlife habitat.

Adaptive Management Plan

Scientists and resource managers have always recognized that, because of the great complexity of even the simplest ecosystems, uncertainty is an integral component of managing those systems. Resource managers must therefore recognize and prepare for the uncertainty that underlies resource management. Adaptive management is one approach for addressing the uncertainty in natural resource management. Kershner (1997) described adaptive management as:

> [T]he process whereby management is initiated, evaluated and refined (Holling 1978, Walters 1986). It differs from traditional management by recognizing and preparing for the uncertainty that underlies resource management decisions. Adaptive management is typically incremental in that it uses information from monitoring and research to continually evaluate and modify management practices. It promotes long-term objectives for ecosystem management and recognizes that the ability to predict results is limited by knowledge of the system. Adaptive management uses information gained from past management experiences to evaluate both success and failure, and to explore new management options.

An adaptive management plan uses information obtained from monitoring activities to continually evaluate the efficacy of management strategies and to make appropriate modifications. Accordingly, successful adaptive management is contingent upon implementation of an effective monitoring program (*see* discussion below).

An adaptive management plan uses information obtained from monitoring activities to continually evaluate the efficacy of management strategies and to make appropriate modifications.

In addition to its obvious virtues for implementing a given wetland conservation plan, adaptive management is a vital tool for increasing the body of knowledge upon which the science of resource management depends. Greater understanding is essential to render effective such areas of endeavor as wetland restoration, watershed management, stream enhancement, water management in controlled systems, grazing intensity, fire management, and individual species management.

While some areas of watershed management have long histories (e.g., management of freshwater marshes as waterfowl habitat), such intricate and varied ecosystems still hold many mysteries. Without a commitment to adaptive management, the success of a regional conservation plan would likely remain an uncertain prospect.

Monitoring Program

Monitoring involves the gathering of information during implementation of a wetland conservation plan. Every plan should include a monitoring program. Monitoring falls into three main categories: compliance monitoring, effects monitoring, and effectiveness monitoring.

Monitoring involves the gathering of information during implementation of a wetland conservation plan.

Wetlands provide habitat for a wide variety of wildlife species.

Compliance monitoring is used to confirm that actions specified in the plan and permits have been conducted as specified. For example, compliance monitoring may confirm that restoration of wetlands has been conducted in keeping with the extent and methodology specified in the plan.

Effects monitoring involves the assessment of impacts that covered activities have on wetlands and waters. Existing conditions and the expected effects of covered activities on aquatic resources must be identified in the regional plan and permit application, but the actual effects may differ from expected effects, especially where the effect of an activity is not well known. Effects monitoring is necessary for long-term regional plans, particularly those that have adopted a process-based approach to the conservation strategy. In such cases, effects must be identified and quantified in order to determine the amount of mitigation necessary as covered activities progress and the conservation plan is implemented over time.

Effectiveness monitoring addresses the expected outcome of conservation/mitigation measures. The plan must provide predictions of specific outcomes of conservation measures, such as maintenance of ecosystem functions in protected wetlands and increased functions in enhanced and restored wetlands. Effectiveness monitoring is conducted to determine if these expectations have been met and, hence, if conservation measures have been successful. Effectiveness monitoring encompasses mitigation performance monitoring (*see* chapter 7, Mitigation Planning); it also develops information to support adaptive management decisions.

Regional wetland conservation plans may need to combine direct and indirect monitoring to measure plan success.

Regional wetland conservation plans may need to combine direct and indirect monitoring to measure plan success. Direct monitoring involves the site-specific compliance, effects, and effectiveness monitoring of impact areas and conservation areas. Indirect monitoring involves the gathering of regional data (e.g., through remote sensing methods) to determine trends in the extent and function of aquatic resources over time. For example, implementation of a regional wetland conservation plan may involve a complete periodic mapping of wetland resources in the planning area to determine overall trends of wetland losses and gains.

The frequency and intensity of monitoring must be determined in light of regulatory requirements, ecological necessity, scientific requirements, and practicability. Regulatory requirements address the need for monitoring to be sufficient to prove the plan is operating successfully. Ecological necessity addresses the need to monitor different types of wetlands/waters in different ways, at different times of year, and with different intensity and frequency. Scientific requirements address the need for statistically and ecologically valid monitoring methods and appropriate experimental designs. Practicability addresses the need to conduct monitoring that is logistically and economically feasible.

As discussed above, a properly designed monitoring plan is crucial to adaptive management. Without appropriate monitoring data, it is impossible to make informed decisions about the degree and causality of either the success or failure of the conservation strategy.

Implementation Costs and Funding Mechanisms

Rivers provide excellent wildlife corridors and create habitat mosaics.

The cost of implementing a regional wetland conservation plan must be estimated and the source of funding identified. Estimating the cost of implementation of a regional wetland conservation plan is not a simple task. Complex implementing entities and processes must often be developed.

Implementation costs typically include:

- Program administration
- Land acquisition through fee title or conservation easement
- Restoration and/or enhancement of wetlands and other waters
- Management and maintenance of conservation areas
- Adaptive management program
- Monitoring program
- Contingency for remedial measures

There are many ways to fund regional wetland conservation plans. Public receptiveness to different funding mechanisms varies. Some examples of funding sources are:

- Impact fees on developers
- Special assessment districts
- Sales or other new taxes
- Federal, state, or private grants
- Specific federal or state legislation
- Land swaps

Wetland conservation typically provides benefits to a wide range of stakeholders. It is, therefore, best to spread implementation funding among several sources so that no single segment of the community bears the full cost.

Alternatives Analysis

Section 404(b)(1) Guidelines require permit applicants to prepare an alternatives analysis in order to receive a standard Section 404 permit. *See* chapter 4, Federal Regulation of Wetlands and Other Waters, and chapter 5, Section 404 Permitting Process. The regulatory language in the Section 404(b)(1) Guidelines is not clear as to the requirement for an alternatives analysis in the process of developing and issuing general permits. USACE does conduct an alternatives analysis when it issues nationwide general permits. The approach to developing alternatives for regional permit applications can be quite different than for project-specific permit applications. Alternatives should be developed to comply with both Section 404(b)(1) Guidelines and NEPA. Examples of alternatives that would be analyzed in a regional Section 404 permit process are:

BA = **Biological assessment**
ESA = **Endangered Species Act**

Parallel Compliance with the Endangered Species Act and Section 404

Wetlands may support habitat for threatened and endangered species; in such cases the goals and objectives of wetland conservation and species conservation overlap. Compliance with Section 404 through a regional wetland conservation plan can be conducted in parallel with regional compliance under the Endangered Species Act (ESA) for take of species listed as threatened and endangered under ESA. Combined conservation planning and parallel regulatory processes can provide tremendous benefits to the resources and much greater efficiency in economic growth and development in a region.

Two regulatory processes are available under ESA:

- The Section 7 consultation process applies to federal agencies where a federal action may affect listed species.
- The Section 10 permitting process applies to nonfederal activities that require incidental take authorization.

Large federal land holdings, where various development and operations actions may have impacts on wetlands and endangered species over extensive areas and an extended time period, are eminently suitable for a combined approach to regional wetland and species conservation planning and parallel Sections 404 and 7 compliance. Military installations, as they conduct ongoing operations or grow to accommodate new missions, are excellent candidates for such an approach. The biological assessment (BA) is the required federal document under Section 7. The BA can be prepared in such a way that it is also suitable for submission to USACE as the mitigation plan for the Section 404 permit application.

Where local or state agencies exert land use authority over large areas, and rapid growth and development threaten wetlands and listed species, combined regional wetland and species conservation planning and parallel Section 404/Section 10 compliance is a recommended approach. Such an approach can be a proactive mechanism to resolve conflicts before they occur. The habitat conservation plan (HCP) is the required regulatory document under Section 10. Like the BA, the HCP can also be prepared for submission to USACE as the mitigation plan for the Section 404 permit application.

Table 8-2 provides a summary comparison of the elements of a regional wetland conservation plan under Section 404 with those of a regional HCP under Section 10 ESA. ■

- No-development alternative
- No-regional-permit alternative
- No-fill alternative
- Different approaches to protection and restoration
- Different amounts of protection and restoration
- Different regulatory approaches

The no-development alternative would be no new development in the region and hence no new impacts on waters of the U.S. This alternative may look at the potential for development to take place entirely outside the planning area. The no-development alternative would not meet the project purpose in most cases because regional wetland conservation plans are typically initiated in areas where growth patterns that do not have alternative regional locations are already resulting from strong economic pressures.

The no-regional-permit alternative evaluates the likely regional development outcome and impacts on wetlands/waters if project-by-project permitting were pursued under existing Section 404 regulations rather than under a regional permit. The no-regional-permit alternative is the USACE no-action alternative.

The no-fill alternative provides a regional plan for growth and development that results in no placement of dredged or fill material in waters of the United States. In most growing regions of the U.S., the density and distribution of wetlands and other waters, and the severe constraints that total avoidance would place on development planning, would preclude a no-fill alternative from being practicable.

Regional plans may analyze alternative approaches to protecting, enhancing, and restoring wetlands and other waters that would achieve the same level of wetland conservation. These alternatives may involve different preserve system locations and designs or different methods of wetland/water restoration.

Alternatives may also be developed that result in different levels of development and wetland conservation. For example, a plan may assess alternatives for:

- The lowest level of development and highest possible level of wetland/water conservation that results in the maximum amount of wetland protection and gains through restoration and enhancement

Table 8-2. Comparison of Elements of a Regional Wetlands Conservation Plan Under CWA Section 404 and a Regional Habitat Conservation Plan Under ESA Section 10

Plan Element	Section 404 Clean Water Act	Section 10 Endangered Species Act
Goals and objectives	Wetlands/waters functions	Species populations and habitat
Existing conditions	Delineation of waters of United States	Threatened and endangered species and habitat occurrences
Geographic scope	Watershed boundaries, political/ownership boundaries	Habitat boundaries, political/ ownership boundaries
Activities covered	Activities resulting in impacts on wetlands/waters	Activities resulting in take of threatened and endangered species
Permit duration	Five years RGP or PGP; set with individual permit	Determined with permit
Impacts	Loss of functions of wetlands/waters	Take of threatened and endangered species
Mitigation measures	Avoid, minimize, compensate (priority sequence)	Avoid, minimize, compensate (priority sequence)
Alternatives analysis	Section 404(b)(1) alternatives analysis	Alternatives to take considered and rejected
Funding	Sources identified in plan	Sources identified in plan (must assure funding)
Monitoring and adaptive management programs	Necessary components of a good plan	Necessary components to a good plan (required)
Authorizations	Placement of dredge or fill into waters of United States	Take of threatened and endangered species
Standards	Least environmentally damaging practicable alternative (LEDPA); no net loss of functions and values	Minimize and mitigate impacts to the maximum extent practicable; not jeopardize the continued existence of species

- An intermediate level of development and intermediate level of wetland/water conservation that exceeds the no-net-loss criterion
- A maximum level of development and minimum level of conservation that addresses only the mitigation of impacts and achieves no net loss

Regional wetland conservation plans may assess alternative methods of permitting. Different forms of regional permits can be analyzed for their anticipated effects on wetlands resources, growth and development, and cost.

As with project-specific permits, regional permits must approach mitigation in the sequence of avoid, minimize, and compensate, and must comply with the LEDPA standard. *See* chapter 4, Federal Regulation of Wetlands and Other Waters, and chapter 5, Section 404 Permitting Process.

LEDPA = Least environmentally damaging practicable alternative

Advantages and Disadvantages of Regional Wetland Conservation Plans

Regional wetland conservation planning is typically a proactive effort to combine the conservation of wetlands with land use planning for growth and development. The existing project-by-project permitting process typically remains an option within regional plan areas. The regional approach to conservation of resources and permitting has both disadvantages and advantages in comparison to the project-level approach.

Disadvantages of Regional Plans

Regional planning is usually a complex process that requires greater knowledge, foresight, and creativity than the project-level permitting process. It also requires the involvement of many more stakeholders for a longer time than does project permitting. Consequently, maintaining momentum, building consensus, and obtaining adequate funding through a potentially arduous process can be a tremendous challenge, requiring finesse and perseverance. The stakeholder desire and political will must be strong, or the regional effort will not succeed. Moreover, developing a regional plan can entail large initial costs, while the benefits may not be realized for many years.

Advantages of Regional Plans

Regional plans offer a number of advantages over project-by-project mitigation and permitting. Regional plans are more efficient and effective in conserving wetlands and other aquatic resources, largely because they can address entire watersheds and the ecosystem processes vital to maintaining wetland functions. Moreover, because of their scope, regional plans are more flexible than the project-specific permitting process in identifying conservation and development areas.

Because of the economy of scale, regional planning is more cost effective per unit of both development and conserved resources than the project-specific approach.

Because of the economy of scale, regional planning is more cost effective per unit of both development and conserved resources than the project-specific approach. Regional planning reduces many of the constraints to land use planning that are attendant upon the project-specific approach; additional savings consequently accrue as long-term development is undertaken. Successfully implemented regional plans typically offer much faster project approvals, which translate into additional cost savings.

Approved and functioning regional plans offer greater certainty in the permitting process and greater certainty in the conservation of wetlands and other waters. By clearly defining the extent and character of aquatic resources that are to be conserved, regional plans offer project proponents a high degree of security as they undertake individual project planning. Finally, while project-by-project permitting is plagued with the necessity to address the same issues repeatedly for each new permit, a regional wetland conservation plan resolves conservation and permitting issues in a single, organized effort.

CHAPTER NINE

Epilogue

In this book, we have shared our collective knowledge and experience about how wetlands, streams, and other waters are protected and regulated across the United States, describing in detail the scope of existing federal government programs, summarizing state programs, and offering suggestions on how to negotiate the permitting process successfully.

What the future holds for the regulation of aquatic resources is not certain, but we offer these predictions.

- While the rate of net loss of wetlands and other waters is expected to slow, new studies will likely uncover gaps in our present knowledge of how ecological functions of aquatic systems have been affected by human activities.
- With continued scientific research and adaptive management programs, new knowledge in aquatic ecology and habitat restoration techniques will improve our ability to avoid, minimize, and compensate for impacts.
- The full ramifications of the Supreme Court's decision in the *SWANCC* case regarding the regulation of isolated waters and their nexus to commerce and navigable waters have not been realized. We expect that there will be continued refinement of the methods used for determining jurisdiction over isolated waters. States may decide to provide protection to waters that USACE is required to remove from Section 404 jurisdiction as a result of *SWANCC*.
- New "regulatory takings" challenges to USACE's jurisdiction under Section 404 of the Clean Water Act will likely be brought to the courts and, depending on the outcome, may result in profound changes in regulation.
- The use of the HGM approach and functional assessments of wetlands and other waters will continue to increase and will become

Alpine Lake

HGM	=	Hydrogeomorphic
SWANCC	=	Solid Waste Agency of Northern Cook County
USACE	=	United States Army Corps of Engineers

a regular part of jurisdictional delineations, impact analyses, and mitigation planning.

- Regional conservation planning for wetlands, streams, and other waters will increase nationwide, resulting in more streamlined and more creative approaches to regulation and conservation of aquatic resources.
- States will continue to add stronger regulations, exceeding or supplementing federal regulations, to their state codes for the protection of wetlands and other waters.
- Federal and state agencies will continue proactive wetland conservation efforts through various programs (e.g., USDA NRCS Wetlands Reserve Program), independent of regulatory programs.
- Judicial decisions, including Supreme Court decisions, will continue to modify the interpretation of various aspects of the Clean Water Act.

NRCS = Natural Resources Conservation Service
USDA = United States Department of Agriculture

Much has changed in regulatory programs since the publication of our earlier book on the subject, *Wetlands Regulation: A Complete Guide to Federal and California Programs,* in 1995. We expect that change will only continue, and recommend use of the web sites listed in the text, tables, and appendix L to supplement and update the information and knowledge acquired here.

Appendices

In the appendices that follow Solano Press has made every effort to ensure that no changes have been made to the contents of the documents as a result of reformatting and reprinting. Any omissions or changes to the contrary are the responsibility of Solano Press Books and not the governmental agencies cited.

APPENDIX A

Clean Water Act

Sections 301, 303, 401, 402, and 404

Section 301, Clean Water Act
33 USC 1311

Effluent limitations

(a) Illegality of pollutant discharges except in compliance with law

Except as in compliance with this section and sections 1312, 1316, 1317, 1328, 1342, and 1344 of this title, the discharge of any pollutant by any person shall be unlawful.

(b) Timetable for achievement of objectives

In order to carry out the objective of this chapter there shall be achieved–

(1) (A) not later than July 1, 1977, effluent limitations for point sources, other than publicly owned treatment works,

(i) which shall require the application of the best practicable control technology currently available as defined by the Administrator pursuant to section 1314(b) of this title, or

(ii) in the case of a discharge into a publicly owned treatment works which meets the requirements of subparagraph (B) of this paragraph, which shall require compliance with any applicable pretreatment requirements and any requirements under section 1317 of this title; and

(B) for publicly owned treatment works in existence on July 1, 1977, or approved pursuant to section 1283 of this title prior to June 30, 1974 (for which construction must be completed within four years of approval), effluent limitations based upon secondary treatment as defined by the Administrator pursuant to section 1314(d)(1) of this title; or,

(C) not later than July 1, 1977, any more stringent limitation, including those necessary to meet water quality standards, treatment standards, or schedules of compliance, established pursuant to any State law or regulations (under authority preserved by section 1370 of this title) or any other Federal law or regulation, or required to implement any applicable water quality standard established pursuant to this chapter.

(2) (A) for pollutants identified in subparagraphs (C), (D), and (F) of this paragraph, effluent limitations for categories and classes of point sources, other than publicly owned treatment works, which

(i) shall require application of the best available technology economically achievable for such category or class, which will result in reasonable further progress toward the national goal of eliminating the discharge of all pollutants, as determined in accordance with regulations issued by the Administrator pursuant to section 1314(b)(2) of this title, which such effluent limitations shall require the elimination of discharges of all pollutants if the Administrator finds, on the basis of information available to him (including information developed pursuant to section 1325 of this title), that such elimination is technologically and economically achievable for a category or class of point sources as determined in accordance with regulations issued by the Administrator pursuant to section 1314(b)(2) of this title, or

(ii) in the case of the introduction of a pollutant into a publicly owned treatment works which meets the requirements of subparagraph (B) of this paragraph, shall require compliance with any applicable pretreatment requirements and any other requirement under section 1317 of this title;

(B) Repealed. Pub. L. 97-117, Sec. 21(b), Dec. 29, 1981, 95 Stat. 1632.

(C) with respect to all toxic pollutants referred to in table 1 of Committee Print Numbered 95-30 of the Committee on Public Works and Transportation of the House of Representatives compliance with effluent limitations in accordance with subparagraph (A) of this paragraph as expeditiously as practicable but in no case later than three years after the date such limitations are promulgated under section 1314(b) of this title, and in no case later than March 31, 1989;

(D) for all toxic pollutants listed under paragraph (1) of subsection (a) of section 1317 of this title which are not referred to in subparagraph (C) of this paragraph compliance with effluent limitations in accordance with subparagraph (A) of this paragraph as expeditiously as practicable, but in no case later than three years after the date such limitations are promulgated under section 1314(b) of this title, and in no case later than March 31, 1989;

(E) as expeditiously as practicable but in no case later than three years after the date such limitations are promulgated under section 1314(b) of this title, and in no case later than March 31, 1989, compliance with effluent limitations for categories and classes of point sources, other than publicly owned treatment works, which in the case of pollutants identified pursuant to section 1314(a)(4) of this title shall require application of the best conventional pollutant control technology as determined in accordance with regulations issued by the Administrator pursuant to section 1314(b)(4) of this title; and

(F) for all pollutants (other than those subject to subparagraphs (C), (D), or (E) of this paragraph) compliance with effluent limitations in accordance with subparagraph (A) of this paragraph as expeditiously as practicable but in no case later than 3 years after the date such limitations are established, and in no case later than March 31, 1989.

(3) (A) for effluent limitations under paragraph (1)(A)(i) of this subsection promulgated after January 1, 1982, and requiring a level of control substantially greater or based on fundamentally different control technology than under permits for an industrial category issued before such date, compliance as expeditiously as practicable but in no case later than three years after the date such limitations are promulgated under section 1314(b) of this title, and in no case later than March 31, 1989; and

(B) for any effluent limitation in accordance with paragraph (1)(A)(i), (2)(A)(i), or (2)(E) of this subsection established only on the basis of section 1342(a)(1) of this title in a permit issued after February 4, 1987, compliance as expeditiously as practicable but in no case later than three years after the date such limitations are established, and in no case later than March 31, 1989.

(c) Modification of timetable

The Administrator may modify the requirements of subsection (b)(2)(A) of this section with respect to any point source for which a permit application is filed after July 1, 1977, upon a showing by the owner or operator of such point source satisfactory to the Administrator that such modified requirements

(1) will represent the maximum use of technology within the economic capability of the owner or operator; and

(2) will result in reasonable further progress toward the elimination of the discharge of pollutants.

(d) Review and revision of effluent limitations

Any effluent limitation required by paragraph (2) of subsection (b) of this section shall be reviewed at least every five years and, if appropriate, revised pursuant to the procedure established under such paragraph.

(e) All point discharge source application of effluent limitations

Effluent limitations established pursuant to this section or section 1312 of this title shall be applied to all point sources of discharge of pollutants in accordance with the provisions of this chapter.

(f) Illegality of discharge of radiological, chemical, or biological warfare agents, high-level radioactive waste, or medical waste

Notwithstanding any other provisions of this chapter it shall be unlawful to discharge any radiological, chemical, or biological warfare agent, any high-level radioactive waste, or any medical waste, into the navigable waters.

(g) Modifications for certain nonconventional pollutants

(1) General authority

The Administrator, with the concurrence of the State, may modify the requirements of subsection (b)(2)(A) of this section with respect to the discharge from any point source of ammonia, chlorine, color, iron, and total phenols (4AAP) (when determined by the Administrator to be a pollutant covered by subsection (b)(2)(F) of this section) and any other pollutant which the Administrator lists under paragraph (4) of this subsection.

(2) Requirements for granting modifications

A modification under this subsection shall be granted only upon a showing by the owner or operator of a point source satisfactory to the Administrator that-

(A) such modified requirements will result at a minimum in compliance with the requirements of subsection (b)(1)(A) or (C) of this section, whichever is applicable;

(B) such modified requirements will not result in any additional requirements on any other point or nonpoint source; and

(C) such modification will not interfere with the attainment or maintenance of that water quality which shall assure protection of public water supplies, and the protection and propagation of a balanced population of shellfish, fish, and wildlife, and allow recreational activities, in and on the water and such modification will not result in the discharge of pollutants in quantities which may reasonably be anticipated to pose an unacceptable risk to human health or the environment because of bioaccumulation, persistency in the environment, acute toxicity, chronic toxicity (including carcinogenicity, mutagenicity or teratogenicity), or synergistic propensities.

(3) Limitation on authority to apply for subsection (c) modification

If an owner or operator of a point source applies for a modification under this subsection with respect to the discharge of any pollutant, such owner or operator shall be eligible to apply for modification under subsection (c) of this section with respect to such pollutant only during the same time period as he is eligible to apply for a modification under this subsection.

(4) Procedures for listing additional pollutants

(A) General authority

Upon petition of any person, the Administrator may add any pollutant to the list of pollutants for which modification under this section is authorized (except for pollutants identified pursuant to section 1314(a)(4) of this title, toxic pollutants subject to section 1317(a) of this title, and the thermal component of discharges) in accordance with the provisions of this paragraph.

(B) Requirements for listing

(i) Sufficient information

The person petitioning for listing of an additional pollutant under this subsection shall submit to the Administrator sufficient information to make the determinations required by this subparagraph.

(ii) Toxic criteria determination.

The Administrator shall determine whether or not the pollutant meets the criteria for listing as a toxic pollutant under section 1317(a) of this title.

(iii) Listing as toxic pollutant

If the Administrator determines that the pollutant meets the criteria for listing as a toxic pollutant under section 1317(a) of this title, the Administrator shall list the pollutant as a toxic pollutant under section 1317(a) of this title.

(iv) Nonconventional criteria determination

If the Administrator determines that the pollutant does not meet the criteria for listing as a toxic pollutant under such section and determines that adequate test methods and sufficient data are available to make the determinations required by paragraph (2) of this subsection with respect to the pollutant, the Administrator shall add the pollutant to the list of pollutants specified in paragraph (1) of this subsection for which modifications are authorized under this subsection.

(C) Requirements for filing of petitions

A petition for listing of a pollutant under this paragraph–

(i) must be filed not later than 270 days after the date of promulgation of an applicable effluent guideline under section 1314 of this title;

(ii) may be filed before promulgation of such guideline; and

(iii) may be filed with an application for a modification under paragraph (1) with respect to the discharge of such pollutant.

(D) Deadline for approval of petition

A decision to add a pollutant to the list of pollutants for which modifications under this subsection are authorized must be made within 270 days after the date of promulgation of an applicable effluent guideline under section 1314 of this title.

(E) Burden of proof

The burden of proof for making the determinations under subparagraph (B) shall be on the petitioner.

(5) Removal of pollutants

The Administrator may remove any pollutant from the list of pollutants for which modifications are authorized under this subsection if the Administrator determines that adequate test methods and sufficient data are no longer available for determining whether or not modifications may be granted with respect to such pollutant under paragraph (2) of this subsection.

(h) Modification of secondary treatment requirements

The Administrator, with the concurrence of the State, may issue a permit under section 1342 of this title which modifies the requirements of subsection (b)(1)(B) of this section with respect to the discharge of any pollutant from a publicly owned treatment works into marine waters, if the applicant demonstrates to the satisfaction of the Administrator that–

(1) there is an applicable water quality standard specific to the pollutant for which the modification is requested, which has been identified under section 1314(a)(6) of this title;

(2) the discharge of pollutants in accordance with such modified requirements will not interfere, alone or in combination with pollutants from other sources, with the attainment or maintenance of that water quality which assures protection of public water supplies and the protection and propagation of a balanced, indigenous population of shellfish, fish, and wildlife, and allows recreational activities, in and on the water;

(3) the applicant has established a system for monitoring the impact of such discharge on a representative sample of aquatic biota, to the extent practicable, and the scope of such monitoring is limited to include only those scientific investigations which are necessary to study the effects of the proposed discharge;

(4) such modified requirements will not result in any additional requirements on any other point or nonpoint source;

(5) all applicable pretreatment requirements for sources introducing waste into such treatment works will be enforced;

(6) in the case of any treatment works serving a population of 50,000 or more, with respect to any toxic pollutant introduced into such works by an industrial discharger for which pollutant there is no applicable pretreatment requirement in effect, sources introducing waste into such works are in compliance with all applicable pretreatment requirements, the applicant will enforce such requirements, and the applicant has in effect a pretreatment program which, in combination with the treatment of discharges from such works, removes the same amount of such pollutant as would be removed if such works were to apply secondary treatment to discharges and if such works had no pretreatment program with respect to such pollutant;

(7) to the extent practicable, the applicant has established a schedule of activities designed to eliminate the entrance of toxic pollutants from nonindustrial sources into such treatment works;

(8) there will be no new or substantially increased discharges from the point source of the pollutant to which the modification applies above that volume of discharge specified in the permit;

(9) the applicant at the time such modification becomes effective will be discharging effluent which has received at least primary or equivalent treatment and which meets the criteria established under section 1314(a)(1) of this title after initial mixing in the waters surrounding or adjacent to the point at which such effluent is discharged.

For the purposes of this subsection the phrase "the discharge of any pollutant into marine waters" refers to a discharge into deep waters of the territorial sea or the waters of the contiguous zone, or into saline estuarine waters where there is strong tidal movement and other hydrological and geological characteristics which the Administrator determines necessary to allow compliance with paragraph (2) of this subsection, and section 1251(a)(2) of this title. For the purposes of paragraph (9), "primary or equivalent treatment" means treatment by screening, sedimentation, and skimming adequate to remove at least 30 percent of the biological oxygen demanding material and of the suspended solids in the treatment works influent, and disinfection, where appropriate. A municipality which applies secondary treatment shall be eligible to receive a permit pursuant to this subsection which modifies the requirements of subsection (b)(1)(B) of this section with respect to the discharge of any pollutant from any treatment works owned by such municipality into marine waters. No permit issued under this subsection shall authorize the discharge of sewage sludge into marine waters. In order for a permit to be issued under this subsection for the discharge of a pollutant into marine waters, such marine waters must exhibit characteristics assuring that water providing dilution does not contain significant amounts of previously discharged effluent from such treatment works. No permit issued under this subsection shall authorize the discharge of any pollutant into saline estuarine waters which at the time of application do not support a balanced indigenous population of shellfish, fish and wildlife, or allow recreation in and on the waters or which exhibit ambient water quality below applicable water quality standards adopted for the protection of public water supplies, shellfish, fish and wildlife or recreational activities or such other standards necessary to assure support and protection of such uses. The prohibition contained in the preceding sentence shall apply without regard to the presence or absence of a causal relationship between such characteristics and the applicant's current or proposed discharge. Notwithstanding any other provisions of this subsection, no permit may be issued under this subsection for discharge of a pollutant into the New York Bight Apex consisting of the ocean waters of the Atlantic Ocean westward of 73 degrees 30 minutes west longitude and northward of 40 degrees 10 minutes north latitude.

(i) Municipal time extensions

(1) Where construction is required in order for a planned or existing publicly owned treatment works to achieve limitations under subsection (b)(1)(B) or (b)(1)(C) of this section, but

(A) construction cannot be completed within the time required in such subsection, or

(B) the United States has failed to make financial assistance under this chapter available in time to achieve such limitations by the time specified in such subsection, the owner or operator of such treatment works may request the Administrator (or if appropriate the State) to issue a permit pursuant to section 1342 of this title or to modify a permit issued pursuant to that section to extend such time for compliance. Any such request shall be filed with the Administrator (or if appropriate the State) within 180 days after February 4, 1987. The Administrator (or if appropriate the State) may grant such request and issue or modify such a permit, which shall contain a schedule of compliance for the publicly owned treatment works based on the earliest date by which such financial assistance will be available from the United States and construction can be completed, but in no event later than July 1, 1988, and shall contain such other terms and conditions, including those necessary to carry out subsections (b) through (g) of section 1281 of this title, section 1317 of this title, and such interim effluent limitations applicable to that treatment works as the Administrator determines are necessary to carry out the provisions of this chapter.

(2) (A) Where a point source (other than a publicly owned treatment works) will not achieve the requirements of subsections (b)(1)(A) and (b)(1)(C) of this section and–

(i) if a permit issued prior to July 1, 1977, to such point source is based upon a discharge into a publicly owned treatment works; or

(ii) if such point source (other than a publicly owned treatment works) had before July 1, 1977, a contract (enforceable against such point source) to discharge into a publicly owned treatment works; or

(iii) if either an application made before July 1, 1977, for a construction grant under this chapter for a publicly owned treatment works, or engineering or architectural plans or working drawings made before July 1, 1977, for a publicly owned treatment works, show that such point source was to discharge into such publicly owned treatment works, and such publicly owned treatment works is presently unable to accept such discharge without construction, and in the case of a discharge to an existing publicly owned treatment works, such treatment works has an extension pursuant to paragraph (1) of this subsection, the owner or operator of such point source may request the Administrator (or if appropriate the State) to issue or modify such a permit pursuant to such section 1342 of this title to extend such time for compliance. Any such request shall be filed with the Administrator (or if appropriate the State) within 180 days after December 27, 1977, or the filing of a request by the appropriate publicly owned treatment works under paragraph (1) of this subsection, whichever is later. If the Administrator (or if appropriate the State) finds that the owner or operator of such point source has acted in good faith, he may grant such request and issue or modify such a permit, which shall contain a schedule of compliance for the point source to achieve the requirements of subsections (b)(1)(A) and (C) of this section and shall contain such other terms and conditions, including pretreatment and interim effluent limitations and water conservation requirements applicable to that point source, as the Administrator determines are necessary to carry out the provisions of this chapter.

(B) No time modification granted by the Administrator (or if appropriate the State) pursuant to paragraph (2)(A) of this subsection shall extend beyond the earliest date practicable for compliance or beyond the date of any extension granted to the appropriate publicly owned treatment works pursuant to paragraph (1) of this subsection, but in no event shall it extend beyond July 1, 1988; and no such time modification shall be granted unless

(i) the publicly owned treatment works will be in operation and available to the point source before July 1, 1988, and will meet the requirements of subsections (b)(1)(B) and (C) of this section after receiving the discharge from that point source; and

(ii) the point source and the publicly owned treatment works have entered into an enforceable contract requiring the point source to discharge into the publicly owned treatment works, the owner or operator of such point source to pay the costs required under section 1284 of this title, and the publicly owned treatment works to accept the discharge from the point source; and

(iii) the permit for such point source requires that point source to meet all requirements under section 1317(a) and (b) of this title during the period of such time modification.

(j) Modification procedures

(1) Any application filed under this section for a modification of the provisions of–

(A) subsection (b)(1)(B) of this section under subsection (h) of this section shall be filed not later that[1] the 365th day which begins after December 29, 1981, except that a publicly owned treatment works which prior to December 31, 1982, had a contractual arrangement to use a portion of the capacity of an ocean outfall operated by another publicly owned treatment works which has applied for or received modification under subsection (h) of this section, may apply for a modification of subsection (h) of this section in its own right not later than 30 days after February 4, 1987, and except as provided in paragraph (5).

(B) subsection (b)(2)(A) of this section as it applies to pollutants identified in subsection (b)(2)(F) of this section shall be filed not later than 270 days after the date of promulgation of an applicable effluent guideline under section 1314 of this title or not later than 270 days after December 27, 1977, whichever is later.

(2) Subject to paragraph (3) of this section, any application for a modification filed under subsection (g) of this section shall not operate to stay any requirement under this chapter, unless in the judgment of the Administrator such a stay or the modification sought will not result in the discharge of pollutants in quantities which may reasonably be anticipated to pose an unacceptable risk to human health or the environment because of bioaccumulation, persistency in the environment, acute toxicity, chronic toxicity (including carcinogenicity, mutagenicity, or teratogenicity), or synergistic propensities, and that there is a substantial likelihood that the applicant will succeed on the merits of such application. In the case of an application filed under subsection (g) of this section, the Administrator may condition any stay granted under this paragraph on requiring the filing of a bond or other appropriate security to assure timely compliance with the requirements from which a modification is sought.

(3) Compliance requirements under subsection (g).–

(A) Effect of filing.–

An application for a modification under subsection (g) of this section and a petition for listing of a pollutant as a pollutant for which modifications are authorized under such subsection shall not stay the requirement that the person seeking such modification or listing comply with effluent limitations under this chapter for all pollutants not the subject of such application or petition.

(B) Effect of disapproval.–

Disapproval of an application for a modification under subsection (g) of this section shall not stay the requirement that the person seeking such modification comply with all applicable effluent limitations under this chapter.

(4) Deadline for subsection (g) decision.–

An application for a modification with respect to a pollutant filed under subsection (g) of this section must be approved or disapproved not later than 365 days after the date of such filing; except that in any case in which a petition for listing such pollutant as a pollutant for which modifications are authorized under such subsection is approved, such application must be approved or disapproved not later than 365 days after the date of approval of such petition.

[1] So in original. Probably should be "than".

(5) Extension of application deadline.–

(A) In general.–

In the 180-day period beginning on October 31, 1994, the city of San Diego, California, may apply for a modification pursuant to subsection (h) of this section of the requirements of subsection (b)(1)(B) of this section with respect to biological oxygen demand and total suspended solids in the effluent discharged into marine waters.

(B) Application.–

An application under this paragraph shall include a commitment by the applicant to implement a waste water reclamation program that, at a minimum, will–

(i) achieve a system capacity of 45,000,000 gallons of reclaimed waste water per day by January 1, 2010; and

(ii) result in a reduction in the quantity of suspended solids discharged by the applicant into the marine environment during the period of the modification.

(C) Additional conditions.–

The Administrator may not grant a modification pursuant to an application submitted under this paragraph unless the Administrator determines that such modification will result in removal of not less than 58 percent of the biological oxygen demand (on an annual average) and not less than 80 percent of total suspended solids (on a monthly average) in the discharge to which the application applies.

(D) Preliminary decision deadline.–

The Administrator shall announce a preliminary decision on an application submitted under this paragraph not later than 1 year after the date the application is submitted.

(k) Innovative technology

In the case of any facility subject to a permit under section 1342 of this title which proposes to comply with the requirements of subsection (b)(2)(A) or (b)(2)(E) of this section by replacing existing production capacity with an innovative production process which will result in an effluent reduction significantly greater than that required by the limitation otherwise applicable to such facility and moves toward the national goal of eliminating the discharge of all pollutants, or with the installation of an innovative control technique that has a substantial likelihood for enabling the facility to comply with the applicable effluent limitation by achieving a significantly greater effluent reduction than that required by the applicable effluent limitation and moves toward the national goal of eliminating the discharge of all pollutants, or by achieving the required reduction with an innovative system that has the potential for significantly lower costs than the systems which have been determined by the Administrator to be economically achievable, the Administrator (or the State with an approved program under section 1342 of this title, in consultation with the Administrator) may establish a date for compliance under subsection (b)(2)(A) or (b)(2)(E) of this section no later than two years after the date for compliance with such effluent limitation which would otherwise be applicable under such subsection, if it is also determined that such innovative system has the potential for industrywide application.

(l) Toxic pollutants

Other than as provided in subsection (n) of this section, the Administrator may not modify any requirement of this section as it applies to any specific pollutant which is on the toxic pollutant list under section 1317(a)(1) of this title.

(m) Modification of effluent limitation requirements for point sources

(1) The Administrator, with the concurrence of the State, may issue a permit under section 1342 of this title which modifies the requirements of subsections (b)(1)(A) and (b)(2)(E) of this section, and of section 1343 of this title, with respect to effluent limitations to the extent such limitations relate to biochemical oxygen demand and pH from discharges by an industrial discharger in such State into deep waters of the territorial seas, if the applicant demonstrates and the Administrator finds that–

(A) the facility for which modification is sought is covered at the time of the enactment of this subsection by National Pollutant Discharge Elimination System permit number CA0005894 or CA0005282;

(B) the energy and environmental costs of meeting such requirements of subsections (b) (1) (A) and (b)(2)(E) of this section and section 1343 of this title exceed by an unreasonable amount the benefits to be obtained, including the objectives of this chapter;

(C) the applicant has established a system for monitoring the impact of such discharges on a representative sample of aquatic biota;

(D) such modified requirements will not result in any additional requirements on any other point or nonpoint source;

(E) there will be no new or substantially increased discharges from the point source of the pollutant to which the modification applies above that volume of discharge specified in the permit;

(F) the discharge is into waters where there is strong tidal movement and other hydrological and geological characteristics which are necessary to allow compliance with this subsection and section 1251(a)(2) of this title;

(G) the applicant accepts as a condition to the permit a contractural [2] obligation to use funds in the amount required (but not less than $250,000 per year for ten years) for research and development of water pollution control technology, including but not limited to closed cycle technology; "contractual".

(H) the facts and circumstances present a unique situation which, if relief is granted, will not establish a precedent or the relaxation of the requirements of this chapter applicable to similarly situated discharges; and

(I) no owner or operator of a facility comparable to that of the applicant situated in the United States has demonstrated that it would be put at a competitive disadvantage to the applicant (or the parent company or any subsidiary thereof) as a result of the issuance of a permit under this subsection.

(2) The effluent limitations established under a permit issued under paragraph (1) shall be sufficient to implement the applicable State water quality standards, to assure the protection of public water supplies and protection and propagation of a balanced, indigenous population of shellfish, fish, fauna, wildlife, and other aquatic organisms, and to allow recreational activities in and on the water. In setting such limitations, the Administrator shall take into account any seasonal variations and the need for an adequate margin of safety, considering the lack of essential knowledge concerning the relationship between effluent limitations and water quality and the lack of essential knowledge of the effects of discharges on beneficial uses of the receiving waters.

(3) A permit under this subsection may be issued for a period not to exceed five years, and such a permit may be renewed for one additional period not to exceed five years upon a demonstration by the applicant and a finding by the Administrator at the time of application for any such renewal that the provisions of this subsection are met.

(4) The Administrator may terminate a permit issued under this subsection if the Administrator determines that there has been

[2] So in original. Probably should be contractual.

a decline in ambient water quality of the receiving waters during the period of the permit even if a direct cause and effect relationship cannot be shown: Provided, That if the effluent from a source with a permit issued under this subsection is contributing to a decline in ambient water quality of the receiving waters, the Administrator shall terminate such permit.

(n) Fundamentally different factors

(1) General rule

The Administrator, with the concurrence of the State, may establish an alternative requirement under subsection (b)(2) of this section or section 1317(b) of this title for a facility that modifies the requirements of national effluent limitation guidelines or categorical pretreatment standards that would otherwise be applicable to such facility, if the owner or operator of such facility demonstrates to the satisfaction of the Administrator that–

(A) the facility is fundamentally different with respect to the factors (other than cost) specified in section 1314(b) or 1314(g) of this title and considered by the Administrator in establishing such national effluent limitation guidelines or categorical pretreatment standards;

(B) the application–

(i) is based solely on information and supporting data submitted to the Administrator during the rulemaking for establishment of the applicable national effluent limitation guidelines or categorical pretreatment standard specifically raising the factors that are fundamentally different for such facility; or

(ii) is based on information and supporting data referred to in clause (i) and information and supporting data the applicant did not have a reasonable opportunity to submit during such rulemaking;

(C) the alternative requirement is no less stringent than justified by the fundamental difference; and

(D) the alternative requirement will not result in a non-water quality environmental impact which is markedly more adverse than the impact considered by the Administrator in establishing such national effluent limitation guideline or categorical pretreatment standard.

(2) Time limit for applications

An application for an alternative requirement which modifies the requirements of an effluent limitation or pretreatment standard under this subsection must be submitted to the Administrator within 180 days after the date on which such limitation or standard is established or revised, as the case may be.

(3) Time limit for decision

The Administrator shall approve or deny by final agency action an application submitted under this subsection within 180 days after the date such application is filed with the Administrator.

(4) Submission of information

The Administrator may allow an applicant under this subsection to submit information and supporting data until the earlier of the date the application is approved or denied or the last day that the Administrator has to approve or deny such application.

(5) Treatment of pending applications

For the purposes of this subsection, an application for an alternative requirement based on fundamentally different factors which is pending on February 4, 1987, shall be treated as having been submitted to the Administrator on the 180th day following February 4, 1987. The applicant may amend the application to take into account the provisions of this subsection.

(6) Effect of submission of application

An application for an alternative requirement under this subsection shall not stay the applicant's obligation to comply with the effluent limitation guideline or categorical pretreatment standard which is the subject of the application.

(7) Effect of denial

If an application for an alternative requirement which modifies the requirements of an effluent limitation or pretreatment standard under this subsection is denied by the Administrator, the applicant must comply with such limitation or standard as established or revised, as the case may be.

(8) Reports

By January 1, 1997, and January 1 of every odd-numbered year thereafter, the Administrator shall submit to the Committee on Environment and Public Works of the Senate and the Committee on Transportation and Infrastructure of the House of Representatives a report on the status of applications for alternative requirements which modify the requirements of effluent limitations under section 1311 or 1314 of this title or any national categorical pretreatment standard under section 1317(b) of this title filed before, on, or after February 4, 1987.

(o) Application fees

The Administrator shall prescribe and collect from each applicant fees reflecting the reasonable administrative costs incurred in reviewing and processing applications for modifications submitted to the Administrator pursuant to subsections (c), (g), (i), (k), (m), and (n) of this section, section 1314(d)(4) of this title, and section 1326(a) of this title. All amounts collected by the Administrator under this subsection shall be deposited into a special fund of the Treasury entitled "Water Permits and Related Services" which shall thereafter be available for appropriation to carry out activities of the Environmental Protection Agency for which such fees were collected.

(p) Modified permit for coal remining operations

(1) In general

Subject to paragraphs (2) through (4) of this subsection, the Administrator, or the State in any case which the State has an approved permit program under section 1342(b) of this title, may issue a permit under section 1342 of this title which modifies the requirements of subsection (b)(2)(A) of this section with respect to the pH level of any pre-existing discharge, and with respect to pre-existing discharges of iron and manganese from the remined area of any coal remining operation or with respect to the pH level or level of iron or manganese in any pre-existing discharge affected by the remining operation. Such modified requirements shall apply the best available technology economically achievable on a case-by-case basis, using best professional judgment, to set specific numerical effluent limitations in each permit.

(2) Limitations

The Administrator or the State may only issue a permit pursuant to paragraph (1) if the applicant demonstrates to the satisfaction of the Administrator or the State, as the case may be, that the coal remining operation will result in the potential for improved water quality from the remining operation but in no event shall such a permit allow the pH level of any discharge, and in no event shall such a permit allow the discharges of iron and manganese, to exceed the levels being discharged from the remined area before the coal remining operation begins. No discharge from, or affected by, the remining operation shall exceed State water quality standards established under section 1313 of this title.

(3) Definitions

For purposes of this subsection–

(A) Coal remining operation

The term "coal remining operation" means a coal mining operation which begins after February 4, 1987 at a site on which coal mining was conducted before August 3, 1977.

(B) Remined area

The term "remined area" means only that area of any coal remining operation on which coal mining was conducted before August 3, 1977.

(C) Pre-existing discharge

The term "pre-existing discharge" means any discharge at the time of permit application under this subsection.

(4) Applicability of strip mining laws

Nothing in this subsection shall affect the application of the Surface Mining Control and Reclamation Act of 1977 (30 U.S.C. 1201 et seq.) to any coal remining operation, including the application of such Act to suspended solids

Section 303, Clean Water Act 33 USC 1313

Water quality standards and implementation plans

(a) Existing water quality standards

(1) In order to carry out the purpose of this chapter, any water quality standard applicable to interstate waters which was adopted by any State and submitted to, and approved by, or is a waiting approval by, the Administrator pursuant to this Act as in effect immediately prior to October 18, 1972, shall remain in effect unless the Administrator determined that such standard is not consistent with the applicable requirements of this Act as in effect immediately prior to October 18, 1972. If the Administrator makes such a determination he shall, within three months after October 18, 1972, notify the State and specify the changes needed to meet such requirements. If such changes are not adopted by the State within ninety days after the date of such notification, the Administrator shall promulgate such changes in accordance with subsection (b) of this section.

(2) Any State which, before October 18, 1972, has adopted, pursuant to its own law, water quality standards applicable to intrastate waters shall submit such standards to the Administrator within thirty days after October 18, 1972. Each such standard shall remain in effect, in the same manner and to the same extent as any other water quality standard established under this chapter unless the Administrator determines that such standard is inconsistent with the applicable requirements of this Act as in effect immediately prior to October 18, 1972. If the Administrator makes such a determination he shall not later than the one hundred and twentieth day after the date of submission of such standards, notify the State and specify the changes needed to meet such requirements. If such changes are not adopted by the State within ninety days after such notification, the Administrator shall promulgate such changes in accordance with subsection (b) of this section.

(3) (A) Any State which prior to October 18, 1972, has not adopted pursuant to its own laws water quality standards applicable to intrastate waters shall, not later than one hundred and eighty days after October 18, 1972, adopt and submit such standards to the Administrator.

(B) If the Administrator determines that any such standards are consistent with the applicable requirements of this Act as in effect immediately prior to October 18, 1972, he shall approve such standards.

(C) If the Administrator determines that any such standards are not consistent with the applicable requirements of this Act as in effect immediately prior to October 18, 1972, he shall, not later than the ninetieth day after the date of submission of such standards, notify the State and specify the changes to meet such requirements. If such changes are not adopted by the State within ninety days after the date of notification, the Administrator shall promulgate such standards pursuant to subsection (b) of this section.

(b) Proposed regulations

(1) The Administrator shall promptly prepare and publish proposed regulations setting forth water quality standards for a State in accordance with the applicable requirements of this Act as in effect immediately prior to October 18, 1972, if–

(A) the State fails to submit water quality standards within the times prescribed in subsection (a) of this section.

(B) a water quality standard submitted by such State under subsection (a) of this section is determined by the Administrator not to be consistent with the applicable requirements of subsection (a) of this section.

(2) The Administrator shall promulgate any water quality standard published in a proposed regulation not later than one hundred and ninety days after the date he publishes any such proposed standard, unless prior to such promulgation, such State has adopted a water quality standard which the Administrator determines to be in accordance with subsection (a) of this section.

(c) Review; revised standards; publication

(1) The Governor of a State or the State water pollution control agency of such State shall from time to time (but at least once each three year period beginning with October 18, 1972) hold public hearings for the purpose of reviewing applicable water quality standards and, as appropriate, modifying and adopting standards. Results of such review shall be made available to the Administrator.

(2) (A) Whenever the State revises or adopts a new standard, such revised or new standard shall be submitted to the Administrator. Such revised or new water quality standard shall consist of the designated uses of the navigable waters involved and the water quality criteria for such waters based upon such uses. Such standards shall be such as to protect the public health or welfare, enhance the quality of water and serve the purposes of this chapter. Such standards shall be established taking into consideration their use and value for public water supplies, propagation of fish and wildlife, recreational purposes, and agricultural, industrial, and other purposes, and also taking into consideration their use and value for navigation.

(B) Whenever a State reviews water quality standards pursuant to paragraph (1) of this subsection, or revises or adopts new standards pursuant to this paragraph, such State shall adopt criteria for all toxic pollutants listed pursuant to section 1317(a)(1) of this title for which criteria have been published under section 1314(a) of this title, the discharge or presence of which in the affected waters could reasonably be expected to interfere with those designated uses adopted by the State, as necessary to support such designated uses. Such criteria shall be specific numerical criteria for such toxic pollutants. Where such numerical criteria are not available, whenever a State reviews water quality standards pursuant to paragraph (1), or revises or adopts new standards pursuant to this paragraph, such State shall adopt criteria based on biological monitoring or assessment methods consistent with information published pursuant to section 1314(a)(8) of this title. Nothing in this section shall be construed to limit or delay the use of effluent limitations or other permit conditions based on or involving biological monitoring or assessment methods or previously adopted numerical criteria.

(3) If the Administrator, within sixty days after the date of submission of the revised or new standard, determines that such standard meets the requirements of this chapter, such standard shall thereafter be the water quality standard for the applicable waters of that State. If the Administrator determines that any such revised or new standard is not

consistent with the applicable requirements of this chapter, he shall not later than the ninetieth day after the date of submission of such standard notify the State and specify the changes to meet such requirements. If such changes are not adopted by the State within ninety days after the date of notification, the Administrator shall promulgate such standard pursuant to paragraph (4) of this subsection.

(4) The Administrator shall promptly prepare and publish proposed regulations setting forth a revised or new water quality standard for the navigable waters involved–

(A) if a revised or new water quality standard submitted by such State under paragraph (3) of this subsection for such waters is determined by the Administrator not to be consistent with the applicable requirements of this chapter, or

(B) in any case where the Administrator determines that a revised or new standard is necessary to meet the requirements of this chapter.

The Administrator shall promulgate any revised or new standard under this paragraph not later than ninety days after he publishes such proposed standards, unless prior to such promulgation, such State has adopted a revised or new water quality standard which the Administrator determines to be in accordance with this chapter.

(d) Identification of areas with insufficient controls; maximum daily load; certain effluent limitations revision

(1) (A) Each State shall identify those waters within its boundaries for which the effluent limitations required by section 1311(b)(1)(A) and section 1311(b)(1)(B) of this title are not stringent enough to implement any water quality standard applicable to such waters. The State shall establish a priority ranking for such waters, taking into account the severity of the pollution and the uses to be made of such waters.

(B) Each State shall identify those waters or parts thereof within its boundaries for which controls on thermal discharges under section 1311 of this title are not stringent enough to assure protection and propagation of a balanced indigenous population of shellfish, fish, and wildlife.

(C) Each State shall establish for the waters identified in paragraph (1)(A) of this subsection, and in accordance with the priority ranking, the total maximum daily load, for those pollutants which the Administrator identifies under section 1314(a)(2) of this title as suitable for such calculation. Such load shall be established at a level necessary to implement the applicable water quality standards with seasonal variations and a margin of safety which takes into account any lack of knowledge concerning the relationship between effluent limitations and water quality.

(D) Each State shall estimate for the waters identified in paragraph (1)(B) of this subsection the total maximum daily thermal load required to assure protection and propagation of a balanced, indigenous population of shellfish, fish, and wildlife. Such estimates shall take into account the normal water temperatures, flow rates, seasonal variations, existing sources of heat input, and the dissipative capacity of the identified waters or parts thereof. Such estimates shall include a calculation of the maximum heat input that can be made into each such part and shall include a margin of safety which takes into account any lack of knowledge concerning the development of thermal water quality criteria for such protection and propagation in the identified waters or parts thereof.

(2) Each State shall submit to the Administrator from time to time, with the first such submission not later than one hundred and eighty days after the date of publication of the first identification of pollutants under section 1314(a)(2)(D) of this title, for his approval the waters identified and the loads established under paragraphs (1)(A), (1)(B), (1)(C), and (1)(D) of this subsection. The Administrator shall either approve or disapprove such identification and load not later than thirty days after the date of submission. If the Administrator approves such identification and load, such State shall incorporate them into its current plan under subsection (e) of this section. If the Administrator disapproves such identification and load, he shall not later than thirty days after the date of such disapproval identify such waters in such State and establish such loads for such waters as he determines necessary to implement the water quality standards applicable to such waters and upon such identification and establishment the State shall incorporate them into its current plan under subsection (e) of this section.

(3) For the specific purpose of developing information, each State shall identify all waters within its boundaries which it has not identified under paragraph (1)(A) and (1)(B) of this subsection and estimate for such waters the total maximum daily load with seasonal variations and margins of safety, for those pollutants which the Administrator identifies under section 1314(a)(2) of this title as suitable for such calculation and for thermal discharges, at a level that would assure protection and propagation of a balanced indigenous population of fish, shellfish, and wildlife.

(4) Limitations on revision of certain effluent limitations.–

(A) Standard not attained.–

For waters identified under paragraph (1)(A) where the applicable water quality standard has not yet been attained, any effluent limitation based on a total maximum daily load or other waste load allocation established under this section may be revised only if

(i) the cumulative effect of all such revised effluent limitations based on such total maximum daily load or waste load allocation will assure the attainment of such water quality standard, or

(ii) the designated use which is not being attained is removed in accordance with regulations established under this section.

(B) Standard attained.–

For waters identified under paragraph (1)(A) where the quality of such waters equals or exceeds levels necessary to protect the designated use for such waters or otherwise required by applicable water quality standards, any effluent limitation based on a total maximum daily load or other waste load allocation established under this section, or any water quality standard established under this section, or any other permitting standard may be revised only if such revision is subject to and consistent with the antidegradation policy established under this section.

(e) Continuing planning process

(1) Each State shall have a continuing planning process approved under paragraph (2) of this subsection which is consistent with this chapter.

(2) Each State shall submit not later than 120 days after October 18, 1972, to the Administrator for his approval a proposed continuing planning process which is consistent with this chapter. Not later than thirty days after the date of submission of such a process the Administrator shall either approve or disapprove such process. The Administrator shall from time to time review each State's approved planning process for the purpose of insuring that such planning process is at all times consistent with this chapter. The Administrator shall not approve any State permit program under subchapter IV of this chapter for any State which does not have an approved continuing planning process under this section.

(3) The Administrator shall approve any continuing planning process submitted to him under this section which will result in plans for all navigable waters within such State, which include, but are not limited to, the following:

(A) effluent limitations and schedules of compliance at least as stringent as those required by section 1311(b)(1), section

1311(b)(2), section 1316, and section 1317 of this title, and at least as stringent as any requirements contained in any applicable water quality standard in effect under authority of this section;

(B) the incorporation of all elements of any applicable area-wide waste management plans under section 1288 of this title, and applicable basin plans under section 1289 of this title;

(C) total maximum daily load for pollutants in accordance with subsection (d) of this section;

(D) procedures for revision;

(E) adequate authority for intergovernmental cooperation;

(F) adequate implementation, including schedules of compliance, for revised or new water quality standards, under subsection (c) of this section;

(G) controls over the disposition of all residual waste from any water treatment processing;

(H) an inventory and ranking, in order of priority, of needs for construction of waste treatment works required to meet the applicable requirements of sections 1311 and 1312 of this title.

(f) Earlier compliance

Nothing in this section shall be construed to affect any effluent limitation, or schedule of compliance required by any State to be implemented prior to the dates set forth in sections 1311(b)(1) and 1311(b)(2) of this title nor to preclude any State from requiring compliance with any effluent limitation or schedule of compliance at dates earlier than such dates.

(g) Heat standards

Water quality standards relating to heat shall be consistent with the requirements of section 1326 of this title.

(h) Thermal water quality standards

For the purposes of this chapter the term "water quality standards" includes thermal water quality standards.

(i) Coastal recreation water quality criteria

(1) Adoption by States

(A) Initial criteria and standards

Not later than 42 months after October 10, 2000, each State having coastal recreation waters shall adopt and submit to the Administrator water quality criteria and standards for the coastal recreation waters of the State for those pathogens and pathogen indicators for which the Administrator has published criteria under section 1314(a) of this title.

(B) New or revised criteria and standards

Not later than 36 months after the date of publication by the Administrator of new or revised water quality criteria under section 1314(a)(9) of this title, each State having coastal recreation waters shall adopt and submit to the Administrator new or revised water quality standards for the coastal recreation waters of the State for all pathogens and pathogen indicators to which the new or revised water quality criteria are applicable.

(2) Failure of States to adopt

(A) In general

If a State fails to adopt water quality criteria and standards in accordance with paragraph (1)(A) that are as protective of human health as the criteria for pathogens and pathogen indicators for coastal recreation waters published by the Administrator, the Administrator shall promptly propose regulations for the State setting forth revised or new water quality standards for pathogens and pathogen indicators described in paragraph (1)(A) for coastal recreation waters of the State.

(B) Exception

If the Administrator proposes regulations for a State described in subparagraph (A) under subsection (c)(4)(B) of this section, the Administrator shall publish any revised or new standard under this subsection not later than 42 months after October 10, 2000.

(3) Applicability

Except as expressly provided by this subsection, the requirements and procedures of subsection (c) of this section apply to this subsection, including the requirement in subsection (c)(2)(A) of this section that the criteria protect public health and welfare.

Revised water quality standards

The review, revision, and adoption or promulgation of revised or new water quality standards pursuant to section 303(c) of the Federal Water Pollution Control Act (33 U.S.C. 1313(c)) shall be completed by the date three years after December 29, 1981. No grant shall be made under title II of the Federal Water Pollution Control Act (33 U.S.C. 1281 et seq.) after such date until water quality standards are reviewed and revised pursuant to section 303(c), except where the State has in good faith submitted such revised water quality standards and the Administrator has not acted to approve or disapprove such submission within one hundred and twenty days of receipt.

Section 401, Clean Water Act 33 USC 1341

Certification

(a) Compliance with applicable requirements; application; procedures; license suspension

(1) Any applicant for a Federal license or permit to conduct any activity including, but not limited to, the construction or operation of facilities, which may result in any discharge into the navigable waters, shall provide the licensing or permitting agency a certification from the State in which the discharge originates or will originate, or, if appropriate, from the interstate water pollution control agency having jurisdiction over the navigable waters at the point where the discharge originates or will originate, that any such discharge will comply with the applicable provisions of sections 1311, 1312, 1313, 1316, and 1317 of this title. In the case of any such activity for which there is not an applicable effluent limitation or other limitation under sections 1311(b) and 1312 of this title, and there is not an applicable standard under sections 1316 and 1317 of this title, the State shall so certify, except that any such certification shall not be deemed to satisfy section 1371(c) of this title. Such State or interstate agency shall establish procedures for public notice in the case of all applications for certification by it and, to the extent it deems appropriate, procedures for public hearings in connection with specific applications. In any case where a State or interstate agency has no authority to give such a certification, such certification shall be from the Administrator. If the State, interstate agency, or Administrator, as the case may be, fails or refuses to act on a request for certification, within a reasonable period of time (which shall not exceed one year) after receipt of such request, the certification requirements of this subsection shall be waived with respect to such Federal application. No license or permit shall be granted until the certification required by this section has been obtained or has been waived as provided in the preceding sentence. No license or permit shall be granted if certification has been denied by the State, interstate agency, or the Administrator, as the case may be.

(2) Upon receipt of such application and certification the licensing or permitting agency shall immediately notify the Administrator of such application and certification. Whenever such a discharge may affect, as

determined by the Administrator, the quality of the waters of any other State, the Administrator within thirty days of the date of notice of application for such Federal license or permit shall so notify such other State, the licensing or permitting agency, and the applicant. If, within sixty days after receipt of such notification, such other State determines that such discharge will affect the quality of its waters so as to violate any water quality requirements in such State, and within such sixty-day period notifies the Administrator and the licensing or permitting agency in writing of its objection to the issuance of such license or permit and requests a public hearing on such objection, the licensing or permitting agency shall hold such a hearing. The Administrator shall at such hearing submit his evaluation and recommendations with respect to any such objection to the licensing or permitting agency. Such agency, based upon the recommendations of such State, the Administrator, and upon any additional evidence, if any, presented to the agency at the hearing, shall condition such license or permit in such manner as may be necessary to insure compliance with applicable water quality requirements. If the imposition of conditions cannot insure such compliance such agency shall not issue such license or permit.

(3) The certification obtained pursuant to paragraph (1) of this subsection with respect to the construction of any facility shall fulfill the requirements of this subsection with respect to certification in connection with any other Federal license or permit required for the operation of such facility unless, after notice to the certifying State, agency, or Administrator, as the case may be, which shall be given by the Federal agency to whom application is made for such operating license or permit, the State, or if appropriate, the interstate agency or the Administrator, notifies such agency within sixty days after receipt of such notice that there is no longer reasonable assurance that there will be compliance with the applicable provisions of sections 1311, 1312, 1313, 1316, and 1317 of this title because of changes since the construction license or permit certification was issued in

(A) the construction or operation of the facility,

(B) the characteristics of the waters into which such discharge is made,

(C) the water quality criteria applicable to such waters or

(D) applicable effluent limitations or other requirements. This paragraph shall be inapplicable in any case where the applicant for such operating license or permit has failed to provide the certifying State, or, if appropriate, the interstate agency or the Administrator, with notice of any proposed changes in the construction or operation of the facility with respect to which a construction license or permit has been granted, which changes may result in violation of section 1311, 1312, 1313, 1316, or 1317 of this title.

(4) Prior to the initial operation of any federally licensed or permitted facility or activity which may result in any discharge into the navigable waters and with respect to which a certification has been obtained pursuant to paragraph (1) of this subsection, which facility or activity is not subject to a Federal operating license or permit, the licensee or permittee shall provide an opportunity for such certifying State, or, if appropriate, the interstate agency or the Administrator to review the manner in which the facility or activity shall be operated or conducted for the purposes of assuring that applicable effluent limitations or other limitations or other applicable water quality requirements will not be violated. Upon notification by the certifying State, or if appropriate, the interstate agency or the Administrator that the operation of any such federally licensed or permitted facility or activity will violate applicable effluent limitations or other limitations or other water quality requirements such Federal agency may, after public hearing, suspend such license or permit. If such license or permit is suspended, it shall remain suspended until notification is received from the certifying State, agency, or Administrator, as the case may be, that there is reasonable assurance that such facility or activity will not violate the applicable provisions of section 1311, 1312, 1313, 1316, or 1317 of this title.

(5) Any Federal license or permit with respect to which a certification has been obtained under paragraph (1) of this subsection may be suspended or revoked by the Federal agency issuing such license or permit upon the entering of a judgment under this chapter that such facility or activity has been operated in violation of the applicable provisions of section 1311, 1312, 1313, 1316, or 1317 of this title.

(6) Except with respect to a permit issued under section 1342 of this title, in any case where actual construction of a facility has been lawfully commenced prior to April 3, 1970, no certification shall be required under this subsection for a license or permit issued after April 3, 1970, to operate such facility, except that any such license or permit issued without certification shall terminate April 3, 1973, unless prior to such termination date the person having such license or permit submits to the Federal agency which issued such license or permit a certification and otherwise meets the requirements of this section.

(b) Compliance with other provisions of law setting applicable water quality requirements

Nothing in this section shall be construed to limit the authority of any department or agency pursuant to any other provision of law to require compliance with any applicable water quality requirements. The Administrator shall, upon the request of any Federal department or agency, or State or interstate agency, or applicant, provide, for the purpose of this section, any relevant information on applicable effluent limitations, or other limitations, standards, regulations, or requirements, or water quality criteria, and shall, when requested by any such department or agency or State or interstate agency, or applicant, comment on any methods to comply with such limitations, standards, regulations, requirements, or criteria.

(c) Authority of Secretary of the Army to permit use of spoil disposal areas by Federal licensees or permittees

In order to implement the provisions of this section, the Secretary of the Army, acting through the Chief of Engineers, is authorized, if he deems it to be in the public interest, to permit the use of spoil disposal areas under his jurisdiction by Federal licensees or permittees, and to make an appropriate charge for such use. Moneys received from such licensees or permittees shall be deposited in the Treasury as miscellaneous receipts.

(d) Limitations and monitoring requirements of certification

Any certification provided under this section shall set forth any effluent limitations and other limitations, and monitoring requirements necessary to assure that any applicant for a Federal license or permit will comply with any applicable effluent limitations and other limitations, under section 1311 or 1312 of this title, standard of performance under section 1316 of this title, or prohibition, effluent standard, or pretreatment standard under section 1317 of this title, and with any other appropriate requirement of State law set forth in such certification, and shall become a condition on any Federal license or permit subject to the provisions of this section.

Section 402, Clean Water Act 33 USC 1342

National pollutant discharge elimination system

(a) Permits for discharge of pollutants

(1) Except as provided in sections 1328 and 1344 of this title, the Administrator may, after opportunity for public hearing

issue a permit for the discharge of any pollutant, or combination of pollutants, notwithstanding section 1311(a) of this title, upon condition that such discharge will meet either

(A) all applicable requirements under sections 1311, 1312, 1316, 1317, 1318, and 1343 of this title, or

(B) prior to the taking of necessary implementing actions relating to all such requirements, such conditions as the Administrator determines are necessary to carry out the provisions of this chapter.

(2) The Administrator shall prescribe conditions for such permits to assure compliance with the requirements of paragraph (1) of this subsection, including conditions on data and information collection, reporting, and such other requirements as he deems appropriate.

(3) The permit program of the Administrator under paragraph (1) of this subsection, and permits issued thereunder, shall be subject to the same terms, conditions, and requirements as apply to a State permit program and permits issued thereunder under subsection (b) of this section.

(4) All permits for discharges into the navigable waters issued pursuant to section 407 of this title shall be deemed to be permits issued under this subchapter, and permits issued under this subchapter shall be deemed to be permits issued under section 407 of this title, and shall continue in force and effect for their term unless revoked, modified, or suspended in accordance with the provisions of this chapter.

(5) No permit for a discharge into the navigable waters shall be issued under section 407 of this title after October 18, 1972. Each application for a permit under section 407 of this title, pending on October 18, 1972, shall be deemed to be an application for a permit under this section. The Administrator shall authorize a State, which he determines has the capability of administering a permit program which will carry out the objectives of this chapter to issue permits for discharges into the navigable waters within the jurisdiction of such State. The Administrator may exercise the authority granted him by the preceding sentence only during the period which begins on October 18, 1972, and ends either on the ninetieth day after the date of the first promulgation of guidelines required by section 1314(i)(2) of this title, or the date of approval by the Administrator of a permit program for such State under subsection (b) of this section, whichever date first occurs, and no such authorization to a State shall extend beyond the last day of such period. Each such permit shall be subject to such conditions as the Administrator determines are necessary to carry out the provisions of this chapter. No such permit shall issue if the Administrator objects to such issuance.

(b) State permit programs

At any time after the promulgation of the guidelines required by subsection (i)(2) of section 1314 of this title, the Governor of each State desiring to administer its own permit program for discharges into navigable waters within its jurisdiction may submit to the Administrator a full and complete description of the program it proposes to establish and administer under State law or under an interstate compact. In addition, such State shall submit a statement from the attorney general (or the attorney for those State water pollution control agencies which have independent legal counsel), or from the chief legal officer in the case of an interstate agency, that the laws of such State, or the interstate compact, as the case may be, provide adequate authority to carry out the described program. The Administrator shall approve each submitted program unless he determines that adequate authority does not exist:

(1) To issue permits which–

(A) apply, and insure compliance with, any applicable requirements of sections 1311, 1312, 1316, 1317, and 1343 of this title;

(B) are for fixed terms not exceeding five years; and

(C) can be terminated or modified for cause including, but not limited to, the following:

(i) violation of any condition of the permit;

(ii) obtaining a permit by misrepresentation, or failure to disclose fully all relevant facts;

(iii) change in any condition that requires either a temporary or permanent reduction or elimination of the permitted discharge;

(D) control the disposal of pollutants into wells;

(2) (A) To issue permits which apply, and insure compliance with, all applicable requirements of section 1318 of this title; or

(B) To inspect, monitor, enter, and require reports to at least the same extent as required in section 1318 of this title;

(3) To insure that the public, and any other State the waters of which may be affected, receive notice of each application for a permit and to provide an opportunity for public hearing before a ruling on each such application;

(4) To insure that the Administrator receives notice of each application (including a copy thereof) for a permit;

(5) To insure that any State (other than the permitting State), whose waters may be affected by the issuance of a permit may submit written recommendations to the permitting State (and the Administrator) with respect to any permit application and, if any part of such written recommendations are not accepted by the permitting State, that the permitting State will notify such affected State (and the Administrator) in writing of its failure to so accept such recommendations together with its reasons for so doing;

(6) To insure that no permit will be issued if, in the judgment of the Secretary of the Army acting through the Chief of Engineers, after consultation with the Secretary of the department in which the Coast Guard is operating, anchorage and navigation of any of the navigable waters would be substantially impaired thereby;

(7) To abate violations of the permit or the permit program, including civil and criminal penalties and other ways and means of enforcement;

(8) To insure that any permit for a discharge from a publicly owned treatment works includes conditions to require the identification in terms of character and volume of pollutants of any significant source introducing pollutants subject to pretreatment standards under section 1317(b) of this title into such works and a program to assure compliance with such pretreatment standards by each such source, in addition to adequate notice to the permitting agency of

(A) new introductions into such works of pollutants from any source which would be a new source as defined in section 1316 of this title if such source were discharging pollutants,

(B) new introductions of pollutants into such works from a source which would be subject to section 1311 of this title if it were discharging such pollutants, or

(C) a substantial change in volume or character of pollutants being introduced into such works by a source introducing pollutants into such works at the time of issuance of the permit. Such notice shall include information on the quality and quantity of effluent to be introduced into such treatment works and any anticipated impact of such change in the quantity or quality of effluent to be discharged from such publicly owned treatment works; and

(9) To insure that any industrial user of any publicly owned treatment works will

comply with sections 1284(b), 1317, and 1318 of this title.

(c) Suspension of Federal program upon submission of State program; withdrawal of approval of State program; return of State program to Administrator

(1) Not later than ninety days after the date on which a State has submitted a program (or revision thereof) pursuant to subsection (b) of this section, the Administrator shall suspend the issuance of permits under subsection (a) of this section as to those discharges subject to such program unless he determines that the State permit program does not meet the requirements of subsection (b) of this section or does not conform to the guidelines issued under section 1314(i)(2) of this title. If the Administrator so determines, he shall notify the State of any revisions or modifications necessary to conform to such requirements or guidelines.

(2) Any State permit program under this section shall at all times be in accordance with this section and guidelines promulgated pursuant to section 1314(i)(2) of this title.

(3) Whenever the Administrator determines after public hearing that a State is not administering a program approved under this section in accordance with requirements of this section, he shall so notify the State and, if appropriate corrective action is not taken within a reasonable time, not to exceed ninety days, the Administrator shall withdraw approval of such program. The Administrator shall not withdraw approval of any such program unless he shall first have notified the State, and made public, in writing, the reasons for such withdrawal.

(4) Limitations on partial permit program returns and withdrawals.–

A State may return to the Administrator administration, and the Administrator may withdraw under paragraph (3) of this subsection approval, of–

(A) a State partial permit program approved under subsection (n)(3) of this section only if the entire permit program being administered by the State department or agency at the time is returned or withdrawn; and

(B) a State partial permit program approved under subsection (n)(4) of this section only if an entire phased component of the permit program being administered by the State at the time is returned or withdrawn.

(d) Notification of Administrator

(1) Each State shall transmit to the Administrator a copy of each permit application received by such State and provide notice to the Administrator of every action related to the consideration of such permit application, including each permit proposed to be issued by such State.

(2) No permit shall issue

(A) if the Administrator within ninety days of the date of his notification under subsection (b)(5) of this section objects in writing to the issuance of such permit, or

(B) if the Administrator within ninety days of the date of transmittal of the proposed permit by the State objects in writing to the issuance of such permit as being outside the guidelines and requirements of this chapter. Whenever the Adminis trator objects to the issuance of a permit under this paragraph such written objection shall contain a statement of the reasons for such objection and the effluent limitations and conditions which such permit would include if it were issued by the Administrator.

(3) The Administrator may, as to any permit application, waive paragraph (2) of this subsection.

(4) In any case where, after December 27, 1977, the Administrator, pursuant to paragraph (2) of this subsection, objects to the issuance of a permit, on request of the State, a public hearing shall be held by the Administrator on such objection. If the State does not resubmit such permit revised to meet such objection within 30 days after completion of the hearing, or, if no hearing is requested within 90 days after the date of such objection, the Administrator may issue the permit pursuant to subsection (a) of this section for such source in accordance with the guidelines and requirements of this chapter.

(e) Waiver of notification requirement

In accordance with guidelines promulgated pursuant to subsection (i)(2) of section 1314 of this title, the Administrator is authorized to waive the requirements of subsection (d) of this section at the time he approves a program pursuant to subsection (b) of this section for any category (including any class, type, or size within such category) of point sources within the State submitting such program.

(f) Point source categories

The Administrator shall promulgate regulations establishing categories of point sources which he determines shall not be subject to the requirements of subsection (d) of this section in any State with a program approved pursuant to subsection (b) of this section. The Administrator may distinguish among classes, types, and sizes within any category of point sources.

(g) Other regulations for safe transportation, handling, carriage, storage, and stowage of pollutants

Any permit issued under this section for the discharge of pollutants into the navigable waters from a vessel or other floating craft shall be subject to any applicable regulations promulgated by the Secretary of the department in which the Coast Guard is operating, establishing specifications for safe transportation, handling, carriage, storage, and stowage of pollutants.

(h) Violation of permit conditions; restriction or prohibition upon introduction of pollutant by source not previously utilizing treatment works

In the event any condition of a permit for discharges from a treatment works (as defined in section 1292 of this title) which is publicly owned is violated, a State with a program approved under subsection (b) of this section or the Administrator, where no State program is approved or where the Administrator determines pursuant to section 1319(a) of this title that a State with an approved program has not commenced appropriate enforcement action with respect to such permit, may proceed in a court of competent jurisdiction to restrict or prohibit the introduction of any pollutant into such treatment works by a source not utilizing such treatment works prior to the finding that such condition was violated.

(i) Federal enforcement not limited

Nothing in this section shall be construed to limit the authority of the Administrator to take action pursuant to section 1319 of this title.

(j) Public information

A copy of each permit application and each permit issued under this section shall be available to the public. Such permit application or permit, or portion thereof, shall further be available on request for the purpose of reproduction.

(k) Compliance with permits

Compliance with a permit issued pursuant to this section shall be deemed compliance, for purposes of sections 1319 and 1365 of this title, with sections 1311, 1312, 1316, 1317, and 1343 of this title, except any standard imposed under section 1317 of this title for a toxic pollutant injurious to human health. Until December 31, 1974, in any case where a permit for discharge has been applied for pursuant to this section, but final administrative disposition of such application has not been made, such discharge shall not be a violation of

(1) section 1311, 1316, or 1342 of this title, or

(2) section 407 of this title, unless the Administrator or other plaintiff proves

that final administrative disposition of such application has not been made because of the failure of the applicant to furnish information reasonably required or requested in order to process the application. For the 180-day period beginning on October 18, 1972, in the case of any point source discharging any pollutant or combination of pollutants immediately prior to such date which source is not subject to section 407 of this title, the discharge by such source shall not be a violation of this chapter if such a source applies for a permit for discharge pursuant to this section within such 180-day period.

(l) Limitation on permit requirement

(1) Agricultural return flows

The Administrator shall not require a permit under this section for discharges composed entirely of return flows from irrigated agriculture, nor shall the Administrator directly or indirectly, require any State to require such a permit.

(2) Stormwater runoff from oil, gas, and mining operations

The Administrator shall not require a permit under this section, nor shall the Administrator directly or indirectly require any State to require a permit, for discharges of stormwater runoff from mining operations or oil and gas exploration, production, processing, or treatment operations or transmission facilities, composed entirely of flows which are from conveyances or systems of conveyances (including but not limited to pipes, conduits, ditches, and channels) used for collecting and conveying precipitation runoff and which are not contaminated by contact with, or do not come into contact with, any overburden, raw material, intermediate products, finished product, byproduct, or waste products located on the site of such operations.

(m) Additional pretreatment of conventional pollutants not required

To the extent a treatment works (as defined in section 1292 of this title) which is publicly owned is not meeting the requirements of a permit issued under this section for such treatment works as a result of inadequate design or operation of such treatment works, the Administrator, in issuing a permit under this section, shall not require pretreatment by a person introducing conventional pollutants identified pursuant to section 1314(a)(4) of this title into such treatment works other than pretreatment required to assure compliance with pretreatment standards under subsection (b)(8) of this section and section 1317(b)(1) of this title. Nothing in this subsection shall affect the Administrator's authority under sections 1317 and 1319 of this title, affect State and local authority under sections 1317(b)(4) and 1370 of this title, relieve such treatment works of its obligations to meet requirements established under this chapter, or otherwise preclude such works from pursuing whatever feasible options are available to meet its responsibility to comply with its permit under this section.

(n) Partial permit program

(1) State submission

The Governor of a State may submit under subsection (b) of this section a permit program for a portion of the discharges into the navigable waters in such State.

(2) Minimum coverage

A partial permit program under this subsection shall cover, at a minimum, administration of a major category of the discharges into the navigable waters of the State or a major component of the permit program required by subsection (b) of this section.

(3) Approval of major category partial permit programs

The Administrator may approve a partial permit program covering administration of a major category of discharges under this subsection if–

(A) such program represents a complete permit program and covers all of the discharges under the jurisdiction of a department or agency of the State; and

(B) the Administrator determines that the partial program represents a significant and identifiable part of the State program required by subsection (b) of this section.

(4) Approval of major component partial permit programs

The Administrator may approve under this subsection a partial and phased permit program covering administration of a major component (including discharge categories) of a State permit program required by subsection (b) of this section if–

(A) the Administrator determines that the partial program represents a significant and identifiable part of the State program required by subsection (b) of this section; and

(B) the State submits, and the Administrator approves, a plan for the State to assume administration by phases of the remainder of the State program required by subsection (b) of this section by a specified date not more than 5 years after submission of the partial program under this subsection and agrees to make all reasonable efforts to assume such administration by such date.

(o) Anti-backsliding

(1) General prohibition

In the case of effluent limitations established on the basis of subsection (a)(1)(B) of this section, a permit may not be renewed, reissued, or modified on the basis of effluent guidelines promulgated under section 1314(b) of this title subsequent to the original issuance of such permit, to contain effluent limitations which are less stringent than the comparable effluent limitations in the previous permit. In the case of effluent limitations established on the basis of section 1311(b)(1)(C) or section 1313(d) or (e) of this title, a permit may not be renewed, reissued, or modified to contain effluent limitations which are less stringent than the comparable effluent limitations in the previous permit except in compliance with section 1313(d)(4) of this title.

(2) Exceptions

A permit with respect to which paragraph (1) applies may be renewed, reissued, or modified to contain a less stringent effluent limitation applicable to a pollutant if–

(A) material and substantial alterations or additions to the permitted facility occurred after permit issuance which justify the application of a less stringent effluent limitation;

(B) (i) information is available which was not available at the time of permit issuance (other than revised regulations, guidance, or test methods) and which would have justified the application of a less stringent effluent limitation at the time of permit issuance; or

(ii) the Administrator determines that technical mistakes or mistaken interpretations of law were made in issuing the permit under subsection (a)(1)(B) of this section;

(C) a less stringent effluent limitation is necessary because of events over which the permittee has no control and for which there is no reasonably available remedy;

(D) the permittee has received a permit modification under section 1311(c), 1311(g), 1311(h), 1311(i), 1311(k), 1311(n), or 1326(a) of this title; or

(E) the permittee has installed the treatment facilities required to meet the effluent limitations in the previous permit and has properly operated and maintained the facilities but has nevertheless been unable to achieve the previous effluent limitations, in which case the limitations in the reviewed, reissued, or modified permit may reflect the level of pollutant control actually achieved (but shall not be less stringent than required by effluent guidelines in effect at the time of permit renewal, reissuance, or modification).

Subparagraph (B) shall not apply to any revised waste load allocations or any alternative grounds for translating water quality standards into effluent limitations, except where the cumulative effect of such revised allocations results in a decrease in the amount of pollutants discharged into the concerned waters, and such revised allocations are not the result of a discharger eliminating or substantially reducing its discharge of pollutants due to complying with the requirements of this chapter or for reasons otherwise unrelated to water quality.

(3) Limitations

In no event may a permit with respect to which paragraph (1) applies be renewed, reissued, or modified to contain an effluent limitation which is less stringent than required by effluent guidelines in effect at the time the permit is renewed, reissued, or modified. In no event may such a permit to discharge into waters be renewed, reissued, or modified to contain a less stringent effluent limitation if the implementation of such limitation would result in a violation of a water quality standard under section 1313 of this title applicable to such waters.

(p) Municipal and industrial stormwater discharges

(1) General rule

Prior to October 1, 1994, the Administrator or the State (in the case of a permit program approved under this section) shall not require a permit under this section for discharges composed entirely of stormwater.

(2) Exceptions

Paragraph (1) shall not apply with respect to the following stormwater discharges:

(A) A discharge with respect to which a permit has been issued under this section before February 4, 1987.

(B) A discharge associated with industrial activity.

(C) A discharge from a municipal separate storm sewer system serving a population of 250,000 or more.

(D) A discharge from a municipal separate storm sewer system serving a population of 100,000 or more but less than 250,000.

(E) A discharge for which the Administrator or the State, as the case may be, determines that the stormwater discharge contributes to a violation of a water quality standard or is a significant contributor of pollutants to waters of the United States.

(3) Permit requirements

(A) Industrial discharges

Permits for discharges associated with industrial activity shall meet all applicable provisions of this section and section 1311 of this title.

(B) Municipal discharge

Permits for discharges from municipal storm sewers–

(i) may be issued on a system- or jurisdiction-wide basis;

(ii) shall include a requirement to effectively prohibit non-stormwater discharges into the storm sewers; and

(iii) shall require controls to reduce the discharge of pollutants to the maximum extent practicable, including management practices, control techniques and system, design and engineering methods, and such other provisions as the Administrator or the State determines appropriate for the control of such pollutants.

(4) Permit application requirements

(A) Industrial and large municipal discharges

Not later than 2 years after February 4, 1987, the Administrator shall establish regulations setting forth the permit application requirements for stormwater discharges described in paragraphs (2)(B) and (2)(C). Applications for permits for such discharges shall be filed no later than 3 years after February 4, 1987. Not later than 4 years after February 4, 1987, the Administrator or the State, as the case may be, shall issue or deny each such permit. Any such permit shall provide for compliance as expeditiously as practicable, but in no event later than 3 years after the date of issuance of such permit.

(B) Other municipal discharges

Not later than 4 years after February 4, 1987, the Administrator shall establish regulations setting forth the permit application requirements for stormwater discharges described in paragraph (2)(D). Applications for permits for such discharges shall be filed no later than 5 years after February 4, 1987. Not later than 6 years after February 4, 1987, the Administrator or the State, as the case may be, shall issue or deny each such permit. Any such permit shall provide for compliance as expeditiously as practicable, but in no event later than 3 years after the date of issuance of such permit.

(5) Studies

The Administrator, in consultation with the States, shall conduct a study for the purposes of–

(A) identifying those stormwater discharges or classes of stormwater discharges for which permits are not required pursuant to paragraphs (1) and (2) of this subsection;

(B) determining, to the maximum extent practicable, the nature and extent of pollutants in such discharges; and

(C) establishing procedures and methods to control stormwater discharges to the extent necessary to mitigate impacts on water quality.

Not later than October 1, 1988, the Administrator shall submit to Congress a report on the results of the study described in subparagraphs (A) and (B). Not later than October 1, 1989, the Administrator shall submit to Congress a report on the results of the study described in subparagraph (C).

(6) Regulations

Not later than October 1, 1993, the Administrator, in consultation with State and local officials, shall issue regulations (based on the results of the studies conducted under paragraph (5)) which designate stormwater discharges, other than those discharges described in paragraph (2), to be regulated to protect water quality and shall establish a comprehensive program to regulate such designated sources. The program shall, at a minimum,

(A) establish priorities,

(B) establish requirements for State stormwater management programs, and

(C) establish expeditious deadlines. The program may include performance standards, guidelines, guidance, and management practices and treatment requirements, as appropriate.

Section 404, Clean Water Act 33 USC 1344

Permits for dredged or fill material

(a) Discharge into navigable waters at specified disposal sites

The Secretary may issue permits, after notice and opportunity for public hearings for the discharge of dredged or fill material into the navigable waters at specified disposal sites. Not later than the fifteenth day after the date an applicant submits all the information required to complete an application for a permit under this subsection, the Secretary shall publish the notice required by this subsection.

(b) Specification for disposal sites

Subject to subsection (c) of this section, each such disposal site shall be specified for each such permit by the Secretary

(1) through the application of guidelines developed by the Administrator, in conjunction with the Secretary, which guidelines shall be based upon criteria comparable to the criteria applicable to the territorial seas, the contiguous zone, and the ocean under section 1343(c) of this title, and

(2) in any case where such guidelines under clause (1) alone would prohibit the specification of a site, through the application additionally of the economic impact of the site on navigation and anchorage.

(c) Denial or restriction of use of defined areas as disposal sites

The Administrator is authorized to prohibit the specification (including the withdrawal of specification) of any defined area as a disposal site, and he is authorized to deny or restrict the use of any defined area for specification (including the withdrawal of specification) as a disposal site, whenever he determines, after notice and opportunity for public hearings, that the discharge of such materials into such area will have an unacceptable adverse effect on municipal water supplies, shellfish beds and fishery areas (including spawning and breeding areas), wildlife, or recreational areas. Before making such determination, the Administrator shall consult with the Secretary. The Administrator shall set forth in writing and make public his findings and his reasons for making any determination under this subsection.

(d) "Secretary" defined

The term "Secretary" as used in this section means the Secretary of the Army, acting through the Chief of Engineers.

(e) General permits on State, regional, or nationwide basis

(1) In carrying out his functions relating to the discharge of dredged or fill material under this section, the Secretary may, after notice and opportunity for public hearing, issue general permits on a State, regional, or nationwide basis for any category of activities involving discharges of dredged or fill material if the Secretary determines that the activities in such category are similar in nature, will cause only minimal adverse environmental effects when performed separately, and will have only minimal cumulative adverse effect on the environment. Any general permit issued under this subsection shall

(A) be based on the guidelines described in subsection (b)(1) of this section, and

(B) set forth the requirements and standards which shall apply to any activity authorized by such general permit.

(2) No general permit issued under this subsection shall be for a period of more than five years after the date of its issuance and such general permit may be revoked or modified by the Secretary if, after opportunity for public hearing, the Secretary determines that the activities authorized by such general permit have an adverse impact on the environment or such activities are more appropriately authorized by individual permits.

(f) Non-prohibited discharge of dredged or fill material

(1) Except as provided in paragraph (2) of this subsection, the discharge of dredged or fill material–

(A) from normal farming, silviculture, and ranching activities such as plowing, seeding, cultivating, minor drainage, harvesting for the production of food, fiber, and forest products, or upland soil and water conservation practices;

(B) for the purpose of maintenance, including emergency reconstruction of recently damaged parts, of currently serviceable structures such as dikes, dams, levees, groins, riprap, breakwaters, causeways, and bridge abutments or approaches, and transportation structures;

(C) for the purpose of construction or maintenance of farm or stock ponds or irrigation ditches, or the maintenance of drainage ditches;

(D) for the purpose of construction of temporary sedimentation basins on a construction site which does not include placement of fill material into the navigable waters;

(E) for the purpose of construction or maintenance of farm roads or forest roads, or temporary roads for moving mining equipment, where such roads are constructed and maintained, in accordance with best management practices, to assure that flow and circulation patterns and chemical and biological characteristics of the navigable waters are not impaired, that the reach of the navigable waters is not reduced, and that any adverse effect on the aquatic environment will be otherwise minimized;

(F) resulting from any activity with respect to which a State has an approved program under section 1288(b)(4) of this title which meets the requirements of subparagraphs (B) and (C) of such section,

is not prohibited by or otherwise subject to regulation under this section or section 1311(a) or 1342 of this title (except for effluent standards or prohibitions under section 1317 of this title).

(2) Any discharge of dredged or fill material into the navigable waters incidental to any activity having as its purpose bringing an area of the navigable waters into a use to which it was not previously subject, where the flow or circulation of navigable waters may be impaired or the reach of such waters be reduced, shall be required to have a permit under this section.

(g) State administration

(1) The Governor of any State desiring to administer its own individual and general permit program for the discharge of dredged or fill material into the navigable waters (other than those waters which are presently used, or are susceptible to use in their natural condition or by reasonable improvement as a means to transport interstate or foreign commerce shoreward to their ordinary high water mark, including all waters which are subject to the ebb and flow of the tide shoreward to their mean high water mark, or mean higher high water mark on the west coast, including wetlands adjacent thereto) within its jurisdiction may submit to the Administrator a full and complete description of the program it proposes to establish and administer under State law or under an interstate compact. In addition, such State shall submit a statement from the attorney general (or the attorney for those State agencies which have independent legal counsel), or from the chief legal officer in the case of an interstate agency, that the laws of such State, or the interstate compact, as the case may be, provide adequate authority to carry out the described program.

(2) Not later than the tenth day after the date of the receipt of the program and statement submitted by any State under paragraph (1) of this subsection, the Administrator shall provide copies of such program and statement to the Secretary and the Secretary of the Interior, acting through the Director of the United States Fish and Wildlife Service.

(3) Not later than the ninetieth day after the date of the receipt by the Administrator of the program and statement submitted by any State, under paragraph (1) of this subsection, the Secretary and the Secretary of the Interior, acting through the Director of the United States Fish and Wildlife Service, shall submit any comments with respect to such program and statement to the Administrator in writing.

(h) Determination of State's authority to issue permits under State program; approval; notification; transfers to State program

(1) Not later than the one-hundred-twentieth day after the date of the receipt by the Administrator of a program and statement submitted by any State under paragraph (1) of this subsection, the Administrator shall determine, taking into account any comments submitted by the Secretary and the Secretary of the Interior, acting through the Director of the United States Fish and Wildlife Service, pursuant to subsection (g) of this section, whether such State has the following authority with respect to the issuance of permits pursuant to such program:

(A) To issue permits which–

(i) apply, and assure compliance with, any applicable requirements of this section, including, but not limited to, the guidelines established under subsection (b)(1) of this section, and sections 1317 and 1343 of this title;

(ii) are for fixed terms not exceeding five years; and

(iii) can be terminated or modified for cause including, but not limited to, the following:

(I) violation of any condition of the permit;

(II) obtaining a permit by misrepresentation, or failure to disclose fully all relevant facts;

(III) change in any condition that requires either a temporary or permanent reduction or elimination of the permitted discharge.

(B) To issue permits which apply, and assure compliance with, all applicable requirements of section 1318 of this title, or to inspect, monitor, enter, and require reports to at least the same extent as required in section 1318 of this title.

(C) To assure that the public, and any other State the waters of which may be affected, receive notice of each application for a permit and to provide an opportunity for public hearing before a ruling on each such application.

(D) To assure that the Administrator receives notice of each application (including a copy thereof) for a permit.

(E) To assure that any State (other than the permitting State), whose waters may be affected by the issuance of a permit may submit written recommendations to the permitting State (and the Administrator) with respect to any permit application and, if any part of such written recommendations are not accepted by the permitting State, that the permitting State will notify such affected State (and the Administrator) in writing of its failure to so accept such recommendations together with its reasons for so doing.

(F) To assure that no permit will be issued if, in the judgment of the Secretary, after consultation with the Secretary of the department in which the Coast Guard is operating, anchorage and navigation of any of the navigable waters would be substantially impaired thereby.

(G) To abate violations of the permit or the permit program, including civil and criminal penalties and other ways and means of enforcement.

(H) To assure continued coordination with Federal and Federal-State water-related planning and review processes.

(2) If, with respect to a State program submitted under subsection (g)(1) of this section, the Administrator determines that such State–

(A) has the authority set forth in paragraph (1) of this subsection, the Administrator shall approve the program and so notify

(i) such State and

(ii) the Secretary, who upon subsequent notification from such State that it is administering such program, shall suspend the issuance of permits under subsections (a) and (e) of this section for activities with respect to which a permit may be issued pursuant to such State program; or

(B) does not have the authority set forth in paragraph (1) of this subsection, the Administrator shall so notify such State, which notification shall also describe the revisions or modifications necessary so that such State may resubmit such program for a determination by the Administrator under this subsection.

(3) If the Administrator fails to make a determination with respect to any program submitted by a State under subsection (g)(1) of this section within one-hundred-twenty days after the date of the receipt of such program, such program shall be deemed approved pursuant to paragraph (2)(A) of this subsection and the Administrator shall so notify such State and the Secretary who, upon subsequent notification from such State that it is administering such program, shall suspend the issuance of permits under subsection (a) and (e) of this section for activities with respect to which a permit may be issued by such State.

(4) After the Secretary receives notification from the Administrator under paragraph (2) or (3) of this subsection that a State permit program has been approved, the Secretary shall transfer any applications for permits pending before the Secretary for activities with respect to which a permit may be issued pursuant to such State program to such State for appropriate action.

(5) Upon notification from a State with a permit program approved under this subsection that such State intends to administer and enforce the terms and conditions of a general permit issued by the Secretary under subsection (e) of this section with respect to activities in such State to which such general permit applies, the Secretary shall suspend the administration and enforcement of such general permit with respect to such activities.

(i) Withdrawal of approval

Whenever the Administrator determines after public hearing that a State is not administering a program approved under subsection (h)(2)(A) of this section, in accordance with this section, including, but not limited to, the guidelines established under subsection (b)(1) of this section, the Administrator shall so notify the State, and, if appropriate corrective action is not taken within a reasonable time, not to exceed ninety days after the date of the receipt of such notification, the Administrator shall

(1) withdraw approval of such program until the Administrator determines such corrective action has been taken, and

(2) notify the Secretary that the Secretary shall resume the program for the issuance of permits under subsections (a) and (e) of this section for activities with respect to which the State was issuing permits and that such authority of the Secretary shall continue in effect until such time as the Administrator makes the determination described in clause (1) of this subsection and such State again has an approved program.

(j) Copies of applications for State permits and proposed general permits to be transmitted to Administrator

Each State which is administering a permit program pursuant to this section shall transmit to the Administrator

(1) a copy of each permit application received by such State and provide notice to the Administrator of every action related to the consideration of such permit application, including each permit proposed to be issued by such State, and

(2) a copy of each proposed general permit which such State intends to issue. Not later than the tenth day after the date of the receipt of such permit application or such proposed general permit, the Administrator shall provide copies of such permit application or such proposed general permit to the Secretary and the Secretary of the Interior,

acting through the Director of the United States Fish and Wildlife Service. If the Administrator intends to provide written comments to such State with respect to such permit application or such proposed general permit, he shall so notify such State not later than the thirtieth day after the date of the receipt of such application or such proposed general permit and provide such written comments to such State, after consideration of any comments made in writing with respect to such application or such proposed general permit by the Secretary and the Secretary of the Interior, acting through the Director of the United States Fish and Wildlife Service, not later than the ninetieth day after the date of such receipt. If such State is so notified by the Administrator, it shall not issue the proposed permit until after the receipt of such comments from the Administrator, or after such ninetieth day, whichever first occurs. Such State shall not issue such proposed permit after such ninetieth day if it has received such written comments in which the Administrator objects

(A) to the issuance of such proposed permit and such proposed permit is one that has been submitted to the Administrator pursuant to subsection (h)(1)(E) of this section, or

(B) to the issuance of such proposed permit as being outside the requirements of this section, including, but not limited to, the guidelines developed under subsection (b)(1) of this section unless it modifies such proposed permit in accordance with such comments. Whenever the Administrator objects to the issuance of a permit under the preceding sentence such written objection shall contain a statement of the reasons for such objection and the conditions which such permit would include if it were issued by the Administrator. In any case where the Administrator objects to the issuance of a permit, on request of the State, a public hearing shall be held by the Administrator on such objection. If the State does not resubmit such permit revised to meet such objection within 30 days after completion of the hearing or, if no hearing is requested within 90 days after the date of such objection, the Secretary may issue the permit pursuant to subsection (a) or (e) of this section, as the case may be, for such source in accordance with the guidelines and requirements of this chapter.

(k) Waiver

In accordance with guidelines promulgated pursuant to subsection (i)(2) of section 1314 of this title, the Administrator is authorized to waive the requirements of subsection (j) of this section at the time of the approval of a program pursuant to subsection (h)(2)(A) of this section for any category (including any class, type, or size within such category) of discharge within the State submitting such program.

(l) Categories of discharges not subject to requirements

The Administrator shall promulgate regulations establishing categories of discharges which he determines shall not be subject to the requirements of subsection (j) of this section in any State with a program approved pursuant to subsection (h)(2)(A) of this section. The Administrator may distinguish among classes, types, and sizes within any category of discharges.

(m) Comments on permit applications or proposed general permits by Secretary of the Interior acting through Director of United States Fish and Wildlife Service

Not later than the ninetieth day after the date on which the Secretary notifies the Secretary of the Interior, acting through the Director of the United States Fish and Wildlife Service that

(1) an application for a permit under subsection (a) of this section has been received by the Secretary, or

(2) the Secretary proposes to issue a general permit under subsection (e) of this section, the Secretary of the Interior, acting through the Director of the United States Fish and Wildlife Service, shall submit any comments with respect to such application or such proposed general permit in writing to the Secretary.

(n) Enforcement authority not limited

Nothing in this section shall be construed to limit the authority of the Administrator to take action pursuant to section 1319 of this title.

(o) Public availability of permits and permit applications

A copy of each permit application and each permit issued under this section shall be available to the public. Such permit application or portion thereof, shall further be available on request for the purpose of reproduction.

(p) Compliance

Compliance with a permit issued pursuant to this section, including any activity carried out pursuant to a general permit issued under this section, shall be deemed compliance, for purposes of sections 1319 and 1365 of this title, with sections 1311, 1317, and 1343 of this title.

(q) Minimization of duplication, needless paperwork, and delays in issuance; agreements

Not later than the one-hundred-eightieth day after December 27, 1977, the Secretary shall enter into agreements with the Administrator, the Secretaries of the Departments of Agriculture, Commerce, Interior, and Transportation, and the heads of other appropriate Federal agencies to minimize, to the maximum extent practicable, duplication, needless paperwork, and delays in the issuance of permits under this section. Such agreements shall be developed to assure that, to the maximum extent practicable, a decision with respect to an application for a permit under subsection (a) of this section will be made not later than the ninetieth day after the date the notice for such application is published under subsection (a) of this section.

(r) Federal projects specifically authorized by Congress

The discharge of dredged or fill material as part of the construction of a Federal project specifically authorized by Congress, whether prior to or on or after December 27, 1977, is not prohibited by or otherwise subject to regulation under this section, or a State program approved under this section, or section 1311(a) or 1342 of this title (except for effluent standards or prohibitions under section 1317 of this title), if information on the effects of such discharge, including consideration of the guidelines developed under subsection (b)(1) of this section, is included in an environmental impact statement for such project pursuant to the National Environmental Policy Act of 1969 (42 U.S.C. 4321 et seq.) and such environmental impact statement has been submitted to Congress before the actual discharge of dredged or fill material in connection with the construction of such project and prior to either authorization of such project or an appropriation of funds for such construction.

(s) Violation of permits

(1) Whenever on the basis of any information available to him the Secretary finds that any person is in violation of any condition or limitation set forth in a permit issued by the Secretary under this section, the Secretary shall issue an order requiring such person to comply with such condition or limitation, or the Secretary shall bring a civil action in accordance with paragraph (3) of this subsection.

(2) A copy of any order issued under this subsection shall be sent immediately by the Secretary to the State in which the violation occurs and other affected States. Any order issued under this subsection shall be by personal service and shall state with reasonable specificity the nature of the violation, specify a time for compliance, not to exceed thirty days, which the Secretary determines

is reasonable, taking into account the seriousness of the violation and any good faith efforts to comply with applicable requirements. In any case in which an order under this subsection is issued to a corporation, a copy of such order shall be served on any appropriate corporate officers.

(3) The Secretary is authorized to commence a civil action for appropriate relief, including a permanent or temporary injunction for any violation for which he is authorized to issue a compliance order under paragraph (1) of this subsection. Any action under this paragraph may be brought in the district court of the United States for the district in which the defendant is located or resides or is doing business, and such court shall have jurisdiction to restrain such violation and to require compliance. Notice of the commencement of such acton[1] shall be given immediately to the appropriate State.

(4) Any person who violates any condition or limitation in a permit issued by the Secretary under this section, and any person who violates any order issued by the Secretary under paragraph (1) of this subsection, shall be subject to a civil penalty not to exceed $25,000 per day for each violation. In determining the amount of a civil penalty the court shall consider the seriousness of the violation or violations, the economic benefit (if any) resulting from the violation, any history of such violations, any good-faith efforts to comply with the applicable requirements, the economic impact of the penalty on the violator, and such other matters as justice may require.

[1] So in original. Probably should be "action".

(t) Navigable waters within State jurisdiction

Nothing in this section shall preclude or deny the right of any State or interstate agency to control the discharge of dredged or fill material in any portion of the navigable waters within the jurisdiction of such State, including any activity of any Federal agency, and each such agency shall comply with such State or interstate requirements both substantive and procedural to control the discharge of dredged or fill material to the same extent that any person is subject to such requirements. This section shall not be construed as affecting or impairing the authority of the Secretary to maintain navigation.

APPENDIX B

River and Harbors Act of 1899

Sections 9 and 10

Section 9, Rivers and Harbors Appropriations Act of 1899 33 USC 401 (1994)

Construction of bridges, causeways, dams or dikes generally; exemptions

It shall not be lawful to construct or commence the construction of any bridge, causeway, dam, or dike over or in any port, roadstead, haven, harbor, canal, navigable river, or other navigable water of the United States until the consent of Congress to the building of such structures shall have been obtained and until the plans for (1) the bridge or causeway shall have been submitted to and approved by the Secretary of Transportation, or (2) the dam or dike shall have been submitted to and approved by the Chief of Engineers and Secretary of the Army. However, such structures may be built under authority of the legislature of a State across rivers and other waterways the navigable portions of which lie wholly within the limits of a single State, provided the location and plans thereof are submitted to and approved by the Secretary of Transportation or by the Chief of Engineers and Secretary of the Army before construction is commenced. When plans for any bridge or other structure have been approved by the Secretary of Transportation or by the Chief of Engineers and Secretary of the Army, it shall not be lawful to deviate from such plans either before or after completion of the structure unless modification of said plans has previously been submitted to and received the approval of the Secretary of Transportation or the Chief of Engineers and the Secretary of the Army. The approval required by this section of the location and plans or any modification of plans of any bridge or causeway does not apply to any bridge or causeway over waters that are not subject to the ebb and flow of the tide and that are not used and are not susceptible to use in their natural condition or by reasonable improvement as a means to transport interstate or foreign commerce.

Section 10, Rivers and Harbors Appropriations Act of 1899 33 USC 403 (1994)

Obstruction of navigable waters generally; wharves; piers, etc.; excavations and filling in

The creation of any obstruction not affirmatively authorized by Congress, to the navigable capacity of any of the waters of the United States is prohibited; and it shall not be lawful to build or commence the building of any wharf, pier, dolphin, boom, weir, breakwater, bulkhead, jetty, or other structures in any port, roadstead, haven, harbor, canal, navigable river, or other water of the United States, outside established harbor lines, or where no harbor lines have been established, except on plans recommended by the Chief of Engineers and authorized by the Secretary of the Army; and it shall not be lawful to excavate or fill, or in any manner to alter or modify the course, location, condition, or capacity of, any port, roadstead, haven, harbor, canal, lake, harbor or refuge, or inclosure within the limits of any breakwater, or of the channel of any navigable water of the United States, unless the work has been recommended by the Chief of Engineers and authorized by the Secretary of the Army prior to beginning the same.

APPENDIX C

USACE Regulations

Section 404 and Section 10

Revised Definitions of Fill Material and Discharge

Section 404 and Section 10

U.S. Army Corps of Engineers 33 CFR 320–331

PART 320 GENERAL REGULATORY POLICIES

AUTHORITY: 33 U.S.C. 401 et seq.; 33 U.S.C. 1344; 33 U.S.C. 1413.

Section 320.1 Purpose and scope

(a) *Regulatory approach of the Corps of Engineers.*

(1) The U.S. Army Corps of Engineers has been involved in regulating certain activities in the nation's waters since 1890. Until 1968, the primary thrust of the Corps' regulatory program was the protection of navigation. As a result of several new laws and judicial decisions, the program has evolved to one involving the consideration of the full public interest by balancing the favorable impacts against the detrimental impacts. This is known as the "public interest review." The program is one which reflects the national concerns for both the protection and utilization of important resources.

(2) The Corps is a highly decentralized organization. Most of the authority for administering the regulatory program has been delegated to the thirty-six district engineers and eleven division engineers. A district engineer's decision on an approved jurisdictional determination, a permit denial, or a declined individual permit is subject to an administrative appeal by the affected party in accordance with the procedures and authorities contained in 33 CFR Part 331. Such administrative appeal must meet the criteria in 33 CFR 331.5; otherwise, no administrative appeal of that decision is allowed. The terms "approved jurisdictional determination," "permit denial," and "declined permit" are defined at 33 CFR 331.2. There shall be no administrative appeal of any issued individual permit that an applicant has accepted, unless the authorized work has not started in waters of the United States, and that issued permit is subsequently modified by the district engineer pursuant to 33 CFR 325.7 (see 33 CFR 331.5(b)(1)). An affected party must exhaust any administrative appeal available pursuant to 33 CFR Part 331 and receive a final Corps decision on the appealed action prior to filing a lawsuit in the Federal courts (see 33 CFR 331.12).

(3) The Corps seeks to avoid unnecessary regulatory controls. The general permit program described in 33 CFR parts 325 and 330 is the primary method of eliminating unnecessary federal control over activities which do not justify individual control or which are adequately regulated by another agency.

(4) The Corps is neither a proponent nor opponent of any permit proposal. However, the Corps believes that applicants are due a timely decision. Reducing unnecessary paperwork and delays is a continuing Corps goal.

(5) The Corps believes that state and federal regulatory programs should complement rather than duplicate one another. The Corps uses general permits, joint processing procedures, interagency review, coordination, and authority transfers (where authorized by law) to reduce duplication.

(6) The Corps has authorized its district engineers to issue formal determinations concerning the applicability of the Clean Water Act or the Rivers and Harbors Act of 1899 to activities or tracts of land and the applicability of general permits or statutory exemptions to proposed activities. A determination pursuant to this authorization shall constitute a Corps final agency action. Nothing contained in this section is intended to affect any authority EPA has under the Clean Water Act.

(b) *Types of activities regulated.* This part and the parts that follow (33 CFR parts 321 through 330) prescribe the statutory authorities, and general and special policies and procedures applicable to the review of applications for Department of the Army (DA) permits for controlling certain activities in waters of the United States or the oceans. This part identifies the various federal statutes which require that DA permits be issued before these activities can be lawfully undertaken; and related Federal laws and the general policies applicable to the review of those activities. Parts 321 through 324 and 330 address special policies and procedures applicable to the following specific classes of activities:

(1) Dams or dikes in navigable waters of the United States (part 321);

(2) Other structures or work including excavation, dredging, and/or disposal activities, in navigable waters of the United States (part 322);

(3) Activities that alter or modify the course, condition, location, or capacity of a navigable water of the United States (part 322);

(4) Construction of artificial islands, installations, and other devices on the outer continental shelf (part 322);

(5) Discharges of dredged or fill material into waters of the United States (part 323);

(6) Activities involving the transportation of dredged material for the purpose of disposal in ocean waters (part 324); and

(7) Nationwide general permits for certain categories of activities (part 330).

(c) *Forms of authorization.* DA permits for the above described activities are issued under various forms of authorization. These include individual permits that are issued following a review of individual applications and general permits that authorize a category or categories of activities in specific geographical regions or nationwide. The term "general permit" as used in these regulations (33 CFR parts 320 through 330) refers to both those regional permits issued by district or division engineers on a regional basis and to nationwide permits which are issued by the Chief of Engineers through publication in the Federal Register and are applicable throughout the nation. The nationwide permits are found in 33 CFR part 330. If an activity is covered by a general permit, an application for a DA permit does not have to be made. In such cases, a person must only comply with the conditions contained in the general permit to satisfy requirements of law for a DA permit. In certain cases pre-notification may be required before initiating construction. (See 33 CFR 330.7)

(d) *General instructions.* General policies for evaluating permit applications are found in this part. Special policies that relate to particular activities are found in parts 321 through 324. The procedures for processing individual permits and general permits are contained in 33 CFR part 325. The terms "navigable waters of the United States" and "waters of the United States" are used frequently throughout these regulations, and it is important from the outset that the reader understand the difference between the two. "Navigable waters of the United States" are defined in 33 CFR part 329. These are waters that are navigable in the traditional sense where permits are required for certain work or structures pursuant to Sections 9 and 10 of the Rivers and Harbors Act of 1899. "Waters of the United States" are defined in 33 CFR part 328. These waters include more than navigable waters of the United States and are the waters where permits are required for the discharge of dredged or fill material pursuant to section 404 of the Clean Water Act.

[51 FR 41220, Nov. 13, 1986, as amended at 64 FR 11714, Mar. 9, 1999; 65 FR 16492, Mar. 28, 2000]

Section 320.2 Authorities to issue permits

(a) Section 9 of the Rivers and Harbors Act, approved March 3, 1899 (33 U.S.C. 401) (hereinafter referred to as section 9), prohibits the construction of any dam or dike across any navigable water of the United States in the absence of Congressional consent and approval of the plans by the Chief of Engineers and the Secretary of the Army. Where the navigable portions of the waterbody lie wholly within the limits of a single state, the structure may be built under authority of the legislature of that state if the location and plans or any modification thereof are approved by the Chief of Engineers and by the Secretary of the Army. The instrument of authorization is designated a permit (See 33 CFR part 321.) Section 9 also pertains to bridges and causeways but the authority of the Secretary of the Army and Chief of Engineers with respect to bridges and causeways was transferred to the Secretary of Transportation under the Department of Transportation Act of October 15, 1966 (49 U.S.C. 1155g(6)(A)). A DA permit pursuant to section 404 of the Clean Water Act is required for the discharge of dredged or fill material into waters of the United States associated with bridges and causeways. (See 33 CFR part 323.)

(b) Section 10 of the Rivers and Harbors Act approved March 3, 1899, (33 U.S.C. 403) (hereinafter referred to as section 10), prohibits the unauthorized obstruction or alteration of any navigable water of the United States. The construction of any structure in or over any navigable water of the United States, the excavating from or depositing of material in such waters, or the accomplishment of any other work affecting the course, location, condition, or capacity of such waters is unlawful unless the work has been recommended by the Chief of Engineers and authorized by the Secretary of the Army. The instrument of authorization is designated a permit. The authority of the Secretary of the Army to prevent obstructions to navigation in navigable waters of the United States was extended to artificial islands, installations, and other devices located on the seabed, to the seaward limit of the outer continental shelf, by section 4(f) of the Outer Continental Shelf Lands Act of 1953 as amended (43 U.S.C. 1333(e)). (See 33 CFR part 322.)

(c) Section 11 of the Rivers and Harbors Act approved March 3, 1899, (33 U.S.C. 404), authorizes the Secretary of the Army to establish harbor lines channelward of which no piers, wharves, bulkheads, or other works may be extended or deposits made without approval of the Secretary of the Army. Effective May 27, 1970, permits for work shoreward of those lines must be obtained in accordance with section 10 and, if applicable, section 404 of the Clean Water Act (see § 320.4(o) of this part).

(d) Section 13 of the Rivers and Harbors Act approved March 3, 1899, (33 U.S.C. 407), provides that the Secretary of the Army, whenever the Chief of Engineers determines that anchorage and navigation will not be injured thereby, may permit the discharge of refuse into navigable waters. In the absence of a permit, such discharge of refuse is prohibited. While the prohibition of this section, known as the Refuse Act, is still in effect, the permit authority of the Secretary of the Army has been superseded by the permit authority provided the Administrator, Environmental Protection Agency (EPA), and the states under sections 402 and 405 of the Clean Water Act, (33 U.S.C. 1342 and 1345). (See 40 CFR parts 124 and 125.)

(e) Section 14 of the Rivers and Harbors Act approved March 3, 1899, (33 U.S.C. 408), provides that the Secretary of the Army, on the recommendation of the Chief of Engineers, may grant permission for the temporary occupation or use of any sea wall, bulkhead, jetty, dike, levee, wharf, pier, or other work built by the United States. This permission will be granted by an appropriate real estate instrument in accordance with existing real estate regulations.

(f) Section 404 of the Clean Water Act (33 U.S.C. 1344) (hereinafter referred to as section 404) authorizes the Secretary of the Army, acting through the Chief of Engineers, to issue permits, after notice and opportunity for public hearing, for the discharge of dredged or fill material into the waters of the United States at specified disposal sites. (See 33 CFR part 323.) The selection and use of disposal sites will be in accordance with guidelines developed by the Administrator of EPA in conjunction with the Secretary of the Army and published in 40 CFR part 230. If these guidelines prohibit the selection or use of a disposal site, the Chief of Engineers shall consider the economic impact on navigation and anchorage of such a prohibition in reaching his decision. Furthermore, the Administrator can deny, prohibit, restrict or withdraw the use of any defined area as a disposal site whenever he determines, after notice and opportunity for public hearing and after consultation with the Secretary of the Army, that the discharge of such materials into such areas will have an unacceptable adverse effect on municipal water supplies, shellfish beds and fishery areas, wildlife, or recreational areas. (See 40 CFR part 230).

(g) Section 103 of the Marine Protection, Research and Sanctuaries Act of 1972, as amended (33 U.S.C. 1413) (hereinafter referred to as section 103), authorizes the Secretary of the Army, acting through the Chief of Engineers, to issue permits, after notice and opportunity for public hearing, for the transportation of dredged material for the purpose of disposal in the ocean where it is determined that the disposal will not unreasonably degrade or endanger human health, welfare, or amenities, or the marine environment, ecological systems, or economic potentialities. The selection of disposal sites will be in accordance with criteria developed by the Administrator of the EPA in consultation with the Secretary of the Army and published in 40 CFR parts 220 through 229. However, similar to the EPA Administrator's limiting authority cited in paragraph (f) of this section, the Administrator can prevent the issuance of a permit under this authority if he finds that the disposal of the material will result in an unacceptable adverse impact on municipal water supplies, shellfish beds, wildlife, fisheries, or recreational areas. (See 33 CFR part 324).

Section 320.3 Related laws

(a) Section 401 of the Clean Water Act (33 U.S.C. 1341) requires any applicant for a federal license or permit to conduct any activity that may result in a discharge of a pollutant into waters of the United States to obtain a certification from the State in which the discharge originates or would originate, or, if appropriate, from the interstate water pollution control agency having jurisdiction over the affected waters at the point where the discharge originates or would originate, that the discharge will comply with the applicable effluent limitations and water quality standards. A certification obtained for the construction of any facility must also pertain to the subsequent operation of the facility.

(b) Section 307(c) of the Coastal Zone Management Act of 1972, as amended (16 U.S.C. 1456(c)), requires federal agencies conducting activities, including development projects, directly affecting a state's coastal zone, to comply to the maximum extent practicable with an approved state coastal zone management program. Indian tribes doing work on federal lands will be treated as a federal agency for the purpose of the Coastal Zone Management Act. The Act also requires any non-federal applicant for a federal license or permit to conduct an activity affecting land or water uses in the state's coastal zone to furnish a certification that the proposed activity will comply with the state's coastal zone management program. Generally, no permit will be issued until the state has concurred with the non-federal applicant's certification. This provision becomes effective upon approval by the Secretary of Commerce of the state's coastal zone management program. (See 15 CFR part 930.)

(c) Section 302 of the Marine Protection, Research and Sanctuaries Act of 1972, as amended (16 U.S.C. 1432), authorizes the Secretary of Commerce, after consultation with other interested federal agencies and with the approval of the President, to designate as marine sanctuaries those areas of the ocean waters, of the Great Lakes and their connecting waters, or of other coastal waters which he determines necessary for the purpose of preserving or restoring such areas for their conservation, recreational, ecological, or aesthetic values. After designating such an area, the Secretary of Commerce shall issue regulations to control any activities within the area. Activities in the sanctuary authorized under other authorities are valid only if the Secretary of Commerce certifies that the activities are consistent with the purposes of Title III of the Act and can be carried out within the regulations for the sanctuary.

(d) The National Environmental Policy Act of 1969 (42 U.S.C. 4321-4347) declares the national policy to encourage a productive and enjoyable harmony between man and his environment. Section 102 of that Act directs that "to the fullest extent possible:

(1) The policies, regulations, and public laws of the United States shall be interpreted and administered in accordance with the policies set forth in this Act, and

(2) All agencies of the Federal Government shall * * * insure that presently unquantified environmental amenities and values may be given appropriate consideration in decision-making along with economic and technical considerations * * *". (See Appendix B of 33 CFR part 325.)

(e) The Fish and Wildlife Act of 1956 (16 U.S.C. 742a, *et seq.*), the Migratory Marine Game-Fish Act (16 U.S.C. 760c–760g), the Fish and Wildlife Coordination Act (16 U.S.C. 661–666c) and other acts express the will of Congress to protect the quality of the aquatic environment as it affects the conservation, improvement and enjoyment of fish and wildlife resources. Reorganization Plan No. 4 of 1970 transferred certain functions, including certain fish and wildlife-water resources coordination responsibilities, from the Secretary of the Interior to the Secretary of Commerce. Under the Fish and Wildlife Coordination Act and Reorganization Plan No. 4, any federal agency that proposes to control or modify any body of water must first consult with the United States Fish and Wildlife Service or the National Marine Fisheries Service, as appropriate, and with the head of the appropriate state agency exercising administration over the wildlife resources of the affected state.

(f) The Federal Power Act of 1920 (16 U.S.C. 791a *et seq.*), as amended, authorizes the Federal Energy Regulatory Agency (FERC) to issue licenses for the construction and the operation and maintenance of dams, water conduits, reservoirs, power houses, transmission lines, and other physical structures of a hydro-power project. However, where such structures will affect the navigable capacity of any navigable water of the United States (as defined in 16 U.S.C. 796), the plans for the dam or other physical structures affecting navigation must be approved by the Chief of Engineers and the Secretary of the Army. In such cases, the interests of navigation should normally be protected by a DA recommendation to FERC for the inclusion of appropriate provisions in the FERC license rather than the issuance of a separate DA permit under 33 U.S.C. 401 *et seq.* As to any other activities in navigable waters not constituting construction and the operation and maintenance of physical structures licensed by FERC under the Federal Power Act of 1920, as amended, the provisions of 33 U.S.C. 401 *et seq.* remain fully applicable. In all cases involving the discharge of dredged or fill material into waters of the United States or the transportation of dredged material for the purpose of disposal in ocean waters, section 404 or section 103 will be applicable.

(g) The National Historic Preservation Act of 1966 (16 U.S.C. 470) created the Advisory Council on Historic Preservation to advise the President and Congress on matters involving historic preservation. In performing its function the Council is authorized to review and comment upon activities licensed by the Federal Government which will have an effect upon properties listed in the National Register of Historic Places, or eligible for such listing. The concern of Congress for the preservation of significant historical sites is also expressed in the Preservation of Historical and Archeological Data Act of 1974 (16 U.S.C. 469 *et seq.*), which amends the Act of June 27, 1960. By this Act, whenever a federal construction project or federally licensed project, activity, or program alters any terrain such that significant historical or archeological data is threatened, the Secretary of the Interior may take action necessary to recover and preserve the data prior to the commencement of the project.

(h) The Interstate Land Sales Full Disclosure Act (15 U.S.C. 1701 *et seq.*) prohibits any developer or agent from selling or leasing any lot in a subdivision (as defined in 15 U.S.C. 1701(3)) unless the purchaser is furnished in advance a printed property report containing information which the Secretary of Housing and Urban Development may, by rules or regulations, require for the protection of purchasers. In the event the lot in question is part of a project that requires DA authorization, the property report is required by Housing and Urban Development regulation to state whether or not a permit for the development has been applied for, issued, or denied by the Corps of Engineers under section 10 or section 404. The property report is also required to state whether or not any enforcement action has been taken as a consequence of non-application for or denial of such permit.

(i) The Endangered Species Act (16 U.S.C. 1531 *et seq.*) declares the intention of the Congress to conserve threatened and endangered species and the ecosystems on which those species depend. The Act requires that federal agencies, in consultation with the U.S. Fish and Wildlife Service and the National Marine Fisheries Service, use their authorities in furtherance of its purposes by carrying out programs for the conservation of endangered or threatened species, and by taking such action necessary to insure that any action authorized, funded, or carried out by the Agency is not likely to jeopardize the continued existence of such endangered or threatened species or result in the destruction or adverse modification of habitat of such species which is determined by the Secretary of the Interior or Commerce, as appropriate, to be critical. (See 50 CFR part 17 and 50 CFR part 402.)

(j) The Deepwater Port Act of 1974 (33 U.S.C. 1501 *et seq.*) prohibits the ownership, construction, or operation of a deepwater port beyond the territorial seas without a license issued by the Secretary of Transportation. The Secretary of Transportation may issue such a license to an applicant if he determines, among other things, that the construction and operation of the deepwater port is in the national interest and consistent with national security and other national policy goals and objectives. An application for a deepwater port license constitutes an application for all federal authorizations required for the ownership, construction, and operation of a deepwater port, including applications for section 10, section 404 and section 103 permits which may also be required pursuant to the authorities listed in § 320.2 and the policies specified in § 320.4 of this part.

(k) The Marine Mammal Protection Act of 1972 (16 U.S.C. 1361 *et seq.*) expresses the intent of Congress that marine mammals be protected and encouraged to develop in order to maintain the health and stability of the marine ecosystem. The Act imposes a perpetual moratorium on the harassment, hunting, capturing, or killing of marine mammals and on the importation of marine mammals and marine mammal products without a permit from either the Secretary of the Interior or the Secretary of Commerce, depending upon the species of marine mammal involved. Such permits may be issued only for purposes of scientific research and for public display if the purpose is consistent with the policies of the Act. The appropriate Secretary is also empowered in certain restricted circumstances to waive the requirements of the Act.

(l) Section 7(a) of the Wild and Scenic Rivers Act (16 U.S.C. 1278 *et seq.*) provides that no department or agency of the United States shall assist by loan, grant, license, or otherwise in the construction of any water resources project that would have a direct and adverse effect on the values for which such river was established, as determined by the Secretary charged with its administration.

(m) The Ocean Thermal Energy Conversion Act of 1980, (42 U.S.C. section 9101 *et seq.*) establishes a licensing regime administered by the Administrator of NOAA for the ownership, construction, location, and operation of ocean thermal energy conversion (OTEC) facilities and plantships. An application for an OTEC license filed with the Administrator constitutes an application for all federal authorizations required for ownership, construction, location, and operation of an OTEC facility or plantship, except for certain activities within the jurisdiction of the Coast Guard. This includes applications for section 10, section 404, section 103 and other DA authorizations which may be required.

(n) Section 402 of the Clean Water Act authorizes EPA to issue permits under procedures established to implement the National Pollutant Discharge Elimination System (NPDES) program. The administration of this program can be, and in most cases has been, delegated to individual states. Section 402(b)(6) states that no NPDES permit will be issued if the Chief of Engineers, acting for the Secretary of the Army and after consulting with the U.S. Coast Guard, determines that navigation and anchorage in any navigable water will be substantially impaired as a result of a proposed activity.

(o) The National Fishing Enhancement Act of 1984 (Pub. L. 98-623) provides for the development of a National Artificial Reef Plan to promote and facilitate responsible and effective efforts to establish artificial reefs. The Act establishes procedures to be followed by the Corps in issuing DA permits for artificial reefs. The Act also establishes the liability of the permittee and the United States. The Act further creates a civil penalty for violation of any provision of a permit issued for an artificial reef.

Section 320.4 General policies for evaluating permit applications

The following policies shall be applicable to the review of all applications for DA permits. Additional policies specifically applicable to certain types of activities are identified in 33 CFR parts 321 through 324.

(a) *Public Interest Review*

(1) The decision whether to issue a permit will be based on an evaluation of the probable impacts, including cumulative impacts, of the proposed activity and its intended use on the public interest. Evaluation of the probable impact which the proposed activity may have on the public interest requires a careful weighing of all those factors which become relevant in each particular case. The benefits which reasonably may be expected to accrue from the proposal must be balanced against its reasonably foreseeable detriments. The decision whether to authorize a proposal, and if so, the conditions under which it will be allowed to occur, are therefore determined by the outcome of this general balancing process. That decision should reflect the national concern for both protection and utilization of important resources. All factors which may be relevant to the proposal must be considered including the cumulative effects thereof: among those are conservation, economics, aesthetics, general environmental concerns, wetlands, historic properties, fish and wildlife values, flood hazards, floodplain values, land use, navigation, shore erosion and accretion, recreation, water supply and conservation, water quality, energy needs, safety, food and fiber production, mineral needs, considerations of property ownership and, in general, the needs and welfare of the people. For activities involving 404 discharges, a permit will be denied if the discharge that would be authorized by such permit would not comply with the Environmental Protection Agency's 404(b)(1) guidelines. Subject to the preceding sentence and any other applicable guidelines and criteria (see §§ 320.2 and 320.3), a permit will be granted unless the district engineer determines that it would be contrary to the public interest.

(2) The following general criteria will be considered in the evaluation of every application:

(i) The relative extent of the public and private need for the proposed structure or work:

(ii) Where there are unresolved conflicts as to resource use, the practicability of using reasonable alternative locations and methods to accomplish the objective of the proposed structure or work; and

(iii) The extent and permanence of the beneficial and/or detrimental effects which the proposed structure or work is likely to have on the public and private uses to which the area is suited.

(3) The specific weight of each factor is determined by its importance and relevance to the particular proposal. Accordingly, how important a factor is and how much consideration it deserves will vary with each proposal. A specific factor may be given great weight on one proposal, while it may not be present or as important on another. However, full consideration and appropriate weight will be given to all comments, including those of federal, state, and local agencies, and other experts on matters within their expertise.

(b) *Effect on wetlands*

(1) Most wetlands constitute a productive and valuable public resource, the unnecessary alteration or destruction of which should be discouraged as contrary to the public interest. For projects to be undertaken or partially or entirely funded by a federal, state, or local agency, additional requirements on wetlands considerations are stated in Executive Order 11990, dated 24 May 1977.

(2) Wetlands considered to perform functions important to the public interest include:

(i) Wetlands which serve significant natural biological functions, including food chain production, general habitat and nesting, spawning, rearing and resting sites for aquatic or land species;

(ii) Wetlands set aside for study of the aquatic environment or as sanctuaries or refuges;

(iii) Wetlands the destruction or alteration of which would affect detrimentally natural drainage characteristics, sedimentation patterns, salinity distribution, flushing characteristics, current patterns, or other environmental characteristics;

(iv) Wetlands which are significant in shielding other areas from wave action, erosion, or storm damage. Such wetlands are often associated with barrier beaches, islands, reefs and bars;

(v) Wetlands which serve as valuable storage areas for storm and flood waters;

(vi) Wetlands which are ground water discharge areas that maintain minimum baseflows important to aquatic resources and those which are prime natural recharge areas;

(vii) Wetlands which serve significant water purification functions; and

(viii) Wetlands which are unique in nature or scarce in quantity to the region or local area.

(3) Although a particular alteration of a wetland may constitute a minor change, the cumulative effect of numerous piecemeal changes can result in a major impairment of wetland resources. Thus, the particular wetland site for which an application is made will be evaluated with the recognition that it may be part of a complete and interrelated wetland area. In addition, the district engineer may undertake, where appropriate, reviews of particular wetland areas in consultation with the Regional Director of the U. S. Fish and Wildlife Service, the Regional Director of the National Marine Fisheries Service of the National Oceanic and Atmospheric Administration, the Regional Administrator of the Environmental Protection Agency, the local representative of the Soil Conservation Service of the Department of Agriculture, and the head of the appropriate state agency to assess the cumulative effect of activities in such areas.

(4) No permit will be granted which involves the alteration of wetlands identified as important by paragraph (b)(2) of this section or because of provisions of paragraph (b)(3), of this section unless the district engineer concludes, on the basis of the analysis required in paragraph (a) of this section, that the benefits of the proposed alteration outweigh the damage to the wetlands resource. In evaluating whether a particular discharge activity should be permitted, the district engineer shall apply the section 404(b)(1) guidelines (40 CFR part 230.10(a) (1), (2), (3)).

(5) In addition to the policies expressed in this subpart, the Congressional policy expressed in the Estuary Protection Act, Pub. L. 90-454, and state regulatory laws or programs for classification and protection of wetlands will be considered.

(c) *Fish and wildlife.* In accordance with the Fish and Wildlife Coordination Act (paragraph 320.3(e) of this section) district engineers will consult with the Regional Director, U.S. Fish and Wildlife Service, the Regional Director, National Marine Fisheries Service, and the head of the agency responsible for fish and wildlife for the state in which work is to be performed, with a view to the conservation of wildlife resources by prevention of their direct and indirect loss and damage due to the activity proposed in a permit application. The Army will give full consideration to the views of those agencies on fish and wildlife matters in deciding on the issuance, denial, or conditioning of individual or general permits.

(d) *Water quality.* Applications for permits for activities which may adversely affect the quality of waters of the United States will be evaluated for compliance with applicable effluent limitations and water quality standards, during the construction and subsequent operation of the proposed activity. The evaluation should include the consideration of both point and non-point sources of pollution. It should be noted, however, that the Clean Water Act assigns responsibility for control of non-point sources of pollution to the states. Certification of compliance with applicable effluent limitations and water quality standards required under provisions of section 401 of the Clean Water Act will be considered conclusive with respect to water quality considerations unless the Regional Administrator, Environmental Protection Agency (EPA), advises of other water quality aspects to be taken into consideration.

(e) *Historic, cultural, scenic, and recreational values.* Applications for DA permits may involve areas which possess recognized historic, cultural, scenic, conservation, recreational or similar values. Full evaluation of the general public interest requires that due consideration be given to the effect which the proposed structure or activity may have on values such as those associated with wild and scenic rivers, historic properties and National Landmarks, National Rivers, National Wilderness Areas, National Seashores, National Recreation Areas, National Lakeshores, National Parks, National Monuments, estuarine and marine sanctuaries, archeological resources, including Indian religious or cultural sites, and such other areas as may be established under federal or state law for similar and related purposes. Recognition of those values is often reflected by state, regional, or local land use classifications, or by similar federal controls or policies. Action on permit applications should, insofar as possible, be consistent with, and avoid significant adverse effects on the values or purposes for which those classifications, controls, or policies were established.

(f) *Effects on limits of the territorial sea.* Structures or work affecting coastal waters may modify the coast line or base line from which the territorial sea is measured for purposes of the Submerged Lands Act and international law. Generally, the coast line or base line is the line of ordinary low water on the mainland; however, there are exceptions where there are islands or lowtide elevations offshore (the Submerged Lands Act, 43 U.S.C. 1301(a) and *United States* v. *California*, 381 U.S.C. 139 (1965), 382 U.S. 448 (1966)). Applications for structures or work affecting coastal waters will therefore be reviewed specifically to determine whether the coast line or base line might be altered. If it is determined that such a change might occur, coordination with the Attorney General and the Solicitor of the Department of the Interior is required before final action is taken. The district engineer will submit a description of the proposed work and a copy of the plans to the Solicitor, Department of the Interior, Washington, DC 20240, and request his comments concerning the effects of the proposed work on the outer continental rights of the United States. These comments will be included in the administrative record of the application. After completion of standard processing procedures, the record will be forwarded to the Chief of Engineers. The decision on the application will be made by the Secretary of the Army after coordination with the Attorney General.

(g) *Consideration of property ownership.* Authorization of work or structures by DA does not convey a property right, nor authorize any injury to property or invasion of other rights.

(1) An inherent aspect of property ownership is a right to reasonable private use. However, this right is subject to the rights and interests of the public in the navigable and other waters of the United States, including the federal navigation servitude and federal regulation for environmental protection.

(2) Because a landowner has the general right to protect property from erosion, applications to erect protective structures will usually receive favorable consideration. However, if the protective structure may cause damage to the property of others, adversely affect public health and safety, adversely impact floodplain or wetland values, or otherwise appears contrary to the public interest, the district engineer will so advise the applicant and inform him of possible alternative methods of protecting his property. Such advice will be given in terms of general guidance only so as not to compete with private engineering firms nor require undue use of government resources.

(3) A riparian landowner's general right of access to navigable waters of the United States is subject to the similar rights of access held by nearby riparian land owners and to the general public's right of navigation on the water surface. In the case of proposals which create undue interference with access to, or use of, navigable waters, the authorization will generally be denied.

(4) Where it is found that the work for which a permit is desired is in navigable waters of the United States (see 33 CFR part 329) and may interfere with an authorized federal project, the applicant should be apprised in writing of the fact and of the possibility that a federal project which may be constructed in the vicinity of the proposed work might necessitate its removal or reconstruction. The applicant should also be informed that the United States will in no case be liable for any damage or injury to the structures or work authorized by Sections 9 or 10 of the Rivers and Harbors Act of 1899 or by section 404 of the Clean Water Act which may be caused by, or result from, future operations undertaken by the Government for the conservation or improvement of navigation or for other purposes, and no claims or right to compensation will accrue from any such damage.

(5) Proposed activities in the area of a federal project which exists or is under construction will be evaluated to insure that they are compatible with the purposes of the project.

(6) A DA permit does not convey any property rights, either in real estate or material, or any exclusive privileges. Furthermore, a DA permit does not authorize any injury to property or invasion of rights or any infringement of Federal, state or local laws or regulations. The applicant's signature on an application is an affirmation that the applicant possesses or will possess the requisite property interest to undertake the activity proposed in the application. The district engineer will not enter into disputes but will remind the applicant of the above. The dispute over property ownership will not be a factor in the Corps public interest decision.

(h) *Activities affecting coastal zones.* Applications for DA permits for activities affecting the coastal zones of those states having a coastal zone management program approved by the Secretary of Commerce will be evaluated with respect to compliance with that program. No permit will be issued to a non-federal applicant until certification has been provided that the proposed activity complies with the coastal zone management program and the appropriate state agency has concurred with the certification or has waived its right to do so. However, a permit may be issued to a non-federal applicant if the Secretary of Commerce, on his own initiative or upon appeal by the applicant, finds that the proposed activity is consistent with the objectives of the Coastal Zone Management Act of 1972 or is otherwise necessary in the interest of national security. Federal agency and Indian tribe applicants for DA permits are responsible for complying with the Coastal Zone Management Act's directives for assuring that their activities directly affecting the coastal zone are consistent, to the maximum extent practicable, with approved state coastal zone management programs.

(i) *Activities in marine sanctuaries.* Applications for DA authorization for activities in a marine sanctuary established by the Secretary of Commerce under authority of section 302 of the Marine Protection, Research and Sanctuaries Act of 1972, as amended, will be evaluated for impact on the marine sanctuary. No permit will be issued until the applicant provides a certification from the Secretary of Commerce that the proposed activity is consistent with the purposes of Title III of the Marine Protection, Research and Sanctuaries Act of 1972, as amended, and can be carried out within the regulations promulgated by the Secretary of Commerce to control activities within the marine sanctuary.

(j) *Other Federal, state, or local requirements*

(1) Processing of an application for a DA permit normally will proceed concurrently with the processing of other required Federal, state, and/or local authorizations or certifications. Final action on the DA permit will normally not be delayed pending action by another Federal, state or local agency (See 33 CFR 325.2 (d)(4)). However, where the required Federal, state and/or local authorization and/or certification has been denied for activities which also require a Department of the Army permit before final action has been taken on the Army permit application, the district engineer will, after considering the likelihood of subsequent approval of the other authorization and/or certification and the time and effort remaining to complete processing the Army permit application, either immediately deny the Army permit without prejudice or continue processing the application to a conclusion. If the district engineer continues processing the application, he will conclude by either denying the permit as contrary to the public interest, or denying it without prejudice indicating that except for the other Federal, state or local denial the Army permit could, under appropriate conditions, be issued. Denial without prejudice means that there is no prejudice to the right of the applicant to reinstate processing of the Army permit application if subsequent approval is received from the appropriate Federal, state and/or local agency on a previously denied authorization and/or certification. Even if official certification and/or authorization is not required by state or federal law, but a state, regional, or local agency having jurisdiction or interest over the particular activity comments on the application, due consideration shall be given to those official views as a reflection of local factors of the public interest.

(2) The primary responsibility for determining zoning and land use matters rests with state, local and tribal governments. The district engineer will normally accept

decisions by such governments on those matters unless there are significant issues of overriding national importance. Such issues would include but are not necessarily limited to national security, navigation, national economic development, water quality, preservation of special aquatic areas, including wetlands, with significant interstate importance, and national energy needs. Whether a factor has overriding importance will depend on the degree of impact in an individual case.

(3) A proposed activity may result in conflicting comments from several agencies within the same state. Where a state has not designated a single responsible coordinating agency, district engineers will ask the Governor to express his views or to designate one state agency to represent the official state position in the particular case.

(4) In the absence of overriding national factors of the public interest that may be revealed during the evaluation of the permit application, a permit will generally be issued following receipt of a favorable state determination provided the concerns, policies, goals, and requirements as expressed in 33 CFR parts 320–324, and the applicable statutes have been considered and followed: e.g., the National Environmental Policy Act; the Fish and Wildlife Coordination Act; the Historical and Archeological Preservation Act; the National Historic Preservation Act; the Endangered Species Act; the Coastal Zone Management Act; the Marine Protection, Research and Sanctuaries Act of 1972, as amended; the Clean Water Act, the Archeological Resources Act, and the American Indian Religious Freedom Act. Similarly, a permit will generally be issued for Federal and Federally-authorized activities; another federal agency's determination to proceed is entitled to substantial consideration in the Corps' public interest review.

(5) Where general permits to avoid duplication are not practical, district engineers shall develop joint procedures with those local, state, and other Federal agencies having ongoing permit programs for activities also regulated by the Department of the Army. In such cases, applications for DA permits may be processed jointly with the state or other federal applications to an independent conclusion and decision by the district engineer and the appropriate Federal or state agency. (See 33 CFR 325.2(e).)

(6) The district engineer shall develop operating procedures for establishing official communications with Indian Tribes within the district. The procedures shall provide for appointment of a tribal representative who will receive all pertinent public notices, and respond to such notices with the official tribal position on the proposed activity. This procedure shall apply only to those tribes which accept this option. Any adopted operating procedures shall be distributed by public notice to inform the tribes of this option.

(k) *Safety of impoundment structures.* To insure that all impoundment structures are designed for safety, non-Federal applicants may be required to demonstrate that the structures comply with established state dam safety criteria or have been designed by qualified persons and, in appropriate cases, that the design has been independently reviewed (and modified as the review would indicate) by similarly qualified persons.

(l) *Floodplain management*

(1) Floodplains possess significant natural values and carry out numerous functions important to the public interest. These include:

(i) Water resources values (natural moderation of floods, water quality maintenance, and groundwater recharge);

(ii) Living resource values (fish, wildlife, and plant resources);

(iii) Cultural resource values (open space, natural beauty, scientific study, outdoor education, and recreation); and

(iv) Cultivated resource values (agriculture, aquaculture, and forestry).

(2) Although a particular alteration to a floodplain may constitute a minor change, the cumulative impact of such changes may result in a significant degradation of floodplain values and functions and in increased potential for harm to upstream and downstream activities. In accordance with the requirements of Executive Order 11988, district engineers, as part of their public interest review, should avoid to the extent practicable, long and short term significant adverse impacts associated with the occupancy and modification of floodplains, as well as the direct and indirect support of floodplain development whenever there is a practicable alternative. For those activities which in the public interest must occur in or impact upon floodplains, the district engineer shall ensure, to the maximum extent practicable, that the impacts of potential flooding on human health, safety, and welfare are minimized, the risks of flood losses are minimized, and, whenever practicable the natural and beneficial values served by floodplains are restored and preserved.

(3) In accordance with Executive Order 11988, the district engineer should avoid authorizing floodplain developments whenever practicable alternatives exist outside the floodplain. If there are no such practicable alternatives, the district engineer shall consider, as a means of mitigation, alternatives within the floodplain which will lessen any significant adverse impact to the floodplain.

(m) *Water supply and conservation.* Water is an essential resource, basic to human survival, economic growth, and the natural environment. Water conservation requires the efficient use of water resources in all actions which involve the significant use of water or that significantly affect the availability of water for alternative uses including opportunities to reduce demand and improve efficiency in order to minimize new supply requirements. Actions affecting water quantities are subject to Congressional policy as stated in section 101(g) of the Clean Water Act which provides that the authority of states to allocate water quantities shall not be superseded, abrogated, or otherwise impaired.

(n) *Energy conservation and development.* Energy conservation and development are major national objectives. District engineers will give high priority to the processing of permit actions involving energy projects.

(o) *Navigation*

(1) Section 11 of the Rivers and Harbors Act of 1899 authorized establishment of harbor lines shoreward of which no individual permits were required. Because harbor lines were established on the basis of navigation impacts only, the Corps of Engineers published a regulation on 27 May 1970 (33 CFR 209.150) which declared that permits would thereafter be required for activities shoreward of the harbor lines. Review of applications would be based on a full public interest evaluation and harbor lines would serve as guidance for assessing navigation impacts. Accordingly, activities constructed shoreward of harbor lines prior to 27 May 1970 do not require specific authorization.

(2) The policy of considering harbor lines as guidance for assessing impacts on navigation continues.

(3) Protection of navigation in all navigable waters of the United States continues to be a primary concern of the federal government.

(4) District engineers should protect navigational and anchorage interests in connection with the NPDES program by recommending to EPA or to the state, if the program has been delegated, that a permit be denied unless appropriate conditions can be included to avoid any substantial impairment of navigation and anchorage.

(p) *Environmental benefits.* Some activities that require Department of the Army permits result in beneficial effects to the quality of the environment. The district engineer will weigh these benefits as well as environmental detriments along with other factors of the public interest.

(q) *Economics.* When private enterprise makes application for a permit, it will generally be assumed that appropriate economic evaluations have been completed, the proposal is economically viable, and is needed in the market place. However, the district engineer in appropriate cases, may make an independent review of the need for the project from the perspective of the overall public interest. The economic benefits of many projects are important to the local community and contribute to needed improvements in the local economic base, affecting such factors as employment, tax revenues, community cohesion, community services, and property values. Many projects also contribute to the National Economic Development (NED), (i.e., the increase in the net value of the national output of goods and services).

(r) *Mitigation.* (FOOTNOTE 1: This is a general statement of mitigation policy which applies to all Corps of Engineers regulatory authorities covered by these regulations (33 CFR parts 320–330). It is not a substitute for the mitigation requirements necessary to ensure that a permit action under section 404 of the Clean Water Act complies with the section 404(b)(1) Guidelines. There is currently an interagency Working Group formed to develop guidance on implementing mitigation requirements of the Guidelines.)

(1) Mitigation is an important aspect of the review and balancing process on many Department of the Army permit applications. Consideration of mitigation will occur throughout the permit application review process and includes avoiding, minimizing, rectifying, reducing, or compensating for resource losses. Losses will be avoided to the extent practicable. Compensation may occur on-site or at an off-site location. Mitigation requirements generally fall into three categories.

(i) Project modifications to minimize adverse project impacts should be discussed with the applicant at pre-application meetings and during application processing. As a result of these discussions and as the district engineer's evaluation proceeds, the district engineer may require minor project modifications. Minor project modifications are those that are considered feasible (cost, constructability, etc.) to the applicant and that, if adopted, will result in a project that generally meets the applicant's purpose and need. Such modifications can include reductions in scope and size; changes in construction methods, materials or timing; and operation and maintenance practices or other similar modifications that reflect a sensitivity to environmental quality within the context of the work proposed. For example, erosion control features could be required on a fill project to reduce sedimentation impacts or a pier could be reoriented to minimize navigational problems even though those projects may satisfy all legal requirements (paragraph (r)(1)(ii) of this section) and the public interest review test (paragraph (r)(1)(iii) of this section) without such modifications.

(ii) Further mitigation measures may be required to satisfy legal requirements. For Section 404 applications, mitigation shall be required to ensure that the project complies with the 404(b)(1) Guidelines. Some mitigation measures are enumerated at 40 CFR 230.70 through 40 CFR 230.77 (Subpart H of the 404(b)(1) Guidelines).

(iii) Mitigation measures in addition to those under paragraphs (r)(1) (i) and (ii) of this section may be required as a result of the public interest review process. (See 33 CFR 325.4(a).) Mitigation should be developed and incorporated within the public interest review process to the extent that the mitigation is found by the district engineer to be reasonable and justified. Only those measures required to ensure that the project is not contrary to the public interest may be required under this subparagraph.

(2) All compensatory mitigation will be for significant resource losses which are specifically identifiable, reasonably likely to occur, and of importance to the human or aquatic environment. Also, all mitigation will be directly related to the impacts of the proposal, appropriate to the scope and degree of those impacts, and reasonably enforceable. District engineers will require all forms of mitigation, including compensatory mitigation, only as provided in paragraphs (r)(1) (i) through (iii) of this section. Additional mitigation may be added at the applicants' request.

PART 321 PERMITS FOR DAMS AND DIKES IN NAVIGABLE WATERS OF THE UNITED STATES

AUTHORITY: 33 U.S.C. 401

Section 321.1 General

This regulation prescribes, in addition to the general policies of 33 CFR part 320 and procedures of 33 CFR part 325, those special policies, practices, and procedures to be followed by the Corps of Engineers in connection with the review of applications for Department of the Army (DA) permits to authorize the construction of a dike or dam in a navigable water of the United States pursuant to section 9 of the Rivers and Harbors Act of 1899 (33 U.S.C. 401). See 33 CFR 320.2(a). Dams and dikes in navigable waters of the United States also require DA permits under section 404 of the Clean Water Act, as amended (33 U.S.C. 1344). Applicants for DA permits under this part should also refer to 33 CFR part 323 to satisfy the requirements of section 404.

Section 321.2 Definitions

For the purpose of this regulation, the following terms are defined:

(a) The term *navigable waters of the United States* means those waters of the United States that are subject to the ebb and flow of the tide shoreward to the mean high water mark and/or are presently used, or have been used in the past, or may be susceptible to use to transport interstate or foreign commerce. See 33 CFR part 329 for a more complete definition of this term.

(b) The term *dike or dam* means, for the purposes of section 9, any impoundment structure that completely spans a navigable water of the United States and that may obstruct interstate waterborne commerce. The term does not include a weir. Weirs are regulated pursuant to section 10 of the Rivers and Harbors Act of 1899. (See 33 CFR part 322.)

Section 321.3 Special policies and procedures

The following additional special policies and procedures shall be applicable to the evaluation of permit applications under this regulation:

(a) The Assistant Secretary of the Army (Civil Works) will decide whether DA authorization for a dam or dike in an interstate navigable water of the United States will be issued, since this authority has not been delegated to the Chief of Engineers. The conditions to be imposed in any instrument of authorization will be recommended by the district engineer when forwarding the report to the Assistant Secretary of the Army (Civil Works), through the Chief of Engineers.

(b) District engineers are authorized to decide whether DA authorization for a dam or dike in an intrastate navigable water of the United States will be issued (see 33 CFR 325.8).

(c) Processing a DA application under section 9 will not be completed until the approval of the United States Congress has been obtained if the navigable water of the United States is an interstate waterbody, or until the approval of the appropriate state legislature has been obtained if the navigable water of the United States is an intrastate waterbody (i.e., the navigable portion of the navigable water of the United States is solely within the boundaries of one state). The district engineer, upon receipt of such an application, will notify the applicant that the consent of Congress or the state legislature must be obtained before a permit can be issued.

PART 322
PERMITS FOR STRUCTURES OR WORK IN OR AFFECTING NAVIGABLE WATERS OF THE UNITED STATES

AUTHORITY: 33 U.S.C. 403

Section 322.1 General

This regulation prescribes, in addition to the general policies of 33 CFR part 320 and procedures of 33 CFR part 325, those special policies, practices, and procedures to be followed by the Corps of Engineers in connection with the review of applications for Department of the Army (DA) permits to authorize certain structures or work in or affecting navigable waters of the United States pursuant to section 10 of the Rivers and Harbors Act of 1899 (33 U.S.C. 403) (hereinafter referred to as section 10). See 33 CFR 320.2(b). Certain structures or work in or affecting navigable waters of the United States are also regulated under other authorities of the DA. These include discharges of dredged or fill material into waters of the United States, including the territorial seas, pursuant to section 404 of the Clean Water Act (33 U.S.C. 1344; see 33 CFR part 323) and the transportation of dredged material by vessel for purposes of dumping in ocean waters, including the territorial seas, pursuant to section 103 of the Marine Protection, Research and Sanctuaries Act of 1972, as amended (33 U.S.C. 1413; see 33 CFR part 324). A DA permit will also be required under these additional authorities if they are applicable to structures or work in or affecting navigable waters of the United States. Applicants for DA permits under this part should refer to the other cited authorities and implementing regulations for these additional permit requirements to determine whether they also are applicable to their proposed activities.

Section 322.2 Definitions

For the purpose of this regulation, the following terms are defined:

(a) The term *navigable waters of the United States* and all other terms relating to the geographic scope of jurisdiction are defined at 33 CFR part 329. Generally, they are those waters of the United States that are subject to the ebb and flow of the tide shoreward to the mean high water mark, and/or are presently used, or have been used in the past, or may be susceptible to use to transport interstate or foreign commerce.

(b) The term *structure* shall include, without limitation, any pier, boat dock, boat ramp, wharf, dolphin, weir, boom, breakwater, bulkhead, revetment, riprap, jetty, artificial island, artificial reef, permanent mooring structure, power transmission line, permanently moored floating vessel, piling, aid to navigation, or any other obstacle or obstruction.

(c) The term *work* shall include, without limitation, any dredging or disposal of dredged material, excavation, filling, or other modification of a navigable water of the United States.

(d) The term *letter of permission* means a type of individual permit issued in accordance with the abbreviated procedures of 33 CFR 325.2(e).

(e) The term *individual permit* means a DA authorization that is issued following a case-by-case evaluation of a specific structure or work in accordance with the procedures of this regulation and 33 CFR part 325, and a determination that the proposed structure or work is in the public interest pursuant to 33 CFR part 320.

(f) The term *general permit* means a DA authorization that is issued on a nationwide or regional basis for a category or categories of activities when:

(1) Those activities are substantially similar in nature and cause only minimal individual and cumulative environmental impacts; or

(2) The general permit would result in avoiding unnecessary duplication of the regulatory control exercised by another Federal, state, or local agency provided it has been determined that the environmental consequences of the action are individually and cumulatively minimal. (See 33 CFR 325.2(e) and 33 CFR part 330.)

(g) The term *artificial reef* means a structure which is constructed or placed in the navigable waters of the United States or in the waters overlying the outer continental shelf for the purpose of enhancing fishery resources and commercial and recreational fishing opportunities. The term does not include activities or structures such as wing deflectors, bank stabilization, grade stabilization structures, or low flow key ways, all of which may be useful to enhance fisheries resources.

Section 322.3 Activities requiring permits

(a) *General.* DA permits are required under section 10 for structures and/or work in or affecting navigable waters of the United States except as otherwise provided in § 322.4 below. Certain activities specified in 33 CFR part 330 are permitted by that regulation ("nationwide general permits"). Other activities may be authorized by district or division engineers on a regional basis ("regional general permits"). If an activity is not exempted by § 322.4 of this part or authorized by a general permit, an individual section 10 permit will be required for the proposed activity. Structures or work are in navigable waters of the United States if they are within limits defined in 33 CFR part 329. Structures or work outside these limits are subject to the provisions of law cited in paragraph (a) of this section, if these structures or work affect the course, location, or condition of the waterbody in such a manner as to impact on its navigable capacity. For purposes of a section 10 permit, a tunnel or other structure or work under or over a navigable water of the United States is considered to have an impact on the navigable capacity of the waterbody.

(b) *Outer continental shelf.* DA permits are required for the construction of artificial islands, installations, and other devices on the seabed, to the seaward limit of the outer continental shelf, pursuant to section 4(f) of the Outer Continental Shelf Lands Act as amended. (See 33 CFR 320.2(b).)

(c) *Activities of Federal agencies*

(1) Except as specifically provided in this paragraph, activities of the type described in paragraphs (a) and (b) of this section, done by or on behalf of any Federal agency are subject to the authorization procedures of these regulations. Work or structures in or affecting navigable waters of the United States that are part of the civil works activities of the Corps of Engineers, unless covered by a nationwide or regional general permit issued pursuant to these regulations, are subject to the procedures of separate regulations. Agreement for construction or engineering services performed for other agencies by the Corps of Engineers does not constitute authorization under this regulation. Division and district engineers will therefore advise Federal agencies accordingly, and cooperate to the fullest extent in expediting the processing of their applications.

(2) Congress has delegated to the Secretary of the Army in section 10 the duty to authorize or prohibit certain work or structures in navigable waters of the United States, upon recommendation of the Chief of Engineers. The general legislation by which Federal agencies are enpowered to act generally is not considered to be sufficient authorization by Congress to satisfy the purposes of section 10. If an agency asserts that it has Congressional authorization meeting the test of section 10 or would otherwise be exempt from the provisions of section 10, the legislative history and/or provisions of the Act should clearly demonstrate that Congress was approving the exact location and plans from which Congress could have considered the effect on navigable waters of the United States or that Congress intended to exempt that agency from the

requirements of section 10. Very often such legislation reserves final approval of plans or construction for the Chief of Engineers. In such cases evaluation and authorization under this regulation are limited by the intent of the statutory language involved.

(3) The policy provisions set out in 33 CFR 320.4(j) relating to state or local certifications and/or authorizations, do not apply to work or structures undertaken by Federal agencies, except where compliance with non-Federal authorization is required by Federal law or Executive policy, e.g., section 313 and section 401 of the Clean Water Act.

Section 322.4 Activities not requiring permits

(a) Activities that were commenced or completed shoreward of established Federal harbor lines before May 27, 1970 (see 33 CFR 320.4(o)) do not require section 10 permits; however, if those activities involve the discharge of dredged or fill material into waters of the United States after October 18, 1972, a section 404 permit is required. (See 33 CFR part 323.)

(b) Pursuant to section 154 of the Water Resource Development Act of 1976 (Pub. L. 94-587), Department of the Army permits are not required under section 10 to construct wharves and piers in any waterbody, located entirely within one state, that is a navigable water of the United States solely on the basis of its historical use to transport interstate commerce.

Section 322.5 Special policies

The Secretary of the Army has delegated to the Chief of Engineers the authority to issue or deny section 10 permits. The following additional special policies and procedures shall also be applicable to the evaluation of permit applications under this regulation.

(a) *General.* DA permits are required for structures or work in or affecting navigable waters of the United States. However, certain structures or work specified in 33 CFR part 330 are permitted by that regulation. If a structure or work is not permitted by that regulation, an individual or regional section 10 permit will be required.

(b) *Artificial Reefs*

(1) When considering an application for an artificial reef, as defined in 33 CFR 322.2(g), the district engineer will review the applicant's provisions for siting, constructing, monitoring, operating, maintaining, and managing the proposed artificial reef and shall determine if those provisions are consistent with the following standards:

(i) The enhancement of fishery resources to the maximum extent practicable;

(ii) The facilitation of access and utilization by United States recreational and commercial fishermen;

(iii) The minimization of conflicts among competing uses of the navigable waters or waters overlying the outer continental shelf and of the resources in such waters;

(iv) The minimization of environmental risks and risks to personal health and property;

(v) Generally accepted principles of international law; and

(vi) the prevention of any unreasonable obstructions to navigation. If the district engineer decides that the applicant's provisions are not consistent with these standards, he shall deny the permit. If the district engineer decides that the provisions are consistent with these standards, and if he decides to issue the permit after the public interest review, he shall make the provisions part of the permit.

(2) In addition, the district engineer will consider the National Artificial Reef Plan developed pursuant to section 204 of the National Fishing Enhancement Act of 1984, and if he decides to issue the permit, will notify the Secretary of Commerce of any need to deviate from that plan.

(3) The district engineer will comply with all coordination provisions required by a written agreement between the DOD and the Federal agencies relative to artificial reefs. In addition, if the district engineer decides that further consultation beyond the normal public commenting process is required to evaluate fully the proposed artificial reef, he may initiate such consultation with any Federal agency, state or local government, or other interested party.

(4) The district engineer will issue a permit for the proposed artificial reef only if the applicant demonstrates, to the district engineer's satisfaction, that the title to the artificial reef construction material is unambiguous, that responsibility for maintenance of the reef is clearly established, and that he has the financial ability to assume liability for all damages that may arise with respect to the proposed artificial reef. A demonstration of financial responsibility might include evidence of insurance, sponsorship, or available assets.

(i) A person to whom a permit is issued in accordance with these regulations and any insurer of that person shall not be liable for damages caused by activities required to be undertaken under any terms and conditions of the permit, if the permittee is in compliance with such terms and conditions.

(ii) A person to whom a permit is issued in accordance with these regulations and any insurer of that person shall be liable, to the extent determined under applicable law, for damages to which paragraph (i) does not apply.

(iii) Any person who has transferred title to artificial reef construction materials to a person to whom a permit is issued in accordance with these regulations shall not be liable for damages arising from the use of such materials in an artificial reef, if such materials meet applicable requirements of the plan published under section 204 of the National Artificial Reef Plan, and are not otherwise defective at the time title is transferred.

(c) *Non-Federal dredging for navigation*

(1) The benefits which an authorized Federal navigation project are intended to produce will often require similar and related operations by non-Federal agencies (e.g., dredging access channels to docks and berthing facilities or deepening such channels to correspond to the Federal project depth). These non-Federal activities will be considered by Corps of Engineers officials in planning the construction and maintenance of Federal navigation projects and, to the maximum practical extent, will be coordinated with interested Federal, state, regional and local agencies and the general public simultaneously with the associated Federal projects. Non-Federal activities which are not so coordinated will be individually evaluated in accordance with these regulations. In evaluating the public interest in connection with applications for permits for such coordinated operations, equal treatment will be accorded to the fullest extent possible to both Federal and non-Federal operations. Permits for non-Federal dredging operations will normally contain conditions requiring the permittee to comply with the same practices or requirements utilized in connection with related Federal dredging operations with respect to such matters as turbidity, water quality, containment of material, nature and location of approved spoil disposal areas (non-Federal use of Federal contained disposal areas will be in accordance with laws authorizing such areas and regulations governing their use), extent and period of dredging, and other factors relating to protection of environmental and ecological values.

(2) A permit for the dredging of a channel, slip, or other such project for navigation may also authorize the periodic maintenance dredging of the project. Authorization procedures and limitations for maintenance dredging shall be as prescribed in 33 CFR 325.6(e). The permit will require the permittee to give advance notice to the district engineer each time maintenance dredging is to be performed. Where the maintenance dredging involves the discharge of dredged material

into waters of the United States or the transportation of dredged material for the purpose of dumping it in ocean waters, the procedures in 33 CFR parts 323 and 324 respectively shall also be followed.

(d) *Structures for small boats*

(1) In the absence of overriding public interest, favorable consideration will generally be given to applications from riparian owners for permits for piers, boat docks, moorings, platforms and similar structures for small boats. Particular attention will be given to the location and general design of such structures to prevent possible obstructions to navigation with respect to both the public's use of the waterway and the neighboring proprietors' access to the waterway. Obstructions can result from both the existence of the structure, particularly in conjunction with other similar facilities in the immediate vicinity, and from its inability to withstand wave action or other forces which can be expected. District engineers will inform applicants of the hazards involved and encourage safety in location, design, and operation. District engineers will encourage cooperative or group use facilities in lieu of individual proprietary use facilities.

(2) Floating structures for small recreational boats or other recreational purposes in lakes controlled by the Corps of Engineers under a resource manager are normally subject to permit authorities cited in § 322.3, of this section, when those waters are regarded as navigable waters of the United States. However, such structures will not be authorized under this regulation but will be regulated under applicable regulations of the Chief of Engineers published in 36 CFR 327.19 if the land surrounding those lakes is under complete Federal ownership. District engineers will delineate those portions of the navigable waters of the United States where this provision is applicable and post notices of this designation in the vicinity of the lake resource manager's office.

(e) *Aids to navigation.* The placing of fixed and floating aids to navigation in a navigable water of the United States is within the purview of Section 10 of the Rivers and Harbors Act of 1899. Furthermore, these aids are of particular interest to the U.S. Coast Guard because of its control of marking, lighting and standardization of such navigation aids. A Section 10 nationwide permit has been issued for such aids provided they are approved by, and installed in accordance with the requirements of the U.S. Coast Guard (33 CFR 330.5(a)(1)). Electrical service cables to such aids are not included in the nationwide permit (an individual or regional Section 10 permit will be required).

(f) *Outer continental shelf.* Artificial islands, installations, and other devices located on the seabed, to the seaward limit of the outer continental shelf, are subject to the standard permit procedures of this regulation. Where the islands, installations and other devices are to be constructed on lands which are under mineral lease from the Mineral Management Service, Department of the Interior, that agency, in cooperation with other federal agencies, fully evaluates the potential effect of the leasing program on the total environment. Accordingly, the decision whether to issue a permit on lands which are under mineral lease from the Department of the Interior will be limited to an evaluation of the impact of the proposed work on navigation and national security. The public notice will so identify the criteria.

(g) *Canals and other artificial waterways connected to navigable waters of the United States.* A canal or similar artificial waterway is subject to the regulatory authorities discussed in § 322.3, of this part, if it constitutes a navigable water of the United States, or if it is connected to navigable waters of the United States in a manner which affects their course, location, condition, or capacity, or if at some point in its construction or operation it results in an effect on the course, location, condition, or capacity of navigable waters of the United States. In all cases the connection to navigable waters of the United States requires a permit. Where the canal itself constitutes a navigable water of the United States, evaluation of the permit application and further exercise of regulatory authority will be in accordance with the standard procedures of these regulations. For all other canals, the exercise of regulatory authority is restricted to those activities which affect the course, location, condition, or capacity of the navigable waters of the United States. The district engineer will consider, for applications for canal work, a proposed plan of the entire development and the location and description of anticipated docks, piers and other similar structures which will be placed in the canal.

(h) *Facilities at the borders of the United States*

(1) The construction, operation, maintenance, or connection of facilities at the borders of the United States are subject to Executive control and must be authorized by the President, Secretary of State, or other delegated official.

(2) Applications for permits for the construction, operation, maintenance, or connection at the borders of the United States of facilities for the transmission of electric energy between the United States and a foreign country, or for the exportation or importation of natural gas to or from a foreign country, must be made to the Secretary of Energy. (Executive Order 10485, September 3, 1953, 16 U.S.C. 824(a)(e), 15 U.S.C. 717(b), as amended by Executive Order 12038, February 3, 1978, and 18 CFR parts 32 and 153).

(3) Applications for the landing or operation of submarine cables must be made to the Federal Communications Commission. (Executive Order 10530, May 10, 1954, 47 U.S.C. 34 to 39, and 47 CFR 1.766).

(4) The Secretary of State is to receive applications for permits for the construction, connection, operation, or maintenance, at the borders of the United States, of pipelines, conveyor belts, and similar facilities for the exportation or importation of petroleum products, coals, minerals, or other products to or from a foreign country; facilities for the exportation or importation of water or sewage to or from a foreign country; and monorails, aerial cable cars, aerial tramways, and similar facilities for the transportation of persons and/or things, to or from a foreign country. (Executive Order 11423, August 16, 1968).

(5) A DA permit under section 10 of the Rivers and Harbors Act of 1899 is also required for all of the above facilities which affect the navigable waters of the United States, but in each case in which a permit has been issued as provided above, the district engineer, in evaluating the general public interest, may consider the basic existence and operation of the facility to have been primarily examined and permitted as provided by the Executive Orders. Furthermore, in those cases where the construction, maintenance, or operation at the above facilities involves the discharge of dredged or fill material in waters of the United States or the transportation of dredged material for the purpose of dumping it into ocean waters, appropriate DA authorizations under section 404 of the Clean Water Act or under section 103 of the Marine Protection, Research and Sanctuaries Act of 1972, as amended, are also required. (See 33 CFR parts 323 and 324.)

(i) *Power transmission lines*

(1) Permits under section 10 of the Rivers and Harbors Act of 1899 are required for power transmission lines crossing navigable waters of the United States unless those lines are part of a water power project subject to the regulatory authorities of the Department of Energy under the Federal Power Act of 1920. If an application is received for a permit for lines which are part of such a water power project, the applicant will be instructed to submit the application to the Department of Energy. If the lines are not

part of such a water power project, the application will be processed in accordance with the procedures of these regulations.

(2) The following minimum clearances are required for aerial electric power transmission lines crossing navigable waters of the United States. These clearances are related to the clearances over the navigable channel provided by existing fixed bridges, or the clearances which would be required by the U.S. Coast Guard for new fixed bridges, in the vicinity of the proposed power line crossing. The clearances are based on the low point of the line under conditions which produce the greatest sag, taking into consideration temperature, load, wind, length or span, and type of supports as outlined in the National Electrical Safety Code.

Nominal System Voltage (kV)	Minimum Additional Clearance (ft.) Above Clearance Required for Bridges
115 and below	20
138	22
161	24
230	26
350	30
500	35
700	42
750–765	45

(3) Clearances for communication lines, stream gaging cables, ferry cables, and other aerial crossings are usually required to be a minimum of ten feet above clearances required for bridges. Greater clearances will be required if the public interest so indicates.

(4) Corps of Engineer regulation ER 1110-2-4401 prescribes minimum vertical clearances for power and communication lines over Corps lake projects. In instances where both this regulation and ER 1110-2-4401 apply, the greater minimum clearance is required.

(j) *Seaplane operations*

(1) Structures in navigable waters of the United States associated with seaplane operations require DA permits, but close coordination with the Federal Aviation Administration (FAA), Department of Transportation, is required on such applications.

(2) The FAA must be notified by an applicant whenever he proposes to establish or operate a seaplane base. The FAA will study the proposal and advise the applicant, district engineer, and other interested parties as to the effects of the proposal on the use of airspace. The district engineer will, therefore, refer any objections regarding the effect of the proposal on the use of airspace to the FAA, and give due consideration to its recommendations when evaluating the general public interest.

(3) If the seaplane base would serve air carriers licensed by the Department of Transportation, the applicant must receive an airport operating certificate from the FAA. That certificate reflects a determination and conditions relating to the installation, operation, and maintenance of adequate air navigation facilities and safety equipment. Accordingly, the district engineer may, in evaluating the general public interest, consider such matters to have been primarily evaluated by the FAA.

(4) For regulations pertaining to seaplane landings at Corps of Engineers projects, see 36 CFR 327.4.

(k) *Foreign trade zones.* The Foreign Trade Zones Act (48 Stat. 998-1003, 19 U.S.C. 81a to 81u, as amended) authorizes the establishment of foreign-trade zones in or adjacent to United States ports of entry under terms of a grant and regulations prescribed by the Foreign-Trade Zones Board. Pertinent regulations are published at Title 15 of the Code of Federal Regulations, part 400. The Secretary of the Army is a member of the Board, and construction of a zone is under the supervision of the district engineer. Laws governing the navigable waters of the United States remain applicable to foreign-trade zones, including the general requirements of these regulations. Evaluation by a district engineer of a permit application may give recognition to the consideration by the Board of the general economic effects of the zone on local and foreign commerce, general location of wharves and facilities, and other factors pertinent to construction, operation, and maintenance of the zone.

(l) *Shipping safety fairways and anchorage areas.* DA permits are required for structures located within shipping safety fairways and anchorage areas established by the U.S. Coast Guard.

(1) The Department of the Army will grant no permits for the erection of structures in areas designated as fairways, except that district engineers may permit anchors and attendant cables or chains for floating or semisubmersible drilling rigs to be placed within a fairway provided the following conditions are met:

(i) The purpose of such anchors and attendant cables or chains as used in this section is to stabilize floating production facilities or semisubmersible drilling rigs which are located outside the boundaries of the fairway.

(ii) In water depths of 600 feet or less, the installation of anchors and attendant cables or chains within fairways must be temporary and shall be allowed to remain only 120 days. This period may be extended by the district engineer provided reasonable cause for such extension can be shown and the extension is otherwise justified. In water depths greater than 600 feet, time restrictions on anchors and attendant cables or chains located within a fairway, whether temporary or permanent, shall not apply.

(iii) Drilling rigs must be at least 500 feet from any fairway boundary or whatever distance necessary to insure that minimum clearance over an anchor line within a fairway will be 125 feet.

(iv) No anchor buoys or floats or related rigging will be allowed on the surface of the water or to a depth of 125 feet from the surface, within the fairway.

(v) Drilling rigs may not be placed closer than 2 nautical miles of any other drilling rig situated along a fairway boundary, and not closer than 3 nautical miles to any drilling rig located on the opposite side of the fairway.

(vi) The permittee must notify the district engineer, Bureau of Land Management, Mineral Management Service, U.S. Coast Guard, National Oceanic and Atmospheric Administration and the U.S. Navy Hydrographic Office of the approximate dates (commencement and completion) the anchors will be in place to insure maximum notification to mariners.

(vii) Navigation aids or danger markings must be installed as required by the U.S. Coast Guard.

(2) District engineers may grant permits for the erection of structures within an area designated as an anchorage area, but the number of structures will be limited by spacing, as follows: The center of a structure to be erected shall be not less than two (2) nautical miles from the center of any existing structure. In a drilling or production complex, associated structures shall be as close together as practicable having due consideration for the safety factors involved. A complex of associated structures, when connected by walkways, shall be considered one structure for the purpose of spacing. A vessel fixed in place by moorings and used in conjunction with the associated structures of a drilling or production complex, shall be considered an attendant vessel and its extent shall include its moorings. When a drilling or production complex includes an attendant vessel and the complex extends more than five hundred (500) yards from the center or the complex, a structure to be erected shall be not closer

than two (2) nautical miles from the near outer limit of the complex. An underwater completion installation in and anchorage area shall be considered a structure and shall be marked with a lighted buoy as approved by the United States Coast Guard.

[51 FR 41228, Nov. 13, 1986, as amended at 60 FR 44761, Aug. 29, 1995]

PART 323 PERMITS FOR DISCHARGES OF DREDGED OR FILL MATERIAL INTO WATERS OF THE UNITED STATES

AUTHORITY: 33 U.S.C. 1344

Section 323.1 General

This regulation prescribes, in addition to the general policies of 33 CFR part 320 and procedures of 33 CFR part 325, those special policies, practices, and procedures to be followed by the Corps of Engineers in connection with the review of applications for DA permits to authorize the discharge of dredged or fill material into waters of the United States pursuant to section 404 of the Clean Water Act (CWA) (33 U.S.C. 1344) (hereinafter referred to as section 404). (See 33 CFR 320.2(g).) Certain discharges of dredged or fill material into waters of the United States are also regulated under other authorities of the Department of the Army. These include dams and dikes in navigable waters of the United States pursuant to section 9 of the Rivers and Harbors Act of 1899 (33 U.S.C. 401; see 33 CFR part 321) and certain structures or work in or affecting navigable waters of the United States pursuant to section 10 of the Rivers and Harbors Act of 1899 (33 U.S.C. 403; see 33 CFR part 322). A DA permit will also be required under these additional authorities if they are applicable to activities involving discharges of dredged or fill material into waters of the United States. Applicants for DA permits under this part should refer to the other cited authorities and implementing regulations for these additional permit requirements to determine whether they also are applicable to their proposed activities.

Section 323.2 Definitions

For the purpose of this part, the following terms are defined:

(a) The term *waters of the United States* and all other terms relating to the geographic scope of jurisdiction are defined at 33 CFR part 328.

(b) The term *lake* means a standing body of open water that occurs in a natural depression fed by one or more streams from which a stream may flow, that occurs due to the widening or natural blockage or cutoff of a river or stream, or that occurs in an isolated natural depression that is not a part of a surface river or stream. The term also includes a standing body of open water created by artificially blocking or restricting the flow of a river, stream, or tidal area. As used in this regulation, the term does not include artificial lakes or ponds created by excavating and/or diking dry land to collect and retain water for such purposes as stock watering, irrigation, settling basins, cooling, or rice growing.

(c) The term *dredged material* means material that is excavated or dredged from waters of the United States.

(d) (1) Except as provided below in paragraph (d)(3), the term *discharge of dredged material* means any addition of dredged material into, including redeposit of dredged material other than incidental fallback within, the waters of the United States. The term includes, but is not limited to, the following:

(i) The addition of dredged material to a specified discharge site located in waters of the United States;

(ii) The runoff or overflow from a contained land or water disposal area; and

(iii) Any addition, including redeposit other than incidental fallback, of dredged material, including excavated material, into waters of the United States which is incidental to any activity, including mechanized landclearing, ditching, channelization, or other excavation.

(2) (i) The Corps and EPA regard the use of mechanized earth-moving equipment to conduct landclearing, ditching, channelization, in-stream mining or other earth-moving activity in waters of the United States as resulting in a discharge of dredged material unless project-specific evidence shows that the activity results in only incidental fallback. This paragraph (i) does not and is not intended to shift any burden in any administrative or judicial proceeding under the CWA.

(ii) *Incidental fallback* is the redeposit of small volumes of dredged material that is incidental to excavation activity in waters of the United States when such material falls back to substantially the same place as the initial removal. Examples of incidental fallback include soil that is disturbed when dirt is shoveled and the back-spill that comes off a bucket when such small volume of soil or dirt falls into substantially the same place from which it was initially removed.

(3) The term *discharge of dredged material* does not include the following:

(i) Discharges of pollutants into waters of the United States resulting from the onshore subsequent processing of dredged material that is extracted for any commercial use (other than fill). These discharges are subject to section 402 of the Clean Water Act even though the extraction and deposit of such material may require a permit from the Corps or applicable State section 404 program.

(ii) Activities that involve only the cutting or removing of vegetation above the ground (e.g., mowing, rotary cutting, and chainsawing) where the activity neither substantially disturbs the root system nor involves mechanized pushing, dragging, or other similar activities that redeposit excavated soil material.

(iii) Incidental fallback.

(4) Section 404 authorization is not required for the following:

(i) Any incidental addition, including redeposit, of dredged material associated with any activity that does not have or would not have the effect of destroying or degrading an area of waters of the United States as defined in paragraphs (d)(5) and (d)(6) of this section; however, this exception does not apply to any person preparing to undertake mechanized landclearing, ditching, channelization and other excavation activity in a water of the United States, which would result in a redeposit of dredged material, unless the person demonstrates to the satisfaction of the Corps, or EPA as appropriate, prior to commencing the activity involving the discharge, that the activity would not have the effect of destroying or degrading any area of waters of the United States, as defined in paragraphs (d)(5) and (d)(6) of this section. The person proposing to undertake mechanized landclearing, ditching, channelization or other excavation activity bears the burden of demonstrating that such activity would not destroy or degrade any area of waters of the United States.

(ii) Incidental movement of dredged material occurring during normal dredging operations, defined as dredging for navigation in *navigable waters of the United States,* as that term is defined in part 329 of this chapter, with proper authorization from the Congress and/or the Corps pursuant to part 322 of this Chapter; however, this exception is not applicable to dredging activities in wetlands, as that term is defined at § 328.3 of this Chapter.

(iii) Certain discharges, such as those associated with normal farming, silviculture, and ranching activities, are not prohibited by or otherwise subject to regulation under section 404. See 33 CFR 323.4 for discharges that do not required permits.

(5) For purposes of this section, an activity associated with a discharge of dredged material destroys an area of waters of the United States if it alters the area in such a way that it would no longer be a water of the United States.

NOTE: Unauthorized discharges into waters of the United States do not eliminate Clean Water Act jurisdiction, even where such unauthorized discharges have the effect of destroying waters of the United States.

(6) For purposes of this section, an activity associated with a discharge of dredged material degrades an area of waters of the United States if it has more than a *de minimis* (i.e., inconsequential) effect on the area by causing an identifiable individual or cumulative adverse effect on any aquatic function.

(e) The term *fill material* means any material used for the primary purpose of replacing an aquatic area with dry land or of changing the bottom elevation of an waterbody. The term does not include any pollutant discharged into the water primarily to dispose of waste, as that activity is regulated under section 402 of the Clean Water Act. See § 323.3(c) concerning the regulation of the placement of pilings in waters of the United States.

(f) The term *discharge of fill material* means the addition of fill material into waters of the United States. The term generally includes, without limitation, the following activities: Placement of fill that is necessary for the construction of any structure in a water of the United States; the building of any structure or impoundment requiring rock, sand, dirt, or other material for its construction; site-development fills for recreational, industrial, commercial, residential, and other uses; causeways or road fills; dams and dikes; artificial islands; property protection and/or reclamation devices such as riprap, groins, seawalls, breakwaters, and revetments; beach nourishment; levees; fill for structures such as sewage treatment facilities, intake and outfall pipes associated with power plants and subaqueous utility lines; and artificial reefs. The term does not include plowing, cultivating, seeding and harvesting for the production of food, fiber, and forest products (See § 323.4 for the definition of these terms). See § 323.3(c) concerning the regulation of the placement of pilings in waters of the United States.

(g) The term *individual permit* means a Department of the Army authorization that is issued following a case-by-case evaluation of a specific project involving the proposed discharge(s) in accordance with the procedures of this part and 33 CFR part 325 and a determination that the proposed discharge is in the public interest pursuant to 33 CFR part 320.

(h) The term *general permit* means a Department of the Army authorization that is issued on a nationwide or regional basis for a category or categories of activities when:

(1) Those activities are substantially similar in nature and cause only minimal individual and cumulative environmental impacts; or

(2) The general permit would result in avoiding unnecessary duplication of regulatory control exercised by another Federal, State, or local agency provided it has been determined that the environmental consequences of the action are individually and cumulatively minimal. (See 33 CFR 325.2(e) and 33 CFR part 330.)

[51 FR 41232, Nov. 13, 1986, as amended at 58 FR 45035, Aug. 25, 1993; 58 FR 48424, Sept. 15, 1993; 63 FR 25123, May 10, 1999; 66 FR 4574, Jan. 17, 2001; 66 FR 10367, Feb. 15, 2001]

Section 323.3 Discharges requiring permits

(a) *General.* Except as provided in § 323.4 of this part, DA permits will be required for the discharge of dredged or fill material into waters of the United States. Certain discharges specified in 33 CFR part 330 are permitted by that regulation ("nationwide permits"). Other discharges may be authorized by district or division engineers on a regional basis ("regional permits"). If a discharge of dredged or fill material is not exempted by § 323.4 of this part or permitted by 33 CFR part 330, an individual or regional section 404 permit will be required for the discharge of dredged or fill material into waters of the United States.

(b) *Activities of Federal agencies.* Discharges of dredged or fill material into waters of the United States done by or on behalf of any Federal agency, other than the Corps of Engineers (see 33 CFR 209.145), are subject to the authorization procedures of these regulations. Agreement for construction or engineering services performed for other agencies by the Corps of Engineers does not constitute authorization under the regulations. Division and district engineers will therefore advise Federal agencies and instrumentalities accordingly and cooperate to the fullest extent in expediting the processing of their applications.

(c) *Pilings*

(1) Placement of pilings in waters of the United States constitutes a discharge of fill material and requires a section 404 permit when such placement has or would have the effect of a discharge of fill material. Examples of such activities that have the effect of a discharge of fill material include, but are not limited to, the following: Projects where the pilings are so closely spaced that sedimentation rates would be increased; projects in which the pilings themselves effectively would replace the bottom of a waterbody; projects involving the placement of pilings that would reduce the reach or impair the flow or circulation of waters of the United States; and projects involving the placement of pilings which would result in the adverse alteration or elimination of aquatic functions.

(2) Placement of pilings in waters of the United States that does not have or would not have the effect of a discharge of fill material shall not require a section 404 permit. Placement of pilings for linear projects, such as bridges, elevated walkways, and powerline structures, generally does not have the effect of a discharge of fill material. Furthermore, placement of pilings in waters of the United States for piers, wharves, and an individual house on stilts generally does not have the effect of a discharge of fill material. All pilings, however, placed in the *navigable waters of the United States,* as that term is defined in part 329 of this chapter, require authorization under section 10 of the Rivers and Harbors Act of 1899 (see part 322 of this chapter).

[51 FR 41232, Nov. 13, 1986, as amended at 58 FR 45036, Aug. 25, 1993]

Section 323.4 Discharges not requiring permits

(a) *General.* Except as specified in paragraphs (b) and (c) of this section, any discharge of dredged or fill material that may result from any of the following activities is not prohibited by or otherwise subject to regulation under section 404:

(1) (i) Normal farming, silviculture and ranching activities such as plowing, seeding, cultivating, minor drainage, and harvesting for the production of food, fiber, and forest products, or upland soil and water conservation practices, as defined in paragraph (a)(1)(iii) of this section.

(ii) To fall under this exemption, the activities specified in paragraph (a)(1)(i) of this section must be part of an established (i.e., on-going) farming, silviculture, or ranching operation and must be in accordance with definitions in § 323.4(a)(1)(iii). Activities on areas lying fallow as part of a conventional rotational cycle are part of an established operation. Activities which bring an area into farming, silviculture, or ranching use are not part of an established operation. An operation ceases to be established when the area on which it was conducted has been converted to another use or has lain idle so long that modifications to the hydrological regime are necessary to resume operations. If an activity takes place outside the waters of the United

States, or if it does not involve a discharge, it does not need a section 404 permit, whether or not it is part of an established farming, silviculture, or ranching operation.

(iii) (A) *Cultivating* means physical methods of soil treatment employed within established farming, ranching and silviculture lands on farm, ranch, or forest crops to aid and improve their growth, quality or yield.

(B) *Harvesting* means physical measures employed directly upon farm, forest, or ranch crops within established agricultural and silvicultural lands to bring about their removal from farm, forest, or ranch land, but does not include the construction of farm, forest, or ranch roads.

(C) (1) *Minor Drainage* means:

(*i*) The discharge of dredged or fill material incidental to connecting upland drainage facilities to waters of the United States, adequate to effect the removal of excess soil moisture from upland croplands. (Construction and maintenance of upland (dryland) facilities, such as ditching and tiling, incidental to the planting, cultivating, protecting, or harvesting of crops, involve no discharge of dredged or fill material into waters of the United States, and as such never require a section 404 permit.);

(*ii*) The discharge of dredged or fill material for the purpose of installing ditching or other such water control facilities incidental to planting, cultivating, protecting, or harvesting of rice, cranberries or other wetland crop species, where these activities and the discharge occur in waters of the United States which are in established use for such agricultural and silvicultural wetland crop production;

(*iii*) The discharge of dredged or fill material for the purpose of manipulating the water levels of, or regulating the flow or distribution of water within, existing impoundments which have been constructed in accordance with applicable requirements of CWA, and which are in established use for the production of rice, cranberries, or other wetland crop species. (The provisions of paragraphs (a)(1)(iii) (C)(*1*)(*ii*) and (*iii*) of this section apply to areas that are in established use exclusively for wetland crop production as well as areas in established use for conventional wetland/non-wetland crop rotation (e.g., the rotations of rice and soybeans) where such rotation results in the cyclical or intermittent temporary dewatering of such areas.)

(*iv*) The discharges of dredged or fill material incidental to the emergency removal of sandbars, gravel bars, or other similar blockages which are formed during flood flows or other events, where such blockages close or constrict previously existing drainageways and, if not promptly removed, would result in damage to or loss of existing crops or would impair or prevent the plowing, seeding, harvesting or cultivating of crops on land in established use for crop production. Such removal does not include enlarging or extending the dimensions of, or changing the bottom elevations of, the affected drainageway as it existed prior to the formation of the blockage. Removal must be accomplished within one year of discovery of such blockages in order to be eligible for exemption.

(*2*) Minor drainage in waters of the U.S. is limited to drainage within areas that are part of an established farming or silviculture operation. It does not include drainage associated with the immediate or gradual conversion of a wetland to a non-wetland (e.g., wetland species to upland species not typically adapted to life in saturated soil conditions), or conversion from one wetland use to another (for example, silviculture to farming). In addition, minor drainage does not include the construction of any canal, ditch, dike or other waterway or structure which drains or otherwise significantly modifies a stream, lake, swamp, bog or any other wetland or aquatic area constituting waters of the United States. Any discharge of dredged or fill material into the waters of the United States incidental to the construction of any such structure or waterway requires a permit.

(D) *Plowing* means all forms of primary tillage, including moldboard, chisel, or wide-blade plowing, discing, harrowing and similar physical means utilized on farm, forest or ranch land for the breaking up, cutting, turning over, or stirring of soil to prepare it for the planting of crops. The term does not include the redistribution of soil, rock, sand, or other surficial materials in a manner which changes any area of the waters of the United States to dry land. For example, the redistribution of surface materials by blading, grading, or other means to fill in wetland areas is not plowing. Rock crushing activities which result in the loss of natural drainage characteristics, the reduction of water storage and recharge capabilities, or the overburden of natural water filtration capacities do not constitute plowing. Plowing as described above will never involve a discharge of dredged or fill material.

(E) *Seeding* means the sowing of seed and placement of seedlings to produce farm, ranch, or forest crops and includes the placement of soil beds for seeds or seedlings on established farm and forest lands.

(2) Maintenance, including emergency reconstruction of recently damaged parts, of currently serviceable structures such as dikes, dams, levees, groins, riprap, breakwaters, causeways, bridge abutments or approaches, and transportation structures. Maintenance does not include any modification that changes the character, scope, or size of the original fill design. Emergency reconstruction must occur within a reasonable period of time after damage occurs in order to qualify for this exemption.

(3) Construction or maintenance of farm or stock ponds or irrigation ditches, or the maintenance (but not construction) of drainage ditches. Discharges associated with siphons, pumps, headgates, wingwalls, weirs, diversion structures, and such other facilities as are appurtenant and functionally related to irrigation ditches are included in this exemption.

(4) Construction of temporary sedimentation basins on a construction site which does not include placement of fill material into waters of the U.S. The term "construction site" refers to any site involving the erection of buildings, roads, and other discrete structures and the installation of support facilities necessary for construction and utilization of such structures. The term also includes any other land areas which involve land-disturbing excavation activities, including quarrying or other mining activities, where an increase in the runoff of sediment is controlled through the use of temporary sedimentation basins.

(5) Any activity with respect to which a State has an approved program under section 208(b)(4) of the CWA which meets the requirements of sections 208(b)(4) (B) and (C).

(6) Construction or maintenance of farm roads, forest roads, or temporary roads for moving mining equipment, where such roads are constructed and maintained in accordance with best management practices (BMPs) to assure that flow and circulation patterns and chemical and biological characteristics of waters of the United States are not impaired, that the reach of the waters of the United States is not reduced, and that any adverse effect on the aquatic environment will be otherwise minimized. These BMPs which must be applied to satisfy this provision shall include those detailed BMPs described in the State's approved program description pursuant to the requirements of 40 CFR 233.22(i), and shall also include the following baseline provisions:

(i) Permanent roads (for farming or forestry activities), temporary access roads (for mining, forestry, or farm purposes) and skid trails (for logging) in waters of the U.S. shall be held to the minimum feasible number, width, and total length consistent with the purpose of specific farming, silvicultural or mining operations, and local topographic and climatic conditions;

(ii) All roads, temporary or permanent, shall be located sufficiently far from streams or other water bodies (except for portions of such roads which must cross water bodies) to minimize discharges of dredged or fill material into waters of the U.S.;

(iii) The road fill shall be bridged, culverted, or otherwise designed to prevent the restriction of expected flood flows;

(iv) The fill shall be properly stabilized and maintained during and following construction to prevent erosion;

(v) Discharges of dredged or fill material into waters of the United States to construct a road fill shall be made in a manner that minimizes the encroachment of trucks, tractors, bulldozers, or other heavy equipment within waters of the United States (including adjacent wetlands) that lie outside the lateral boundaries of the fill itself;

(vi) In designing, constructing, and maintaining roads, vegetative disturbance in the waters of the U.S. shall be kept to a minimum;

(vii) The design, construction and maintenance of the road crossing shall not disrupt the migration or other movement of those species of aquatic life inhabiting the water body;

(viii) Borrow material shall be taken from upland sources whenever feasible;

(ix) The discharge shall not take, or jeopardize the continued existence of, a threatened or endangered species as defined under the Endangered Species Act, or adversely modify or destroy the critical habitat of such species;

(x) Discharges into breeding and nesting areas for migratory waterfowl, spawning areas, and wetlands shall be avoided if practical alternatives exist;

(xi) The discharge shall not be located in the proximity of a public water supply intake;

(xii) The discharge shall not occur in areas of concentrated shellfish production;

(xiii) The discharge shall not occur in a component of the National Wild and Scenic River System;

(xiv) The discharge of material shall consist of suitable material free from toxic pollutants in toxic amounts; and

(xv) All temporary fills shall be removed in their entirety and the area restored to its original elevation.

(b) If any discharge of dredged or fill material resulting from the activities listed in paragraphs (a) (1) through (6) of this section contains any toxic pollutant listed under section 307 of the CWA such discharge shall be subject to any applicable toxic effluent standard or prohibition, and shall require a section 404 permit.

(c) Any discharge of dredged or fill material into waters of the United States incidental to any of the activities identified in paragraphs (a) (1) through (6) of this section must have a permit if it is part of an activity whose purpose is to convert an area of the waters of the United States into a use to which it was not previously subject, where the flow or circulation of waters of the United States may be impaired or the reach of such waters reduced. Where the proposed discharge will result in significant discernible alterations to flow or circulation, the presumption is that flow or circulation may be impaired by such alteration. For example, a permit will be required for the conversion of a cypress swamp to some other use or the conversion of a wetland from silvicultural to agricultural use when there is a discharge of dredged or fill material into waters of the United States in conjunction with construction of dikes, drainage ditches or other works or structures used to effect such conversion. A conversion of a section 404 wetland to a non-wetland is a change in use of an area of waters of the United States. A discharge which elevates the bottom of waters of the United States without converting it to dry land does not thereby reduce the reach of, but may alter the flow or circulation of, waters of the United States.

(d) Federal projects which qualify under the criteria contained in section 404(r) of the CWA are exempt from section 404 permit requirements, but may be subject to other State or Federal requirements.

Section 323.5 Program transfer to States

Section 404(h) of the CWA allows the Administrator of the Environmental Protection Agency (EPA) to transfer administration of the section 404 permit program for discharges into certain waters of the United States to qualified States. (The program cannot be transferred for those waters which are presently used, or are susceptible to use in their natural condition or by reasonable improvement as a means to transport interstate or foreign commerce shoreward to their ordinary high water mark, including all waters which are subject to the ebb and flow of the tide shoreward to the high tide line, including wetlands adjacent thereto). See 40 CFR parts 233 and 124 for procedural regulations for transferring section 404 programs to States. Once a State's 404 program is approved and in effect, the Corps of Engineers will suspend processing of section 404 applications in the applicable waters and will transfer pending applications to the State agency responsible for administering the program. District engineers will assist EPA and the States in any way practicable to effect transfer and will develop appropriate procedures to ensure orderly and expeditious transfer.

Section 323.6 Special policies and procedures

(a) The Secretary of the Army has delegated to the Chief of Engineers the authority to issue or deny section 404 permits. The district engineer will review applications for permits for the discharge of dredged or fill material into waters of the United States in accordance with guidelines promulgated by the Administrator, EPA, under authority of section 404(b)(1) of the CWA. (See 40 CFR part 230.) Subject to consideration of any economic impact on navigation and anchorage pursuant to section 404(b)(2), a permit will be denied if the discharge that would be authorized by such a permit would not comply with the 404(b)(1) guidelines. If the district engineer determines that the proposed discharge would comply with the 404(b)(1) guidelines, he will grant the permit unless issuance would be contrary to the public interest.

(b) The Corps will not issue a permit where the regional administrator of EPA has notified the district engineer and applicant in writing pursuant to 40 CFR 231.3(a)(1) that he intends to issue a public notice of a proposed determination to prohibit or withdraw the specification, or to deny, restrict or withdraw the use for specification, of any defined area as a disposal site in accordance with section 404(c) of the Clean Water Act. However the Corps will continue to complete the administrative processing of the application while the section 404(c) procedures are underway including completion of final coordination with EPA under 33 CFR part 325.

PART 324 PERMITS FOR OCEAN DUMPING OF DREDGED MATERIAL

AUTHORITY: 33 U.S.C. 1413

Section 324.1 General

This regulation prescribes in addition to the general policies of 33 CFR part 320 and procedures of 33 CFR part 325, those special policies, practices and procedures to be followed by the Corps of Engineers in connection with the review of applications for Department of the Army (DA) permits to authorize the transportation of dredged material by vessel or other vehicle for the purpose of dumping it in ocean waters at dumping sites designated under 40 CFR part 228 pursuant to section 103 of the Marine Protection, Research and Sanctuaries Act of 1972, as amended (33 U.S.C. 1413) (hereinafter referred to as section 103). See 33 CFR 320.2(h). Activities involving the transportation of dredged material for the purpose of dumping in the ocean waters also require DA permits under Section 10 of the Rivers and Harbors Act of 1899 (33 U.S.C. 403) for the dredging in navigable waters of the United States. Applicants for DA permits under this part should also refer to 33 CFR part 322 to satisfy the requirements of Section 10.

Section 324.2 Definitions

For the purpose of this regulation, the following terms are defined:

(a) The term *ocean waters* means those waters of the open seas lying seaward of the base line from which the territorial sea is measured, as provided for in the Convention on the Territorial Sea and the Contiguous Zone (15 UST 1606: TIAS 5639).

(b) The term *dredged material* means any material excavated or dredged from navigable waters of the United States.

(c) The term *transport* or *transportation* refers to the conveyance and related handling of dredged material by a vessel or other vehicle.

Section 324.3 Activities requiring permits

(a) *General.* DA permits are required for the transportation of dredged material for the purpose of dumping it in ocean waters.

(b) *Activities of Federal agencies.*

(1) The transportation of dredged material for the purpose of disposal in ocean waters done by or on behalf of any Federal agency other than the activities of the Corps of Engineers is subject to the procedures of this regulation. Agreement for construction or engineering services performed for other agencies by the Corps of Engineers does not constitute authorization under these regulations. Division and district engineers will therefore advise Federal agencies accordingly and cooperate to the fullest extent in the expeditious processing of their applications. The activities of the Corps of Engineers that involve the transportation of dredged material for disposal in ocean waters are regulated by 33 CFR 209.145.

(2) The policy provisions set out in 33 CFR 320.4(j) relating to state or local authorizations do not apply to work or structures undertaken by Federal agencies, except where compliance with non-Federal authorization is required by Federal law or Executive policy. Federal agencies are responsible for conformance with such laws and policies. (See EO 12088, October 18, 1978.) Federal agencies are not required to obtain and provide certification of compliance with effluent limitations and water quality standards from state or interstate water pollution control agencies in connection with activities involving the transport of dredged material for dumping into ocean waters beyond the territorial sea.

Section 324.4 Special procedures

The Secretary of the Army has delegated to the Chief of Engineers the authority to issue or deny section 103 permits. The following additional procedures shall also be applicable under this regulation.

(a) *Public notice.* For all applications for section 103 permits, the district engineer will issue a public notice which shall contain the information specified in 33 CFR 325.3.

(b) *Evaluation.* Applications for permits for the transportation of dredged material for the purpose of dumping it in ocean waters will be evaluated to determine whether the proposed dumping will unreasonably degrade or endanger human health, welfare, amenities, or the marine environment, ecological systems or economic potentialities. District engineers will apply the criteria established by the Administrator of EPA pursuant to section 102 of the Marine Protection, Research and Sanctuaries Act of 1972 in making this evaluation. (See 40 CFR parts 220–229) Where ocean dumping is determined to be necessary, the district engineer will, to the extent feasible, specify disposal sites using the recommendations of the Administrator pursuant to section 102(c) of the Act.

(c) *EPA review.* When the Regional Administrator, EPA, in accordance with 40 CFR 225.2(b), advises the district engineer, in writing, that the proposed dumping will comply with the criteria, the district engineer will complete his evaluation of the application under this part and 33 CFR parts 320 and 325. If, however, the Regional Administrator advises the district engineer, in writing, that the proposed dumping does not comply with the criteria, the district engineer will proceed as follows:

(1) The district engineer will determine whether there is an economically feasible alternative method or site available other than the proposed ocean disposal site. If there are other feasible alternative methods or sites available, the district engineer will evaluate them in accordance with 33 CFR parts 320, 322, 323, and 325 and this part, as appropriate.

(2) If the district engineer determines that there is no economically feasible alternative method or site available, and the proposed project is otherwise found to be not contrary to the public interest, he will so advise the Regional Administrator setting forth his reasons for such determination. If the Regional Administrator has not removed his objection within 15 days, the district engineer will submit a report of his determination to the Chief of Engineers for further coordination with the Administrator, EPA, and decision. The report forwarding the case will contain the analysis of whether there are other economically feasible methods or sites available to dispose of the dredged material.

(d) *Chief of Engineers review.* The Chief of Engineers shall evaluate the permit application and make a decision to deny the permit or recommend its issuance. If the decision of the Chief of Engineers is that ocean dumping at the proposed disposal site is required because of the unavailability of economically feasible alternatives, he shall so certify and request that the Secretary of the Army seek a waiver from the Administrator, EPA, of the criteria or of the critical site designation in accordance with 40 CFR 225.4.

PART 325 PROCESSING OF DEPARTMENT OF THE ARMY PERMITS

AUTHORITY: 33 U.S.C. 401 et seq.; 33 U.S.C. 1344; 33 U.S.C. 1413

Section 325.1 Applications for permits

(a) *General.* The processing procedures of this part apply to any Department of the Army (DA) permit. Special procedures and additional information are contained in 33 CFR parts 320 through 324, 327 and part 330. This part is arranged in the basic timing sequence used by the Corps of Engineers in processing applications for DA permits.

(b) *Pre-application consultation for major applications.* The district staff element having responsibility for administering, processing, and enforcing federal laws and regulations relating to the Corps of Engineers regulatory program shall be available to advise potential applicants of studies or other information foreseeably required for later federal action. The district engineer will establish local procedures and policies including appropriate publicity programs which will allow potential applicants to contact the district engineer or the regulatory staff element to request pre-application consultation. Upon receipt of such request, the district engineer will assure the conduct of an orderly process which may involve other staff elements and affected agencies (Federal, state, or local) and the public. This early process should be brief but thorough so that the potential applicant may begin to assess the viability of some of the more obvious potential alternatives in the application. The district engineer will endeavor, at this stage, to provide the potential applicant with all helpful information necessary in pursuing the application, including factors which the Corps must consider in its permit decision making process. Whenever the district engineer becomes aware of planning for work which may require a DA permit and which may involve the preparation of an environmental document, he shall contact the principals involved to advise them of the requirement for the permit(s) and the attendant public interest review including the development of an environmental document. Whenever a potential applicant indicates the intent to submit an application for work which may require the preparation of an environmental document, a single point of contact shall be designated within the district's regulatory staff to effectively coordinate the regulatory process, including the National Environmental Policy Act (NEPA) procedures and all attendant reviews, meetings, hearings, and other actions, including the scoping process if appropriate, leading to a decision by the district engineer. Effort devoted to this process should be commensurate with the likelihood of a permit application actually being submitted to the Corps. The regulatory staff coordinator shall maintain an open relationship with each potential applicant or his consultants so as to assure that the potential applicant is fully aware of the substance (both quantitative and qualitative) of the data required by the district engineer for use in preparing an environmental assessment or an environmental impact statement (EIS) in accordance with 33 CFR part 230, Appendix B.

(c) *Application form.* Applicants for all individual DA permits must use the standard application form (ENG Form 4345, OMB Approval No. OMB 49-R0420). Local variations of the application form for purposes of facilitating coordination with federal, state and local agencies may be used. The appropriate form may be obtained from the district office having jurisdiction over the waters in which the activity is proposed to be located. Certain activities have been authorized by general permits and do not require submission of an application form but may require a separate notification.

(d) *Content of application.*

(1) The application must include a complete description of the proposed activity including necessary drawings, sketches, or plans sufficient for public notice (detailed engineering plans and specifications are not required); the location, purpose and need for the proposed activity; scheduling of the activity; the names and addresses of adjoining property owners; the location and dimensions of adjacent structures; and a list of authorizations required by other federal, interstate, state, or local agencies for the work, including all approvals received or denials already made. See § 325.3 for information required to be in public notices. District and division engineers are not authorized to develop additional information forms but may request specific information on a case-by-case basis. (See § 325.1(e)).

(2) All activities which the applicant plans to undertake which are reasonably related to the same project and for which a DA permit would be required should be included in the same permit application. District engineers should reject, as incomplete, any permit application which fails to comply with this requirement. For example, a permit application for a marina will include dredging required for access as well as any fill associated with construction of the marina.

(3) If the activity would involve dredging in navigable waters of the United States, the application must include a description of the type, composition and quantity of the material to be dredged, the method of dredging, and the site and plans for disposal of the dredged material.

(4) If the activity would include the discharge of dredged or fill material into the waters of the United States or the transportation of dredged material for the purpose of disposing of it in ocean waters the application must include the source of the material; the purpose of the discharge, a description of the type, composition and quantity of the material; the method of transportation and disposal of the material; and the location of the disposal site. Certification under section 401 of the Clean Water Act is required for such discharges into waters of the United States.

(5) If the activity would include the construction of a filled area or pile or float-supported platform the project description must include the use of, and specific structures to be erected on, the fill or platform.

(6) If the activity would involve the construction of an impoundment structure, the applicant may be required to demonstrate that the structure complies with established state dam safety criteria or that the structure has been designed by qualified persons and, in appropriate cases, independently reviewed (and modified as the review would indicate) by similarly qualified persons. No specific design criteria are to be prescribed nor is an independent detailed engineering review to be made by the district engineer.

(7) *Signature on application.* The application must be signed by the person who desires to undertake the proposed activity (i.e. the applicant) or by a duly authorized agent. When the applicant is represented by an agent, that information will be included in the space provided on the application or by a separate written statement. The signature of the applicant or the agent will be an affirmation that the applicant possesses or will possess the requisite property interest to undertake the activity proposed in the application, except where the lands are under the control of the Corps of Engineers, in which cases the district engineer will coordinate the transfer of the real estate and the permit action. An application may include the activity of more than one owner provided the character of the activity of each owner is similar and in the same general area and each owner submits a statement designating the same agent.

(8) If the activity would involve the construction or placement of an artificial reef, as defined in 33 CFR 322.2(g), in the navigable waters of the United States or in the waters overlying the outer continental shelf, the application must include provisions for siting, constructing, monitoring, and managing the artificial reef.

(9) *Complete application.* An application will be determined to be complete when sufficient information is received to issue a public notice (See 33 CFR 325.1(d) and 325.3(a).) The issuance of a public notice will not be delayed to obtain information necessary to evaluate an application.

(e) *Additional information.* In addition to the information indicated in paragraph (d) of this section, the applicant will be required to furnish only such additional information as the district engineer deems essential to make a public interest determination including, where applicable, a determination of compliance with the section 404(b)(1) guidelines or ocean dumping criteria. Such additional information may include environmental data and information on alternate methods and sites as may be necessary for the preparation of the required environmental documentation.

(f) *Fees.* Fees are required for permits under section 404 of the Clean Water Act, section 103 of the Marine Protection, Research and Sanctuaries Act of 1972, as amended, and sections 9 and 10 of the Rivers and Harbors Act of 1899. A fee of $100.00 will be charged when the planned or ultimate purpose of the project is commercial or industrial in nature and is in support of operations that charge for the production, distribution or sale of goods or services. A $10.00 fee will be charged for permit applications when the proposed work is non-commercial in nature and would provide personal benefits that have no connection with a commercial enterprise. The final decision as to the basis for a fee (commercial vs. non-commercial) shall be solely the responsibility of the district engineer. No fee will be charged if the applicant withdraws the application at any time prior to issuance of the permit or if the permit is denied. Collection of the fee will be deferred until the proposed activity has been determined to be not contrary to the public interest. Multiple fees are not to be charged if more than one law is applicable. Any modification significant enough to require publication of a public notice will also require a fee. No fee will be assessed when a permit is transferred from one property owner to another. No fees will be charged for time extensions, general permits or letters of permission. Agencies or instrumentalities of federal, state or local governments will not be required to pay any fee in connection with permits.

Section 325.2 Processing of applications

(a) *Standard procedures*

(1) When an application for a permit is received the district engineer shall immediately assign it a number for identification, acknowledge receipt thereof, and advise the applicant of the number assigned to it. He shall review the application for completeness, and if the application is incomplete, request from the applicant within 15 days of receipt of the application any additional information necessary for further processing.

(2) Within 15 days of receipt of an application the district engineer will either determine that the application is complete (see 33 CFR 325.1(d)(9) and issue a public notice as described in § 325.3 of this part, unless specifically exempted by other provisions of this regulation or that it is incomplete and notify the applicant of the information necessary for a complete application. The district engineer will issue a supplemental, revised, or corrected public notice if in his view there is a change in the application data that would affect the public's review of the proposal.

(3) The district engineer will consider all comments received in response to the public notice in his subsequent actions on the permit application. Receipt of the comments will be acknowledged, if appropriate, and they will be made a part of the administrative record of the application. Comments received as form letters or petitions may be acknowledged as a group to the person or organization responsible for the form letter or petition. If comments relate to matters within the special expertise of another federal agency, the district engineer may seek the advice of that agency. If the district engineer determines, based on comments received, that he must have the views of the applicant on a particular issue to make a public interest determination, the applicant will be given the opportunity to furnish his views on such issue to the district engineer (see § 325.2(d)(5)). At the earliest practicable time other substantive comments will be furnished to the applicant for his information and any views he may wish to offer. A summary of the comments, the actual letters or portions thereof, or representative comment letters may be furnished to the applicant. The applicant may voluntarily elect to contact objectors in an attempt to resolve objections but will not be required to do so. District engineers will ensure that all parties are informed that the Corps alone is responsible for reaching a decision on the merits of any application. The district engineer may also offer Corps regulatory staff to be present at meetings between applicants and objectors, where appropriate, to provide information on the process, to mediate differences, or to gather information to aid in the decision process. The district engineer should not delay processing of the application unless the applicant requests a reasonable delay, normally not to exceed 30 days, to provide additional information or comments.

(4) The district engineer will follow Appendix B of 33 CFR part 230 for environmental procedures and documentation required by the National Environmental Policy Act of 1969. A decision on a permit application will require either an environmental assessment or an environmental impact statement unless it is included within a categorical exclusion.

(5) The district engineer will also evaluate the application to determine the need for a public hearing pursuant to 33 CFR part 327.

(6) After all above actions have been completed, the district engineer will determine in accordance with the record and applicable regulations whether or not the permit should be issued. He shall prepare a statement of findings (SOF) or, where an EIS has been prepared, a record of decision (ROD), on all permit decisions. The SOF or ROD shall include the district engineer's views on the probable effect of the proposed work on the public interest including conformity with the guidelines published for the discharge of dredged or fill material into waters of the United States (40 CFR part 230) or with the criteria for dumping of dredged material in ocean waters (40 CFR parts 220 to 229), if applicable, and the conclusions of the district engineer. The SOF or ROD shall be dated, signed, and included in the record prior to final action on the application. Where the district engineer has delegated authority to sign permits for and in his behalf, he may similarly delegate the signing of the SOF or ROD. If a district engineer makes a decision on a permit application which is contrary to state or local decisions (33 CFR 320.4(j) (2) & (4)), the district engineer will include in the decision document the significant national issues and explain how they are overriding in importance. If a permit is warranted, the district engineer will determine the special conditions, if any, and duration which should be incorporated into the permit. In accordance with the authorities specified in § 325.8 of this part, the district engineer will take final action or forward the application with all pertinent comments, records, and studies, including the final EIS or environmental assessment, through channels to the official authorized to make the final decision. The report forwarding the application for decision will be in a format prescribed by the Chief of Engineers. District and division engineers will notify the applicant and interested federal and state agencies that the application has been forwarded to higher headquarters. The district or division engineer may, at his option, disclose his recommendation to the news media and other interested parties, with the caution that it is only a recommendation and not a final decision. Such disclosure is encouraged in permit cases which have become controversial and have been the subject of stories in the media or have generated strong public interest. In those cases where the application is forwarded for decision in the format prescribed by the Chief of Engineers, the report will serve as the SOF or ROD. District engineers will generally combine the SOF, environmental assessment, and findings of no significant impact (FONSI), 404(b)(1) guideline analysis, and/or the criteria for dumping of dredged material in ocean waters into a single document.

(7) If the final decision is to deny the permit, the applicant will be advised in writing of the reason(s) for denial. If the final decision is to issue the permit and a standard individual permit form will be used, the issuing official will forward the permit to the applicant for signature accepting the conditions of the permit. The permit is not valid until signed by the issuing official. Letters of permission require only the signature of the issuing official. Final action on the permit application is the signature on the letter notifying the applicant of the denial of the permit or signature of the issuing official on the authorizing document.

(8) The district engineer will publish monthly a list of permits issued or denied during the previous month. The list will identify each action by public notice number, name of applicant, and brief description of activity involved. It will also note that relevant environmental documents and the SOF's or

ROD's are available upon written request and, where applicable, upon the payment of administrative fees. This list will be distributed to all persons who may have an interest in any of the public notices listed.

(9) Copies of permits will be furnished to other agencies in appropriate cases as follows:

(i) If the activity involves the construction of artificial islands, installations or other devices on the outer continental shelf, to the Director, Defense Mapping Agency, Hydrographic Center, Washington, DC 20390 Attention, Code NS12, and to the National Ocean Service, Office of Coast Survey, N/CS261, 1315 East West Highway, Silver Spring, Maryland 20910-3282.

(ii) If the activity involves the construction of structures to enhance fish propagation (e.g., fishing reefs) along the coasts of the United States, to the Defense Mapping Agency, Hydrographic Center and National Ocean Service as in paragraph (a)(9)(i) of this section and to the Director, Office of Marine Recreational Fisheries, National Marine Fisheries Service, Washington, DC 20235.

(iii) If the activity involves the erection of an aerial transmission line, submerged cable, or submerged pipeline across a navigable water of the United States, to the National Ocean Service, Office of Coast Survey, N/CS261, 1315 East West Highway, Silver Spring, Maryland 20910-3282.

(iv) If the activity is listed in paragraphs (a)(9) (i), (ii), or (iii) of this section, or involves the transportation of dredged material for the purpose of dumping it in ocean waters, to the appropriate District Commander, U.S. Coast Guard.

(b) *Procedures for particular types of permit situations–*

(1) *Section 401 Water Quality Certification.* If the district engineer determines that water quality certification for the proposed activity is necessary under the provisions of section 401 of the Clean Water Act, he shall so notify the applicant and obtain from him or the certifying agency a copy of such certification.

(i) The public notice for such activity, which will contain a statement on certification requirements (see § 325.3(a)(8)), will serve as the notification to the Administrator of the Environmental Protection Agency (EPA) pursuant to section 401(a)(2) of the Clean Water Act. If EPA determines that the proposed discharge may affect the quality of the waters of any state other than the state in which the discharge will originate, it will so notify such other state, the district engineer, and the applicant. If such notice or a request for supplemental information is not received within 30 days of issuance of the public notice, the district engineer will assume EPA has made a negative determination with respect to section 401(a)(2). If EPA determines another state's waters may be affected, such state has 60 days from receipt of EPA's notice to determine if the proposed discharge will affect the quality of its waters so as to violate any water quality requirement in such state, to notify EPA and the district engineer in writing of its objection to permit issuance, and to request a public hearing. If such occurs, the district engineer will hold a public hearing in the objecting state. Except as stated below, the hearing will be conducted in accordance with 33 CFR part 327. The issues to be considered at the public hearing will be limited to water quality impacts. EPA will submit its evaluation and recommendations at the hearing with respect to the state's objection to permit issuance. Based upon the recommendations of the objecting state, EPA, and any additional evidence presented at the hearing, the district engineer will condition the permit, if issued, in such a manner as may be necessary to insure compliance with applicable water quality requirements. If the imposition of conditions cannot, in the district engineer's opinion, insure such compliance, he will deny the permit.

(ii) No permit will be granted until required certification has been obtained or has been waived. A waiver may be explicit, or will be deemed to occur if the certifying agency fails or refuses to act on a request for certification within sixty days after receipt of such a request unless the district engineer determines a shorter or longer period is reasonable for the state to act. In determining whether or not a waiver period has commenced or waiver has occurred, the district engineer will verify that the certifying agency has received a valid request for certification. If, however, special circumstances identified by the district engineer require that action on an application be taken within a more limited period of time, the district engineer shall determine a reasonable lesser period of time, advise the certifying agency of the need for action by a particular date, and that, if certification is not received by that date, it will be considered that the requirement for certification has been waived. Similarly, if it appears that circumstances may reasonably require a period of time longer than sixty days, the district engineer, based on information provided by the certifying agency, will determine a longer reasonable period of time, not to exceed one year, at which time a waiver will be deemed to occur.

(2) *Coastal Zone Management Consistency.* If the proposed activity is to be undertaken in a state operating under a coastal zone management program approved by the Secretary of Commerce pursuant to the Coastal Zone Management (CZM) Act (see 33 CFR 320.3(b)), the district engineer shall proceed as follows:

(i) If the applicant is a federal agency, and the application involves a federal activity in or affecting the coastal zone, the district engineer shall forward a copy of the public notice to the agency of the state responsible for reviewing the consistency of federal activities. The federal agency applicant shall be responsible for complying with the CZM Act's directive for ensuring that federal agency activities are undertaken in a manner which is consistent, to the maximum extent practicable, with approved CZM Programs. (See 15 CFR part 930.) If the state coastal zone agency objects to the proposed federal activity on the basis of its inconsistency with the state's approved CZM Program, the district engineer shall not make a final decision on the application until the disagreeing parties have had an opportunity to utilize the procedures specified by the CZM Act for resolving such disagreements.

(ii) If the applicant is not a federal agency and the application involves an activity affecting the coastal zone, the district engineer shall obtain from the applicant a certification that his proposed activity complies with and will be conducted in a manner that is consistent with the approved state CZM Program. Upon receipt of the certification, the district engineer will forward a copy of the public notice (which will include the applicant's certification statement) to the state coastal zone agency and request its concurrence or objection. If the state agency objects to the certification or issues a decision indicating that the proposed activity requires further review, the district engineer shall not issue the permit until the state concurs with the certification statement or the Secretary of Commerce determines that the proposed activity is consistent with the purposes of the CZM Act or is necessary in the interest of national security. If the state agency fails to concur or object to a certification statement within six months of the state agency's receipt of the certification statement, state agency concurrence with the certification statement shall be conclusively presumed. District engineers will seek agreements with state CZM agencies that the agency's failure to provide comments during the public notice comment period will be considered as a concurrence with the certification or waiver of the right to concur or non-concur.

(iii) If the applicant is requesting a permit for work on Indian reservation lands which are in the coastal zone, the district engineer shall treat the application in the same manner as prescribed for a Federal applicant in paragraph (b)(2)(i) of this section. However, if the applicant is requesting a permit on non-trust Indian lands, and the state CZM agency has decided to assert jurisdiction over such lands, the district engineer shall treat the application in the same manner as prescribed for a non-Federal applicant in paragraph (b)(2)(ii) of this section.

(3) *Historic Properties.* If the proposed activity would involve any property listed or eligible for listing in the National Register of Historic Places, the district engineer will proceed in accordance with Corps National Historic Preservation Act implementing regulations.

(4) *Activities Associated with Federal Projects.* If the proposed activity would consist of the dredging of an access channel and/or berthing facility associated with an authorized federal navigation project, the activity will be included in the planning and coordination of the construction or maintenance of the federal project to the maximum extent feasible. Separate notice, hearing, and environmental documentation will not be required for activities so included and coordinated, and the public notice issued by the district engineer for these federal and associated non-federal activities will be the notice of intent to issue permits for those included non-federal dredging activities. The decision whether to issue or deny such a permit will be consistent with the decision on the federal project unless special considerations applicable to the proposed activity are identified. (See § 322.5(c).)

(5) *Endangered Species.* Applications will be reviewed for the potential impact on threatened or endangered species pursuant to section 7 of the Endangered Species Act as amended. The district engineer will include a statement in the public notice of his current knowledge of endangered species based on his initial review of the application (see 33 CFR 325.2(a)(2)). If the district engineer determines that the proposed activity would not affect listed species or their critical habitat, he will include a statement to this effect in the public notice. If he finds the proposed activity may affect an endangered or threatened species or their critical habitat, he will initiate formal consultation procedures with the U.S. Fish and Wildlife Service or National Marine Fisheries Service. Public notices forwarded to the U.S. Fish and Wildlife Service or National Marine Fisheries Service will serve as the request for information on whether any listed or proposed to be listed endangered or threatened species may be present in the area which would be affected by the proposed activity, pursuant to section 7(c) of the Act. References, definitions, and consultation procedures are found in 50 CFR part 402.

(c) [Reserved]

(d) *Timing of processing of applications.* The district engineer will be guided by the following time limits for the indicated steps in the evaluation process:

(1) The public notice will be issued within 15 days of receipt of all information required to be submitted by the applicant in accordance with paragraph 325.1.(d) of this part.

(2) The comment period on the public notice should be for a reasonable period of time within which interested parties may express their views concerning the permit. The comment period should not be more than 30 days nor less than 15 days from the date of the notice. Before designating comment periods less than 30 days, the district engineer will consider:

(i) Whether the proposal is routine or noncontroversial,

(ii) Mail time and need for comments from remote areas,

(iii) Comments from similar proposals, and

(iv) The need for a site visit. After considering the length of the original comment period, paragraphs (a)(2) (i) through (iv) of this section, and other pertinent factors, the district engineer may extend the comment period up to an additional 30 days if warranted.

(3) District engineers will decide on all applications not later than 60 days after receipt of a complete application, unless

(i) precluded as a matter of law or procedures required by law (see below),

(ii) The case must be referred to higher authority (see § 325.8 of this part),

(iii) The comment period is extended,

(iv) A timely submittal of information or comments is not received from the applicant,

(v) The processing is suspended at the request of the applicant, or

(vi) Information needed by the district engineer for a decision on the application cannot reasonably be obtained within the 60-day period. Once the cause for preventing the decision from being made within the normal 60-day period has been satisfied or eliminated, the 60-day clock will start running again from where it was suspended. For example, if the comment period is extended by 30 days, the district engineer will, absent other restraints, decide on the application within 90 days of receipt of a complete application. Certain laws (e.g., the Clean Water Act, the CZM Act, the National Environmental Policy Act, the National Historic Preservation Act, the Preservation of Historical and Archeological Data Act, the Endangered Species Act, the Wild and Scenic Rivers Act, and the Marine Protection, Research and Sanctuaries Act) require procedures such as state or other federal agency certifications, public hearings, environmental impact statements, consultation, special studies, and testing which may prevent district engineers from being able to decide certain applications within 60 days.

(4) Once the district engineer has sufficient information to make his public interest determination, he should decide the permit application even though other agencies which may have regulatory jurisdiction have not yet granted their authorizations, except where such authorizations are, by federal law, a prerequisite to making a decision on the DA permit application. Permits granted prior to other (non-prerequisite) authorizations by other agencies should, where appropriate, be conditioned in such manner as to give those other authorities an opportunity to undertake their review without the applicant biasing such review by making substantial resource commitments on the basis of the DA permit. In unusual cases the district engineer may decide that due to the nature or scope of a specific proposal, it would be prudent to defer taking final action until another agency has acted on its authorization. In such cases, he may advise the other agency of his position on the DA permit while deferring his final decision.

(5) The applicant will be given a reasonable time, not to exceed 30 days, to respond to requests of the district engineer. The district engineer may make such requests by certified letter and clearly inform the applicant that if he does not respond with the requested information or a justification why additional time is necessary, then his application will be considered withdrawn or a final decision will be made, whichever is appropriate. If additional time is requested, the district engineer will either grant the time, make a final decision, or consider the application as withdrawn.

(6) The time requirements in these regulations are in terms of calendar days rather than in terms of working days.

(e) *Alternative procedures.* Division and district engineers are authorized to use alternative procedures as follows:

(1) *Letters of permission.* Letters of permission are a type of permit issued through an abbreviated processing procedure which includes coordination with Federal and state fish and wildlife agencies, as required by the Fish and Wildlife Coordination Act, and a public interest evaluation, but without the

publishing of an individual public notice. The letter of permission will not be used to authorize the transportation of dredged material for the purpose of dumping it in ocean waters. Letters of permission may be used:

(i) In those cases subject to section 10 of the Rivers and Harbors Act of 1899 when, in the opinion of the district engineer, the proposed work would be minor, would not have significant individual or cumulative impacts on environmental values, and should encounter no appreciable opposition.

(ii) In those cases subject to section 404 of the Clean Water Act after:

(A) The district engineer, through consultation with Federal and state fish and wildlife agencies, the Regional Administrator, Environmental Protection Agency, the state water quality certifying agency, and, if appropriate, the state Coastal Zone Management Agency, develops a list of categories of activities proposed for authorization under LOP procedures;

(B) The district engineer issues a public notice advertising the proposed list and the LOP procedures, requesting comments and offering an opportunity for public hearing; and

(C) A 401 certification has been issued or waived and, if appropriate, CZM consistency concurrence obtained or presumed either on a generic or individual basis.

(2) *Regional permits.* Regional permits are a type of general permit as defined in 33 CFR 322.2(f) and 33 CFR 323.2(n). They may be issued by a division or district engineer after compliance with the other procedures of this regulation. After a regional permit has been issued, individual activities falling within those categories that are authorized by such regional permits do not have to be further authorized by the procedures of this regulation. The issuing authority will determine and add appropriate conditions to protect the public interest. When the issuing authority determines on a case-by-case basis that the concerns for the aquatic environment so indicate, he may exercise discretionary authority to override the regional permit and require an individual application and review. A regional permit may be revoked by the issuing authority if it is determined that it is contrary to the public interest provided the procedures of § 325.7 of this part are followed. Following revocation, applications for future activities in areas covered by the regional permit shall be processed as applications for individual permits. No regional permit shall be issued for a period of more than five years.

(3) *Joint procedures.* Division and district engineers are authorized and encouraged to develop joint procedures with states and other Federal agencies with ongoing permit programs for activities also regulated by the Department of the Army. Such procedures may be substituted for the procedures in paragraphs (a)(1) through (a)(5) of this section provided that the substantive requirements of those sections are maintained. Division and district engineers are also encouraged to develop management techniques such as joint agency review meetings to expedite the decision-making process. However, in doing so, the applicant's rights to a full public interest review and independent decision by the district or division engineer must be strictly observed.

(4) *Emergency procedures.* Division engineers are authorized to approve special processing procedures in emergency situations. An "emergency" is a situation which would result in an unacceptable hazard to life, a significant loss of property, or an immediate, unforeseen, and significant economic hardship if corrective action requiring a permit is not undertaken within a time period less than the normal time needed to process the application under standard procedures. In emergency situations, the district engineer will explain the circumstances and recommend special procedures to the division engineer who will instruct the district engineer as to further processing of the application. Even in an emergency situation, reasonable efforts will be made to receive comments from interested Federal, state, and local agencies and the affected public. Also, notice of any special procedures authorized and their rationale is to be appropriately published as soon as practicable.

[51 FR 41236, Nov. 13, 1986, as amended at 62 FR 26230, May 13, 1997]

Section 325.3 Public notice

(a) *General.* The public notice is the primary method of advising all interested parties of the proposed activity for which a permit is sought and of soliciting comments and information necessary to evaluate the probable impact on the public interest. The notice must, therefore, include sufficient information to give a clear understanding of the nature and magnitude of the activity to generate meaningful comment. The notice should include the following items of information:

(1) Applicable statutory authority or authorities;

(2) The name and address of the applicant;

(3) The name or title, address and telephone number of the Corps employee from whom additional information concerning the application may be obtained;

(4) The location of the proposed activity;

(5) A brief description of the proposed activity, its purpose and intended use, so as to provide sufficient information concerning the nature of the activity to generate meaningful comments, including a description of the type of structures, if any, to be erected on fills or pile or float-supported platforms, and a description of the type, composition, and quantity of materials to be discharged or disposed of in the ocean;

(6) A plan and elevation drawing showing the general and specific site location and character of all proposed activities, including the size relationship of the proposed structures to the size of the impacted waterway and depth of water in the area;

(7) If the proposed activity would occur in the territorial seas or ocean waters, a description of the activity's relationship to the baseline from which the territorial sea is measured;

(8) A list of other government authorizations obtained or requested by the applicant, including required certifications relative to water quality, coastal zone management, or marine sanctuaries;

(9) If appropriate, a statement that the activity is a categorical exclusion for purposes of NEPA (see paragraph 7 of Appendix B to 33 CFR part 230);

(10) A statement of the district engineer's current knowledge on historic properties;

(11) A statement of the district engineer's current knowledge on endangered species (see § 325.2(b)(5));

(12) A statement(s) on evaluation factors (see § 325.3(c));

(13) Any other available information which may assist interested parties in evaluating the likely impact of the proposed activity, if any, on factors affecting the public interest;

(14) The comment period based on § 325.2(d)(2);

(15) A statement that any person may request, in writing, within the comment period specified in the notice, that a public hearing be held to consider the application. Requests for public hearings shall state, with particularity, the reasons for holding a public hearing;

(16) For non-federal applications in states with an approved CZM Plan, a statement on compliance with the approved Plan; and

(17) In addition, for section 103 (ocean dumping) activities:

(i) The specific location of the proposed disposal site and its physical boundaries;

(ii) A statement as to whether the proposed disposal site has been designated for use by the Administrator, EPA, pursuant to section 102(c) of the Act;

(iii) If the proposed disposal site has not been designated by the Administrator, EPA, a description of the characteristics of the proposed disposal site and an explanation as to why no previously designated disposal site is feasible;

(iv) A brief description of known dredged material discharges at the proposed disposal site;

(v) Existence and documented effects of other authorized disposals that have been made in the disposal area (e.g., heavy metal background reading and organic carbon content);

(vi) An estimate of the length of time during which disposal would continue at the proposed site; and

(vii) Information on the characteristics and composition of the dredged material.

(b) *Public notice for general permits.* District engineers will publish a public notice for all proposed regional general permits and for significant modifications to, or reissuance of, existing regional permits within their area of jurisdiction. Public notices for statewide regional permits may be issued jointly by the affected Corps districts. The notice will include all applicable information necessary to provide a clear understanding of the proposal. In addition, the notice will state the availability of information at the district office which reveals the Corps' provisional determination that the proposed activities comply with the requirements for issuance of general permits. District engineers will publish a public notice for nationwide permits in accordance with 33 CFR 330.4.

(c) *Evaluation factors.* A paragraph describing the various evaluation factors on which decisions are based shall be included in every public notice.

(1) Except as provided in paragraph (c)(3) of this section, the following will be included:

"The decision whether to issue a permit will be based on an evaluation of the probable impact including cumulative impacts of the proposed activity on the public interest. That decision will reflect the national concern for both protection and utilization of important resources. The benefit which reasonably may be expected to accrue from the proposal must be balanced against its reasonably foreseeable detriments. All factors which may be relevant to the proposal will be considered including the cumulative effects thereof; among those are conservation, economics, aesthetics, general environmental concerns, wetlands, historic properties, fish and wildlife values, flood hazards, floodplain values, land use, navigation, shoreline erosion and accretion, recreation, water supply and conservation, water quality, energy needs, safety, food and fiber production, mineral needs, considerations of property ownership and, in general, the needs and welfare of the people."

(2) If the activity would involve the discharge of dredged or fill material into the waters of the United States or the transportation of dredged material for the purpose of disposing of it in ocean waters, the public notice shall also indicate that the evaluation of the impact of the activity on the public interest will include application of the guidelines promulgated by the Administrator, EPA, (40 CFR part 230) or of the criteria established under authority of section 102(a) of the Marine Protection, Research and Sanctuaries Act of 1972, as amended (40 CFR parts 220 to 229), as appropriate. (See 33 CFR parts 323 and 324).

(3) In cases involving construction of artificial islands, installations and other devices on outer continental shelf lands which are under mineral lease from the Department of the Interior, the notice will contain the following statement: "The decision as to whether a permit will be issued will be based on an evaluation of the impact of the proposed work on navigation and national security."

(d) *Distribution of public notices*

(1) Public notices will be distributed for posting in post offices or other appropriate public places in the vicinity of the site of the proposed work and will be sent to the applicant, to appropriate city and county officials, to adjoining property owners, to appropriate state agencies, to appropriate Indian Tribes or tribal representatives, to concerned Federal agencies, to local, regional and national shipping and other concerned business and conservation organizations, to appropriate River Basin Commissions, to appropriate state and areawide clearing houses as prescribed by OMB Circular A-95, to local news media and to any other interested party. Copies of public notices will be sent to all parties who have specifically requested copies of public notices, to the U.S. Senators and Representatives for the area where the work is to be performed, the field representative of the Secretary of the Interior, the Regional Director of the Fish and Wildlife Service, the Regional Director of the National Park Service, the Regional Administrator of the Environmental Protection Agency (EPA), the Regional Director of the National Marine Fisheries Service of the National Oceanic and Atmospheric Administration (NOAA), the head of the state agency responsible for fish and wildlife resources, the State Historic Preservation Officer, and the District Commander, U.S. Coast Guard.

(2) In addition to the general distribution of public notices cited above, notices will be sent to other addressees in appropriate cases as follows:

(i) If the activity would involve structures or dredging along the shores of the seas or Great Lakes, to the Coastal Engineering Research Center, Washington, DC 20016.

(ii) If the activity would involve construction of fixed structures or artificial islands on the outer continental shelf or in the territorial seas, to the Assistant Secretary of Defense (Manpower, Installations, and Logistics (ASD(MI&L)), Washington, DC 20310; the Director, Defense Mapping Agency (Hydrographic Center) Washington, DC 20390, Attention, Code NS12; and the National Ocean Service, Office of Coast Survey, N/CS261, 1315 East West Highway, Silver Spring, Maryland 20910-3282, and to affected military installations and activities.

(iii) If the activity involves the construction of structures to enhance fish propagation (e.g., fishing reefs) along the coasts of the United States, to the Director, Office of Marine Recreational Fisheries, National Marine Fisheries Service, Washington, DC 20235.

(iv) If the activity involves the construction of structures which may affect aircraft operations or for purposes associated with seaplane operations, to the Regional Director of the Federal Aviation Administration.

(v) If the activity would be in connection with a foreign-trade zone, to the Executive Secretary, Foreign-Trade Zones Board, Department of Commerce, Washington, DC 20230 and to the appropriate District Director of Customs as Resident Representative, Foreign-Trade Zones Board.

(3) It is presumed that all interested parties and agencies will wish to respond to public notices; therefore, a lack of response will be interpreted as meaning that there is no objection to the proposed project. A copy of the public notice with the list of the addresses to whom the notice was sent will be included in the record. If a question develops with respect to an activity for which another agency has responsibility and that other agency has not responded to the public notice, the district engineer may request its comments. Whenever a response to a public notice has been received from a member of Congress, either in behalf of a constituent or himself, the district engineer will inform the member of Congress of the final decision.

(4) District engineers will update public notice mailing lists at least once every two years.

Section 325.4 Conditioning of permits

(a) District engineers will add special conditions to Department of the Army permits when such conditions are necessary to satisfy legal requirements or to otherwise satisfy the public interest requirement. Permit conditions will be directly related to the impacts of the proposal, appropriate to the scope and degree of those impacts, and reasonably enforceable.

(1) Legal requirements which may be satisfied by means of Corps permit conditions include compliance with the 404(b)(1) guidelines, the EPA ocean dumping criteria, the Endangered Species Act, and requirements imposed by conditions on state section 401 water quality certifications.

(2) Where appropriate, the district engineer may take into account the existence of controls imposed under other federal, state, or local programs which would achieve the objective of the desired condition, or the existence of an enforceable agreement between the applicant and another party concerned with the resource in question, in determining whether a proposal complies with the 404(b)(1) guidelines, ocean dumping criteria, and other applicable statutes, and is not contrary to the public interest. In such cases, the Department of the Army permit will be conditioned to state that material changes in, or a failure to implement and enforce such program or agreement, will be grounds for modifying, suspending, or revoking the permit.

(3) Such conditions may be accomplished on-site, or may be accomplished off-site for mitigation of significant losses which are specifically identifiable, reasonably likely to occur, and of importance to the human or aquatic environment.

(b) District engineers are authorized to add special conditions, exclusive of paragraph (a) of this section, at the applicant's request or to clarify the permit application.

(c) If the district engineer determines that special conditions are necessary to insure the proposal will not be contrary to the public interest, but those conditions would not be reasonably implementable or enforceable, he will deny the permit.

(d) *Bonds.* If the district engineer has reason to consider that the permittee might be prevented from completing work which is necessary to protect the public interest, he may require the permittee to post a bond of sufficient amount to indemnify the government against any loss as a result of corrective action it might take.

Section 325.5 Forms of permits

(a) *General discussion*

(1) DA permits under this regulation will be in the form of individual permits or general permits. The basic format shall be ENG Form 1721, DA Permit (Appendix A).

(2) The general conditions included in ENG Form 1721 are normally applicable to all permits; however, some conditions may not apply to certain permits and may be deleted by the issuing officer. Special conditions applicable to the specific activity will be included in the permit as necessary to protect the public interest in accordance with § 325.4 of this part.

(b) *Individual permits* –

(1) *Standard permits.* A standard permit is one which has been processed through the public interest review procedures, including public notice and receipt of comments, described throughout this part. The standard individual permit shall be issued using ENG Form 1721.

(2) *Letters of permission.* A letter of permission will be issued where procedures of § 325.2(e)(1) have been followed. It will be in letter form and will identify the permittee, the authorized work and location of the work, the statutory authority, any limitations on the work, a construction time limit and a requirement for a report of completed work. A copy of the relevant general conditions from ENG Form 1721 will be attached and will be incorporated by reference into the letter of permission.

(c) *General permits* –

(1) *Regional permits.* Regional permits are a type of general permit. They may be issued by a division or district engineer after compliance with the other procedures of this regulation. If the public interest so requires, the issuing authority may condition the regional permit to require a case-by-case reporting and acknowledgment system. However, no separate applications or other authorization documents will be required.

(2) *Nationwide permits.* Nationwide permits are a type of general permit and represent DA authorizations that have been issued by the regulation (33 CFR part 330) for certain specified activities nationwide. If certain conditions are met, the specified activities can take place without the need for an individual or regional permit.

(3) *Programmatic permits.* Programmatic permits are a type of general permit founded on an existing state, local or other Federal agency program and designed to avoid duplication with that program.

(d) *Section 9 permits.* Permits for structures in interstate navigable waters of the United States under section 9 of the Rivers and Harbors Act of 1899 will be drafted at DA level.

Section 325.6 Duration of permits

(a) *General.* DA permits may authorize both the work and the resulting use. Permits continue in effect until they automatically expire or are modified, suspended, or revoked.

(b) *Structures.* Permits for the existence of a structure or other activity of a permanent nature are usually for an indefinite duration with no expiration date cited. However, where a temporary structure is authorized, or where restoration of a waterway is contemplated, the permit will be of limited duration with a definite expiration date.

(c) *Works.* Permits for construction work, discharge of dredged or fill material, or other activity and any construction period for a structure with a permit of indefinite duration under paragraph (b) of this section will specify time limits for completing the work or activity. The permit may also specify a date by which the work must be started, normally within one year from the date of issuance. The date will be established by the issuing official and will provide reasonable times based on the scope and nature of the work involved. Permits issued for the transport of dredged material for the purpose of disposing of it in ocean waters will specify a completion date for the disposal not to exceed three years from the date of permit issuance.

(d) *Extensions of time.* An authorization or construction period will automatically expire if the permittee fails to request and receive an extension of time. Extensions of time may be granted by the district engineer. The permittee must request the extension and explain the basis of the request, which will be granted unless the district engineer determines that an extension would be contrary to the public interest. Requests for extensions will be processed in accordance with the regular procedures of § 325.2 of this part, including issuance of a public notice, except that such processing is not required where the district engineer determines that there have been no significant changes in the attendant circumstances since the authorization was issued.

(e) *Maintenance dredging.* If the authorized work includes periodic maintenance dredging, an expiration date for the authorization of that maintenance dredging will be included in the permit. The expiration date, which in no event is to exceed ten years from the date of issuance of the permit, will be established by the issuing official after evaluation of the proposed method of dredging and disposal of the dredged material in accordance with the requirements of 33 CFR parts 320 to 325. In such cases, the district engineer shall require notification of the maintenance dredging prior to actual performance to insure continued compliance with the requirements of this regulation and 33 CFR parts 320 to 324. If the permittee desires to continue maintenance dredging beyond the expiration date, he must request a new permit. The permittee should be advised to apply for the new permit six months prior to the time he wishes to do the maintenance work.

Section 325.7 Modification, suspension, or revocation of permits

(a) *General.* The district engineer may reevaluate the circumstances and conditions of any permit, including regional permits, either on his own motion, at the request of the permittee, or a third party, or as the result of periodic progress inspections, and initiate action to modify, suspend, or revoke a permit as may be made necessary by considerations of the public interest. In the case of regional permits, this reevaluation may cover individual activities, categories of activities, or geographic areas. Among the factors to be considered are the extent of the permittee's compliance with the terms and conditions of the permit; whether or not circumstances relating to the authorized activity have changed since the permit was issued or extended, and the continuing adequacy of or need for the permit conditions; any significant objections to the authorized activity which were not earlier considered; revisions to applicable statutory and/or regulatory authorities; and the extent to which modification, suspension, or other action would adversely affect plans, investments and actions the permittee has reasonably made or taken in reliance on the permit. Significant increases in scope of a permitted activity will be processed as new applications for permits in accordance with § 325.2 of this part, and not as modifications under this section.

(b) *Modification.* Upon request by the permittee or, as a result of reevaluation of the circumstances and conditions of a permit, the district engineer may determine that the public interest requires a modification of the terms or conditions of the permit. In such cases, the district engineer will hold informal consultations with the permittee to ascertain whether the terms and conditions can be modified by mutual agreement. If a mutual agreement is reached on modification of the terms and conditions of the permit, the district engineer will give the permittee written notice of the modification, which will then become effective on such date as the district engineer may establish. In the event a mutual agreement cannot be reached by the district engineer and the permittee, the district engineer will proceed in accordance with paragraph (c) of this section if immediate suspension is warranted. In cases where immediate suspension is not warranted but the district engineer determines that the permit should be modified, he will notify the permittee of the proposed modification and reasons therefor, and that he may request a meeting with the district engineer and/or a public hearing. The modification will become effective on the date set by the district engineer which shall be at least ten days after receipt of the notice by the permittee unless a hearing or meeting is requested within that period. If the permittee fails or refuses to comply with the modification, the district engineer will proceed in accordance with 33 CFR part 326. The district engineer shall consult with resource agencies before modifying any permit terms or conditions, that would result in greater impacts, for a project about which that agency expressed a significant interest in the term, condition, or feature being modified prior to permit issuance.

(c) *Suspension.* The district engineer may suspend a permit after preparing a written determination and finding that immediate suspension would be in the public interest. The district engineer will notify the permittee in writing by the most expeditious means available that the permit has been suspended with the reasons therefor, and order the permittee to stop those activities previously authorized by the suspended permit. The permittee will also be advised that following this suspension a decision will be made to either reinstate, modify, or revoke the permit, and that he may within 10 days of receipt of notice of the suspension, request a meeting with the district engineer and/or a public hearing to present information in this matter. If a hearing is requested, the procedures prescribed in 33 CFR part 327 will be followed. After the completion of the meeting or hearing (or within a reasonable period of time after issuance of the notice to the permittee that the permit has been suspended if no hearing or meeting is requested), the district engineer will take action to reinstate, modify, or revoke the permit.

(d) *Revocation.* Following completion of the suspension procedures in paragraph (c) of this section, if revocation of the permit is found to be in the public interest, the authority who made the decision on the original permit may revoke it. The permittee will be advised in writing of the final decision.

(e) *Regional permits.* The issuing official may, by following the procedures of this section, revoke regional permits for individual activities, categories of activities, or geographic areas. Where groups of permittees are involved, such as for categories of activities or geographic areas, the informal discussions provided in paragraph (b) of this section may be waived and any written notification nay be made through the general public notice procedures of this regulation. If a regional permit is revoked, any permittee may then apply for an individual permit which shall be processed in accordance with these regulations.

Section 325.8 Authority to issue or deny permits

(a) *General.* Except as otherwise provided in this regulation, the Secretary of the Army, subject to such conditions as he or his authorized representative may from time to time impose, has authorized the Chief of Engineers and his authorized representatives to issue or deny permits for dams or dikes in intrastate waters of the United States pursuant to section 9 of the Rivers and Harbors Act of 1899; for construction or other work in or affecting navigable waters of the United States pursuant to section 10 of the Rivers and Harbors Act of 1899; for the discharge of dredged or fill material into waters of the United States pursuant to section 404 of the Clean Water Act; or for the transportation of dredged material for the purpose of disposing of it into ocean waters pursuant to section 103 of the Marine Protection, Research and Sanctuaries Act of 1972, as amended. The authority to issue or deny permits in interstate navigable waters of the United States pursuant to section 9 of the Rivers and Harbors Act of March 3, 1899 has not been delegated to the Chief of Engineers or his authorized representatives.

(b) *District engineer's authority.* District engineers are authorized to issue or deny permits in accordance with these regulations pursuant to sections 9 and 10 of the Rivers and Harbors Act of 1899; section 404 of the Clean Water Act; and section 103 of the Marine Protection, Research and Sanctuaries Act of 1972, as amended, in all cases not required to be referred to higher authority (see below). It is essential to the legality of a permit that it contain the name of the district engineer as the issuing officer. However, the permit need not be signed by the district engineer in person but may be signed for and in behalf of him by whomever he designates. In cases where permits are denied for reasons other than navigation or failure to obtain required local, state, or other federal approvals or certifications, the Statement of Findings must conclusively justify a denial decision. District engineers are authorized to deny permits without issuing a public notice or taking other procedural steps where required local, state, or other federal permits for the proposed activity have been denied or where he determines that the activity will clearly interfere with navigation except in all cases required to be referred to higher authority (see below). District engineers are also authorized to add, modify, or delete special conditions in permits in accordance with § 325.4 of this part, except for those conditions which may have been imposed by higher authority, and to modify, suspend and revoke permits according to the procedures of § 325.7 of this part. District engineers will refer the following applications to the division engineer for resolution:

(1) When a referral is required by a written agreement between the head of a Federal agency and the Secretary of the Army;

(2) When the recommended decision is contrary to the written position of the Governor of the state in which the work would be performed;

(3) When there is substantial doubt as to authority, law, regulations, or policies applicable to the proposed activity;

(4) When higher authority requests the application be forwarded for decision; or

(5) When the district engineer is precluded by law or procedures required by law from taking final action on the application (e.g. section 9 of the Rivers and Harbors Act of 1899, or territorial sea baseline changes).

(c) *Division engineer's authority.* Division engineers will review and evaluate all permit applications referred by district engineers. Division engineers may authorize the issuance or denial of permits pursuant to section 10 of the Rivers and Harbors Act of 1899; section 404 of the Clean Water Act; and section 103 of the Marine Protection, Research and Sanctuaries Act of 1972, as amended; and the inclusion of conditions in accordance with § 325.4 of this part in all cases not required to be referred to the Chief of Engineers. Division engineers will refer the following applications to the Chief of Engineers for resolution:

(1) When a referral is required by a written agreement between the head of a Federal agency and the Secretary of the Army;

(2) When there is substantial doubt as to authority, law, regulations, or policies applicable to the proposed activity;

(3) When higher authority requests the application be forwarded for decision; or

(4) When the division engineer is precluded by law or procedures required by law from taking final action on the application.

Section 325.9 Authority to determine jurisdiction

District engineers are authorized to determine the area defined by the terms "navigable waters of the United States" and "waters of the United States" except:

(a) When a determination of navigability is made pursuant to 33 CFR 329.14 (division engineers have this authority); or

(b) When EPA makes a Section 404 jurisdiction determination under its authority.

Section 325.10 Publicity

The district engineer will establish and maintain a program to assure that potential applicants for permits are informed of the requirements of this regulation and of the steps required to obtain permits for activities in waters of the United States or ocean waters. Whenever the district engineer becomes aware of plans being developed by either private or public entities which might require permits for implementation, he should advise the potential applicant in writing of the statutory requirements and the provisions of this regulation. Whenever the district engineer is aware of changes in Corps of Engineers regulatory jurisdiction, he will issue appropriate public notices.

Appendix A to Part 325 Permit Form and Special Conditions

A. Permit Form

Department of the Army Permit

Permittee____________________________

Permit No. ____________________________

Issuing Office __________________________

NOTE–The term "you"' and its derivatives, as used in this permit, means the permittee or any future transferee. The term "this office" refers to the appropriate district or division office of the Corps of Engineers having jurisdiction over the permitted activity or the appropriate official of that office acting under the authority of the commanding officer.

You are authorized to perform work in accordance with the terms and conditions specified below.

Project Description: (Describe the permitted activity and its intended use with references to any attached plans or drawings that are considered to be a part of the project description. Include a description of the types and quantities of dredged or fill materials to be discharged in jurisdictional waters.)

Project Location: (Where appropriate, provide the names of and the locations on the waters where the permitted activity and any off-site disposals will take place. Also, using name, distance, and direction, locate the permitted activity in reference to a nearby landmark such as a town or city.)

Permit Conditions:

General Conditions:

1. The time limit for completing the work authorized ends on ____________. If you find that you need more time to complete the authorized activity, submit your request for a time extension to this office for consideration at least one month before the above date is reached.

2. You must maintain the activity authorized by this permit in good condition and in conformance with the terms and conditions of this permit. You are not relieved of this requirement if you abandon the permitted activity, although you may make a good faith transfer to a third party in compliance with General Condition 4 below. Should you wish to cease to maintain the authorized activity or should you desire to abandon it without a good faith transfer, you must obtain a modification of this permit from this office, which may require restoration of the area.

3. If you discover any previously unknown historic or archeological remains while accomplishing the activity authorized by this permit, you must immediately notify this office of what you have found. We will initiate the Federal and state coordination required to determine if the remains warrant a recovery effort or if the site is eligible for listing in the National Register of Historic Places.

4. If you sell the property associated with this permit, you must obtain the signature of the new owner in the space provided and forward a copy of the permit to this office to validate the transfer of this authorization.

5. If a conditioned water quality certification has been issued for your project, you must comply with the conditions specified in the certification as special conditions to this permit. For your convenience, a copy of the certification is attached if it contains such conditions.

6. You must allow representatives from this office to inspect the authorized activity at any time deemed necessary to ensure that it is being or has been accomplished in accordance with the terms and conditions of your permit.

Special Conditions: (Add special conditions as required in this space with reference to a continuation sheet if necessary.)

Further Information:

1. Congressional Authorities: You have been authorized to undertake the activity described above pursuant to:

( ) Section 10 of the Rivers and Harbors Act of 1899 (33 U.S.C. 403).

( ) Section 404 of the Clean Water Act (33 U.S.C. 1344).

( ) Section 103 of the Marine Protection, Research and Sanctuaries Act of 1972 (33 U.S.C. 1413).

2. Limits of this authorization.

a. This permit does not obviate the need to obtain other Federal, state, or local authorizations required by law.

b. This permit does not grant any property rights or exclusive privileges.

c. This permit does not authorize any injury to the property or rights of others.

d. This permit does not authorize interference with any existing or proposed Federal project.

3. Limits of Federal Liability. In issuing this permit, the Federal Government does not assume any liability for the following:

a. Damages to the permitted project or uses thereof as a result of other permitted or unpermitted activities or from natural causes.

b. Damages to the permitted project or uses thereof as a result of current or future activities undertaken by or on behalf of the United States in the public interest.

c. Damages to persons, property, or to other permitted or unpermitted activities or structures caused by the activity authorized by this permit.

d. Design or construction deficiencies associated with the permitted work.

e. Damage claims associated with any future modification, suspension, or revocation of this permit.

4. Reliance on Applicant's Data: The determination of this office that issuance of this permit is not contrary to the public interest was made in reliance on the information you provided.

5. Reevaluation of Permit Decision. This office may reevaluate its decision on this permit at any time the circumstances warrant. Circumstances that could require a reevaluation include, but are not limited to, the following:

a. You fail to comply with the terms and conditions of this permit.

b. The information provided by you in support of your permit application proves to have been false, incomplete, or inaccurate (See 4 above).

c. Significant new information surfaces which this office did not consider in reaching the original public interest decision. Such a reevaluation may result in a determination that it is appropriate to use the suspension, modification, and revocation procedures contained in 33 CFR 325.7 or enforcement procedures such as those contained in 33 CFR 326.4 and 326.5. The referenced enforcement procedures provide for the issuance of an administrative order requiring you to comply with the terms and conditions of your permit and for the initiation of legal action where appropriate. You will be required to pay for any corrective measures ordered by this office, and if you fail to comply with such directive, this office may in certain situations (such as those specified in 33 CFR 209.170) accomplish the corrective measures by contract or otherwise and bill you for the cost.

6. Extensions. General condition 1 establishes a time limit for the completion of the activity authorized by this permit. Unless there are circumstances requiring either a prompt completion of the authorized activity or a reevaluation of the public interest decision, the Corps will normally give favorable consideration to a request for an extension of this time limit. Your signature below, as permittee, indicates that you accept and agree to comply with the terms and conditions of this permit.

____

(Permittee)

____

(Date)

This permit becomes effective when the Federal official, designated to act for the Secretary of the Army, has signed below.

____

(District Engineer)

____

(Date)

When the structures or work authorized by this permit are still in existence at the time the property is transferred, the terms and conditions of this permit will continue to be binding on the new owner(s) of the property. To validate the transfer of this permit and the associated liabilities associated with compliance with its terms and conditions, have the transferee sign and date below.

____

(Transferee)

____

(Date)

B. Special Conditions

No special conditions will be preprinted on the permit form. The following and other special conditions should be added, as appropriate, in the space provided after the general conditions or on a referenced continuation sheet:

1. Your use of the permitted activity must not interfere with the public's right to free navigation on all navigable waters of the United States.

2. You must have a copy of this permit available on the vessel used for the authorized transportation and disposal of dredged material.

3. You must advise this office in writing, at least two weeks before you start maintenance dredging activities under the authority of this permit.

4. You must install and maintain, at your expense, any safety lights and signals prescribed by the United States Coast Guard (USCG), through regulations or otherwise, on your authorized facilities. The USCG may be reached at the following address and telephone number:

____

____

5. The condition below will be used when a Corps permit authorizes an artificial reef, an aerial transmission line, a submerged cable or pipeline, or a structure on the outer continental shelf. National Ocean Service (NOS) has been notified of this authorization. You must notify NOS and this office in writing, at least two weeks before you begin work and upon completion of the activity authorized by this permit. Your notification of completion must include a drawing which certifies the location and configuration of the completed activity (a certified permit drawing may be used). Notifications to NOS will be sent to the following address: National Ocean Service, Office of Coast Survey, N/CS261, 1315 East West Highway, Silver Spring, Maryland 20910-3282.

6. The following condition should be used for every permit where legal recordation of the permit would be reasonably practicable and recordation could put a subsequent purchaser or owner of property on notice of permit conditions.

You must take the actions required to record this permit with the Registrar of Deeds or other appropriate official charged with the responsibility for maintaining records of title to or interest in real property.

[51 FR 41236, Nov. 13, 1986, as amended at 62 FR 26230, May 13, 1997]

Appendix B to Part 325 NEPA Implementation Procedures for the Regulatory Program

1. Introduction
2. General
3. Development of Information and Data
4. Elimination of Duplication with State and Local Procedures
5. Public Involvement
6. Categorical Exclusions
7. EA/FONSI Document
8. Environmental Impact Statement–General
9. Organization and Content of Draft EISs
10. Notice of Intent
11. Public Hearing
12. Organization and Content of Final EIS
13. Comments Received on the Final EIS
14. EIS Supplement
15. Filing Requirements
16. Timing
17. Expedited Filing
18. Record of Decision
19. Predecision Referrals by Other Agencies
20. Review of Other Agencies' EISs
21. Monitoring

1. *Introduction.* In keeping with Executive Order 12291 and 40 CFR 1500.2, where interpretive problems arise in implementing this regulation, and consideration of all other factors do not give a clear indication of a reasonable interpretation, the interpretation (consistent with the spirit and intent of NEPA) which results in the least paperwork and delay will be used. Specific examples of ways to reduce paperwork in the NEPA process are found at 40 CFR 1500.4. Maximum advantage of these recommendations should be taken.

2. *General.* This Appendix sets forth implementing procedures for the Corps regulatory program. For additional guidance, see the Corps NEPA regulation 33 CFR part 230 and for general policy guidance, see the CEQ regulations 40 CFR 1500–1508.

3. *Development of Information and Data.* See 40 CFR 1506.5. The district engineer may require the applicant to furnish appropriate information that the district engineer considers necessary for the preparation of an Environmental Assessment (EA) or Environmental Impact Statement (EIS). See also 40 CFR 1502.22 regarding incomplete or unavailable information.

4. *Elimination of Duplication with State and Local Procedures.* See 40 CFR 1506.2.

5. *Public Involvement.* Several paragraphs of this appendix (paragraphs 7, 8, 11, 13, and 19) provide information on the requirements for district engineers to make available to the public certain environmental documents in accordance with 40 CFR 1506.6.

6. *Categorical Exclusions–*

a. *General.* Even though an EA or EIS is not legally mandated for any Federal action falling within one of the "categorical exclusions," that fact does not exempt any Federal action from procedural or substantive compliance with any other Federal law. For example, compliance with the Endangered Species Act, the Clean Water Act, etc., is always mandatory, even for actions not requiring an EA or EIS. The following activities are not considered to be major Federal actions significantly affecting the quality of the human environment and are therefore categorically excluded from NEPA documentation:

(1) Fixed or floating small private piers, small docks, boat hoists and boathouses.

(2) Minor utility distribution and collection lines including irrigation;

(3) Minor maintenance dredging using existing disposal sites;

(4) Boat launching ramps;

(5) All applications which qualify as letters of permission (as described at 33 CFR 325.5(b)(2)).

b. *Extraordinary Circumstances.* District engineers should be alert for extraordinary circumstances where normally excluded actions could have substantial environmental effects and thus require an EA or EIS. For a period of one year from the effective date of these regulations, district engineers should maintain an information list on the type and number of categorical exclusion actions which, due to extraordinary circumstances, triggered the need for an EA/FONSI or EIS. If a district engineer determines that a categorical exclusion should be modified, the information will be furnished to the division engineer who will review and analyze the actions and circumstances to determine if there is a basis for recommending a modification to the list of categorical exclusions. HQUSACE (CECW-OR) will review recommended changes for Corps-wide consistency and revise the list accordingly.

7. *EA/FONSI Document.* (See 40 CFR 1508.9 and 1508.13 for definitions)–

a. *Environmental Assessment (EA) and Findings of No Significant Impact (FONSI).* The EA should normally be combined with other required documents (EA/404 (b)(1)/SOF/FONSI). "EA" as used throughout this Appendix normally refers to this combined document. The district engineer should complete an EA as soon as practicable after all relevant information is available (i.e., after the comment period for the public notice of the permit application has expired) and when the EA is a separate document it must be completed prior to completion of the statement of finding (SOF). When the EA confirms that the impact of the applicant's proposal is not significant and there are no "unresolved conflicts concerning alternative uses of available resources * * *" (section 102(2)(E) of NEPA), and the proposed activity is a "water dependent" activity as defined in 40 CFR 230.10(a)(3), the EA need not include a discussion on alternatives. In all other cases where the district engineer determines that there are unresolved conflicts concerning alternative uses of available resources, the EA shall include a discussion of the reasonable alternatives which are to be considered by the ultimate decision-maker. The decision options available to the Corps, which embrace all of the applicant's alternatives, are issue the permit, issue with modifications or deny the permit. Modifications are limited to those project modifications within the scope of established permit conditioning policy (See 33 CFR 325.4). The decision option to deny the permit results in the "no action" alternative (i.e., no activity requiring a Corps permit). The combined document normally should not exceed 15 pages and shall conclude with a FONSI (See 40 CFR 1508.13) or a determination that an EIS is required. The district engineer may delegate the signing of the NEPA document. Should the EA demonstrate that an EIS is necessary, the district engineer shall follow the procedures outlined in paragraph 8 of this Appendix. In those cases where it is obvious an EIS is required, an EA is not required. However, the district engineer should document his reasons for requiring an EIS.

b. *Scope of Analysis.*

(1) In some situations, a permit applicant may propose to conduct a specific activity requiring a Department of the Army (DA) permit (e.g., construction of a pier in a navigable water of the United States) which is merely one component of a larger project (e.g., construction of an oil refinery on an upland area). The district engineer should establish the scope of the NEPA document (e.g., the EA or EIS) to address the impacts of the specific activity requiring a DA permit and those portions of the entire project over which the district engineer has sufficient control and responsibility to warrant Federal review.

(2) The district engineer is considered to have control and responsibility for portions of the project beyond the limits of Corps jurisdiction where the Federal involvement is sufficient to turn an essentially private action into a Federal action. These are cases where the environmental consequences of the larger project are essentially products of the Corps permit action.

Typical factors to be considered in determining whether sufficient "control and responsibility" exists include:

(i) Whether or not the regulated activity comprises "merely a link" in a corridor type project (e.g., a transportation or utility transmission project).

(ii) Whether there are aspects of the upland facility in the immediate vicinity of the regulated activity which affect the location and configuration of the regulated activity.

(iii) The extent to which the entire project will be within Corps jurisdiction.

(iv) The extent of cumulative Federal control and responsibility.

A. Federal control and responsibility will include the portions of the project beyond the limits of Corps jurisdiction where the cumulative Federal involvement of the Corps and other Federal agencies is sufficient to grant legal control over such additional portions of the project. These are cases where the environmental consequences of the additional portions of the projects are essentially products of Federal financing, assistance, direction, regulation, or approval (not including funding assistance solely in the form of general revenue sharing funds, with no Federal agency control over the subsequent use of such funds, and not including judicial or administrative civil or criminal enforcement actions).

B. In determining whether sufficient cumulative Federal involvement exists to expand the scope of Federal action the district engineer should consider whether other Federal agencies are required to take Federal action under the Fish and Wildlife Coordination Act (16 U.S.C. 661 *et seq.*), the National Historic Preservation Act of 1966 (16 U.S.C. 470 *et seq.*), the Endangered Species Act of 1973 (16 U.S.C. 1531 *et seq.*), Executive Order 11990, Protection of Wetlands, (42 U.S.C. 4321 91977), and other environmental review laws and executive orders.

C. The district engineer should also refer to paragraphs 8(b) and 8(c) of this appendix for guidance on determining whether it should be the lead or a cooperating agency in these situations.

These factors will be added to or modified through guidance as additional field experience develops.

(3) *Examples:* If a non-Federal oil refinery, electric generating plant, or industrial facility is proposed to be built on an upland site and the only DA permit requirement relates to a connecting pipeline, supply loading terminal or fill road, that pipeline, terminal or fill road permit, in and of itself, normally would not constitute sufficient overall Federal involvement with the project to justify expanding the scope of a Corps NEPA document to cover upland portions of the facility beyond the structures in the immediate vicinity of the regulated activity that would effect the location and configuration of the regulated activity.

Similarly, if an applicant seeks a DA permit to fill waters or wetlands on which other construction or work is proposed, the control and responsibility of the Corps, as well as its overall Federal involvement would extend to the portions of the project to be located on the permitted fill. However, the NEPA review would be extended to the entire project, including portions outside waters of the United States, only if sufficient Federal control and responsibility over the entire project is determined to exist; that is, if the regulated activities, and those activities involving regulation, funding, etc. by other Federal agencies, comprise a substantial portion of the overall project. In any case, once the scope of analysis has been defined, the NEPA analysis for that action should include direct, indirect and cumulative impacts on all Federal interests within the purview of the NEPA statute. The district engineer should, whenever practicable, incorporate by reference and rely upon the reviews of other Federal and State agencies.

For those regulated activities that comprise merely a link in a transportation or utility transmission project, the scope of analysis should address the Federal action, i.e., the specific activity requiring a DA permit and any other portion of the project that is within the control or responsibility of the Corps of Engineers (or other Federal agencies).

For example, a 50-mile electrical transmission cable crossing a 1 1/4 mile wide river that is a navigable water of the United States requires a DA permit. Neither the origin and destination of the cable nor its route to and from the navigable water, except as the route applies to the location and configuration of the crossing, are within the control or responsibility of the Corps of Engineers. Those matters would not be included in the scope of analysis which, in this case, would address the impacts of the specific cable crossing.

Conversely, for those activities that require a DA permit for a major portion of a transportation or utility transmission project, so that the Corps permit bears upon the origin and destination as well as the route of the project outside the Corps regulatory boundaries, the scope of analysis should include those portions of the project outside the boundaries of the Corps section 10/404 regulatory jurisdiction. To use the same example, if 30 miles of the 50-mile transmission line crossed wetlands or other "waters of the United States," the scope of analysis should reflect impacts of the whole 50-mile transmission line.

For those activities that require a DA permit for a major portion of a shoreside facility, the scope of analysis should extend to upland portions of the facility. For example, a shipping terminal normally requires dredging, wharves, bulkheads, berthing areas and disposal of dredged material in order to function. Permits for such activities are normally considered sufficient Federal control and responsibility to warrant extending the scope of analysis to include the upland portions of the facility.

In all cases, the scope of analysis used for analyzing both impacts and alternatives should be the same scope of analysis used for analyzing the benefits of a proposal.

8. *Environmental Impact Statement–General–*

a. *Determination of Lead and Cooperating Agencies.* When the district engineer determines that an EIS is required, he will contact all appropriate Federal agencies to determine their respective role(s), i.e., that of lead agency or cooperating agency.

b. *Corps as Lead Agency.* When the Corps is lead agency, it will be responsible for managing the EIS process, including those portions which come under the jurisdiction of other Federal agencies. The district engineer is authorized to require the applicant to furnish appropriate information as discussed in paragraph 3 of this appendix. It is permissible for the Corps to reimburse, under agreement, staff support from other Federal agencies beyond the immediate jurisdiction of those agencies.

c. *Corps as Cooperating Agency.* If another agency is the lead agency as set forth by the CEQ regulations (40 CFR 1501.5 and 1501.6(a) and 1508.16), the district engineer will coordinate with that agency as a cooperating agency under 40 CFR 1501.6(b) and 1508.5 to insure that agency's resulting EIS may be adopted by the Corps for purposes of exercising its regulatory authority. As a cooperating agency the Corps will be responsible to the lead agency for providing environmental information which is directly related to the regulatory matter involved and which is required for the preparation of an EIS. This in no way shall be construed as lessening the district engineer's ability to request the applicant to furnish appropriate information as discussed in paragraph 3 of this appendix.

When the Corps is a cooperating agency because of a regulatory responsibility, the district engineer should, in accordance with 40 CFR 1501.6(b)(4), "make available staff support at the lead agency's request" to enhance the latter's interdisciplinary capability provided the request pertains to the Corps regulatory action covered by the EIS, to the extent this is practicable. Beyond this, Corps staff support will generally be made available to the lead agency to the extent practicable within its own responsibility and available resources.

Any assistance to a lead agency beyond this will normally be by written agreement with the lead agency providing for the Corps expenses on a cost reimbursable basis. If the district engineer believes a public hearing should be held and another agency is lead agency, the district engineer should request such a hearing and provide his reasoning for the request. The district engineer should suggest a joint hearing and offer to take an active part in the hearing and ensure coverage of the Corps concerns.

d. *Scope of Analysis.* See paragraph 7b.

e. *Scoping Process.* Refer to 40 CFR 1501.7 and 33 CFR 230.12.

f. *Contracting.* See 40 CFR 1506.5.

(1) The district engineer may prepare an EIS, or may obtain information needed to prepare an EIS, either with his

own staff or by contract. In choosing a contractor who reports directly to the district engineer, the procedures of 40 CFR 1506.5(c) will be followed.

(2) Information required for an EIS also may be furnished by the applicant or a consultant employed by the applicant. Where this approach is followed, the district engineer will (i) advise the applicant and/or his consultant of the Corps information requirements, and (ii) meet with the applicant and/or his consultant from time to time and provide him with the district engineer's views regarding adequacy of the data that are being developed (including how the district engineer will view such data in light of any possible conflicts of interest).

The applicant and/or his consultant may accept or reject the district engineer's guidance. The district engineer, however, may after specifying the information in contention, require the applicant to resubmit any previously submitted data which the district engineer considers inadequate or inaccurate. In all cases, the district engineer should document in the record the Corps independent evaluation of the information and its accuracy, as required by 40 CFR 1506.5(a).

g. *Change in EIS Determination.* If it is determined that an EIS is not required after a notice of intent has been published, the district engineer shall terminate the EIS preparation and withdraw the notice of intent. The district engineer shall notify in writing the appropriate division engineer; HQUSACE (CECW-OR); the appropriate EPA regional administrator, the Director, Office of Federal Activities (A-104), EPA, 401 M Street SW., Washington, DC 20460 and the public of the determination.

h. *Time Limits.* For regulatory actions, the district engineer will follow 33 CFR 230.17(a) unless unusual delays caused by applicant inaction or compliance with other statutes require longer time frames for EIS preparation. At the outset of the EIS effort, schedule milestones will be developed and made available to the applicant and the public. If the milestone dates are not met the district engineer will notify the applicant and explain the reason for delay.

9. *Organization and Content of Draft EISs–*

a. *General.* This section gives detailed information for preparing draft EISs. When the Corps is the lead agency, this draft EIS format and these procedures will be followed. When the Corps is one of the joint lead agencies, the joint lead agencies will mutually decide which agency's format and procedures will be followed.

b. *Format –*

(1) *Cover Sheet*

(a) Ref. 40 CFR 1502.11

(b) The "person at the agency who can supply further information" (40 CFR 1502.11(c) is the project manager handling that permit application.

(c) The cover sheet should identify the EIS as a Corps permit action and state the authorities (sections 9, 10, 404, 103, etc.) under which the Corps is exerting its jurisdiction.

(2) *Summary.* In addition to the requirements of 40 CFR 1502.12, this section should identify the proposed action as a Corps permit action stating the authorities (sections 9, 10, 404, 103, etc.) under which the Corps is exerting its jurisdiction. It shall also summarize the purpose and need for the proposed action and shall briefly state the beneficial/adverse impacts of the proposed action.

(3) *Table of Contents*

(4) *Purpose and Need.* See 40 CFR 1502.13. If the scope of analysis for the NEPA document (see paragraph 7b) covers only the proposed specific activity requiring a Department of the Army permit, then the underlying purpose and need for that specific activity should be stated. (For example, "The purpose and need for the pipe is to obtain cooling water from the river for the electric generating plant.") If the scope of analysis covers a more extensive project, only part of which may require a DA permit, then the underlying purpose and need for the entire project should be stated. (For example, "The purpose and need for the electric generating plant is to provide increased supplies of electricity to the (named) geographic area.") Normally, the applicant should be encouraged to provide a statement of his proposed activity's purpose and need from his perspective (for example, "to construct an electric generating plant"). However, whenever the NEPA document's scope of analysis renders it appropriate, the Corps also should consider and express that activity's underlying purpose and need from a public interest perspective (to use that same example, "to meet the public's need for electric energy"). Also, while generally focusing on the applicant's statement, the Corps, will in all cases, exercise independent judgment in defining the purpose and need for the project from both the applicant's and the public's perspective.

(5) *Alternatives.* See 40 CFR 1502.14. The Corps is neither an opponent nor a proponent of the applicant's proposal; therefore, the applicant's final proposal will be identified as the "applicant's preferred alternative" in the final EIS. Decision options available to the district engineer, which embrace all of the applicant's alternatives, are issue the permit, issue with modifications or conditions or deny the permit.

(a) Only reasonable alternatives need be considered in detail, as specified in 40 CFR 1502.14(a). Reasonable alternatives must be those that are feasible and such feasibility must focus on the accomplishment of the underlying purpose and need (of the applicant or the public) that would be satisfied by the proposed Federal action (permit issuance). The alternatives analysis should be thorough enough to use for both the public interest review and the 404(b)(1) guidelines (40 CFR part 230) where applicable. Those alternatives that are unavailable to the applicant, whether or not they require Federal action (permits), should normally be included in the analysis of the no-Federal-action (denial) alternative. Such alternatives should be evaluated only to the extent necessary to allow a complete and objective evaluation of the public interest and a fully informed decision regarding the permit application.

(b) The "no-action" alternative is one which results in no construction requiring a Corps permit. It may be brought by (1) the applicant electing to modify his proposal to eliminate work under the jurisdiction of the Corps or (2) by the denial of the permit. District engineers, when evaluating this alternative, should discuss, when appropriate, the consequences of other likely uses of a project site, should the permit be denied.

(c) The EIS should discuss geographic alternatives, e.g., changes in location and other site specific variables, and functional alternatives, e.g., project substitutes and design modifications.

(d) The Corps shall not prepare a cost-benefit analysis for projects requiring a Corps permit. 40 CFR 1502.23 states that the weighing of the various alternatives need not be displayed in a cost-benefit analysis and "* * * should not be when there are important qualitative considerations." The EIS should, however, indicate any cost considerations that are likely to be relevant to a decision.

(e) Mitigation is defined in 40 CFR 1508.20, and Federal action agencies are directed in 40 CFR 1502.14 to include appropriate mitigation measures. Guidance on the conditioning of permits to require mitigation is in 33 CFR 320.4(r) and 325.4. The nature and extent of mitigation conditions are dependent on the results of the public interest review in 33 CFR 320.4.

(6) *Affected Environment.* See Ref. 40 CFR 1502.15.

(7) *Environmental Consequences.* See Ref. 40 CFR 1502.16.

(8) *List of Preparers*. See Ref. 40 CFR 1502.17.

(9) *Public Involvement*. This section should list the dates and nature of all public notices, scoping meetings and public hearings and include a list of all parties notified.

(10) *Appendices*. See 40 CFR 1502.18. Appendices should be used to the maximum extent practicable to minimize the length of the main text of the EIS. Appendices normally should not be circulated with every copy of the EIS, but appropriate appendices should be provided routinely to parties with special interest and expertise in the particular subject.

(11) *Index*. The Index of an EIS, at the end of the document, should be designed to provide for easy reference to items discussed in the main text of the EIS.

10. *Notice of Intent*. The district engineer shall follow the guidance in 33 CFR part 230, Appendix C in preparing a notice of intent to prepare a draft EIS for publication in the Federal Register.

11. *Public Hearing*. If a public hearing is to be held pursuant to 33 CFR part 327 for a permit application requiring an EIS, the actions analyzed by the draft EIS should be considered at the public hearing. The district engineer should make the draft EIS available to the public at least 15 days in advance of the hearing. If a hearing request is received from another agency having jurisdiction as provided in 40 CFR 1506.6(c)(2), the district engineer should coordinate a joint hearing with that agency whenever appropriate.

12. *Organization and Content of Final EIS*. The organization and content of the final EIS including the abbreviated final EIS procedures shall follow the guidance in 33 CFR 230.14(a).

13. *Comments Received on the Final EIS*. For permit cases to be decided at the district level, the district engineer should consider all incoming comments and provide responses when substantive issues are raised which have not been addressed in the final EIS. For permit cases decided at higher authority, the district engineer shall forward the final EIS comment letters together with appropriate responses to higher authority along with the case. In the case of a letter recommending a referral under 40 CFR part 1504, the district engineer will follow the guidance in paragraph 19 of this appendix.

14. *EIS Supplement*. See 33 CFR 230.13(b).

15. *Filing Requirements*. See 40 CFR 1506.9. Five (5) copies of EISs shall be sent to Director, Office of Federal Activities (A-104), Environmental Protection Agency, 401 M Street SW, Washington, DC 20460. The official review periods commence with EPA's publication of a notice of availability of the draft or final EISs in the Federal Register. Generally, this notice appears on Friday of each week. At the same time they are mailed to EPA for filing, one copy of each draft or final EIS, or EIS supplement should be mailed to HQUSACE (CECW-OR) WASH DC 20314-1000.

16. *Timing*. 40 CFR 1506.10 describes the timing of an agency action when an EIS is involved.

17. *Expedited Filing*. 40 CFR 1506.10 provides information on allowable time reductions and time extensions associated with the EIS process. The district engineer will provide the necessary information and facts to HQUSACE (CECW-RE) WASH DC 20314-1000 (with copy to CECW-OR) for consultation with EPA for a reduction in the prescribed review periods.

18. *Record of Decision*. In those cases involving an EIS, the statement of findings will be called the record of decision and shall incorporate the requirements of 40 CFR 1505.2. The record of decision is not to be included when filing a final EIS and may not be signed until 30 days after the notice of availability of the final EIS is published in the Federal Register. To avoid duplication, the record of decision may reference the EIS.

19. *Predecision Referrals by Other Agencies*. See 40 CFR part 1504. The decisionmaker should notify any potential referring Federal agency and CEQ of a final decision if it is contrary to the announced position of a potential referring agency. (This pertains to a NEPA referral, not a 404(q) referral under the Clean Water Act. The procedures for a 404(q) referral are outlined in the 404(q) Memoranda of Agreement. The potential referring agency will then have 25 calendar days to refer the case to CEQ under 40 CFR part 1504. Referrals will be transmitted through division to CECW-RE for further guidance with an information copy to CECW-OR.

20. *Review of Other Agencies' EISs*. District engineers should provide comments directly to the requesting agency specifically related to the Corps jurisdiction by law or special expertise as defined in 40 CFR 1508.15 and 1508.26 and identified in Appendix II of CEQ regulations (49 FR 49750, December 21, 1984). If the district engineer determines that another agency's draft EIS which involves a Corps permit action is inadequate with respect to the Corps permit action, the district engineer should attempt to resolve the differences concerning the Corps permit action prior to the filing of the final EIS by the other agency. If the district engineer finds that the final EIS is inadequate with respect to the Corps permit action, the district engineer should incorporate the other agency's final EIS or a portion thereof and prepare an appropriate and adequate NEPA document to address the Corps involvement with the proposed action. See 33 CFR 230.21 for guidance. The agency which prepared the original EIS should be given the opportunity to provide additional information to that contained in the EIS in order for the Corps to have all relevant information available for a sound decision on the permit.

21. *Monitoring*. Monitoring compliance with permit requirements should be carried out in accordance with 33 CFR 230.15 and with 33 CFR part 325.

[53 FR 3134, Feb. 3, 1988]

Appendix C to Part 325 Procedures for the Protection of Historic Properties

1. Definitions
2. General Policy
3. Initial Review
4. Public Notice
5. Investigations
6. Eligibility Determinations
7. Assessing Effects
8. Consultation
9. ACHP Review and Comment
10. District Engineer Decision
11. Historic Properties Discovered During Construction
12. Regional General Permits
13. Nationwide General Permits
14. Emergency Procedures
15. Criteria of Effect and Adverse Effect

1. *Definitions*

a. *Designated historic property* is a historic property listed in the National Register of Historic Places (National Register) or which has been determined eligible for listing in the National Register pursuant to 36 CFR part 63. A historic property that, in both the opinion of the SHPO and the district engineer, appears to meet the criteria for inclusion in the National Register will be treated as a "designated historic property."

b. *Historic property* is a property which has historical importance to any person or group. This term includes the types of districts, sites, buildings, structures or objects eligible for inclusion, but not necessarily listed, on the National Register.

c. *Certified local government* is a local government certified in accordance with section 101(c)(1) of the NHPA (See 36 CFR part 61).

d. The term "criteria for inclusion in the National Register" refers to the criteria published by the Department of Interior at 36 CFR 60.4.

e. An "effect" on a "designated historic property" occurs when the undertaking may alter the characteristics of the property that qualified the property for inclusion in the National Register. Consideration of effects on "designated historic properties" includes indirect effects of the undertaking. The criteria for effect and adverse effect are described in Paragraph 15 of this appendix.

f. The term "undertaking" as used in this appendix means the work, structure or discharge that requires a Department of the Army permit pursuant to the Corps regulations at 33 CFR 320–334.

g. Permit area

(1) The term "permit area" as used in this appendix means those areas comprising the waters of the United States that will be directly affected by the proposed work or structures and uplands directly affected as a result of authorizing the work or structures. The following three tests must all be satisfied for an activity undertaken outside the waters of the United States to be included within the "permit area":

(i) Such activity would not occur but for the authorization of the work or structures within the waters of the United States;

(ii) Such activity must be integrally related to the work or structures to be authorized within waters of the United States. Or, conversely, the work or structures to be authorized must be essential to the completeness of the overall project or program; and

(iii) Such activity must be directly associated (first order impact) with the work or structures to be authorized.

(2) For example, consider an application for a permit to construct a pier and dredge an access channel so that an industry may be established and operated on an upland area.

(i) Assume that the industry requires the access channel and the pier and that without such channel and pier the project would not be feasible. Clearly then, the industrial site, even though upland, would be within the "permit area." It would not be established "but for" the access channel and pier; it also is integrally related to the work and structure to be authorized; and finally it is directly associated with the work and structure to be authorized. Similarly, all three tests are satisfied for the dredged material disposal site and it too is in the "permit area" even if located on uplands.

(ii) Consider further that the industry, if established, would cause local agencies to extend water and sewer lines to service the area of the industrial site. Assume that the extension would not itself involve the waters of the United States and is not solely the result of the industrial facility. The extensions would not be within the "permit area" because they would not be directly associated with the work or structure to be authorized.

(iii) Now consider that the industry, if established, would require increased housing for its employees, but that a private developer would develop the housing. Again, even if the housing would not be developed but for the authorized work and structure, the housing would not be within the permit area because it would not be directly associated with or integrally related to the work or structure to be authorized.

(3) Consider a different example. This time an industry will be established that requires no access to the navigable waters for its operation. The plans for the facility, however, call for a recreational pier with an access channel. The pier and channel will be used for the company-owned yacht and employee recreation. In the example, the industrial site is not included within the permit area. Only areas of dredging, dredged material disposal, and pier construction would be within the permit area.

(4) Lastly, consider a linear crossing of the waters of the United States; for example, by a transmission line, pipeline, or highway.

(i) Such projects almost always *can* be undertaken without Corps authorization, if they are designed to avoid affecting the waters of the United States. Corps authorization is sought because it is less expensive or more convenient for the applicant to do so than to avoid affecting the waters of the United States. Thus the "but for" test is not met by the entire project right-of-way. The "same undertaking" and "integral relationship" tests are met, but this is not sufficient to make the whole right-of-way part of the permit area. Typically, however, some portion of the right-of-way, approaching the crossing, would not occur in its given configuration "but for" the authorized activity. This portion of the right-of-way, whose location is determined by the location of the crossing, meets all three tests and hence is part of the permit area.

(ii) Accordingly, in the case of the linear crossing, the permit area shall extend in either direction from the crossing to that point at which alternative alignments leading to reasonable alternative locations for the crossing can be considered and evaluated. Such a point may often coincide with the physical feature of the waterbody to be crossed, for example, a bluff, the limit of the flood plain, a vegetational change, etc., or with a jurisdictional feature associated with the waterbody, for example, a zoning change, easement limit, etc., although such features should not be controlling in selecting the limits of the permit area.

2. *General Policy*

This appendix establishes the procedures to be followed by the U.S. Army Corps of Engineers (Corps) to fulfill the requirements set forth in the National Historic Preservation Act (NHPA), other applicable historic preservation laws, and Presidential directives as they relate to the regulatory program of the Corps of Engineers (33 CFR parts 320–334).

a. The district engineer will take into account the effects, if any, of proposed undertakings on historic properties both within and beyond the waters of the U.S. Pursuant to section 110(f) of the NHPA, the district engineer, where the undertaking that is the subject of a permit action may directly and adversely affect any National Historic Landmark, shall, to the maximum extent possible, condition any issued permit as may be necessary to minimize harm to such landmark.

b. In addition to the requirements of the NHPA, all historic properties are subject to consideration under the National Environmental Policy Act, (33 CFR part 325, appendix B), and the Corps' public interest review requirements contained in 33 CFR 320.4. Therefore, historic properties will be included as a factor in the district engineer's decision on a permit application.

c. In processing a permit application, the district engineer will generally accept for Federal or Federally assisted projects the Federal agency's or Federal lead agency's compliance with the requirements of the NHPA.

d. If a permit application requires the preparation of an Environmental Impact Statement (EIS) pursuant to the National Environmental Policy Act, the draft EIS will contain the information required by paragraph 9.a. below. Furthermore, the SHPO and the ACHP will be given the opportunity to participate in the scoping process and to comment on the Draft and Final EIS.

e. During pre-application consultations with a prospective applicant the district engineer will encourage the consideration of historic properties at the earliest practical time in the planning process.

f. This appendix is organized to follow the Corps standard permit process and to indicate how historic property considerations are to be addressed during the processing and evaluating of permit applications. The procedures of this Appendix are not intended to diminish the full consideration of historic properties in the Corps regulatory program. Rather, this appendix is intended to provide for the maximum consideration of historic properties within the time and jurisdictional constraints of the Corps regulatory program. The Corps will make every effort to provide information on historic properties and the effects of proposed undertakings on them to the public by the public notice within the time constraints required by the Clean Water Act. Within the time constraints of applicable laws, executive orders, and regulations, the Corps will provide the maximum coordination and comment opportunities to interested parties especially the SHPO and ACHP. The Corps will discuss with and encourage the applicant to avoid or minimize effects on historic properties. In reaching its decisions on permits, the Corps will adhere to the goals of the NHPA and other applicable laws dealing with historic properties.

3. *Initial Review*

a. Upon receipt of a completed permit application, the district engineer will consult district files and records, the latest published version(s) of the National Register, lists of properties determined eligible, and other appropriate sources of information to determine if there are any designated historic properties which may be affected by the proposed undertaking. The district engineer will also consult with other appropriate sources of information for knowledge of undesignated historic properties which may be affected by the proposed undertaking. The district engineer will establish procedures (e.g., telephone calls) to obtain supplemental information from the SHPO and other appropriate sources. Such procedures shall be accomplished within the time limits specified in this appendix and 33 CFR part 325.

b. In certain instances, the nature, scope, and magnitude of the work, and/or structures to be permitted may be such that there is little likelihood that a historic property exists or may be affected. Where the district engineer determines that such a situation exists, he will include a statement to this effect in the public notice. Three such situations are:

(1) Areas that have been extensively modified by previous work. In such areas, historic properties that may have at one time existed within the permit area may be presumed to have been lost unless specific information indicates the presence of such a property (e.g., a shipwreck).

(2) Areas which have been created in modern times. Some recently created areas, such as dredged material disposal islands, have had no human habitation. In such cases, it may be presumed that there is no potential for the existence of historic properties unless specific information indicates the presence of such a property.

(3) Certain types of work or structures that are of such limited nature and scope that there is little likelihood of impinging upon a historic property even if such properties were to be present within the affected area.

c. If, when using the pre-application procedures of 33 CFR 325.1(b), the district engineer believes that a designated historic property may be affected, he will inform the prospective applicant for consideration during project planning of the potential applicability of the Secretary of the Interior's Standards and Guidelines for Archeology and Historic Preservation (48 FR 44716). The district engineer will also inform the prospective applicant that the Corps will consider any effects on historic properties in accordance with this appendix.

d. At the earliest practical time the district engineer will discuss with the applicant measures or alternatives to avoid or minimize effects on historic properties.

4. *Public Notice*

a. Except as specified in subparagraph 4.c., the district engineer's current knowledge of the presence or absence of historic properties and the effects of the undertaking upon these properties will be included in the public notice. The public notice will be sent to the SHPO, the regional office of the National Park Service (NPS), certified local governments (see paragraph (1.c.) and Indian tribes, and interested citizens. If there are designated historic properties which reasonably may be affected by the undertaking or if there are undesignated historic properties within the affected area which the district engineer reasonably expects to be affected by the undertaking and which he believes meet the criteria for inclusion in the National Register, the public notice will also be sent to the ACHP.

b. During permit evaluation for newly designated historic properties or undesignated historic properties which reasonably may be affected by the undertaking and which have been newly identified through the public interest review process, the district engineer will immediately inform the applicant, the SHPO, the appropriate certified local government and the ACHP of the district engineer's current knowledge of the effects of the undertaking upon these properties. Commencing from the date of the district engineer's letter, these entities will be given 30 days to submit their comments.

c. Locational and sensitive information related to archeological sites is excluded from the Freedom of Information Act (Section 304 of the NHPA and Section 9 of ARPA). If the district engineer or the Secretary of the Interior determine that the disclosure of information to the public relating to the location or character of sensitive historic resources may create a substantial risk of harm, theft, or destruction to such resources or to the area or place where such resources are located, then the district engineer will not include such information in the public notice nor otherwise make it available to the public. Therefore, the district engineer will furnish such information to the ACHP and the SHPO by separate notice.

5. *Investigations*

a. When initial review, addition submissions by the applicant, or response to the public notice indicates the existence of a potentially eligible property, the district engineer shall examine the pertinent evidence to determine the need for further investigation. The evidence must set forth specific reasons for the need to further investigate within the permit area and may consist of:

(1) Specific information concerning properties which may be eligible for inclusion in the National Register and which are known to exist in the vicinity of the project; and

(2) Specific information concerning known sensitive areas which are likely to yield resources eligible for inclusion in the National Register, particularly where such sensitive area determinations are based upon data collected from other, similar areas within the general vicinity.

b. Where the scope and type of work proposed by the applicant or the evidence presented leads the district engineer to conclude that the chance of disturbance by the undertaking to any potentially eligible historic property is too remote to justify further investigation, he shall so advise the reporting party and the SHPO.

c. If the district engineer's review indicates that an investigation for the presence of potentially eligible historic properties on the upland locations of the permit area (see paragraph 1.g.) is justified, the district engineer will conduct or cause to be conducted such an investigation. Additionally, if the notification indicates that a potentially eligible historic property may exist within waters of the U.S., the district engineer will conduct or cause to be conducted an investigation to determine whether this property may be eligible for inclusion in the National Register.

Comments or information of a general nature will not be considered as sufficient evidence to warrant an investigation.

d. In addition to any investigations conducted in accordance with paragraph 6.a. above, the district engineeer may conduct or cause to be conducted additional investigations which the district engineer determines are essential to reach the public interest decision. As part of any site visit, Corps personnel will examine the permit area for the presence of potentially eligible historic properties. The Corps will notify the SHPO, if any evidence is found which indicates the presence of potentially eligible historic properties.

e. As determined by the district engineer, investigations may consist of any of the following: further consultations with the SHPO, the State Archeologist, local governments, Indian tribes, local historical and archeological societies, university archeologists, and others with knowledge and expertise in the identification of historical, archeological, cultural and scientific resources; field examinations; and archeological testing. In most cases, the district engineer will require, in accordance with 33 CFR 325.1(e), that the applicant conduct the investigation at his expense and usually by third party contract.

f. The Corps of Engineers' responsibilities to seek eligibility determinations for potentially eligible historic properties is limited to resources located within waters of the U.S. that are directly affected by the undertaking. The Corps responsibilities to identify potentially eligible historic properties is limited to resources located within the permit area that are directly affected by related upland activities. The Corps is not responsible for identifying or assessing potentially eligible historic properties outside the permit area, but will consider the effects of undertakings on any known historic properties that may occur outside the permit area.

6. *Eligibility determinations*

a. For a historic property within waters of the U.S. that will be directly affected by the undertaking the district engineer will, for the purposes of this Appendix and compliance with the NHPA:

(1) Treat the historic property as a "designated historic property," if both the SHPO and the district engineer agree that it is eligible for inclusion in the National Register; or

(2) Treat the historic property as not eligible, if both the SHPO and the district engineer agree that it is not eligible for inclusion in the National Register; or

(3) Request a determination of eligibility from the Keeper of the National Register in accordance with applicable National Park Service regulations and notify the applicant, if the SHPO and the district engineer disagree or the ACHP or the Secretary of the Interior so request. If the Keeper of the National Register determines that the resources are not eligible for listing in the National Register or fails to respond within 45 days of receipt of the request, the district engineer may proceed to conclude his action on the permit application.

b. For a historic property outside of waters of the U.S. that will be directly affected by the undertaking the district engineer will, for the purposes of this appendix and compliance with the NHPA:

(1) Treat the historic property as a "designated historic property," if both the SHPO and the district engineer agree that it is eligible for inclusion in the National Register; or

(2) Treat the historic property as not eligible, if both the SHPO and the district engineer agree that it is not eligible for inclusion in the National Register; or

(3) Treat the historic property as not eligible unless the Keeper of the National Register determines it is eligible for or lists it on the National Register. (See paragraph 6.c. below.)

c. If the district engineer and the SHPO do not agree pursuant to paragraph 6.b.(1) and the SHPO notifies the district engineer that it is nominating a potentially eligible historic property for the National Register that may be affected by the undertaking, the district engineer will wait a reasonable period of time for that determination to be made before concluding his action on the permit. Such a reasonable period of time would normally be 30 days for the SHPO to nominate the historic property plus 45 days for the Keeper of the National Register to make such determination. The district engineer will encourage the applicant to cooperate with the SHPO in obtaining the information necessary to nominate the historic property.

7. *Assessing Effects*

a. *Applying the Criteria of Effect and Adverse Effect.* During the public notice comment period or within 30 days after the determination or discovery of a designated history property the district engineer will coordinate with the SHPO and determine if there is an effect and if so, assess the effect. (See Paragraph 15.)

b. *No Effect.* If the SHPO concurs with the district engineer's determination of no effect or fails to respond within 15 days of the district engineer's notice to the SHPO of a no effect determination, then the district engineer may proceed with the final decision.

c. *No Adverse Effect.* If the district engineer, based on his coordination with the SHPO (see paragraph 7.a.), determines that an effect is not adverse, the district engineer will notify the ACHP and request the comments of the ACHP. The district engineer's notice will include a description of both the project and the designated historic property; both the district engineer's and the SHPO's views, as well as any views of affected local governments, Indian tribes, Federal agencies, and the public, on the no adverse effect determination; and a description of the efforts to identify historic properties and solicit the views of those above. The district engineer may conclude the permit decision if the ACHP does not object to the district engineer's determination or if the district engineer accepts any conditions requested by the ACHP for a no adverse effect determination, or the ACHP fails to respond within 30 days of the district engineer's notice to the ACHP. If the ACHP objects or the district engineer does not accept the conditions proposed by the ACHP, then the effect shall be considered as adverse.

d. *Adverse Effect.* If an adverse effect on designated historic properties is found, the district engineer will notify the ACHP and coordinate with the SHPO to seek ways to avoid or reduce effects on designated historic properties. Either the district engineer or the SHPO may request the ACHP to participate. At its discretion, the ACHP may participate without such a request. The district engineer, the SHPO or the ACHP may state that further coordination will not be productive. The district engineer shall then request the ACHP's comments in accordance with paragraph 9.

8. *Consultation*

At any time during permit processing, the district engineer may consult with the involved parties to discuss and consider possible alternatives or measures to avoid or minimize the adverse effects of a proposed activity. The district engineer will terminate any consultation immediately upon determining that further consultation is not productive and will immediately notify the consulting parties. If the consultation results in a mutual agreement among the SHPO, ACHP, applicant and the district engineer regarding the treatment of designated historic properties, then the district engineer may formalize that agreement either through permit conditioning or by signing a Memorandum of Agreement (MOA) with these parties. Such MOA will constitute the comments of the ACHP and the SHPO, and the district engineer may proceed with the permit decision. Consultation shall not continue beyond the comment period provided in paragraph 9.b.

9. *ACHP Review and Comment*

a. If: (i) The district engineer determines that coordination with the SHPO is unproductive; or (ii) the ACHP, within the appropriate comment period, requests additional information in order to provide its comments; or (iii) the ACHP objects to any agreed resolution of impacts on designated historic properties; the district engineer, normally within 30 days, shall provide the ACHP with:

(1) A project description, including, as appropriate, photographs, maps, drawings, and specifications (such as, dimensions of structures, fills, or excavations; types of materials and quantity of material);

(2) A listing and description of the designated historic properties that will be affected, including the reports from any surveys or investigations;

(3) A description of the anticipated adverse effects of the undertaking on the designated historic properties and of the proposed mitigation measures and alternatives considered, if any; and

(4) The views of any commenting parties regarding designated historic properties.

In developing this information, the district engineer may coordinate with the applicant, the SHPO, and any appropriate Indian tribe or certified local government. Copies of the above information also should be forwarded to the applicant, the SHPO, and any appropriate Indian tribe or certified local government. The district engineer will not delay his decision but will consider any comments these parties may wish to provide.

b. The district engineer will provide the ACHP 60 days from the date of the district engineer's letter forwarding the information in paragraph 9.a., to provide its comments. If the ACHP does not comment by the end of this comment period, the district engineer will complete processing of the permit application. When the permit decision is otherwise delayed as provided in 33 CFR 325.2(d)(3) & (4), the district engineer will provide additional time for the ACHP to comment consistent with, but not extending beyond that delay.

10. *District Engineer Decision*

a. In making the public interest decision on a permit application, in accordance with 33 CFR 320.4, the district engineer shall weigh all factors, including the effects of the undertaking on historic properties and any comments of the ACHP and the SHPO, and any views of other interested parties. The district engineer will add permit conditions to avoid or reduce effects on historic properties which he determines are necessary in accordance with 33 CFR 325.4. In reaching his determination, the district engineer will consider the Secretary of the Interior's Standards and Guidelines for Archeology and Historic Preservation (48 FR 44716).

b. If the district engineer concludes that permitting the activity would result in the irrevocable loss of important scientific, prehistoric, historical, or archeological data, the district engineer, in accordance with the Archeological and Historic Preservation Act of 1974, will advise the Secretary of the Interior (by notifying the National Park Service (NPS)) of the extent to which the data may be lost if the undertaking is permitted, any plans to mitigate such loss that will be implemented, and the permit conditions that will be included to ensure that any required mitigation occurs.

11. *Historic Properties Discovered During Construction*

After the permit has been issued, if the district engineer finds or is notified that the permit area contains a previously unknown potentially eligible historic property which he reasonably expects will be affected by the undertaking, he shall immediately inform the Department of the Interior Departmental Consulting Archeologist and the regional office of the NPS of the current knowledge of the potentially eligible historic property and the expected effects, if any, of the undertaking on that property. The district engineer will seek voluntary avoidance of construction activities that could affect the historic property pending a recommendation from the National Park Service pursuant to the Archeological and Historic Preservation Act of 1974. Based on the circumstances of the discovery, equity to all parties, and considerations of the public interest, the district engineer may modify, suspend or revoke a permit in accordance with 33 CFR 325.7.

12. *Regional General Permits*

Potential impacts on historic properties will be considered in development and evaluation of general permits. However, many of the specific procedures contained in this appendix are not normally applicable to general permits. In developing general permits, the district engineer will seek the views of the SHPO and, the ACHP and other organizations and/or individuals with expertise or interest in historic properties. Where designated historic properties are reasonably likely to be affected, general permits shall be conditioned to protect such properties or to limit the applicability of the permit coverage.

13. *Nationwide General Permit*

a. The criteria at paragraph 15 of this Appendix will be used for determining compliance with the nationwide permit condition at 33 CFR 330.5(b)(9) regarding the effect on designated historic properties. When making this determination the district engineer may consult with the SHPO, the ACHP or other interested parties.

b. If the district engineer is notified of a potentially eligible historic property in accordance with nationwide permit regulations and conditions, he will immediately notify the SHPO. If the district engineer believes that the potentially eligible historic property meets the criteria for inclusion in the National Register and that it may be affected by the proposed undertaking then he may suspend authorization of the nationwide permit until he provides the ACHP and the SHPO the opportunity to comment in accordance with the provisions of this Appendix. Once these provisions have been satisfied, the district engineer may notify the general permittee that the activity is authorized including any special activity specific conditions identified or that an individual permit is required.

14. *Emergency Procedures*

The procedures for processing permits in emergency situations are described at 33 CFR 325.2(e)(4). In an emergency situation the district engineer will make every reasonable effort to receive comments from the SHPO and the ACHP, when the proposed undertaking can reasonably be expected to affect a potentially eligible or designated historic property and will comply with the provisions of this Appendix to the extent time and the emergency situation allows.

15. *Criteria of Effect and Adverse Effect*

(a) An undertaking has an effect on a designated historic property when the undertaking may alter characteristics of the property that qualified the property for inclusion in the National Register. For the purpose of determining effect, alteration to features of a property's location, setting, or use may be relevant, and depending on a property's important characteristics, should be considered.

(b) An undertaking is considered to have an adverse effect when the effect on a designated historic property may diminish the integrity of the property's location, design, setting, materials, workmanship, feeling, or association. Adverse effects on designated historic properties include, but are not limited to:

(1) Physical destruction, damage, or alteration of all or part of the property;

(2) Isolation of the property from or alteration of the character of the property's setting when that character contributes to the property's qualification for the National Register;

(3) Introduction of visual, audible, or atmospheric elements that are out of character with the property or alter its setting;

(4) Neglect of a property resulting in its deterioration or destruction; and

(5) Transfer, lease, or sale of the property.

(c) Effects of an undertaking that would otherwise be found to be adverse may be considered as being not adverse for the purpose of this appendix:

(1) When the designated historic property is of value only for its potential contribution to archeological, historical, or architectural research, and when such value can be substantially preserved through the conduct of appropriate research, and such research is conducted in accordance with applicable professional standards and guidelines;

(2) When the undertaking is limited to the rehabilitation of buildings and structures and is conducted in a manner that preserves the historical and architectural value of affected designated historic properties through conformance with the Secretary's "Standards for Rehabilitation and Guidelines for Rehabilitating Historic Buildings", or

(3) When the undertaking is limited to the transfer, lease, or sale of a designated historic property, and adequate restrictions or conditions are included to ensure preservation of the property's important historic features.

[55 FR 27003, June 29, 1990]

PART 326 ENFORCEMENT

AUTHORITY: 33 U.S.C. 401 et seq.; 33 U.S.C. 1344; 33 U.S.C. 1413; 33 U.S.C. 2101

Section 326.1 Purpose

This part prescribes enforcement policies (§ 326.2) and procedures applicable to activities performed without required Department of the Army permits (§ 326.3) and to activities not in compliance with the terms and conditions of issued Department of the Army permits (§ 326.4). Procedures for initiating legal actions are prescribed in § 326.5. Nothing contained in this part shall establish a non-discretionary duty on the part of district engineers nor shall deviation from these procedures give rise to a private right of action against a district engineer.

Section 326.2 Policy

Enforcement, as part of the overall regulatory program of the Corps, is based on a policy of regulating the waters of the United States by discouraging activities that have not been properly authorized and by requiring corrective measures, where appropriate, to ensure those waters are not misused and to maintain the integrity of the program. There are several methods discussed in the remainder of this part which can be used either singly or in combination to implement this policy, while making the most effective use of the enforcement resources available. As EPA has independent enforcement authority under the Clean Water Act for unauthorized discharges, the district engineer should normally coordinate with EPA to determine the most effective and efficient manner by which resolution of a section 404 violation can be achieved.

Section 326.3 Unauthorized activities

(a) *Surveillance.* To detect unauthorized activities requiring permits, district engineers should make the best use of all available resources. Corps employees; members of the public; and representatives of state, local, and other Federal agencies should be encouraged to report suspected violations. Additionally, district engineers should consider developing joint surveillance procedures with Federal, state, or local agencies having similar regulatory responsibilities, special expertise, or interest.

(b) *Initial investigation.* District engineers should take steps to investigate suspected violations in a timely manner. The scheduling of investigations will reflect the nature and location of the suspected violations, the anticipated impacts, and the most effective use of inspection resources available to the district engineer. These investigations should confirm whether a violation exists, and if so, will identify the extent of the violation and the parties responsible.

(c) *Formal notifications to parties responsible for violations.* Once the district engineer has determined that a violation exists, he should take appropriate steps to notify the responsible parties.

(1) If the violation involves a project that is not complete, the district engineer's notification should be in the form of a cease and desist order prohibiting any further work pending resolution of the violation in accordance with the procedures contained in this part. See paragraph (c)(4) of this section for exception to this procedure.

(2) If the violation involves a completed project, a cease and desist order should not be necessary. However, the district engineer should still notify the responsible parties of the violation.

(3) All notifications, pursuant to paragraphs (c) (1) and (2) of this section, should identify the relevant statutory authorities, indicate potential enforcement consequences, and direct the responsible parties to submit any additional information that the district engineer may need at that time to determine what course of action he should pursue in resolving the violation; further information may be requested, as needed, in the future.

(4) In situations which would, if a violation were not involved, qualify for emergency procedures pursuant to 33 CFR part 325.2(e)(4), the district engineer may decide it would not be appropriate to direct that the unauthorized work be stopped. Therefore, in such situations, the district engineer may, at his discretion, allow the work to continue, subject to appropriate limitations and conditions as he may prescribe, while the violation is being resolved in accordance with the procedures contained in this part.

(5) When an unauthorized activity requiring a permit has been undertaken by American Indians (including Alaskan natives, Eskimos, and Aleuts, but not including Native Hawaiians) on reservation lands or in pursuit of specific treaty rights, the district engineer should use appropriate means to coordinate proposed directives and orders with the Assistant Chief Counsel for Indian Affairs (DAEN-CCI).

(6) When an unauthorized activity requiring a permit has been undertaken by an official acting on behalf of a foreign government, the district engineer should use appropriate means to coordinate proposed directives and orders with the Office, Chief of Engineers, ATTN: DAEN-CCK.

(d) *Initial corrective measures*

(1) The district engineer should, in appropriate cases, depending upon the nature of the impacts associated with the unauthorized, completed work, solicit the views of the Environmental Protection Agency; the U.S. Fish and Wildlife Service; the National Marine Fisheries Service, and other Federal, state, and local agencies to facilitate his decision on what initial corrective measures are required. If the district engineer determines as a result of his investigation, coordination, and preliminary evaluation that initial corrective measures are required, he should issue an appropriate order to the parties responsible for the violation. In determining what initial corrective measures are required, the district engineer should consider whether serious jeopardy to life, property, or important public resources (see 33 CFR 320.4) may be reasonably anticipated to occur during the period required for the ultimate resolution of the violation. In his order, the district engineer will specify the initial corrective measures required and the time limits for completing this work. In unusual cases where initial corrective measures substantially eliminate all current and future detrimental impacts resulting from the unauthorized work, further enforcement actions

should normally be unnecessary. For all other cases, the district engineer's order should normally specify that compliance with the order will not foreclose the Government's options to initiate appropriate legal action or to later require the submission of a permit application.

(2) An order requiring initial corrective measures that resolve the violation may also be issued by the district engineer in situations where the acceptance or processing of an after-the-fact permit application is prohibited or considered not appropriate pursuant to § 326.3(e)(1) (iii) through (iv) below. However, such orders will be issued only when the district engineer has reached an independent determination that such measures are necessary and appropriate.

(3) It will not be necessary to issue a Corps permit in connection with initial corrective measures undertaken at the direction of the district engineer.

(e) *After-the-fact permit applications*

(1) Following the completion of any required initial corrective measures, the district engineer will accept an after-the-fact permit application unless he determines that one of the exceptions listed in subparagraphs i–iv below is applicable. Applications for after-the-fact permits will be processed in accordance with the applicable procedures in 33 CFR parts 320 through 325. Situations where no permit application will be processed or where the acceptance of a permit application must be deferred are as follows:

(i) No permit application will be processed when restoration of the waters of the United States has been completed that eliminates current and future detrimental impacts to the satisfaction of the district engineer.

(ii) No permit application will be accepted in connection with a violation where the district engineer determines that legal action is appropriate (§ 326.5(a)) until such legal action has been completed.

(iii) No permit application will be accepted where a Federal, state, or local authorization or certification, required by Federal law, has already been denied.

(iv) No permit application will be accepted nor will the processing of an application be continued when the district engineer is aware of enforcement litigation that has been initiated by other Federal, state, or local regulatory agencies, unless he determines that concurrent processing of an after-the-fact permit application is clearly appropriate.

(v) No appeal of an approved jurisdictional determination (JD) associated with an unauthorized activity or after-the-fact permit application will be accepted unless and until the applicant has furnished a signed statute of limitations tolling agreement to the district engineer. A separate statute of limitations tolling agreement will be prepared for each unauthorized activity. Any person who appeals an approved JD associated with an unauthorized activity or applies for an after-the-fact permit, where the application is accepted and evaluated by the Corps, thereby agrees that the statute of limitations regarding any violation associated with that application is suspended until one year after the final Corps decision, as defined at 33 CFR 331.10. Moreover, the recipient of an approved JD associated with an unauthorized activity or an application for an after-the-fact permit must also memorialize that agreement to toll the statute of limitations, by signing an agreement to that effect, in exchange for the Corps acceptance of the after-the-fact permit application, and/or any administrative appeal. Such agreement will state that, in exchange for the Corps acceptance of any after-the-fact permit application and/or any administrative appeal associated with the unauthorized activity, the responsible party agrees that the statute of limitations will be suspended (i.e., tolled) until one year after the final Corps decision on the after-the-fact permit application or, if there is an administrative appeal, one year after the final Corps decision as defined at 33 CFR 331.10, whichever date is later.

(2) Upon completion of his review in accordance with 33 CFR parts 320 through 325, the district engineer will determine if a permit should be issued, with special conditions if appropriate, or denied. In reaching a decision to issue, he must determine that the work involved is not contrary to the public interest, and if section 404 is applicable, that the work also complies with the Environmental Protection Agency's section 404(b)(1) guidelines. If he determines that a denial is warranted, his notification of denial should prescribe any final corrective actions required. His notification should also establish a reasonable period of time for the applicant to complete such actions unless he determines that further information is required before the corrective measures can be specified. If further information is required, the final corrective measures may be specified at a later date. If an applicant refuses to undertake prescribed corrective actions ordered subsequent to permit denial or refuses to accept a conditioned permit, the district engineer may initiate legal action in accordance with § 326.5.

(f) *Combining steps.* The procedural steps in this section are in the normal sequence. However, these regulations do not prohibit the streamlining of the enforcement process through the combining of steps.

(g) *Coordination with EPA.* In all cases where the district engineer is aware that EPA is considering enforcement action, he should coordinate with EPA to attempt to avoid conflict or duplication. Such coordination applies to interim protective measures and after-the-fact permitting, as well as to appropriate legal enforcement actions.

[51 FR 41246, Nov. 13, 1986, as amended at 64 FR 11714, Mar. 9, 1999; 65 FR 16493, Mar. 28, 2000]

Section 326.4 Supervision of authorized activities

(a) *Inspections.* District engineers will, at their discretion, take reasonable measures to inspect permitted activities, as required, to ensure that these activities comply with specified terms and conditions. To supplement inspections by their enforcement personnel, district engineers should encourage their other personnel; members of the public; and interested state, local, and other Federal agency representatives to report suspected violations of Corps permits. To facilitate inspections, district engineers will, in appropriate cases, require that copies of ENG Form 4336 be posted conspicuously at the sites of authorized activities and will make available to all interested persons information on the terms and conditions of issued permits. The U.S. Coast Guard will inspect permitted ocean dumping activities pursuant to section 107(c) of the Marine Protection, Research and Sanctuaries Act of 1972, as amended.

(b) *Inspection limitations.* § 326.4 does not establish a non-discretionary duty to inspect permitted activities for safety, sound engineering practices, or interference with other permitted or unpermitted structures or uses in the area. Further, the regulations implementing the Corps regulatory program do not establish a non-discretionary duty to inspect permitted activities for any other purpose.

(c) *Inspection expenses.* The expenses incurred in connection with the inspection of permitted activities will normally be paid by the Federal Government unless daily supervision or other unusual expenses are involved. In such unusual cases, the district engineer may condition permits to require permittees to pay inspection expenses pursuant to the authority contained in section 9701 of Pub L. 97-258 (33 U.S.C. 9701). The collection and disposition of inspection expense funds obtained from applicants will be administered in accordance with the relevant Corps regulations governing such funds.

(d) *Non-compliance.* If a district engineer determines that a permittee has violated the terms or conditions of the permit and that the violation is sufficiently serious to require an enforcement action, then he should, unless at his discretion he deems it inappropriate:

(1) First contact the permittee;

(2) Request corrected plans reflecting actual work, if needed; and

(3) Attempt to resolve the violation. Resolution of the violation may take the form of the permitted project being voluntarily brought into compliance or of a permit modification (33 CFR 325.7(b)).

If a mutually agreeable solution cannot be reached, a written order requiring compliance should normally be issued and delivered by personal service. Issuance of an order is not, however, a prerequisite to legal action. If an order is issued, it will specify a time period of not more than 30 days for bringing the permitted project into compliance, and a copy will be sent to the appropriate state official pursuant to section 404(s)(2) of the Clean Water Act. If the permittee fails to comply with the order within the specified period of time, the district engineer may consider using the suspension/ revocation procedures in 33 CFR 325.7(c) and/or he may recommend legal action in accordance with § 326.5.

Section 326.5 Legal action

(a) *General.* For cases the district engineer determines to be appropriate, he will recommend criminal or civil actions to obtain penalties for violations, compliance with the orders and directives he has issued pursuant to §§ 326.3 and 326.4, or other relief as appropriate. Appropriate cases for criminal or civil action include, but are not limited to, violations which, in the district engineer's opinion, are willful, repeated, flagrant, or of substantial impact.

(b) *Preparation of case.* If the district engineer determines that legal action is appropriate, he will prepare a litigation report or such other documentation that he and the local U.S. Attorney have mutually agreed to, which contains an analysis of the information obtained during his investigation of the violation or during the processing of a permit application and a recommendation of appropriate legal action. The litigation report or alternative documentation will also recommend what, if any, restoration or mitigative measures are required and will provide the rationale for any such recommendation.

(c) *Referral to the local U.S. Attorney.* Except as provided in paragraph (d) of this section, district engineers are authorized to refer cases directly to the U.S. Attorney. Because of the unique legal system in the Trust Territories, all cases over which the Department of Justice has no authority will be referred to the Attorney General for the trust Territories. Information copies of all letters of referral shall be forwarded to the appropriate division counsel, the Office, Chief of Engineers, ATTN: DAEN-CCK, the Office of the Assistant Secretary of the Army (Civil Works), and the Chief of the Environmental Defense Section, Lands and Natural Resources Division, U.S. Department of Justice.

(d) *Referral to the Office, Chief of Engineers.* District engineers will forward litigation reports with recommendations through division offices to the Office, Chief of Engineers, ATTN: DAEN-CCK, for all cases that qualify under the following criteria:

(1) Significant precedential or controversial questions of law or fact;

(2) Requests for elevation to the Washington level by the Department of Justice;

(3) Violations of section 9 of the Rivers and Harbors Act of 1899;

(4) Violations of section 103 the Marine Protection, Research and Sanctuaries Act of 1972;

(5) All cases involving violations by American Indians (original of litigation report to DAEN-CCI with copy to DAEN-CCK) on reservation lands or in pursuit of specific treaty rights;

(6) All cases involving violations by officials acting on behalf of foreign governments; and

(7) Cases requiring action pursuant to paragraph (e) of this section.

(e) *Legal option not available.* In cases where the local U.S. Attorney declines to take legal action, it would be appropriate for the district engineer to close the enforcement case record unless he believes that the case warrants special attention. In that situation, he is encouraged to forward a litigation report to the Office, Chief of Engineers, ATTN: DAEN-CCK, for direct coordination through the Office of the Assistant Secretary of the Army (Civil Works) with the Department of Justice. Further, the case record should not be closed if the district engineer anticipates that further administrative enforcement actions, taken in accordance with the procedures prescribed in this part, will identify remedial measures which, if not complied with by the parties responsible for the violation, will result in appropriate legal action at a later date.

Section 326.6 Class I administrative penalties

(a) *Introduction*

(1) This section sets forth procedures for initiation and administration of Class I administrative penalty orders under section 309(g) of the Clean Water Act, and section 205 of the National Fishing Enhancement Act. Section 309(g)(2)(A) specifies that Class I civil penalties may not exceed $10,000 per violation, except that the maximum amount of any Class I civil penalty shall not exceed $25,000. The National Fishing Enhancement Act, section 205(e), provides that penalties for violations of permits issued in accordance with that Act shall not exceed $10,000 for each violation.

(2) These procedures supplement the existing enforcement procedures at §§ 326.1 through 326.5. However, as a matter of Corps enforcement discretion once the Corps decides to proceed with an administrative penalty under these procedures it shall not subsequently pursue judicial action pursuant to § 326.5. Therefore, an administrative penalty should not be pursued if a subsequent judicial action for civil penalties is desired. An administrative civil penalty may be pursued in conjunction with a compliance order; request for restoration and/or request for mitigation issued under § 326.4.

(3) *Definitions.* For the purposes of this section of the regulation:

(i) *Corps* means the Secretary of the Army, acting through the U.S. Army Corps of Engineers, with respect to the matters covered by this regulation.

(ii) *Interested person outside the Corps* includes the permittee, any person who filed written comments on the proposed penalty order, and any other person not employed by the Corps with an interest in the subject of proposed penalty order, and any attorney of record for those persons.

(iii) *Interested Corps staff* means those Corps employees, whether temporary or permanent, who may investigate, litigate, or present evidence, arguments, or the position of the Corps in the hearing or who participated in the preparation, investigation or deliberations concerning the proposed penalty order, including any employee, contractor, or consultant who may be called as a witness.

(iv) *Permittee* means the person to whom the Corps issued a permit under section 404 of the Clean Water Act, (or section 10 of the Rivers and Harbors Act for an Artificial Reef) the conditions and limitations of which permit have allegedly been violated.

(v) *Presiding Officer* means a member of Corps Counsel staff or any other qualified person designated by the District Engineer (DE), to hold a hearing on a proposed administrative civil penalty order (hereinafter referred to as "proposed order") in accordance with the rules set forth in this regulation and to make such recommendations to the DE as prescribed in this regulation.

(vi) *Ex parte communication* means any communication, written or oral, relating to the merits of the proceeding, between the Presiding Officer and an interested person outside the Corps or the interested Corps staff, which was not originally filed or stated in the administrative record or in the hearing. Such communication is not an "ex parte communication" if all parties have received prior written notice of the proposed communication and have been given the opportunity to participate herein.

(b) *Initiation of action*

(1) If the DE or a delegatee of the DE finds that a recipient of a Department of the Army permit (hereinafter referred to as "the permittee") has violated any permit condition or limitation contained in that permit, the DE is authorized to prepare and process a proposed order in accordance with these procedures. The proposed order shall specify the amount of the penalty which the permittee may be assessed and shall describe with reasonable specificity the nature of the violation.

(2) The permittee will be provided actual notice, in writing, of the DE's proposal to issue an administrative civil penalty and will be advised of the right to request a hearing and to present evidence on the alleged violation. Notice to the permittee will be provided by certified mail, return receipt requested, or other notice, at the discretion of the DE when he determines justice so requires. This notice will be accompanied by a copy of the proposed order, and will include the following information:

(i) A description of the alleged violation and copies of the applicable law and regulations;

(ii) An explanation of the authority to initiate the proceeding;

(iii) An explanation, in general terms, of the procedure for assessing civil penalties, including opportunities for public participation;

(iv) A statement of the amount of the penalty that is proposed and a statement of the maximum amount of the penalty which the DE is authorized to assess for the violations alleged;

(v) A statement that the permittee may within 30 calendar days of receipt of the notice provided under this subparagraph, request a hearing prior to issuance of any final order. Further, that the permittee must request a hearing within 30 calendar days of receipt of the notice provided under this subparagraph in order to be entitled to receive such a hearing;

(vi) The name and address of the person to whom the permittee must send a request for hearing;

(vii) Notification that the DE may issue the final order on or after 30 calendar days following receipt of the notice provided under these rules, if the permittee does not request a hearing; and

(viii) An explanation that any final order issued under this section shall become effective 30 calendar days following its issuance unless a petition to set aside the order and to hold a hearing is filed by a person who commented on the proposed order and such petition is granted or an appeal is taken under section 309(g)(8) of the Clean Water Act.

(3) At the same time that actual notice is provided to the permittee, the DE shall give public notice of the proposed order, and provide reasonable opportunity for public comment on the proposed order, prior to issuing a final order assessing an administrative civil penalty. Procedures for giving public notice and providing the opportunity for public comment are contained in § 326.6(c).

(4) At the same time that actual notice is provided to the permittee, the DE shall provide actual notice, in writing, to the appropriate state agency for the state in which the violation occurred. Procedures for providing actual notice to and consulting with the appropriate state agency are contained in § 326.6(d).

(c) *Public notice and comment*

(1) At the same time the permittee and the appropriate state agency are provided actual notice, the DE shall provide public notice of and a reasonable opportunity to comment on the DE's proposal to issue an administrative civil penalty against the permittee.

(2) A 30 day public comment period shall be provided. Any person may submit written comments on the proposed administrative penalty order. The DE shall include all written comments in an administrative record relating to the proposed order. Any person who comments on a proposed order shall be given notice of any hearing held on the proposed order. Such persons shall have a reasonable opportunity to be heard and to present evidence in such hearings.

(3) If no hearing is requested by the permittee, any person who has submitted comments on the proposed order shall be given notice by the DE of any final order issued, and will be given 30 calendar days in which to petition the DE to set aside the order and to provide a hearing on the penalty. The DE shall set aside the order and provide a hearing in accordance with these rules if the evidence presented by the commenter in support of the commenter's petition for a hearing is material and was not considered when the order was issued. If the DE denies a hearing, the DE shall provide notice to the commenter filing the petition for the hearing, together with the reasons for the denial. Notice of the denial and the reasons for the denial shall be published in the Federal Register by the DE.

(4) The DE shall give public notice by mailing a copy of the information listed in paragraph (c)(5), of this section to:

(i) Any person who requests notice;

(ii) Other persons on a mailing list developed to include some or all of the following sources:

(A) Persons who request in writing to be on the list;

(B) Persons on "area lists" developed from lists of participants in past similar proceedings in that area, including hearings or other actions related to section 404 permit issuance as required by § 325.3(d)(1).

The DE may update the mailing list from time to time by requesting written indication of continued interest from those listed. The DE may delete from the list the name of any person who fails to respond to such a request.

(5) All public notices under this subpart shall contain at a minimum the information provided to the permittee as described in § 326.6(b)(2) and:

(i) A statement of the opportunity to submit written comments on the proposed order and the deadline for submission of such comments;

(ii) Any procedures through which the public may comment on or participate in proceedings to reach a final decision on the order;

(iii) The location of the administrative record referenced in § 326.6(e), the times at which the administrative record will be available for public inspection, and a statement that all information submitted by the permittee and persons commenting on the proposed order is available as part of the administrative record, subject to provisions of law restricting the public disclosure of confidential information.

(d) *State consultation*

(1) At the same time that the permittee is provided actual notice, the DE shall send the appropriate state agency written notice of proposal to issue an administrative civil penalty order. This notice will include the same information required pursuant to § 326.6(c)(5).

(2) For the purposes of this regulation, the appropriate State agency will be the agency administering the 401 certification

program, unless another state agency is agreed to by the District and the respective state through formal/informal agreement with the state.

(3) The appropriate state agency will be provided the same opportunity to comment on the proposed order and participate in any hearing that is provided pursuant to § 326.6(c).

(e) *Availability of the administrative record*

(1) At any time after the public notice of a proposed penalty order is given under § 326.6(c), the DE shall make available the administrative record at reasonable times for inspection and copying by any interested person, subject to provisions of law restricting the public disclosure of confidential information. Any person requesting copies of the administrative record or portions of the administrative record may be required by the DE to pay reasonable charges for reproducing the information requested.

(2) The administrative record shall include the following:

(i) Documentation relied on by the DE to support the violations alleged in the proposed penalty order with a summary of violations, if a summary has been prepared;

(ii) Proposed penalty order or assessment notice;

(iii) Public notice of the proposed order with evidence of notice to the permittee and to the public;

(iv) Comments by the permittee and/or the public on the proposed penalty order, including any requests for a hearing;

(v) All orders or notices of the Presiding Officer;

(vi) Subpoenas issued, if any, for the attendance and testimony of witnesses and the production of relevant papers, books, or documents in connection with any hearings;

(vii) All submittals or responses of any persons or comments to the proceeding, including exhibits, if any;

(viii) A complete and accurate record or transcription of any hearing;

(ix) The recommended decision of the Presiding Officer and final decision and/or order of the Corps issued by the DE; and

(x) Any other appropriate documents related to the administrative proceeding;

(f) *Counsel.* A permittee may be represented at all stages of the proceeding by counsel. After receiving notification that a permittee or any other party or commenter is represented by counsel, the Presiding Officer and DE shall direct all further communications to that counsel.

(g) *Opportunity for hearing*

(1) The permittee may request a hearing and may provide written comments on the proposed administrative penalty order at any time within 30 calendar days after receipt of the notice set forth in § 326.6(b)(2). The permittee must request the hearing in writing, specifying in summary form the factual and legal issues which are in dispute and the specific factual and legal grounds for the permittee's defense.

(2) The permittee waives the right to a hearing to present evidence on the alleged violation or violations if the permittee does not submit the request for the hearing to the official designated in the notice of the proposed order within 30 calendar days of receipt of the notice. The DE shall determine the date of receipt of notice by permittee's signed and dated return receipt or such other evidence that constitutes proof of actual notice on a certain date.

(3) The DE shall promptly schedule requested hearings and provide reasonable notice of the hearing schedule to all participants, except that no hearing shall be scheduled prior to the end of the thirty day public comment period provided in § 326.6(c)(2). The DE may grant any delays or continuances necessary or desirable to resolve the case fairly.

(4) The hearing shall be held at the district office or a location chosen by the DE, except the permittee may request in writing upon a showing of good cause that the hearing be held at an alternative location. Action on such request is at the discretion of the DE.

(h) *Hearing*

(1) Hearings shall afford permittees with an opportunity to present evidence on alleged violations and shall be informal, adjudicatory hearings and shall not be subject to section 554 or 556 of the Administrative Procedure Act. Permittees may present evidence either orally or in written form in accordance with the hearing procedures specified in § 326.6(i).

(2) The DE shall give written notice of any hearing to be held under these rules to any person who commented on the proposed administrative penalty order under § 326.6(c). This notice shall specify a reasonable time prior to the hearing within which the commenter may request an opportunity to be heard and to present oral evidence or to make comments in writing in any such hearing. The notice shall require that any such request specify the facts or issues which the commenter wishes to address. Any commenter who files comments pursuant to § 326.6(c)(2) shall have a right to be heard and to present evidence at the hearing in conformance with these procedures.

(3) The DE shall select a member of the Corps counsel staff or other qualified person to serve as Presiding Officer of the hearing. The Presiding Officer shall exercise no other responsibility, direct or supervisory, for the investigation or prosecution of any case before him. The Presiding Officer shall conduct hearings as specified by these rules and make a recommended decision to the DE.

(4) The Presiding Officer shall consider each case on the basis of the evidence presented, and must have no prior connection with the case. The Presiding Officer is solely responsible for the recommended decision in each case.

(5) *Ex Parte Communications*

(i) No interested person outside the Corps or member of the interested Corps staff shall make, or knowingly cause to be made, any ex parte communication on the merits of the proceeding.

(ii) The Presiding Officer shall not make, or knowingly cause to be made, any ex parte communication on the proceeding to any interested person outside the Corps or to any member of the interested Corps staff.

(iii) The DE may replace the Presiding Officer in any proceeding in which it is demonstrated to the DE's satisfaction that the Presiding Officer has engaged in prohibited ex parte communications to the prejudice of any participant.

(iv) Whenever an ex parte communication in violation of this section is received by the Presiding Officer or made known to the Presiding Officer, the Presiding Officer shall immediately notify all participants in the proceeding of the circumstances and substance of the communication and may require the person who made the communication or caused it to be made, or the party whose representative made the communication or caused it to be made, to the extent consistent with justice and the policies of the Clean Water Act, to show cause why that person or party's claim or interest in the proceedings should not be dismissed, denied, disregarded, or otherwise adversely affected on account of such violation.

(v) The prohibitions of this paragraph apply upon designation of the Presiding Officer and terminate on the date of final action or the final order.

(i) *Hearing Procedures*

(1) The Presiding Officer shall conduct a fair and impartial proceeding in which the participants are given a reasonable opportunity to present evidence.

(2) The Presiding Officer may subpoena witnesses and issue subpoenas for documents pursuant to the provisions of the Clean Water Act.

(3) The Presiding Officer shall provide interested parties a reasonable opportunity to be heard and to present evidence. Interested parties include the permittee, any person who filed a request to participate under 33 CFR 326.6(c), and any other person attending the hearing. The Presiding Officer may establish reasonable time limits for oral testimony.

(4) The permittee may not challenge the permit condition or limitation which is the subject matter of the administrative penalty order.

(5) Prior to the commencement of the hearing, the DE shall provide to the Presiding Officer the complete administrative record as of that date. During the hearing, the DE, or an authorized representative of the DE may summarize the basis for the proposed administrative order. Thereafter, the administrative record shall be admitted into evidence and the Presiding Officer shall maintain the administrative record of the proceedings and shall include in that record all documentary evidence, written statements, correspondence, the record of hearing, and any other relevant matter.

(6) The Presiding Officer shall cause a tape recording, written transcript or other permanent, verbatim record of the hearing to be made, which shall be included in the administrative record, and shall, upon written request, be made available, for inspection or copying, to the permittee or any person, subject to provisions of law restricting the public disclosure of confidential information. Any person making a request may be required to pay reasonable charges for copies of the administrative record or portions thereof.

(7) In receiving evidence, the Presiding Officer is not bound by strict rules of evidence. The Presiding Officer may determine the weight to be accorded the evidence.

(8) The permittee has the right to examine, and to respond to the administrative record. The permittee may offer into evidence, in written form or through oral testimony, a response to the administrative record including, any facts, statements, explanations, documents, testimony, or other exculpatory items which bear on any appropriate issues. The Presiding Officer may question the permittee and require the authentication of any written exhibit or statement. The Presiding Officer may exclude any repetitive or irrelevant matter.

(9) At the close of the permittee's presentation of evidence, the Presiding Officer should allow the introduction of rebuttal evidence. The Presiding Officer may allow the permittee to respond to any such rebuttal evidence submitted and to cross-examine any witness.

(10) The Presiding Officer may take official notice of matters that are not reasonably in dispute and are commonly known in the community or are ascertainable from readily available sources of known accuracy. Prior to taking official notice of a matter, the Presiding Officer shall give the Corps and the permittee an opportunity to show why such notice should not be taken. In any case in which official notice is taken, the Presiding Officer shall place a written statement of the matters as to which such notice was taken in the record, including the basis for such notice and a statement that the Corps or permittee consented to such notice being taken or a summary of the objections of the Corps or the permittee.

(11) After all evidence has been presented, any participant may present argument on any relevant issue, subject to reasonable time limitations set at the discretion of the Presiding Officer.

(12) The hearing record shall remain open for a period of 10 business days from the date of the hearing so that the permittee or any person who has submitted comments on the proposed order may examine and submit responses for the record.

(13) At the close of this 10 business day period, the Presiding Officer may allow the introduction of rebuttal evidence. The Presiding Officer may hold the record open for an additional 10 business days to allow the presentation of such rebuttal evidence.

(j) *The decision*

(1) Within a reasonable time following the close of the hearing and receipt of any statements following the hearing and after consultation with the state pursuant to § 326.6(d), the Presiding Officer shall forward a recommended decision accompanied by a written statement of reasons to the DE. The decision shall recommend that the DE withdraw, issue, or modify and issue the proposed order as a final order. The recommended decision shall be based on a preponderance of the evidence in the administrative record. If the Presiding Officer finds that there is not a preponderance of evidence in the record to support the penalty or the amount of the penalty in a proposed order, the Presiding Officer may recommend that the order be withdrawn or modified and then issued on terms that are supported by a preponderance of evidence on the record. The Presiding Officer also shall make the complete administrative record available to the DE for review.

(2) The Presiding Officer's recommended decision to the DE shall become part of the administrative record and shall be made available to the parties to the proceeding at the time the DE's decision is released pursuant to § 326.6(j)(5). The Presiding Officer's recommended decision shall not become part of the administrative record until the DE's final decision is issued, and shall not be made available to the permittee or public prior to that time.

(3) The rules applicable to Presiding Officers under § 326.6(h)(5) regarding ex parte communications are also applicable to the DE and to any person who advises the DE on the decision or the order, except that communications between the DE and the Presiding Officer do not constitute ex parte communications, nor do communications between the DE and his staff prior to issuance of the proposed order.

(4) The DE may request additional information on specified issues from the participants, in whatever form the DE designates, giving all participants a fair opportunity to be heard on such additional matters. The DE shall include this additional information in the administrative record.

(5) Within a reasonable time following receipt of the Presiding Officer's recommended decision, the DE shall withdraw, issue, or modify and issue the proposed order as a final order. The DE's decision shall be based on a preponderance of the evidence in the administrative record, shall consider the penalty factors set out in section 309(g)(3) of the CWA, shall be in writing, shall include a clear and concise statement of reasons for the decision, and shall include any final order assessing a penalty. The DE's decision, once issued, shall constitute final Corps action for purposes of judicial review.

(6) The DE shall issue the final order by sending the order, or written notice of its withdrawal, to the permittee by certified mail. Issuance of the order under this subparagraph constitutes final Corps action for purposes of judicial review.

(7) The DE shall provide written notice of the issuance, modification and issuance, or withdrawal of the proposed order to every person who submitted written comments on the proposed order.

(8) The notice shall include a statement of the right to judicial review and of the procedures and deadlines for obtaining judicial review. The notice shall also note the right of a commenter to petition for a hearing pursuant to 33 CFR 326.6(c)(3) if no hearing was previously held.

(k) *Effective date of order*

(1) Any final order issued under this subpart shall become effective 30 calendar

days following its issuance unless an appeal is taken pursuant to section 309(g)(8) of the Clean Water Act, or in the case where no hearing was held prior to the final order, and a petition for hearing is filed by a prior commenter.

(2) If a petition for hearing is received within 30 days after the final order is issued, the DE shall:

(i) Review the evidence presented by the petitioner.

(ii) If the evidence is material and was not considered in the issuance of the order, the DE shall immediately set aside the final order and schedule a hearing. In that case, a hearing will be held, a new recommendation will be made by the Presiding Officer to the DE and a new final decision issued by the DE.

(iii) If the DE denies a hearing under this subparagraph, the DE shall provide to the petitioner, and publish in the Federal Register, notice of, and the reasons for, such denial.

(l) *Judicial review*

(1) Any permittee against whom a final order assessing a civil penalty under these regulations or any person who provided written comments on a proposed order may obtain judicial review of the final order.

(2) In order to obtain judicial review, the permittee or commenter must file a notice of appeal in the United States District Court for either the District of Columbia, or the district in which the violation was alleged to have occurred, within 30 calendar days after the date of issuance of the final order.

(3) Simultaneously with the filing of the notice of appeal, the permittee or commenter must send a copy of such notice by certified mail to the DE and the Attorney General.

[54 FR 50709, Dec. 8, 1989]

PART 327
PUBLIC HEARINGS

AUTHORITY: 33 U.S.C. 1344; 33 U.S.C. 1413

Section 327.1 Purpose

This regulation prescribes the policy, practice and procedures to be followed by the U.S. Army Corps of Engineers in the conduct of public hearings conducted in the evaluation of a proposed DA permit action or Federal project as defined in § 327.3 of this part including those held pursuant to section 404 of the Clean Water Act (33 U.S.C. 1344) and section 103 of the Marine Protection, Research and Sanctuaries Act (MPRSA), as amended (33 U.S.C. 1413).

Section 327.2 Applicability

This regulation is applicable to all divisions and districts responsible for the conduct of public hearings.

Section 327.3 Definitions

(a) *Public hearing* means a public proceeding conducted for the purpose of acquiring information or evidence which will be considered in evaluating a proposed DA permit action, or Federal project, and which affords the public an opportunity to present their views, opinions, and information on such permit actions or Federal projects.

(b) *Permit action*, as used herein means the evaluation of and decision on an application for a DA permit pursuant to sections 9 or 10 of the Rivers and Harbors Act of 1899, section 404 of the Clean Water Act, or section 103 of the MPRSA, as amended, or the modification, suspension or revocation of any DA permit (see 33 CFR 325.7).

(c) *Federal project* means a Corps of Engineers project (work or activity of any nature for any purpose which is to be performed by the Chief of Engineers pursuant to Congressional authorizations) involving the discharge of dredged or fill material into waters of the United States or the transportation of dredged material for the purpose of dumping it in ocean waters subject to section 404 of the Clean Water Act, or section 103 of the MPRSA.

Section 327.4 General policies

(a) A public hearing will be held in connection with the consideration of a DA permit application or a Federal project whenever a public hearing is needed for making a decision on such permit application or Federal project. In addition, a public hearing may be held when it is proposed to modify or revoke a permit. (See 33 CFR 325.7.)

(b) Unless the public notice specifies that a public hearing will be held, any person may request, in writing, within the comment period specified in the public notice on a DA permit application or on a Federal project, that a public hearing be held to consider the material matters at issue in the permit application or with respect to Federal project. Upon receipt of any such request, stating with particularity the reasons for holding a public hearing, the district engineer may expeditiously attempt to resolve the issues informally. Otherwise, he shall promptly set a time and place for the public hearing, and give due notice thereof, as prescribed in § 327.11 of this part. Requests for a public hearing under this paragraph shall be granted, unless the district engineer determines that the issues raised are insubstantial or there is otherwise no valid interest to be served by a hearing. The district engineer will make such a determination in writing, and communicate his reasons therefor to all requesting parties. Comments received as form letters or petitions may be acknowledged as a group to the person or organization responsible for the form letter or petition.

(c) In case of doubt, a public hearing shall be held. HQDA has the discretionary power to require hearings in any case.

(d) In fixing the time and place for a hearing, the convenience and necessity of the interested public will be duly considered.

Section 327.5 Presiding officer

(a) The district engineer, in whose district a matter arises, shall normally serve as the presiding officer. When the district engineer is unable to serve, he may designate the deputy district engineer or other qualified person as presiding officer. In cases of unusual interest, the Chief of Engineers or the division engineer may appoint such person as he deems appropriate to serve as the presiding officer.

(b) The presiding officer shall include in the administrative record of the permit action the request or requests for the hearing and any data or material submitted in justification thereof, materials submitted in opposition to or in support of the proposed action, the hearing transcript, and such other material as may be relevant or pertinent to the subject matter of the hearing. The administrative record shall be available for public inspection with the exception of material exempt from disclosure under the Freedom of Information Act.

Section 327.6 Legal adviser

At each public hearing, the district counsel or his designee may serve as legal advisor to the presiding officer. In appropriate circumstances, the district engineer may waive the requirement for a legal advisor to be present.

Section 327.7 Representation

At the public hearing, any person may appear on his own behalf, or may be represented by counsel, or by other representatives.

Section 327.8 Conduct of hearings

(a) The presiding officer shall make an opening statement outlining the purpose of the hearing and prescribing the general procedures to be followed.

(b) Hearings shall be conducted by the presiding officer in an orderly but expeditious manner. Any person shall be permitted to submit oral or written statements concerning the subject matter of the hearing, to call witnesses who may present oral or written statements, and to present recommendations as to an appropriate decision. Any person may present written statements for the hearing record prior to the time the hearing record is closed to public submissions, and may present proposed findings and recommendations. The presiding officer shall afford participants a reasonable opportunity for rebuttal.

(c) The presiding officer shall have discretion to establish reasonable limits upon the time allowed for statements of witnesses, for arguments of parties or their counsel or representatives, and upon the number of rebuttals.

(d) Cross-examination of witnesses shall not be permitted.

(e) All public hearings shall be reported verbatim. Copies of the transcripts of proceedings may be purchased by any person from the Corps of Engineers or the reporter of such hearing. A copy will be available for public inspection at the office of the appropriate district engineer.

(f) All written statements, charts, tabulations, and similar data offered in evidence at the hearing shall, subject to exclusion by the presiding officer for reasons of redundancy, be received in evidence and shall constitute a part of the record.

(g) The presiding officer shall allow a period of not less than 10 days after the close of the public hearing for submission of written comments.

(h) In appropriate cases, the district engineer may participate in joint public hearings with other Federal or state agencies, provided the procedures of those hearings meet the requirements of this regulation. In those cases in which the other Federal or state agency allows a cross-examination in its public hearing, the district engineer may still participate in the joint public hearing but shall not require cross examination as a part of his participation.

Section 327.9 Filing of the transcript of the public hearing

Where the presiding officer is the initial action authority, the transcript of the public hearing, together with all evidence introduced at the public hearing, shall be made a part of the administrative record of the permit action or Federal project. The initial action authority shall fully consider the matters discussed at the public hearing in arriving at his initial decision or recommendation and shall address, in his decision or recommendation, all substantial and valid issues presented at the hearing. Where a person other than the initial action authority serves as presiding officer, such person shall forward the transcript of the public hearing and all evidence received in connection therewith to the initial action authority together with a report summarizing the issues covered at the hearing. The report of the presiding officer and the transcript of the public hearing and evidence submitted thereat shall in such cases be fully considered by the initial action authority in making his decision or recommendation to higher authority as to such permit action or Federal project.

Section 327.10 Authority of the presiding officer

Presiding officers shall have the following authority:

(a) To regulate the course of the hearing including the order of all sessions and the scheduling thereof, after any initial session, and the recessing, reconvening, and adjournment thereof; and

(b) To take any other action necessary or appropriate to the discharge of the duties vested in them, consistent with the statutory or other authority under which the Chief of Engineers functions, and with the policies and directives of the Chief of Engineers and the Secretary of the Army.

Section 327.11 Public notice

(a) Public notice shall be given of any public hearing to be held pursuant to this regulation. Such notice should normally provide for a period of not less than 30 days following the date of public notice during which time interested parties may prepare themselves for the hearing. Notice shall also be given to all Federal agencies affected by the proposed action, and to state and local agencies and other parties having an interest in the subject matter of the hearing. Notice shall be sent to all persons requesting a hearing and shall be posted in appropriate government buildings and provided to newspapers of general circulation for publication. Comments received as form letters or petitions may be acknowledged as a group to the person or organization responsible for the form letter or petition.

(b) The notice shall contain time, place, and nature of hearing; the legal authority and jurisdiction under which the hearing is held; and location of and availability of the draft environmental impact statement or environmental assessment.

PART 328 DEFINITION OF WATERS OF THE UNITED STATES

AUTHORITY: 33 U.S.C. 1344

Section 328.1 Purpose

This section defines the term "waters of the United States" as it applies to the jurisdictional limits of the authority of the Corps of Engineers under the Clean Water Act. It prescribes the policy, practice, and procedures to be used in determining the extent of jurisdiction of the Corps of Engineers concerning "waters of the United States." The terminology used by section 404 of the Clean Water Act includes "navigable waters" which is defined at section 502(7) of the Act as "waters of the United States including the territorial seas." To provide clarity and to avoid confusion with other Corps of Engineer regulatory programs, the term "waters of the United States" is used throughout 33 CFR parts 320 through 330. This section does not apply to authorities under the Rivers and Harbors Act of 1899 except that some of the same waters may be regulated under both statutes (see 33 CFR parts 322 and 329).

Section 328.2 General scope

Waters of the United States include those waters listed in § 328.3(a). The lateral limits of jurisdiction in those waters may be divided into three categories. The categories include the territorial seas, tidal waters, and non-tidal waters (see 33 CFR 328.4 (a), (b), and (c), respectively).

Section 328.3 Definitions

For the purpose of this regulation these terms are defined as follows:

(a) The term *waters of the United States* means

(1) All waters which are currently used, or were used in the past, or may be susceptible to use in interstate or foreign commerce, including all waters which are subject to the ebb and flow of the tide;

(2) All interstate waters including interstate wetlands;

(3) All other waters such as intrastate lakes, rivers, streams (including intermittent streams), mudflats, sandflats, wetlands, sloughs, prairie potholes, wet meadows, playa lakes, or natural ponds, the use, degradation or destruction of which could affect interstate or foreign commerce including any such waters:

(i) Which are or could be used by interstate or foreign travelers for recreational or other purposes; or

(ii) From which fish or shellfish are or could be taken and sold in interstate or foreign commerce; or

(iii) Which are used or could be used for industrial purpose by industries in interstate commerce;

(4) All impoundments of waters otherwise defined as waters of the United States under the definition;

(5) Tributaries of waters identified in paragraphs (a) (1) through (4) of this section;

(6) The territorial seas;

(7) Wetlands adjacent to waters (other than waters that are themselves wetlands) identified in paragraphs (a)(1) through (6) of this section.

(8) Waters of the United States do not include prior converted cropland. Notwithstanding the determination of an area's status as prior converted cropland by any other Federal agency, for the purposes of the Clean Water Act, the final authority regarding Clean Water Act jurisdiction remains with EPA. Waste treatment systems, including treatment ponds or lagoons designed to meet the requirements of CWA (other than cooling ponds as defined in 40 CFR 423.11(m) which also meet the criteria of this definition) are not waters of the United States.

(b) The term *wetlands* means those areas that are inundated or saturated by surface or ground water at a frequency and duration sufficient to support, and that under normal circumstances do support, a prevalence of vegetation typically adapted for life in saturated soil conditions. Wetlands generally include swamps, marshes, bogs, and similar areas.

(c) The term *adjacent* means bordering, contiguous, or neighboring. Wetlands separated from other waters of the United States by man-made dikes or barriers, natural river berms, beach dunes and the like are "adjacent wetlands."

(d) The term *high tide line* means the line of intersection of the land with the water's surface at the maximum height reached by a rising tide. The high tide line may be determined, in the absence of actual data, by a line of oil or scum along shore objects, a more or less continuous deposit of fine shell or debris on the foreshore or berm, other physical markings or characteristics, vegetation lines, tidal gages, or other suitable means that delineate the general height reached by a rising tide. The line encompasses spring high tides and other high tides that occur with periodic frequency but does not include storm surges in which there is a departure from the normal or predicted reach of the tide due to the piling up of water against a coast by strong winds such as those accompanying a hurricane or other intense storm.

(e) The term *ordinary high water mark* means that line on the shore established by the fluctuations of water and indicated by physical characteristics such as clear, natural line impressed on the bank, shelving, changes in the character of soil, destruction of terrestrial vegetation, the presence of litter and debris, or other appropriate means that consider the characteristics of the surrounding areas.

(f) The term *tidal waters* means those waters that rise and fall in a predictable and measurable rhythm or cycle due to the gravitational pulls of the moon and sun. Tidal waters end where the rise and fall of the water surface can no longer be practically measured in a predictable rhythm due to masking by hydrologic, wind, or other effects.

[51 FR 41250, Nov. 13, 1986, as amended at 58 FR 45036, Aug. 25, 1993]

Section 328.4 Limits of jurisdiction

(a) *Territorial Seas.* The limit of jurisdiction in the territorial seas is measured from the baseline in a seaward direction a distance of three nautical miles. (See 33 CFR 329.12)

(b) *Tidal Waters of the United States.* The landward limits of jurisdiction in tidal waters:

(1) Extends to the high tide line, or

(2) When adjacent non-tidal waters of the United States are present, the jurisdiction extends to the limits identified in paragraph (c) of this section.

(c) *Non-Tidal Waters of the United States.* The limits of jurisdiction in non-tidal waters:

(1) In the absence of adjacent wetlands, the jurisdiction extends to the ordinary high water mark, or

(2) When adjacent wetlands are present, the jurisdiction extends beyond the ordinary high water mark to the limit of the adjacent wetlands.

(3) When the water of the United States consists only of wetlands the jurisdiction extends to the limit of the wetland.

Section 328.5 Changes in limits of waters of the United States

Permanent changes of the shoreline configuration result in similar alterations of the boundaries of waters of the United States. Gradual changes which are due to natural causes and are perceptible only over some period of time constitute changes in the bed of a waterway which also change the boundaries of the waters of the United States. For example, changing sea levels or subsidence of land may cause some areas to become waters of the United States while siltation or a change in drainage may remove an area from waters of the United States. Man-made changes may affect the limits of waters of the United States; however, permanent changes should not be presumed until the particular circumstances have been examined and verified by the district engineer. Verification of changes to the lateral limits of jurisdiction may be obtained from the district engineer.

PART 329 DEFINITION OF NAVIGABLE WATERS OF THE UNITED STATES

AUTHORITY: 33 U.S.C. 401 et seq.

Section 329.1 Purpose

This regulation defines the term "navigable waters of the United States" as it is used to define authorities of the Corps of Engineers. It also prescribes the policy, practice and procedure to be used in determining the extent of the jurisdiction of the Corps of Engineers and in answering inquiries concerning "navigable waters of the United States." This definition does not apply to authorities under the Clean Water Act which definitions are described under 33 CFR parts 323 and 328.

Section 329.2 Applicability

This regulation is applicable to all Corps of Engineers districts and divisions having civil works responsibilities.

Section 329.3 General policies

Precise definitions of "navigable waters of the United States" or "navigability" are ultimately dependent on judicial interpretation and cannot be made conclusively by administrative agencies. However, the policies and criteria contained in this regulation are in close conformance with the tests used by Federal courts and determinations made under this regulation are considered binding in regard to the activities of the Corps of Engineers.

Section 329.4 General definition

Navigable waters of the United States are those waters that are subject to the ebb and flow of the tide and/or are presently used, or have been used in the past, or may be susceptible for use to transport interstate or foreign commerce. A determination of navigability, once made, applies laterally over the entire surface of the waterbody, and is not extinguished by later actions or events which impede or destroy navigable capacity.

Section 329.5 General scope of determination

The several factors which must be examined when making a determination whether a waterbody is a navigable water of the United States are discussed in detail below. Generally, the following conditions must be satisfied:

(a) Past, present, or potential presence of interstate or foreign commerce;

(b) Physical capabilities for use by commerce as in paragraph (a) of this section; and

(c) Defined geographic limits of the waterbody.

Section 329.6 Interstate or foreign commerce

(a) *Nature of commerce: type, means, and extent of use.* The types of commercial use of a waterway are extremely varied and will depend on the character of the region, its products, and the difficulties or dangers of navigation. It is the waterbody's capability of use by the public for purposes of transportation of commerce which is the determinative factor, and not the time, extent or manner of that use. As discussed in § 329.9 of this part, it is sufficient to establish the potential for commercial use at any past, present, or future time. Thus, sufficient commerce may be shown by historical use of canoes, bateaux, or other frontier craft, as long as that type of boat was common or well-suited to the place and period. Similarly, the particular items of commerce may vary widely, depending again on the region and period. The goods involved might be grain, furs, or other commerce of the time. Logs are a common example; transportation of logs has been a substantial and well-recognized commercial use of many navigable waters of the United States. Note, however, that the mere presence of floating logs will not of itself make the river "navigable"; the logs must have been related to a commercial venture. Similarly, the presence of recreational craft may indicate that a waterbody is capable of bearing some forms of commerce, either presently, in the future, or at a past point in time.

(b) *Nature of commerce: interstate and intrastate.* Interstate commerce may of course be existent on an intrastate voyage which occurs only between places within the same state. It is only necessary that goods may be brought from, or eventually be destined to go to, another state. (For purposes of this regulation, the term "interstate commerce" hereinafter includes "foreign commerce" as well.)

Section 329.7 Intrastate or interstate nature of waterway

A waterbody may be entirely within a state, yet still be capable of carrying interstate commerce. This is especially clear when it physically connects with a generally acknowledged avenue of interstate commerce, such as the ocean or one of the Great Lakes, and is yet wholly within one state. Nor is it necessary that there be a physically navigable connection across a state boundary. Where a waterbody extends through one or more states, but substantial portions, which are capable of bearing interstate commerce, are located in only one of the states, the entirety of the waterway up to the head (upper limit) of navigation is subject to Federal jurisdiction.

Section 329.8 Improved or natural conditions of the waterbody

Determinations are not limited to the natural or original condition of the waterbody. Navigability may also be found where artificial aids have been or may be used to make the waterbody suitable for use in navigation.

(a) *Existing improvements: artificial waterbodies*

(1) An artificial channel may often constitute a navigable water of the United States, even though it has been privately developed and maintained, or passes through private property. The test is generally as developed above, that is, whether the waterbody is capable of use to transport interstate commerce. Canals which connect two navigable waters of the United States and which are used for commerce clearly fall within the test, and themselves become navigable. A canal open to navigable waters of the United States on only one end is itself navigable where it in fact supports interstate commerce. A canal or other artificial waterbody that is subject to ebb and flow of the tide is also a navigable water of the United States.

(2) The artificial waterbody may be a major portion of a river or harbor area or merely a minor backwash, slip, or turning area (see § 329.12(b) of this part).

(3) Private ownership of the lands underlying the waterbody, or of the lands through which it runs, does not preclude a finding of navigability. Ownership does become a controlling factor if a privately constructed and operated canal is not used to transport interstate commerce nor used by the public; it is then not considered to be a navigable water of the United States. However, a private waterbody, even though not itself navigable, may so affect the navigable capacity of nearby waters as to nevertheless be subject to certain regulatory authorities.

(b) *Non-existing improvements, past or potential.* A waterbody may also be considered navigable depending on the feasibility of use to transport interstate commerce after the construction of whatever "reasonable" improvements may potentially be made. The improvement need not exist, be planned, nor even authorized; it is enough that potentially they could be made. What is a "reasonable" improvement is always a matter of degree; there must be a balance between cost and need at a time when the improvement would be (or would have been) useful. Thus, if an improvement were "reasonable" at a time of past use, the water was therefore navigable in law from that time forward. The changes in engineering practices or the coming of new industries with varying classes of freight may affect the type of the improvement; those which may be entirely reasonable in a thickly populated, highly developed industrial region may have been entirely too costly for the same region in the days of the pioneers. The determination of reasonable improvement is often similar to the cost analyses presently made in Corps of Engineers studies.

Section 329.9 Time at which commerce exists or determination is made

(a) *Past use.* A waterbody which was navigable in its natural or improved state, or which was susceptible of reasonable improvement (as discussed in § 329.8(b) of this part) retains its character as "navigable in law" even though it is not presently used for commerce, or is presently incapable of such use because of changed conditions or the presence of obstructions. Nor does absence of use because of changed economic conditions affect the legal character of the waterbody. Once having attained the character of "navigable in law," the Federal authority remains in existence, and cannot be abandoned by administrative officers or court action. Nor is mere inattention or ambiguous action by Congress an abandonment of Federal control. However, express statutory declarations by Congress that described portions of a waterbody are non-navigable, or have been abandoned, are binding upon the Department of the Army. Each statute must be carefully examined, since Congress often reserves the power to amend the Act, or assigns special duties of supervision and control to the Secretary of the Army or Chief of Engineers.

(b) *Future or potential use.* Navigability may also be found in a waterbody's susceptibility for use in its ordinary condition or by reasonable improvement to transport interstate commerce. This may be either in its natural or improved condition, and may thus be existent although there has been no actual use to date. Non-use in the past therefore does not prevent recognition of the potential for future use.

Section 329.10 Existence of obstructions

A stream may be navigable despite the existence of falls, rapids, sand bars, bridges, portages, shifting currents, or similar obstructions. Thus, a waterway in its original condition might have had substantial obstructions which were overcome by frontier boats and/or portages, and nevertheless be a "channel" of commerce, even though boats had to be removed from the water in some stretches, or logs be brought around an obstruction by means of artificial chutes. However, the question is ultimately a matter of degree, and it must be recognized that there is some point beyond which navigability could not be established.

Section 329.11 Geographic and jurisdictional limits of rivers and lakes

(a) *Jurisdiction over entire bed.* Federal regulatory jurisdiction, and powers of improvement for navigation, extend laterally to the entire water surface and bed of a navigable waterbody, which includes all the land and waters below the ordinary high water mark. Jurisdiction thus extends to the edge (as determined above) of all such waterbodies, even though portions of the waterbody may be extremely shallow, or obstructed by shoals, vegetation or other barriers. Marshlands and similar areas are thus considered navigable in law, but only so far as the area is subject to inundation by the ordinary high waters.

(1) The "ordinary high water mark" on non-tidal rivers is the line on the shore established by the fluctuations of water and indicated by physical characteristics such as a clear, natural line impressed on the bank; shelving; changes in the character of soil; destruction of terrestrial vegetation; the presence of litter and debris; or other appropriate means that consider the characteristics of the surrounding areas.

(2) Ownership of a river or lake bed or of the lands between high and low water marks will vary according to state law; however, private ownership of the underlying lands has no bearing on the existence or extent of the dominant Federal jurisdiction over a navigable waterbody.

(b) *Upper limit of navigability.* The character of a river will, at some point along its length, change from navigable to non-navigable. Very often that point will be at a major fall or rapids, or other place where there is a marked decrease in the navigable capacity of the river. The upper limit will therefore often be the same point traditionally recognized as the head of navigation, but may, under some of the tests described above, be at some point yet farther upstream.

Section 329.12 Geographic and jurisdictional limits of oceanic and tidal waters

(a) *Ocean and coastal waters.* The navigable waters of the United States over which Corps of Engineers regulatory jurisdiction extends include all ocean and coastal waters within a zone three geographic (nautical) miles seaward from the baseline (The Territorial Seas). Wider zones are recognized for special regulatory powers exercised over the outer continental shelf. (See 33 CFR 322.3(b)).

(1) *Baseline defined.* Generally, where the shore directly contacts the open sea, the line on the shore reached by the ordinary low tides comprises the baseline from which the distance of three geographic miles is measured. The baseline has significance for both domestic and international law and is subject to precise definitions. Special problems arise when offshore rocks, islands, or other bodies exist, and the baseline may have to be drawn seaward of such bodies.

(2) *Shoreward limit of jurisdiction.* Regulatory jurisdiction in coastal areas extends to the line on the shore reached by the plane of the mean (average) high water. Where precise determination of the actual location of the line becomes necessary, it must be established by survey with reference to the available tidal datum, preferably averaged over a period of 18.6 years. Less precise methods, such as observation of the "apparent shoreline" which is determined by reference to physical markings, lines of vegetation, or changes in type of vegetation, may be used only where an estimate is needed of the line reached by the mean high water.

(b) *Bays and estuaries.* Regulatory jurisdiction extends to the entire surface and bed of all waterbodies subject to tidal action. Jurisdiction thus extends to the edge (as determined by paragraph (a)(2) of this section) of all such waterbodies, even though portions of the waterbody may be extremely shallow, or obstructed by shoals, vegetation, or other barriers. Marshlands and similar areas are thus considered "navigable in law," but only so far as the area is subject to inundation by the mean high waters. The relevant test is therefore the presence of the mean high tidal waters, and not the general test described above, which generally applies to inland rivers and lakes.

Section 329.13 Geographic limits: Shifting boundaries

Permanent changes of the shoreline configuration result in similar alterations of the boundaries of the navigable waters of the United States. Thus, gradual changes which are due to natural causes and are perceptible only over some period of time constitute changes in the bed of a waterbody which also change the shoreline boundaries of the navigable waters of the United States. However, an area will remain "navigable in law," even though no longer covered with water, whenever the change has occurred suddenly, or was caused by artificial forces intended to produce that change. For example, shifting sand bars within a river or estuary remain part of the navigable water of the United States, regardless that they may be dry at a particular point in time.

Section 329.14 Determination of navigability

(a) *Effect on determinations.* Although conclusive determinations of navigability can be made only by federal Courts, those made by federal agencies are nevertheless accorded substantial weight by the courts. It is therefore necessary that when jurisdictional questions arise, district personnel carefully investigate those waters which may be subject to Federal regulatory jurisdiction under guidelines set out above, as the resulting determination may have substantial impact upon a judicial body. Official determinations by an agency made in the past can be revised or reversed as necessary to reflect changed rules or interpretations of the law.

(b) *Procedures of determination.* A determination whether a waterbody is a navigable water of the United States will be made by the division engineer, and will be based on a report of findings prepared at the district level in accordance with the criteria set out in this regulation. Each report of findings will be prepared by the district engineer, accompanied by an opinion of the district counsel, and forwarded to the division engineer for final determination. Each report of findings will be based substantially on applicable portions of the format in paragraph (c) of this section.

(c) *Suggested format of report of findings:*

(1) Name of waterbody:

(2) Tributary to:

(3) Physical characteristics:

(i) Type: (river, bay, slough, estuary, etc.)

(ii) Length:

(iii) Approximate discharge volumes: Maximum, Minimum, Mean:

(iv) Fall per mile:

(v) Extent of tidal influence:

(vi) Range between ordinary high and ordinary low water:

(vii) Description of improvements to navigation not listed in paragraph (c)(5) of this section:

(4) Nature and location of significant obstructions to navigation in portions of the waterbody used or potentially capable of use in interstate commerce:

(5) Authorized projects:

(i) Nature, condition and location of any improvements made under projects authorized by Congress:

(ii) Description of projects authorized but not constructed:

(iii) List of known survey documents or reports describing the waterbody:

(6) Past or present interstate commerce:

(i) General types, extent, and period in time:

(ii) Documentation if necessary:

(7) Potential use for interstate commerce, if applicable:

(i) If in natural condition:

(ii) If improved:

(8) Nature of jurisdiction known to have been exercised by Federal agencies if any:

(9) State or Federal court decisions relating to navigability of the waterbody, if any:

(10) Remarks:

(11) Finding of navigability (with date) and recommendation for determination:

Section 329.15 Inquiries regarding determinations

(a) Findings and determinations should be made whenever a question arises regarding the navigability of a waterbody. Where no determination has been made, a report of findings will be prepared and forwarded to the division engineer, as described above. Inquiries may be answered by an interim reply which indicates that a final agency determination must be made by the division engineer. If a need develops for an emergency determination, district engineers may act in reliance on a finding prepared as in § 329.14 of this part. The report of findings should then be forwarded to the division engineer on an expedited basis.

(b) Where determinations have been made by the division engineer, inquiries regarding the *navigability* of specific portions of waterbodies covered by these determinations may be answered as follows:

This Department, in the administration of the laws enacted by Congress for the protection and preservation of the navigable waters of the United States, has determined that ______ (River) (Bay) (Lake, etc.) is a navigable water of the United States from ______ to ______. Actions which modify or otherwise affect those waters are subject to the jurisdiction of this Department, whether such actions occur within or outside the navigable areas.

(c) Specific inquiries regarding the *jurisdiction* of the Corps of Engineers can be answered only after a determination whether

(1) the waters are navigable waters of the United States or

(2) If not navigable, whether the proposed type of activity may nevertheless so affect the navigable waters of the United States that the assertion of regulatory jurisdiction is deemed necessary.

Section 329.16 Use and maintenance of lists of determinations

(a) Tabulated lists of final determinations of navigability are to be maintained in each district office, and be updated as necessitated by court decisions, jurisdictional inquiries, or other changed conditions.

(b) It should be noted that the lists represent only those waterbodies for which determinations have been made; absence from that list should not be taken as an indication that the waterbody is not navigable.

(c) Deletions from the list are not authorized. If a change in status of a waterbody from navigable to non-navigable is deemed necessary, an updated finding should be forwarded to the division engineer; changes are not considered final until a determination has been made by the division engineer.

PART 330 NATIONWIDE PERMIT PROGRAM

AUTHORITY: 33 U.S.C. 401 et seq.; 33 U.S.C. 1344; 33 U.S.C. 1413

Section 330.1 Purpose and policy

(a) *Purpose.* This part describes the policy and procedures used in the Department of the Army's nationwide permit program to issue, modify, suspend, or revoke nationwide permits; to identify conditions, limitations, and restrictions on the nationwide permits; and, to identify any procedures, whether required or optional, for authorization by nationwide permits.

(b) *Nationwide permits.* Nationwide permits (NWPs) are a type of general permit issued by the Chief of Engineers and are designed to regulate with little, if any, delay or paperwork certain activities having minimal impacts. The NWPs are proposed, issued, modified, reissued (extended), and revoked from time to time after an opportunity for public notice and comment. Proposed NWPs or modifications to or reissuance of existing NWPs will be adopted only after the Corps gives notice and allows the public an opportunity to comment on and request a public hearing regarding the proposals. The Corps will give full consideration to all comments received prior to reaching a final decision.

(c) *Terms and conditions.* An activity is authorized under an NWP only if that activity and the permittee satisfy all of the NWP's terms and conditions. Activities that do not qualify for authorization under an NWP still may be authorized by an individual or regional general permit. The Corps will consider unauthorized any activity requiring Corps authorization if that activity is under construction or completed and does not comply with all of the terms and conditions of an NWP, regional general permit, or an individual permit. The Corps will evaluate unauthorized activities for enforcement action under 33 CFR part 326. The district engineer (DE) may elect to suspend enforcement proceedings if the permittee modifies his project to comply with an NWP or a regional general permit. After considering whether a violation was knowing or intentional, and other indications of the need for a penalty, the DE can elect to terminate an enforcement proceeding with an after-the-fact authorization under an NWP, if all terms and conditions of the NWP have been satisfied, either before or after the activity has been accomplished.

(d) *Discretionary authority.* District and division engineers have been delegated a discretionary authority to suspend, modify, or revoke authorizations under an NWP. This discretionary authority may be used by district and division engineers only to further condition or restrict the applicability of an NWP for cases where they have concerns for the aquatic environment under the Clean Water Act section 404(b)(1) Guidelines or for any factor of the public interest. Because of the nature of most activities authorized by NWP, district and division engineers will not have to review every such activity to decide whether to exercise discretionary authority. The terms and conditions of certain NWPs require the DE to review the proposed activity before the NWP authorizes its construction. However, the DE has the discretionary authority to review any activity authorized by NWP to determine whether the activity complies with the NWP. If the DE finds that the proposed activity would have more than minimal individual or cumulative net adverse effects on the environment or otherwise may be contrary to the public interest, he shall modify the NWP authorization to reduce or eliminate those adverse effects, or he shall instruct the prospective permittee to apply for a regional general permit or an individual permit. Discretionary authority is also discussed at 33 CFR 330.4(e) and 330.5.

(e) *Notifications*

(1) In most cases, permittees may proceed with activities authorized by NWPs without notifying the DE. However, the prospective permittee should carefully review the language of the NWP to ascertain whether he must notify the DE prior to commencing the authorized activity. For NWPs requiring advance notification, such notification must be made in writing as early as possible prior to commencing the proposed activity. The permittee may presume that his project qualifies for the NWP unless he is otherwise notified by the DE within a 30-day period. The 30-day

period starts on the date of receipt of the notification in the Corps district office and ends 30 calendar days later regardless of weekends or holidays. If the DE notifies the prospective permittee that the notification is incomplete, a new 30-day period will commence upon receipt of the revised notification. The prospective permittee may not proceed with the proposed activity before expiration of the 30-day period unless otherwise notified by the DE. If the DE fails to act within the 30-day period, he must use the procedures of 33 CFR 330.5 in order to modify, suspend, or revoke the NWP authorization.

(2) The DE will review the notification and may add activity-specific conditions to ensure that the activity complies with the terms and conditions of the NWP and that the adverse impacts on the aquatic environment and other aspects of the public interest are individually and cumulatively minimal.

(3) For some NWPs involving discharges into wetlands, the notification must include a wetland delineation. The DE will review the notification and determine if the individual and cumulative adverse environmental effects are more than minimal. If the adverse effects are more than minimal the DE will notify the prospective permittee that an individual permit is required or that the prospective permittee may propose measures to mitigate the loss of special aquatic sites, including wetlands, to reduce the adverse impacts to minimal. The prospective permittee may elect to propose mitigation with the original notification. The DE will consider that proposed mitigation when deciding if the impacts are minimal. The DE shall add activity-specific conditions to ensure that the mitigation will be accomplished. If sufficient mitigation cannot be developed to reduce the adverse environmental effects to the minimal level, the DE will not allow authorization under the NWP and will instruct the prospective permittee on procedures to seek authorization under an individual permit.

(f) *Individual Applications.* DEs should review all incoming applications for individual permits for possible eligibility under regional general permits or NWPs. If the activity complies with the terms and conditions of one or more NWP, he should verify the authorization and so notify the applicant. If the DE determines that the activity could comply after reasonable project modifications and/or activity-specific conditions, he should notify the applicant of such modifications and conditions. If such modifications and conditions are accepted by the applicant, verbally or in writing, the DE will verify the authorization with the modifications and conditions in accordance with 33 CFR 330.6(a). However, the DE will proceed with processing the application as an individual permit and take the appropriate action within 15 calendar days of receipt, in accordance with 33 CFR 325.2(a)(2), unless the applicant indicates that he will accept the modifications or conditions.

(g) *Authority.* NWPs can be issued to satisfy the permit requirements of section 10 of the Rivers and Harbors Act of 1899, section 404 of the Clean Water Act, section 103 of the Marine Protection, Research, and Sanctuaries Act, or some combination thereof. The applicable authority will be indicated at the end of each NWP. NWPs and their conditions previously published at 33 CFR 330.5 and 330.6 will remain in effect until they expire or are modified or revoked in accordance with the procedures of this part.

Section 330.2 Definitions

(a) The definitions found in 33 CFR parts 320–329 are applicable to the terms used in this part.

(b) *Nationwide* permit refers to a type of general permit which authorizes activities on a nationwide basis unless specifically limited. (Another type of general permit is a "regional permit" which is issued by division or district engineers on a regional basis in accordance with 33 CFR part 325). (See 33 CFR 322.2(f) and 323.2(h) for the definition of a general permit.)

(c) *Authorization* means that specific activities that qualify for an NWP may proceed, provided that the terms and conditions of the NWP are met. After determining that the activity complies with all applicable terms and conditions, the prospective permittee may assume an authorization under an NWP. This assumption is subject to the DE's authority to determine if an activity complies with the terms and conditions of an NWP. If requested by the permittee in writing, the DE will verify in writing that the permittee's proposed activity complies with the terms and conditions of the NWP. A written verification may contain activity-specific conditions and regional conditions which a permittee must satisfy for the authorization to be valid.

(d) *Headwaters* means non-tidal rivers, streams, and their lakes and impoundments, including adjacent wetlands, that are part of a surface tributary system to an interstate or navigable water of the United States upstream of the point on the river or stream at which the average annual flow is less than five cubic feet per second. The DE may estimate this point from available data by using the mean annual area precipitation, area drainage basin maps, and the average runoff coefficient, or by similar means. For streams that are dry for long periods of the year, DEs may establish the point where headwaters begin as that point on the stream where a flow of five cubic feet per second is equaled or exceeded 50 percent of the time.

(e) *Isolated waters* means those non-tidal waters of the United States that are:

(1) Not part of a surface tributary system to interstate or navigable waters of the United States; and

(2) Not adjacent to such tributary waterbodies.

(f) *Filled area* means the area within jurisdictional waters which is eliminated or covered as a direct result of the discharge (i.e., the area actually covered by the discharged material). It does not include areas excavated nor areas impacted as an indirect effect of the fill.

(g) *Discretionary authority* means the authority described in §§ 330.1(d) and 330.4(e) which the Chief of Engineers delegates to division or district engineers to modify an NWP authorization by adding conditions, to suspend an NWP authorization, or to revoke an NWP authorization and thus require individual permit authorization.

(h) *Terms and conditions.* The "terms" of an NWP are the limitations and provisions included in the description of the NWP itself. The "conditions" of NWPs are additional provisions which place restrictions or limitations on all of the NWPs. These are published with the NWPs. Other conditions may be imposed by district or division engineers on a geographic, category-of-activity, or activity-specific basis (See 33 CFR 330.4(e)).

(i) *Single and complete project* means the total project proposed or accomplished by one owner/developer or partnership or other association of owners/developers. For example, if construction of a residential development affects several different areas of a headwater or isolated water, or several different headwaters or isolated waters, the cumulative total of all filled areas should be the basis for deciding whether or not the project will be covered by an NWP. For linear projects, the "single and complete project" (i.e. single and complete crossing) will apply to each crossing of a separate water of the United States (i.e. single waterbody) at that location; except that for linear projects crossing a single waterbody several times at separate and distant locations, each crossing is considered a single and complete project. However, individual channels in a braided stream or river, or individual arms of a large, irregularly-shaped wetland or lake, etc., are not separate waterbodies.

(j) *Special aquatic sites* means wetlands, mudflats, vegetated shallows, coral reefs, riffle and pool complexes, sanctuaries, and refuges as defined at 40 CFR 230.40 through 230.45.

Section 330.3 Activities occurring before certain dates

The following activities were permitted by NWPs issued on July 19, 1977, and, unless the activities are modified, they do not require further permitting:

(a) Discharges of dredged or fill material into waters of the United States outside the limits of navigable waters of the United States that occurred before the phase-in dates which extended Section 404 jurisdiction to all waters of the United States. The phase-in dates were: After July 25, 1975, discharges into navigable waters of the United States and adjacent wetlands; after September 1, 1976, discharges into navigable waters of the United States and their primary tributaries, including adjacent wetlands, and into natural lakes, greater than 5 acres in surface area; and after July 1, 1977, discharges into all waters of the United States, including wetlands. (section 404)

(b) Structures or work completed before December 18, 1968, or in waterbodies over which the DE had not asserted jurisdiction at the time the activity occurred, provided in both instances, there is no interference with navigation. Activities completed shoreward of applicable Federal Harbor lines before May 27, 1970 do not require specific authorization. (section 10)

Section 330.4 Conditions, limitations, and restrictions

(a) *General.* A prospective permittee must satisfy all terms and conditions of an NWP for a valid authorization to occur. Some conditions identify a "threshold" that, if met, requires additional procedures or provisions contained in other paragraphs in this section. It is important to remember that the NWPs only authorize activities from the perspective of the Corps regulatory authorities and that other Federal, state, and local permits, approvals, or authorizations may also be required.

(b) *Further information*

(1) DEs have authority to determine if an activity complies with the terms and conditions of an NWP.

(2) NWPs do not obviate the need to obtain other Federal, state, or local permits, approvals, or authorizations required by law.

(3) NWPs do not grant any property rights or exclusive privileges.

(4) NWPs do not authorize any injury to the property or rights of others.

(5) NWPs do not authorize interference with any existing or proposed Federal project.

(c) *State 401 water quality certification*

(1) State 401 water quality certification pursuant to section 401 of the Clean Water Act, or waiver thereof, is required prior to the issuance or reissuance of NWPs authorizing activities which may result in a discharge into waters of the United States.

(2) If, prior to the issuance or reissuance of such NWPs, a state issues a 401 water quality certification which includes special conditions, the division engineer will make these special conditions regional conditions of the NWP for activities which may result in a discharge into waters of United States in that state, unless he determines that such conditions do not comply with the provisions of 33 CFR 325.4. In the latter case, the conditioned 401 water quality certification will be considered a denial of the certification (see paragraph (c)(3) of this section).

(3) If a state denies a required 401 water quality certification for an activity otherwise meeting the terms and conditions of a particular NWP, that NWP's authorization for all such activities within that state is denied without prejudice until the state issues an individual 401 water quality certification or waives its right to do so. State denial of 401 water quality certification for any specific NWP affects only those activities which may result in a discharge. That NWP continues to authorize activities which could not reasonably be expected to result in discharges into waters of the United States.[1]

(4) DEs will take appropriate measures to inform the public of which activities, waterbodies, or regions require an individual 401 water quality certification before authorization by NWP.

(5) The DE will not require or process an individual permit application for an activity which may result in a discharge and otherwise qualifies for an NWP solely on the basis that the 401 water quality certification has been denied for that NWP. However, the district or division engineer may consider water quality, among other appropriate factors, in determining whether to exercise his discretionary authority and require a regional general permit or an individual permit.

1. NWPs numbered 1, 2, 8, 9, 10, 11, 19, 24, 28, and 35, do not require 401 water quality certification since they would authorize activities which, in the opinion of the Corps, could not reasonably be expected to result in a discharge and in the case of NWP 8 is seaward of the territorial seas. NWPs numbered 3, 4, 5, 6, 7, 13, 14, 18, 20, 21, 22, 23, 27, 32, 36, 37, and 38, involve various activities, some of which may result in a discharge and require 401 water quality certification, and others of which do not. State denial of 401 water quality certification for any specific NWP in this category affects only those activities which may result in a discharge. For those activities not involving discharges, the NWP remains in effect. NWPs numbered 12, 15, 16, 17, 25, 26, and 40 involve activities which would result in discharges and therefore 401 water quality certification is required.

(6) In instances where a state has denied the 401 water quality certification for discharges under a particular NWP, permittees must furnish the DE with an individual 401 water quality certification or a copy of the application to the state for such certification. For NWPs for which a state has denied the 401 water quality certification, the DE will determine a reasonable period of time after receipt of the request for an activity-specific 401 water quality certification (generally 60 days), upon the expiration of which the DE will presume state waiver of the certification for the individual activity covered by the NWP's. However, the DE and the state may negotiate for additional time for the 401 water quality certification, but in no event shall the period exceed one (1) year (see 33 CFR 325.2(b)(1)(ii)). Upon receipt of an individual 401 water quality certification, or if the prospective permittee demonstrates to the DE state waiver of such certification, the proposed work can be authorized under the NWP. For NWPs requiring a 30-day predischarge notification the district engineer will immediately begin, and complete, his review prior to the state action on the individual section 401 water quality certification. If a state issues a conditioned individual 401 water quality certification for an individual activity, the DE will include those conditions as activity-specific conditions of the NWP.

(7) Where a state, after issuing a 401 water quality certification for an NWP, subsequently attempts to withdraw it for substantive reasons after the effective date of the NWP, the division engineer will review those reasons and consider whether there is substantial basis for suspension, modification, or revocation of the NWP authorization as outlined in § 330.5. Otherwise, such attempted state withdrawal is not effective and the Corps will consider the state certification to be valid for the NWP authorizations until such time as the NWP is modified or reissued.

(d) *Coastal zone management consistency determination*

(1) Section 307(c)(1) of the Coastal Zone Management Act (CZMA) requires the Corps to provide a consistency determination and receive state agreement prior to the issuance, reissuance, or expansion of activities authorized by an NWP that authorizes activities within a state with a Federally-approved Coastal Management Program when activities that would occur within, or outside, that state's coastal zone will affect land or water uses or natural resources of the state's coastal zone.

(2) If, prior to the issuance, reissuance, or expansion of activities authorized by an NWP, a state indicates that additional conditions are necessary for the state to agree with

the Corps consistency determination, the division engineer will make such conditions regional conditions for the NWP in that state, unless he determines that the conditions do not comply with the provisions of 33 CFR 325.4 or believes for some other specific reason it would be inappropriate to include the conditions. In this case, the state's failure to agree with the Corps consistency determination without the conditions will be considered to be a disagreement with the Corps consistency determination.

(3) When a state has disagreed with the Corps consistency determination, authorization for all such activities occurring within or outside the state's coastal zone that affect land or water uses or natural resources of the state's coastal zone is denied without prejudice until the prospective permittee furnishes the DE an individual consistency certification pursuant to section 307(c)(3) of the CZMA and demonstrates that the state has concurred in it (either on an individual or generic basis), or that concurrence should be presumed (see paragraph (d)(6) of this section).

(4) DEs will take appropriate measures, such as public notices, to inform the public of which activities, waterbodies, or regions require prospective permittees to make an individual consistency determination and seek concurrence from the state.

(5) DEs will not require or process an individual permit application for an activity otherwise qualifying for an NWP solely on the basis that the activity has not received CZMA consistency agreement from the state. However, the district or division engineer may consider that factor, among other appropriate factors, in determining whether to exercise his discretionary authority and require a regional general permit or an individual permit application.

(6) In instances where a state has disagreed with the Corps consistency determination for activities under a particular NWP, permittees must furnish the DE with an individual consistency concurrence or a copy of the consistency certification provided to the state for concurrence. If a state fails to act on a permittee's consistency certification within six months after receipt by the state, concurrence will be presumed. Upon receipt of an individual consistency concurrence or upon presumed consistency, the proposed work is authorized if it complies with all terms and conditions of the NWP. For NWPs requiring a 30-day predischarge notification the DE will immediately begin, and may complete, his review prior to the state action on the individual consistency certification. If a state indicates that individual conditions are necessary for consistency with the state's Federally-approved coastal management program for that individual activity, the DE will include those conditions as activity-specific conditions of the NWP unless he determines that such conditions do not comply with the provisions of 33 CFR 325.4. In the latter case the DE will consider the conditioned concurrence as a nonconcurrence unless the permittee chooses to comply voluntarily with all the conditions in the conditioned concurrence.

(7) Where a state, after agreeing with the Corps consistency determination, subsequently attempts to reverse its agreement for substantive reasons after the effective date of the NWP, the division engineer will review those reasons and consider whether there is substantial basis for suspension, modification, or revocation as outlined in 33 CFR 330.5. Otherwise, such attempted reversal is not effective and the Corps will consider the state CZMA consistency agreement to be valid for the NWP authorization until such time as the NWP is modified or reissued.

(8) Federal activities must be consistent with a state's Federally-approved coastal management program to the maximum extent practicable. Federal agencies should follow their own procedures and the Department of Commerce regulations appearing at 15 CFR part 930 to meet the requirements of the CZMA. Therefore, the provisions of 33 CFR 330.4(d)(1)–(7) do not apply to Federal activities. Indian tribes doing work on Indian Reservation lands shall be treated in the same manner as Federal applicants.

(e) *Discretionary authority.* The Corps reserves the right (i.e., discretion) to modify, suspend, or revoke NWP authorizations. Modification means the imposition of additional or revised terms or conditions on the authorization. Suspension means the temporary cancellation of the authorization while a decision is made to either modify, revoke, or reinstate the authorization. Revocation means the cancellation of the authorization. The procedures for modifying, suspending, or revoking NWP authorizations are detailed in § 330.5.

(1) A division engineer may assert discretionary authority by modifying, suspending, or revoking NWP authorizations for a specific geographic area, class of activity, or class of waters within his division, including on a statewide basis, whenever he determines sufficient concerns for the environment under the section 404(b)(1) Guidelines or any other factor of the public interest so requires, or if he otherwise determines that the NWP would result in more than minimal adverse environmental effects either individually or cumulatively.

(2) A DE may assert discretionary authority by modifying, suspending, or revoking NWP authorization for a specific activity whenever he determines sufficient concerns for the environment or any other factor of the public interest so requires. Whenever the DE determines that a proposed specific activity covered by an NWP would have more than minimal individual or cumulative adverse effects on the environment or otherwise may be contrary to the public interest, he must either modify the NWP authorization to reduce or eliminate the adverse impacts, or notify the prospective permittee that the proposed activity is not authorized by NWP and provide instructions on how to seek authorization under a regional general or individual permit.

(3) The division or district engineer will restore authorization under the NWPs at any time he determines that his reason for asserting discretionary authority has been satisfied by a condition, project modification, or new information.

(4) When the Chief of Engineers modifies or reissues an NWP, division engineers must use the procedures of § 330.5 to reassert discretionary authority to reinstate regional conditions or revocation of NWP authorizations for specific geographic areas, class of activities, or class of waters. Division engineers will update existing documentation for each NWP. Upon modification or reissuance of NWPs, previous activity-specific conditions or revocations of NWP authorization will remain in effect unless the DE specifically removes the activity-specific conditions or revocations.

(f) *Endangered species.* No activity is authorized by any NWP if that activity is likely to jeopardize the continued existence of a threatened or endangered species as listed or proposed for listing under the Federal Endangered Species Act (ESA), or to destroy or adversely modify the critical habitat of such species.

(1) Federal agencies should follow their own procedures for complying with the requirements of the ESA.

(2) Non-federal permittees shall notify the DE if any Federally listed (or proposed for listing) endangered or threatened species or critical habitat might be affected or is in the vicinity of the project. In such cases, the prospective permittee will not begin work under authority of the NWP until notified by the district engineer that the requirements of the Endangered Species Act have been satisfied and that the activity is authorized. If the DE determines that the activity may affect any Federally listed species or critical habitat, the DE must initiate section 7 consultation in accordance with the ESA. In such cases, the DE may:

(i) Initiate section 7 consultation and then, upon completion, authorize the activity under the NWP by adding, if appropriate, activity-specific conditions; or

(ii) Prior to or concurrent with section 7 consultation, assert discretionary authority (see 33 CFR 330.4(e)) and require an individual permit (see 33 CFR 330.5(d)).

(3) Prospective permittees are encouraged to obtain information on the location of threatened or endangered species and their critical habitats from the U.S. Fish and Wildlife Service, Endangered Species Office, and the National Marine Fisheries Service.

(g) *Historic properties.* No activity which may affect properties listed or properties eligible for listing in the National Register of Historic Places, is authorized until the DE has complied with the provisions of 33 CFR part 325, appendix C.

(1) Federal permittees should follow their own procedures for compliance with the requirements of the National Historic Preservation Act and other Federal historic preservation laws.

(2) Non-federal permittees will notify the DE if the activity may affect historic properties which the National Park Service has listed, determined eligible for listing, or which the prospective permittee has reason to believe may be eligible for listing, on the National Register of Historic Places. In such cases, the prospective permittee will not begin the proposed activity until notified by the DE that the requirements of the National Historic Preservation Act have been satisfied and that the activity is authorized. If a property in the permit area of the activity is determined to be an historic property in accordance with 33 CFR part 325, appendix C, the DE will take into account the effects on such properties in accordance with 33 CFR part 325, appendix C. In such cases, the district engineer may:

(i) After complying with the requirements of 33 CFR part 325, appendix C, authorize the activity under the NWP by adding, if appropriate, activity-specific conditions; or

(ii) Prior to or concurrent with complying with the requirements of 33 CFR part 325, appendix C, he may assert discretionary authority (see 33 CFR 330.4(e)) and instruct the prospective permittee of procedures to seek authorization under a regional general permit or an individual permit. (See 33 CFR 330.5(d).)

(3) The permittee shall immediately notify the DE if, before or during prosecution of the work authorized, he encounters an historic property that has not been listed or determined eligible for listing on the National Register, but which the prospective permittee has reason to believe may be eligible for listing on the National Register.

(4) Prospective permittees are encouraged to obtain information on the location of historic properties from the State Historic Preservation Officer and the National Register of Historic Places.

Section 330.5 Issuing, modifying, suspending, or revoking nationwide permits and authorizations

(a) *General.* This section sets forth the procedures for issuing and reissuing NWPs and for modifying, suspending, or revoking NWPs and authorizations under NWPs.

(b) *Chief of Engineers*

(1) Anyone may, at any time, suggest to the Chief of Engineers, (ATTN: CECW-OR), any new NWPs or conditions for issuance, or changes to existing NWPs, which he believes to be appropriate for consideration. From time-to-time new NWPs and revocations of or modifications to existing NWPs will be evaluated by the Chief of Engineers following the procedures specified in this section. Within five years of issuance of the NWPs, the Chief of Engineers will review the NWPs and propose modification, revocation, or reissuance.

(2) *Public notice*

(i) Upon proposed issuance of new NWPs or modification, suspension, revocation, or reissuance of existing NWPs, the Chief of Engineers will publish a document seeking public comments, including the opportunity to request a public hearing. This document will also state that the information supporting the Corps' provisional determination that proposed activities comply with the requirements for issuance under general permit authority is available at the Office of the Chief of Engineers and at all district offices. The Chief of Engineers will prepare this information which will be supplemented, if appropriate, by division engineers.

(ii) Concurrent with the Chief of Engineers' notification of proposed, modified, reissued, or revoked NWPs, DEs will notify the known interested public by a notice issued at the district level. The notice will include proposed regional conditions or proposed revocations of NWP authorizations for specific geographic areas, classes of activities, or classes of waters, if any, developed by the division engineer.

(3) *Documentation.* The Chief of Engineers will prepare appropriate NEPA documents and, if applicable, section 404(b)(1) Guidelines compliance analyses for proposed NWPs. Documentation for existing NWPs will be modified to reflect any changes in these permits and to reflect the Chief of Engineers' evaluation of the use of the permit since the last issuance. Copies of all comments received on the document will be included in the administrative record. The Chief of Engineers will consider these comments in making his decision on the NWPs, and will prepare a statement of findings outlining his views regarding each NWP and discussing how substantive comments were considered. The Chief of Engineers will also determine the need to hold a public hearing for the proposed NWPs.

(4) *Effective dates.* The Chief of Engineers will advise the public of the effective date of any issuance, modification, or revocation of an NWP.

(c) *Division Engineer*

(1) A division engineer may use his discretionary authority to modify, suspend, or revoke NWP authorizations for any specific geographic area, class of activities, or class of waters within his division, including on a statewide basis, by issuing a public notice or notifying the individuals involved. The notice will state his concerns regarding the environment or the other relevant factors of the public interest. Before using his discretionary authority to modify or revoke such NWP authorizations, division engineers will:

(i) Give an opportunity for interested parties to express their views on the proposed action (the DE will publish and circulate a notice to the known interested public to solicit comments and provide the opportunity to request a public hearing);

(ii) Consider fully the views of affected parties;

(iii) Prepare supplemental documentation for any modifications or revocations that may result through assertion of discretionary authority. Such documentation will include comments received on the district public notices and a statement of findings showing how substantive comments were considered;

(iv) Provide, if appropriate, a grandfathering period as specified in § 330.6(b) for those who have commenced work or are under contract to commence in reliance on the NWP authorization; and

(v) Notify affected parties of the modification, suspension, or revocation, including the effective date (the DE will publish and circulate a notice to the known interested public and to anyone who commented on the proposed action).

(2) The modification, suspension, or revocation of authorizations under an NWP by the division engineer will become effective by issuance of public notice or a notification to the individuals involved.

(3) A copy of all regional conditions imposed by division engineers on activities authorized by NWPs will be forwarded to the Office of the Chief of Engineers, ATTN: CECW-OR.

(d) *District Engineer*

(1) When deciding whether to exercise his discretionary authority to modify, suspend, or revoke a case specific activity's authorization under an NWP, the DE should consider to the extent relevant and appropriate: Changes in circumstances relating to the authorized activity since the NWP itself was issued or since the DE confirmed authorization under the NWP by written verification; the continuing need for, or adequacy of, the specific conditions of the authorization; any significant objections to the authorization not previously considered; progress inspections of individual activities occurring under an NWP; cumulative adverse environmental effects resulting from activities occurring under the NWP; the extent of the permittee's compliance with the terms and conditions of the NWPs; revisions to applicable statutory or regulatory authorities; and, the extent to which asserting discretionary authority would adversely affect plans, investments, and actions the permittee has made or taken in reliance on the permit; and, other concerns for the environment, including the aquatic environment under the section 404(b)(1) Guidelines, and other relevant factors of the public interest.

(2) *Procedures*

(i) When considering whether to modify or revoke a specific authorization under an NWP, whenever practicable, the DE will initially hold informal consultations with the permittee to determine whether special conditions to modify the authorization would be mutually agreeable or to allow the permittee to furnish information which satisfies the DE's concerns. If a mutual agreement is reached, the DE will give the permittee written verification of the authorization, including the special conditions. If the permittee furnishes information which satisfies the DE's concerns, the permittee may proceed. If appropriate, the DE may suspend the NWP authorization while holding informal consultations with the permittee.

(ii) If the DE's concerns remain after the informal consultation, the DE may suspend a specific authorization under an NWP by notifying the permittee in writing by the most expeditious means available that the authorization has been suspended, stating the reasons for the suspension, and ordering the permittee to stop any activities being done in reliance upon the authorization under the NWP. The permittee will be advised that a decision will be made either to reinstate or revoke the authorization under the NWP; or, if appropriate, that the authorization under the NWP may be modified by mutual agreement. The permittee will also be advised that within 10 days of receipt of the notice of suspension, he may request a meeting with the DE, or his designated representative, to present information in this matter. After completion of the meeting (or within a reasonable period of time after suspending the authorization if no meeting is requested), the DE will take action to reinstate, modify, or revoke the authorization.

(iii) Following completion of the suspension procedures, if the DE determines that sufficient concerns for the environment, including the aquatic environment under the section 404(b)(1) Guidelines, or other relevant factors of the public interest so require, he will revoke authorization under the NWP. The DE will provide the permittee a written final decision and instruct him on the procedures to seek authorization under a regional general permit or an individual permit.

(3) The DE need not issue a public notice when asserting discretionary authority over a specific activity. The modification, suspension, or revocation will become effective by notification to the prospective permittee.

Section 330.6 Authorization by nationwide permit

(a) *Nationwide permit verification*

(1) Nationwide permittees may, and in some cases must, request from a DE confirmation that an activity complies with the terms and conditions of an NWP. DEs should respond as promptly as practicable to such requests.

(2) If the DE decides that an activity does not comply with the terms or conditions of an NWP, he will notify the person desiring to do the work and instruct him on the procedures to seek authorization under a regional general permit or individual permit.

(3) If the DE decides that an activity does not comply with the terms or conditions of an NWP, he will notify the nationwide permittee.

(i) The DE may add conditions on a case-by-case basis to clarify compliance with the terms and conditions of an NWP or to ensure that the activity will have only minimal individual and cumulative adverse effects on the environment, and will not be contrary to the public interest.

(ii) The DE's response will state that the verification is valid for a specific period of time (generally but no more than two years) unless the NWP authorization is modified, suspended, or revoked. The response should also include a statement that the verification will remain valid for the specified period of time, if during that time period, the NWP authorization is reissued without modification or the activity complies with any subsequent modification of the NWP authorization. Furthermore, the response should include a statement that the provisions of § 330.6(b) will apply, if during that period of time, the NWP authorization expires, or is suspended or revoked, or is modified, such that the activity would no longer comply with the terms and conditions of an NWP. Finally, the response should include any known expiration date that would occur during the specified period of time. A period of time less than two years may be used if deemed appropriate.

(iii) For activities where a state has denied 401 water quality certification and/or did not agree with the Corps consistency determination for an NWP the DE's response will state that the proposed activity meets the terms and conditions for authorization under the NWP with the exception of a state 401 water quality certification and/or CZM consistency concurrence. The response will also indicate the activity is denied without prejudice and cannot be authorized until the requirements of §§ 330.4(c)(3), 330.4(c)(6), 330.4(d)(3), and 330.4(d)(6) are satisfied. The response will also indicate that work may only proceed subject to the terms and conditions of the state 401 water quality certification and/or CZM concurrence.

(iv) Once the DE has provided such verification, he must use the procedures of 33 CFR 330.5 in order to modify, suspend, or revoke the authorization.

(b) *Expiration of nationwide permits.* The Chief of Engineers will periodically review NWPs and their conditions and will decide to either modify, reissue, or revoke the permits. If an NWP is not modified or reissued within five years of its effective date, it automatically expires and becomes null and void. Activities which have commenced (i.e., are under construction) or are under contract to commence in reliance upon an NWP will remain authorized provided the activity is completed within twelve months of the date of an NWP's expiration, modification, or revocation, unless discretionary authority has been exercised on a case-by-case basis to modify, suspend, or revoke the authorization in accordance with 33 CFR 330.4(e) and 33 CFR 330.5 (c) or (d). Activities completed under the authorization of an NWP which was in effect at the time the activity was completed continue to be authorized by that NWP.

(c) *Multiple use of nationwide permits.* Two or more different NWPs can be combined to authorize a "single and complete project" as defined at 33 CFR 330.2(i). However, the same NWP cannot be used more than once for a single and complete project.

(d) *Combining nationwide permits with individual permits.* Subject to the following qualifications, portions of a larger project may proceed under the authority of the NWPs while the DE evaluates an individual permit application for other portions of the same project, but only if the portions of the project qualifying for NWP authorization would have independent utility and are able to function or meet their purpose independent of the total project. When the functioning or usefulness of a portion of the total project qualifying for an NWP is dependent on the remainder of the project, such that its construction and use would not be fully justified even if the Corps were to deny the individual permit, the NWP does not apply and all portions of the project must be evaluated as part of the individual permit process.

(1) When a portion of a larger project is authorized to proceed under an NWP, it is with the understanding that its construction will in no way prejudice the decision on the individual permit for the rest of the project. Furthermore, the individual permit documentation must include an analysis of the impacts of the entire project, including related activities authorized by NWP.

(2) NWPs do not apply, even if a portion of the project is not dependent on the rest of the project, when any portion of the project is subject to an enforcement action by the Corps or EPA.

(e) *After-the-fact authorizations.* These authorizations often play an important part in the resolution of violations. In appropriate cases where the activity complies with the terms and conditions of an NWP, the DE can elect to use the NWP for resolution of an after-the-fact permit situation following a consideration of whether the violation being resolved was knowing or intentional and other indications of the need for a penalty. For example, where an unauthorized fill meets the terms and conditions of NWP 13, the DE can consider the appropriateness of allowing the residual fill to remain, in situations where said fill would normally have been permitted under NWP 13. A knowing, intentional, willful violation should be the subject of an enforcement action leading to a penalty, rather than an after-the-fact authorization. Use of after-the-fact NWP authorization must be consistent with the terms of the Army/EPA Memorandum of Agreement on Enforcement. Copies are available from each district engineer.

PART 331 ADMINISTRATIVE APPEALS PROCESS

Authority: 33 U.S.C. 401 et seq.; 33 U.S.C. 1344; 33 U.S.C. 1413

Section 331.1 Purpose and policy

(a) *General.* The purpose of this Part is to establish policies and procedures to be used for the administrative appeal of approved jurisdictional determinations (JDs), permit applications denied with prejudice, and declined permits. The appeal process will allow the affected party to pursue an administrative appeal of certain Corps of Engineers decisions with which they disagree. The basis for an appeal and the specific policies and procedures of the appeal process are described in the following sections. It shall be the policy of the Corps of Engineers to promote and maintain an administrative appeal process that is independent, objective, fair, prompt, and efficient.

(b) *Level of decision maker.* Appealable actions decided by a division engineer or higher authority may be appealed to an Army official at least one level higher than the decision maker. This higher Army official shall make the decision on the merits of the appeal, and may appoint a qualified individual to act as a review officer (as defined in § 331.2). References to the division engineer in this Part shall be understood as also referring to a higher level Army official when such official is conducting an administrative appeal.

Section 331.2 Definitions

The terms and definitions contained in 33 CFR Parts 320 through 330 are applicable to this part. In addition, the following terms are defined for the purposes of this part:

Affected party means a permit applicant, landowner, a lease, easement or option holder (*i.e.*, an individual who has an identifiable and substantial legal interest in the property) who has received an approved JD, permit denial, or has declined a proffered individual permit.

Agent(s) means the affected party's business partner, attorney, consultant, engineer, planner, or any individual with legal authority to represent the appellant's interests.

Appealable action means an approved JD, a permit denial, or a declined permit, as these terms are defined in this section.

Appellant means an affected party who has filed an appeal of an approved JD, a permit denial or declined permit under the criteria and procedures of this part.

Approved jurisdictional determination means a Corps document stating the presence or absence of waters of the United States on a parcel or a written statement and map identifying the limits of waters of the United States on a parcel. Approved JDs are clearly designated appealable actions and will include a basis of JD with the document.

Basis of jurisdictional determination is a summary of the indicators that support the Corps approved JD. Indicators supporting the Corps approved JD can include, but are not limited to: indicators of wetland hydrology, hydric soils, and hydrophytic plant communities; indicators of ordinary high water marks, high tide lines, or mean high water marks; indicators of adjacency to navigable or interstate waters; indicators that the wetland or waterbody is of part of a tributary system; or indicators of linkages between isolated water bodies and interstate or foreign commerce.

Declined permit means a proffered individual permit, including a letter of permission, that an applicant has refused to accept, because he has objections to the terms and special conditions therein. A declined permit can also be an individual permit that the applicant originally accepted, but where such permit was subsequently modified by the district engineer, pursuant to 33 CFR 325.7, in such a manner that the resulting permit contains terms and special conditions that lead the applicant to decline the modified permit, provided that the applicant has not started work in waters of the United States authorized by such permit. Where an applicant declines a permit (either initial or modified), the applicant does not have a valid permit to conduct regulated activities in waters of the United States, and must not begin construction of the work requiring a Corps permit unless and until the applicant receives and accepts a valid Corps permit.

Denial determination means a letter from the district engineer detailing the reasons a permit was denied with prejudice. The decision document for the project will be attached to the denial determination in all cases.

Jurisdictional determination (JD) means a written Corps determination that a wetland and/or waterbody is subject to regulatory jurisdiction under Section 404 of the Clean Water Act (33 U.S.C. 1344) or a written determination that a waterbody is subject to regulatory jurisdiction under Section 9 or 10 of the Rivers and Harbors Act of 1899 (33 U.S.C. 401 *et seq.*). Additionally, the term includes a written reverification of expired JDs and a written reverification of JDs where new information has become available that may affect the previously written determination. For example, such geographic JDs may include, but are not limited to, one or more of the following determinations: the presence or absence of wetlands; the location(s) of the wetland boundary, ordinary high water mark, mean high water mark, and/or high tide line; interstate commerce nexus for isolated waters; and adjacency of wetlands to other waters of the United States. All JDs will be in writing and will be identified as either preliminary or approved. JDs do not include determinations that a particular activity requires a DA permit.

Notification of Appeal Process (NAP) means a fact sheet that explains the criteria and procedures of the administrative appeal process. Every approved JD, permit denial, and every proffered individual permit returned for reconsideration after review by the district engineer in accordance with § 331.6(b) will have an NAP form attached.

Notification of Applicant Options (NAO) means a fact sheet explaining an applicant's options with a proffered individual permit under the administrative appeal process.

Permit denial means a written denial with prejudice (see 33 CFR 320.4(j)) of an individual permit application as defined in 33 CFR 325.5(b).

Preliminary JDs are written indications that there may be waters of the United States on a parcel or indications of the approximate location(s) of waters of the United States on a parcel. Preliminary JDs are advisory in nature and may not be appealed. Preliminary JDs include compliance orders that have an implicit JD, but no approved JD.

Proffered permit means a permit that is sent to an applicant that is in the proper format for the applicant to sign (for a standard permit) or accept (for a letter of permission). The term "initial proffered permit" as used in this part refers to the first time a permit is sent to the applicant. The initial proffered permit is not an appealable action. However, the applicant may object to the terms or conditions of the initial proffered permit and, if so, a second reconsidered permit will be sent to the applicant. The term "proffered permit" as used in this part refers to the second permit that is sent to the applicant. Such proffered permit is an appealable action.

Request for appeal (RFA) means the affected party's official request to initiate the appeal process. The RFA must include the name of the affected party, the Corps file number of the approved JD, denied permit, or declined permit, the reason(s) for the appeal, and any supporting data and information. No new information may be submitted. A grant of right of entry for the Corps to the project site is a condition of the RFA to allow the RO to clarify elements of the record or to conduct field tests or sampling for purposes directly related to the appeal. A standard RFA form will be provided to the affected party with the NAP form. For appeals of decisions related to unauthorized activities a signed tolling agreement, as required by 33 CFR 326.3(e)(1)(v), must be included with the RFA, unless a signed tolling agreement has previously been furnished to the Corps district office. The affected party initiates the administrative appeal process by providing an acceptable RFA to the appropriate Corps of Engineers division office. An acceptable RFA contains all the required information and provides reasons for appeal that meets the criteria identified in § 331.5.

Review officer (RO) means the Corps official responsible for assisting the division engineer or higher authority responsible for rendering the final decision on the merits of an appeal.

Tolling agreement refers to a document signed by any person who appeals an approved JD associated with an unauthorized activity or applies for an after-the-fact (ATF) permit, where the application is accepted and evaluated by the Corps. The agreement states that the affected party agrees to have the statute of limitations regarding any violation associated with that approved JD or application "tolled" or temporarily set aside until one year after the final Corps decision, as defined at § 331.10. No ATF permit application or administrative appeal associated with an unauthorized activity will be accepted until a tolling agreement is furnished to the district engineer.

Section 331.3 Review officer

(a) *Authority*

(1) The division engineer has the authority and responsibility for administering a fair, reasonable, prompt, and effective administrative appeal process. The division engineer may act as the review officer (RO), or may delegate, either generically or on a case-by-case basis, any authority or responsibility described in this part as that of the RO. With the exception of JDs, as described in this paragraph (a)(1), the division engineer may not delegate any authority or responsibility described in this part as that of the division engineer. For approved JDs only, the division engineer may delegate any authority or responsibility described in this part as that of the division engineer, including the final appeal decision. In such cases, any delegated authority must be granted to an official that is at the same or higher grade level than the grade level of the official that signed the approved JD. Regardless of any delegation of authority or responsibility for ROs or for final appeal decisions for approved JDs, the division engineer retains overall responsibility for the administrative appeal process.

(2) The RO will assist the division engineer in reaching and documenting the division engineer's decision on the merits of an appeal, if the division engineer has delegated this responsibility as explained in paragraph (a)(1) of this section. The division engineer has the authority to make the final decision on the merits of the appeal. Neither the RO nor the division engineer has the authority to make a final decision to issue or deny any particular permit nor to make an approved JD, pursuant to the administrative appeal process established by this part. The authority to issue or deny permits remains with the district engineer. However, the division engineer may exercise the authority at 33 CFR 325.8(c) to elevate any permit application, and subsequently make the final permit decision. In such a case, any appeal process of the district engineer's initial decision is terminated. If a particular permit application is elevated to the division engineer pursuant to 33 CFR 325.8(c), and the division engineer's decision on the permit application is a permit denial or results in a declined permit, that permit denial or declined permit would be subject to an administrative appeal to the Chief of Engineers.

(3) *Qualifications.* The RO will be a Corps employee with extensive knowledge of the Corps regulatory program. Where the permit decision being appealed was made by the division engineer or higher authority, a Corps official at least one level higher than the decision maker shall make the decision on the merits of the RFA, and this Corps official shall appoint a qualified individual as the RO to conduct the appeal process.

(b) *General–*

(1) *Independence.* The RO will not perform, or have been involved with, the preparation, review, or decision making of the action being appealed. The RO will be independent and impartial in reviewing any appeal, and when assisting the division engineer to make a decision on the merits of the appeal.

(2) *Review.* The RO will conduct an independent review of the administrative record to address the reasons for the appeal cited by the applicant in the RFA. In addition, to the extent that it is practicable and feasible, the RO will also conduct an independent review of the administrative record to verify that the record provides an adequate and reasonable basis supporting the district engineer's decision, that facts or analysis essential to the district engineer's decision have not been omitted from the administrative record, and that all relevant requirements of law, regulations, and officially promulgated Corps policy guidance have been satisfied. Should the RO require expert advice regarding any subject, he may seek such advice from any employee of the Corps or of another Federal or state agency, or from any recognized expert, so long as that person had not been previously involved in the action under review.

Section 331.4 Notification of appealable actions

Affected parties will be notified in writing of a Corps decision on those activities that are eligible for an appeal. For approved JDs, the notification must include an NAP fact sheet, an RFA form, and a basis of JD. For permit denials, the notification must include a copy of the decision document for the permit application, an NAP fact sheet and an RFA form. For proffered individual permits, when the initial proffered permit is sent to the applicant, the notification must include an NAO fact sheet. For declined permits (i.e., proffered individual permits that the applicant refuses to accept and sends back to the Corps), the notification must include an NAP fact sheet and an RFA form. Additionally, an affected party has the right to obtain a copy of the administrative record.

Section 331.5 Criteria

(a) *Criteria for appeal–*

(1) *Submission of RFA.* The appellant must submit a completed RFA (as defined at § 331.2) to the appropriate division office in order to appeal an approved JD, a permit denial, or a declined permit. An individual permit that has been signed by the applicant, and subsequently unilaterally modified by the district engineer pursuant to 33 CFR 325.7, may be appealed under this process, provided that the applicant has not started work in waters of the United States authorized by the permit. The RFA must be received by the division engineer within 60 days of the date of the NAP.

(2) *Reasons for appeal.* The reason(s) for requesting an appeal of an approved JD, a permit denial, or a declined permit must be specifically stated in the RFA and must be more than a simple request for appeal because the affected party did not like the approved JD, permit decision, or the permit conditions. Examples of reasons for appeals include, but are not limited to, the following: A procedural error; an incorrect application of law, regulation or officially promulgated policy; omission of material fact; incorrect application of the current regulatory criteria and associated guidance for identifying and delineating wetlands; incorrect application of the Section 404(b)(1) Guidelines (see 40 CFR Part 230); or use of incorrect data. The reasons for appealing a permit denial or a declined permit may include jurisdiction issues, whether or not a previous approved JD was appealed.

(b) *Actions not appealable.* An action or decision is not subject to an administrative appeal under this part if it falls into one or more of the following categories:

(1) An individual permit decision (including a letter of permission or a standard permit with special conditions), where the permit has been accepted and signed by the permittee. By signing the permit, the applicant waives all rights to appeal the terms and conditions of the permit, unless the authorized work has not started in waters of the United States and that issued permit is subsequently modified by the district engineer pursuant to 33 CFR 325.7;

(2) Any site-specific matter that has been the subject of a final decision of the Federal courts;

(3) A final Corps decision that has resulted from additional analysis and evaluation, as directed by a final appeal decision;

(4) A permit denial without prejudice or a declined permit, where the controlling factor cannot be changed by the Corps decision maker (*e.g.*, the requirements of a binding statute, regulation, state Section 401 water quality certification, state coastal zone management disapproval, *etc.* (See 33 CFR 320.4(j));

(5) A permit denial case where the applicant has subsequently modified the proposed project, because this would constitute an amended application that would require a new public interest review, rather than an appeal of the existing record and decision;

(6) Any request for the appeal of an approved JD, a denied permit, or a declined permit where the RFA has not been received by the division engineer within 60 days of the date of the NAP;

(7) A previously approved JD that has been superceded by another approved JD based on new information or data submitted by the applicant. The new approved JD is an appealable action;

(8) An approved JD associated with an individual permit where the permit has been accepted and signed by the permittee;

(9) A preliminary JD; or

(10) A JD associated with unauthorized activities except as provided in § 331.11.

Section 331.6 Filing an appeal

(a) An affected party appealing an approved JD, permit denial or declined permit must submit an RFA that is received by the division engineer within 60 days of the date of the NAP. Flow charts illustrating the appeal process are in the Appendices of this part.

(b) In the case where an applicant objects to an initial proffered individual permit, the appeal process proceeds as follows. To initiate the appeal process regarding the terms and special conditions of the permit, the applicant must write a letter to the district engineer explaining his objections to the permit. The district engineer, upon evaluation of the applicant's objections, may: Modify the permit to address all of the applicant's objections or modify the permit to address some, but not all, of the applicant's objections, or not modify the permit, having determined that the permit should be issued as previously written. In the event that the district engineer agrees to modify the initial proffered individual permit to address all of the applicant's objections, the district engineer will proffer such modified permit to the applicant, enclosing an NAP fact sheet and an RFA form as well. Should the district engineer modify the initial proffered individual permit to address some, but not all, of the applicant's objections, the district engineer will proffer such modified permit to the applicant, enclosing an NAP fact sheet, RFA form, and a copy of the decision document for the project. If the district engineer does not modify the initial proffered individual permit, the district engineer will proffer the unmodified permit to the applicant a second time, enclosing an NAP fact sheet, an RFA form, and a copy of the decision document. If the applicant still has objections, after receiving the second proffered permit (modified or unmodified), the applicant may decline such proffered permit; this declined permit may be appealed to the division engineer upon submittal of a complete RFA form. The completed RFA must be received by the division engineer within 60 days of the NAP. A flow chart of an applicant's options for an initial proffered individual permit is shown in appendix B of this part. A flow chart of the appeal process for a permit denial or a declined permit (*i.e.*, a proffered permit declined after the Corps decision on the applicant's objections to the initial proffered permit) is shown in appendix A of this part. A flow chart of the appeal process for an approved jurisdictional determination is shown in appendix C of this part. A flow chart of the process for when an unacceptable request for appeal is returned to an applicant is shown in appendix D of this part.

(c) An approved JD will be reconsidered by the district engineer if the affected party submits new information or data to the district engineer within 60 days of the date of the NAP. (An RFA that contains new information will either be returned to the district engineer for reconsideration or the appeal will be processed if the applicant withdraws the new information.) The district engineer has 60 days from the receipt of such new information or data to review the new information or data, consider whether or not that information changes the previously approved JD, and, reissue the approved JD or issue a new approved JD. The reconsideration of an approved JD by the district engineer does not commence the administrative appeal process.

The affected party may appeal the district engineer's reissued or new approved JD.

(d) The district engineer may not delegate his signature authority to deny the permit with prejudice or to return an individual permit to the applicant with unresolved objections. The district engineer may delegate signature authority for JDs, including approved JDs.

(e) Affected parties may appeal approved JDs where the determination was dated after March 28, 2000, but may not appeal approved JDs dated on or before March 28, 2000. The Corps will begin processing JD appeals no later than May 30, 2000. All appeals must meet the criteria set forth in § 331.5. If work is authorized by either general or individual permit, and the affected party wishes to request an appeal of the JD associated with the general permit authorization or individual permit or the special conditions of the proffered individual permit, the appeal must be received by the Corps and the appeal process concluded prior to the commencement of any work in waters of the United States and prior to any work that could alter the hydrology of waters of the United States.

Section 331.7 Review procedures

(a) *General.* The administrative appeal process for approved JDs, permit denials, and declined permits is a one level appeal, normally to the division engineer. The appeal process will normally be conducted by the RO. The RO will document the appeal process, and assist the division engineer in making a decision on the merits of the appeal. The division engineer may participate in the appeal process as the division engineer deems appropriate. The division engineer will make the decision on the merits of the appeal, and provide any instructions, as appropriate, to the district engineer.

(b) *Requests for the appeal of approved JDs, permit denials, or declined permits.* Upon receipt of an RFA, the RO shall review the RFA to determine whether the RFA is acceptable (*i.e.*, complete and meets the criteria for appeal). If the RFA is acceptable, the RO will so notify the appellant in writing within 30 days of the receipt of the acceptable RFA. If the RO determines that the RFA is not complete the RO will so notify the appellant in writing within 30 days of the receipt of the RFA detailing the reason(s) why the RFA is not complete. If the RO believes that the RFA does not meet the criteria for appeal (see § 331.5), the RO will make a recommendation on the RFA to the division engineer. If the division engineer determines that the RFA is not acceptable, the division engineer will notify the appellant of this determination by a certified letter detailing the reason(s) why the appeal failed to meet the criteria for appeal. No further administrative appeal is available, unless the appellant revises the RFA to correct the deficiencies noted in the division engineer's letter or the RO's letter. The revised RFA must be received by the division engineer within 30 days of the date of the Corps letter indicating that the initial RFA is not acceptable. If the RO determines that the revised RFA is still not complete, the RO will again so notify the appellant in writing within 30 days of the receipt of the RFA detailing the reason(s) why the RFA is not complete. If the division engineer determines that the revised RFA is still not acceptable, the division engineer will notify the appellant of this determination by a certified letter within 30 days of the date of the receipt of the revised RFA, and will advise the appellant that the matter is not eligible for appeal. No further RFAs will be accepted after this point.

(c) *Site investigations.* Within 30 days of receipt of an acceptable RFA, the RO should determine if a site investigation is needed to clarify the administrative record. The RO should normally conduct any such site investigation within 60 days of receipt of an acceptable RFA. The RO may also conduct a site investigation at the request of the appellant, provided the RO has determined that such an investigation would be of benefit in interpreting the administrative record. The appellant and the appellant's authorized agent(s) must be provided an opportunity to participate in any site investigation, and will be given 15 days notice of any site investigation. The RO will attempt to schedule any site investigation at the earliest practicable time acceptable to both the RO and the appellant. The RO, the appellant, the appellant's agent(s) and the Corps district staff are authorized participants at any site investigation. The RO may also invite any other party the RO has determined to be appropriate, such as any technical experts consulted by the Corps. For permit denials and declined permit appeals, any site investigation should be scheduled in conjunction with the appeal review conference, where practicable. If extenuating circumstances occur at the site that preclude the appellant and/or the RO from conducting any required site visit within 60 days, the RO may extend the time period for review. Examples of extenuating circumstances may include seasonal hydrologic conditions, winter weather, or disturbed site conditions. The site visit must be conducted as soon as practicable as allowed by the extenuating circumstances, however, in no case shall any site visit extend the total appeals process beyond twelve months from the date of receipt of the RFA. If any site visit delay is necessary, the RO will notify the appellant in writing.

(d) *Approved JD appeal meeting.* The RO may schedule an informal meeting moderated by the RO or conference call with the appellant, his authorized agent, or both, and appropriate Corps regulatory personnel to review and discuss issues directly related to the appeal for the purpose of clarifying the administrative record. If a meeting is held, the appellant will bear his own costs associated with necessary arrangements, exhibits, travel, and representatives. The approved JD appeal meeting should be held at a location of reasonable convenience to the appellant and near the site where the approved JD was conducted.

(e) *Permit denials and declined permits appeal conference.* Conferences held in accordance with this part will be informal, and will be chaired by the RO. The purpose of the appeal conference is to provide a forum that allows the participants to discuss freely all relevant issues and material facts associated with the appeal. An appeal conference will be held for every appeal of a permit denial or a declined permit, unless the RO and the appellant mutually agree to forego a conference. The conference will take place within 60 days of receipt of an acceptable RFA, unless the RO determines that unforeseen or unusual circumstances require scheduling the conference for a later date. The purpose of the conference will be to allow the appellant and the Corps district representatives to discuss supporting data and information on issues previously identified in the administrative record, and to allow the RO the opportunity to clarify elements of the administrative record. Presentations by the appellant and the Corps district representatives may include interpretation, clarification, or explanation of the legal, policy, and factual bases for their positions. The conference will be governed by the following guidelines:

(1) *Notification.* The RO will set a date, time, and location for the conference. The RO will notify the appellant and the Corps district office in writing within 30 days of receipt of the RFA, and not less than 15 days before the date of the conference.

(2) *Facilities.* The conference will be held at a location that has suitable facilities and that is reasonably convenient to the appellant, preferably in the proximity of the project site. Public facilities available at no expense are preferred. If a free facility is not available, the Corps will pay the cost for the facility.

(3) *Participants.* The RO, the appellant, the appellant's agent(s) and the Corps district staff are authorized participants in the conference. The RO may also invite any other

party the RO has determined to be appropriate, such as any technical experts consulted by the Corps, adjacent property owners or Federal or state agency personnel to clarify elements of the administrative record. The division engineer and/or the district engineer may attend the conference at their discretion. If the appellant or his authorized agent(s) fail to attend the appeal conference, the appeal process is terminated, unless the RO excuses the appellant for a justifiable reason. Furthermore, should the process be terminated in such a manner, the district engineer's original decision on the appealed action will be sustained.

(4) *The role of the RO.* The RO shall be in charge of conducting the conference. The RO shall open the conference with a summary of the policies and procedures for conducting the conference. The RO will conduct a fair and impartial conference, hear and fully consider all relevant issues and facts, and seek clarification of any issues of the administrative record, as needed, to allow the division engineer to make a final determination on the merits of the appeal. The RO will also be responsible for documenting the appeal conference.

(5) *Appellant rights.* The appellant, and/or the appellant's authorized agent(s), will be given a reasonable opportunity to present the appellant's views regarding the subject permit denial or declined permit.

(6) *Subject matter.* The purpose of the appeal conference will be to discuss the reasons for appeal contained in the RFA. Any material in the administrative record may be discussed during the conference, but the discussion should be focused on relevant issues needed to address the reasons for appeal contained in the RFA. The RO may question the appellant or the Corps representatives with respect to interpretation of particular issues in the record, or otherwise to clarify elements of the administrative record. Issues not identified in the administrative record by the date of the NAP for the application may not be raised or discussed, because substantive new information or project modifications would be treated as a new permit application (see § 331.5(b)(5)).

(7) *Documentation of the appeal conference.* The appeal conference is an informal proceeding, intended to provide clarifications and explanations of the administrative record for the RO and the division engineer; it is not intended to supplement the administrative record. Consequently, the proceedings of the conference will not be recorded verbatim by the Corps or any other party attending the conference, and no verbatim transcripts of the conference will be made. However, after the conference, the RO will write a memorandum for the record (MFR) summarizing the presentations made at the conference, and will provide a copy of that MFR to the division engineer, the appellant, and the district engineer.

(8) *Appellant costs.* The appellant will be responsible for his own expenses for attending the appeal conference.

(f) *Basis of decision and communication with the RO.* The appeal of an approved JD, a permit denial, or a declined permit is limited to the information contained in the administrative record by the date of the NAP for the application or approved JD, the proceedings of the appeal conference, and any relevant information gathered by the RO as described in § 331.5. Neither the appellant nor the Corps may present new information not already contained in the administrative record, but both parties may interpret, clarify or explain issues and information contained in the record.

(g) *Applicability of appeal decisions.* Because a decision to determine geographic jurisdiction, deny a permit, or condition a permit depends on the facts, circumstances, and physical conditions particular to the specific project and/or site being evaluated, appeal decisions would be of little or no precedential utility. Therefore, an appeal decision of the division engineer is applicable only to the instant appeal, and has no other precedential effect. Such a decision may not be cited in any other administrative appeal, and may not be used as precedent for the evaluation of any other jurisdictional determination or permit application. While administrative appeal decisions lack precedential value and may not be cited by an appellant or a district engineer in any other appeal proceeding, the Corps goal is to have the Corps regulatory program operate as consistently as possible, particularly with respect to interpretations of law, regulation, an Executive Order, and officially-promulgated policy. Therefore, a copy of each appeal decision will be forwarded to Corps Headquarters; those decisions will be periodically reviewed at the headquarters level for consistency with law, Executive Orders, and policy. Additional official guidance will be issued as necessary to maintain or improve the consistency of the Corps' appellate and permit decisions.

Section 331.8 Timeframes for final appeal decisions

The Division Engineer will make a final decision on the merits of the appeal at the earliest practicable time, in accordance with the following time limits. The administrative appeal process is initiated by the receipt of an RFA by the division engineer. The Corps will review the RFA to determine whether the RFA is acceptable. The Corps will notify the appellant accordingly within 30 days of the receipt of the RFA in accordance with § 331.7(b). If the Corps determines that the RFA is acceptable, the RO will immediately request the administrative record from the district engineer. The division engineer will normally make a final decision on the merits of the appeal within 90 days of the receipt of an acceptable RFA unless any site visit is delayed pursuant to § 331.7(c). In such case, the RO will complete the appeal review and the division engineer will make a final appeal decision within 30 days of the site visit. In no case will a site visit delay extend the total appeal process beyond twelve months from the date of receipt of an acceptable RFA.

Section 331.9 Final appeal decision

(a) In accordance with the authorities contained in § 331.3(a), the division engineer will make a decision on the merits of the appeal. While reviewing an appeal and reaching a decision on the merits of an appeal, the division engineer can consult with or seek information from any person, including the district engineer.

(b) The division engineer will disapprove the entirety of or any part of the district engineer's decision only if he determines that the decision on some relevant matter was arbitrary, capricious, an abuse of discretion, not supported by substantial evidence in the administrative record, or plainly contrary to a requirement of law, regulation, an Executive Order, or officially promulgated Corps policy guidance. The division engineer will not attempt to substitute his judgment for that of the district engineer regarding a matter of fact, so long as the district engineer's determination was supported by substantial evidence in the administrative record, or regarding any other matter if the district engineer's determination was reasonable and within the zone of discretion delegated to the district engineer by Corps regulations. The division engineer may instruct the district engineer on how to correct any procedural error that was prejudicial to the appellant (i.e., that was not a "harmless" procedural error), or to reconsider the decision where any essential part of the district engineer's decision was not supported by accurate or sufficient information, or analysis, in the administrative record. The division engineer will document his decision on the merits of the appeal in writing, and provide a copy of this decision to the applicant (using certified mail) and the district engineer.

(c) The final decision of the division engineer on the merits of the appeal will conclude the administrative appeal process, and this decision will be filed in the administrative record for the project.

Section 331.10 Final Corps decision

The final Corps decision on a permit application is the initial decision to issue or deny a permit, unless the applicant submits an RFA, and the division engineer accepts the RFA, pursuant to this Part. The final Corps decision on an appealed action is as follows:

(a) If the division engineer determines that the appeal is without merit, the final Corps decision is the district engineer's letter advising the applicant that the division engineer has decided that the appeal is without merit, confirming the district engineer's initial decision, and sending the permit denial or the proffered permit for signature to the appellant; or

(b) If the division engineer determines that the appeal has merit, the final Corps decision is the district engineer's decision made pursuant to the division engineer's remand of the appealed action. The division engineer will remand the decision to the district engineer with specific instructions to review the administrative record, and to further analyze or evaluate specific issues. If the district engineer determines that the effects of the district engineer's reconsideration of the administrative record would be narrow in scope and impact, the district engineer must provide notification only to those parties who commented or participated in the original review, and would allow 15 days for the submission of supplemental comments. For permit decisions, where the district engineer determines that the effect of the district engineer's reconsideration of the administrative record would be substantial in scope and impact, the district engineer's review process will include issuance of a new public notice, and/or preparation of a supplemental environmental analysis and decision document (see 33 CFR 325.7). Subsequently, the district engineer's decision made pursuant to the division engineer's remand of the appealed action becomes the final Corps permit decision. Nothing in this part precludes the agencies' authorities pursuant to Section 404(q) of the Clean Water Act.

Section 331.11 Unauthorized activities

Approved JDs, permit denials, and declined permits associated with after-the-fact permit applications are appealable actions for the purposes of this part. If the Corps accepts an after-the-fact permit application, an administrative appeal of an approved JD, permit denial, or declined permit may be filed and processed in accordance with these regulations subject to the provisions of paragraphs (a), (b), and (c) of this section. An appeal of an approved JD associated with unauthorized activities will normally not be accepted unless the Corps accepts an after-the-fact permit application. However, in rare cases, the district engineer may accept an appeal of such an approved JD, if the district engineer determines that the interests of justice, fairness, and administrative efficiency would be served thereby. Furthermore, no such appeal will be accepted if the unauthorized activity is the subject of a referral to the Department of Justice or the EPA, or for which the EPA has the lead enforcement authority or has requested lead enforcement authority.

(a) *Initial corrective measures.* If the district engineer determines that initial corrective measures are necessary pursuant to 33 CFR 326.3(d), an RFA for an appealable action will not be accepted by the Corps, until the initial corrective measures have been completed to the satisfaction of the district engineer.

(b) *Penalties.* If an affected party requests, under this Section, an administrative appeal of an appealable action prior to the resolution of the unauthorized activity, and the division engineer determines that the appeal has no merit, the responsible party remains subject to any civil, criminal, and administrative penalties as provided by law.

(c) *Tolling of statute of limitations.* Any person who appeals an approved JD associated with an unauthorized activity or applies for an after-the-fact permit, where the application is accepted and processed by the Corps, thereby agrees that the statute of limitations regarding any violation associated with that approved JD or application is tolled until one year after the final Corps decision, as defined at § 331.10. Moreover, the recipient of an approved JD associated with an unauthorized activity or applicant for an after-the-fact permit must also memorialize that agreement to toll the statute of limitations, by signing an agreement to that effect, in exchange for the Corps acceptance of the after-the-fact permit application, and/or any administrative appeal (See 33 CFR 326.3(e)(1)(v)). No administrative appeal associated with an unauthorized activity or after-the-fact permit application will be accepted until such signed tolling agreement is furnished to the district engineer.

Section 331.12 Exhaustion of administrative remedies

No affected party may file a legal action in the Federal courts based on a permit denial or a proffered permit until after a final Corps decision has been made and the appellant has exhausted all applicable administrative remedies under this part. The appellant is considered to have exhausted all administrative remedies when a final Corps permit decision is made in accordance with § 331.10.

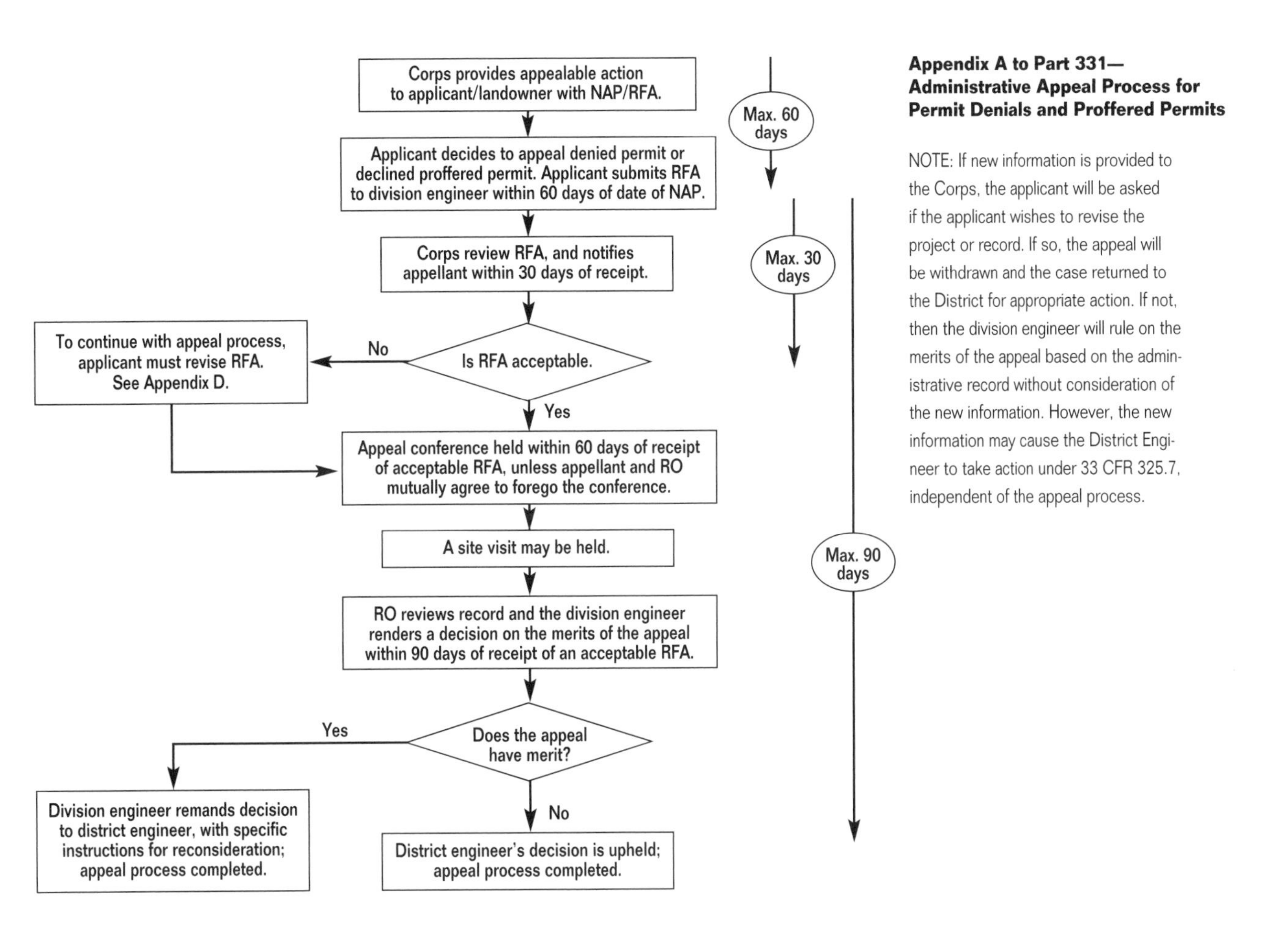

Appendix A to Part 331—Administrative Appeal Process for Permit Denials and Proffered Permits

NOTE: If new information is provided to the Corps, the applicant will be asked if the applicant wishes to revise the project or record. If so, the appeal will be withdrawn and the case returned to the District for appropriate action. If not, then the division engineer will rule on the merits of the appeal based on the administrative record without consideration of the new information. However, the new information may cause the District Engineer to take action under 33 CFR 325.7, independent of the appeal process.

Appendix B to Part 331—Applicant Options With Initial Proffered Permit

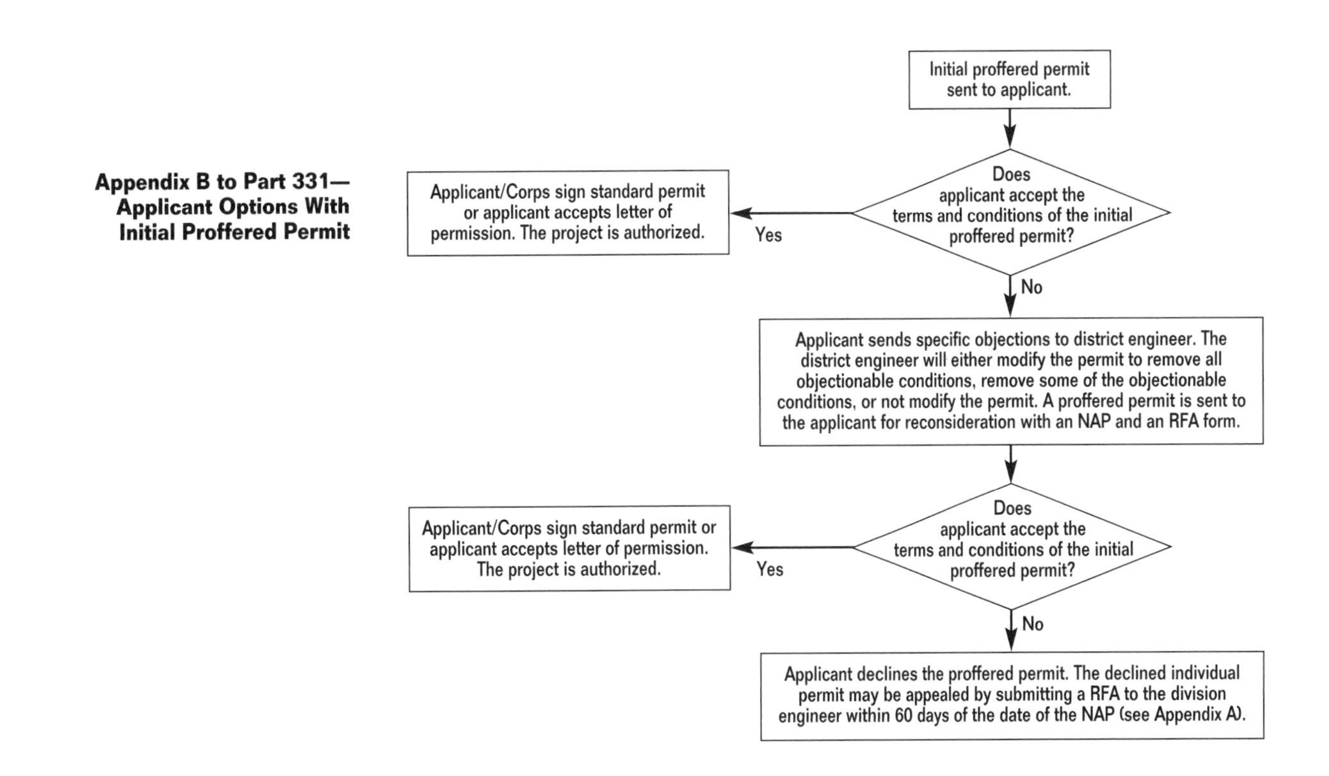

Appendix C to Part 331—Administrative Appeal Process for Approved Jurisdictional Determinations

Appendix D to Part 331—Process for Unacceptable Request for Appeal

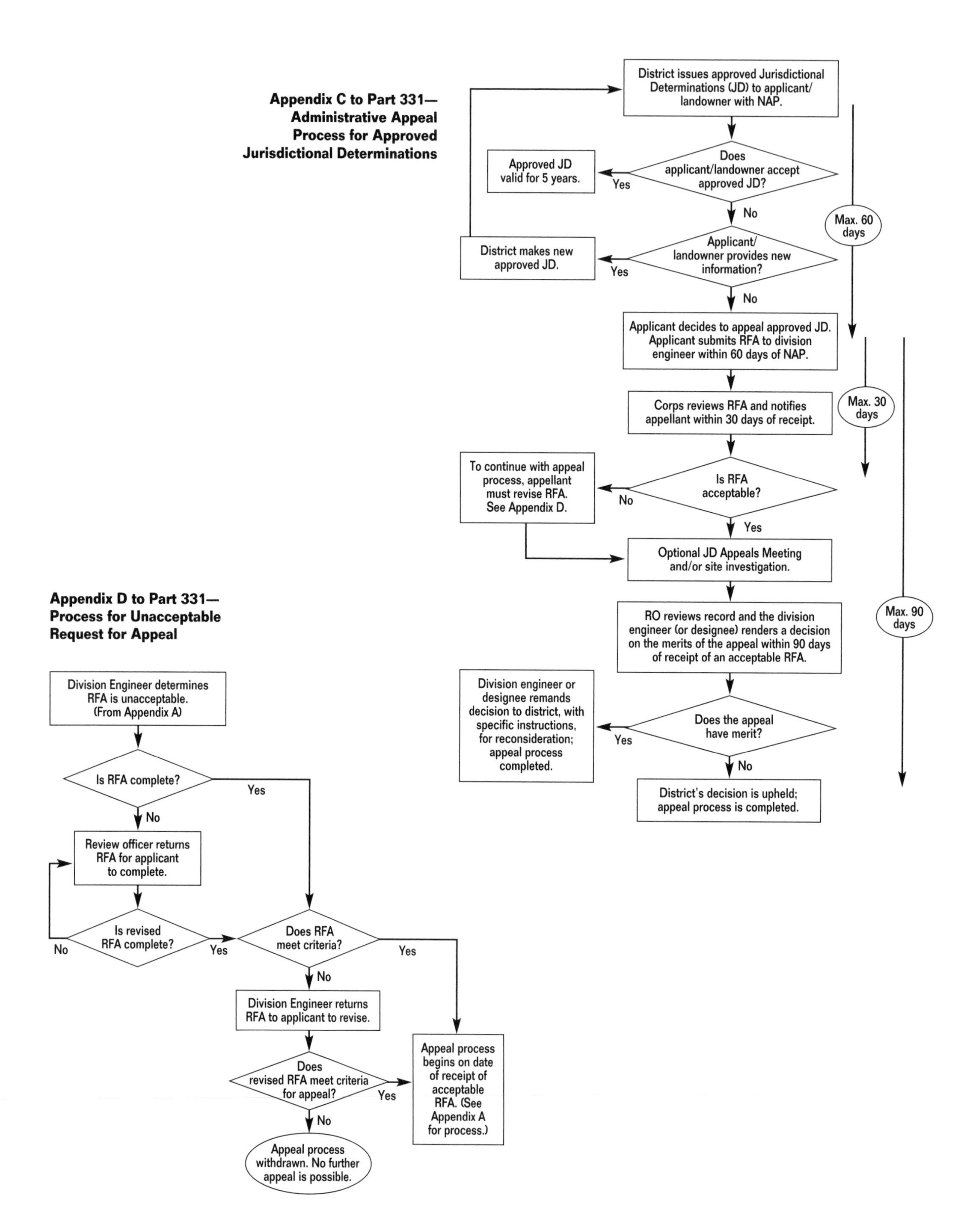

Revised Definitions of Fill Material and Discharge

Federal Register
Vol. 67, No. 90
Pages 31129–31143
Thursday, May 9, 2002

Department of the Army, Corps of Engineers
33 CFR Part 323

Environmental Protection Agency
40 CFR Part 232

Final Revisions to the Clean Water Act Regulatory Definitions of "Fill Material" and "Discharge of Fill Material"

ACTION: Final rule

SUMMARY: The U.S. Army Corps of Engineers (Corps) and the Environmental Protection Agency (EPA) are promulgating a final rule to reconcile our Clean Water Act (CWA) section 404 regulations defining the term "fill material" and to amend our definitions of "discharge of fill material." Today's final rule completes the rulemaking process initiated by the April 20, 2000, proposal in which we jointly proposed to amend our respective regulations so that both agencies would have identical definitions of these key terms. The proposal was intended to clarify the Section 404 regulatory framework and generally to be consistent with existing regulatory practice. Today's final rule satisfies those goals.

Today's final rule defines "fill material" in both the Corps' and EPA's regulations as material placed in waters of the U.S. where the material has the effect of either replacing any portion of a water of the United States with dry land or changing the bottom elevation of any portion of a water. The examples of "fill material" identified in today's rule include rock, sand, soil, clay, plastics, construction debris, wood chips, overburden from mining or other excavation activities, and materials used to create any structure or infrastructure in waters of the U.S. This rule retains the effects-based approach of the April 2000 proposal and reflects the approach in EPA's longstanding regulations.Today's final rule, however, includes an explicit exclusion from the definition of "fill material" for trash or garbage.

Today's final rule also includes several clarifying changes to the term "discharge of fill material." Specifically, the term "infrastructure" has been added in several places following the term "structure" to further define the situations where the placement of fill material is considered a "discharge of fill material." In addition, the phrases "placement of fill material for construction or maintenance of any liner, berm, or other infrastructure associated with solid waste landfills" and "placement of overburden, slurry, or tailings or similar mining-related materials" have been added to the definition of "discharge of fill material" to provide further clarification of the types of activities regulated under section 404.

As indicated in the proposal, as a general matter, this final rule will not modify existing regulatory practice. Today's final rule, which establishes uniform language for the Corps' and EPA's definitions of "fill material" and "discharge of fill material," will enhance the agencies' ability to protect aquatic resources by ensuring more consistent and effective implementation of CWA requirements.

EFFECTIVE DATE: June 10, 2002.

FOR FURTHER INFORMATION CONTACT: For information on today's rule, contact either Mr. Thaddeus J. Rugiel, U.S. Army Corps of Engineers, ATTN CECW-OR, 441 "G" Street, NW, Washington, DC 20314-1000, phone: (202) 761-4595, e-mail address: *thaddeus.j. rugiel@hq02. usace.army.mil,* or Ms. Brenda Mallory, U.S. Environmental Protection Agency, EPA West, Office of Wetlands, Oceans and Watersheds (4502T), 1200 Pennsylvania Avenue, NW, Washington, DC 20460, phone: (202) 566–1368, e-mail address: *mallory.brenda @epa.gov.*

SUPPLEMENTARY INFORMATION

I. Background

A. Potentially Regulated Entities

Persons or entities that discharge material to waters of the U.S. that has the effect of replacing any portion of a water of the U.S. with dry land or changing the bottom elevation of any portion of a water of the U.S. could be regulated by today's rule. The CWA generally prohibits the discharge of pollutants into waters of the U.S. without a permit issued by EPA, or a State or Tribe approved by EPA under section 402 of the Act, or, in the case of dredged or fill material, by the Corps or an approved State or Tribe under section 404 of the Act. Today's final rule addresses the CWA section 404 program's definitions of "fill material" and "discharge of fill material," which are important for determining whether a particular discharge is subject to regulation under CWA section 404. Today's final rule reconciles EPA's and the Corps' differing definitions of "fill material" and provides further clarification for the regulated public on what constitutes a "discharge of fill material." Examples of entities potentially regulated include:

CATEGORY	EXAMPLES OF POTENTIALLY REGULATED ENTITIES
State/Tribal governments or instrumentalities	State/Tribal agencies or instrumentalities that discharge material that has the effect of replacing any portion of a water of the U.S. with dry land or changing the bottom elevation of a water of the U.S.
Local governments or instrumentalities	Local governments or instrumentalities that discharge material that has the effect of replacing any portion of a water of the U.S. with dry land or changing the bottom elevation of a water of the U.S.
Federal government agencies or instrumentalities	Federal government agencies or instrumentalities that discharge material that has the effect of replacing any portion of a water of the U.S. with dry land or changing the bottom elevation of a water of the U.S.
Industrial, commercial, or agricultural entities	Industrial, commercial, or agricultural entities that discharge material that has the effect of replacing any portion of a water of the U.S. with dry land or changing the bottom elevation of a water of the U.S.
Land developers and landowners	Land developers and landowners that discharge material that has the effect of replacing any portion of a water of the U.S. with dry land or changing the bottom elevation of a water of the U.S.

This table is not intended to be exhaustive, but rather provides a guide for readers regarding entities that are likely to be regulated by this action. This table lists the types of entities that we are now aware of that could potentially be regulated by this action. Other types of entities not listed in the table also could be regulated. To determine whether your organization or its activities are regulated by this action, you should carefully examine the applicability criteria in sections 230.2 of Title 40 and 323.2 of Title 33 of the Code of Federal Regulations, as well as the preamble discussion in Section II of today's final rule. If you have questions regarding the applicability of this action to a particular entity, consult the persons listed in the preceding section entitled FOR FURTHER INFORMATION, CONTACT.

B. Summary of Regulatory History Leading to Final Rule and Related Litigation

The CWA governs the "discharge" of "pollutants" into "navigable waters," which are defined as "waters of the United States." Specifically, Section 301 of the CWA generally prohibits the discharge of pollutants into waters of the U.S., except in accordance with the requirements of one of the two permitting programs established under the CWA: Section 404, which regulates the discharge of dredged or fill material, or section 402, which regulates all other pollutants under the National Pollutant Discharge Elimination System (NPDES) program. Section 404 is primarily administered by the Corps, or States/Tribes that have assumed the program pursuant to section 404(g), with input and oversight by EPA. In contrast, Section 402 and the remainder of the CWA are administered by EPA or approved States or Tribes. The CWA defines the term "pollutant" to include materials such as rock, sand, and cellar dirt that often serve as "fill material." The CWA, however, does not define the terms "fill material" and "discharge of fill material," leaving it to the agencies to adopt definitions consistent with the statutory framework of the CWA.

Prior to 1977, both the Corps and EPA had defined "fill material" as "any pollutant used to create fill in the traditional sense of replacing an aquatic area with dry land or of changing the bottom elevation of a water body for any purpose. * * *" 40 FR 31325 (July 25, 1975); 40 FR 41291 (September 5, 1975). In 1977, the Corps amended its definition of "fill material" to add a "primary purpose test," and specifically excluded from that definition material that was discharged primarily to dispose of waste. 42 FR 37130 (July 19, 1977). This change was adopted by the Corps because it recognized that some discharges of solid waste materials technically fit the definition of fill material; however, the Corps believed that such waste materials should not be subject to regulation under the CWA section 404 program. Specifically, the Corps' definition of "fill material" adopted in 1977 reads as follows:

> (e) The term "fill material means any material used *for the primary purpose* of replacing an aquatic area with dry land or of changing the bottom elevation of an [sic] water body. The term does not include any pollutant discharged into the water primarily to dispose of waste, *as that activity is regulated under section 402 of the Clean Water Act.*" 33 CFR 323.2(e) (2001) (emphasis added).

EPA did not amend its regulations to adopt a "primary purpose test" similar to that used by the Corps. Instead, the EPA regulations at 40 CFR 232.2 defined "fill material" as "any 'pollutant' which replaces portions of the 'waters of the United States' with dry land or which changes the bottom elevation of a water body *for any purpose*" (emphasis added). EPA's definition focused on the effect of the material (an effects-based test), rather than the purpose of the discharge in determining whether it would be regulated by section 404 or section 402.

C. April 2000 Proposal

These differing definitions of "fill material" have resulted in some confusion for some members of the regulated community which has not promoted effective implementation of the CWA. *See* 65 FR at 21294. As a result, in April 2000, the agencies proposed revisions to their respective definitions of "fill material" and "discharge of fill material," adopting a single effects-based definition similar to that in EPA's regulations. The April 2000 proposed rule defined "fill material" as material that has the effect of replacing any portion of a water of the U.S. with dry land, or changing the bottom elevation of any portion of a water of the U.S. The agencies believe that an effects-based definition is, as a general matter, the most effective approach for identifying discharges that are regulated as "fill material" under section 404. Thus, the proposal removed from the Corps' definition the "primary purpose" test and the provision excluding pollutants discharged into water primarily to dispose of waste.

The April 2000 proposal also would have excluded from the definition discharges subject to an EPA proposed or promulgated effluent limitation guideline or standard under CWA sections 301, 304, 306, or discharges covered under a NPDES permit under CWA section 402. Finally, the April 2000 proposal solicited comments on the idea of the agencies creating an "unsuitable fill" category in the regulations that would identify materials that the Corps District Engineer could determine were not appropriate as fill material and, consequently, refuse to process an application seeking authorization to discharge such material.

In the preamble for the April 2000 proposal, the agencies discussed the need to address the confusion created by the agencies' differing definitions. While in practice some Corps Districts and EPA Regions have developed consistent approaches for determining whether proposed activities would result in a discharge of fill material, national uniformity will ensure better environmental results. Moreover, two judicial decisions discussed in the April 2000 proposal, *Resource Investments Incorporated* v. *U.S. Army Corps of Engineers*, 151 F. 3d 1162 (9th Cir. 1998) ("*RII*") and *Bragg* v. *Robertson* (Civil Action No. 2:98-636, S.D. W. Va.), *vacated on other grounds*, 248 F.3d 275 (4th Cir. 2001) ("*Bragg*"), indicate that the differing EPA and Corps definitions can result in judicial decisions that further confuse the regulatory context. *See* 65 FR at 21294–95. The clarification in the April 2000 proposal was intended to promote clearer understanding and application of our regulatory programs.

With respect to the term "discharge of fill material," the April 2000 proposal also included several clarifying changes. Unlike the definition of "fill material," EPA's and the Corps' then-existing regulations defining the term "discharge of fill material" were substantively identical. The proposed changes to the term were intended to provide further clarification of the issue. Specifically, the proposal provided for adding two phrases to the definition: (1) "Placement of fill material for construction or maintenance of liners, berms, and other infrastructure associated with solid waste landfills;" and (2) "placement of coal mining overburden."

As summarized in more detail in the U.S. Army Corps of Engineers' and Environmental Protection Agency's Response to Comments on the April 20, 2000, Proposed Rule Revising the Clean Water Act Regulatory Definitions of "Fill Material" and "Discharge of Fill Material," dated May 3, 2002 ("Response to Comments"), we received a number of comments addressing these proposed changes. The comments and the above-referenced document are part of the administrative record for this rule and are available from either agency. *See* the section entitled FOR FURTHER INFORMATION CONTACT.

II. Discussion of Final Rule

A. Overall Summary of Comments

We received over 17,200 comments on the proposed rule, including several hundred late comments, most of which consisted of identical or substantially identical e-mails, letters, and postcards opposing the rule. (In April 2002, an additional several thousand letters and e-mails were sent opposing the adoption of a rule similar to the proposal.) Approximately 500 of the original comments consisted of more individualized letters, with a mixture of those comments supporting and opposing the rule. The comments of environmental groups and the various form letters were strongly opposed to the proposal, in particular, the elimination of the waste exclusion and the discussion in the preamble regarding treatment of unsuitable fill material. Except for several landfill representatives, comments from the regulated community generally supported the proposal, in particular, the fact that the rule would create uniform definitions of "fill material" for the Corps' and EPA's rules and maintain regulation of certain discharges under section 404 as opposed to section 402 of the CWA. A detailed discussion of the issues raised in the comments and the agencies' responses can be found in the Response to Comments document.

The April 2000 proposal would have achieved four major outcomes and these were the focus of many of the comments. These outcomes were (1) Conforming the EPA and Corps definitions of "fill material" to one another; (2) adopting an effects-based test, as opposed to the Corps' primary purpose test, for defining "fill material;" (3) eliminating the waste exclusion from the Corps' regulation; and (4) soliciting comments on whether to develop a definition for "unsuitable fill material." A summary of comments relating to these four issues and our responses are discussed in section II.B of this preamble, which describes today's final rule.

In addition, comments asserted the need for the agencies to prepare an environmental impact statement (EIS) in order to comply with the National Environmental Policy Act; and questioned the consistency of the April 2000 proposal with the CWA, existing judicial decisions, and agency guidance documents. These comments are addressed in this section of the preamble.

With respect to the need for an EIS, many of the comments opposing the adoption of the rule argued that an EIS should have been prepared, particularly to address the impacts of eliminating the waste exclusion. Supporters of an EIS rejected the notion that the issues will be addressed in the individual permit situations. First, they pointed out that many of the mining activities have historically been permitted under the nationwide permit program where truncated environmental review occurs and no individual NEPA analysis is undertaken. Second, they argued that the cumulative impacts often are not appropriately addressed in this context. As described in section III.J of this final preamble and in the Response to Comments document, the agencies have concluded that preparation of an EIS is not required for this rule pursuant to NEPA. While supporters of an EIS suggest that finalizing this rule will result in significant new discharges that previously would not have occurred, that is not the case. Although the rule will clarify the appropriate regulatory framework, we do not expect there to be any significant change in the nature and scope of discharges that will occur.

Finally, a number of comments asserted that the proposal should not be finalized because it violated the then-existing law (*e.g.*, CWA, *Bragg*, and *RII*). Other comments argued that the proposal was consistent with the CWA and current regulatory practice. We do not agree that the proposal or today's final rule violate the CWA or any other law. Moreover, we believe that agencies have an obligation to take whatever steps may be necessary, including making revisions to their regulations, to ensure that their programs are appropriately implementing statutory mandates. As indicated, the Corps and EPA believe that the current inconsistency between their respective definitions of "fill material" is impeding the effective implementation of the section 404 program. Under those circumstances, we believe that a change in the regulatory language is justified and that by adopting the substance of EPA's longstanding definition, we are minimizing potential confusion and disruption to the program, while remaining consistent with the CWA. We agree with those comments that recognize the consistency of our action with the CWA and current practice. As described in more detail in the Response to Comments document and sections II. B and D of this preamble, today's final rule clarifies the governing regulatory framework in a manner consistent with the CWA and existing practice.

B. Discussion of the Final Rule

1. Definition of "Fill Material"

Today's final rule modifies both the EPA's and Corps' existing definitions of "fill material" and has retained the effects-based approach set forth in the proposal. The final rule defines "fill material" as material placed in waters of the U.S. where the material has the effect of either replacing any portion of a water of the United States with dry land or changing the bottom elevation of any portion of a water. The examples of "fill material" identified in today's rule include rock, sand, soil, clay, plastics, construction debris, wood chips, overburden from mining or other excavation activities, and materials used to create any structure or infrastructure in waters of the U.S. The proposed rule only specifically identified rock, earth and sand as examples, but the preamble made it clear that these were merely illustrative. In addition, in the preamble to the proposal, we indicated that wood chips, coal mining overburden, and similar materials would also constitute "fill material" if they had the effect of fill. As a result of questions raised in the comments about the scope of the term "fill material," we have included additional examples in the final rule, several of which were discussed in the proposed preamble. We believe that these additional examples will further clarify the rule.

Although today's final rule adopts a general effects-based approach for defining "fill material," it specifically excludes trash or garbage. Today's final rule does not modify any other Section 404 jurisdictional terms or alter any procedures governing the individual or general permit processes for Section 404 authorizations, requirements under Section 402, or the governing permit programs. Following is a summary of the actions that the agencies have taken in response to public comments.

a. Reconciling Agencies' Definitions. The majority of the comments from both the environmental and industry perspectives addressing the issue of whether the agencies should have identical definitions expressed the general view that the agencies should have the same definitions for the key jurisdictional terms "fill material" and "discharge of fill material." Many of the comments also noted that the differences between the Corps' and EPA's rules have historically caused confusion for the regulated community. Several asserted that despite differences in the regulatory language, some Corps Districts have been applying an effects-based test for some time. As described in the Response to Comments document, the agencies agree with those comments supporting the promulgation in both the Corps' and EPA's regulations of a uniform definition for the terms "fill material" and "discharge of fill material." Today's final rule achieves this result.

b. Effects-Based Test. Most of the comments supported the proposed rule's use of an effects based test similar to EPA's longstanding definition for defining "fill material" and the elimination of the "primary purpose" test from the Corps regulations. Those disagreeing with such an approach gave a variety of reasons including, the lack of any demonstrated justification that eliminating the primary purpose test from the Corps' regulation was necessary; the existence of similar purpose tests in other statutes involving waste materials as well as in the Section 404(b)(1) Guidelines as demonstrating that such tests need not be unwieldy; the existence of alternative ways of addressing the issues of concern without resorting to this rule change; and concerns about the inappropriate expansion of section 404 jurisdiction. As will be explained, the agencies are not persuaded by these arguments.

First, we believe that the objective standard created by the effects-based test will yield more consistent results in determining what is "fill material" and will provide greater certainty in the implementation of the program. We believe that these benefits provide sufficient justification for today's rule change. In addition, although similar "purpose" tests may be used under other statutes and even under the section 404 program, this does not negate the difficulties we have faced in applying the primary purpose test, as well as some confusion that has resulted from the use of the subjective primary purpose test in the section 404 jurisdictional context. An objective, effects-based standard also helps ensure that discharges with similar environmental effects will be treated in a similar manner under the regulatory program. The subjective, purpose-based standard led in some cases to inconsistent treatment of similar discharges, a result which hampers effective implementation of the statute.

Moreover, we believe there is an important distinction between the use of a purpose test here, where it determines the basic jurisdiction of the section 404 versus the section 402 program, and its use in the other contexts, such as in the evaluation of whether alternatives to a discharge of dredged material are "practicable" within the meaning of the section 404(b)(1) Guidelines. See 40 CFR 230.10(a)(2). The use of project purpose in the latter case is appropriate because it would make no sense to consider an alternative "practicable" if it did not satisfy the basic or overall purpose of the project proposed by the applicant. The definition of fill material, on the other hand, determines which legal requirements must be met for a discharge to be authorized under the statute. In that circumstance, we believe it is important to use an objective, effects-based test that ensures consistent treatment of like discharges, and prevents uncertainty for the regulated community as to what regulatory program applies to particular discharges. Moreover, we disagree that alternatives other than a rulemaking could have adequately addressed the agencies' concerns since the facial differences in our regulations could only be completely reconciled by revising the rules. In addition, them agencies previously had attempted to clarify their interpretation of the rules in a 1986 Memorandum of Agreement (MOA). Nevertheless, issues persisted.

Finally, we disagree that the rule causes an inappropriate expansion of section 404 jurisdiction. The CWA does not limit section 404 jurisdiction over fill material to materials meeting the primary purpose test. The "primary purpose test" is a regulatory definition and within the agencies province to modify as long as the modification is consistent with the CWA. In sum, as described in the Response to Comments document, the final rule, just as the proposal, adopts an effects-based approach to defining fill material. We believe the clarity and consistency created by the agencies relying on a more objective test for defining these key jurisdictional terms will result in more effective regulation under the CWA.

c. *Elimination of Waste Exclusion.* Many comments opposed the proposal to eliminate the waste exclusion from the Corps' regulation. Some of these comments recommended that, in addition to the effects-based test, the agencies should include a general exclusion from the definition of "fill material" for any discharge of "waste." These comments asserted that such an approach provides the advantages of EPA's effects-based approach while more effectively implementing the Corps' exclusion of waste material from regulation under section 404. Some of the comments argued that the proposed rule's deletion of the waste exclusion language from the Corps' regulations violates the CWA. According to these comments, while waste material can permissibly be covered by section 404 when it is placed in waters for a beneficial purpose, the CWA categorically prohibits authorizing such discharges under section 404 when their purpose is waste disposal. These comments pointed to the decisions in *RII* and *Bragg* to argue that all waste material is outside the scope of section 404.

These comments do not object to, nor claim that the CWA prohibits, issuance of a section 404 permit for waste material discharged into waters of the U.S. under all circumstances. Where waste is discharged for a purpose other than waste disposal (e.g., to create fast land for development), these comments acknowledged that the Corps' issuance of a section 404 permit in accordance with the section 404(b)(1) Guidelines adequately protects the environment and is consistent with the CWA. On this point, we agree.

However, where the identical material–with identical environmental effects–is discharged into waters for purposes of waste disposal, the comments contend that issuance of a section 404 permit in accordance with the Guidelines would neither protect the environment nor be allowed by the CWA. Here, we disagree. Simply because a material is disposed of for purposes of waste disposal does not, in our view, justify excluding it categorically from the definition of fill. Some waste (e.g., mine overburden) consists of material such as soil, rock and earth, that is similar to "traditional" fill material used for purposes of creating fast land for development. In addition, other kinds of waste having the effect of fill (e.g., certain other mining wastes) can, unlike trash or garbage, be indistinguishable either upon discharge or over time from structures created for purposes of creating fast land. Given the similarities of some discharges of waste to "traditional" fill, we believe that a categorical exclusion for waste would be over-broad. Instead, where a waste has the effect of fill, we believe that regulation under the section 404 program is appropriate.

This does not mean, however, that today's rule opens up waters of the U.S. to be filled for any waste disposal purposes. As explained previously, today's rule is generally consistent with current agency practice and so it does not expand the types of discharges that will be covered under section 404. The section 404(b)(1) Guidelines provide for a demonstration that there are no less damaging alternatives to the discharge, and that all appropriate and practicable steps have been taken to avoid, minimize and compensate for any effects on the waters. We recognize that, some fill material may exhibit characteristics, such as chemical contamination, which may be of environmental concern in certain circumstances. This is true under either a primary purpose or effects based definition of fill material. The section 404 permitting process, however, is expressly designed to address the entire range of environmental concerns arising from discharges of dredged or fill material. See 40 CFR Part 230, subparts C–G (containing comprehensive provisions for addressing physical, chemical and biological impacts of discharges).

The 404(b)(1) guidelines provide a comprehensive means of evaluating whether any discharge of fill material, regardless of its purpose, is environmentally acceptable and therefore may be discharged in accordance with the CWA. Where the practicable alternatives test has been satisfied and all practicable steps have been taken both to minimize effects on the aquatic environment and to compensate for the loss of aquatic functions and values, we believe the section 404 permitting process is adequate to ensure protection of the aquatic ecosystem for any pollutant that fills waters. There is no environmental basis for contending that the sufficiency of the permitting process to protect waters of the U.S. depends on the purpose of the discharge.

The position reflected in some of the comments appears to be based on the contention that Congress did not intend for waste disposal to be a permissible purpose of discharging pollutants into waters of the U.S. While we agree that Congress wanted to prevent utilization of waters as unlicensed dumping grounds for waste material, the Act as a whole is focused primarily on discharges of waste material, as shown by the Act's definition of pollutant, which includes solid waste, sewage, garbage, discarded equipment, industrial, municipal and agricultural waste. *See* CWA section 502(6). While the elimination of all discharges is an important goal of the Act (*see* CWA section 101(a)(1)), the Act seeks to meet that goal not by banning discharges of waste outright, but by imposing carefully tailored restrictions on discharges of pollutants based on factors such as the impact of the discharge on the receiving water, availability of treatment technologies, cost, and the availability of alternatives to the discharge. *See, e.g.*, CWA sections 301(b), 304(b) (requiring discharges to meet technology-based effluent limitations guidelines and standards); section 306(a)(1) (defining new source performance standard to include no discharge of pollutants "where practicable"); section 301(b)(1)(C) (requiring dischargers to comply with any more stringent limitations necessary to meet water quality standards); sections 404(b)(1) and 403(c)(1)(F) (requiring that 404(b)(1) Guidelines be based on section 403(c) criteria, which include consideration of "other possible locations" of disposal).

Nor do we think that there is any indication that Congress intended to exclude discharges for purposes of waste disposal entirely from coverage under section 404. For example, section 404 applies to "dredged material" (referred to as dredged "spoil" in the definition of pollutant in section 502(6)), which is typically discharged not for any beneficial purpose, but as a waste product from a dredging operation. Moreover, section 404(a) authorizes the Corps to issue permits for discharges of dredged or fill material at specified "disposal" sites. Congress' use of the word "disposal" supports the reasonableness of our view that regulating waste material having the effect of fill under section 404 is consistent with the Act.

We also disagree with the interpretation of some of the comments on the *RII* and *Bragg* decisions as mandating that the Corps retain the current exclusion of waste disposal in the definition of fill material. We note first that the decision of the district court in *Bragg* has been vacated by the Fourth Circuit on 11th amendment grounds. *Bragg* v. *Robertson*, 72 F. Supp. 2d 642 (S.D. W. Va. 1999), *rev'd*, 248 F. 3d 275 (4th Cir. 2001). In any event, both *Bragg* and *RII* applied the Corps' then-existing definition of fill material to conclude that certain discharges were not covered by section 404. Nothing in those decisions suggests that the Act itself precluded the regulation of waste materials with the effect of fill under section 404. See section II.D. of this preamble for further discussion of the *RII* decision. While we agree that trash or garbage generally should be excluded from the definition of fill material (for the reasons explained in section II.B.1d of this preamble), we do not agree that an exclusion for all waste is appropriate and have not included such a provision in today's rule. These issues are discussed in section II.B.1d of the preamble and are addressed more fully in the Response to Comments document.

d. Trash or Garbage. The agencies have added an exclusion for trash or garbage to the definition of "fill material" for several reasons. First, the preamble to the proposed rule and many of the comments recognized that trash or garbage, such as debris, junk cars, used tires, discarded kitchen appliances, and similar materials, are not appropriately used, as a general matter, for fill material in waters of the U.S. In particular, we agree that the discharge of trash or garbage often results in adverse environmental impacts to waters of the U.S. by creating physical obstructions that alter the natural hydrology of waters and may cause physical hazards as well as other environmental effects. We also agree that these impacts are generally avoidable because there are alternative clean and safe forms of fill material that can be used to accomplish project objectives and because there are widely available landfills and other approved facilities for disposal of trash or garbage.

Accordingly, a party may not obtain a section 404 permit to dispose of trash or garbage in regulated waters. Because the discharge of any pollutant into jurisdictional waters is prohibited under CWA section 301 except in accordance with a permit issued under sections 404 or 402, section 402 would govern such discharges. For many of the reasons identified in this preamble, such as the physical obstruction and hazards that such materials would create in waters of the U.S., we would emphasize that trash or garbage are unlikely to be eligible to receive a permit under the section 402 regulatory program. We also note that where such materials are placed in waters of the U.S. without a permit, EPA or an approved State/Tribal agency with permitting authority, remains the lead enforcement agency. Today's rule does not affect the application of section 402 of the CWA to discharges of pollutants other than fill material that may be associated with such things as solid waste landfill structures and mine impoundments. Where such structures release pollutants into waters of the U.S., a permit under section 402 of the CWA is required that will ensure protection of any downstream waters, including compliance with State water quality standards.

While the agencies have generally excluded materials characterized as trash or garbage from the definition of "fill material," we agree that there are very specific circumstances where certain types of material that might otherwise be considered trash or garbage may be appropriate for use in a particular project to create a structure or infrastructure in waters of the U.S. In such situations, this material would be regulated as fill material. Such material would have to be suitably cleaned up and not include constituents that would cause significant environmental degradation. An example would be where recycled porcelain fixtures are cleaned and placed in waters of the U.S. to create environmentally beneficial artificial reefs. Such material would not be considered trash or garbage and thus would not be subject to the exclusion. The agencies believe that this is appropriate, and even environmentally beneficial, in situations where (1) the otherwise excluded materials are being placed in waters of the U.S. in a manner consistent with traditional uses of fill material to create a structure or infrastructure, (2) the material's characteristics are suitable to the project purpose, and (3) the review under section 404 can effectively ensure that the material will not cause or contribute to significant environmental degradation.

We also note that as stated in the preamble to the proposal, it is important to draw a clear distinction between solid waste discharged directly into waters of the U.S. and sanitary solid waste landfills. With respect to solid waste landfills, the liners, berms, and other infrastructure that are constructed of fill materials in waters of the U.S. are regulated under section 404 of the CWA. In the case of a landfill that has received a section 404 permit for the placement of berms, dikes, liners and similar activities needed to construct the facility, the subsequent disposal of solid waste into the landfill, while subject to regulation under the RCRA, would not be subject to regulation under the CWA because the constructed facility is not waters of the U.S. As with current practice, discharges of leachate from landfills into waters of the U.S. would remain subject to CWA section 402. Today's final rule does not change this general regulatory framework for landfills. *See* section II D of this preamble for further discussion.

e. Unsuitable Fill Material. With respect to developing a potential definition of "unsuitable fill material," there was almost unanimous opposition to the unsuitable fill concept as discussed in the preamble. Some comments viewed it as an inadequate substitute for the elimination of the waste exclusion. Others argued that having an unsuitable fill provision would be a good idea but that it would need to be much broader and to specifically include mining-related wastes. These commenters also objected to leaving the question of whether something was "unsuitable fill material" to the discretion of the District

Engineer. Some comments expressed concern that the definition of unsuitable fill material focused on materials that have a potential to leach or that have toxic constituents in toxic amounts. They argued that the definition could result in prohibiting activities that with appropriate permit terms and conditions potentially are allowable under section 404. They also argued that such issues should be addressed in the context of the permitting process and should not result in the permit application being rejected. As described in the Response to Comments document, the agencies have not included an unsuitable fill category in the final rule but, as discussed, the final rule does narrow the scope of "fill material" by excluding trash or garbage.

f. Effluent Guideline Limitations and 402 Permits. In addition to the changes already discussed in this preamble, today's final rule also deletes the exclusion contained in the proposal for discharges covered by effluent limitation guidelines or standards or NPDES permits. Several of the comments raised concerns that the exclusion included in the proposed definition for discharges covered by proposed or existing effluent limitation guidelines or standards or NPDES permits was vague and would result in uncertainty with respect to the regulation of certain discharges. Other comments stated that it was inappropriate for rule language to allow reliance on proposed effluent limitation guidelines or standards before they are promulgated as a final rule. In addition, including the language in the actual rule could raise questions as to whether the reference to effluent guidelines was meant to refer only to those in existence at the time today's rule was promulgated or whether the reference was prospective.

In light of the concerns and confusion associated with the proposed provision, we have decided to delete it from the rule. However, although we have removed the language in question from the rule itself, we emphasize that today's rule generally is intended to maintain our existing approach to regulating pollutants under either section 402 or 404 of the CWA. Effluent limitation guidelines and new source performance standards ("effluent guidelines") promulgated under section 304 and 306 of the CWA establish limitations and standards for specified wastestreams from industrial categories, and those limitations and standards are incorporated into permits issued under section 402 of the Act. EPA has never sought to regulate fill material under effluent guidelines. Rather, effluent guidelines restrict discharges of pollutants from identified wastestreams based upon the pollutant reduction capabilities of available treatment technologies. Recognizing that some discharges (such as suspended or settleable solids) can have the associated effect, over time, of raising the bottom elevation of a water due to settling of waterborne pollutants, we do not consider such pollutants to be "fill material," and nothing in today's rule changes that view. Nor does today's rule change any determination we have made regarding discharges that are subject to an effluent limitation guideline and standards, which will continue to be regulated under section 402 of the CWA. Similarly, this rule does not alter the manner in which water quality standards currently apply under the section 402 or the section 404 programs.

2. Definition of "Discharge of Fill Material"

Most of the comments addressing "discharge of fill material" supported the inclusion of items related to solid waste landfills, although several asserted that the regulation of discharges associated with solid waste landfills was inconsistent with the court's decision in *Resource Investments Inc. v. U.S. Army Corps of Engineers,* 151 F. 3d 1162 (9th Cir. 1998). See detailed discussion in section II.D of this final preamble. With respect to the placement of coal mining overburden, two diametrically opposed views were reflected in the comments. Many of the comments argued that coal overburden was "waste" material and that allowing such discharges was a violation of the CWA. In contrast, other comments argued that focusing on "coal mining overburden" was confusing, because it created the impression that the overburden or similar materials from other mining processes may not be regulated as "discharges of fill material."

Today's final rule responds to the comments in the following ways. First, the agencies continue to agree with those comments that supported including the placement of material associated with construction and maintenance of solid waste landfills and related facilities in the discharge of fill material. For the reasons discussed in section II.D of this final preamble and in the Response to Comments document, we do not agree that we are precluded by the *RII* decision from issuing a rule that defines "fill material" or the "discharge of fill material" as encompassing discharges associated with the construction of solid waste landfill infrastructures. Second, the agencies have modified the "placement of coal mining overburden" to read "placement of overburden, slurry, or tailings or similar mining-related materials." The language in today's final rule will clarify that any mining-related material that has the effect of fill when discharged will be regulated as "fill material." We made this clarification because it was clear from the comments that some were reading the examples we identified as an exclusive list. The general intent of this rule is to cover materials that have the effect of fill, not simply to focus on any one industrial activity. We believe that the additional mining related examples will address the confusion reflected in the comments. Finally, as discussed in section II.B.1.c of this preamble, we do not agree that the CWA contains a blanket prohibition precluding discharges of "waste" materials in to waters of the U.S. Instead, the Act establishes the framework for regulating discharges into waters and we believe the section 404 program is the most appropriate vehicle for regulating overburden and other mining-related materials. Several other minor changes, editorial in nature, have also been made in today's final rule.

C. Appropriate Reliance on the Environmental Reviews Conducted by Other Federal or State Programs

As indicated, today's rule is designed to improve the effective implementation of the section 404 program by having the Corps and EPA adopt a single, uniform definition for these key jurisdictional terms. We also believe that we can improve the effective implementation of the program by placing greater emphasis on coordination among the Federal agencies and with relevant State and Tribal programs. There are numerous examples of where the agencies can effectively work together and with other State, Tribal and Federal programs in the review of proposed projects that involve a section 404 discharge to jointly develop information that is relevant and reliable. Projects involving discharges to waters of the U.S. are often subject to review under other Federal and State permit programs, including the RCRA, the Surface Mining Control and Reclamation Act (SMCRA), the Coastal Zone Management Act (CZMA), CWA Section 402 NPDES, and others. Examples where closer coordination may be beneficial include the review of proposed solid waste landfills under the CWA and RCRA, proposed highway projects under the CWA and NEPA, proposed mining projects under the CWA and SMCRA, and proposed coastal restoration projects under the CWA and CZMA.

As EPA and the Corps implement today's rule, we will be placing even greater emphasis on effective coordination with other relevant State, Tribal and Federal programs and, consistent with our legal responsibilities, on reliance, as appropriate, on the information developed and conclusions reached by other agencies to support the decisions required under these programs and ours. We are confident that this coordination will serve to make the implementation of today's rule and, more broadly, the CWA section 404 program, more effective, consistent and environmentally protective.

Some comments expressed concern that an effects-based approach to the definition of "fill material" would result in a duplication of effort among Federal programs and an increased workload for the Corps. We believe that more effective coordination among the State, Tribal and Federal agencies and appropriate reliance on the analyses of other agencies will help significantly to address these concerns.

First, it is important to note that EPA and Corps regulations encourage coordination and allow for appropriate reliance on relevant information and analyses developed under other programs to help satisfy section 404 program requirements. In the most effective circumstances, the Corps is able to coordinate with other relevant State, Tribal and Federal agencies before and during project review to identify the most efficient and effective role for each agency and ensure mutual reliance on information and analyses, particularly where that reliance is consistent with individual agency expertise and experience. For example, for many years, subject to advice from EPA, the Corps has relied on State determinations regarding water quality matters, as those State determinations are reflected in State CWA section 401 water quality certifications (see 33 CFR 320.4(d)). Such Corps reliance on State water quality determinations will continue for discharges associated with activities such as mining and solid waste landfills. In regulating discharges associated with mining, close coordination with the State, Tribal and Federal entities responsible for implementation of SMCRA, CWA section 401 and section 402 will enable the Corps to take advantage of the specialized expertise of the agencies as the Corps completes the section 404 review. Such coordination also helps to reduce the costs associated with project reviews, promotes consistent and predictable decision-making, and ultimately ensures the most effective protection for human health and the environment. EPA and the Corps anticipate that Corps District offices will rely on State/Federal site selection under SMCRA regarding the siting of coal mining related discharges to the extent allowed under current law and regulations. Similarly, the Corps will make full use of State RCRA information regarding the siting, design and construction of solid waste landfills, and will defer to those State decisions to the extent allowed by current law and regulation.

Both agencies recognize, however, that the Corps is ultimately responsible under the CWA for making the required determinations that support each permit decision based on the Corps' independent evaluation of the record. The Corps itself determines the extent of deference to information generated from other programs including, for example, site selection under SMCRA and RCRA, that is appropriate on a case-by-case basis. Ultimately the Corps is relying on, rather than relinquishing to, these other sources of information as a record is developed and the Corps makes the determinations required by the Section 404 regulatory program. For example, the Corps will make full use of State site selection decisions under SMCRA (*e.g.*, coal slurry impoundments) and RCRA (*e.g.*, solid waste landfills), but the Corps will independently review those decisions and the State processes that generated them, to ensure that any Corps permit decision for a discharge site will fully comply with NEPA, the section 404(b)(1) Guidelines, and other relevant legal requirements. The Corps and EPA believe that effective coordination with other State and Federal agencies and the information they develop will help the Corps continue to make more timely, consistent and environmentally protective permit decisions.

D. The Final Rule and the Resource Investments Decision

In *Resource Investments Inc. v. Corps,* 151 F.3d 1162 (9th Cir. 1998), the Ninth Circuit held that the Corps lacked the authority to regulate a solid waste landfill in waters of the U.S. The court found that: (1) Neither the solid waste itself nor the liner consisting of layers of gravel and low-permeability soil constituted "fill material" under Corps regulations; and (2) because of the potential for inconsistent results if landfills were regulated under both section 404 of the CWA and Subtitle D of RCRA, requiring these facilities to be subject solely to RCRA would "harmonize" the statutes.

We discussed this decision in the preamble to the proposed rule as an example of some of the confusion engendered by the "primary purpose" test. The court found in *RII* that the liner was not fill material because its primary purpose was not to replace an aquatic area with dry land or change the bottom elevation of a waterbody, "but rather to serve as a leak detection and collection system." 151 F. 3d at 1168. We explained in the proposal that fills typically serve some other purpose than just creating dry land or raising a water's bottom elevation and that, if the court's reasoning were taken to its logical conclusion, many traditional fills in waters of the U.S. would not be subject to section 404.

Some commenters objected to our proposal not to follow the decision in *RII* in this rulemaking. They criticized the proposal as an improper attempt to "override" or "overrule" the Ninth Circuit's decision, particularly within the Ninth Circuit where the decision is binding. They also argued that the proposed rule failed to address the potential for duplication and inconsistency in decision-making by State and Federal agencies identified in *RII.*

In our view, these comments raise two distinct issues. The first is whether we should follow the *RII* decision outside the Ninth Circuit and cease regulating discharges associated with the construction of solid waste landfills under section 404. The second issue is whether *RII* precludes us from regulating discharges associated with construction of solid waste landfill structures within the Ninth Circuit, even after today's rule. We address each of these issues in turn.

Regarding the first question, we note first that, after *RII* was decided, we chose not to acquiesce in the decision outside the Ninth Circuit. While we agreed that the solid waste disposal placed in a landfill is not fill material (and such waste continues to be excluded under today's rule), we believed that the court misapplied the primary purpose test in the Corps' regulations, and that the court's conclusion that RCRA supplanted CWA regulation was contrary to Congressional intent. See *Resource Investments Inc. et al.* v. *Corps,* No. 97-35934 (Government's Petition for Rehearing and Suggestion for Rehearing En Banc, September 30, 1998). Thus, after the court decided *RII*, the Corps has continued to issue section 404 permits for the construction of solid waste landfill infrastructures outside the Ninth Circuit.

After considering public comments, we continue to decline to follow *RII* outside the Ninth Circuit and have, therefore, maintained the approach in the proposed rule to the regulation of solid waste landfills. The revisions to the Corps' definition of fill material in today's rule address the basis for the court's holding that the landfill did not involve the discharge of fill material under section 404. For the reasons explained elsewhere in today's notice, we believe that an effects-based test is the appropriate means of evaluating whether a pollutant is "fill material" and should be regulated under section 404 as opposed to section 402 of the CWA. The placement of berms, liners and other infrastructure (such as roads) associated with construction of a solid waste landfill in waters of the U.S. has the effect of replacing water with dry land or raising the bottom elevation of a water. Therefore, under today's rule, they constitute fill material. Such discharges are indistinguishable from similar discharges associated with other construction activity, which the Corps has always regulated as fill under section 404. *See* 40 CFR 232.2; 33 CFR 323.2 (defining "discharge of fill material," to include "fill that is necessary for the construction of any structure in a water of the U.S.; the building of any structure or impoundment requiring rock, sand, dirt or other material for its construction; site-development fills for recreational, industrial, commercial, residential and other uses; causeways or road fills; * * *"). We have amended our definition of this term to include the "placement of fill material for construction or maintenance of any liner, berm, or other infrastructure associated with solid waste landfills." That amendment does not change substantively the prior definition, but merely adds solid waste landfills as an example to make clear that it constitutes a "discharge of fill material." Thus, under our new regulations, discharges associated with the creation of solid waste landfill structures clearly constitute "fill material."

To the extent some commenters asserted that revising our regulation was an improper attempt to "overrule" or "override" this holding in *RII*, we disagree. The court's analysis of the "fill material" in *RII* was based entirely on the Corps regulations as they existed at that time, and not upon the interpretation of the CWA itself. Moreover, the CWA does not define "fill material." Therefore, both the statute and the Ninth Circuit's decision leave us the discretion to adopt a reasonable definition consistent with the statutory scheme. We have explained elsewhere why we believe today's definition of fill is reasonable and appropriate under the CWA. To the extent today's rule has the practical effect of "overriding" this aspect of the court's decision in *RII*, that is neither remarkable nor inappropriate, since it is entirely proper for agencies to consider and, if appropriate, revise their regulations in light of judicial interpretation of them.

For purposes of deciding whether to apply the *RII* decision outside the Ninth Circuit, we have also evaluated the second basis for the court's decision–that regulation solely under Subtitle D of RCRA instead of section 404 would "harmonize" the statutes and avoid necessary duplication. We decline to follow that holding both on legal and policy grounds. First, we believe, notwithstanding *RII*, that eliminating the CWA permitting requirement on the grounds that an activity is regulated under RCRA is contrary to Congressional intent in both statutes. Second, we do not agree with the court that regulation under Subtitle D and section 404 would constitute unnecessary duplication, in light of the distinct purposes served by these authorities, the differing Federal roles under the two statutes, and our clarification in today's rulemaking of our intent to give all appropriate deference to State RCRA decisionmaking in the section 404 permitting process.

We first do not agree with the court's legal reasons for concluding that regulation under Subtitle D of RCRA supplants CWA regulation. The CWA prohibits the discharge of any pollutant into waters of the U.S. without a permit under the Act. *See* CWA section 301(a). Even though an activity associated with a discharge may be regulated under other Federal or State authorities, we believe there is not any basis to conclude that such regulation by itself makes section 301(a) of the Act inapplicable to a discharge of a pollutant into waters of the U.S. In effect, the court concluded that enactment of a regulatory scheme under Subtitle D of RCRA impliedly repealed the statutory permit requirement under the CWA. But "the intention of the legislature to repeal must be clear and manifest." *Radzanower v. Touche Ross & Co.*, 426 U.S. 148, 154 (1976), and the court must conclude that the two acts are in irreconcilable conflict or that the later act covers the whole subject of the earlier one and is clearly intended as a substitute. *Id.* The court in *RII* did not, and could not, make these findings. In fact, Congress itself made precisely the opposite findings when it enacted RCRA. Section 1006(a) states:

> Nothing in this chapter shall be construed to apply to (or to authorize any State, interstate, or local authority to regulate) any activity or substance which is subject to the [CWA] except to the extent such application (or regulation) is not inconsistent with the requirements of (the CWA).

This provision precludes regulation of solid waste landfills under Subtitle D in a manner inconsistent with the requirements of the CWA. In our view, it is plainly "inconsistent" with the requirements of the CWA to hold that regulation under RCRA eliminates CWA permitting requirement altogether. Instead, the court relied upon certain Corps regulations, statements by Corps officials and a 1986 interagency MOA. The court first stated that applying section 404 to solid waste landfills was "unreasonable" because there would be "potentially inconsistent results" where both the State and the Corps were applying the same criteria in regulating solid waste landfills. 151 F. 3d at 1169. The court held that this "regulatory overlap is inconsistent with" Corps regulations stating that "the Corps believes that State and Federal regulatory programs should complement rather than duplicate one another." 33 CFR 320.1(a)(5). In addition, the court cited statements by the Corps in a 1984 letter to EPA stating that EPA was in a better position than the Corps to regulate solid waste landfills. Finally, the court cited the 1986 MOA between the Corps and EPA.

However, none of these "authorities" purport to modify the statutory permitting requirements of the CWA, nor could they. The Corps' regulation cited by the court is simply a statement of the Corps' policy objective of working in concert with State regulatory programs, an important and continuing Corps objective that was discussed previously. The Corps' letter and the MOA reflected our efforts to manage our programs in light of our differing definitions of fill material, but did not speak to the CWA statutory permitting requirement. The court also misconstrued the 1986 MOA entered into by EPA and the Corps as indicating we intended to make the regulation of solid waste facilities within "the sole purview of the EPA and affected states" after EPA promulgated certain Subtitle D regulations. 151 F.3d at 1169. In fact, we stated,

> EPA and Army agree that consideration given to the control of discharges of solid waste both in waters of the United States and upland should take into account the results of studies being implemented under the 1984 Hazardous and Solid Waste Amendments (HSWA) to the Resource Conservation and Recovery Act (RCRA), signed into law on November 8, 1984....

Unless extended by mutual agreement, the agreement will expire at such time as EPA has accomplished specified steps in its implementation of RCRA, at which time the results of the study of the adequacy of the existing Subtitle D criteria and proposed revisions to the Subtitle D criteria for solid waste disposal facilities, including those that may receive hazardous household wastes and small quantity generator waste, will be known. In addition, data resulting from actions under the interim agreement can be considered at that time.

It should be noted that this MOA is about the regulation of solid waste disposal, not about the construction of infrastructure, including solid waste landfill infrastructure, that involves discharges of fill material to waters of the U.S. We did not address in the MOA how solid waste landfills would be regulated after EPA completed its study and certain RCRA regulations, but said only that these developments would "be taken into account" as we decided how to address these discharges in the future. Thus, in addition to the inability of the agencies as a legal matter to modify the CWA statutory permitting requirement through an MOA, we expressly reserved any judgment about the appropriate regulatory approach to be taken after certain actions were taken under RCRA. The court appears to have assumed that the MOA expired after we completed the specified steps under RCRA, and that regulatory authority over solid waste landfills thereafter became the sole purview of RCRA. In fact, the MOA did not expire, and it has continued to provide the framework for regulation of solid waste landfills under section 404 of the CWA. See Memorandum of John F. Studt, U.S. Army Corps of Engineers, May 17, 1993 (stating "the subject MOA remains effective in its entirety until further notice" and noting that this position was coordinated with EPA).

We conclude, therefore, that it would be contrary to the language and intent of both the CWA and RCRA to conclude that RCRA subtitle D supplants the CWA permitting requirement for discharges into waters of the U.S. associated with the construction of solid waste landfills. The different Federal roles in the permitting schemes in these statutes supports this conclusion. Subtitle D provides that each State will "adopt and implement a permit program or other system of prior approval and conditions" to assure that each solid waste management facility within the State "will comply" with criteria established by EPA for the siting, design, construction, operation and closure of solid waste landfills. RCRA section 4005(c)(1)(B). States are required to submit permit programs for EPA to review and EPA is required to "determine whether each State has developed an adequate program" to ensure compliance with EPA's Subtitle D regulations. RCRA section 4005(c)(1)(B) and (C). However, RCRA does not grant to EPA authority to issue permits for solid waste landfills, review State permitting decisions or enforce Subtitle D requirements in States with approved programs. The court in *RII* appeared to misunderstand EPA's authorities under Subtitle D of RCRA when it stated that EPA would be the permitting authority in the absence of an approved State program. *See* 151 F.3d 1169 ("we hold that when a proposed project affecting a wetlands area is a solid waste landfill, the *EPA* (or the approved State program) . . . will have the permit authority under RCRA.") (Emphasis added); 151 F.3d at 1167 ("RCRA gives the EPA authority to issue permits for the disposal of solid waste, but allows states to substitute their own permit programs for the Federal program if the State program is approved by EPA."). While this authority exists with regard to disposal of hazardous waste under Subtitle C of RCRA, EPA does not have this authority with regard to disposal of nonhazardous solid waste under Subtitle D.

In contrast, the CWA requires either a Federal permit for discharges of pollutants into waters of the U.S., or issuance of a permit by a State/Tribe with an approved program, subject to EPA's authority to object to a permit where EPA finds it fails to meet the guidelines and requirements of the CWA. CWA sections 402(d); 404(j). EPA also has authority under the CWA to enforce conditions in Federal or State permits under the Act. CWA section 309.

These contrasting statutory schemes support the conclusion that eliminating CWA authority over discharges of fill material associated with construction of solid waste landfills would mean a significant departure from the statutory structure created by Congress in the CWA, a scheme which Congress expressly sought to preserve when it adopted RCRA. *See* RCRA section 1006(a). This does not mean that we view the Federal role as one of secondguessing every decision made by State regulatory authorities under RCRA. To the contrary, both RCRA and the CWA reflect a strong presumption in favor of State-administered regulatory programs. As discussed elsewhere, we intend to rely on State decision-making under RCRA to the extent allowed under current law and regulations. However, we believe that eliminating a Federal role entirely on these matters is neither appropriate nor consistent with Congressional intent under RCRA or the CWA.

Thus, we decline to follow the decision in *RII* outside the Ninth Circuit because we conclude there is not an adequate legal basis on which to conclude that discharges of pollutants associated with solid waste landfills no longer need to be authorized by a CWA permit solely because the project receives a permit under Subtitle D of RCRA.

We nonetheless share the basic policy perspective expressed by the court in *RII* about the need to avoid unnecessary duplication and potential inconsistent application of regulatory programs under the CWA and RCRA. In fact, RCRA expressly vests EPA with the responsibility to "integrate all provisions of (RCRA) for purposes of administration and enforcement and (to) avoid duplication, to the maximum extent practicable, with the appropriate provisions of the * * * (CWA). * * * Such integration shall be effected only to the extent that it can be done in a manner consistent with the goals and policies of this chapter and the CWA. * * *" RCRA section 1006(b). EPA has sought such integration first by promulgating location restrictions for landfills that are consistent with the criteria for issuance of section 404 permits. See 40 CFR 258.12; 230.10. Among other requirements, a landfill may not be located in wetlands unless it is demonstrated to the State that there are not less environmentally damaging practicable alternatives, the facility will not cause significant degradation of wetlands, and that appropriate and practicable steps have been taken to mitigate the loss of wetlands from the facility. However, EPA never purported to substitute Subtitle D regulation for the CWA permitting requirement, a result that would violate both section 1006(a) and (b). Instead, the Subtitle D RCRA regulations make clear that owners or operators of municipal solid waste landfills "must comply with any other applicable Federal rules, laws, regulations, or other requirements." 40 CFR 258.3. At the time EPA promulgated this regulation, the agency expressly noted that such requirements include those arising under the CWA. *See* 56 FR 51042 (October 9, 1991).

We do not believe, however, that the Subtitle D and section 404 programs are redundant. Rather, each program has a distinct focus. The State RCRA permitting process addresses a much broader range of issues, including technical operating and design criteria, ground water monitoring, corrective action, closure and post-closure care and financial assurances. In contrast, the section 404 process is focused exclusively on the impacts of discharges of dredged or fill material on the aquatic ecosystem, and ways of ensuring that those impacts are avoided, minimized and compensated. Because of the Corps' expertise in protecting aquatic ecosystems, we have found that State RCRA permitting agencies often incorporate by reference the requirements of section 404 permits. (For example, the State RCRA permit for the *RII* landfill required the applicant to implement the wetlands and mitigation plan to be approved by the Corps through the 404 permit process.) We believe that, in these and other ways, State and Federal permitting authorities can create efficiencies by relying on each other's expertise in making regulatory decisions.

We intend to make additional efforts to avoid unnecessary duplication in the Federal and State permitting process. As explained in section II.C of this final preamble, we intend that the Corps will rely on decisions by the State RCRA authority about the siting, design and construction of solid waste landfills in waters of the U.S. to the extent allowed by law and regulations. Appropriate deference to State decision-making will help avoid duplication, while still ensuring that the Corps fulfills its responsibilities to authorize discharges of fill material associated with solid waste landfills in accordance with CWA requirements.

This does not mean that, in every single case, State and Federal decisionmakers will agree on whether a particular project or configuration is environmentally acceptable. Nevertheless, instances of disagreement have been rare. We intend to further enhance our efforts to ensure effective coordination between State and Federal officials. However, we do not agree with the court in *RII* that the only way to avoid unnecessary duplication is to eliminate the CWA permitting requirement altogether.

We next address commenters' assertions that the decision in *RII* continues to preclude us from regulating solid waste landfills under section 404 within the Ninth Circuit. These comments also argue that, given the "statutory" basis for the court's decision, we cannot change the result in the Ninth Circuit through this rulemaking.

As noted in this preamble, the court construed administrative materials of the Corps and EPA as supporting the conclusion that the agencies did not intend to regulate solid waste landfills under section 404 of the CWA. In light of this agency intent, the court concluded that subjecting landfills to regulation solely under RCRA would "harmonize" the statutes and "give effect to each [statute] while preserving their sense and purpose." 151 F.3d at 1169. The court found that this harmonization "is consistent with the sense of the CWA that discharges of solid waste materials are beyond the scope of section 404...and avoids unnecessary duplication of Federal and State efforts in the area of wetlands protection." *Id.*

We again emphasize the distinction between "discharges of solid waste material," as referenced by the court and discharges of fill material associated with the construction of infrastructure. In this rulemaking, we have clarified that discharges having the effect of raising the bottom elevation of a water or replacing water with dry land, including fill used to create landfills such as liners, berms and other infrastructure associated with solid waste landfills are discharges of fill material subject to the section 404 program. Therefore, we have altered the landscape as understood by the court in *RII* (i.e., that these facilities were entirely outside the intended purview of section 404). We do not agree with commenters who argued that there was a "statutory" basis to the court's decision in the sense that the holding of the decision turned on an interpretation of Congressional intent in the CWA or RCRA. The court did not cite any provision of the CWA or RCRA to support its conclusions. Rather, the court derived the "sense and purpose" of the CWA based on agency regulations, guidance and correspondence. By clarifying the scope of section 404 authorities in this rulemaking, we have altered the "sense and purpose" of the CWA underlying the court's conclusion that regulation solely under RCRA would "harmonize" the statutes. Because the premises before the court have changed, we do not view the court's decision as continuing to bar the regulation under section 404 of discharges associated with solid waste landfills within the Ninth Circuit. At a minimum, today's rule calls into question the continuing vitality of the court's reasoning and conclusions and, should a case be brought within the Ninth Circuit challenging our authority to regulate solid waste landfills, we would ask the court to address the question anew in light of the clarification of our authorities in today's rule.

III. Administrative Requirements

A. Plain Language

In compliance with the principle in Executive Order 12866 regarding plain language, this preamble is written using plain language. Thus, the use of "we" in this notice refers to EPA and the Corps, and the use of "you" refers to the reader. We have also used active voice, short sentences, and common every day terms except for necessary technical terms.

B. Paperwork Reduction Act

This action does not impose any new information collection burden under the provisions of the Paperwork Production Act, 44 U.S.C. 3501 *et seq.* This rule merely reconciles EPA and Corps CWA section 404 regulations defining the term "fill material" and amends our definitions of "discharge of fill material." Thus, this action is not subject to the Paperwork Reduction Act.

Burden means the total time, effort, or financial resources expended by persons to generate, maintain, retain, or disclose or provide information to or for a Federal agency. This includes the time needed to review instructions; develop, acquire, install, and utilize technology and systems for the purposes of collecting, validating, and verifying information, processing and maintaining information, and disclosing and providing information; adjust the existing ways to comply with any previously applicable instructions and requirements; train personnel to be able to respond to a collection of information; search data sources; complete and review the collection of information; and transmit or otherwise disclose the information.

An Agency may not conduct or sponsor, and a person is not required to respond to, a collection of information unless it displays a currently valid OMB control number. The OMB control numbers for EPA's regulations are displayed in 40 CFR part 9 and 48 CFR chapter 15. For the CWA section regulatory 404 program, the current OMB approval number for information requirements is maintained by the Corps of Engineers (OMB approval number 0710-0003, expires December 31, 2004).

C. Executive Order 12866

Under Executive Order 12866 (58 FR 51735, October 4, 1993), EPA and the Corps must determine whether the regulatory action is "significant" and therefore subject to review by the Office of Management and Budget (OMB) and the requirements of the Executive Order. The Order defines "significant regulatory action" as one that is likely to result in a rule that may:

(1) Have an annual effect on the economy of $100 million or more or adversely affect in a material way the economy, a sector of the economy, productivity, competition, jobs, the environment, public health or safety, or State, local, or Tribal governments or communities;

(2) Create a serious inconsistency or otherwise interfere with an action taken or planned by another agency;

(3) Materially alter the budgetary impact of entitlements, grants, user fees, or loan programs or the rights and obligations of recipients thereof; or

(4) Raise novel legal or policy issues arising out of legal mandates, the President's priorities, or the principles set forth in the Executive Order.

Pursuant to the terms of Executive Order 12866, it has been determined that this rule is a "significant regulatory action" in light of the provisions of paragraph (4) above. As such, this action was submitted to OMB for review. Changes made in response to OMB suggestions or recommendations will be documented in the public record.

D. Executive Order 13132

Executive Order 13132, entitled "Federalism" (64 FR 43255, August 10, 1999), requires EPA and the Corps to develop an accountable process to ensure "meaningful and timely input by State and local officials in the development of regulatory policies that have Federalism implications." "Policies that have Federalism implications" is defined in the Executive Order to include regulations that have "substantial direct effects on the States, on the relationship between the national government and the States, or on the distribution of power and responsibilities among the various levels of government."

This final rule does not have Federalism implications. It will not have substantial direct effects on the States, on the relationship between the national government and the States, or on the distribution of power and responsibilities among the various levels of government, as specified in Executive Order 13132. Currently, under the CWA, any discharge of pollutants into waters of the U.S. requires a permit under either section 402 or 404 of the CWA. Today's rule conforms our two regulatory definitions of "fill material" and thereby clarifies whether a particular discharge is subject to regulation under section 402 or Section 404. It is generally consistent with current agency practice and does not impose new substantive requirements. Within California, Oregon, Washington, Idaho, Wyoming, Nevada, Arizona, Hawaii, Guam, and the Northern Mariana Islands, after today's rule, the Corps will again be issuing Section 404 permits for the construction of solid waste landfills in waters of the U.S., which the Corps had ceased doing after the decision in *RII* (the decision did not affect the permitting requirement outside these states). *See* section II.D. of this preamble. However, resuming the issuance of section 404 permits for construction of solid waste landfills in waters of the U.S. in these areas does not have Federalism implications. None of the States within the Ninth Circuit will incur administrative costs as a result of today's rule, because none currently administer the section 404 program and, in any event, the administrative costs of permitting solid waste landfills are minimal in the context of the overall section 404 permitting program. In addition, this change does not impose any additional substantive obligations on State or local governments seeking to construct solid waste landfills in waters of the U.S. since Subtitle D of RCRA currently requires such facilities to meet comparable conditions for receiving a section 404 permit. *See* section II. D of this preamble. Finally, we do not believe that requiring any State or local governments seeking to construct solid waste landfills in waters of the U.S. to undergo the Section 404 permitting process itself will have substantial direct effects on the States, on the relationship between the national government and the States, or on the distribution of power and responsibilities among the various levels of government. Thus, Executive Order 13132 does not apply to this rule.

E. Regulatory Flexibility Act (RFA), as Amended by the Small Business Regulatory Enforcement Fairness Act of 1996 (SBREFA), 5 U.S.C. 601 et seq.

The RFA generally requires an agency to prepare a regulatory flexibility analysis of any rule subject to notice-and-comment rulemaking requirements under the Administrative Procedure Act or any other statute unless the agency certifies that the rule will not have a significant economic impact on a substantial number of small entities. Small entities include small businesses, small organizations and small governmental jurisdictions.

For purposes of assessing the impacts of today's rule on small entities, a small entity is defined as : (1) A small business based on SBA size standards; (2) a small governmental jurisdiction that is a government of a city, county, town, school district, or special district with a population of less than 50,000; and (3) a small organization that is any not-for-profit enterprise which is independently owned and operated and is not dominant in its field.

After considering the economic impacts of today's final rule on small entities, we certify that this action will not have a significant economic impact on a substantial number of small entities. Currently, under the CWA, any discharge of pollutants into waters of the U.S. requires a permit under either section 402 or 404 of the CWA. Today's rule conforms our two regulatory definitions of "fill material" and thereby clarifies whether a particular discharge is subject to regulation under section 402 or section 404. Today's rule is generally consistent with current agency practice, does not impose new substantive requirements and therefore would not have a significant economic impact on a substantial number of small entities.

F. Unfunded Mandates Reform Act

Title II of the Unfunded Mandates Reform Act of 1995 (UMRA), Public Law 104-4, establishes requirements for Federal agencies to assess the effects of their regulatory actions on State, local, and Tribal governments and the private sector. Under section 202 of the UMRA, the agencies generally must prepare a written statement, including a cost benefit analysis, for proposed and final rules with "Federal mandates" that may result in expenditures to State, local, and Tribal governments, in the aggregate, or to the private sector, of $100 million or more in any one year. Before promulgating an EPA or Corps rule for which a written statement is needed, section 205 of the UMRA generally requires the agencies to identify and consider a reasonable number of regulatory alternatives and adopt the least costly, most cost-effective or least burdensome alternative that achieves the objectives of the rule. The provisions of section 205 do not apply when they are inconsistent with applicable law. Moreover, section 205 allows EPA and the Corps to adopt an alternative other than the least costly, most cost-effective or least burdensome alternative if the Administrator and Secretary of the Army publish with the final rule an explanation why that alternative was not adopted. Before EPA or the Corps establishes any regulatory requirements that may significantly or uniquely affect small governments, including Tribal governments, they must have developed under section 203 of the UMRA a small government agency plan. The plan must provide for notifying potentially affected small governments, enabling officials of affected small governments to have meaningful and timely input in the development of EPA or Corps regulatory proposals with significant Federal intergovernmental mandates, and informing, educating, and advising small governments on compliance with the regulatory requirements.

We have determined that this rule does not contain a Federal mandate that may result in expenditures of $100 million or more for State, local, and Tribal governments, in the aggregate, or the private sector in any one year. Currently, under the CWA, any discharge of pollutants into waters of the U.S. requires a permit under either section 402 or 404 of the CWA. Today's rule conforms our two regulatory definitions of "fill material" and thereby clarifies whether a particular discharge is subject to regulation under section 402 or section 404. Today's rule is generally consistent with current agency practice, does not impose new substantive requirements and therefore does not contain a Federal mandate that may result in expenditures of $100 million or more for State, local, and Tribal governments, in the aggregate, or the private sector in any one year. Thus, today's rule is not subject to the requirements of sections 202 and 205 of the UMRA. For the same reasons, we have determined that this rule contains no regulatory requirements that might significantly or uniquely affect small governments. Thus today's rule is not subject to the requirements of section 203 of UMRA.

G. National Technology Transfer and Advancement Act

As noted in the proposed rule, Section 12(d) of the National Technology Transfer and Advancement Act of 1995 (the NTTAA), Public Law 104-113, section 12(d) (15 U.S.C. 272 note) directs us to use voluntary consensus standards in our regulatory activities unless to do so would be inconsistent with applicable law or otherwise impractical. Voluntary consensus standards are technical standards (*e.g.*, materials specifications, test methods, sampling procedures, and business practices) that are developed or adopted by voluntary consensus standards bodies. The NTTAA directs us to provide Congress, through OMB, explanations when we decide not to use available and applicable voluntary consensus standards.

This rule does not involve technical standards. Therefore, we did not consider the use of any voluntary consensus standards.

H. Executive Order 13045

Executive Order 13045: "Protection of Children from Environmental Health Risks and Safety Risks" (62 FR 19885, April 23, 1997) applies to any rule that: (1) Is determined to be "economically significant" as defined under Executive Order 12866, and (2) concerns an environmental health or safety risk that we have reason to believe may have a disproportionate effect on children. If the regulatory action meets both criteria, we must evaluate the environmental health or safety effects of the planned rule on children, and explain why the planned regulation is preferable to other potentially effective and reasonably feasible alternatives considered by us.

This final rule is not subject to the Executive Order because it is not economically significant as defined in Executive Order 12866. In addition, it does not concern an environmental or safety risk that we have reason to believe may have a disproportionate effect on children.

I. Executive Order 13175

Executive Order 13175, entitled "Consultation and Coordination with Indian Tribal Governments" (65 FR 67249, November 6, 2000), requires the agencies to develop an accountable process to ensure "meaningful and timely input by tribal officials in the development of regulatory policies that have tribal implications." "Policies that have tribal implications" is defined in the Executive Order to include regulations that have "substantial direct effects on one or more Indian tribes, on the relationship between the Federal government and the Indian tribes, or on the distribution of power and responsibilities between the Federal government and Indian tribes."

Today's rule does not have tribal implications. It will not have substantial direct effects on tribal governments, on the relationship between the Federal government and the Indian tribes, or on the distribution of power and responsibilities between the Federal gov ernment and Indian tribes, as specified in Executive Order 13175. Currently, under the CWA, any discharge of pollutants into waters of the U.S. requires a permit under either section 402 or 404 of the CWA. Today's rule conforms our two regulatory definitions of "fill material" and thereby clarifies whether a particular discharge is subject to regulation under section 402 or section 404. It is generally consistent with current agency practice and does not impose new substantive requirements. Within California, Oregon, Washington, Idaho, Wyoming, Nevada, Arizona, Hawaii, Guam, and the Northern Mariana Islands, after today's rule, the Corps will again be issuing Section 404 permits for the construction of solid waste landfills in waters of the U.S., which the Corps had ceased doing after the decision in *RII* (the decision did not affect the permitting requirement outside these states). *See* section II. D. of this preamble. However, resuming the issuance of section 404 permits for construction of solid waste landfills in waters of the U.S. in these areas does not have tribal implications. No tribes within the Ninth Circuit will incur administrative costs as a result of today's rule, because none currently administer the section 404 program and, in any event, the administrative costs of permitting solid waste landfills are minimal in the context of the overall section 404 permitting program. In addition, this change does not impose any additional substantive obligations on any Tribe seeking to construct solid waste landfills in waters of the U.S. since Subtitle D of RCRA currently requires such facilities to meet comparable conditions for receiving a section 404 permit. *See* section II.D. of this preamble. Finally, we do not believe that requiring any tribal government seeking to construct solid waste landfills in waters of the U.S. to undergo the Section 404 permitting process itself will have substantial direct effects on one or more Indian tribes, on the relationship between the Federal government and the Indian tribes, or on the distribution of power and responsibilities between the Federal government and Indian tribes. Thus, Executive Order 13175 does not apply to this rule.

J. Environmental Documentation

As required by the NEPA, the Corps prepares appropriate environmental documentation for its activities affecting the quality of the human environment. The Corps has prepared an environmental assessment (EA) of the final rule. The Corps' EA ultimately concludes that, since the adoption of this rule will not significantly affect the quality of the human environment, the preparation and coordination of an EIS is not required. The EA, included in the administrative record for today's rule, explains the rationale for the Corps' conclusion.

K. Congressional Review Act

The Congressional Review Act, 5 U.S.C. 801 *et seq.*, as added by the Small Business Regulatory Enforcement Fairness Act of 1996, generally provides that before a rule may take effect, the agency promulgating the rule must submit a rule report, which includes a copy of the rule, to each House of the Congress and to the Comptroller General of the United States. We will submit a report containing this rule and other required information to the U.S. Senate, the U.S. House of Representatives, and the Comptroller General of the United States prior to publication of the rule in the **Federal Register**. A major rule cannot take effect until 60 days after it is published in the **Federal Register**. This rule is not a "major rule" as defined by 5 U.S.C. section 804(2). This rule will be effective June 10, 2002.

L. Executive Order 12898

Executive Order 12898 requires that, to the greatest extent practicable and permitted by law, each Federal agency must make achieving environmental justice part of its mission. Executive Order 12898 provides that each Federal agency conduct its programs, policies, and activities that substantially affect human health or the environment in a manner that ensures that such programs, policies, and activities do not have the effect of excluding persons (including populations) from participation in, denying persons (including populations) the benefits of, or subjecting persons (including populations) to discrimination under such programs, policies, and activities because of their race, color, or national origin.

Today's rule is not expected to negatively impact any community, and therefore is not expected to cause any disproportionately high and adverse impacts to minority or low-income communities. Today's rule relates solely to whether a particular discharge is appropriately authorized under section 402 or section 404 of the Clean Water Act. Moreover, the proposed allocation of authority between these programs is generally consistent with existing agency practice.

M. Executive Order 13211

This rule is not a "significant energy action" as defined in Executive Order 13211, "Actions Concerning Regulations That Significantly Affect Energy Supply, Distribution, or Use" (66 FR 28355 (May 22, 2001)) because it is not likely to have a significant adverse effect on the supply, distribution, or use of energy. Today's rule conforms our two regulatory definitions of "fill material" and thereby clarifies whether a particular discharge is subject to regulation under section 402 or section 404. Today's rule is generally consistent with current agency practice, does not impose new substantive requirements and therefore will not have a significant adverse effect on the supply, distribution, or use of energy.

List of Subjects

33 CFR Part 323
Water pollution control, Waterways.

40 CFR Part 232
Environmental protection, Intergovernmental relations, Water pollution control.

Corps of Engineers

33 CFR Chapter II

Accordingly, as set forth in the preamble 33 CFR part 323 is amended as set forth below:

PART 323 [AMENDED]

1. The authority citation for part 323 continues to read as follows:

Authority: 33 U.S.C. 1344

2. Amend § 323.2 as follows:

a. Paragraph (e) is revised.

b. In paragraph (f), in the second sentence: add the words "or infrastructure" after the words "for the construction of any structure"; add the word ", infrastructure," after the words "building of any structure"; remove the words "residential, and" and add in their place the words "residential, or"; and add the words "placement of fill material for construction or maintenance of any liner, berm, or other infrastructure associated with solid waste landfills; placement of overburden, slurry, or tailings or similar mining-related materials;" after the words "utility lines;".

The revision reads as follows:

Section 323.2 Definitions

* * * * *

(e) (1) Except as specified in paragraph (e)(3) of this section, the term fill material means material placed in waters of the United States where the material has the effect of:

(i) Replacing any portion of a water of the United States with dry land; or

(ii) Changing the bottom elevation of any portion of a water of the United States.

(2) Examples of such fill material include, but are not limited to: rock, sand, soil, clay, plastics, construction debris, wood chips, overburden from mining or other excavation activities, and materials used to create any structure or infrastructure in the waters of the United States.

(3) The term fill material does not include trash or garbage.

* * * * *

Dated: May 3, 2002

Dominic Izzo,

Principal Deputy Assistant Secretary of the Army (Civil Works), Department of the Army.

Environmental Protection Agency

40 CFR Chapter I

Accordingly, as set forth in the preamble 40 CFR part 232 is amended as set forth below:

PART 232 [AMENDED]

1. The authority citation for part 232 continues to read as follows:

Authority: 33 U.S.C. 1344

2. Amend § 232.2 as follows:

a. The definition of "Fill material" is revised.

b. In the definition of "Discharge of fill material", in paragraph (1): add the words "or infrastructure" after the words "for the construction of any structure"; add the word ", infrastructure," after the words "building of any structure"; remove the words "residential, and" and add in their place the words "residential, or"; and add the words "placement of fill material for construction or maintenance of any liner, berm, or other infrastructure associated with solid waste landfills; placement of overburden, slurry, or tailings or similar mining-related materials;" after the words "utility lines;".

The revision reads as follows:

Section 232.2 Definitions

* * * * *

Fill material. (1) Except as specified in paragraph (3) of this definition, the term fill material means material placed in waters of the United States where the material has the effect of:

(i) Replacing any portion of a water of the United States with dry land; or

(ii) Changing the bottom elevation of any portion of a water of the United States.

(2) Examples of such fill material include, but are not limited to: rock, sand, soil, clay, plastics, construction debris, wood chips, overburden from mining or other excavation activities, and materials used to create any structure or infrastructure in the waters of the United States.

(3) The term fill material does not include trash or garbage.

* * * * *

Dated: May 3, 2002.

Christine Todd Whitman,
Administrator, Environmental Protection Agency.

[FR Doc. 02-11547 Filed 5-8-02; 8:45 am]

BILLING CODE 3710-92-P

APPENDIX D

USEPA Regulations

Section 404 (Section 404[B][1] Guidelines)

U.S. Environmental Protection Agency 40 CFR 230-233

PART 230 SECTION 404(B)(1) GUIDELINES FOR SPECIFICATION OF DISPOSAL SITES FOR DREDGED OR FILL MATERIAL

Authority: Secs. 404(b) and 501(a) of the Clean Water Act of 1977 33 U.S.C. 1344(b) and 1361(a)

Source: 45 FR 85344, Dec. 24, 1980, unless otherwise noted.

Subpart A—General

Section 230.1 Purpose and policy

(a) The purpose of these Guidelines is to restore and maintain the chemical, physical, and biological integrity of waters of the United States through the control of discharges of dredged or fill material.

(b) Congress has expressed a number of policies in the Clean Water Act. These Guidelines are intended to be consistent with and to implement those policies.

(c) Fundamental to these Guidelines is the precept that dredged or fill material should not be discharged into the aquatic ecosystem, unless it can be demonstrated that such a discharge will not have an unacceptable adverse impact either individually or in combination with known and/or probable impacts of other activities affecting the ecosystems of concern.

(d) From a national perspective, the degradation or destruction of special aquatic sites, such as filling operations in wetlands, is considered to be among the most severe environmental impacts covered by these Guidelines. The guiding principle should be that degradation or destruction of special sites may represent an irreversible loss of valuable aquatic resources.

Section 230.2 Applicability

(a) These Guidelines have been developed by the Administrator of the Environmental Protection Agency in conjunction with the Secretary of the Army acting through the Chief of Engineers under section 404(b)(1) of the Clean Water Act (33 U.S.C. 1344). The Guidelines are applicable to the specification of disposal sites for discharges of dredged or fill material into waters of the United States. Sites may be specified through:

(1) The regulatory program of the U.S. Army Corps of Engineers under sections 404(a) and (e) of the Act (see 33 CFR Parts 320, 323 and 325);

(2) The civil works program of the U.S. Army Corps of Engineers (see 33 CFR 209.145 and section 150 of Pub. L. 94-587, Water Resources Development Act of 1976);

(3) Permit programs of States approved by the Administrator of the Environmental Protection Agency in accordance with section 404(g) and (h) of the Act (see 40 CFR parts 122, 123 and 124);

(4) Statewide dredged or fill material regulatory programs with best management practices approved under section 208(b)(4)(B) and (C) of the Act (see 40 CFR 35.1560);

(5) Federal construction projects which meet criteria specified in section 404(r) of the Act.

(b) These Guidelines will be applied in the review of proposed discharges of dredged or fill material into navigable waters which lie inside the baseline from which the territorial sea is measured, and the discharge of fill material into the territorial sea, pursuant to the procedures referred to in paragraphs (a)(1) and (2) of this section. The discharge of dredged material into the territorial sea is governed by the Marine Protection, Research, and Sanctuaries Act of 1972, Pub. L. 92-532, and regulations and criteria issued pursuant thereto (40 CFR parts 220 through 228).

(c) Guidance on interpreting and implementing these Guidelines may be prepared jointly by EPA and the Corps at the national or regional level from time to time. No modifications to the basic application, meaning, or intent of these Guidelines will be made without rulemaking by the Administrator under the Administrative Procedure Act (5 U.S.C. 551 *et seq.*).

Section 230.3 Definitions

For purposes of this part, the following terms shall have the meanings indicated:

(a) The term *Act* means the Clean Water Act (also known as the Federal Water Pollution Control Act or FWPCA) Pub. L. 92-500, as amended by Pub. L. 95-217, 33 U.S.C. 1251, *et seq.*

(b) The term *adjacent* means bordering, contiguous, or neighboring. Wetlands separated from other waters of the United States by man-made dikes or barriers, natural river berms, beach dunes, and the like are "adjacent wetlands."

(c) The terms *aquatic environment* and *aquatic ecosystem* mean waters of the United

States, including wetlands, that serve as habitat for interrelated and interacting communities and populations of plants and animals.

(d) The term *carrier of contaminant* means dredged or fill material that contains contaminants.

(e) The term *contaminant* means a chemical or biological substance in a form that can be incorporated into, onto or be ingested by and that harms aquatic organisms, consumers of aquatic organisms, or users of the aquatic environment, and includes but is not limited to the substances on the 307(a)(1) list of toxic pollutants promulgated on January 31, 1978 (43 FR 4109).

(f)-(g) [Reserved]

(h) The term *discharge point* means the point within the disposal site at which the dredged or fill material is released.

(i) The term *disposal site* means that portion of the "waters of the United States" where specific disposal activities are permitted and consist of a bottom surface area and any overlying volume of water. In the case of wetlands on which surface water is not present, the disposal site consists of the wetland surface area.

(j) [Reserved]

(k) The term *extraction site* means the place from which the dredged or fill material proposed for discharge is to be removed.

(l) [Reserved]

(m) The term *mixing zone* means a limited volume of water serving as a zone of initial dilution in the immediate vicinity of a discharge point where receiving water quality may not meet quality standards or other requirements otherwise applicable to the receiving water. The mixing zone should be considered as a place where wastes and water mix and not as a place where effluents are treated.

(n) The term *permitting authority* means the District Engineer of the U.S. Army Corps of Engineers or such other individual as may be designated by the Secretary of the Army to issue or deny permits under section 404 of the Act; or the State Director of a permit program approved by EPA under section 404(g) and section 404(h) or his delegated representative.

(o) The term *pollutant* means dredged spoil, solid waste, incinerator residue, sewage, garbage, sewage sludge, munitions, chemical wastes, biological materials, radioactive materials not covered by the Atomic Energy Act, heat, wrecked or discarded equipment, rock, sand, cellar dirt, and industrial, municipal, and agricultural waste discharged into water. The legislative history of the Act reflects that "radioactive materials" as included within the definition of "pollutant" in section 502 of the Act means only radioactive materials which are not encompassed in the definition of source, byproduct, or special nuclear materials as defined by the Atomic Energy Act of 1954, as amended, and regulated under the Atomic Energy Act. Examples of radioactive materials not covered by the Atomic Energy Act and, therefore, included within the term "pollutant", are radium and accelerator produced isotopes. See *Train* v. *Colorado Public Interest Research Group, Inc.*, 426 U.S. 1 (1976).

(p) The term *pollution* means the man-made or man-induced alteration of the chemical, physical, biological or radiological integrity of an aquatic ecosystem.

(q) The term *practicable* means available and capable of being done after taking into consideration cost, existing technology, and logistics in light of overall project purposes.

(q-1) *Special aquatic sites* means those sites identified in subpart E. They are geographic areas, large or small, possessing special ecological characteristics of productivity, habitat, wildlife protection, or other important and easily disrupted ecological values. These areas are generally recognized as significantly influencing or positively contributing to the general overall environmental health or vitality of the entire ecosystem of a region. (See § 230.10(a)(3))

(r) The term *territorial sea* means the belt of the sea measured from the baseline as determined in accordance with the Convention on the Territorial Sea and the Contiguous Zone and extending seaward a distance of three miles.

(s) The term *waters of the United States* means:

(1) All waters which are currently used, or were used in the past, or may be susceptible to use in interstate or foreign commerce, including all waters which are subject to the ebb and flow of the tide;

(2) All interstate waters including interstate wetlands;

(3) All other waters such as intrastate lakes, rivers, streams (including intermittent streams), mudflats, sandflats, wetlands, sloughs, prairie potholes, wet meadows, playa lakes, or natural ponds, the use, degradation or destruction of which could affect interstate or foreign commerce including any such waters:

(i) Which are or could be used by interstate or foreign travelers for recreational or other purposes; or

(ii) From which fish or shellfish are or could be taken and sold in interstate or foreign commerce; or

(iii) Which are used or could be used for industrial purposes by industries in interstate commerce;

(4) All impoundments of waters otherwise defined as waters of the United States under this definition;

(5) Tributaries of waters identified in paragraphs (s)(1) through (4) of this section;

(6) The territorial sea;

(7) Wetlands adjacent to waters (other than waters that are themselves wetlands) identified in paragraphs (s)(1) through (6) of this section; waste treatment systems, including treatment ponds or lagoons designed to meet the requirements of CWA (other than cooling ponds as defined in 40 CFR 423.11(m) which also meet the criteria of this definition) are not waters of the United States. Waters of the United States do not include prior converted cropland. Notwithstanding the determination of an area's status as prior converted cropland by any other federal agency, for the purposes of the Clean Water Act, the final authority regarding Clean Water Act jurisdiction remains with EPA.

(t) The term *wetlands* means those areas that are inundated or saturated by surface or ground water at a frequency and duration sufficient to support, and that under normal circumstances do support, a prevalence of vegetation typically adapted for life in saturated soil conditions. Wetlands generally include swamps, marshes, bogs and similar areas.

[45 FR 85344, Dec. 24, 1980, as amended at 58 FR 45037, Aug. 25, 1993]

Section 230.4 Organization

The Guidelines are divided into eight subparts. Subpart A presents those provisions of general applicability, such as purpose and definitions. Subpart B establishes the four conditions which must be satisfied in order to make a finding that a proposed discharge of dredged or fill material complies with the Guidelines. § 230.11 of subpart B, sets forth factual determinations which are to be considered in determining whether or not a proposed discharge satisfies the subpart B conditions of compliance. Subpart C describes the physical and chemical components of a site and provides guidance as to how proposed discharges of dredged or fill material may affect these components. Subparts D through F detail the special characteristics of particular aquatic ecosystems in terms of their values, and the possible loss of these values due to discharges of dredged or fill material. Subpart G prescribes a number of physical, chemical, and biological evaluations and testing procedures to be used in reaching the required factual determinations. Subpart H details the means to prevent or minimize adverse effects. Subpart I concerns advanced identification of disposal areas.

Section 230.5 General procedures to be followed

In evaluating whether a particular discharge site may be specified, the permitting authority should use these Guidelines in the following sequence:

(a) In order to obtain an overview of the principal regulatory provisions of the Guidelines, review the restrictions on discharge in § 230.10(a) through (d), the measures to minimize adverse impact of subpart H, and the required factual determinations of § 230.11.

(b) Determine if a General permit (§230.7) is applicable; if so, the applicant needs merely to comply with its terms, and no further action by the permitting authority is necessary. Special conditions for evaluation of proposed General permits are contained in §230.7. If the discharge is not covered by a General permit:

(c) Examine practicable alternatives to the proposed discharge, that is, not discharging into the waters of the U.S. or discharging into an alternative aquatic site with potentially less damaging consequences (§ 230.10(a)).

(d) Delineate the candidate disposal site consistent with the criteria and evaluations of § 230.11(f).

(e) Evaluate the various physical and chemical components which characterize the non-living environment of the candidate site, the substrate and the water including its dynamic characteristics (subpart C).

(f) Identify and evaluate any special or critical characteristics of the candidate disposal site, and surrounding areas which might be affected by use of such site, related to their living communities or human uses (subparts D, E, and F).

(g) Review Factual Determinations in § 230.11 to determine whether the information in the project file is sufficient to provide the documentation required by § 230.11 or to perform the pre-testing evaluation described in § 230.60, or other information is necessary.

(h) Evaluate the material to be discharged to determine the possibility of chemical contamination or physical incompatibility of the material to be discharged (§ 230.60).

(i) If there is a reasonable probability of chemical contamination, conduct the appropriate tests according to the section on Evaluation and Testing (§ 230.61).

(j) Identify appropriate and practicable changes to the project plan to minimize the environmental impact of the discharge, based upon the specialized methods of minimization of impacts in subpart H.

(k) Make and document Factual Determinations in § 230.11.

(l) Make and document Findings of Compliance (§ 230.12) by comparing Factual Determinations with the requirements for discharge of § 230.10. This outline of the steps to follow in using the Guidelines is simplified for purposes of illustration. The actual process followed may be iterative, with the results of one step leading to a reexamination of previous steps. The permitting authority must address all of the relevant provisions of the Guidelines in reaching a Finding of Compliance in an individual case.

Section 230.6 Adaptability

(a) The manner in which these Guidelines are used depends on the physical, biological, and chemical nature of the proposed extraction site, the material to be discharged, and the candidate disposal site, including any other important components of the ecosystem being evaluated. Documentation to demonstrate knowledge about the extraction site, materials to be extracted, and the candidate disposal site is an essential component of guideline application. These Guidelines allow evaluation and documentation for a variety of activities, ranging from those with large, complex impacts on the aquatic environment to those for which the impact is likely to be innocuous. It is unlikely that the Guidelines will apply in their entirety to any one activity, no matter how complex. It is anticipated that substantial numbers of permit applications will be for minor, routine activities that have little, if any, potential for significant degradation of the aquatic environment. It generally is not intended or expected that extensive testing, evaluation or analysis will be needed to make findings of compliance in such routine cases. Where the conditions for General permits are met, and where numerous applications for similar activities are likely, the use of General permits will eliminate repetitive evaluation and documentation for individual discharges.

(b) The Guidelines user, including the agency or agencies responsible for implementing the Guidelines, must recognize the different levels of effort that should be associated with varying degrees of impact and require or prepare commensurate documentation. The level of documentation should reflect the significance and complexity of the discharge activity.

(c) An essential part of the evaluation process involves making determinations as to the relevance of any portion(s) of the Guidelines and conducting further evaluation only as needed. However, where portions of the Guidelines review procedure are "short form" evaluations, there still must be sufficient information (including consideration of both individual and cumulative impacts) to support the decision of whether to specify the site for disposal of dredged or fill material and to support the decision to curtail or abbreviate the evaluation process. The presumption against the discharge in § 230.1 applies to this decision-making.

(d) In the case of activities covered by General permits or section 208(b)(4)(B) and (C) Best Management Practices, the analysis and documentation required by the Guidelines will be performed at the time of General permit issuance or section 208(b)(4)(B) and (C) Best Management Practices promulgation and will not be repeated when activities are conducted under a General permit or section 208(b)(4)(B) and (C) Best Management Practices control. These Guidelines do not require reporting or formal written communication at the time individual activities are initiated under a General permit or section 208(b)(4)(B) and (C) Best Management Practices. However, a particular General permit may require appropriate reporting.

Section 230.7 General permits

(a) *Conditions for the issuance of General permits.* A General permit for a category of activities involving the discharge of dredged or fill material complies with the Guidelines if it meets the applicable restrictions on the discharge in § 230.10 and if the permitting authority determines that:

(1) The activities in such category are similar in nature and similar in their impact upon water quality and the aquatic environment;

(2) The activities in such category will have only minimal adverse effects when performed separately; and

(3) The activities in such category will have only minimal cumulative adverse effects on water quality and the aquatic environment.

(b) *Evaluation process.* To reach the determinations required in paragraph (a) of this section, the permitting authority shall set forth in writing an evaluation of the potential individual and cumulative impacts of the category of activities to be regulated under the General permit. While some of the information necessary for this evaluation can be obtained from potential permittees and others through the proposal of General permits for public review, the evaluation must be completed before any General permit is issued, and the results must be published with the final permit.

(1) This evaluation shall be based upon consideration of the prohibitions listed

in § 230.10(b) and the factors listed in § 230.10(c), and shall include documented information supporting each factual determination in § 230.11 of the Guidelines (consideration of alternatives in § 230.10(a) are not directly applicable to General permits);

(2) The evaluation shall include a precise description of the activities to be permitted under the General permit, explaining why they are sufficiently similar in nature and in environmental impact to warrant regulation under a single General permit based on subparts C through F of the Guidelines. Allowable differences between activities which will be regulated under the same General permit shall be specified. Activities otherwise similar in nature may differ in environmental impact due to their location in or near ecologically sensitive areas, areas with unique chemical or physical characteristics, areas containing concentrations of toxic substances, or areas regulated for specific human uses or by specific land or water management plans (e.g., areas regulated under an approved Coastal Zone Management Plan). If there are specific geographic areas within the purview of a proposed General permit (called a draft General permit under a State 404 program), which are more appropriately regulated by individual permit due to the considerations cited in this paragraph, they shall be clearly delineated in the evaluation and excluded from the permit. In addition, the permitting authority may require an individual permit for any proposed activity under a General permit where the nature or location of the activity makes an individual permit more appropriate.

(3) To predict cumulative effects, the evaluation shall include the number of individual discharge activities likely to be regulated under a General permit until its expiration, including repetitions of individual discharge activities at a single location.

Subpart B—Compliance with the Guidelines

Section 230.10 Restrictions on discharge

Note: Because other laws may apply to particular discharges and because the Corps of Engineers or State 404 agency may have additional procedural and substantive requirements, a discharge complying with the requirement of these Guidelines will not automatically receive a permit.

Although all requirements in § 230.10 must be met, the compliance evaluation procedures will vary to reflect the seriousness of the potential for adverse impacts on the aquatic ecosystems posed by specific dredged or fill material discharge activities.

(a) Except as provided under section 404(b)(2), no discharge of dredged or fill material shall be permitted if there is a practicable alternative to the proposed discharge which would have less adverse impact on the aquatic ecosystem, so long as the alternative does not have other significant adverse environmental consequences.

(1) For the purpose of this requirement, practicable alternatives include, but are not limited to:

(i) Activities which do not involve a discharge of dredged or fill material into the waters of the United States or ocean waters;

(ii) Discharges of dredged or fill material at other locations in waters of the United States or ocean waters;

(2) An alternative is practicable if it is available and capable of being done after taking into consideration cost, existing technology, and logistics in light of overall project purposes. If it is otherwise a practicable alternative, an area not presently owned by the applicant which could reasonably be obtained, utilized, expanded or managed in order to fulfill the basic purpose of the proposed activity may be considered.

(3) Where the activity associated with a discharge which is proposed for a special aquatic site (as defined in subpart E) does not require access or proximity to or siting within the special aquatic site in question to fulfill its basic purpose (i.e., is not "water dependent"), practicable alternatives that do not involve special aquatic sites are presumed to be available, unless clearly demonstrated otherwise. In addition, where a discharge is proposed for a special aquatic site, all practicable alternatives to the proposed discharge which do not involve a discharge into a special aquatic site are presumed to have less adverse impact on the aquatic ecosystem, unless clearly demonstrated otherwise.

(4) For actions subject to NEPA, where the Corps of Engineers is the permitting agency, the analysis of alternatives required for NEPA environmental documents, including supplemental Corps NEPA documents, will in most cases provide the information for the evaluation of alternatives under these Guidelines. On occasion, these NEPA documents may address a broader range of alternatives than required to be considered under this paragraph or may not have considered the alternatives in sufficient detail to respond to the requirements of these Guidelines. In the latter case, it may be necessary to supplement these NEPA documents with this additional information.

(5) To the extent that practicable alternatives have been identified and evaluated under a Coastal Zone Management program, a section 208 program, or other planning process, such evaluation shall be considered by the permitting authority as part of the consideration of alternatives under the Guidelines. Where such evaluation is less complete than that contemplated under this subsection, it must be supplemented accordingly.

(b) No discharge of dredged or fill material shall be permitted if it:

(1) Causes or contributes, after consideration of disposal site dilution and dispersion, to violations of any applicable State water quality standard;

(2) Violates any applicable toxic effluent standard or prohibition under section 307 of the Act;

(3) Jeopardizes the continued existence of species listed as endangered or threatened under the Endangered Species Act of 1973, as amended, or results in likelihood of the destruction or adverse modification of a habitat which is determined by the Secretary of Interior or Commerce, as appropriate, to be a critical habitat under the Endangered Species Act of 1973, as amended. If an exemption has been granted by the Endangered Species Committee, the terms of such exemption shall apply in lieu of this subparagraph;

(4) Violates any requirement imposed by the Secretary of Commerce to protect any marine sanctuary designated under title III of the Marine Protection, Research, and Sanctuaries Act of 1972.

(c) Except as provided under section 404(b)(2), no discharge of dredged or fill material shall be permitted which will cause or contribute to significant degradation of the waters of the United States. Findings of significant degradation related to the proposed discharge shall be based upon appropriate factual determinations, evaluations, and tests required by subparts B and G, after consideration of subparts C through F, with special emphasis on the persistence and permanence of the effects outlined in those subparts. Under these Guidelines, effects contributing to significant degradation considered individually or collectively, include:

(1) Significantly adverse effects of the discharge of pollutants on human health or welfare, including but not limited to effects on municipal water supplies, plankton, fish, shellfish, wildlife, and special aquatic sites.

(2) Significantly adverse effects of the discharge of pollutants on life stages of aquatic life and other wildlife dependent on aquatic ecosystems, including the transfer, concentration, and spread of pollutants or their byproducts outside of the disposal site through biological, physical, and chemical processes;

(3) Significantly adverse effects of the discharge of pollutants on aquatic ecosystem diversity, productivity, and stability. Such effects may include, but are not limited to, loss of fish and wildlife habitat or loss of the capacity of a wetland to assimilate nutrients, purify water, or reduce wave energy; or

(4) Significantly adverse effects of discharge of pollutants on recreational, aesthetic, and economic values.

(d) Except as provided under section 404(b)(2), no discharge of dredged or fill material shall be permitted unless appropriate and practicable steps have been taken which will minimize potential adverse impacts of the discharge on the aquatic ecosystem. Subpart H identifies such possible steps.

Section 230.11 Factual determinations

The permitting authority shall determine in writing the potential short-term or long-term effects of a proposed discharge of dredged or fill material on the physical, chemical, and biological components of the aquatic environment in light of subparts C through F. Such factual determinations shall be used in § 230.12 in making findings of compliance or non-compliance with the restrictions on discharge in § 230.10. The evaluation and testing procedures described in § 230.60 and § 230.61 of subpart G shall be used as necessary to make, and shall be described in, such determination. The determinations of effects of each proposed discharge shall include the following:

(a) *Physical substrate determinations.* Determine the nature and degree of effect that the proposed discharge will have, individually and cumulatively, on the characteristics of the substrate at the proposed disposal site. Consideration shall be given to the similarity in particle size, shape, and degree of compaction of the material proposed for discharge and the material constituting the substrate at the disposal site, and any potential changes in substrate elevation and bottom contours, including changes outside of the disposal site which may occur as a result of erosion, slumpage, or other movement of the discharged material. The duration and physical extent of substrate changes shall also be considered. The possible loss of environmental values (§ 230.20) and actions to minimize impact (subpart H) shall also be considered in making these determinations. Potential changes in substrate elevation and bottom contours shall be predicted on the basis of the proposed method, volume, location, and rate of discharge, as well as on the individual and combined effects of current patterns, water circulation, wind and wave action, and other physical factors that may affect the movement of the discharged material.

(b) *Water circulation, fluctuation, and salinity determinations.* Determine the nature and degree of effect that the proposed discharge will have individually and cumulatively on water, current patterns, circulation including downstream flows, and normal water fluctuation. Consideration shall be given to water chemistry, salinity, clarity, color, odor, taste, dissolved gas levels, temperature, nutrients, and eutrophication plus other appropriate characteristics. Consideration shall also be given to the potential diversion or obstruction of flow, alterations of bottom contours, or other significant changes in the hydrologic regime. Additional consideration of the possible loss of environmental values (§§ 230.23 through 230.25) and actions to minimize impacts (subpart H), shall be used in making these determinations. Potential significant effects on the current patterns, water circulation, normal water fluctuation and salinity shall be evaluated on the basis of the proposed method, volume, location, and rate of discharge.

(c) *Suspended particulate/turbidity determinations.* Determine the nature and degree of effect that the proposed discharge will have, individually and cumulatively, in terms of potential changes in the kinds and concentrations of suspended particulate/turbidity in the vicinity of the disposal site. Consideration shall be given to the grain size of the material proposed for discharge, the shape and size of the plume of suspended particulates, the duration of the discharge and resulting plume and whether or not the potential changes will cause violations of applicable water quality standards. Consideration should also be given to the possible loss of environmental values (§ 30.21) and to actions for minimizing impacts (subpart H). Consideration shall include the proposed method, volume, location, and rate of discharge, as well as the individual and combined effects of current patterns, water circulation and fluctuations, wind and wave action, and other physical factors on the movement of suspended particulates.

(d) *Contaminant determinations.* Determine the degree to which the material proposed for discharge will introduce, relocate, or increase contaminants. This determination shall consider the material to be discharged, the aquatic environment at the proposed disposal site, and the availability of contaminants.

(e) *Aquatic ecosystem and organism determinations.* Determine the nature and degree of effect that the proposed discharge will have, both individually and cumulatively, on the structure and function of the aquatic ecosystem and organisms. Consideration shall be given to the effect at the proposed disposal site of potential changes in substrate characteristics and elevation, water or substrate chemistry, nutrients, currents, circulation, fluctuation, and salinity, on the recolonization and existence of indigenous aquatic organisms or communities. Possible loss of environmental values (§ 230.31), and actions to minimize impacts (subpart H) shall be examined. Tests as described in § 230.61 (Evaluation and Testing), may be required to provide information on the effect of the discharge material on communities or populations of organisms expected to be exposed to it.

(f) *Proposed disposal site determinations.*

(1) Each disposal site shall be specified through the application of these Guidelines. The mixing zone shall be confined to the smallest practicable zone within each specified disposal site that is consistent with the type of dispersion determined to be appropriate by the application of these Guidelines. In a few special cases under unique environmental conditions, where there is adequate justification to show that widespread dispersion by natural means will result in no significantly adverse environmental effects, the discharged material may be intended to be spread naturally in a very thin layer over a large area of the substrate rather than be contained within the disposal site.

(2) The permitting authority and the Regional Administrator shall consider the following factors in determining the acceptability of a proposed mixing zone:

(i) Depth of water at the disposal site;

(ii) Current velocity, direction, and variability at the disposal site;

(iii) Degree of turbulence;

(iv) Stratification attributable to causes such as obstructions, salinity or density profiles at the disposal site;

(v) Discharge vessel speed and direction, if appropriate;

(vi) Rate of discharge;

(vii)Ambient concentration of constituents of interest;

(viii) Dredged material characteristics, particularly concentrations of constituents, amount of material, type of material (sand, silt, clay, etc.) and settling velocities;

(ix) Number of discharge actions per unit of time;

(x) Other factors of the disposal site that affect the rates and patterns of mixing.

(g) *Determination of cumulative effects on the aquatic ecosystem.*

(1) Cumulative impacts are the changes in an aquatic ecosystem that are attributable to the collective effect of a number of individual discharges of dredged or fill

material. Although the impact of a particular discharge may constitute a minor change in itself, the cumulative effect of numerous such piecemeal changes can result in a major impairment of the water resources and interfere with the productivity and water quality of existing aquatic ecosystems.

(2) Cumulative effects attributable to the discharge of dredged or fill material in waters of the United States should be predicted to the extent reasonable and practical. The permitting authority shall collect information and solicit information from other sources about the cumulative impacts on the aquatic ecosystem. This information shall be documented and considered during the decision-making process concerning the evaluation of individual permit applications, the issuance of a General permit, and monitoring and enforcement of existing permits.

(h) *Determination of secondary effects on the aquatic ecosystem.*

(1) Secondary effects are effects on an aquatic ecosystem that are associated with a discharge of dredged or fill materials, but do not result from the actual placement of the dredged or fill material. Information about secondary effects on aquatic ecosystems shall be considered prior to the time final section 404 action is taken by permitting authorities.

(2) Some examples of secondary effects on an aquatic ecosystem are fluctuating water levels in an impoundment and downstream associated with the operation of a dam, septic tank leaching and surface runoff from residential or commercial developments on fill, and leachate and runoff from a sanitary landfill located in waters of the U.S. Activities to be conducted on fast land created by the discharge of dredged or fill material in waters of the United States may have secondary impacts within those waters which should be considered in evaluating the impact of creating those fast lands.

Section 230.12 Findings of compliance or non-compliance with the restrictions on discharge

(a) On the basis of these Guidelines (subparts C through G) the proposed disposal sites for the discharge of dredged or fill material must be:

(1) Specified as complying with the requirements of these Guidelines; or

(2) Specified as complying with the requirements of these Guidelines with the inclusion of appropriate and practicable discharge conditions (see subpart H) to minimize pollution or adverse effects to the affected aquatic ecosystems; or

(3) Specified as failing to comply with the requirements of these Guidelines where:

(i) There is a practicable alternative to the proposed discharge that would have less adverse effect on the aquatic ecosystem, so long as such alternative does not have other significant adverse environmental consequences; or

(ii) The proposed discharge will result in significant degradation of the aquatic ecosystem under § 230.10(b) or (c); or

(iii) The proposed discharge does not include all appropriate and practicable measures to minimize potential harm to the aquatic ecosystem; or

(iv) There does not exist sufficient information to make a reasonable judgment as to whether the proposed discharge will comply with these Guidelines.

(b) Findings under this section shall be set forth in writing by the permitting authority for each proposed discharge and made available to the permit applicant. These findings shall include the factual determinations required by § 230.11, and a brief explanation of any adaptation of these Guidelines to the activity under consideration. In the case of a General permit, such findings shall be prepared at the time of issuance of that permit rather than for each subsequent discharge under the authority of that permit.

Subpart C—Potential Impacts on Physical and Chemical Characteristics of the Aquatic Ecosystem

Note: The effects described in this subpart should be considered in making the factual determinations and the findings of compliance or non-compliance in subpart B.

Section 230.20 Substrate

(a) The substrate of the aquatic ecosystem underlies open waters of the United States and constitutes the surface of wetlands. It consists of organic and inorganic solid materials and includes water and other liquids or gases that fill the spaces between solid particles.

(b) Possible loss of environmental characteristics and values: The discharge of dredged or fill material can result in varying degrees of change in the complex physical, chemical, and biological characteristics of the substrate. Discharges which alter substrate elevation or contours can result in changes in water circulation, depth, current pattern, water fluctuation and water temperature. Discharges may adversely affect bottom-dwelling organisms at the site by smothering immobile forms or forcing mobile forms to migrate. Benthic forms present prior to a discharge are unlikely to recolonize on the discharged material if it is very dissimilar from that of the discharge site. Erosion, slumping, or lateral displacement of surrounding bottom of such deposits can adversely affect areas of the substrate outside the perimeters of the disposal site by changing or destroying habitat. The bulk and composition of the discharged material and the location, method, and timing of discharges may all influence the degree of impact on the substrate.

Section 230.21 Suspended particulates/turbidity

(a) Suspended particulates in the aquatic ecosystem consist of fine-grained mineral particles, usually smaller than silt, and organic particles. Suspended particulates may enter water bodies as a result of land runoff, flooding, vegetative and planktonic breakdown, resuspension of bottom sediments, and man's activities including dredging and filling. Particulates may remain suspended in the water column for variable periods of time as a result of such factors as agitation of the water mass, particulate specific gravity, particle shape, and physical and chemical properties of particle surfaces.

(b) Possible loss of environmental characteristics and values: The discharge of dredged or fill material can result in greatly elevated levels of suspended particulates in the water column for varying lengths of time. These new levels may reduce light penetration and lower the rate of photosynthesis and the primary productivity of an aquatic area if they last long enough. Sight-dependent species may suffer reduced feeding ability leading to limited growth and lowered resistance to disease if high levels of suspended particulates persist. The biological and the chemical content of the suspended material may react with the dissolved oxygen in the water, which can result in oxygen depletion. Toxic metals and organics, pathogens, and viruses absorbed or adsorbed to fine-grained particulates in the material may become biologically available to organisms either in the water column or on the substrate. Significant increases in suspended particulate levels create turbid plumes which are highly visible and aesthetically displeasing. The extent and persistence of these adverse impacts caused by discharges depend upon the relative increase in suspended particulates above the amount occurring naturally, the duration of the higher levels, the current patterns, water level, and fluctuations present when such discharges occur, the volume, rate, and duration of the discharge, particulate deposition, and the seasonal timing of the discharge.

Section 230.22 Water

(a) Water is the part of the aquatic ecosystem in which organic and inorganic constituents are dissolved and suspended. It constitutes part of the liquid phase and is contained by the substrate. Water forms part of a dynamic aquatic life-supporting system. Water clarity, nutrients and chemical content, physical and biological content, dissolved gas levels, pH, and temperature contribute to its life-sustaining capabilities.

(b) Possible loss of environmental characteristics and values: The discharge of dredged or fill material can change the chemistry and the physical characteristics of the receiving water at a disposal site through the introduction of chemical constituents in suspended or dissolved form. Changes in the clarity, color, odor, and taste of water and the addition of contaminants can reduce or eliminate the suitability of water bodies for populations of aquatic organisms, and for human consumption, recreation, and aesthetics. The introduction of nutrients or organic material to the water column as a result of the discharge can lead to a high biochemical oxygen demand (BOD), which in turn can lead to reduced dissolved oxygen, thereby potentially affecting the survival of many aquatic organisms. Increases in nutrients can favor one group of organisms such as algae to the detriment of other more desirable types such as submerged aquatic vegetation, potentially causing adverse health effects, objectionable tastes and odors, and other problems.

Section 230.23 Current patterns and water circulation

(a) Current patterns and water circulation are the physical movements of water in the aquatic ecosystem. Currents and circulation respond to natural forces as modified by basin shape and cover, physical and chemical characteristics of water strata and masses, and energy dissipating factors.

(b) Possible loss of environmental characteristics and values: The discharge of dredged or fill material can modify current patterns and water circulation by obstructing flow, changing the direction or velocity of water flow, changing the direction or velocity of water flow and circulation, or otherwise changing the dimensions of a water body. As a result, adverse changes can occur in: Location, structure, and dynamics of aquatic communities; shoreline and substrate erosion and deposition rates; the deposition of suspended particulates; the rate and extent of mixing of dissolved and suspended components of the water body; and water stratification.

Section 230.24 Normal water fluctuations

(a) Normal water fluctuations in a natural aquatic system consist of daily, seasonal, and annual tidal and flood fluctuations in water level. Biological and physical components of such a system are either attuned to or characterized by these periodic water fluctuations.

(b) Possible loss of environmental characteristics and values: The discharge of dredged or fill material can alter the normal water-level fluctuation pattern of an area, resulting in prolonged periods of inundation, exaggerated extremes of high and low water, or a static, nonfluctuating water level. Such water level modifications may change salinity patterns, alter erosion or sedimentation rates, aggravate water temperature extremes, and upset the nutrient and dissolved oxygen balance of the aquatic ecosystem. In addition, these modifications can alter or destroy communities and populations of aquatic animals and vegetation, induce populations of nuisance organisms, modify habitat, reduce food supplies, restrict movement of aquatic fauna, destroy spawning areas, and change adjacent, upstream, and downstream areas.

Section 230.25 Salinity gradients

(a) Salinity gradients form where salt water from the ocean meets and mixes with fresh water from land.

(b) Possible loss of environmental characteristics and values: Obstructions which divert or restrict flow of either fresh or salt water may change existing salinity gradients. For example, partial blocking of the entrance to an estuary or river mouth that significantly restricts the movement of the salt water into and out of that area can effectively lower the volume of salt water available for mixing within that estuary. The downstream migration of the salinity gradient can occur, displacing the maximum sedimentation zone and requiring salinity-dependent aquatic biota to adjust to the new conditions, move to new locations if possible, or perish. In the freshwater zone, discharge operations in the upstream regions can have equally adverse impacts. A significant reduction in the volume of fresh water moving into an estuary below that which is considered normal can affect the location and type of mixing thereby changing the characteristic salinity patterns. The resulting changed circulation pattern can cause the upstream migration of the salinity gradient displacing the maximum sedimentation zone. This migration may affect those organisms that are adapted to freshwater environments. It may also affect municipal water supplies.

Note: Possible actions to minimize adverse impacts regarding site characteristics can be found in subpart H.

Subpart D—Potential Impacts on Biological Characteristics of the Aquatic Ecosystem

Note: The impacts described in this subpart should be considered in making the factual determinations and the findings of compliance or non-compliance in subpart B.

Section 230.30 Threatened and endangered species

(a) An endangered species is a plant or animal in danger of extinction throughout all or a significant portion of its range. A threatened species is one in danger of becoming an endangered species in the foreseeable future throughout all or a significant portion of its range. Listings of threatened and endangered species as well as critical habitats are maintained by some individual States and by the U.S. Fish and Wildlife Service of the Department of the Interior (codified annually at 50 CFR 17.11). The Department of Commerce has authority over some threatened and endangered marine mammals, fish and reptiles.

(b) Possible loss of values: The major potential impacts on threatened or endangered species from the discharge of dredged or fill material include:

(1) Covering or otherwise directly killing species;

(2) The impairment or destruction of habitat to which these species are limited. Elements of the aquatic habitat which are particularly crucial to the continued survival of some threatened or endangered species include adequate good quality water, spawning and maturation areas, nesting areas, protective cover, adequate and reliable food supply, and resting areas for migratory species. Each of these elements can be adversely affected by changes in either the normal water conditions for clarity, chemical content, nutrient balance, dissolved oxygen, pH, temperature, salinity, current patterns, circulation and fluctuation, or the physical removal of habitat; and

(3) Facilitating incompatible activities.

(c) Where consultation with the Secretary of the Interior occurs under section 7 of the Endangered Species Act, the conclusions of the Secretary concerning the impact(s) of the discharge on threatened and endangered species and their habitat shall be considered final.

Section 230.31 Fish, crustaceans, mollusks, and other aquatic organisms in the food web

(a) Aquatic organisms in the food web include, but are not limited to, finfish, crustaceans, mollusks, insects, annelids, planktonic organisms, and the plants and animals on which they feed and depend upon for their needs. All forms and life stages of an organism, throughout its geographic range, are included in this category.

(b) Possible loss of values: The discharge of dredged or fill material can variously affect populations of fish, crustaceans, mollusks and other food web organisms through the release of contaminants which adversely affect adults, juveniles, larvae, or eggs, or result in the establishment or proliferation of an undesirable competitive species of plant or animal at the expense of the desired resident species. Suspended particulates settling on attached or buried eggs can smother the eggs by limiting or sealing off their exposure to oxygenated water. Discharge of dredged and fill material may result in the debilitation or death of sedentary organisms by smothering, exposure to chemical contaminants in dissolved or suspended form, exposure to high levels of suspended particulates, reduction in food supply, or alteration of the substrate upon which they are dependent. Mollusks are particularly sensitive to the discharge of material during periods of reproduction and growth and development due primarily to their limited mobility. They can be rendered unfit for human consumption by tainting, by production and accumulation of toxins, or by ingestion and retention of pathogenic organisms, viruses, heavy metals or persistent synthetic organic chemicals. The discharge of dredged or fill material can redirect, delay, or stop the reproductive and feeding movements of some species of fish and crustacea, thus preventing their aggregation in accustomed places such as spawning or nursery grounds and potentially leading to reduced populations. Reduction of detrital feeding species or other representatives of lower trophic levels can impair the flow of energy from primary consumers to higher trophic levels. The reduction or potential elimination of food chain organism populations decreases the overall productivity and nutrient export capability of the ecosystem.

Section 230.32 Other wildlife

(a) Wildlife associated with aquatic ecosystems are resident and transient mammals, birds, reptiles, and amphibians.

(b) Possible loss of values: The discharge of dredged or fill material can result in the loss or change of breeding and nesting areas, escape cover, travel corridors, and preferred food sources for resident and transient wildlife species associated with the aquatic ecosystem. These adverse impacts upon wildlife habitat may result from changes in water levels, water flow and circulation, salinity, chemical content, and substrate characteristics and elevation. Increased water turbidity can adversely affect wildlife species which rely upon sight to feed, and disrupt the respiration and feeding of certain aquatic wildlife and food chain organisms. The availability of contaminants from the discharge of dredged or fill material may lead to the bioaccumulation of such contaminants in wildlife. Changes in such physical and chemical factors of the environment may favor the introduction of undesirable plant and animal species at the expense of resident species and communities. In some aquatic environments lowering plant and animal species diversity may disrupt the normal functions of the ecosystem and lead to reductions in overall biological productivity.

Note: Possible actions to minimize adverse impacts regarding characteristics of biological components of the aquatic ecosystem can be found in subpart H.

Subpart E—Potential Impacts on Special Aquatic Sites

Note: The impacts described in this subpart should be considered in making the factual determinations and the findings of compliance or non-compliance in subpart B. The definition of special aquatic sites is found in § 230.3(q–1).

Section 230.40 Sanctuaries and refuges

(a) Sanctuaries and refuges consist of areas designated under State and Federal laws or local ordinances to be managed principally for the preservation and use of fish and wildlife resources.

(b) Possible loss of values: Sanctuaries and refuges may be affected by discharges of dredged or fill material which will:

(1) Disrupt the breeding, spawning, migratory movements or other critical life requirements of resident or transient fish and wildlife resources;

(2) Create unplanned, easy and incompatible human access to remote aquatic areas;

(3) Create the need for frequent maintenance activity;

(4) Result in the establishment of undesirable competitive species of plants and animals;

(5) Change the balance of water and land areas needed to provide cover, food, and other fish and wildlife habitat requirements in a way that modifies sanctuary or refuge management practices;

(6) Result in any of the other adverse impacts discussed in subparts C and D as they relate to a particular sanctuary or refuge.

Section 230.41 Wetlands

(a) (1) Wetlands consist of areas that are inundated or saturated by surface or ground water at a frequency and duration sufficient to support, and that under normal circumstances do support, a prevalence of vegetation typically adapted for life in saturated soil conditions.

(2) Where wetlands are adjacent to open water, they generally constitute the transition to upland. The margin between wetland and open water can best be established by specialists familiar with the local environment, particularly where emergent vegetation merges with submerged vegetation over a broad area in such places as the lateral margins of open water, headwaters, rainwater catch basins, and groundwater seeps. The landward margin of wetlands also can best be identified by specialists familiar with the local environment when vegetation from the two regions merges over a broad area.

(3) Wetland vegetation consists of plants that require saturated soils to survive (obligate wetland plants) as well as plants, including certain trees, that gain a competitive advantage over others because they can tolerate prolonged wet soil conditions and their competitors cannot. In addition to plant populations and communities, wetlands are delimited by hydrological and physical characteristics of the environment. These characteristics should be considered when information about them is needed to supplement information available about vegetation, or where wetland vegetation has been removed or is dormant.

(b) Possible loss of values: The discharge of dredged or fill material in wetlands is likely to damage or destroy habitat and adversely affect the biological productivity of wetlands ecosystems by smothering, by dewatering, by permanently flooding, or by altering substrate elevation or periodicity of water movement. The addition of dredged or fill material may destroy wetland vegetation or result in advancement of succession to dry land species. It may reduce or eliminate nutrient exchange by a reduction of the system's productivity, or by altering current patterns and velocities. Disruption or elimination of the wetland system can degrade water quality by obstructing circulation patterns that flush large expanses of wetland systems, by interfering with the filtration function of wetlands, or by changing the aquifer recharge capability of a wetland. Discharges can also change the wetland habitat value for fish and wildlife as discussed in subpart D. When disruptions in flow and circulation patterns occur, apparently minor loss of wetland acreage may result in major losses through secondary impacts. Discharging fill material in wetlands as part of municipal, industrial or recreational development may modify the capacity of wetlands to retain and store floodwaters and to serve as a buffer zone shielding upland areas from wave actions, storm damage and erosion.

Section 230.42 Mud flats

(a) Mud flats are broad flat areas along the sea coast and in coastal rivers to the head of tidal influence and in inland lakes, ponds, and riverine systems. When mud flats are inundated, wind and wave action may resuspend bottom sediments. Coastal mud flats are exposed at extremely low tides and inundated at high tides with the water table at or near the surface of the substrate. The substrate of mud flats contains organic material and particles smaller in size than sand. They are either unvegetated or vegetated only by algal mats.

(b) Possible loss of values: The discharge of dredged or fill material can cause changes in water circulation patterns which may permanently flood or dewater the mud flat or disrupt periodic inundation, resulting in an increase in the rate of erosion or accretion. Such changes can deplete or eliminate mud flat biota, foraging areas, and nursery areas. Changes in inundation patterns can affect the chemical and biological exchange and decomposition process occurring on the mud flat and change the deposition of suspended material affecting the productivity of the area. Changes may reduce the mud flat's capacity to dissipate storm surge runoff.

Section 230.43 Vegetated shallows

(a) Vegetated shallows are permanently inundated areas that under normal circumstances support communities of rooted aquatic vegetation, such as turtle grass and eelgrass in estuarine or marine systems as well as a number of freshwater species in rivers and lakes.

(b) Possible loss of values: The discharge of dredged or fill material can smother vegetation and benthic organisms. It may also create unsuitable conditions for their continued vigor by:

(1) Changing water circulation patterns;

(2) releasing nutrients that increase undesirable algal populations;

(3) releasing chemicals that adversely affect plants and animals;

(4) increasing turbidity levels, thereby reducing light penetration and hence photosynthesis; and

(5) changing the capacity of a vegetated shallow to stabilize bottom materials and decrease channel shoaling. The discharge of dredged or fill material may reduce the value of vegetated shallows as nesting, spawning, nursery, cover, and forage areas, as well as their value in protecting shorelines from erosion and wave actions. It may also encourage the growth of nuisance vegetation.

Section 230.44 Coral reefs

(a) Coral reefs consist of the skeletal deposit, usually of calcareous or silicaceous materials, produced by the vital activities of anthozoan polyps or other invertebrate organisms present in growing portions of the reef.

(b) Possible loss of values: The discharge of dredged or fill material can adversely affect colonies of reef building organisms by burying them, by releasing contaminants such as hydrocarbons into the water column, by reducing light penetration through the water, and by increasing the level of suspended particulates. Coral organisms are extremely sensitive to even slight reductions in light penetration or increases in suspended particulates. These adverse effects will cause a loss of productive colonies which in turn provide habitat for many species of highly specialized aquatic organisms.

Section 230.45 Riffle and pool complexes

(a) Steep gradient sections of streams are sometimes characterized by riffle and pool complexes. Such stream sections are recognizable by their hydraulic characteristics. The rapid movement of water over a coarse substrate in riffles results in a rough flow, a turbulent surface, and high dissolved oxygen levels in the water. Pools are deeper areas associated with riffles. Pools are characterized by a slower stream velocity, a steaming flow, a smooth surface, and a finer substrate. Riffle and pool complexes are particularly valuable habitat for fish and wildlife.

(b) Possible loss of values: Discharge of dredged or fill material can eliminate riffle and pool areas by displacement, hydrologic modification, or sedimentation. Activities which affect riffle and pool areas and especially riffle/pool ratios, may reduce the aeration and filtration capabilities at the discharge site and downstream, may reduce stream habitat diversity, and may retard repopulation of the disposal site and downstream waters through sedimentation and the creation of unsuitable habitat. The discharge of dredged or fill material which alters stream hydrology may cause scouring or sedimentation of riffles and pools. Sedimentation induced through hydrological modification or as a direct result of the deposition of unconsolidated dredged or fill material may clog riffle and pool areas, destroy habitats, and create anaerobic conditions. Eliminating pools and meanders by the discharge of dredged or fill material can reduce water holding capacity of streams and cause rapid runoff from a watershed. Rapid runoff can deliver large quantities of flood water in a short time to downstream areas resulting in the destruction of natural habitat, high property loss, and the need for further hydraulic modification.

Note: Possible actions to minimize adverse impacts on site or material characteristics can be found in subpart H.

Subpart F—Potential Effects on Human Use Characteristics

Note: The effects described in this subpart should be considered in making the factual determinations and the findings of compliance or non-compliance in subpart B.

Section 230.50 Municipal and private water supplies

(a) Municipal and private water supplies consist of surface water or ground water which is directed to the intake of a municipal or private water supply system.

(b) Possible loss of values: Discharges can affect the quality of water supplies with respect to color, taste, odor, chemical content and suspended particulate concentration, in such a way as to reduce the fitness of the water for consumption. Water can be rendered unpalatable or unhealthy by the addition of suspended particulates, viruses and pathogenic organisms, and dissolved materials. The expense of removing such substances before the water is delivered for consumption can be high. Discharges may also affect the quantity of water available for municipal and private water supplies. In addition, certain commonly used water treatment chemicals have the potential for combining with some suspended or dissolved substances from dredged or fill material to form other products that can have a toxic effect on consumers.

Section 230.51 Recreational and commercial fisheries

(a) Recreational and commercial fisheries consist of harvestable fish, crustaceans, shellfish, and other aquatic organisms used by man.

(b) Possible loss of values: The discharge of dredged or fill materials can affect the suitability of recreational and commercial fishing grounds as habitat for populations of consumable aquatic organisms. Discharges can result in the chemical contamination of recreational or commercial fisheries. They may also interfere with the reproductive success of recreational and commercially important aquatic species through disruption of migration and spawning areas. The introduction of pollutants at critical times in their life cycle may directly reduce populations of commercially important aquatic organisms or indirectly reduce them by reducing organisms upon which they depend for food. Any of these impacts can be of short duration or prolonged, depending upon the physical and chemical impacts of the discharge and the biological availability of contaminants to aquatic organisms.

Section 230.52 Water-related recreation

(a) Water-related recreation encompasses activities undertaken for amusement and relaxation. Activities encompass two broad categories of use: consumptive, e.g., harvesting resources by hunting and fishing; and non-consumptive, e.g. canoeing and sightseeing.

(b) Possible loss of values: One of the more important direct impacts of dredged or fill disposal is to impair or destroy the resources which support recreation activities. The disposal of dredged or fill material may adversely modify or destroy water use for recreation by changing turbidity, suspended particulates, temperature, dissolved oxygen, dissolved materials, toxic materials, pathogenic organisms, quality of habitat, and the aesthetic qualities of sight, taste, odor, and color.

Section 230.53 Aesthetics

(a) Aesthetics associated with the aquatic ecosystem consist of the perception of beauty by one or a combination of the senses of sight, hearing, touch, and smell. Aesthetics of aquatic ecosystems apply to the quality of life enjoyed by the general public and property owners.

(b) Possible loss of values: The discharge of dredged or fill material can mar the beauty of natural aquatic ecosystems by degrading water quality, creating distracting disposal sites, inducing inappropriate development, encouraging unplanned and incompatible human access, and by destroying vital elements that contribute to the compositional harmony or unity, visual distinctiveness, or diversity of an area. The discharge of dredged or fill material can adversely affect the particular features, traits, or characteristics of an aquatic area which make it valuable to property owners. Activities which degrade water quality, disrupt natural substrate and vegetational characteristics, deny access to or visibility of the resource, or result in changes in odor, air quality, or noise levels may reduce the value of an aquatic area to private property owners.

Section 230.54 Parks, national and historical monuments, national seashores, wilderness areas, research sites, and similar preserves

(a) These preserves consist of areas designated under Federal and State laws or local ordinances to be managed for their aesthetic, educational, historical, recreational, or scientific value.

(b) Possible loss of values: The discharge of dredged or fill material into such areas may modify the aesthetic, educational, historical, recreational and/or scientific qualities thereby reducing or eliminating the uses for which such sites are set aside and managed.

Note: Possible actions to minimize adverse impacts regarding site or material characteristics can be found in subpart H.

Subpart G—Evaluation and Testing

Section 230.60 General evaluation of dredged or fill material

The purpose of these evaluation procedures and the chemical and biological testing sequence outlined in § 230.61 is to provide information to reach the determinations required by § 230.11. Where the results of prior evaluations, chemical and biological tests, scientific research, and experience can provide information helpful in making a determination, these should be used. Such prior results may make new testing unnecessary. The information used shall be documented. Where the same information applies to more than one determination, it may be documented once and referenced in later determinations.

(a) If the evaluation under paragraph (b) indicates the dredged or fill material is not a carrier of contaminants, then the required determinations pertaining to the presence and effects of contaminants can be made without testing. Dredged or fill material is most likely to be free from chemical, biological, or other pollutants where it is composed primarily of sand, gravel, or other naturally occurring inert material. Dredged material so composed is generally found in areas of high current or wave energy such as streams with large bed loads or coastal areas with shifting bars and channels. However, when such material is discolored or contains other indications that contam inants may be present, further inquiry should be made.

(b) The extraction site shall be examined in order to assess whether it is sufficiently removed from sources of pollution to provide reasonable assurance that the proposed discharge material is not a carrier of contaminants. Factors to be considered include but are not limited to:

(1) Potential routes of contaminants or contaminated sediments to the extraction site, based on hydrographic or other maps, aerial photography, or other materials that show watercourses, surface relief, proximity to tidal movement, private and public roads, location of buildings, municipal and industrial areas, and agricultural or forest lands.

(2) Pertinent results from tests previously carried out on the material at the extraction site, or carried out on similar material for other permitted projects in the vicinity. Materials shall be considered similar if the sources of contamination, the physical configuration of the sites and the sediment composition of the materials are comparable, in light of water circulation and stratification, sediment accumulation and general sediment characteristics. Tests from other sites may be relied on only if no changes have occurred at the extraction sites to render the results irrelevant.

(3) Any potential for significant introduction of persistent pesticides from land runoff or percolation;

(4) Any records of spills or disposal of petroleum products or substances designated as hazardous under section 311 of the Clean Water Act (See 40 CFR part 116);

(5) Information in Federal, State and local records indicating significant introduction of pollutants from industries, municipalities, or other sources, including types and amounts of waste materials discharged along the potential routes of contaminants to the extraction site; and

(6) Any possibility of the presence of substantial natural deposits of minerals or other substances which could be released to the aquatic environment in harmful quantities by man-induced discharge activities.

(c) To reach the determinations in § 230.11 involving potential effects of the discharge on the characteristics of the disposal site, the narrative guidance in subparts C through F shall be used along with the general evaluation procedure in § 230.60 and, if necessary, the chemical and biological testing sequence in § 230.61. Where the discharge site is adjacent to the extraction site and subject to the same sources of contaminants, and materials at the two sites are substantially similar, the fact that the material to be discharged may be a carrier of contaminants is not likely to result in degradation of the disposal site. In such circumstances, when dissolved material and suspended particulates can be controlled to prevent carrying pollutants to less contaminated areas, testing will not be required.

(d) Even if the § 230.60(b) evaluation (previous tests, the presence of polluting industries and information about their discharge or runoff into waters of the U.S., bioinventories, etc.) leads to the conclusion that there is a high probability that the material proposed for discharge is a carrier of contaminants, testing may not be necessary if constraints are available to reduce contamination to acceptable levels within the disposal site and to prevent contaminants from being transported beyond the boundaries of the disposal site, if such constraints are acceptable to the permitting authority and the Regional Administrator, and if the potential discharger is willing and able to implement such constraints. However, even if tests are not performed, the permitting authority must still determine the probable impact of the operation on the receiving aquatic ecosystem. Any decision not to test must be explained in the determinations made under § 230.11.

Section 230.61 Chemical, biological, and physical evaluation and testing

Note: The Agency is today proposing revised testing guidelines. The evaluation and testing procedures in this section are based on the 1975 section 404(b)(1) interim final Guidelines and shall remain in effect until the revised testing guidelines are published as final regulations.

(a) No single test or approach can be applied in all cases to evaluate the effects of proposed discharges of dredged or fill materials. This section provides some guidance in determining which test and/or evaluation procedures are appropriate in a given case. Interim guidance to applicants concerning the applicability of specific approaches or procedures will be furnished by the permitting authority.

(b) *Chemical-biological interactive effects.* The principal concerns of discharge of dredged or fill material that contain contaminants are the potential effects on the water column and on communities of aquatic organisms.

(1) *Evaluation of chemical-biological interactive effects.* Dredged or fill material may be excluded from the evaluation procedures specified in paragraphs (b) (2) and (3) of this section if it is determined, on the basis of the evaluation in § 230.60, that the likelihood of contamination by contaminants is acceptably low, unless the permitting authority, after evaluating and considering any comments received from the Regional Administrator, determines that these procedures are necessary. The Regional Administrator may require, on a case-by-case basis, testing approaches and procedures by stating what additional information is needed through further analyses and how the results of the analyses will be of value in evaluating potential environmental effects. If the General Evaluation indicates the presence of a sufficiently large number of chemicals to render impractical the identification of all contaminants by chemical testing, information may be obtained from bioassays in lieu of chemical tests.

(2) *Water column effects.*

(i) Sediments normally contain constituents that exist in various chemical forms and in various concentrations in several locations within the sediment. An elutriate test may be used to predict the effect on water quality due to release of contaminants from the sediment to the water column. However, in the case of fill material originating on land which may be a carrier of contaminants, a water leachate test is appropriate.

(ii) Major constituents to be analyzed in the elutriate are those deemed critical by the permitting authority, after evaluating and considering any comments received from the Regional Administrator, and considering results of the evaluation in § 230.60. Elutriate concentrations should be compared to concentrations of the same constituents in water from the disposal site. Results should be evaluated in light of the volume and rate of the intended discharge, the type of discharge, the hydrodynamic regime at the disposal site, and other information relevant to the impact on water quality. The permitting authority should consider the mixing zone in evaluating water column effects. The permitting authority may specify bioassays when such procedures will be of value.

(3) *Effects on benthos.* The permitting authority may use an appropriate benthic bioassay (including bioaccumulation tests) when such procedures will be of value in assessing ecological effects and in establishing discharge conditions.

(c) *Procedure for comparison of sites.*

(1) When an inventory of the total concentration of contaminants would be of value in comparing sediment at the dredging site with sediment at the disposal site, the permitting authority may require a sediment chemical analysis. Markedly different concentrations of contaminants between the excavation and disposal sites may aid in making an environmental assessment of the proposed disposal operation. Such differences should be interpreted in terms of the potential for harm as supported by any pertinent scientific literature.

(2) When an analysis of biological community structure will be of value to assess the potential for adverse environmental impact at the proposed disposal site, a comparison of the biological characteristics between the excavation and disposal sites may be required by the permitting authority. Biological indicator species may be useful in evaluating the existing degree of stress at both sites. Sensitive species representing community components colonizing various substrate types within the sites should be identified as possible bioassay organisms if tests for toxicity are required. Community structure studies should be performed only when they will be of value in determining discharge conditions. This is particularly applicable to large quantities of dredged material known to contain adverse quantities of toxic materials. Community studies should include benthic organisms such as microbiota and harvestable shellfish and finfish. Abundance, diversity, and distribution should be documented and correlated with substrate type and other appropriate physical and chemical environmental characteristics.

(d) *Physical tests and evaluation.* The effect of a discharge of dredged or fill material on physical substrate characteristics at the disposal site, as well as on the water circulation, fluctuation, salinity, and suspended particulates content there, is important in making factual determinations in § 230.11. Where information on such effects is not otherwise available to make these factual determinations, the permitting authority shall require appropriate physical tests and evaluations as are justified and deemed necessary. Such tests may include sieve tests, settleability tests, compaction tests, mixing zone and suspended particulate plume determinations, and site assessments of water flow, circulation, and salinity characteristics.

Subpart H—Actions to Minimize Adverse Effects

Note: There are many actions which can be undertaken in response to § 203.10(d) to minimize the adverse effects of discharges of dredged or fill material. Some of these, grouped by type of activity, are listed in this subpart.

Section 230.70 Actions concerning the location of the discharge

The effects of the discharge can be minimized by the choice of the disposal site. Some of the ways to accomplish this are by:

(a) Locating and confining the discharge to minimize smothering of organisms;

(b) Designing the discharge to avoid a disruption of periodic water inundation patterns;

(c) Selecting a disposal site that has been used previously for dredged material discharge;

(d) Selecting a disposal site at which the substrate is composed of material similar to that being discharged, such as discharging sand on sand or mud on mud;

(e) Selecting the disposal site, the discharge point, and the method of discharge to minimize the extent of any plume;

(f) Designing the discharge of dredged or fill material to minimize or prevent the creation of standing bodies of water in areas of normally fluctuating water levels, and minimize or prevent the drainage of areas subject to such fluctuations.

Section 230.71 Actions concerning the material to be discharged

The effects of a discharge can be minimized by treatment of, or limitations on the material itself, such as:

(a) Disposal of dredged material in such a manner that physiochemical conditions are maintained and the potency and availability of pollutants are reduced.

(b) Limiting the solid, liquid, and gaseous components of material to be discharged at a particular site;

(c) Adding treatment substances to the discharge material;

(d) Utilizing chemical flocculants to enhance the deposition of suspended particulates in diked disposal areas.

Section 230.72 Actions controlling the material after discharge

The effects of the dredged or fill material after discharge may be controlled by:

(a) Selecting discharge methods and disposal sites where the potential for erosion, slumping or leaching of materials into the surrounding aquatic ecosystem will be reduced. These sites or methods include, but are not limited to:

(1) Using containment levees, sediment basins, and cover crops to reduce erosion;

(2) Using lined containment areas to reduce leaching where leaching of chemical constituents from the discharged material is expected to be a problem;

(b) Capping in-place contaminated material with clean material or selectively discharging the most contaminated material first to be capped with the remaining material;

(c) Maintaining and containing discharged material properly to prevent point and nonpoint sources of pollution;

(d) Timing the discharge to minimize impact, for instance during periods of unusual high water flows, wind, wave, and tidal actions.

Section 230.73 Actions affecting the method of dispersion

The effects of a discharge can be minimized by the manner in which it is dispersed, such as:

(a) Where environmentally desirable, distributing the dredged material widely in a thin layer at the disposal site to maintain natural substrate contours and elevation;

(b) Orienting a dredged or fill material mound to minimize undesirable obstruction to the water current or circulation pattern, and utilizing natural bottom contours to minimize the size of the mound;

(c) Using silt screens or other appropriate methods to confine suspended particulate/turbidity to a small area where settling or removal can occur;

(d) Making use of currents and circulation patterns to mix, disperse and dilute the discharge;

(e) Minimizing water column turbidity by using a submerged diffuser system. A similar effect can be accomplished by submerging pipeline discharges or otherwise releasing materials near the bottom;

(f) Selecting sites or managing discharges to confine and minimize the release of suspended particulates to give decreased turbidity levels and to maintain light penetration for organisms;

(g) Setting limitations on the amount of material to be discharged per unit of time or volume of receiving water.

Section 230.74 Actions related to technology

Discharge technology should be adapted to the needs of each site. In determining whether the discharge operation sufficiently minimizes adverse environmental impacts, the applicant should consider:

(a) Using appropriate equipment or machinery, including protective devices, and the use of such equipment or machinery in activities related to the discharge of dredged or fill material;

(b) Employing appropriate maintenance and operation on equipment or machinery, including adequate training, staffing, and working procedures;

(c) Using machinery and techniques that are especially designed to reduce damage to wetlands. This may include machines equipped with devices that scatter rather than mound excavated materials, machines with specially designed wheels or tracks, and the use of mats under heavy machines to reduce wetland surface compaction and rutting;

(d) Designing access roads and channel spanning structures using culverts, open channels, and diversions that will pass both low and high water flows, accommodate fluctuating water levels, and maintain circulation and faunal movement;

(e) Employing appropriate machinery and methods of transport of the material for discharge.

Section 230.75 Actions affecting plant and animal populations

Minimization of adverse effects on populations of plants and animals can be achieved by:

(a) Avoiding changes in water current and circulation patterns which would interfere with the movement of animals;

(b) Selecting sites or managing discharges to prevent or avoid creating habitat conducive to the development of undesirable predators or species which have a competitive edge ecologically over indigenous plants or animals;

(c) Avoiding sites having unique habitat or other value, including habitat of threatened or endangered species;

(d) Using planning and construction practices to institute habitat development and restoration to produce a new or modified environmental state of higher ecological value by displacement of some or all of the existing environmental characteristics. Habitat development and restoration techniques can be used to minimize adverse impacts and to compensate for destroyed habitat. Use techniques that have been demonstrated to be effective in circumstances similar to those under consideration wherever possible. Where proposed development and restoration techniques have not yet advanced to the pilot demonstration stage, initiate their use on a small scale to allow corrective action if unanticipated adverse impacts occur;

(e) Timing discharge to avoid spawning or migration seasons and other biologically critical time periods;

(f) Avoiding the destruction of remnant natural sites within areas already affected by development.

Section 230.76 Actions affecting human use

Minimization of adverse effects on human use potential may be achieved by:

(a) Selecting discharge sites and following discharge procedures to prevent or minimize any potential damage to the aesthetically pleasing features of the aquatic site (e.g., viewscapes), particularly with respect to water quality;

(b) Selecting disposal sites which are not valuable as natural aquatic areas;

(c) Timing the discharge to avoid the seasons or periods when human recreational activity associated with the aquatic site is most important;

(d) Following discharge procedures which avoid or minimize the disturbance of aesthetic features of an aquatic site or ecosystem;

(e) Selecting sites that will not be detrimental or increase incompatible human activity, or require the need for frequent dredge or fill maintenance activity in remote fish and wildlife areas;

(f) Locating the disposal site outside of the vicinity of a public water supply intake.

Section 230.77 Other actions

(a) In the case of fills, controlling runoff and other discharges from activities to be conducted on the fill;

(b) In the case of dams, designing water releases to accommodate the needs of fish and wildlife;

(c) In dredging projects funded by Federal agencies other than the Corps of Engineers, maintain desired water quality of the return discharge through agreement with the Federal funding authority on scientifically defensible pollutant concentration levels in addition to any applicable water quality standards;

(d) When a significant ecological change in the aquatic environment is proposed by the discharge of dredged or fill material, the permitting authority should consider the ecosystem that will be lost as well as the environmental benefits of the new system.

Subpart I—Planning to Shorten Permit Processing Time

Section 230.80 Advanced identification of disposal areas

(a) Consistent with these Guidelines, EPA and the permitting authority, on their own initiative or at the request of any other party and after consultation with any affected State that is not the permitting authority, may identify sites which will be considered as:

(1) Possible future disposal sites, including existing disposal sites and non-sensitive areas; or

(2) Areas generally unsuitable for disposal site specification;

(b) The identification of any area as a possible future disposal site should not be deemed to constitute a permit for the discharge of dredged or fill material within such area or a specification of a disposal site. The identification of areas that generally will not be available for disposal site specification should not be deemed as prohibiting applications for permits to discharge dredged or fill material in such areas. Either type of identification constitutes information to facilitate individual or General permit application and processing.

(c) An appropriate public notice of the proposed identification of such areas shall be issued;

(d) To provide the basis for advanced identification of disposal areas, and areas unsuitable for disposal, EPA and the permitting authority shall consider the likelihood that use of the area in question for dredged or fill material disposal will comply with these Guidelines. To facilitate this analysis, EPA and the permitting authority should review available water resources management data including data available from the public, other Federal and State agencies, and information from approved Coastal Zone Management programs and River Basin Plans;

(e) The permitting authority should maintain a public record of the identified areas and a written statement of the basis for identification.

PART 231
SECTION 404(c) PROCEDURES

Authority: 33 U.S.C. 1344(c)

Source: 44 FR 58082, Oct. 9, 1979, unless otherwise noted.

Section 231.1 Purpose and scope

(a) The Regulations of this part include the procedures to be followed by the Environmental Protection agency in prohibiting or withdrawing the specification, or denying, restricting, or withdrawing the use for specification, of any defined area as a disposal site for dredged or fill material pursuant to section 404(c) of the Clean Water Act ("CWA"), 33 U.S.C. 1344(c). The U.S. Army Corps of Engineers or a state with a 404 program which has been approved under section 404(h) may grant permits specifying disposal sites for dredged or fill material by determining that the section 404(b)(1) Guidelines (40 CFR Part 230) allow specification of a particular site to receive dredged or fill material. The Corps may also grant permits by determining that the discharge of dredged or fill material is necessary under the economic impact provision of section 404(b)(2). Under section 404(c), the Administrator may exercise a veto over the specification by the U.S. Army Corps of Engineers or by a state of a site for the discharge of dredged or fill material. The Administrator may also prohibit the specification of a site under section 404(c) with regard to any existing or potential disposal site before a permit application has been submitted to or approved by the Corps or a state. The Administrator is authorized to prohibit or otherwise restrict a site whenever he determines that the discharge of dredged or fill material is having or will have an "unacceptable adverse effect" on municipal water supplies, shellfish beds and fishery areas (including spawning and breeding areas), wildlife, or recreational areas. In making this determination, the Administrator will take into account all information available to him, including any written determination of compliance with the section 404(b)(1) Guidelines made in 40 CFR part 230, and will consult with the Chief of Engineers or with the state.

(b) These regulations establish procedures for the following steps:

(1) The Regional Administrator's proposed determinations to prohibit or withdraw the specification of a defined area as a disposal site, or to deny, restrict or withdraw the use of any defined area for the discharge of any particular dredged or fill material;

(2) The Regional Administrator's recommendation to the Administrator for determination as to the specification of a defined area as a disposal site.

(3) The Administrator's final determination to affirm, modify or rescind the recommended determination after consultation with the Chief of Engineers or with the state.

(c) Applicability: The regulations set forth in this part are applicable whenever the Administrator is considering whether the specification of any defined area as a disposal site should be prohibited, denied, restricted, or withdrawn. These regulations apply to all existing, proposed or potential disposal sites for discharges of dredged or fill material into waters of the United States, as defined in 40 CFR 230.2.

Section 231.2 Definitions

For the purposes of this part, the definitions of terms in 40 CFR 230.2 shall apply. In addition, the term:

(a) *Withdraw specification* means to remove from designation any area already specified as a disposal site by the U.S. Army Corps of Engineers or by a state which has assumed the section 404 program, or any portion of such area.

(b) *Prohibit specification* means to prevent the designation of an area as a present or future disposal site.

(c) *Deny or restrict the use of any defined area for specification* is to deny or restrict the use of any area for the present or future discharge of any dredged or fill material.

(d) *Person* means an individual, corporation, partnership, association, Federal agency, state, municipality, or commission, or political subdivision of a state, or any interstate body.

(e) *Unacceptable adverse effect* means impact on an aquatic or wetland ecosystem which is likely to result in significant degradation of municipal water supplies (including surface or ground water) or significant loss of or damage to fisheries, shellfishing, or wildlife habitat or recreation areas. In evaluating the unacceptability of such impacts, consideration should be given to the relevant portions of the section 404(b)(1) guidelines (40 CFR part 230).

(f) *State* means any state agency administering a 404 program which has been approved under section 404(h).

Section 231.3 Procedures for proposed determinations

(a) If the Regional Administrator has reason to believe after evaluating the information available to him, including any record developed under the section 404 referral process specified in 33 CFR 323.5(b), that an "unacceptable adverse effect" could result from the specification or use for specification of a defined area for the disposal of dredged or fill material, he may initiate the following actions:

(1) The Regional Administrator will notify the District Engineer or the state, if the site is covered by an approved state program, the owner of record of the site, and the applicant, if any, in writing that the Regional Administrator intends to issue a public notice of a proposed determination to prohibit or withdraw the specification, or to deny, restrict or withdraw the use for specification, whichever the case may be, of any defined area as a disposal site.

(2) If within 15 days of receipt of the Regional Administrator's notice under paragraph (a)(1) of this section, it has not been demonstrated to the satisfaction of the Regional Administrator that no unacceptable adverse effect(s) will occur or the District Engineer or state does not notify the Regional Administrator of his intent to take corrective action to prevent an unacceptable adverse effect satisfactory to the Regional Administrator, the Regional Administrator shall publish notice of a proposed determination in accordance with the procedures of this section. Where the Regional Administrator has notified the District Engineer under paragraph (a)(1) of this section that he is considering exercising section 404(c) authority with respect to a particular disposal site for which a permit application is pending but for which no permit has been issued, the District Engineer, in accordance with 33 CFR 325.8, shall not issue the permit until final action is taken under this part.

Comment: In cases involving a proposed disposal site for which a permit application is pending, it is anticipated that the procedures of the section 404 referral process will normally be exhausted prior to any final decision of whether to initiate a 404(c) proceeding.

(b) Public notice of every proposed determination and notice of all public hearings shall be given by the Regional Administrator. Every public notice shall contain, at a minimum:

(1) An announcement that the Regional Administrator has proposed a determination to prohibit or withdraw specification, or to deny, restrict, or withdraw the use for specification, of an area as a disposal site, including a summary of the facts on which the proposed determination is based;

(2) The location of the existing, proposed or potential disposal site, and a summary of its characteristics;

(3) A summary of information concerning the nature of the proposed discharge, where applicable;

(4) The identity of the permit applicant, if any;

(5) A brief description of the right to, and procedures for requesting, a public hearing; and

(6) The address and telephone number of the office where interested persons may obtain additional information, including copies of the proposed determination; and

(7) Such additional statements, representations, or information as the Regional Administrator considers necessary or proper.

(c) In addition to the information required under paragraph (b) of this section, public notice of a public hearing held under § 231.4 shall contain the following information:

(1) Reference to the date of public notice of the proposed determination;

(2) Date, time and place of the hearing; and

(3) A brief description of the nature and purpose of the hearing including the applicable rules and procedures.

(d) The following procedures for giving public notice of the proposed determination or of a public hearing shall be followed:

(1) Publication at least once in a daily or weekly newspaper of general circulation in the area in which the defined area is located. In addition the Regional Administrator may (i) post a copy of the notice at the principal office of the municipality in which the defined area is located, or if the defined area is not located near a sizeable community, at the principal office of the political subdivision (State, county or local, whichever is appropriate) with general jurisdiction over the area in which the disposal site is located, and (ii) post a copy of the notice at the United States Post Office serving that area.

(2) A copy of the notice shall be mailed to the owner of record of the site, to the permit applicant or permit holder, if any, to the U.S. Fish and Wildlife Service, National Marine Fisheries Service and any other interested Federal and State water pollution control and resource agencies, and to any person who has filed a written request with the Regional Administrator to receive copies of notices relating to section 404(c) determinations;

(3) A copy of the notice shall be mailed to the appropriate District and Division Engineer(s) and state;

(4) The notice will also be published in the Federal Register.

Section 231.4 Public comments and hearings

(a) The Regional Administrator shall provide a comment period of not less than 30 or more than 60 days following the date of public notice of the proposed determination. During this period any interested persons may submit written comments on the proposed determination. Comments should be directed to whether the proposed determination should become the final determination and corrective action that could be taken to reduce the adverse impact of the discharge. All such comments shall be considered by the Regional Administrator or his designee in preparing his recommended determination in § 231.5.

(b) Where the Regional Administrator finds a significant degree of public interest in a proposed determination or that it would be otherwise in the public interest to hold a hearing, or if an affected landowner or permit applicant or holder requests a hearing, he or his designee shall hold a public hearing. Public notice of that hearing shall be given as specified in § 231.3(c). No hearing may be held prior to 21 days after the date of the public notice. The hearing may be scheduled either by the Regional Administrator at his own initiative, or in response to a request received during the comment period provided for in paragraph (a) of this section. If no public hearing is held the Regional Administrator shall notify any persons who requested a hearing of the reasons for that decision. Where practicable, hearings shall be conducted in the vicinity of the affected site.

(c) Hearings held under this section shall be conducted by the Regional Administrator, or his designee, in an orderly and expeditious manner. A record of the proceeding shall be made by either tape recording or verbatim transcript.

(d) Any person may appear at the hearing and submit oral or written statements and data and may be represented by counsel or other authorized representative. Any person may present written statements for the hearing file prior to the time the hearing file is closed to public submissions, and may present proposed findings and recommendations. The Regional Administrator or his designee shall afford the participants an opportunity for rebuttal.

(e) The Regional Administrator, or his designee, shall have discretion to establish reasonable limits on the nature, amount or form of presentation of documentary material and oral presentations. No cross examination of any hearing participant shall be permitted, although the Regional Administrator, or his designee, may make appropriate inquiries of any such participant.

(f) The Regional Administrator or his designee shall allow a reasonable time not to exceed 15 days after the close of the public hearing for submission of written comments. After such time has expired, unless such period is extended by the Regional Administrator or his designee for good cause, the hearing file shall be closed to additional public written comments.

(g) No later than the time a public notice of proposed determination is issued, a Record Clerk shall be designated with responsibility for maintaining the administrative record identified in § 231.5(e). Copying of any documents in the record shall be permitted under appropriate arrangements to prevent their loss. The charge for such copies shall be in accordance with the written schedule contained in part 2 of this chapter.

Section 231.5 Recommended determination

(a) The Regional Administrator or his designee shall, within 30 days after the conclusion of the public hearing (but not before the end of the comment period), or, if no hearing is held, within 15 days after the expiration of the comment period on the public notice of the proposed determination, either withdraw the proposed determination or prepare a recommended determination to prohibit or withdraw specification, or to deny, restrict, or withdraw the use for specification, of the disposal site because the discharge of dredged or fill material at such site would be likely to have an unacceptable adverse effect.

(b) Where a recommended determination is prepared, the Regional Administrator or his designee shall promptly forward the recommended determination and administrative record to the Administrator for review, with a copy of the recommended determination to the Assistant Administrator for Water and Waste Management.

(c) Where the Regional Administrator, or his designee, decides to withdraw the proposed determination, he shall promptly notify the Administrator by mail, with a copy to the Assistant Administrator for Water and Waste Management, who shall have 10 days from receipt of such notice to notify the Regional Administrator of his intent to review such withdrawal. Copies of the notification shall be sent to all persons who commented on the proposed determination or participated at the hearing. Such persons may submit timely written recommendations concerning review.

(1) If the Administrator does not notify him, the Regional Administrator shall give notice at the withdrawal of the proposed determination as provided in § 231.3(d). Such notice shall constitute final agency action.

(2) If the Administrator does decide to review, the Regional Administrator or his designee shall forward the administrative record to the Administrator for a final determination under § 231.6. Where there is review of a withdrawal of proposed determination or review of a recommended determination under § 231.6, final agency action does not occur until the Administrator makes a final determination.

(d) Any recommended determination under paragraph (b) of this section shall include the following:

(1) A summary of the unacceptable adverse effects that could occur from use of the disposal site for the proposed discharge;

(2) Recommendations regarding a final determination to prohibit, deny, restrict, or withdraw, which shall confirm or modify the proposed determination, with a statement of reasons.

(e) The administrative record shall consist of the following:

(1) A copy of the proposed determination, public notice, written comments on the public notice and written submissions in the hearing file;

(2) A transcript or recording of the public hearing, where a hearing was held;

(3) The recommended determination;

(4) Where possible a copy of the record of the Corps or the state pertaining to the site in question;

(5) Any other information considered by the Regional Administrator or his designee.

Section 231.6 Administrator's final determinations

After reviewing the recommendations of the Regional Administrator or his designee, the Administrator shall within 30 days of receipt of the recommendations and administrative record initiate consultation with the Chief of Engineers, the owner of record, and, where applicable, the State and the applicant, if any. They shall have 15 days to notify the Administrator of their intent to take corrective action to prevent an unacceptable adverse effect(s), satisfactory to the Administrator. Within 60 days of receipt of the recommendations and record, the Administrator shall make a final determination affirming, modifying, or rescinding the recommended determination. The final determination shall describe the satisfactory corrective action, if any, make findings, and state the reasons for the final determination. Notice of such final determination shall be published as provided in § 231.3, and shall be given to all persons who participated in the public hearing. Notice of the Administrator's final determination shall also be published in the Federal Register. For purposes of judicial review, a final determination constitutes final agency action under section 404(c) of the Act.

Section 231.7 Emergency procedure

Where a permit has already been issued, and the Administrator has reason to believe that a discharge under the permit presents an imminent danger of irreparable harm to municipal water supplies, shellfish beds and fishery areas (including spawning and breeding areas) wildlife, or recreational areas, and that the public health, interest, or safety requires, the Administrator may ask the Chief of Engineers to suspend the permit under 33 CFR 325.7, or the state, pending completion of proceedings under Part 231. The Administrator may also take appropriate action as authorized under section 504 of the Clean Water Act. If a permit is suspended, the Administrator and Regional Administrator (or his designee) may, where appropriate, shorten the times allowed by these regulations to take particular actions.

Section 231.8 Extension of time

The Administrator or the Regional Administrator may, upon a showing of good cause, extend the time requirements in these regulations. Notice of any such extension shall be published in the Federal Register and, as appropriate, through other forms of notice.

PART 232 404 PROGRAM DEFINITIONS; EXEMPT ACTIVITIES NOT REQUIRING 404 PERMITS

Authority: 33 U.S.C. 1344

Source: 53 FR 20773, June 6, 1988, unless otherwise noted.

Section 232.1 Purpose and scope of this part

Part 232 contains definitions applicable to the section 404 program for discharges of dredged or fill material. These definitions apply to both the federally operated program and State administered programs after program approval. This part also describes those activities which are exempted from regulation. Regulations prescribing the substantive environmental criteria for issuance of section 404 permits appear at 40 CFR part 230. Regulations establishing procedures to be followed by the EPA in denying or restricting a disposal site appear at 40 CFR part 231. Regulations containing the procedures and policies used by the Corps in administering the 404 program appear at 33 CFR parts 320–330. Regulations specifying the procedures EPA will follow, and the criteria EPA will apply in approving, monitoring, and withdrawing approval of section 404 State programs appear at 40 CFR part 233.

Section 232.2 Definitions

Administrator means the Administrator of the Environmental Protection Agency or an authorized representative.

Application means a form for applying for a permit to discharge dredged or fill material into waters of the United States.

Approved program means a State program which has been approved by the Regional Administrator under part 233 of this chapter or which is deemed approved under section 404(h)(3), 33 U.S.C. 1344(h)(3).

Best management practices (BMPs) means schedules of activities, prohibitions of practices, maintenance procedures, and other management practices to prevent or reduce the pollution of waters of the United States from discharges of dredged or fill material. BMPs include methods, measures, practices, or design and performance standards which facilitate compliance with the section 404(b)(1) Guidelines (40 CFR part 230), effluent limitations or prohibitions under section 307(a), and applicable water quality standards.

Discharge of dredged material.

(1) Except as provided below in paragraph (3), the term *discharge of dredged material* means any addition of dredged material into, including redeposit of dredged material other than incidental fallback within, the waters of the United States. The term includes, but is not limited to, the following:

(i) The addition of dredged material to a specified discharge site located in waters of the United States;

(ii) The runoff or overflow, associated with a dredging operation, from a contained land or water disposal area; and

(iii) Any addition, including redeposit other than incidental fallback, of dredged material, including excavated material, into waters of the United States which is incidental to any activity, including mechanized landclearing, ditching, channelization, or other excavation.

(2) (i) The Corps and EPA regard the use of mechanized earth-moving equipment to conduct landclearing, ditching, channelization, in-stream mining or other earth-moving activity in waters of the United States as resulting in a discharge of dredged material unless project-specific evidence shows that the activity results in only incidental fallback. This paragraph (i) does not and is not intended to shift any burden in any administrative or judicial proceeding under the CWA.

(ii) *Incidental fallback* is the redeposit of small volumes of dredged material that is incidental to excavation activity in waters of the United States when such material falls back to substantially the same place as the initial removal. Examples of incidental fallback include soil that is disturbed when dirt is shoveled and the back-spill that comes off a bucket when such small volume of soil or dirt falls into substantially the same place from which it was initially removed.

(3) The term *discharge of dredged material* does not include the following:

(i) Discharges of pollutants into waters of the United States resulting from the onshore subsequent processing of dredged material that is extracted for any commercial use (other than fill). These discharges are subject to section 402 of the Clean Water Act even though the extraction and deposit of such material may require a permit from the Corps or applicable state.

(ii) Activities that involve only the cutting or removing of vegetation above the ground (e.g., mowing, rotary cutting, and chainsawing) where the activity neither substantially disturbs the root system nor involves mechanized pushing, dragging, or other similar activities that redeposit excavated soil material.

(iii) Incidental fallback.

(4) Section 404 authorization is not required for the following:

(i) Any incidental addition, including redeposit, of dredged material associated with any activity that does not have or would not have the effect of destroying or degrading an area of waters of the U.S. as defined in paragraphs (5) and (6) of this definition; however, this exception does not apply to any person preparing to undertake mechanized landclearing, ditching, channelization and other excavation activity in a water of the United States, which would result in a redeposit of dredged material, unless the person demonstrates to the satisfaction of the Corps, or EPA as appropriate, prior to commencing the activity involving the discharge, that the activity would not have the effect of destroying or degrading any area of waters of the United States, as defined in paragraphs (5) and (6) of this definition. The person proposing to undertake mechanized landclearing, ditching, channelization or other excavation activity bears the burden of demonstrating that such activity would not destroy or degrade any area of waters of the United States.

(ii) Incidental movement of dredged material occurring during normal dredging operations, defined as dredging for navigation in *navigable waters of the United States,* as that term is defined in 33 CFR part 329, with proper authorization from the Congress or the Corps pursuant to 33 CFR part 322; however, this exception is not applicable to dredging activities in wetlands, as that term is defined at § 232.2(r) of this chapter.

(iii) Certain discharges, such as those associated with normal farming, silviculture, and ranching activities, are not prohibited by or otherwise subject to regulation under Section 404. See 40 CFR 232.3 for discharges that do not require permits.

(5) For purposes of this section, an activity associated with a discharge of dredged material destroys an area of waters of the United States if it alters the area in such a way that it would no longer be a water of the United States.

Note: Unauthorized discharges into waters of the United States do not eliminate Clean Water Act jurisdiction, even where such unauthorized discharges have the effect of destroying waters of the United States.

(6) For purposes of this section, an activity associated with a discharge of dredged material degrades an area of waters of the United States if it has more than a *de minimis* (i.e., inconsequential) effect on the area by causing an identifiable individual or cumulative adverse effect on any aquatic function.

Discharge of fill material

(1) The term *discharge of fill material* means the addition of fill material into waters of the United States. The term generally includes, without limitation, the following activities: Placement of fill that is necessary for the construction of any structure in a water of the United States; the building of any structure or impoundment requiring rock, sand, dirt, or other material for its construction; site-development fills for recreational, industrial, commercial, residential, and other uses; causeways or road fills; dams and dikes; artificial islands; property protection and/or reclamation devices such as riprap, groins, seawalls, breakwaters, and revetments; beach nourishment; levees; fill for structures such as sewage

treatment facilities, intake and outfall pipes associated with power plants and subaqueous utility lines; and artificial reefs.

(2) In addition, placement of pilings in waters of the United States constitutes a discharge of fill material and requires a Section 404 permit when such placement has or would have the effect of a discharge of fill material. Examples of such activities that have the effect of a discharge of fill material include, but are not limited to, the following: Projects where the pilings are so closely spaced that sedimentation rates would be increased; projects in which the pilings themselves effectively would replace the bottom of a waterbody; projects involving the placement of pilings that would reduce the reach or impair the flow or circulation of waters of the United States; and projects involving the placement of pilings which would result in the adverse alteration or elimination of aquatic functions.

(i) Placement of pilings in waters of the United States that does not have or would not have the effect of a discharge of fill material shall not require a Section 404 permit. Placement of pilings for linear projects, such as bridges, elevated walkways, and powerline structures, generally does not have the effect of a discharge of fill material. Furthermore, placement of pilings in waters of the United States for piers, wharves, and an individual house on stilts generally does not have the effect of a discharge of fill material. All pilings, however, placed in the *navigable waters of the United States,* as that term is defined in 33 CFR part 329, require authorization under section 10 of the Rivers and Harbors Act of 1899 (see 33 CFR part 322).

(ii) [Reserved]

Dredged material means material that is excavated or dredged from waters of the United States.

Effluent means dredged material or fill material, including return flow from confined sites.

Federal Indian reservation means all land within the limits of any Indian reservation under the jurisdiction of the United States Government, notwithstanding the issuance of any patent, and including rights-of-way running through the reservation.

Fill material means any "pollutant" which replaces portions of the "waters of the United States" with dry land or which changes the bottom elevation of a water body for any purpose.

General permit means a permit authorizing a category of discharges of dredged or fill material under the Act. General permits are permits for categories of discharge which are similar in nature, will cause only minimal adverse environmental effects when performed separately, and will have only minimal cumulative adverse effect on the environment.

Indian Tribe means any Indian Tribe, band, group, or community recognized by the Secretary of the Interior and exercising governmental authority over a Federal Indian reservation.

Owner or operator means the owner or operator of any activity subject to regulation under the 404 program.

Permit means a written authorization issued by an approved State to implement the requirements of part 233, or by the Corps under 33 CFR parts 320-330. When used in these regulations, "permit" includes "general permit" as well as individual permit.

Person means an individual, association, partnership, corporation, municipality, State or Federal agency, or an agent or employee thereof.

Regional Administrator means the Regional Administrator of the appropriate Regional Office of the Environmental Protection Agency or the authorized representative of the Regional Administrator.

Secretary means the Secretary of the Army acting through the Chief of Engineers.

State means any of the 50 States, the District of Columbia, Guam, the Commonwealth of Puerto Rico, the Virgin Islands, American Samoa, the Commonwealth of the Northern Mariana Islands, the Trust Territory of the Pacific Islands, or an Indian Tribe as defined in this part, which meet the requirements of § 233.60.

State regulated waters means those waters of the United States in which the Corps suspends the issuance of section 404 permits upon approval of a State's section 404 permit program by the Administrator under section 404(h). The program cannot be transferred for those waters which are presently used, or are susceptible to use in their natural condition or by reasonable improvement as a means to transport interstate or foreign commerce shoreward to their ordinary high water mark, including all waters which are subject to the ebb and flow of the tide shoreward to the high tide line, including wetlands adjacent thereto. All other waters of the United States in a State with an approved program shall be under jurisdiction of the State program, and shall be identified in the program description as required by part 233.

Waters of the United States means:

All waters which are currently used, were used in the past, or may be susceptible to us in interstate or foreign commerce, including all waters which are subject to the ebb and flow of the tide.

All interstate waters including interstate wetlands.

All other waters, such as intrastate lakes, rivers, streams (including intermittent streams), mudflats, sandflats, wetlands, sloughs, prairie potholes, wet meadows, playa lakes, or natural ponds, the use, degradation, or destruction of which would or could affect interstate or foreign commerce including any such waters:

Which are or could be used by interstate or foreign travelers for recreational or other purposes; or

From which fish or shellfish are or could be taken and sold in interstate or foreign commerce; or

Which are used or could be used for industrial purposes by industries in interstate commerce.

All impoundments of waters otherwise defined as waters of the United States under this definition;

Tributaries of waters identified in paragraphs (g)(1)–(4) of this section;

The territorial sea; and

Wetlands adjacent to waters (other than waters that are themselves wetlands) identified in paragraphs (q)(1)–(6) of this section.

Waste treatment systems, including treatment ponds or lagoons designed to meet the requirements of the Act (other than cooling ponds as defined in 40 CFR 123.11(m) which also meet the criteria of this definition) are not waters of the United States.

Waters of the United States do not include prior converted cropland. Notwithstanding the determination of an area's status as prior converted cropland by any other federal agency, for the purposes of the Clean Water Act, the final authority regarding Clean Water Act jurisdiction remains with EPA.

Wetlands means those areas that are inundated or saturated by surface or ground water at a frequency and duration sufficient to support, and that under normal circumstances do support, a prevalence of vegetation typically adapted for life in saturated soil conditions. Wetlands generally include swamps, marshes, bogs, and similar areas.

[53 FR 20773, June 6, 1988, as amended at 58 FR 8182, Feb. 11, 1993; 58 FR 45037, Aug. 25, 1993; 64 FR 25123, May 10, 1999; 66 FR 4575, Jan. 17, 2001]

Section 232.3 Activities not requiring permits

Except as specified in paragraphs (a) and (b) of this section, any discharge of dredged or fill material that may result from any of the activities described in paragraph (c) of this section is not prohibited by or otherwise subject to regulation under this part.

(a) If any discharge of dredged or fill material resulting from the activities listed in paragraph (c) of this section contains any toxic pollutant listed under section 307 of the Act, such discharge shall be subject to any applicable toxic effluent standard or prohibition, and shall require a section 404 permit.

(b) Any discharge of dredged or fill material into waters of the United States incidental to any of the activities identified in paragraph (c) of this section must have a permit if it is part of an activity whose purpose is to convert an area of the waters of the United States into a use to which it was not previously subject, where the flow or circulation of waters of the United States may be impaired or the reach of such waters reduced. Where the proposed discharge will result in significant discernable alterations to flow or circulation, the presumption is that flow or circulation may be impaired by such alteration.

Note: For example, a permit will be required for the conversion of a cypress swamp to some other use or the conversion of a wetland from silvicultural to agricultural use when there is a discharge of dredged or fill material into waters of the United States in conjunction with construction of dikes, drainage ditches or other works or structures used to effect such conversion. A conversion of section 404 wetland to a non-wetland is a change in use of an area of waters of the U.S. A discharge which elevates the bottom of waters of the United States without converting it to dry land does not thereby reduce the reach of, but may alter the flow or circulation of, waters of the United States.

(c) The following activities are exempt from section 404 permit requirements, except as specified in paragraphs (a) and (b) of this section:

(1) (i) Normal farming, silviculture and ranching activities such as plowing, seeding, cultivating, minor drainage, and harvesting for the production of food, fiber, and forest products, or upland soil and water conservation practices, as defined in paragraph (d) of this section.

(ii) (A) To fall under this exemption, the activities specified in paragraph (c)(1) of this section must be part of an established (i.e., ongoing) farming, silviculture, or ranching operation, and must be in accordance with definitions in paragraph (d) of this section. Activities on areas lying fallow as part of a conventional rotational cycle are part of an established operation.

(B) Activities which bring an area into farming, silviculture or ranching use are not part of an established operation. An operation ceases to be established when the area in which it was conducted has been converted to another use or has lain idle so long that modifications to the hydrological regime are necessary to resume operation. If an activity takes place outside the waters of the United States, or if it does not involve a discharge, it does not need a section 404 permit whether or not it was part of an established farming, silviculture or ranching operation.

(2) Maintenance, including emergency reconstruction of recently damaged parts, of currently serviceable structures such as dikes, dams, levees, groins, riprap, breakwaters, causeways, bridge abutments or approaches, and transportation structures. Maintenance does not include any modification that changes the character, scope, or size of the original fill design. Emergency reconstruction must occur within a reasonable period of time after damage occurs in order to qualify for this exemption.

(3) Construction or maintenance of farm or stock ponds or irrigation ditches or the maintenance (but not construction) of drainage ditches. Discharge associated with siphons, pumps, headgates, wingwalls, wiers, diversion structures, and such other facilities as are appurtenant and functionally related to irrigation ditches are included in this exemption.

(4) Construction of temporary sedimentation basins on a construction site which does not include placement of fill material into waters of the United States. The term "construction site" refers to any site involving the erection of buildings, roads, and other discrete structures and the installation of support facilities necessary for construction and utilization of such structures. The term also includes any other land areas which involve land-disturbing excavation activities, including quarrying or other mining activities, where an increase in the runoff of sediment is controlled through the use of temporary sedimentation basins.

(5) Any activity with respect to which a State has an approved program under section 208(b)(4) of the Act which meets the requirements of section 208(b)(4)(B) and (C).

(6) Construction or maintenance of farm roads, forest roads, or temporary roads for moving mining equipment, where such roads are constructed and maintained in accordance with best management practices (BMPs) to assure that flow and circulation patterns and chemical and biological characteristics of waters of the United States are not impaired, that the reach of the waters of the United States is not reduced, and that any adverse effect on the aquatic environment will be otherwise minimized. The BMPs which must be applied to satisfy this provision include the following baseline provisions:

(i) Permanent roads (for farming or forestry activities), temporary access roads (for mining, forestry, or farm purposes) and skid trails (for logging) in waters of the United States shall be held to the minimum feasible number, width, and total length consistent with the purpose of specific farming, silvicultural or mining operations, and local topographic and climatic conditions;

(ii) All roads, temporary or permanent, shall be located sufficiently far from streams or other water bodies (except for portions of such roads which must cross water bodies) to minimize discharges of dredged or fill material into waters of the United States;

(iii) The road fill shall be bridged, culverted, or otherwise designed to prevent the restriction of expected flood flows;

(iv) The fill shall be properly stabilized and maintained to prevent erosion during and following construction;

(v) Discharges of dredged or fill material into waters of the United States to construct a road fill shall be made in a manner that minimizes the encroachment of trucks, tractors, bulldozers, or other heavy equipment within the waters of the United States (including adjacent wetlands) that lie outside the lateral boundaries of the fill itself;

(vi) In designing, constructing, and maintaining roads, vegetative disturbance in the waters of the United States shall be kept to a minimum;

(vii) The design, construction and maintenance of the road crossing shall not disrupt the migration or other movement of those species of aquatic life inhabiting the water body;

(viii) Borrow material shall be taken from upland sources whenever feasible;

(ix) The discharge shall not take, or jeopardize the continued existence of, a threatened or endangered species as defined under the Endangered Species Act, or adversely modify or destroy the critical habitat of such species;

(x) Discharges into breeding and nesting areas for migratory waterfowl, spawning areas, and wetlands shall be avoided if practical alternatives exist;

(xi) The discharge shall not be located in the proximity of a public water supply intake;

(xii) The discharge shall not occur in areas of concentrated shellfish production;

(xiii) The discharge shall not occur in a component of the National Wild and Scenic River System;

(xiv) The discharge of material shall consist of suitable material free from toxic pollutants in toxic amounts; and

(xv) All temporary fills shall be removed in their entirety and the area restored to its original elevation.

(d) For purpose of paragraph (c)(1) of this section, cultivating, harvesting, minor drainage, plowing, and seeding are defined as follows:

(1) Cultivating means physical methods of soil treatment employed within established farming, ranching and silviculture lands on farm, ranch, or forest crops to aid and improve their growth, quality, or yield.

(2) Harvesting means physical measures employed directly upon farm, forest, or ranch crops within established agricultural and silvicultural lands to bring about their removal from farm, forest, or ranch land, but does not include the construction of farm, forest, or ranch roads.

(3) (i) Minor drainage means:

(A) The discharge of dredged or fill material incidental to connecting upland drainage facilities to waters of the United States, adequate to effect the removal of excess soil moisture from upland croplands. Construction and maintenance of upland (dryland) facilities, such as ditching and tiling, incidental to the planting, cultivating, protecting, or harvesting of crops, involve no discharge of dredged or fill material into waters of the United States, and as such never require a section 404 permit;

(B) The discharge of dredged or fill material for the purpose of installing ditching or other water control facilities incidental to planting, cultivating, protecting, or harvesting of rice, cranberries or other wetland crop species, where these activities and the discharge occur in waters of the United States which are in established use for such agricultural and silvicultural wetland crop production;

(C) The discharge of dredged or fill material for the purpose of manipulating the water levels of, or regulating the flow or distribution of water within, existing impoundments which have been constructed in accordance with applicable requirements of the Act, and which are in established use for the production or rice, cranberries, or other wetland crop species.

Note: The provisions of paragraphs (d)(3)(i)(B) and (C) of this section apply to areas that are in established use exclusively for wetland crop production as well as areas in established use for conventional wetland/non-wetland crop rotation (e.g., the rotations of rice and soybeans) where such rotation results in the cyclical or intermittent temporary dewatering of such areas.

(D) The discharge of dredged or fill material incidental to the emergency removal of sandbars, gravel bars, or other similar blockages which are formed during flood flows or other events, where such blockages close or constrict previously existing drainageways and, if not promptly removed, would result in damage to or loss of existing crops or would impair or prevent the plowing, seeding, harvesting or cultivating of crops on land in established use for crop production. Such removal does not include enlarging or extending the dimensions of, or changing the bottom elevations of, the affected drainageway as it existed prior to the formation of the blockage. Removal must be accomplished within one year after such blockages are discovered in order to be eligible for exemption.

(ii) Minor drainage in waters of the United States is limited to drainage within areas that are part of an established farming or silviculture operation. It does not include drainage associated with the immediate or gradual conversion of a wetland to a non-wetland (e.g., wetland species to upland species not typically adequate to life in saturated soil conditions), or conversion from one wetland use to another (for example, silviculture to farming). In addition, minor drainage does not include the construction of any canal, ditch, dike or other waterway or structure which drains or otherwise significantly modifies a stream, lake, swamp, bog or any other wetland or aquatic area constituting waters of the United States. Any discharge of dredged or fill material into the waters of the United States incidental to the construction of any such structure or waterway requires a permit.

(4) Plowing means all forms of primary tillage, including moldboard, chisel, or wide-blade plowing, discing, harrowing, and similar physical means used on farm, forest or ranch land for the breaking up, cutting, turning over, or stirring of soil to prepare it for the planting of crops. Plowing does not include the redistribution of soil, rock, sand, or other surficial materials in a manner which changes any area of the waters of the United States to dryland. For example, the redistribution of surface materials by blading, grading, or other means to fill in wetland areas is not plowing. Rock crushing activities which result in the loss of natural drainage characteristics, the reduction of water storage and recharge capabilities, or the overburden of natural water filtration capacities do not constitute plowing. Plowing, as described above, will never involve a discharge of dredged or fill material.

(5) Seeding means the sowing of seed and placement of seedlings to produce farm, ranch, or forest crops and includes the placement of soil beds for seeds or seedlings on established farm and forest lands.

(e) Federal projects which qualify under the criteria contained in section 404(r) of the Act are exempt from section 404 permit requirements, but may be subject to other State or Federal requirements.

PART 233
404 STATE PROGRAM REGULATIONS

Authority: 33 U.S.C. 1251 et seq.

Source: 53 FR 20776, June 1, 1988, unless otherwise noted.

Subpart A—General

Section 233.1 Purpose and scope

(a) This part specifies the procedures EPA will follow, and the criteria EPA will apply, in approving, reviewing, and withdrawing approval of State programs under section 404 of the Act.

(b) Except as provided in § 232.3, a State program must regulate all discharges of dredged or fill material into waters regulated by the State under section 404(g)–(1). Partial State programs are not approvable under section 404. A State's decision not to assume existing Corps' general permits does not constitute a partial program. The discharges previously authorized by general permit will be regulated by State individual permits. However, in many cases, States other than Indian Tribes will lack authority to regulate activities on Indian lands. This lack of authority does not impair that State's ability to obtain full program approval in accordance with this part, i.e., inability of a State which is not an Indian Tribe to regulate activities on Indian lands does not constitute a partial program. The Secretary of the Army acting through the Corps of Engineers will continue to administer the program on Indian lands if a State which is not an Indian Tribe does not seek and have authority to regulate activities on Indian lands.

(c) Nothing in this part precludes a State from adopting or enforcing requirements which are more stringent or from operating a program with greater scope, than required under this part. Where an approved State program has a greater scope than required by Federal law, the additional coverage is not part of the Federally approved program and is not subject to Federal oversight or enforcement.

Note: State assumption of the section 404 program is limited to certain waters, as provided in section 404(g)(1). The Federal program operated by the Corps of Engineers continues to apply to the remaining waters in the State even after program approval. However, this does not restrict States from regulating discharges of dredged or fill material into those waters over which the Secretary retains section 404 jurisdiction.

(d) Any approved State Program shall, at all times, be conducted in accordance with the requirements of the Act and of this part. While States may impose more stringent requirements, they may not impose any less stringent requirements for any purpose.

[53 FR 20776, June 1, 1988, as amended at 58 FR 8183, Feb. 11, 1993]

Section 233.2 Definitions

The definitions in parts 230 and 232 as well as the following definitions apply to this part.

Act means the Clean Water Act (33 U.S.C. 1251 *et seq.*).

Corps means the U.S. Army Corps of Engineers.

Federal Indian reservation means all land within the limits of any Indian reservation under the jurisdiction of the United States Government, notwithstanding the issuance of any patent, and including rights-of-way running through the reservation.

FWS means the U.S. Fish and Wildlife Service.

Indian Tribe means any Indian Tribe, band, group, or community recognized by the Secretary of the Interior and exercising governmental authority over a Federal Indian reservation.

Interstate agency means an agency of two or more States established by or under an agreement or compact approved by the Congress, or any other agency of two or more States having substantial powers or duties pertaining to the control of pollution.

NMFS means the National Marine Fisheries Service.

State means any of the 50 States, the District of Columbia, Guam, the Commonwealth of Puerto Rico, the Virgin Islands, American Samoa, the Commonwealth of the Northern Mariana Islands, the Trust Territory of the Pacific Islands, or an Indian Tribe, as defined in this part, which meet the requirements of § 233.60. For purposes of this part, the word State also includes any interstate agency requesting program approval or administering an approved program.

State Director (*Director*) means the chief administrative officer of any State or interstate agency operating an approved program, or the delegated representative of the Director. If responsibility is divided among two or more State or interstate agencies, Director means the chief administrative officer of the State or interstate agency authorized to perform the particular procedure or function to which reference is made.

State 404 program or *State program* means a State program which has been approved by EPA under section 404 of the Act to regulate the discharge of dredged or fill material into certain waters as defined in § 232.2(p).

[53 FR 20776, June 1, 1988, as amended at 58 FR 8183, Feb. 11, 1993]

Section 233.3 Confidentiality of information

(a) Any information submitted to EPA pursuant to these regulations may be claimed as confidential by the submitter at the time of submittal and a final determination as to that claim will be made in accordance with the procedures of 40 CFR part 2 and paragraph (c) of this section.

(b) Any information submitted to the Director may be claimed as confidential in accordance with State law, subject to paragraphs (a) and (c) of this section.

(c) Claims of confidentiality for the following information will be denied:

(1) The name and address of any permit applicant or permittee,

(2) Effluent data,

(3) Permit application, and

(4) Issued permit.

Section 233.4 Conflict of interest

Any public officer or employee who has a direct personal or pecuniary interest in any matter that is subject to decision by the agency shall make known such interest in the official records of the agency and shall refrain from participating in any manner in such decision.

Subpart B—Program Approval

Section 233.10 Elements of a program submission

Any State that seeks to administer a 404 program under this part shall submit to the Regional Administrator at least three copies of the following:

(a) A letter from the Governor of the State requesting program approval.

(b) A complete program description, as set forth in § 233.11.

(c) An Attorney General's statement, as set forth in § 233.12.

(d) A Memorandum of Agreement with the Regional Administrator, as set forth in § 233.13.

(e) A Memorandum of Agreement with the Secretary, as set forth in § 233.14.

(f) Copies of all applicable State statutes and regulations, including those governing applicable State administrative procedures.

Section 233.11 Program description

The program description as required under § 233.10 shall include:

(a) A description of the scope and structure of the State's program. The description should include extent of State's jurisdiction, scope of activities regulated, anticipated coordination, scope of permit exemptions if any, and permit review criteria;

(b) A description of the State's permitting, administrative, judicial review, and other applicable procedures;

(c) A description of the basic organization and structure of the State agency (agencies) which will have responsibility for administering the program. If more than one State agency is responsible for the administration of the program, the description shall address the responsibilities of each agency and how the agencies intend to coordinate administration and evaluation of the program;

(d) A description of the funding and manpower which will be available for program administration;

(e) An estimate of the anticipated workload, e.g., number of discharges.

(f) Copies of permit application forms, permit forms, and reporting forms;

(g) A description of the State's compliance evaluation and enforcement programs, including a description of how the State will coordinate its enforcement strategy with that of the Corps and EPA;

(h) A description of the waters of the United States within a State over which the State assumes jurisdiction under the approved program; a description of the waters of the United States within a State over which the Secretary retains jurisdiction subsequent to program approval; and a comparison of the State and Federal definitions of wetlands.

Note: States should obtain from the Secretary an identification of those waters of the U.S. within the State over which the Corps retains authority under section 404(g) of the Act.

(i) A description of the specific best management practices proposed to be used to satisfy the exemption provisions of section 404(f)(1)(E) of the Act for construction or maintenance of farm roads, forest roads, or temporary roads for moving mining equipment.

Section 233.12 Attorney General's statement

(a) Any State that seeks to administer a program under this part shall submit a statement from the State Attorney General (or the attorney for those State or interstate agencies which have independence legal counsel), that the laws and regulations of the State, or an interstate compact, provide adequate authority to carry out the program and meet the applicable requirements of this part. This statement shall cite specific statutes and administrative regulations which are lawfully adopted at the time the statement is signed and which shall be fully effective by the time the program is approved, and, where appropriate, judicial decisions which demonstrate adequate authority. The attorney signing the statement required by this section must have authority to represent the State agency in court on all matters pertaining to the State program.

(b) If a State seeks approval of a program covering activities on Indian lands, the statement shall contain an analysis of the State's authority over such activities.

(c) The State Attorney General's statement shall contain a legal analysis of the effect of State law regarding the prohibition on taking private property without just compensation on the successful implementation of the State's program.

(d) In those States where more than one agency has responsibility for administering the State program, the statement must include certification that each agency has full authority to administer the program within its category of jurisdiction and that the State, as a whole, has full authority to administer a complete State section 404 program.

Section 233.13 Memorandum of Agreement with Regional Administrator

(a) Any State that seeks to administer a program under this part shall submit a Memorandum of Agreement executed by the Director and the Regional Administrator. The Memorandum of Agreement shall become effective upon approval of the State program. When more than one agency within a State has responsibility for administering the State program, Directors of each of the responsible State agencies shall be parties to the Memorandum of Agreement.

(b) The Memorandum of Agreement shall set out the State and Federal responsibilities for program administration and enforcement. These shall include, but not be limited to:

(1) Provisions specifying classes and categories of permit applications for which EPA will waive Federal review (as specified in § 233.51).

(2) Provisions specifying the frequency and content of reports, documents and other information which the State may be required to submit to EPA in addition to the annual report, as well as a provision establishing the submission date for the annual report. The State shall also allow EPA routinely to review State records, reports and files relevant to the administration and enforcement of the approved program.

(3) Provisions addressing EPA and State roles and coordination with respect to compliance monitoring and enforcement activities.

(4) Provisions addressing modification of the Memorandum of Agreement.

Section 233.14 Memorandum of Agreement with the Secretary

(a) Before a State program is approved under this part, the Director shall enter into a Memorandum of Agreement with the Secretary. When more than one agency within a State has responsibility for administering the State program, Directors of each of the responsible agencies shall be parties of the Memorandum of Agreement.

(b) The Memorandum of Agreement shall include:

(1) A description of waters of the United States within the State over which the Secretary retains jurisdiction, as identified by the Secretary.

(2) Procedures whereby the Secretary will, upon program approval, transfer to the State pending 404 permit applications for discharges in State regulated waters and other relevant information not already in the possession of the Director.

Note: Where a State permit program includes coverage of those traditionally navigable waters in which only the Secretary may issue section 404 permits, the State is encouraged to establish in this MOA procedures for joint processing of Federal and State permits, including joint public notices and public hearings.

(3) An identification of all general permits issued by the Secretary the terms and conditions of which the State intends to administer and enforce upon receiving approval of its program, and a plan for transferring responsibility for these general permits to the State, including procedures for the prompt transmission from the Secretary to the Director of relevant information not already in the possession of the Director, including support files for permit issuance, compliance reports and records of enforcement actions.

Section 233.15 Procedures for approving State programs

(a) The 120 day statutory review period shall commence on the date of receipt of a complete State program submission as set out in § 233.10 of this part. EPA shall determine whether the submission is complete within 30 days of receipt of the submission and shall notify the State of its determination. If EPA finds that a State's submission is incomplete, the statutory review period shall not begin until all the necessary information is received by EPA.

(b) If EPA determines the State significantly changes its submission during the review period, the statutory review period shall begin again upon the receipt of a revised submission.

(c) The State and EPA may extend the statutory review period by agreement.

(d) Within 10 days of receipt of a complete State section 404 program submission, the Regional Administrator shall provide copies of the State's submission to the Corps, FWS, and NMFS (both Headquarters and appropriate Regional organizations.)

(e) After determining that a State program submission is complete, the Regional Administrator shall publish notice of the State's application in the Federal Register and in enough of the largest newspapers in the State to attract statewide attention. The Regional Administrator shall also mail notice to persons known to be interested in such matters. Existing State, EPA, Corps, FWS, and NMFS mailing lists shall be used as a basis for this mailing. However, failure to mail all such notices shall not be grounds for invalidating approval (or disapproval) of an otherwise acceptable (or unacceptable) program. This notice shall:

(1) Provide for a comment period of not less than 45 days during which interested members of the public may express their views on the State program.

(2) Provide for a public hearing within the State to be held not less than 30 days after notice of hearing is published in the Federal Register;

(3) Indicate where and when the State's submission may be reviewed by the public;

(4) Indicate whom an interested member of the public with questions should contact; and

(5) Briefly outline the fundamental aspects of the State's proposed program and the process for EPA review and decision.

(f) Within 90 days of EPA's receipt of a complete program submission, the Corps, FWS, and NMFS shall submit to EPA any comments on the State's program.

(g) Within 120 days of receipt of a complete program submission (unless an extension is agreed to by the State), the Regional Administrator shall approve or disapprove the program based on whether the State's program fulfills the requirements of this part and the Act, taking into consideration all comments received. The Regional Administrator shall prepare a responsiveness summary of significant comments received and his response to these comments. The Regional Administrator shall respond individually to comments received from the Corps, FWS, and NMFS.

(h) If the Regional Administrator approves the State's section 404 program, he shall notify the State and the Secretary of the decision and publish notice in the Federal Register. Transfer of the program to the State shall not be considered effective until such notice appears in the Federal Register. The Secretary shall suspend the issuance by the Corps of section 404 permits in State regulated waters on such effective date.

(i) If the Regional Administrator disapproves the State's program based on the State not meeting the requirements of the Act and this part, the Regional Administrator shall notify the State of the reasons for the disapproval and of any revisions or modifications to the State's program which are necessary to obtain approval. If the State resubmits a program submission remedying the identified problem areas, the approval procedure and statutory review period shall begin upon receipt of the revised submission.

Section 233.16 Procedures for revision of State programs

(a) The State shall keep the Regional Administrator fully informed of any proposed or actual changes to the State's statutory or regulatory authority or any other modifications which are significant to administration of the program.

(b) Any approved program which requires revision because of a modification to this part or to any other applicable Federal statute or regulation shall be revised within one year of the date of promulgation of such regulation, except that if a State must amend or enact a statute in order to make the required revision, the revision shall take place within two years.

(c) States with approved programs shall notify the Regional Administrator whenever they propose to transfer all or part of any program from the approved State agency to any other State agency. The new agency is not authorized to administer the program until approved by the Regional Administrator under paragraph (d) of this section.

(d) Approval of revision of a State program shall be accomplished as follows:

(1) The Director shall submit a modified program description or other documents which the Regional Administrator determines to be necessary to evaluate whether the program complies with the requirements of the Act and this part.

(2) Notice of approval of program changes which are not substantial revisions may be given by letter from the Regional Administrator to the Governor or his designee.

(3) Whenever the Regional Administrator determines that the proposed revision is substantial, he shall publish and circulate notice to those persons known to be interested in such matters, provide opportunity for a public hearing, and consult with the Corps, FWS, and NMFS. The Regional Administrator shall approve or disapprove program revisions based on whether the program fulfills the requirements of the Act and this part, and shall publish notice of his decision in the Federal Register. For purposes of this paragraph, substantial revisions include, but are not limited to, revisions that affect the area of jurisdiction, scope of activities regulated, criteria for review of permits, public participation, or enforcement capability.

(4) Substantial program changes shall become effective upon approval by the Regional Administrator and publication of notice in the Federal Register.

(e) Whenever the Regional Administrator has reason to believe that circumstances have changed with respect to a State's program, he may request and the State shall provide a supplemental Attorney General's statement, program description, or such other documents or information as are necessary to evaluate the program's compliance with the requirements of the Act and this part.

Subpart C—Permit Requirements

Section 233.20 Prohibitions

No permit shall be issued by the Director in the following circumstances:

(a) When permit does not comply with the requirements of the Act or regulations thereunder, including the section 404(b)(1) Guidelines (part 230 of this chapter).

(b) When the Regional Administrator has objected to issuance of the permit under § 233.50 and the objection has not been resolved.

(c) When the proposed discharges would be in an area which has been prohibited, withdrawn, or denied as a disposal site by the Administrator under section 404(c) of the Act, or when the discharge would fail to comply with a restriction imposed thereunder.

(d) If the Secretary determines, after consultation with the Secretary of the Department in which the Coast Guard is operating, that anchorage and navigation of any of the navigable waters would be substantially impaired.

Section 233.21 General permits

(a) Under section 404(h)(5) of the Act, States may, after program approval, administer and enforce general permits previously issued by the Secretary in State regulated waters.

Note: If States intend to assume existing general permits, they must be able to ensure compliance with existing permit conditions an any reporting monitoring, or prenotification requirements.

(b) The Director may issue a general permit for categories of similar activities if he determines that the regulated activities will cause only minimal adverse environmental effects when performed separately and will have only minimal cumulative adverse effects on the environment. Any general permit issued shall be in compliance with the section 404(b)(1) Guidelines.

(c) In addition to the conditions specified in § 233.23, each general permit shall contain:

(1) A specific description of the type(s) of activities which are authorized, including limitations for any single operation. The description shall be detailed enough to ensure that the requirements of paragraph (b) of this -section are met. (This paragraph supercedes § 233.23(c)(1) for general permits.)

(2) A precise description of the geographic area to which the general permit applies, including limitations on the type(s) of water where operations may be conducted sufficient to ensure that the requirements of paragraph (b) of this section are met.

(d) Predischarge notification or other reporting requirements may be required by the Director on a permit-by-permit basis as appropriate to ensure that the general permit will comply with the requirement (section 404(e) of the Act) that the regulated activities will cause only minimal adverse environmental effects when performed separately and will have only minimal cumulative adverse effects on the environment.

(e) The Director may, without revoking the general permit, require any person authorized under a general permit to apply for an individual permit. This discretionary authority will be based on concerns for the aquatic environment including compliance with paragraph (b) of this section and the 404(b)(1) Guidelines (40 CFR part 230.)

(1) This provision in no way affects the legality of activities undertaken pursuant to the general permit prior to notification by the Director of such requirement.

(2) Once the Director notifies the discharger of his decision to exercise discretionary authority to require an individual permit, the discharger's activity is no longer authorized by the general permit.

Section 233.22 Emergency permits

(a) Notwithstanding any other provision of this part, the Director may issue a temporary emergency permit for a discharge of dredged or fill material if unacceptable harm to life or severe loss of physical property is likely to occur before a permit could be issued or modified under procedures normally required.

(b) Emergency permits shall incorporate, to the extent possible and not inconsistent with the emergency situation, all applicable requirements of § 233.23.

(1) Any emergency permit shall be limited to the duration of time (typically no more than 90 days) required to complete the authorized emergency action.

(2) The emergency permit shall have a condition requiring appropriate restoration of the site.

(c) The emergency permit may be terminated at any time without process (§ 233.36) if the Director determines that termination is necessary to protect human health or the environment.

(d) The Director shall consult in an expeditious manner, such as by telephone, with the Regional Administrator, the Corps, FWS, and NMFS about issuance of an emergency permit.

(e) The emergency permit may be oral or written. If oral, it must be followed within 5 days by a written emergency permit. A copy of the written permit shall be sent to the Regional Administrator.

(f) Notice of the emergency permit shall be published and public comments solicited in accordance with § 233.32 as soon as possible but no later than 10 days after the issuance date.

Section 233.23 Permit conditions

(a) For each permit the Director shall establish conditions which assure compliance with all applicable statutory and regulatory requirements, including the 404(b)(1) Guidelines, applicable section 303 water quality standards, and applicable section 307 effluent standards and prohibitions.

(b) Section 404 permits shall be effective for a fixed term not to exceed 5 years.

(c) Each 404 permit shall include conditions meeting or implementing the following requirements:

(1) A specific identification and complete description of the authorized activity including name and address of permittee, location and purpose of discharge, type and quantity of material to be discharged. (This subsection is not applicable to general permits).

(2) Only the activities specifically described in the permit are authorized.

(3) The permittee shall comply with all conditions of the permit even if that requires halting or reducing the permitted activity to maintain compliance. Any permit violation constitutes a violation of the Act as well as of State statute and/or regulation.

(4) The permittee shall take all reasonable steps to minimize or prevent any discharge in violation of this permit.

(5) The permittee shall inform the Director of any expected or known actual noncompliance.

(6) The permittee shall provide such information to the Director, as the Director requests, to determine compliance status, or whether cause exists–for permit modification, revocation or–termination.

(7) Monitoring, reporting and recordkeeping requirements as needed to safeguard the aquatic environment. (Such requirements will be determined on a case-by-case basis, but at a minimum shall include monitoring and reporting of any expected leachates, reporting of noncompliance, planned changes or transfer of the permit.)

(8) Inspection and entry. The permittee shall allow the Director, or his authorized representative, upon presentation of proper identification, at reasonable times to:

(i) Enter upon the permittee's premises where a regulated activity is located or where records must be kept under the conditions of the permit,

(ii) Have access to and copy any records that must be kept under the conditions of the permit,

(iii) Inspect operations regulated or required under the permit, and

(iv) Sample or monitor, for the purposes of assuring permit compliance or as otherwise authorized by the Act, any substances or parameters at any location.

(9) Conditions assuring that the discharge will be conducted in a manner which minimizes adverse impacts upon the physical, chemical and biological integrity of the waters of the United States, such as requirements for restoration or mitigation.

Subpart D—Program Operation

Section 233.30 Application for a permit

(a) Except when an activity is authorized by a general permit issued pursuant to § 233.21 or is exempt from the requirements to obtain a permit under § 232.3, any person who proposes to discharge dredged or fill material into State regulated waters shall complete, sign and submit a permit application to the Director. Persons proposing to discharge dredged or fill material under the authorization of a general permit must comply with any reporting requirements of the general permit.

(b) A complete application shall include:

(1) Name, address, telephone number of the applicant and name(s) and address(es) of adjoining property owners.

(2) A complete description of the proposed activity including necessary drawings, sketches or plans sufficient for public notice (the applicant is not generally expected to submit detailed engineering plans and specifications); the location, purpose and intended use of the proposed activity; scheduling of the activity; the location and dimensions of adjacent structures; and a list of authorizations required by other Federal, interstate, State or local agencies for the work, including all approvals received or denials already made.

(3) The application must include a description of the type, composition, source and quantity of the material to be discharged, the method of discharge, and the site and plans for disposal of the dredged or fill material.

(4) A certification that all information contained in the application is true and accurate and acknowledging awareness of penalties for submitting false information.

(5) All activities which the applicant plans to undertake which are reasonably related to the same project should be included in the same permit application.

(c) In addition to the information indicated in § 233.30(b), the applicant will be required to furnish such additional information as the Director deems appropriate to assist in the evaluation of the application. Such additional information may include environmental data and information on alternate methods and sites as may be necessary for the preparation of the required environmental documentation.

(d) The level of detail shall be reasonably commensurate with the type and size of discharge, proximity to critical areas, likelihood of long-lived toxic chemical substances, and potential level of environmental degradation.

Note: EPA encourages States to provide permit applicants guidance regarding the level of detail of information and documentation required under this subsection. This guidance can be provided either through the application form or on an individual basis. EPA also encourages the State to maintain a program to inform potential applicants for permits of the requirements of the State program and of the steps required to obtain permits for activities in State regulated waters.

Section 233.31 Coordination requirements

(a) If a proposed discharge may affect the biological, chemical, or physical integrity of the waters of any State(s) other than the State in which the discharge occurs, the Director shall provide an opportunity for such State(s) to submit written comments within the public comment period and to suggest permit conditions. If these recommendations are not accepted by the Director, he shall notify the affected State and the Regional Administrator prior to permit issuance in writing of his failure to accept these recommendations, together with his reasons for so doing. The Regional Administrator shall then have the time provided for in § 233.50(d) to comment upon, object to, or make recommendations.

(b) State section 404 permits shall be coordinated with Federal and Federal-State water related planning and review processes.

Section 233.32 Public notice

(a) Applicability.

(1) The Director shall give public notice of the following actions:

(i) Receipt of a permit application.

(ii) Preparation of a draft general permit.

(iii) Consideration of a major modification to an issued permit.

(iv) Scheduling of a public hearing.

(v) Issuance of an emergency permit.

(2) Public notices may describe more than one permit or action.

(b) Timing.

(1) The public notice shall provide a reasonable period of time, normally at least 30 days, within which interested parties may express their views concerning the permit application.

(2) Public notice of a public hearing shall be given at least 30 days before the hearing.

(3) The Regional Administrator may approve a program with shorter public notice timing if the Regional Administrator determines that sufficient public notice is provided for.

(c) The Director shall give public notice by each of the following methods:

(1) By mailing a copy of the notice to the following persons (any person otherwise entitled to receive notice under this paragraph may waive his rights to receive notice for any classes or categories of permits):

(i) The applicant.

(ii) Any agency with jurisdiction over the activity or the disposal site, whether or not the agency issues a permit.

(iii) Owners of property adjoining the property where the regulated activity will occur.

(iv) All persons who have specifically requested copies of public notices. (The Director may update the mailing list from time to time by requesting written indication of continued interest from those listed. The Director may delete from the list the name of any person who fails to respond to such a request.)

(v) Any State whose waters may be affected by the proposed discharge.

(2) In addition, by providing notice in at least one other way (such as advertisement in a newspaper of sufficient circulation) reasonably calculated to cover the area affected by the activity.

(d) All public notices shall contain at least the following information:

(1) The name and address of the applicant and, if different, the address or location of the activity(ies) regulated by the permit.

(2) The name, address, and telephone number of a person to contact for further information.

(3) A brief description of the comment procedures and procedures to request a public hearing, including deadlines.

(4) A brief description of the proposed activity, its purpose and intended use, so as to provide sufficient information concerning the nature of the activity to generate meaningful comments, including a description of the type of structures, if any, to be erected on fills, and a description of the type, composition and quantity of materials to be discharged.

(5) A plan and elevation drawing showing the general and specific site location and character of all proposed activities, including the size relationship of the proposed structures to the size of the impacted waterway and depth of water in the area.

(6) A paragraph describing the various evaluation factors, including the 404(b)(1) Guidelines or State-equivalent criteria, on which decisions are based.

(7) Any other information which would significantly assist interested parties in evaluating the likely impact of the proposed activity.

(e) Notice of public hearing shall also contain the following information:

(1) Time, date, and place of hearing.

(2) Reference to the date of any previous public notices relating to the permit.

(3) Brief description of the nature and purpose of the hearing.

Section 233.33 Public hearing

(a) Any interested person may request a public hearing during the public comment period as specified in § 233.32. Requests shall be in writing and shall state the nature of the issues proposed to be raised at the hearing.

(b) The Director shall hold a public hearing whenever he determines there is a significant degree of public interest in a permit application or a draft general permit. He may also hold a hearing, at his discretion, whenever he determines a hearing may be useful to a decision on the permit application.

(c) At a hearing, any person may submit oral or written statements or data concerning the permit application or draft general permit. The public comment period shall automatically be extended to the close of any public hearing under this section. The presiding officer may also extend the comment period at the hearing.

(d) All public hearings shall be reported verbatim. Copies of the record of proceedings may be purchased by any person from the Director or the reporter of such hearing. A copy of the transcript (or if none is prepared, a tape of the proceedings) shall be made available for public inspection at an appropriate State office.

Section 233.34 Making a decision on the permit application

(a) The Director will review all applications for compliance with the 404(b)(1) Guidelines and/or equivalent State environmental criteria as well as any other applicable State laws or regulations.

(b) The Director shall consider all comments received in response to the public notice, and public hearing if a hearing is held. All comments, as well as the record of any public hearing, shall be made part of the official record on the application.

(c) After the Director has completed his review of the application and consideration of comments, the Director will determine, in accordance with the record and all applicable regulations, whether or not the permit should be issued. No permit shall be issued by the Director under the circumstances described in § 233.20. The Director shall prepare a written determination on each application outlining his decision and rationale for his decision. The determination shall be dated, signed and included in the official record prior to final action on the application. The official record shall be open to the public.

Section 233.35 Issuance and effective date of permit

(a) If the Regional Administrator comments on a permit application or draft general permit under § 233.50, the Director shall follow the procedures specified in that section in issuing the permit.

(b) If the Regional Administrator does not comment on a permit application or draft general permit, the Director shall make a final permit decision after the close of the public comment period and shall notify the applicant.

(1) If the decision is to issue a permit, the permit becomes effective when it is signed by the Director and the applicant.

(2) If the decision is to deny the permit, the Director will notify the applicant in writing of the reason(s) for denial.

Section 233.36 Modification, suspension or revocation of permits

(a) *General.* The Director may reevaluate the circumstances and conditions of a permit either on his own motion or at the request of the permittee or of a third party and initiate action to modify, suspend, or revoke a permit if he determines that sufficient cause exists. Among the factors to be considered are:

(1) Permittee's noncompliance with any of the terms or conditions of the permit;

(2) Permittee's failure in the application or during the permit issuance process to disclose fully all relevant facts or the permittee's misrepresentation of any relevant facts at the time;

(3) Information that activities authorized by a general permit are having more than minimal individual or cumulative adverse effect on the environment, or that the permitted activities are more appropriately regulated by individual permits;

(4) Circumstances relating to the authorized activity have changed since the permit was issued and justify changed permit conditions or temporary or permanent cessation of any discharge controlled by the permit;

(5) Any significant information relating to the activity authorized by the permit if such information was not available at the time the permit was issued and would have justified the imposition of different permit conditions or denial at the time of issuance;

(6) Revisions to applicable statutory or regulatory authority, including toxic effluent standards or prohibitions or water quality standards.

(b) *Limitations.* Permit modifications shall be in compliance with § 233.20.

(c) *Procedures.*

(1) The Director shall develop procedures to modify, suspend or revoke permits if he determines cause exists for such action (§ 233.36(a)). Such procedures shall provide opportunity for public comment (§ 233.32), coordination with the Federal review agencies (§ 233.50), and opportunity for public hearing (§ 233.33) following notification of the permittee. When permit modification is proposed, only the conditions subject to modification need be reopened.

(2) Minor modification of permits. The Director may, upon the consent of the permittee, use abbreviated procedures to modify a permit to make the following corrections or allowance for changes in the permitted activity:

(i) Correct typographical errors;

(ii) Require more frequent monitoring or reporting by permittee;

(iii) Allow for a change in ownership or operational control of a project or activity where the Director determines that no other change in the permit is necessary, provided that a written agreement containing a specific date for transfer of permit responsibility, coverage, and liability between the current and new permittees has been submitted to the Director;

(iv) Provide for minor modification of project plans that do not significantly change the character, scope, and/or purpose of the project or result in significant change in environmental impact;

(v) Extend the term of a permit, so long as the modification does not extend the term of the permit beyond 5 years from its original effective date and does not result in any increase in the amount of dredged or fill material allowed to be discharged.

Section 233.37 Signatures on permit applications and reports

The application and any required reports must be signed by the person who desires to undertake the proposed activity or by that person's duly authorized agent if accompanied by a statement by that person designating the agent. In either case, the signature of the applicant or the agent will be understood to be an affirmation that he possesses or represents the person who possesses the requisite property interest to undertake the activity proposed in the application.

Section 233.38 Continuation of expiring permits

A Corps 404 permit does not continue in force beyond its expiration date under Federal law if, at that time, a State is the permitting authority. States authorized to administer the 404 Program may continue Corps or State-issued permits until the effective date of the new permits, if State law allows. Otherwise, the discharge is being conducted without a permit from the time of expiration of the old permit to the effective date of a new State-issued permit, if any.

Subpart E—Compliance Evaluation and Enforcement

Section 233.40 Requirements for compliance evaluation programs

(a) In order to abate violations of the permit program, the State shall maintain a program designed to identify persons subject to regulation who have failed to obtain a permit or to comply with permit conditions.

(b) The Director and State officers engaged in compliance evaluation, upon presentation of proper identification, shall have authority to enter any site or premises subject to regulation or in which records relevant to program operation are kept in order to copy any records, inspect, monitor or otherwise investigate compliance with the State program.

(c) The State program shall provide for inspections to be conducted, samples to be taken and other information to be gathered in a manner that will produce evidence admissible in an enforcement proceeding.

(d) The State shall maintain a program for receiving and ensuring proper consideration of information submitted by the public about violations.

Section 233.41 Requirements for enforcement authority

(a) Any State agency administering a program shall have authority:

(1) To restrain immediately and effectively any person from engaging in any unauthorized activity;

(2) To sue to enjoin any threatened or continuing violation of any program requirement;

(3) To assess or sue to recover civil penalties and to seek criminal remedies, as follows:

(i) The agency shall have the authority to assess or recover civil penalties for discharges of dredged or fill material without a required permit or in violation of any section 404 permit condition in an amount of at least $5,000 per day of such violation.

(ii) The agency shall have the authority to seek criminal fines against any person who willfully or with criminal negligence discharges dredged or fill material without a required permit or violates any permit condition issued under section 404 in the amount of at least $10,000 per day of such violation.

(iii) The agency shall have the authority to seek criminal fines against any person who knowingly makes false statements, representation, or certification in any application, record, report, plan, or other document filed or required to be maintained under the Act, these regulations or the approved State program, or who falsifies, tampers with, or knowingly renders inaccurate any monitoring device or method required to be maintained under the permit, in an amount of at least $5,000 for each instance of violation.

(b) (1) The approved maximum civil penalty or criminal fine shall be assessable for each violation and, if the violation is continuous, shall be assessable in that maximum amount for each day of violation.

(2) The burden of proof and degree of knowledge or intent required under State law for establishing violations under paragraph (a)(3) of this section, shall be no greater than the burden of proof or degree of knowledge or intent EPA must bear when it brings an action under the Act.

(c) The civil penalty assessed, sought, or agreed upon by the Director under paragraph (a)(3) of this section shall be appropriate to the violation.

Note: To the extent that State judgments or settlements provide penalties in amounts which EPA believes to be substantially inadequate in comparison to the amounts which EPA would require under similar facts, EPA may, when authorized by section 309 of the Act, commence separate action for penalties.

(d) (1) The Regional Administrator may approve a State program where the State lacks authority to recover penalties of the levels required under paragraphs (a)(3)(i)–(iii) of this section only if the Regional Administrator determines, after evaluating a record of at least one year for an alternative enforcement program, that the State has an alternate, demonstrably effective method of ensuring compliance which has both punitive and deterrence effects.

(2) States whose programs were approved via waiver of monetary penalties shall keep the Regional Administrator informed of all enforcement actions taken under any alternative method approved pursuant to paragraph (d)(1) of this section. The manner of reporting will be established in the Memorandum of Agreement with the Regional Administrator (§ 233.13).

(e) Any State administering a program shall provide for public participation in the State enforcement process by providing either:

(1) Authority which allows intervention of right in any civil or administrative action to obtain remedies specified in paragraph (a)(3) of this section by any citizen having an interest which is or may be adversely affected, or

(2) Assurance that the State agency or enforcement authority will:

(i) Investigate and provide written responses to all citizen complaints submitted pursuant to State procedures;

(ii) Not oppose intervention by any citizen when permissive intervention may be authorized by statute, rule, or regulation; and

(iii) Publish notice of and provide at least 30 days for public comment on any proposed settlement of a State enforcement action.

(f) *Provision for Tribal criminal enforcement authority.* To the extent that an Indian Tribe does not assert or is precluded from asserting criminal enforcement authority (§ 233.41(a)(3) (ii) and (iii)), the Federal government will continue to exercise primary criminal enforcement responsibility. The Tribe, with the EPA Region and Corps District(s) with jurisdiction, shall develop a system where the Tribal agency will refer such a violation to the Regional Administrator or the District Engineer(s), as agreed to by the parties, in an appropriate and timely manner. This agreement shall be incorporated into joint or separate Memorandum of Agreement with the EPA Region and the Corps District(s), as appropriate.

[53 FR 20776, June 1, 1988, as amended at 58 FR 8183, Feb. 11, 1993]

Subpart F—Federal Oversight

Section 233.50 Review of and objection to State permits

(a) The Director shall promptly transmit to the Regional Administrator:

(1) A copy of the public notice for any complete permit applications received by the Director, except those for which permit review has been waived under § 233.51. The State shall supply the Regional Administrator with copies of public notices for permit applications for which permit review has been waived whenever requested by EPA.

(2) A copy of a draft general permit whenever the State intends to issue a general permit.

(3) Notice of every significant action taken by the State agency related to the consideration of any permit application except those for which Federal review has been waived or draft general permit.

(4) A copy of every issued permit.

(5) A copy of the Director's response to another State's comments/recommendations, if the Director does not accept these recommendations (§ 233.32(a)).

(b) Unless review has been waived under § 233.51, the Regional Administrator shall provide a copy of each public notice, each draft general permit, and other information needed for review of the application to the Corps, FWS, and NMFS, within 10 days of receipt. These agencies shall notify the Regional Administrator within 15 days of their receipt if they wish to comment on the public notice or draft general permit. Such agencies should submit their evaluation and comments to the Regional Administrator within 50 days of such receipt. The final decision to comment, object or to require permit conditions shall be made by the Regional Administrator. (These times may be shortened by mutual agreement of the affected Federal agencies and the State.)

(c) If the information provided is inadequate to determine whether the permit application or draft general permit meets the requirements of the Act, these regulations, and the 404(b)(1) Guidelines, the Regional Administrator may, within 30 days of receipt, request the Director to transmit to the Regional Administrator the complete record of the permit proceedings before the State, or any portions of the record, or other information, including a supplemental application, that the Regional Administrator determines necessary for review.

(d) If the Regional Administrator intends to comment upon, object to, or make recommendations with respect to a permit application, draft general permit, or the Director's failure to accept the recommendations of an affected State submitted pursuant to § 233.31(a), he shall notify the Director of his intent within 30 days of receipt. If the Director has been so notified, the permit shall not be issued until after the receipt of such comments or 90 days of the Regional Administrator's receipt of the public notice, draft general permit or Director's response (§ 233.31(a)), whichever comes first. The Regional Administrator may notify the Director within 30 days of receipt that there is no comment but that he reserves the right to object within 90 days of receipt, based on any new information brought out by the public during the comment period or at a hearing.

(e) If the Regional Administrator has given notice to the Director under paragraph (d) of this section, he shall submit to the Director, within 90 days of receipt of the public notice, draft general permit, or Director's response (§ 233.31(a)), a written statement of his comments, objections, or recommendations; the reasons for the comments, objections, or recommendations; and the actions that must be taken by the Director in order to eliminate any objections. Any such objection shall be based on the Regional Administrator's determination that the proposed permit is (1) the subject of an interstate dispute under § 233.31(a) and/or (2) outside requirements of the Act, these regulations, or the 404(b)(1) Guidelines. The Regional Administrator shall make available upon request a copy of any comment, objection, or recommendation on a permit application or draft general permit to the permit applicant or to the public.

(f) When the Director has received an EPA objection or requirement for a permit condition to a permit application or draft general permit under this section, he shall not issue the permit unless he has taken the steps required by the Regional Administrator to eliminate the objection.

(g) Within 90 days of receipt by the Director of an objection or requirement for a permit condition by the Regional Administrator, the State or any interested person may request that the Regional Administrator hold a public hearing on the objection or requirement. The Regional Administrator shall conduct a public hearing whenever requested by the State proposing to issue the permit, or if warranted by significant public interest based on requests received.

(h) If a public hearing is held under paragraph (g) of this section, the Regional Administrator shall, following that hearing, reaffirm, modify or withdraw the objection or requirement for a permit condition, and notify the Director of this decision.

(1) If the Regional Administrator withdraws his objection or requirement for a permit condition, the Director may issue the permit.

(2) If the Regional Administrator does not withdraw the objection or requirement for a permit condition, the Director must issue a permit revised to satisfy the Regional Administrator's objection or requirement for a permit condition or notify EPA of its intent to deny the permit within 30 days of receipt of the Regional Administrator's notification.

(i) If no public hearing is held under paragraph (g) of this section, the Director within 90 days of receipt of the objection or requirement for a permit condition shall either issue the permit revised to satisfy EPA's objections or notify EPA of its intent to deny the permit.

(j) In the event that the Director neither satisfies EPA's objections or requirement for a permit condition nor denies the permit, the Secretary shall process the permit application.

[53 FR 20776, June 1, 1988; 53 FR 41649, Oct. 24, 1988]

Section 233.51 Waiver of review

(a) The MOA with the Regional Administrator shall specify the categories of discharge for which EPA will waive Federal review of State permit applications. After program approval, the MOA may be modified to reflect any additions or deletions of categories of discharge for which EPA will waive review. The Regional Administrator shall consult with the Corps, FWS, and NMFS prior to specifying or modifying such categories.

(b) With the following exceptions, any category of discharge is eligible for consideration for waiver:

(1) Draft general permits;

(2) Discharges with reasonable potential for affecting endangered or threatened species as determined by FWS;

(3) Discharges with reasonable potential for adverse impacts on waters of another State;

(4) Discharges known or suspected to contain toxic pollutants in toxic amounts (section 101(a)(3) of the Act) or hazardous substances in reportable quantities (section 311 of the Act);

(5) Discharges located in proximity of a public water supply intake;

(6) Discharges within critical areas established under State or Federal law, including but not limited to National and State parks, fish and wildlife sanctuaries and refuges, National and historical monuments, wilderness areas and preserves, sites identified or proposed under the National Historic Preservation Act, and components of the National Wild and Scenic Rivers System.

(c) The Regional Administrator retains the right to terminate a waiver as to future permit actions at any time by sending the Director written notice of termination.

Section 233.52 Program reporting

(a) The starting date for the annual period to be covered by reports shall be established in the Memorandum of Agreement with the Regional Administrator (§ 233.13.)

(b) The Director shall submit to the Regional Administrator within 90 days after completion of the annual period, a draft annual report evaluating the State's administration of its program identifying problems the State has encountered in the administration of its program and recommendations for resolving these problems. Items that shall be addressed in the annual report include an assessment of the cumulative impacts of the State's permit program on the integrity of the State regulated waters; identification of areas of particular concern and/or interest within the State; the number and nature of individual and general permits issued, modified, and denied; number of violations identified and number and nature of enforcement actions taken; number of suspected unauthorized activities reported and nature of action taken; an estimate of extent of activities regulated by general permits; and the number of permit applications received but not yet processed.

(c) The State shall make the draft annual report available for public inspection.

(d) Within 60 days of receipt of the draft annual report, the Regional Administrator will complete review of the draft report and transmit comments, questions, and/or requests for additional evaluation and/or information to the Director.

(e) Within 30 days of receipt of the Regional Administrator's comments, the Director will finalize the annual report, incorporating and/or responding to the Regional Administrator's comments, and transmit the final report to the Regional Administrator.

(f) Upon acceptance of the annual report, the Regional Administrator shall publish notice of availability of the final annual report.

Section 233.53 Withdrawal of program approval

(a) A State with a program approved under this part may voluntarily transfer program responsibilities required by Federal law to the Secretary by taking the following actions, or in such other manner as may be agreed upon with the Administrator.

(1) The State shall give the Administrator and the Secretary 180 days notice of the proposed transfer. The State shall also submit a plan for the orderly transfer of all relevant program information not in the possession of the Secretary (such as permits, permit files, reports, permit applications) which are necessary for the Secretary to administer the program.

(2) Within 60 days of receiving the notice and transfer plan, the Administrator and the Secretary shall evaluate the State's transfer plan and shall identify for the State any additional information needed by the Federal government for program administration.

(3) At least 30 days before the transfer is to occur the Administrator shall publish notice of transfer in the Federal Register and in a sufficient number of the largest newspapers in the State to provide statewide

coverage, and shall mail notice to all permit holders, permit applicants, other regulated persons and other interested persons on appropriate EPA, Corps and State mailing lists.

(b) The Administrator may withdraw program approval when a State program no longer complies with the requirements of this part, and the State fails to take corrective action. Such circumstances include the following:

(1) When the State's legal authority no longer meets the requirements of this part, including:

(i) Failure of the State to promulgate or enact new authorities when necessary; or

(ii) Action by a State legislature or court striking down or limiting State authorities.

(2) When the operation of the State program fails to comply with the requirements of this part, including:

(i) Failure to exercise control over activities required to be regulated under this part, including failure to issue permits;

(ii) Issuance of permits which do not conform to the requirements of this part; or

(iii) Failure to comply with the public participation requirements of this part.

(3) When the State's enforcement program fails to comply with the requirements of this part, including:

(i) Failure to act on violations of permits or other program requirements;

(ii) Failure to seek adequate enforcement penalties or to collect administrative fines when imposed, or to implement alternative enforcement methods approved by the Administrator; or

(iii) Failure to inspect and monitor activities subject to regulation.

(4) When the State program fails to comply with the terms of the Memorandum of Agreement required under § 233.13.

(c) The following procedures apply when the Administrator orders the commencement of proceedings to determine whether to withdraw approval of a State program:

(1) *Order.* The Administrator may order the commencement of withdrawal proceedings on the Administrator's initiative or in response to a petition from an interested person alleging failure of the State to comply with the requirements of this part as set forth in subsection (b) of this section. The Administrator shall respond in writing to any petition to commence withdrawal proceedings. He may conduct an informal review of the allegations in the petition to determine whether cause exists to commence proceedings under this paragraph. The Administrator's order commencing proceedings under this paragraph shall fix a time and place for the commencement of the hearing, shall specify the allegations against the State which are to be considered at the hearing, and shall be published in the Federal Register. Within 30 days after publication of the Administrator's order in the Federal Register, the State shall admit or deny these allegations in a written answer. The party seeking withdrawal of the State's program shall have the burden of coming forward with the evidence in a hearing under this paragraph.

(2) *Definitions.* For purposes of this paragraph the definition of *Administrative Law Judge, Hearing Clerk,* and *Presiding Officer* in 40 CFR 22.03 apply in addition to the following:

(i) *Party* means the petitioner, the State, the Agency, and any other person whose request to participate as a party is granted.

(ii) *Person* means the Agency, the State and any individual or organization having an interest in the subject matter of the proceedings.

(iii) *Petitioner* means any person whose petition for commencement of withdrawal proceedings has been granted by the Administrator.

(3) *Procedures.*

(i) The following provisions of 40 CFR Part 22 [Consolidated Rules of Practice] are applicable to proceedings under this paragraph:

(A) Section 22.02–(use of number/gender);

(B) Section 22.04–(authorities of Presiding Officer);

(C) Section 22.06–(filing/service of rulings and orders);

(D) Section 22.09–(examination of filed documents);

(E) Section 22.19 (a), (b) and (c)–(prehearing conference);

(F) Section 22.22–(evidence);

(G) Section 22.23–(objections/offers of proof);

(H) Section 22.25–(filing the transcript; and

(I) Section 22.26–(findings/conclusions).

(ii) The following provisions are also applicable:

(A) Computation and extension of time.

(1) Computation. In computing any period of time prescribed or allowed in these rules of practice, except as otherwise provided, the day of the event from which the designated period begins to run shall not be included. Saturdays, Sundays, and Federal legal holidays shall be included. When a stated time expires on a Saturday, Sunday or Federal legal holiday, the stated time period shall be extended to include the next business day.

(2) Extensions of time. The Administrator, Regional Administrator, or Presiding Officer, as appropriate, may grant an extension of time for the filing of any pleading, document, or motion

(i) upon timely motion of a party to the proceeding, for good cause shown and after consideration of prejudice to other parties, or

(ii) upon his own motion. Such a motion by a party may only be made after notice to all other parties, unless the movant can show good cause why serving notice is impracticable. The motion shall be filed in advance of the date on which the pleading, document or motion is due to be filed, unless the failure of a party to make timely motion for extension of time was the result of excusable neglect.

(3) The time for commencement of the hearing shall not be extended beyond the date set in the Administrator's order without approval of the Administrator.

(B) Ex parte discussion of proceeding.

At no time after the issuance of the order commencing proceedings shall the Administrator, the Regional Administrator, the Regional Judicial Officer, the Presiding Officer, or any other person who is likely to advise these officials in the decisions on the case, discuss ex parte the merits of the proceeding with any interested person outside the Agency, with any Agency staff member who performs a prosecutorial or investigative function in such proceeding or a factually related proceeding, or with any representative of such person. Any ex parte memorandum or other communication addressed to the Administrator, the Regional Administrator, the Regional Judicial Officer, or the Presiding Officer during the pendency of the proceeding and relating to the merits thereof, by or on behalf of any party shall be regarded as argument made in the proceeding and shall be served upon all other parties. The other parties shall be given an opportunity to reply to such memorandum or communication.

(C) Intervention–

(1) Motion. A motion for leave to intervene in any proceeding conducted under these rules of practice must set forth the grounds for the proposed intervention, the position and interest of the movant and the likely impact that intervention will

have on the expeditious progress of the proceeding. Any person already a party to the proceeding may file an answer to a motion to intervene, making specific reference to the factors set forth in the foregoing sentence and paragraph (b)(3)(ii)(C)(*3*) of this section, within ten (10) days after service of the motion for leave to intervene.

(*2*) However, motions to intervene must be filed within 15 days from the date the notice of the Administrator's order is published in the Federal Register.

(*3*) *Disposition.* Leave to intervene may be granted only if the movant demonstrates that

(*i*) his presence in the proceeding would not unduly prolong or otherwise prejudice the adjudication of the rights of the original parties;

(*ii*) the movant will be adversely affected by a final order; and

(*iii*) the interests of the movant are not being adequately represented by the original parties. The intervenor shall become a full party to the proceeding upon the granting of leave to intervene.

(*4*) *Amicus curiae.* Persons not parties to the proceeding who wish to file briefs may so move. The motion shall identify the interest of the applicant and shall state the reasons why the proposed amicus brief is desirable. If the motion is granted, the Presiding Officer or Administrator shall issue an order setting the time for filing such brief. An amicus curiae is eligible to participate in any briefing after his motion is granted, and shall be served with all briefs, reply briefs, motions, and orders relating to issues to be briefed.

(D) Motions–

(*1*) *General.* All motions, except those made orally on the record during a hearing, shall

(*i*) be in writing;

(*ii*) state the grounds therefore with particularity;

(*iii*) set forth the relief or order sought; and

(*iv*) be accompanied by any affidavit, certificate, other evidence, or legal memorandum relied upon. Such motions shall be served as provided by paragraph (b)(4) of this section.

(*2*) *Response to motions.* A party's response to any written motion must be filed within ten (10) days after service of such motion, unless additional time is allowed for such response. The response shall be accompanied by any affidavit, certificate, other evidence, or legal memorandum relied upon. If no response is filed within the designated period, the parties may be deemed to have waived any objection to the granting of the motion. The Presiding Officer, Regional Administrator, or Administrator, as appropriate, may set a shorter time for response, or make such other orders concerning the disposition of motions as they deem appropriate.

(*3*) *Decision.* The Administrator shall rule on all motions filed or made after service of the recommended decision upon the parties. The Presiding Officer shall rule on all other motions. Oral argument on motions will be permitted where the Presiding Officer, Regional Administrator, or the Administrator considers it necessary or desirable.

(4) *Record of proceedings.*

(i) The hearing shall be either stenographically reported verbatim or tape recorded, and thereupon transcribed by an official reporter designated by the Presiding Officer;

(ii) All orders issued by the Presiding Officer, transcripts of testimony, written statements of position, stipulations, exhibits, motions, briefs, and other written material of any kind submitted in the hearing shall be a part of the record and shall be available for inspection or copying in the Office of the Hearing Clerk, upon payment of costs. Inquiries may be made at the Office of the Administrative Law Judges, Hearing Clerk, 1200 Pennsylvania Ave., NW., Washington, DC 20460;

(iii) Upon notice to all parties the Presiding Officer may authorize corrections to the transcript which involve matters of substance;

(iv) An original and two (2) copies of all written submissions to the hearing shall be filed with the Hearing Clerk;

(v) A copy of each such submission shall be served by the person making the submission upon the Presiding Officer and each party of record. Service under this paragraph shall take place by mail or personal delivery;

(vi) Every submission shall be accompanied by acknowledgement of service by the person served or proof of service in the form of a statement of the date, time, and manner of service and the names of the persons served, certified by the person who made service; and

(vii) The Hearing Clerk shall maintain and furnish to any person upon request, a list containing the name, service address, and telephone number of all parties and their attorneys or duly authorized representatives.

(5) *Participation by a person not a party.* A person who is not a party may, in the discretion of the Presiding Officer, be permitted to make a limited appearance by making an oral or written statement of his/her position on the issues within such limits and on such conditions as may be fixed by the Presiding Officer, but he/she may not otherwise participate in the proceeding.

(6) *Rights of parties.*

(i) All parties to the proceeding may:

(A) Appear by counsel or other representative in all hearing and prehearing proceedings;

(B) Agree to stipulations of facts which shall be made a part of the record.

(7) *Recommended decision.*

(i) Within 30 days after the filing of proposed findings and conclusions and reply briefs, the Presiding Officer shall evaluate the record before him/her, the proposed findings and conclusions and any briefs filed by the parties, and shall prepare a recommended decision, and shall certify the entire record, including the recommended decision, to the Administrator.

(ii) Copies of the recommended decision shall be served upon all parties.

(iii) Within 20 days after the certification and filing of the record and recommended decision, all parties may file with the Administrator exceptions to the recommended decision and a supporting brief.

(8) *Decision by Administrator*

(i) Within 60 days after certification of the record and filing of the Presiding Officer's recommended decision, the Administrator shall review the record before him and issue his own decision.

(ii) If the Administrator concludes that the State has administered the program in conformity with the Act and this part, his decision shall constitute "final agency action" within the meaning of 5 U.S.C. 704.

(iii) If the Administrator concludes that the State has not administered the program in conformity with the Act and regulations, he shall list the deficiencies in the program and provide the State a reasonable time, not to exceed 90 days, to take such appropriate corrective action as the Administrator determines necessary.

(iv) Within the time prescribed by the Administrator the State shall take such appropriate corrective action as required by the Administrator and shall file with the Administrator and all parties a statement certified by the State Director that appropriate corrective action has been taken.

(v) The Administrator may require a further showing in addition to the certified statement that corrective action has been taken.

(vi) If the state fails to take appropriate corrective action and file a certified statement thereof within the time prescribed by the Administrator, the Administrator shall issue a supplementary order withdrawing approval of the State program. If the State takes appropriate corrective action, the Administrator shall issue a supplementary order stating that approval of authority is not withdrawn.

(vii) The Administrator's supplementary order shall constitute final Agency action within the meaning of 5 U.S. 704.

(d) Withdrawal of authorization under this section and the Act does not relieve any person from complying with the requirements of State law, nor does it affect the validity of actions taken by the State prior to withdrawal.

[53 FR 20776, June 1, 1988, as amended at 57 FR 5346, Feb. 13, 1992]

Subpart G—Eligible Indian Tribes

Source: 58 FR 8183, Feb. 11, 1993, unless otherwise noted.

Section 233.60 Requirements for eligibility

Section 518(e) of the CWA, 33 U.S.C. 1378(e), authorizes the Administrator to treat an Indian Tribe as eligible to apply for the 404 permit program under section 404(g)(1) if it meets the following criteria:

(a) The Indian Tribe is recognized by the Secretary of the Interior.

(b) The Indian Tribe has a governing body carrying out substantial governmental duties and powers.

(c) The functions to be exercised by the Indian Tribe pertain to the management and protection of water resources which are held by an Indian Tribe, held by the Untied States in trust for the Indians, held by a member of an Indian Tribe if such property interest is subject to a trust restriction an alienation, or otherwise within the borders of the Indian reservation.

(d) The Indian Tribe is reasonably expected to be capable, in the Administrator's judgment, of carrying out the functions to be exercised, in a manner consistent with the terms and purposes of the Act and applicable regulations, of an effective section 404 dredge and fill permit program.

[58 FR 8183, Feb. 11, 1993, as amended at 59 FR 64345, Dec. 14, 1994]

Section 233.61 Determination of Tribal eligibility

An Indian Tribe may apply to the Regional Administrator for a determination that it meets the statutory criteria which authorize EPA to treat the Tribe in a manner similar to that in which it treats a State, for purposes of the section 404 program. The application shall be concise and describe how the Indian Tribe will meet each of the requirements of § 233.60. The application should include the following information:

(a) A statement that the Tribe is recognized by the Secretary of the Interior.

(b) A descriptive statement demonstrating that the Tribal governing body is currently carrying out substantial governmental duties and powers over a defined area. This Statement should:

(1) Describe the form of the Tribal government.

(2) Describe the types of governmental functions currently performed by the Tribal governing body, such as, but not limited to, the exercise of police powers affecting (or relating to) the health, safety, and welfare of the affected population; taxation; and the exercise of the power of eminent domain; and

(3) Identify the source of the Tribal government's authority to carry out the governmental functions currently being performed.

(c) (1) A map or legal description of the area over which the Indian Tribe asserts regulatory authority pursuant to section 518(e)(2) of the CWA and § 233.60(c);

(2) A statement by the Tribal Attorney General (or equivalent official) which describes the basis for the Tribe's assertion under section 518(e)(2) (including the nature or subject matter of the asserted regulatory authority) which may include a copy of documents such as Tribal constitutions, by-laws, charters, executive orders, codes, ordinances, and/or resolutions which support the Tribe's assertion of authority;

(d) A narrative statement describing the capability of the Indian Tribe to administer an effective 404 permit program. The Statement may include:

(1) A description of the Indian Tribe's previous management experience which may include the administration of programs and services authorized by the Indian Self Determination & Education Act (25 U.S.C. 450 *et seq.*), The Indian Mineral Development Act (25 U.S.C. 2101 *et seq.*), or the Indian Sanitation Facility Construction Activity Act (42 U.S.C. 2004a).

(2) A list of existing environmental or public health programs administered by the Tribal governing body, and a copy of related Tribal laws, regulations, and policies;

(3) A description of the entity (or entities) which exercise the executive, legislative, and judicial functions of the Tribal government.

(4) A description of the existing, or proposed, agency of the Indian Tribe which will assume primary responsibility for establishing and administering a section 404 dredge and fill permit program or plan which proposes how the Tribe will acquire additional administrative and technical expertise. The plan must address how the Tribe will obtain the funds to acquire the administrative and technical expertise.

(5) A description of the technical and administrative abilities of the staff to administer and manage an effective, environmentally sound 404 dredge and fill permit program.

(e) The Administrator may, at his discretion, request further documentation necessary to support a Tribal application.

(f) If the Administrator has previously determined that a Tribe has met the requirements for eligibility or for "treatment as a State" for programs authorized under the Safe Drinking Water Act or the Clean Water Act, then that Tribe need only provide additional information unique to the particular statute or program for which the Tribe is seeking additional authorization.

(Approved by the Office of Management and Budget under control number 2040-0140)

[58 FR 8183, Feb. 11, 1993, as amended at 59 FR 64345, Dec. 14, 1994]

Section 233.62 Procedures for processing an Indian Tribe's application

(a) The Regional Administrator shall process an application of an Indian Tribe submitted pursuant to § 233.61 in a timely manner. He shall promptly notify the Indian Tribe of receipt of the application.

(b) The Regional Administrator shall follow the procedures described in § 233.15 in processing a Tribe's request to assume the 404 dredge and fill permit program.

[58 FR 8183, Feb. 11, 1993, as amended at 59 FR 64346, Dec. 14, 1994]

Subpart H—Approved State Programs

Section 233.70 Michigan

The applicable regulatory program for discharges of dredged or fill material into waters of the United States in Michigan that are not presently used, or susceptible for use in their natural condition or by reasonable improvement as a means to transport interstate or foreign commerce shoreward to the ordinary high water mark, including wetlands adjacent thereto, except those on Indian lands, is the program administered by the Michigan Department of Natural Resources, approved by EPA, pursuant to section 404 of the CWA.

Notice of this approval was published in the Federal Register on October 2, 1984; the effective date of this program is October 16, 1984. This program consists of the following elements, as submitted to EPA in the State's program application.

(a) *Incorporation by reference.* The requirements set forth in the State statutes and regulations cited in this paragraph are hereby incorporated by reference and made a part of the applicable 404 Program under the CWA for the State of Michigan. This incorporation by reference was approved by the Director of the Federal Register on October 16, 1984.

(1) The Great Lakes Submerged Lands Act, MCL 322.701 *et seq.*, reprinted in Michigan 1983 Natural Resources Law.

(2) The Water Resources Commission Act, MCL 323.1 *et seq.*, reprinted in Michigan 1983 Natural Resources Law.

(3) The Goemaere-Anderson Wetland Protection Act, MCL 281.701 *et seq.*, reprinted in Michigan 1983 Natural Resources Law.

(4) The Inland Lakes and Stream Act, MCL 281.951 *et seq.*, reprinted in Michigan 1983 Natural Resources Law.

(5) The Michigan Administrative Procedures Act of 1969, MCL 24-201 *et seq.*

(6) An act concerning the Erection of Dams, MCL 281.131 *et seq.*, reprinted in Michigan 1983 Natural Resources Law.

(7) R 281.811 through R 281.819 inclusive, R 281.821, R 281.823, R 281.824, R 281.832 through R 281.839 inclusive, and R 281.841 through R 281.845 inclusive of the Michigan Administrative Code (1979 ed., 1982 supp.).

(b) *Other Laws.* The following statutes and regulations, although not incorporated by reference, also are part of the approved State-administered program:

(1) Administrative Procedures Act, MCLA 24.201 *et seq.*

(2) Freedom of Information Act, MCLA 15.231 *et seq.*

(3) Open Meetings Act, MCLA 15.261 *et seq.*

(4) Michigan Environmental Protection Act, MCLA 691.1201 *et seq.*

(c) *Memoranda of Agreement.*

(1) The Memorandum of Agreement between EPA Region V and the Michigan Department of Natural Resources, signed by the EPA Region V Administrator on December 9, 1983.

(2) The Memorandum of Agreement between the U.S. Army Corps of Engineers and the Michigan Department of Natural Resources, signed by the Commander, North Central Division, on March 27, 1984.

(d) *Statement of Legal Authority.*

(1) "Attorney General Certification section 404/State of Michigan", signed by Attorney General of Michigan, as submitted with the request for approval of "The State of Michigan 404 Program", October 26, 1983.

(e) The Program description and any other materials submitted as part of the original application or supplements thereto.

(33 U.S.C. 13344, CWA 404)

[49 FR 38948, Oct. 2, 1984. Redesignated at 53 FR 20776, June 6, 1988. Redesignated at 58 FR 8183, Feb. 11, 1993]

Section 233.71 New Jersey

The applicable regulatory program for discharges of dredged or fill material into waters of the United States in New Jersey that are not presently used, or susceptible for use in their natural condition or by reasonable improvement as a means to transport interstate or foreign commerce shoreward to the ordinary high water mark, including wetlands adjacent thereto, except those on Indian lands, is the program administered by the New Jersey Department of Environmental Protection and Energy, approved by EPA, pursuant to section 404 of the CWA. The program becomes effective March 2, 1994. This program consists of the following elements, as submitted to EPA in the State's program application:

(a) *Incorporation by reference.* The requirements set forth in the State statutes and regulations cited in paragraph (b) of this section are hereby incorporated by reference and made a part of the applicable 404 Program under the CWA for the State of New Jersey, for incorporation by reference by the Director of the Federal Register in accordance with 552(a) and 1 CFR part 51. Material is incorporated as it exists at 1 p.m. on March 2, 1994 and notice of any change in the material will be published in the Federal Register.

(b) Copies of materials incorporated by reference may be inspected at the Office of the Federal Register, 800 North Capitol Street, NW., suite 700, Washington, DC. Copies of materials incorporated by reference may be obtained or inspected at the EPA UST Docket, located at 1235 Jefferson Davis Highway, First Floor, Arlington, VA 22202 (telephone number: 703-603-9231), or send mail to Mail Code 5305G, 1200 Pennsylvania Ave., NW., Washington, DC 20460, and at the Library of the Region 2 Regional Office, Federal Office Building, 26 Federal Plaza, New York, NY 10278.

(1) New Jersey Statutory Requirements Applicable to the Freshwater Wetlands Program, 1994.

(2) New Jersey Regulatory Requirements Applicable to the Freshwater Wetlands Program, 1994.

(c) *Other laws.* The following statutes and regulations, although not incorporated by reference, also are part of the approved State-administered program:

(1) Administrative Procedure Act, N.J.S.A. 52:14B-1 *et seq.*

(2) New Jersey Uniform Administrative Procedure Rules, N.J.A.C. 1:1-1.1 *et seq.*

(3) Open Public Meetings Act, N.J.S.A. 10:4-6 *et seq.*

(4) Examination and Copies of Public Records, N.J.S.A. 47:1A-1 *et seq.*

(5) Environmental Rights Act, N.J.S.A. 2A:35A-1 *et seq.*

(6) Department of Environmental Protection (and Energy), N.J.S.A. 13:1D-1 *et seq.*

(7) Water Pollution Control Act, N.J.S.A. 58:10A-1 *et seq.*

(d) *Memoranda of agreement.* The following memoranda of agreement, although not incorporated by reference also are part of the approved State administered program:

(1) The Memorandum of Agreement between EPA Region II and the New Jersey Department of Environmental Protection and Energy, signed by the EPA Region II Acting Regional Administrator on June 15, 1993.

(2) The Memorandum of Agreement between the U.S. Army Corps of Engineers and the New Jersey Department of Environmental Protection and Energy, signed by the Division Engineer on March 4, 1993.

(3) The Memorandum of Agreement between EPA Region II, the New Jersey Department of Environmental Protection and Energy, and the U.S. Fish and Wildlife Service, signed by all parties on December 22, 1993.

(e) *Statement of legal authority.* The following documents, although not incorporated by reference, also are part of the approved State administered program:

(1) Attorney General's Statement, signed by the Attorney General of New Jersey, as submitted with the request for approval of The State of New Jersey's 404 Program.

(2) The program description and any other materials submitted as part of the original application or supplements thereto.

[59 FR 9933, Mar. 2, 1994, as amended at 65 FR 47325, Aug. 2, 2000]

APPENDIX E

Nationwide Permits Issued January 15, 2002

Per February 13, 2002 Corrections[1]

Federal Register
Volume 67, Number 10
Pages 2077–2095
Tuesday, January 15, 2002

Department of the Army, Corps of Engineers Issuance of Nationwide Permits; Notice

Nationwide Permits, Conditions, Further Information, and Definitions

A. Index of Nationwide Permits, Conditions, Further Information, and Definitions

Nationwide Permits

1. Aids to Navigation
2. Structures in Artificial Canals
3. Maintenance
4. Fish and Wildlife Harvesting, Enhancement, and Attraction Devices and Activities
5. Scientific Measurement Devices
6. Survey Activities
7. Outfall Structures and Maintenance
8. Oil and Gas Structures
9. Structures in Fleeting and Anchorage Areas
10. Mooring Buoys
11. Temporary Recreational Structures
12. Utility Line Activities
13. Bank Stabilization
14. Linear Transportation Projects
15. U.S. Coast Guard Approved Bridges
16. Return Water from Upland Contained Disposal Areas
17. Hydropower Projects
18. Minor Discharges
19. Minor Dredging
20. Oil Spill Cleanup
21. Surface Coal Mining Activities
22. Removal of Vessels
23. Approved Categorical Exclusions
24. State Administered Section 404 Programs
25. Structural Discharges
26. [Reserved]
27. Stream and Wetland Restoration Activities
28. Modifications of Existing Marinas
29. Single-family Housing
30. Moist Soil Management for Wildlife
31. Maintenance of Existing Flood Control Facilities
32. Completed Enforcement Actions
33. Temporary Construction, Access and Dewatering
34. Cranberry Production Activities
35. Maintenance Dredging of Existing Basins
36. Boat Ramps
37. Emergency Watershed Protection and Rehabilitation
38. Cleanup of Hazardous and Toxic Waste
39. Residential, Commercial, and Institutional Developments
40. Agricultural Activities
41. Reshaping Existing Drainage Ditches
42. Recreational Facilities
43. Stormwater Management Facilities
44. Mining Activities

Nationwide Permit General Conditions

1. Navigation
2. Proper Maintenance
3. Soil Erosion and Sediment Controls
4. Aquatic Life Movements
5. Equipment
6. Regional and Case-by-Case Conditions
7. Wild and Scenic Rivers
8. Tribal Rights
9. Water Quality
10. Coastal Zone Management
11. Endangered Species
12. Historic Properties
13. Notification
14. Compliance Certification
15. Use of Multiple Nationwide Permits
16. Water Supply Intakes
17. Shellfish Beds
18. Suitable Material
19. Mitigation
20. Spawning Areas
21. Management of Water Flows

1. This document contains corrections to the January 15, 2002 NWPs per the Federal Register, volume 67, number 20, pages 6692–6695, Wednesday, February 13, 2002, Department of the Army, Corps of Engineers, Issuance of Nationwide Permits; Notice; Correction. Corrections are presented throughout the text bracketed and underlined. *See* www.usace.army.mil/inet/functions/cw/cecwo/reg/citizen.htm for more recent regulatory announcements and decisions.

22. Adverse Effects from Impoundments
23. Waterfowl Breeding Areas
24. Removal of Temporary Fills
25. Designated Critical Resource Waters
26. Fills Within 100-year Floodplains
27. Construction Period

Further Information

Definitions

Best Management Practices (BMPs)
Compensatory Mitigation
Creation
Enhancement
Ephemeral Stream
Farm Tract
Flood Fringe
Floodway
Independent Utility
Intermittent Stream
Loss of Waters of the US
Non-tidal Wetland
Open Water
Perennial Stream
Permanent Above-grade Fill
Preservation
Restoration
Riffle and Pool Complex
Single and Complete Project
Stormwater Management
Stormwater Management Facilities
Stream Bed
Stream Channelization
Tidal Wetland
Vegetated Buffer
Vegetated Shallows
Waterbody

B. Nationwide Permits

1. Aids to Navigation. The placement of aids to navigation and Regulatory markers which are approved by and installed in accordance with the requirements of the U.S. Coast Guard (USCG) (See 33 CFR, chapter I, subchapter C part 66). (Section 10)

2. Structures in Artificial Canals. Structures constructed in artificial canals within principally residential developments where the connection of the canal to navigable water of the US has been previously authorized (see 33 CFR 322.5(g)). (Section 10)

3. Maintenance. Activities related to:

(i) The repair, rehabilitation, or replacement of any previously authorized, currently serviceable, structure, or fill, or of any currently serviceable structure or fill authorized by 33 CFR 330.3, provided that the structure or fill is not to be put to uses differing from those uses specified or contemplated for it in the original permit or the most recently authorized modification. Minor deviations in the structure's configuration or filled area including those due to changes in materials, construction techniques, or current construction codes or safety standards which are necessary to make repair, rehabilitation, or replacement are permitted, provided the adverse environmental effects resulting from such repair, rehabilitation, or replacement are minimal. Currently serviceable means useable as is or with some maintenance, but not so degraded as to essentially require reconstruction. This NWP authorizes the repair, rehabilitation, or replacement of those structures or fills destroyed or damaged by storms, floods, fire or other discrete events, provided the repair, rehabilitation, or replacement is commenced, or is under contract to commence, within two years of the date of their destruction or damage. In cases of catastrophic events, such as hurricanes or tornadoes, this two-year limit may be waived by the District Engineer, provided the permittee can demonstrate funding, contract, or other similar delays.

(ii) Discharges of dredged or fill material, including excavation, into all waters of the US to remove accumulated sediments and debris in the vicinity of, and within, existing structures (e.g., bridges, culverted road crossings, water intake structures, etc.) and the placement of new or additional riprap to protect the structure, provided the permittee notifies the District Engineer in accordance with General Condition 13. The removal of sediment is limited to the minimum necessary to restore the waterway in the immediate vicinity of the structure to the approximate dimensions that existed when the structure was built, but cannot extend further than 200 feet in any direction from the structure. The placement of rip rap must be the minimum necessary to protect the structure or to ensure the safety of the structure. All excavated materials must be deposited and retained in an upland area unless otherwise specifically approved by the District Engineer under separate authorization. Any bank stabilization measures not directly associated with the structure will require a separate authorization from the District Engineer.

(iii)Discharges of dredged or fill material, including excavation, into all waters of the US for activities associated with the restoration of upland areas damaged by a storm, flood, or other discrete event, including the construction, placement, or installation of upland protection structures and minor dredging to remove obstructions in a water of the US. (Uplands lost as a result of a storm, flood, or other discrete event can be replaced without a Section 404 permit provided the uplands are restored to their original pre-event location. This NWP is for the activities in waters of the US associated with the replacement of the uplands.) The permittee must notify the District Engineer, in accordance with General Condition 13, within 12-months of the date of the damage and the work must commence, or be under contract to commence, within two years of the date of the damage. The permittee should provide evidence, such as a recent topographic survey or photographs, to justify the extent of the proposed restoration. The restoration of the damaged areas cannot exceed the contours, or ordinary high water mark, that existed before the damage. The District Engineer retains the right to determine the extent of the pre-existing conditions and the extent of any restoration work authorized by this permit. Minor dredging to remove obstructions from the adjacent waterbody is limited to 50 cubic yards below the plane of the ordinary high water mark, and is limited to the amount necessary to restore the preexisting bottom contours of the waterbody. The dredging may not be done primarily to obtain fill for any restoration activities. The discharge of dredged or fill material and all related work needed to restore the upland must be part of a single and complete project. This permit cannot be used in conjunction with NWP 18 or NWP 19 to restore damaged upland areas. This permit cannot be used to reclaim historic lands lost, over an extended period, to normal erosion processes. This permit does not authorize maintenance dredging for the primary purpose of navigation and beach restoration.

This permit does not authorize new stream channelization or stream relocation projects. Any work authorized by this permit must not cause more than minimal degradation of water quality, more than minimal changes to the flow characteristics of the stream, or increase flooding (See General Conditions 9 and 21). (Sections 10 and 404)

Note: This NWP authorizes the repair, rehabilitation, or replacement of any previously authorized structure or fill that does not qualify for the Section 404(f) exemption for maintenance.

4. Fish and Wildlife Harvesting, Enhancement, and Attraction Devices and Activities. Fish and wildlife harvesting devices and activities such as pound nets, crab traps, crab dredging, eel pots, lobster traps, duck blinds, clam and oyster digging; and small fish attraction devices such as open water fish concentrators (sea kites, etc.). This NWP authorizes shellfish seeding provided this activity does not occur in wetlands or sites that support submerged aquatic vegetation (including sites where submerged aquatic vegetation is documented to exist, but may not be present in a given year.). This NWP does not authorize artificial reefs or impoundments and semi-impoundments of waters of the US for the culture or holding of motile species such as lobster or the use of covered oyster trays or clam racks. (Sections 10 and 404)

5. Scientific Measurement Devices. Devices, whose purpose is to measure and record scientific data such as staff gages, tide gages, water recording devices, water quality testing and improvement devices and similar structures. Small weirs and flumes constructed primarily to record water quantity and velocity are also authorized provided the discharge is limited to 25 cubic yards and further for discharges of 10 to 25 cubic yards provided the permittee notifies the District Engineer in accordance with the "Notification" General Condition. (Sections 10 and 404)

6. Survey Activities. Survey activities including core sampling, seismic exploratory operations, plugging of seismic shot holes and other exploratory-type bore holes, soil survey, sampling, and historic resources surveys. Discharges and structures associated with the recovery of historic resources are not authorized by this NWP. Drilling and the discharge of excavated material from test wells for oil and gas exploration is not authorized by this NWP; the plugging of such wells is authorized. Fill placed for roads, pads and other similar activities is not authorized by this NWP. The NWP does not authorize any permanent structures. The discharge of drilling mud and cuttings may require a permit under section 402 of the CWA. (Sections 10 and 404)

7. Outfall Structures and Maintenance. Activities related to:

(i) Construction of outfall structures and associated intake structures where the effluent from the outfall is authorized, conditionally authorized, or specifically exempted, or are otherwise in compliance with regulations issued under the National Pollutant Discharge Elimination System Program (Section 402 of the CWA), and

(ii) Maintenance excavation, including dredging, to remove accumulated sediments blocking or restricting outfall and intake structures, accumulated sediments from small impoundments associated with outfall and intake structures, and accumulated sediments from canals associated with outfall and intake structures, provided that the activity meets all of the following criteria:

a. The permittee notifies the District Engineer in accordance with General Condition 13;

b. The amount of excavated or dredged material must be the minimum necessary to restore the outfalls, intakes, small impoundments, and canals to original design capacities and design configurations (i.e., depth and width);

c. The excavated or dredged material is deposited and retained at an upland site, unless otherwise approved by the District Engineer under separate authorization; and

d. Proper soil erosion and sediment control measures are used to minimize reentry of sediments into waters of the US.

The construction of intake structures is not authorized by this NWP, unless they are directly associated with an authorized outfall structure. For maintenance excavation and dredging to remove accumulated sediments, the notification must include information regarding the original design capacities and configurations of the facility and the presence of special aquatic sites (e.g., vegetated shallows) in the vicinity of the proposed work. (Sections 10 and 404)

8. Oil and Gas Structures. Structures for the exploration, production, and transportation of oil, gas, and minerals on the outer continental shelf within areas leased for such purposes by the DOI, Minerals Management Service (MMS). Such structures shall not be placed within the limits of any designated shipping safety fairway or traffic separation scheme, except temporary anchors that comply with the fairway regulations in 33 CFR 322.5(l). (Where such limits have not been designated, or where changes are anticipated, District Engineers will consider asserting discretionary authority in accordance with 33 CFR 330.4(e) and will also review such proposals to ensure they comply with the provisions of the fairway regulations in 33 CFR 322.5(l). Any Corps review under this permit will be limited to the effects on navigation and national security in accordance with 33 CFR 322.5(f)). Such structures will not be placed in established danger zones or restricted areas as designated in 33 CFR part 334: nor will such structures be permitted in EPA or Corps designated dredged material disposal areas. (Section 10)

9. Structures in Fleeting and Anchorage Areas. Structures, buoys, floats and other devices placed within anchorage or fleeting areas to facilitate moorage of vessels where the USCG has established such areas for that purpose. (Section 10)

10. Mooring Buoys. Non-commercial, single-boat, mooring buoys. (Section 10)

11. Temporary Recreational Structures. Temporary buoys, markers, small floating docks, and similar structures placed for recreational use during specific events such as water skiing competitions and boat races or seasonal use provided that such structures are removed within 30 days after use has been discontinued. At Corps of Engineers reservoirs, the reservoir manager must approve each buoy or marker individually. (Section 10)

12. Utility Line Activities. Activities required for the construction, maintenance and repair of utility lines and associated facilities in waters of the US as follows:

(i) Utility lines: The construction, maintenance, or repair of utility lines, including outfall and intake structures and the associated excavation, backfill, or bedding for the utility lines, in all waters of the US, provided there is no change in preconstruction contours. A "utility line" is defined as any pipe or pipeline for the transportation of any gaseous, liquid, liquescent, or slurry substance, for any purpose, and any cable, line, or wire for the transmission for any purpose of electrical energy, telephone, and telegraph messages, and radio and television communication (see Note 1, below). Material resulting from trench excavation may be temporarily sidecast (up to three months) into waters of the US, provided that the material is not placed in such a manner that it is dispersed by currents or other forces. The District Engineer may extend the period of temporary side casting not to exceed a total of 180 days, where appropriate. In wetlands, the top 6_to 12_of the trench should normally be backfilled with topsoil from the trench. Furthermore, the trench cannot be constructed in such a manner as to drain waters of the US (e.g., backfilling with extensive gravel layers, creating a french drain effect). For example, utility line trenches can be backfilled with clay blocks to ensure that the trench does not drain the waters of the US through which the utility line is installed. Any exposed slopes and stream banks must be stabilized immediately upon completion of the utility line crossing of each waterbody.

(ii) Utility line substations: The construction, maintenance, or expansion of a substation facility associated with a power line or utility line in non-tidal waters of the US, excluding non-tidal wetlands adjacent to tidal waters, provided the activity does not result in the loss of greater than 1/2-acre of non-tidal waters of the US.

(iii) Foundations for overhead utility line towers, poles, and anchors: The construction or maintenance of foundations for overhead utility line towers, poles, and anchors in all waters of the US, provided the foundations are the minimum size necessary and separate footings for each tower leg (rather than a larger single pad) are used where feasible.

(iv) Access roads: The construction of access roads for the construction and maintenance of utility lines, including overhead power lines and utility line substations, in non-tidal waters of the US, excluding non-tidal wetlands adjacent to tidal waters, provided the discharges do not cause the loss of greater than 1/2-acre of non-tidal waters of the US. Access roads shall be the minimum width necessary (see Note 2, below). Access roads must be constructed so that the length of the road minimizes the adverse effects on waters of the US and as near as possible to preconstruction contours and elevations (e.g., at grade corduroy roads or geotextile/gravel roads). Access roads constructed above preconstruction contours and elevations in waters of the US must be properly bridged or culverted to maintain surface flows.

The term "utility line" does not include activities which drain a water of the US, such as drainage tile, or french drains; however, it does apply to pipes conveying drainage from another area. For the purposes of this NWP, the loss of waters of the US includes the filled area plus waters of the US that are adversely affected by flooding, excavation, or drainage as a result of the project. Activities authorized by paragraph (i) through (iv) may not exceed a total of 1/2-acre loss of waters of the US. Waters of the US temporarily affected by filling, flooding, excavation, or drainage, where the project area is restored to preconstruction contours and elevation, is not included in the calculation of permanent loss of waters of the US. This includes temporary construction mats (e.g., timber, steel, geotextile) used during construction and removed upon completion of the work. Where certain functions and values of waters of the US are permanently adversely affected, such as the conversion of a forested wetland to a herbaceous wetland in the permanently maintained utility line right-of-way, mitigation will be required to reduce the adverse effects of the project to the minimal level.

Mechanized land clearing necessary for the construction, maintenance, or repair of utility lines and the construction, maintenance and expansion of utility line substations, foundations for overhead utility lines, and access roads is authorized, provided the cleared area is kept to the minimum necessary and preconstruction contours are maintained as near as possible. The area of waters of the US that is filled, excavated, or flooded must be limited to the minimum necessary to construct the utility line, substations, foundations, and access roads. Excess material must be removed to upland areas immediately upon completion of construction. This NWP may authorize utility lines in or affecting navigable waters of the US even if there is no associated discharge of dredged or fill material (See 33 CFR part 322).

Notification: The permittee must notify the District Engineer in accordance with General Condition 13, if any of the following criteria are met:

(a) Mechanized land clearing in a forested wetland for the utility line right-of-way;

(b) A Section 10 permit is required;

(c) The utility line in waters of the US, excluding overhead lines, exceeds 500 feet;

(d) The utility line is placed within a jurisdictional area (i.e., water of the US), and it runs parallel to a stream bed that is within that jurisdictional area;

(e) Discharges associated with the construction of utility line substations that result in the loss of greater than 1/10-acre of waters of the US; or [Delete "or"]

(f) Permanent access roads constructed above grade in waters of the US for a distance of more than 500 feet. [Replace with: "; or"]

(g) Permanent access roads constructed in waters of the US with impervious materials. (Sections 10 and 404)

Note 1: Overhead utility lines constructed over Section 10 waters and utility lines that are routed in or under Section 10 waters without a discharge of dredged or fill material require a Section 10 permit; except for pipes or pipelines used to transport gaseous, liquid, liquescent, or slurry substances over navigable waters of the US, which are considered to be bridges, not utility lines, and may require a permit from the USCG pursuant to section 9 of the Rivers and Harbors Act of 1899. However, any discharges of dredged or fill material associated with such pipelines will require a Corps permit under Section 404.

Note 2: Access roads used for both construction and maintenance may be authorized, provided they meet the terms and conditions of this NWP. Access roads used solely for construction of the utility line must be removed upon completion of the work and the area restored to preconstruction contours, elevations, and wetland conditions. Temporary access roads for construction may be authorized by NWP 33.

Note 3: Where the proposed utility line is constructed or installed in navigable waters of the US (*i.e.,* Section 10 waters), copies of the PCN and NWP verification will be sent by the Corps to the National Oceanic and Atmospheric Administration (NOAA), National Ocean Service (NOS), for charting the utility line to protect navigation.

13. Bank Stabilization. Bank stabilization activities necessary for erosion prevention provided the activity meets all of the following criteria:

a. No material is placed more than the minimum needed for erosion protection; [Replace with: "No material is placed in excess of the minimum needed for erosion protection;"]

b. The bank stabilization activity is less than 500 feet in length;

c. The activity will not exceed an average of one cubic yard per running foot placed along the bank below the plane of the ordinary high water mark or the high tide line;

d. No material is placed in any special aquatic site, including wetlands;

e. No material is of the type, or is placed in any location, or in any manner, to impair surface water flow into or out of any wetland area;

f. No material is placed in a manner that will be eroded by normal or expected high flows (properly anchored trees and treetops may be used in low energy areas); and,

g. The activity is part of a single and complete project.

Bank stabilization activities in excess of 500 feet in length or greater than an average of one cubic yard per running foot may be authorized if the permittee notifies the District Engineer in accordance with the "*Notification*" General Condition 13 and the District Engineer determines the activity complies with the other terms and conditions of the NWP and the adverse environmental effects are minimal both individually and cumulatively. This NWP may not be used for the channelization of waters of the US. (Sections 10 and 404)

14. Linear Transportation Projects. Activities required for the construction, expansion, modification, or improvement of linear transportation crossings (e.g., highways, railways, trails, airport runways, and taxiways) in waters of the US, including wetlands, if the activity meets the following criteria:

a. This NWP is subject to the following acreage limits:

(1) For linear transportation projects in non-tidal waters, provided the discharge does not cause the loss of greater than 1/2-acre of waters of the US; [Insert: "or"]

(2) For linear transportation projects in tidal waters, provided the discharge does not cause the loss of greater than 1/3-acre of waters of the US.

b. The permittee must notify the District Engineer in accordance with General Condition 13 if any of the following criteria are met:

(1) The discharge causes the loss of greater than 1/10-acre of waters of the US; or

(2) There is a discharge in a special aquatic site, including wetlands;

c. The *notification* must include a compensatory mitigation proposal to offset permanent losses of waters of the US to ensure that those losses result only in minimal adverse effects to the aquatic environment and a statement describing how temporary losses will be minimized to the maximum extent practicable;

d. For discharges in special aquatic sites, including wetlands, and stream riffle and pool complexes, the *notification* must include a delineation of the affected special aquatic sites;

e. The width of the fill is limited to the minimum necessary for the crossing;

f. This permit does not authorize stream channelization, and the authorized activities must not cause more than minimal changes to the hydraulic flow characteristics of the stream, increase flooding, or cause more than minimal degradation of water quality of any stream (see General Conditions 9 and 21);

g. This permit cannot be used to authorize non-linear features commonly associated with transportation projects, such as vehicle maintenance or storage buildings, parking lots, train stations, or aircraft hangars; and

h. The crossing is a single and complete project for crossing waters of the US. Where a road segment (*i.e.*, the shortest segment of a road with independent utility that is part of a larger project) has multiple crossings of streams (several single and complete projects) the Corps will consider whether it should use its discretionary authority to require an Individual Permit. (Sections 10 and 404)

Note: Some discharges for the construction of farm roads, forest roads, or temporary roads for moving mining equipment may be eligible for an exemption from the need for a Section 404 permit (see 33 CFR 323.4).

15. U.S. Coast Guard Approved Bridges. Discharges of dredged or fill material incidental to the construction of bridges across navigable waters of the US, including cofferdams, abutments, foundation seals, piers, and temporary construction and access fills provided such discharges have been authorized by the USCG as part of the bridge permit. Causeways and approach fills are not included in this NWP and will require an individual or regional Section 404 permit. (Section 404)

16. Return Water From Upland Contained Disposal Areas. Return water from upland, contained dredged material disposal area. The dredging itself may require a Section 404 permit (33 CFR 323.2(d)), but will require a Section 10 permit if located in navigable waters of the US. The return water from a contained disposal area is administratively defined as a discharge of dredged material by 33 CFR 323.2(d), even though the disposal itself occurs on the upland and does not require a Section 404 permit. This NWP satisfies the technical requirement for a Section 404 permit for the return water where the quality of the return water is controlled by the state through the Section 401 certification procedures. (Section 404)

17. Hydropower Projects. Discharges of dredged or fill material associated with (a) small hydropower projects at existing reservoirs where the project, which includes the fill, are licensed by the Federal Energy Regulatory Commission (FERC) under the Federal Power Act of 1920, as amended; and has a total generating capacity of not more than 5000 kW; and the permittee notifies the District Engineer in accordance with the "Notification" General Condition; or (b) hydropower projects for which the FERC has granted an exemption from licensing pursuant to section 408 of the Energy Security Act of 1980 (16 U.S.C. 2705 and 2708) and section 30 of the Federal Power Act, as amended; provided the permittee notifies the District Engineer in accordance with the "Notification" General Condition. (Section 404)

18. Minor Discharges. Minor discharges of dredged or fill material into all waters of the US if the activity meets all of the following criteria:

a. The quantity of discharged material and the volume of area excavated do not exceed 25 cubic yards below the plane of the ordinary high water mark or the high tide line;

b. The discharge, including any excavated area, will not cause the loss of more than 1/10-acre of a special aquatic site, including wetlands. For the purposes of this NWP, the acreage limitation includes the filled area and excavated area plus special aquatic sites that are adversely affected by flooding and special aquatic sites that are drained so that they would no longer be a water of the US as a result of the project;

c. If the discharge, including any excavated area, exceeds 10 cubic yards below the plane of the ordinary high water mark or the high tide line or if the discharge is in a special aquatic site, including wetlands, the permittee notifies the District Engineer in accordance with the "Notification" General Condition. For discharges in special aquatic sites, including wetlands, the notification must also include a delineation of affected special aquatic sites, including wetlands (also see 33 CFR 330.1(e)); and

d. The discharge, including all attendant features, both temporary and permanent, is part of a single and complete project and is not placed for the purpose of a stream diversion. (Sections 10 and 404)

19. Minor Dredging. Dredging of no more than 25 cubic yards below the plane of the ordinary high water mark or the mean high water mark from navigable waters of the US (i.e., Section 10 waters) as part of a single and complete project. This NWP does not authorize the dredging or degradation through siltation of coral reefs, sites that support submerged aquatic vegetation (including sites where submerged aquatic vegetation is documented to exist, but may not be present in a given year), anadromous fish spawning areas, or wetlands, or the connection of canals or other artificial waterways to navigable waters of the US (see 33 CFR 322.5(g)). (Sections 10 and 404)

20. Oil Spill Cleanup. Activities required for the containment and cleanup of oil and hazardous substances which are subject to the National Oil and Hazardous Substances Pollution Contingency Plan (40 CFR part 300) provided that the work is done in accordance with the Spill Control and Countermeasure Plan required by 40 CFR 112.3 and any existing state contingency plan and provided that the Regional Response Team (if one exists in the area) concurs with the proposed containment and cleanup action. (Sections 10 and 404)

21. Surface Coal Mining Activities. Discharges of dredged or fill material into waters of the US associated with surface coal mining and reclamation operations provided the coal mining activities are authorized by the DOI, Office of Surface Mining (OSM), or by states with approved programs under Title V of the Surface Mining Control and Reclamation Act of 1977 and provided the permittee notifies the District Engineer in accordance with the "Notification" General Condition. In addition, to be authorized by this NWP, the District Engineer must determine that the activity complies with the terms and conditions of the NWP and that the adverse environmental effects are minimal both individually and cumulatively and must notify the project sponsor of this determination in writing. The Corps, at the discretion of the District Engineer, may require a bond to ensure success of the mitigation, if no other Federal or state agency has required one. For discharges in special aquatic sites, including wetlands, and stream riffle and pool complexes, the notification must also include a delineation of affected special aquatic sites, including wetlands. (also, see 33 CFR 330.1(e))

Mitigation: In determining the need for as well as the level and type of mitigation, the

District Engineer will ensure no more than minimal adverse effects to the aquatic environment occur. As such, District Engineers will determine on a case-by-case basis the requirement for adequate mitigation to ensure the effects to aquatic systems are minimal. In cases where OSM or the state has required mitigation for the loss of aquatic habitat, the Corps may consider this in determining appropriate mitigation under Section 404. (Sections 10 and 404)

22. Removal of Vessels. Temporary structures or minor discharges of dredged or fill material required for the removal of wrecked, abandoned, or disabled vessels, or the removal of man-made obstructions to navigation. This NWP does not authorize the removal of vessels listed or determined eligible for listing on the National Register of Historic Places unless the District Engineer is notified and indicates that there is compliance with the "Historic Properties" General Condition. This NWP does not authorize maintenance dredging, shoal removal, or riverbank snagging. Vessel disposal in waters of the US may need a permit from EPA (see 40 CFR 229.3). (Sections 10 and 404)

23. Approved Categorical Exclusions. Activities undertaken, assisted, authorized, regulated, funded, or financed, in whole or in part, by another Federal agency or department where that agency or department has determined, pursuant to the Council on Environmental Quality Regulation for Implementing the Procedural Provisions of the National Environmental Policy Act (NEPA) (40 CFR part 1500 *et seq.*), that the activity, work, or discharge is categorically excluded from environmental documentation, because it is included within a category of actions which neither individually nor cumulatively have a significant effect on the human environment, and the Office of the Chief of Engineers (ATTN: CECWOR) has been furnished notice of the agency's or department's application for the categorical exclusion and concurs with that determination. Before approval for purposes of this NWP of any agency's categorical exclusions, the Chief of Engineers will solicit public comment. In addressing these comments, the Chief of Engineers may require certain conditions for authorization of an agency's categorical exclusions under this NWP. (Sections 10 and 404)

24. State Administered Section 404 Program. Any activity permitted by a state administering its own Section 404 permit program pursuant to 33 U.S.C. 1344(g)–(l) is permitted pursuant to section 10 of the Rivers and Harbors Act of 1899. Those activities that do not involve a Section 404 state permit are not included in this NWP, but certain structures will be exempted by section 154 of Pub. L. 94–587, 90 Stat. 2917 (33 U.S.C. 591) (see 33 CFR 322.3(a)(2)). (Section 10)

25. Structural Discharges. Discharges of material such as concrete, sand, rock, etc., into tightly sealed forms or cells where the material will be used as a structural member for standard pile supported structures, such as bridges, transmission line footings, and walkways or for general navigation, such as mooring cells, including the excavation of bottom material from within the form prior to the discharge of concrete, sand, rock, etc. This NWP does not authorize filled structural members that would support buildings, building pads, homes, house pads, parking areas, storage areas and other such structures. The structure itself may require a Section 10 permit if located in navigable waters of the US. (Section 404)

26. [Reserved]

27. Stream and Wetland Restoration Activities. Activities in waters of the US associated with the restoration of former waters, the enhancement of degraded tidal and non-tidal wetlands and riparian areas, the creation of tidal and non-tidal wetlands and riparian areas, and the restoration and enhancement of non-tidal streams and non-tidal open water areas as follows:

(a) *The activity is conducted on:*

(1) Non-Federal public lands and private lands, in accordance with the terms and conditions of a binding wetland enhancement, restoration, or creation agreement between the landowner and the U.S. Fish and Wildlife Service (FWS) or the Natural Resources Conservation Service (NRCS), the National Marine Fisheries Service, the National Ocean Service, or voluntary wetland restoration, enhancement, and creation actions documented by the NRCS pursuant to NRCS regulations; or

(2) Reclaimed surface coal mine lands, in accordance with a Surface Mining Control and Reclamation Act permit issued by the OSM or the applicable state agency (the future reversion does not apply to streams or wetlands created, restored, or enhanced as mitigation for the mining impacts, nor naturally due to hydrologic or topographic features, nor for a mitigation bank); or

(3) Any other public, private or tribal lands;

(b) *Notification:* For activities on any public or private land that are not described by paragraphs (a)(1) or (a)(2) above, the permittee must notify the District Engineer in accordance with General Condition 13; and

(c) Planting of only native species should occur on the site. Activities authorized by this NWP include, to the extent that a Corps permit is required, but are not limited to: the removal of accumulated sediments; the installation, removal, and maintenance of small water control structures, dikes, and berms; the installation of current deflectors; the enhancement, restoration, or creation of riffle and pool stream structure; the placement of in-stream habitat structures; modifications of the stream bed and/or banks to restore or create stream meanders; the backfilling of artificial channels and drainage ditches; the removal of existing drainage structures; the construction of small nesting islands; the construction of open water areas; the construction of oyster habitat over unvegetated bottom in tidal waters; activities needed to reestablish vegetation, including plowing or discing for seed bed preparation and the planting of appropriate wetland species; mechanized land clearing to remove non-native invasive, exotic or nuisance vegetation; and other related activities. This NWP does not authorize the conversion of a stream to another aquatic use, such as the creation of an impoundment for waterfowl habitat. This NWP does not authorize stream channelization.

This NWP does not authorize the conversion of natural wetlands to another aquatic use, such as creation of waterfowl impoundments where a forested wetland previously existed. However, this NWP authorizes the relocation of non-tidal waters, including non-tidal wetlands, on the project site provided there are net gains in aquatic resource functions and values. For example, this NWP may authorize the creation of an open water impoundment in a non-tidal emergent wetland, provided the non-tidal emergent wetland is replaced by creating that wetland type on the project site. This NWP does not authorize the relocation of tidal waters or the conversion of tidal waters, including tidal wetlands, to other aquatic uses, such as the conversion of tidal wetlands into open water impoundments.

Reversion. For enhancement, restoration, and creation projects conducted under paragraphs (a)(3), this NWP does not authorize any future discharge of dredged or fill material associated with the reversion of the area to its prior condition. In such cases a separate permit would be required for any reversion. For restoration, enhancement, and creation projects conducted under paragraphs (a)(1) and (a)(2), this NWP also authorizes any future discharge of dredged or fill material associated with the reversion of the area to its documented prior condition and use (i.e., prior to the restoration, enhancement, or creation activities). The reversion must occur within five years after expiration of a limited term wetland restoration or creation agreement or permit, even if the discharge occurs after this NWP expires. This NWP also authorizes the reversion of wetlands that were restored, enhanced, or created on prior-converted cropland that has not been abandoned, in accordance with a binding agreement between the landowner and NRCS or FWS (even though the restoration, enhancement, or creation activity did not require a Section 404 permit). The five-year reversion limit does not apply to agreements without time limits reached under paragraph (a)(1). The prior condition will be documented in the original agreement or permit, and the determination of return to prior conditions will be made by the Federal agency or appropriate state agency executing the agreement or permit.

Before any reversion activity the permittee or the appropriate Federal or state agency must notify the District Engineer and include the documentation of the prior condition. Once an area has reverted to its prior physical condition, it will be subject to whatever the Corps Regulatory requirements will be at that future date. (Sections 10 and 404)

Note: Compensatory mitigation is not required for activities authorized by this NWP, provided the authorized work results in a net increase in aquatic resource functions and values in the project area. This NWP can be used to authorize compensatory mitigation projects, including mitigation banks, provided the permittee notifies the District Engineer in accordance with General Condition 13, and the project includes compensatory mitigation for impacts to waters of the US caused by the authorized work. However, this NWP does not authorize the reversion of an area used for a compensatory mitigation project to its prior condition. NWP 27 can be used to authorize impacts at a mitigation bank, but only in circumstances where it has been approved under the Interagency Federal Mitigation Bank Guidelines.

28. Modifications of Existing Marinas. Reconfiguration of existing docking facilities within an authorized marina area. No dredging, additional slips, dock spaces, or expansion of any kind within waters of the US is authorized by this NWP. (Section 10)

29. Single-family Housing. Discharges of dredged or fill material into non-tidal waters of the US, including non-tidal wetlands for the construction or expansion of a single-family home and attendant features (such as a garage, driveway, storage shed, and/or septic field) for an Individual Permittee provided that the activity meets all of the following criteria:

a. The discharge does not cause the loss of more than 1/4-acre of non-tidal waters of the US, including non-tidal wetlands;

b. The permittee notifies the District Engineer in accordance with the "Notification" General Condition;

c. The permittee has taken all practicable actions to minimize the onsite and offsite impacts of the discharge. For example, the location of the home may need to be adjusted onsite to avoid flooding of adjacent property owners;

d. The discharge is part of a single and complete project; furthermore, that for any subdivision created on or after November 22, 1991, the discharges authorized under this NWP may not exceed an aggregate total loss of waters of the US of 1/4-acre for the entire subdivision;

e. An individual may use this NWP only for a single-family home for a personal residence;

f. This NWP may be used only once per parcel;

g. This NWP may not be used in conjunction with NWP 14 or NWP 18, for any parcel; and,

h. Sufficient vegetated buffers must be maintained adjacent to all open water bodies, streams, etc., to preclude water quality degradation due to erosion and sedimentation.

For the purposes of this NWP, the acreage of loss of waters of the US includes the filled area previously permitted, the proposed filled area, and any other waters of the US that are adversely affected by flooding, excavation, or drainage as a result of the project. This NWP authorizes activities only by individuals; for this purpose, the term "individual" refers to a natural person and/or a married couple, but does not include a corporation, partnership, or similar entity. For the purposes of this NWP, a parcel of land is defined as "the entire contiguous quantity of land in possession of, recorded as property of, or owned (in any form of ownership, including land owned as a partner, corporation, joint tenant, etc.) by the same individual (and/or that individual's spouse), and comprises not only the area of wetlands sought to be filled, but also all land contiguous to those wetlands, owned by the individual (and/or that individual's spouse) in any form of ownership." (Sections 10 and 404)

30. Moist Soil Management for Wildlife. Discharges of dredged or fill material and maintenance activities that are associated with moist soil management for wildlife performed on non-tidal Federally-owned or managed, state-owned or managed property, and local government agency-owned or managed property, for the purpose of continuing ongoing, site-specific, wildlife management activities where soil manipulation is used to manage habitat and feeding areas for wildlife. Such activities include, but are not limited to: The repair, maintenance or replacement of existing water control structures; the repair or maintenance of dikes; and plowing or discing to impede succession, prepare seed beds, or establish fire breaks. Sufficient vegetated buffers must be maintained adjacent to all open water bodies, streams, etc., to preclude water quality degradation due to erosion and sedimentation. This NWP does not authorize the construction of new dikes, roads, water control structures, etc. associated with the management areas. This NWP does not authorize converting wetlands to uplands, impoundments or other open water bodies. (Section 404)

31. Maintenance of Existing Flood Control Facilities. Discharge of dredge or fill material resulting from activities associated with the maintenance of existing flood control facilities, including debris basins, retention/detention basins, and channels that

(i) were previously authorized by the Corps by Individual Permit, General Permit, by 33 CFR 330.3, or did not require a permit at the time it was constructed, or

(ii) were constructed by the Corps and transferred to a non-Federal sponsor for operation and maintenance. Activities authorized by this NWP are limited to those resulting from maintenance activities that are conducted within the "maintenance baseline," as described in the definition below. Activities including the discharges of dredged or fill materials, associated with maintenance activities in flood control facilities in any watercourse that has previously been determined to be within the maintenance baseline, are authorized under this NWP. The NWP does not authorize the removal of sediment and associated vegetation from the natural water courses except to the extent that these have been included in the maintenance baseline. All dredged material must be placed in an upland site or an authorized disposal site in waters of the US, and proper siltation controls must be used. (Activities of any kind that result in only incidental fallback, or only the cutting and removing of vegetation above the ground, e.g., mowing, rotary cutting, and chainsawing, where the activity neither substantially disturbs the root system nor involves mechanized pushing, dragging, or other similar activities that redeposit excavated soil material, do not require a Section 404 permit in accordance with 33 CFR 323.2(d)(2)).

Notification: After the maintenance baseline is established, and before any maintenance work is conducted, the permittee must notify the District Engineer in accordance with the "Notification" General Condition. The notification may be for activity-specific maintenance or for maintenance of the entire flood control facility by submitting a five year (or less) maintenance plan.

Maintenance Baseline: The maintenance baseline is a description of the physical characteristics (e.g., depth, width, length, location, configuration, or design flood capacity, etc.) of a flood control project within which maintenance activities are normally authorized by NWP 31, subject to any case-specific conditions required by the District Engineer. The District Engineer will approve the maintenance baseline based on the approved or constructed capacity of the flood control facility, whichever is smaller, including any areas where there are no constructed channels, but which are part of the facility. If no evidence of the constructed capacity exist, the approved constructed capacity will be used. The prospective permittee will provide documentation of the physical characteristics of the flood control facility (which will normally consist of as-built or approved drawings) and documentation of the design capacities of the flood control facility. The documentation will also include BMPs to ensure that the impacts to the aquatic environment are minimal, especially in maintenance areas where there are no constructed channels. (The Corps may request maintenance records in areas where there has not been recent maintenance.) Revocation or modification of the final determination of the maintenance baseline can only be done in accordance with 33 CFR 330.5. Except in emergencies as described below, this NWP cannot be used until the District Engineer approves the maintenance baseline and determines the need for mitigation and any regional or activity-specific conditions. Once determined, the maintenance baseline will remain valid for any subsequent reissuance of this NWP. This permit does not authorize maintenance of a flood control facility that has been abandoned. A flood control facility will be considered abandoned if it has operated at a significantly reduced capacity without needed maintenance being accomplished in a timely manner.

Mitigation: The District Engineer will determine any required mitigation onetime only for impacts associated with maintenance work at the same time that the maintenance baseline is approved. Such one-time mitigation will be required when necessary to ensure that adverse environmental impacts are no more than minimal, both individually and cumulatively. Such mitigation will only be required once for any specific reach of a flood control project. However, if one-time mitigation is required for impacts associated with maintenance activities, the District Engineer will not delay needed maintenance, provided the District Engineer and the permittee establish a schedule for identification, approval, development, construction and completion of any such required mitigation. Once the one-time mitigation described above has been completed, or a determination made that mitigation is not required, no further mitigation will be required for maintenance activities within the maintenance baseline. In determining appropriate mitigation, the District Engineer will give special consideration to natural water courses that have been included in the maintenance baseline and require compensatory mitigation and/or BMPs as appropriate.

Emergency Situations: In emergency situations, this NWP may be used to authorize maintenance activities in flood control facilities for which no maintenance baseline has been approved. Emergency situations are those which would result in an unacceptable hazard to life, a significant loss of property, or an immediate, unforeseen, and significant economic hardship if action is not taken before a maintenance baseline can be approved. In such situations, the determination of mitigation requirements, if any, may be deferred until the emergency has been resolved. Once the emergency has ended, a maintenance baseline must be established expeditiously, and mitigation, including mitigation for maintenance conducted during the emergency, must be required as appropriate. (Sections 10 and 404)

32. Completed Enforcement Actions. Any structure, work or discharge of or fill material, remaining in place, or undertaken for mitigation, restoration, or environmental benefit in compliance with either:

(i) The terms of a final written Corps non-judicial settlement agreement resolving a violation of section 404 of the CWA and/or section 10 of the Rivers and Harbors Act of 1899; or the terms of an EPA 309(a) order on consent resolving a violation of section 404 of the CWA, provided that:

a. The unauthorized activity affected no more than 5 acres of non-tidal wetlands or 1 acre of tidal wetlands;

b. The settlement agreement provides for environmental benefits, to an equal or greater degree, than the environmental detriments caused by the unauthorized activity that is authorized by this NWP; and

c. The District Engineer issues a verification letter authorizing the activity subject to the terms and conditions of this NWP and the settlement agreement, including a specified completion date; or

(ii) The terms of a final Federal court decision, consent decree, or settlement agreement resulting from an enforcement action brought by the U.S. under section 404 of the CWA and/or section 10 of the Rivers and Harbors Act of 1899; or

(iii)The terms of a final court decision, consent decree, settlement agreement, or non-judicial settlement agreement resulting from a natural resource damage claim brought by a trustee or trustees for natural resources (as defined by the National Contingency Plan at 40 CFR subpart G) under section 311 of the Clean Water Act (CWA), section 107 of the Comprehensive Environmental Response, Compensation and Liability Act (CERCLA or Superfund), section 312 of the National Marine Sanctuaries Act (NMSA), section 1002 of the Oil Pollution Act of 1990 (OPA), or the Park System Resource Protection Act at 16 U.S.C. '19jj, to the extent that a Corps permit is required.

For either (i), (ii) or (iii) above, compliance is a condition of the NWP itself. Any authorization under this NWP is automatically revoked if the permittee does not comply with the terms of this NWP or the terms of the court decision, consent decree, or judicial/ non-judicial settlement agreement or fails to complete the work by the specified completion date. This NWP does not apply to any activities occurring after the date of the decision, decree, or agreement that are not for the purpose of mitigation, restoration, or environmental benefit. Before reaching any settlement agreement, the Corps will ensure compliance with the provisions of 33 CFR part 326 and 33 CFR 330.6 (d)(2) and (e). (Sections 10 and 404)

33. Temporary Construction, Access and Dewatering. Temporary structures, work and discharges, including cofferdams, necessary for construction activities or access fills or dewatering of construction sites; provided that the associated primary activity is authorized by the Corps of Engineers or the USCG, or for other construction activities not subject to the Corps or USCG regulations. Appropriate measures must be taken to maintain near normal downstream flows and to minimize flooding. Fill must be of materials, and placed in a manner, that will not be eroded by expected high flows. The use of dredged material may be allowed if it is determined by the District Engineer that it will not cause more than minimal adverse effects on aquatic resources. Temporary fill must be entirely removed to upland areas, or dredged material returned to its original location, following completion of the construction activity, and the affected areas must be restored to the pre-project conditions. Cofferdams cannot be used to dewater wetlands or other aquatic areas to change their use. Structures left in place after cofferdams are removed require a Section 10 permit if located in navigable waters of the U.S. (See 33 CFR part 322). The permittee must notify the District Engineer in accordance with the "Notification" General Condition. The notification must also include a restoration plan of reasonable measures to avoid and minimize adverse effects to aquatic resources. The District Engineer will add Special Conditions, where necessary, to ensure environmental adverse effects is minimal. Such conditions may include: limiting the temporary work to the minimum necessary; requiring seasonal restrictions; modifying the restoration plan; and requiring alternative construction methods (e.g. construction mats in wetlands where practicable.). (Sections 10 and 404)

34. Cranberry Production Activities. Discharges of dredged or fill material for dikes, berms, pumps, water control structures or leveling of cranberry beds associated with expansion, enhancement, or modification activities at existing cranberry production operations provided that the activity meets all of the following criteria:

a. The cumulative total acreage of disturbance per cranberry production operation, including but not limited to, filling, flooding, ditching, or clearing, does not exceed 10 acres of waters of the U.S., including wetlands;

b. The permittee notifies the District Engineer in accordance with the "Notification" General Condition. The notification must include a delineation of affected special aquatic sites, including wetlands; and,

c. The activity does not result in a net loss of wetland acreage. This NWP does not authorize any discharge of dredged or fill material related to other cranberry production activities such as warehouses, processing facilities, or parking areas. For the purposes of this NWP, the cumulative total of 10 acres will be measured over the period that this NWP is valid. (Section 404)

35. Maintenance Dredging of Existing Basins. Excavation and removal of accumulated sediment for maintenance of existing marina basins, access channels to marinas or boat slips, and boat slips to previously authorized depths or controlling depths for ingress/egress, whichever is less, provided the dredged material is disposed of at an upland site and proper siltation controls are used. (Section 10)

36. Boat Ramps. Activities required for the construction of boat ramps provided:

a. The discharge into waters of the U.S. does not exceed 50 cubic yards of concrete, rock, crushed stone or gravel into forms, or placement of pre-cast concrete planks or slabs. (Unsuitable material that causes unacceptable chemical pollution or is structurally unstable is not authorized);

b. The boat ramp does not exceed 20 feet in width;

c. The base material is crushed stone, gravel or other suitable material;

d. The excavation is limited to the area necessary for site preparation and all excavated material is removed to the upland; and,

e. No material is placed in special aquatic sites, including wetlands.

Another NWP, Regional General Permit, or Individual Permit may authorize dredging to provide access to the boat ramp after obtaining a Section 10 if located in navigable waters of the U.S. [Replace with: "Dredging to provide access to the boat ramp may be authorized by another NWP, regional general permit, or individual permit pursuant to Section 10 if located in navigable waters of the United States."] (Sections 10 and 404)

37. Emergency Watershed Protection and Rehabilitation. Work done by or funded by:

a. The NRCS which is a situation requiring immediate action under its emergency Watershed Protection Program (7 CFR part 624); or

b. The USFS under its Burned-Area Emergency Rehabilitation Handbook (FSH 509.13); or

c. The DOI for wildland fire management burned area emergency stabilization and rehabilitation (DOI Manual part 620, Ch. 3).

For all of the above provisions, the District Engineer must be notified in accordance with the General Condition 13. (Also, see 33 CFR 330.1(e)). (Sections 10 and 404)

38. Cleanup of Hazardous and Toxic Waste. Specific activities required to effect the containment, stabilization, or removal of hazardous or toxic waste materials that are performed, ordered, or sponsored by a government agency with established legal or regulatory authority provided the permittee notifies the District Engineer in accordance with the "Notification" General Condition. For discharges in special aquatic sites, including wetlands, the notification must also include a delineation of affected special aquatic sites, including wetlands. Court ordered remedial action plans or related settlements are also authorized by this NWP. This NWP does not authorize the establishment of new disposal sites or the expansion of existing sites used for the disposal of hazardous or toxic waste. Activities undertaken entirely on a Comprehensive Environmental Response, Compensation, and Liability Act (CERCLA) site by authority of CERCLA as approved or required by EPA, are not required to obtain permits under section 404 of the CWA or section 10 of the Rivers and Harbors Act. (Sections 10 and 404)

39. Residential, Commercial, and Institutional Developments. Discharges of dredged or fill material into non-tidal waters of the U.S., excluding non-tidal wetlands adjacent to tidal waters, for the construction or expansion of residential, commercial, and institutional building foundations and building pads and attendant features that are necessary for the use and maintenance of the structures. Attendant features may include, but are not limited to, roads, parking lots, garages, yards, utility lines, stormwater management facilities, and recreation facilities such as playgrounds, playing fields, and golf courses (provided the golf course is an integral part of the residential development). The construction of new ski areas or oil and gas wells is not authorized by this NWP.

Residential developments include multiple and single unit developments. Examples of commercial developments include retail stores, industrial facilities, restaurants, business parks, and shopping centers. Examples of institutional developments include schools, fire stations, government office buildings, judicial buildings, public works buildings, libraries, hospitals, and places of worship. The activities listed above are authorized, provided the activities meet all of the following criteria:

a. The discharge does not cause the loss of greater than 1/2-acre of non-tidal waters of the U.S., excluding non-tidal wetlands adjacent to tidal waters;

b. The discharge does not cause the loss of greater than 300 linear-feet of a stream bed, unless for intermittent stream beds this criterion is waived in writing pursuant to a determination by the District Engineer, as specified below, that the project complies with all terms and conditions of this NWP and that any adverse impacts of the project on the aquatic environment are minimal, both individually and cumulatively;

c. The permittee must notify the District Engineer in accordance with General Condition 13, if any of the following criteria are met:

(1) The discharge causes the loss of greater than 1/10-acre of non-tidal waters of the US, excluding non-tidal wetlands adjacent to tidal waters; or

(2) The discharge causes the loss of any open waters, including perennial or intermittent streams, below the ordinary high water mark (see Note, below); or

(3) The discharge causes the loss of greater than 300 linear feet of intermittent stream bed. In such case, to be authorized the District Engineer must determine that the activity complies with the other terms and conditions of the NWP, determine adverse environmental effects are minimal both individually and cumulatively, and waive the limitation on stream impacts in writing before the permittee may proceed;

d. For discharges in special aquatic sites, including wetlands, the notification must include a delineation of affected special aquatic sites;

e. The discharge is part of a single and complete project;

f. The permittee must avoid and minimize discharges into waters of the US at the project site to the maximum extent practicable. The notification, when required, must include a written statement explaining how avoidance and minimization of losses of waters of the US were achieved on the project site. Compensatory mitigation will normally be required to offset the losses of waters of the US. (See General Condition 19.) The notification must also include a compensatory mitigation proposal for offsetting unavoidable losses of waters of the US. If an applicant asserts that the adverse effects of the project are minimal without mitigation, then the applicant may submit justification explaining why compensatory mitigation should not be required for the District Engineer's consideration;

g. When this NWP is used in conjunction with any other NWP, any combined total permanent loss of waters of the US exceeding 1/10-acre requires that the permittee notify the District Engineer in accordance with General Condition 13;

h. Any work authorized by this NWP must not cause more than minimal degradation of water quality or more than minimal changes to the flow characteristics of any stream (see General Conditions 9 and 21);

i. For discharges causing the loss of 1/10-acre or less of waters of the US, the permittee must submit a report, within 30 days of completion of the work, to the District Engineer that contains the following information: (1) The name, address, and telephone number of the permittee; (2) The location of the work; (3) A description of the work; (4) The type and acreage of the loss of waters of the US (e.g., 1/12-acre of emergent wetlands); and (5) The type and acreage of any compensatory mitigation used to offset the loss of waters of the US (e.g., 1/12-acre of emergent wetlands created on-site);

j. If there are any open waters or streams within the project area, the permittee will establish and maintain, to the maximum extent practicable, wetland or upland vegetated buffers next to those open waters or streams consistent with General Condition 19. Deed restrictions, conservation easements, protective covenants, or other means of land conservation and preservation are required to protect and maintain the vegetated buffers established on the project site.

Only residential, commercial, and institutional activities with structures on the foundation(s) or building pad(s), as well as the attendant features, are authorized by this NWP. The compensatory mitigation proposal that is required in paragraph (e) [Replace with: "paragraph (f)"] of this NWP may be either conceptual or detailed. The wetland or upland vegetated buffer required in paragraph (i) [Replace with: "paragraph (j)"] of this NWP will be determined on a case-by-case basis by the District Engineer for addressing water quality concerns. The required wetland or upland vegetated buffer is part of the overall compensatory mitigation requirement for this NWP. If the project site was previously used for agricultural purposes and the farm owner/operator used NWP 40 to authorize activities in waters of the US to increase production or construct farm buildings, NWP 39 cannot be used by the developer to authorize additional activities. This is more than the acreage limit for NWP 39 impacts to waters of the US (i.e., the combined acreage loss authorized under NWPs 39 and 40 cannot exceed 1/2-acre, see General Condition 15). [Replace with: "If the project site was previously used for agricultural purposes and the farm owner/operator used NWP 40 to authorize activities in waters of the United States to increase production or construct farm buildings, NWP 39 cannot be used by the developer to authorize additional activities in waters of the United States on the project site in excess of the acreage limit for NWP 39 (i.e., the combined acreage loss authorized under NWPs 39 and 40 cannot exceed 1/2-acre)."]

Subdivisions: For residential subdivisions, the aggregate total loss of waters of US authorized by NWP 39 can not exceed 1/2-acre. This includes any loss of waters associated with development of individual subdivision lots. (Sections 10 and 404)

Note: Areas where wetland vegetation is not present should be determined by the presence or absence of an ordinary high water mark or bed and bank. Areas that are waters of the US based on this criterion would require a PCN although water is infrequently present in the stream channel (except for ephemeral waters, which do not require PCNs) [Replace with: "(except for ephemeral waters, which do not require PCNs under paragraph (c)(2), above; however, activities that result in the loss of greater than 1/10 acre of ephemeral waters would require PCNs under paragraph (c)(1), above)."]

40. Agricultural Activities. Discharges of dredged or fill material into non-tidal waters of the US, excluding non-tidal wetlands adjacent to tidal waters, for improving agricultural production and the construction of building pads for farm buildings. Authorized activities include the installation, placement, or construction of drainage tiles, ditches, or levees; mechanized land clearing; land leveling; the relocation of existing serviceable drainage ditches constructed in waters of the US; and similar activities, provided the permittee complies with the following terms and conditions:

a. For discharges into non-tidal wetlands to improve agricultural production, the following criteria must be met if the permittee is an United States Department of Agriculture (USDA) Program participant:

(1) The permittee must obtain a categorical minimal effects exemption, minimal effect exemption, or mitigation exemption from NRCS in accordance with the provisions of the Food Security Act of 1985, as amended (16 U.S.C. 3801 et seq.);

(2) The discharge into non-tidal wetlands does not result in the loss of greater than 1/2-acre of non-tidal wetlands on a farm tract;

(3) The permittee must have NRCS certified wetland delineation;

(4) The permittee must implement an NRCS-approved compensatory mitigation plan that fully offsets wetland losses, if required; and

(5) The permittee must submit a report, within 30 days of completion of the authorized work, to the District Engineer that contains the following information: (a) The name, address, and telephone number of the permittee; (b) The location of the work; (c) A description of the work; (d) The type and acreage (or square feet) of the loss of wetlands (e.g., 1/3-acre of emergent wetlands); and (e) The type, acreage (or square feet), and location of compensatory mitigation (e.g. 1/3-acre of emergent wetland on a farm tract; credits purchased from a mitigation bank); or

b. For discharges into non-tidal wetlands to improve agricultural production, the following criteria must be met if the permittee is not a USDA Program participant (or a USDA Program participant for which the proposed work does not qualify for authorization under paragraph (a) of this NWP):

(1) The discharge into non-tidal wetlands does not result in the loss of greater

than 1/2-acre of non-tidal wetlands on a farm tract;

(2) The permittee must notify the District Engineer in accordance with General Condition 13, if the discharge results in the loss of greater than 1/10-acre of non-tidal wetlands;

(3) The notification must include a delineation of affected wetlands; and

(4) The notification must include a compensatory mitigation proposal to offset losses of waters of the US; or

c. For the construction of building pads for farm buildings, the discharge does not cause the loss of greater than 1/2-acre of non-tidal wetlands that were in agricultural production prior to December 23, 1985, (i.e., farmed wetlands) and the permittee must notify the District Engineer in accordance with General Condition 13; and

d. Any activity in other waters of the US is limited to the relocation of existing serviceable drainage ditches constructed in non-tidal streams. This NWP does not authorize the relocation of greater than 300 linear-feet of existing serviceable drainage ditches constructed in non-tidal streams unless, for drainage ditches constructed in intermittent non-tidal streams, the District Engineer waives this criterion in writing, and the District Engineer has determined that the project complies with all terms and conditions of this NWP, and that any adverse impacts of the project on the aquatic environment are minimal, both individually and cumulatively. For impacts exceeding 300-linear feet of impacts to existing serviceable ditches constructed in intermittent non-tidal streams, the permittee must notify the District Engineer in accordance with the "Notification" General Condition 13; and

e. The term "farm tract" refers to a parcel of land identified by the Farm Service Agency. The Corps will identify other waters of the US on the farm tract. NRCS will determine if a proposed agricultural activity meets the terms and conditions of paragraph a. of this NWP, except as provided below. For those activities that require notification, the District Engineer will determine if a proposed agricultural activity is authorized by paragraphs b., c., and/or d. of this NWP. USDA Program participants requesting authorization for discharges of dredged or fill material into waters of the US authorized by paragraphs (c) or (d) of this NWP, in addition to paragraph (a), must notify the District Engineer in accordance with General Condition 13 and the District Engineer will determine if the entire single and complete project is authorized by this NWP. Discharges of dredged or fill material into waters of the US associated with completing required compensatory mitigation are authorized by this NWP. However, total impacts, including other authorized impacts under this NWP, may not exceed the 1/2-acre limit of this NWP. This NWP does not affect, or otherwise regulate, discharges associated with agricultural activities when the discharge qualifies for an exemption under section 404(f) of the CWA, even though a categorical minimal effects exemption, minimal effect exemption, or mitigation exemption from NRCS pursuant to the Food Security Act of 1985, as amended, may be required. Activities authorized by paragraphs a. through d. may not exceed a total of 1/2- acre on a single farm tract. If the site was used for agricultural purposes and the farm owner/operator used either paragraphs a., b., or c. of this NWP to authorize activities in waters of the US to increase agricultural production or construct farm buildings, and the current landowner wants to use NWP 39 to authorize residential, commercial, or industrial development activities in waters of the US on the site, the combined acreage loss authorized by NWPs 39 and 40 cannot exceed 1/2-acre (see General Condition 15). (Section 404)

41. Reshaping Existing Drainage Ditches. Discharges of dredged or fill material into non-tidal waters of the US, excluding non-tidal wetlands adjacent to tidal waters, to modify the cross-sectional configuration of currently serviceable drainage ditches constructed in waters of the US. The reshaping of the ditch cannot increase drainage capacity beyond the original design capacity. Nor can it expand the area drained by the ditch as originally designed (i.e., the capacity of the ditch must be the same as originally designed and it cannot drain additional wetlands or other waters of the US). Compensatory mitigation is not required because the work is designed to improve water quality (e.g., by regrading the drainage ditch with gentler slopes, which can reduce erosion, increase growth of vegetation, increase uptake of nutrients and other substances by vegetation, etc.).

Notification: The permittee must notify the District Engineer in accordance with General Condition 13 if greater than 500 linear feet of drainage ditch will be reshaped. Material resulting from excavation may not be permanently sidecast into waters but may be temporarily sidecast (up to three months) into waters of the US, provided the material is not placed in such a manner that it is dispersed by currents or other forces. The District Engineer may extend the period of temporary sidecasting not to exceed a total of 180 days, where appropriate. In general, this NWP does not apply to reshaping drainage ditches constructed in uplands, since these areas are generally not waters of the US, and thus no permit from the Corps is required, or to the maintenance of existing drainage ditches to their original dimensions and configuration, which does not require a Section 404 permit (see 33 CFR 323.4(a)(3)). This NWP does not authorize the relocation of drainage ditches constructed in waters of the US; the location of the centerline of the reshaped drainage ditch must be approximately the same as the location of the centerline of the original drainage ditch. This NWP does not authorize stream channelization or stream relocation projects. (Section 404)

42. Recreational Facilities. Discharges of dredged or fill material into non-tidal waters of the US, excluding non-tidal wetlands adjacent to tidal waters, for the construction or expansion of recreational facilities, provided the activity meets all of the following criteria:

a. The discharge does not cause the loss of greater than 1/2-acre of non-tidal waters of the US, excluding non-tidal wetlands adjacent to tidal waters;

b. The discharge does not cause the loss of greater than 300 linear-feet of a stream bed, unless for intermittent stream beds this criterion is waived in writing pursuant to a determination by the District Engineer, as specified below, that the project complies with all terms and conditions of this NWP and that any adverse impacts of the project on the aquatic environment are minimal, both individually and cumulatively;

c. The permittee notifies the District Engineer in accordance with the "Notification" General Condition 13 for discharges exceeding 300 linear feet of impact of intermittent stream beds. In such cases, to be authorized the District Engineer must determine that the activity complies with the other terms and conditions of the NWP, determine the adverse environmental effects are minimal both individually and cumulatively, and waive this limitation in writing before the permittee may proceed;

d. For discharges causing the loss of greater than 1/10-acre of non-tidal waters of the US, the permittee notifies the District Engineer in accordance with General Condition 13;

e. For discharges in special aquatic sites, including wetlands, the notification must include a delineation, of affected special aquatic sites;

f. The discharge is part of a single and complete project; and

g. Compensatory mitigation will normally be required to offset the losses of waters of the US. The notification must also include a compensatory mitigation proposal to offset authorized losses of waters of the US.

For the purposes of this NWP, the term "recreational facility" is defined as a recreational activity that is integrated into the natural landscape and does not substantially change preconstruction grades or deviate from natural landscape contours. For the purpose of this permit, the primary function of recreational facilities does not include the use of motor vehicles, buildings, or impervious surfaces. Examples of recreational facilities that may be authorized by this NWP include hiking trails, bike paths, horse paths, nature centers, and campgrounds (excluding trailer parks). This NWP may authorize the construction or expansion of golf courses and the expansion of ski areas, provided the golf course or ski area does not substantially deviate from natural landscape contours. Additionally, these activities are designed to minimize adverse effects to waters of the US and riparian areas through the use of such practices as integrated pest management, adequate stormwater management facilities, vegetated buffers, reduced fertilizer use, etc. The facility must have an adequate water quality management plan [Replace with: "adequate water quality management measures"] in accordance with General Condition 9, such as a stormwater management facility, to ensure that the recreational facility results in no substantial adverse effects to water quality. This NWP also authorizes the construction or expansion of small support facilities, such as maintenance and storage buildings and stables that are directly related to the recreational activity. This NWP does not authorize other buildings, such as hotels, restaurants, etc. The construction or expansion of playing fields (e.g., baseball, soccer, or football fields), basketball and tennis courts, racetracks, stadiums, arenas, and the construction of new ski areas are not authorized by this NWP. (Section 404)

43. Stormwater Management Facilities. Discharges of dredged or fill material into non-tidal waters of the US, excluding non-tidal wetlands adjacent to tidal waters, for the construction and maintenance of stormwater management facilities, including activities for the excavation of stormwater ponds/facilities, detention basins, and retention basins; the installation and maintenance of water control structures, outfall structures and emergency spillways; and the maintenance dredging of existing stormwater management ponds/facilities and detention and retention basins, provided the activity meets all of the following criteria:

a. The discharge for the construction of new stormwater management facilities does not cause the loss of greater than 1/2-acre of non-tidal waters of the US, excluding non-tidal wetlands adjacent to tidal waters;

b. The discharge does not cause the loss of greater than 300 linear-feet of a stream bed, unless for intermittent stream beds this criterion is waived in writing pursuant to a determination by the District Engineer, as specified below, that the project complies with all terms and conditions of this NWP and that any adverse impacts of the project on the aquatic environment are minimal, both individually and cumulatively;

c. For discharges causing the loss of greater than 300 linear feet of intermittent stream beds, the permittee notifies the District Engineer in accordance with the "Notification" General Condition 13. In such cases, to be authorized the District Engineer must determine that the activity complies with the other terms and conditions of the NWP, determine the adverse environmental effects are minimal both individually and cumulatively, and waive this limitation in writing before the permittee may proceed;

d. The discharges of dredged or fill material for the construction of new stormwater management facilities in perennial streams is not authorized;

e. For discharges or excavation for the construction of new stormwater management facilities or for the maintenance of existing stormwater management facilities causing the loss of greater than 1/10-acre of non-tidal waters, excluding non-tidal wetlands adjacent to tidal waters, provided the permittee notifies the District Engineer in accordance with the "Notification" General Condition 13. In addition, the notification must include:

(1) A maintenance plan. The maintenance plan should be in accordance with state and local requirements, if any such requirements exist;

(2) For discharges in special aquatic sites, including wetlands and submerged aquatic vegetation, the notification must include a delineation of affected areas; and

(3) A compensatory mitigation proposal that offsets the loss of waters of the US. Maintenance in constructed areas will not require mitigation provided such maintenance is accomplished in designated maintenance areas and not within compensatory mitigation areas (i.e., District Engineers may designate nonmaintenance areas, normally at the downstream end of the stormwater management facility, in existing stormwater management facilities). (No mitigation will be required for activities that are exempt from Section 404 permit requirements);

f. The permittee must avoid and minimize discharges into waters of the US at the project site to the maximum extent practicable, and the notification must include a written statement to the District Engineer detailing compliance with this condition (i.e. why the discharge must occur in waters of the US and why additional minimization cannot be achieved);

g. The stormwater management facility must comply with General Condition 21 and be designed using BMPs and watershed protection techniques. Examples may include forebays (deeper areas at the upstream end of the stormwater management facility that would be maintained through excavation), vegetated buffers, and siting considerations to minimize adverse effects to aquatic resources. Another example of a BMP would be bioengineering methods incorporated into the facility design to benefit water quality and minimize adverse effects to aquatic resources from storm flows, especially downstream of the facility, that provide, to the maximum extent practicable, for long term aquatic resource protection and enhancement;

h. Maintenance excavation will be in accordance with an approved maintenance plan and will not exceed the original contours of the facility as approved and constructed; and

i. The discharge is part of a single and complete project. (Section 404)

44. Mining Activities. Discharges of dredged or fill material into:

(i) Isolated waters; streams where the annual average flow is 1 cubic foot per second or less, and non-tidal wetlands adjacent to headwater streams, for aggregate mining (i.e., sand, gravel, and crushed and broken stone) and associated support activities;

(ii) Lower perennial streams, excluding wetlands adjacent to lower perennial streams, for aggregate mining activities (support activities in lower perennial streams or adjacent wetlands are not authorized by this NWP); and/or

(iii)Isolated waters and non-tidal wetlands adjacent to headwater streams, for hard rock/mineral mining activities (i.e., extraction of metalliferous ores from subsurface locations) and associated support activities, provided the discharge meets the following criteria:

a. The mined area within waters of the US, plus the acreage loss of waters of the US resulting from support activities, cannot exceed 1/2-acre;

b. The permittee must avoid and minimize discharges into waters of the US at the project site to the maximum extent practicable, and the notification must include a written statement detailing compliance with this condition (i.e., why the discharge must occur in waters of the US and why additional minimization cannot be achieved);

c. In addition to General Conditions 17 and 20, activities authorized by this permit must not substantially alter the sediment characteristics of areas of concentrated shellfish beds or fish spawning areas. Normally, the mandated water quality management plan should address these impacts; [Replace with: "Normally, the water quality management measures required by General Condition 9 should address these impacts;"]

d. The permittee must implement necessary measures to prevent increases in stream gradient and water velocities and to prevent adverse effects (e.g., head cutting, bank erosion) to upstream and downstream channel conditions;

e. Activities authorized by this permit must not result in adverse effects on the course, capacity, or condition of navigable waters of the US;

f. The permittee must use measures to minimize downstream turbidity;

g. Wetland impacts must be compensated through mitigation approved by the Corps;

h. Beneficiation and mineral processing for hard rock/mineral mining activities may not occur within 200 feet of the ordinary high water mark of any open waterbody. Although the Corps does not regulate discharges from these activities, a CWA section 402 permit may be required;

i. All activities authorized must comply with General Conditions 9 and 21. Further, the District Engineer may require modifications to the required water quality management plan to ensure that the authorized work results in minimal adverse effects to water quality; [Replace with: "Further the District Engineer may require water quality management measures to ensure the authorized work results in minimal adverse effects to water quality;"]

j. Except for aggregate mining activities in lower perennial streams, no aggregate mining can occur within stream beds where the average annual flow is greater than 1 cubic foot per second or in waters of the US within 100 feet of the ordinary high water mark of headwater stream segments where the average annual flow of the stream is greater than 1 cubic foot per second (aggregate mining can occur in areas immediately adjacent to the ordinary high water mark of a stream where the average annual flow is 1 cubic foot per second or less);

k. Single and complete project: The discharge must be for a single and complete project, including support activities. Discharges of dredged or fill material into waters of the US for multiple mining activities on several designated parcels of a single and complete mining operation can be authorized by this NWP provided the 1/2-acre limit is not exceeded; and

l. **Notification:** The permittee must notify the District Engineer in accordance with General Condition 13. The notification must include:

(1) A description of waters of the US adversely affected by the project;

(2) A written statement to the District Engineer detailing compliance with paragraph (b), above (i.e., why the discharge must occur in waters of the US and why additional minimization cannot be achieved);

(3) A description of measures taken to ensure that the proposed work complies with paragraphs (c) through (f), above; and

(4) A reclamation plan (for aggregate mining in isolated waters and nontidal wetlands adjacent to headwaters and hard rock/mineral mining only).

This NWP does not authorize hard rock/mineral mining, including placer mining, in streams. No hard rock/mineral mining can occur in waters of the US within 100 feet of the ordinary high water mark of headwater streams. The term's "headwaters" and "isolated waters" are defined at 33 CFR 330.2(d) and (e), respectively. For the purposes of this NWP, the term "lower perennial stream" is defined as follows: "A stream in which the gradient is low and water velocity is slow, there is no tidal influence, some water flows throughout the year, and the substrate consists mainly of sand and mud." (Sections 10 and 404)

C. Nationwide Permits General Conditions

The following General Conditions must be followed in order for any authorization by an NWP to be valid:

1. Navigation. No activity may cause more than a minimal adverse effect on navigation.

2. Proper Maintenance. Any structure or fill authorized shall be properly maintained, including maintenance to ensure public safety.

3. Soil Erosion and Sediment Controls. Appropriate soil erosion and sediment controls must be used and maintained in effective operating condition during construction, and all exposed soil and other fills, as well as any work below the ordinary high water mark or high tide line, must be permanently stabilized at the earliest practicable date. Permittees are encouraged to perform work within waters of the United States during periods of low-flow or no-flow.

4. Aquatic Life Movements. No activity may substantially disrupt the necessary life-cycle movements of those species of aquatic life indigenous to the waterbody, including those species that normally migrate through the area, unless the activity's primary purpose is to impound water. Culverts placed in streams must be installed to maintain low flow conditions.

5. Equipment. Heavy equipment working in wetlands must be placed on mats, or other measures must be taken to minimize soil disturbance.

6. Regional and Case-by-Case Conditions. The activity must comply with any regional conditions that may have been added by the Division Engineer (see 33 CFR 330.4(e)). Additionally, any case specific conditions added by the Corps or by the state or tribe in its Section 401 Water Quality Certification and Coastal Zone Management Act consistency determination. [Replace with: "The activity must comply with any regional conditions that may have been added by the Division Engineer (see 33 CFR 330.4(e)) and with any case specific conditions added by the Corps or by the state or tribe in its Section 401 Water Quality Certification and Coastal Zone Management Act consistency determination."]

7. Wild and Scenic Rivers. No activity may occur in a component of the National Wild and Scenic River System; or in a river officially designated by Congress as a "study river" for possible inclusion in the system, while the river is in an official study status; unless the appropriate Federal agency, with direct management responsibility for such river, has determined in writing that the proposed activity will not adversely affect the Wild and Scenic River designation, or study status. Information on Wild and Scenic Rivers may be obtained from the appropriate Federal land management agency in the area (e.g., National Park Service, U.S. Forest Service, Bureau of Land Management, U.S. Fish and Wildlife Service).

8. Tribal Rights. No activity or its operation may impair reserved tribal rights, including, but not limited to, reserved water rights and treaty fishing and hunting rights.

9. Water Quality.

(a) In certain states and tribal lands an individual 401 Water Quality Certification must be obtained or waived (See 33 CFR 330.4(c)).

(b) For NWPs 12, 14, 17, 18, 32, 39, 40, 42, 43, and 44, where the state or tribal 401 certification (either generically or individually) does not require or approve water quality management measures, the permittee must provide water quality management measures that will ensure that the authorized work does not result in more than minimal degradation of water quality (or the Corps determines that compliance with state or local standards, where applicable, will ensure no more than minimal adverse effect on water quality). An important component of water quality management includes stormwater management that minimizes degradation of the downstream aquatic system, including water quality (refer to General Condition 21 for stormwater management requirements). Another important component of water quality management is the establishment and maintenance of vegetated buffers next to open waters, including streams (refer to General Condition 19 for vegetated buffer requirements for the NWPs).

This condition is only applicable to projects that have the potential to affect water quality. While appropriate measures must be taken, in most cases it is not necessary to conduct detailed studies to identify such measures or to require monitoring.

10. Coastal Zone Management. In certain states, an individual state coastal zone management consistency concurrence must be obtained or waived (see Section 330.4(d)). [Replace with: "(see 33 CFR 330.4(d))."]

11. Endangered Species.

(a) No activity is authorized under any NWP which is likely to jeopardize the continued existence of a threatened or endangered species or a species proposed for such designation, as identified under the Federal Endangered Species Act (ESA), or which will destroy or adversely modify the critical habitat of such species. Non-federal permittees shall notify the District Engineer if any listed species or designated critical habitat might be affected or is in the vicinity of the project, or is located in the designated critical habitat and shall not begin work on the activity until notified by the District Engineer that the requirements of the ESA have been satisfied and that the activity is authorized. For activities that may affect Federally-listed endangered or threatened species or designated critical habitat, the notification must include the name(s) of the endangered or threatened species that may be affected by the proposed work or that utilize the designated critical habitat that may be affected by the proposed work. As a result of formal or informal consultation with the FWS or NMFS the District Engineer may add species-specific regional endangered species conditions to the NWPs.

(b) Authorization of an activity by a NWP does not authorize the "take" of a threatened or endangered species as defined under the ESA. In the absence of separate authorization (e.g., an ESA Section 10 Permit, a Biological Opinion with "incidental take" provisions, etc.) from the USFWS or the NMFS, both lethal and non-lethal "takes" of protected species are in violation of the ESA. Information on the location of threatened and endangered species and their critical habitat can be obtained directly from the offices of the USFWS and NMFS or their world wide web pages at http://www.fws.gov/r9endspp/endspp.html and http://www.nfms.gov/prot_res/esahome. html [Replace with: "http://www.nmfs.noaa.gov/prot res/overview/es.html"] respectively.

12. Historic Properties. No activity which may affect historic properties listed, or eligible for listing, in the National Register of Historic Places is authorized, until the District Engineer has complied with the provisions of 33 CFR part 325, Appendix C. The prospective permittee must notify the District Engineer if the authorized activity may affect any historic properties listed, determined to be eligible, or which the prospective permittee has reason to believe may be eligible for listing on the National Register of Historic Places, and shall not begin the activity until notified by the District Engineer that the requirements of the National Historic Preservation Act have been satisfied and that the activity is authorized. Information on the location and existence of historic resources can be obtained from the State Historic Preservation Office and the National Register of Historic Places (see 33 CFR 330.4(g)). For activities that may affect historic properties listed in, or eligible for listing in, the National Register of Historic Places, the notification must state which historic property may be affected by the proposed work or include a vicinity map indicating the location of the historic property.

13. Notification.

(a) **Timing;** where required by the terms of the NWP, the prospective permittee must notify the District Engineer with a preconstruction notification (PCN) as early as possible. The District Engineer must determine if the notification is complete within 30 days of the date of receipt and can request additional information necessary to make the PCN complete only once. However, if the prospective permittee does not provide all of the requested information, then the District Engineer will notify the prospective permittee that the notification is still incomplete and the PCN review process will not commence until all of the requested information has been received by the District Engineer. The prospective permittee shall not begin the activity:

(1) Until notified in writing by the District Engineer that the activity may proceed under the NWP with any special conditions imposed by the District or Division Engineer; or

(2) If notified in writing by the District or Division Engineer that an Individual Permit is required; or

(3) Unless 45 days have passed from the District Engineer's receipt of the complete notification and the prospective permittee has not received written notice from the District or Division Engineer. Subsequently, the permittee's right to proceed under the NWP may be modified, suspended, or revoked only in accordance with the procedure set forth in 33 CFR 330.5(d)(2).

(b) **Contents of Notification:** The notification must be in writing and include the following information:

(1) Name, address and telephone numbers of the prospective permittee;

(2) Location of the proposed project;

(3) Brief description of the proposed project; the project's purpose; direct and indirect adverse environmental effects the project would cause; any other NWP(s), Regional General Permit(s), or Individual Permit(s) used or intended to be used to authorize any part of the proposed project or any related activity. Sketches should be provided when necessary to show that the activity complies with the terms of the NWP (Sketches usually clarify the project and when provided result in a quicker decision.);

(4) For NWPs 7, 12, 14, 18, 21, 34, 38, 39, [Insert: "40,"] 41, 42, and 43, the PCN must also include a delineation of affected special aquatic sites, including wetlands, vegetated shallows (e.g., submerged aquatic vegetation, seagrass beds), and riffle and pool complexes (see paragraph 13(f));

(5) For NWP 7 (Outfall Structures and Maintenance), the PCN must include information regarding the original design capacities and configurations of those areas of the facility where maintenance dredging or excavation is proposed;

(6) For NWP 14 (Linear Transportation Crossings Crossings [Replace with: "Projects"]), the PCN must include a compensatory mitigation proposal to offset permanent losses of waters of the US and a statement describing how temporary losses of waters of the US will be minimized to the maximum extent practicable;

(7) For NWP 21 (Surface Coal Mining Activities), the PCN must include an Office of Surface Mining (OSM) or state-approved mitigation plan, if applicable. To

be authorized by this NWP, the District Engineer must determine that the activity complies with the terms and conditions of the NWP and that the adverse environmental effects are minimal both individually and cumulatively and must notify the project sponsor of this determination in writing;

(8) For NWP 27 (Stream and Wetland Restoration [Insert: "Activities"]), the PCN must include documentation of the prior condition of the site that will be reverted by the permittee;

(9) For NWP 29 (Single-Family Housing), the PCN must also include:

(i) Any past use of this NWP by the Individual Permittee and/or the permittee's spouse;

(ii) A statement that the single-family housing activity is for a personal residence of the permittee;

(iii) A description of the entire parcel, including its size, and a delineation of wetlands. For the purpose of this NWP, parcels of land measuring 1/4-acre or less will not require a formal on-site delineation. However, the applicant shall provide an indication of where the wetlands are and the amount of wetlands that exists on the property. For parcels greater than 1/4-acre in size, formal wetland delineation must be prepared in accordance with the current method required by the Corps. (See paragraph 13(f));

(iv) A written description of all land (including, if available, legal descriptions) owned by the prospective permittee and/or the prospective permittee's spouse, within a one mile radius of the parcel, in any form of ownership (including any land owned as a partner, corporation, joint tenant, co-tenant, or as a tenant-by-the-entirety) and any land on which a purchase and sale agreement or other contract for sale or purchase has been executed;

(10) For NWP 31 (Maintenance of Existing Flood Control Projects [Replace with: "Facilities"]), the prospective permittee must either notify the District Engineer with a PCN prior to each maintenance activity or submit a five year (or less) maintenance plan. In addition, the PCN must include all of the following:

(i) Sufficient baseline information identifying the approved channel depths and configurations and existing facilities. Minor deviations are authorized, provided the approved flood control protection or drainage is not increased;

(ii) A delineation of any affected special aquatic sites, including wetlands; and,

(iii) Location of the dredged material disposal site;

(11) For NWP 33 (Temporary Construction, Access, and Dewatering), the PCN must also include a restoration plan of reasonable measures to avoid and minimize adverse effects to aquatic resources;

(12) For NWPs 39, 43 and 44, the PCN must also include a written statement to the District Engineer explaining how avoidance and minimization for losses of waters of the US were achieved on the project site;

(13) For NWP 39 and NWP 42, the PCN must include a compensatory mitigation proposal to offset losses of waters of the US or justification explaining why compensatory mitigation should not be required. For discharges that cause the loss of greater than 300 linear feet of an intermittent stream bed, to be authorized, the District Engineer must determine that the activity complies with the other terms and conditions of the NWP, determine adverse environmental effects are minimal both individually and cumulatively, and waive the limitation on stream impacts in writing before the permittee may proceed;

(14) For NWP 40 (Agricultural Activities), the PCN must include a compensatory mitigation proposal to offset losses of waters of the US. This NWP does not authorize the relocation of greater than 300 linear-feet of existing serviceable drainage ditches constructed in non-tidal streams unless, for drainage ditches constructed in intermittent nontidal streams, the District Engineer waives this criterion in writing, and the District Engineer has determined that the project complies with all terms and conditions of this NWP, and that any adverse impacts of the project on the aquatic environment are minimal, both individually and cumulatively;

(15) For NWP 43 (Stormwater Management Facilities), the PCN must include, for the construction of new stormwater management facilities, a maintenance plan (in accordance with state and local requirements, if applicable) and a compensatory mitigation proposal to offset losses of waters of the US. For discharges that cause the loss of greater than 300 linear feet of an intermittent stream bed, to be authorized, the District Engineer must determine that the activity complies with the other terms and conditions of the NWP, determine adverse environmental effects are minimal both individually and cumulatively, and waive the limitation on stream impacts in writing before the permittee may proceed;

(16) For NWP 44 (Mining Activities), the PCN must include a description of all waters of the US adversely affected by the project, a description of measures taken to minimize adverse effects to waters of the US, a description of measures taken to comply with the criteria of the NWP, and a reclamation plan (for all aggregate mining activities in isolated waters and non tidal wetlands adjacent to headwaters and any hard rock/mineral mining activities);

(17) For activities that may adversely affect Federally-listed endangered or threatened species, the PCN must include the name(s) of those endangered or threatened species that may be affected by the proposed work or utilize the designated critical habitat that may be affected by the proposed work; and

(18) For activities that may affect historic properties listed in, or eligible for listing in, the National Register of Historic Places, the PCN must state which historic property may be affected by the proposed work or include a vicinity map indicating the location of the historic property.

(c) **Form of Notification:** The standard Individual Permit application form (Form ENG 4345) may be used as the notification but must clearly indicate that it is a PCN and must include all of the information required in (b) (1)–(18) of General Condition 13. A letter containing the requisite information may also be used.

(d) **District Engineer's Decision:** In reviewing the PCN for the proposed activity, the District Engineer will determine whether the activity authorized by the NWP will result in more than minimal individual or cumulative adverse environmental effects or may be contrary to the public interest. The prospective permittee may submit a proposed mitigation plan with the PCN to expedite the process. The District Engineer will consider any proposed compensatory mitigation the applicant has included in the proposal in determining whether the net adverse environmental effects to the aquatic environment of the proposed work are minimal. If the District Engineer determines that the activity complies with the terms and conditions of the NWP and that the adverse effects on the aquatic environment are minimal, after considering mitigation, the District Engineer will notify the permittee and include any conditions the District Engineer deems necessary. The District Engineer must approve any compensatory mitigation proposal before the permittee commences work. If the prospective permittee is required to submit a compensatory mitigation proposal with the PCN, the proposal may be either conceptual or detailed. If the prospective permittee elects to submit a compensatory mitigation plan with the PCN, the District Engineer will expeditiously review the proposed compensatory mitigation plan. The District Engineer must review the plan within 45 days of receiving a complete PCN and determine whether the conceptual or specific proposed mitigation would ensure no more than minimal adverse effects on the aquatic environment. If the net adverse effects of the project on the aquatic environment (after consideration of the compensatory mitigation proposal) are determined by the District Engineer to be minimal, the District Engineer will provide a timely written response to the applicant. The response will state that the project can proceed under the terms and conditions of the NWP.

If the District Engineer determines that the adverse effects of the proposed work are more than minimal, then the District Engineer will notify the applicant either:

(1) That the project does not qualify for authorization under the NWP and instruct the applicant on the procedures to seek authorization under an Individual Permit;

(2) that the project is authorized under the NWP subject to the applicant's submission of a mitigation proposal that would reduce the adverse effects on the aquatic environment to the minimal level; or

(3) that the project is authorized under the NWP with specific modifications or conditions. Where the District Engineer determines that mitigation is required to ensure no more than minimal adverse effects occur to the aquatic environment, the activity will be authorized within the 45-day PCN period. The authorization will include the necessary conceptual or specific mitigation or a requirement that the applicant submit a mitigation proposal that would reduce the adverse effects on the aquatic environment to the minimal level. When conceptual mitigation is included, or a mitigation plan is required under item (2) above, no work in waters of the US will occur until the District Engineer has approved a specific mitigation plan.

(e) **Agency Coordination:** The District Engineer will consider any comments from Federal and state agencies concerning the proposed activity's compliance with the terms and conditions of the NWPs and the need for mitigation to reduce the project's adverse environmental effects to a minimal level.

For activities requiring notification to the District Engineer that result in the loss of greater than 1/2-acre of waters of the US, the District Engineer will provide immediately (e.g., via facsimile transmission, overnight mail, or other expeditious manner) a copy to the appropriate Federal or state offices (USFWS, state natural resource or water quality agency, EPA, State Historic Preservation Officer (SHPO), and, if appropriate, the NMFS). With the exception of NWP 37, these agencies will then have 10 calendar days from the date the material is transmitted to telephone or fax the District Engineer notice that they intend to provide substantive, site-specific comments. If so contacted by an agency, the District Engineer will wait an additional 15 calendar days before making a decision on the notification. The District Engineer will fully consider agency comments received within the specified time frame, but will provide no response to the resource agency, except as provided below. The District Engineer will indicate in the administrative record associated with each notification that the resource agencies' concerns were considered. As required by section 305(b)(4)(B) of the Magnuson-Stevens Fishery Conservation and Management Act, the District Engineer will provide a response to NMFS within 30 days of receipt of any Essential Fish Habitat conservation recommendations. Applicants are encouraged to provide the Corps multiple copies of notifications to expedite agency notification.

(f) **Wetland Delineations:** Wetland delineations must be prepared in accordance with the current method required by the Corps (For NWP 29 see paragraph (b)(9)(iii) for parcels less than (1/4-acre in size). The permittee may ask the Corps to delineate the special aquatic site. There may be some delay if the Corps does the delineation. Furthermore, the 45-day period will not start until the wetland delineation has been completed and submitted to the Corps, where appropriate.

14. Compliance Certification. Every permittee who has received NWP verification from the Corps will submit a signed certification regarding the completed work and any required mitigation. The certification will be forwarded by the Corps with the authorization letter and will include:

(a) A statement that the authorized work was done in accordance with the Corps authorization, including any general or specific conditions;

(b) A statement that any required mitigation was completed in accordance with the permit conditions; and

(c) The signature of the permittee certifying the completion of the work and mitigation.

15. Use of Multiple Nationwide Permits. The use of more than one NWP for a single and complete project is prohibited, except when the acreage loss of waters of the US authorized by the NWPs does not exceed the acreage limit of the NWP with the highest specified acreage limit (e.g. if a road crossing over tidal waters is constructed under NWP 14, with associated bank stabilization authorized by NWP 13, the maximum acreage loss of waters of the US for the total project cannot exceed 1/3-acre).

16. Water Supply Intakes. No activity, including structures and work in navigable waters of the US or discharges of dredged or fill material, may occur in the proximity of a public water supply intake except where the activity is for repair of the public water supply intake structures or adjacent bank stabilization.

17. Shellfish Beds. No activity, including structures and work in navigable waters of the US or discharges of dredged or fill material, may occur in areas of concentrated shellfish populations, unless the activity is directly related to a shellfish harvesting activity authorized by NWP 4.

18. Suitable Material. No activity, including structures and work in navigable waters of the US or discharges of dredged or fill material, may consist of unsuitable material (e.g., trash, debris, car bodies, asphalt, etc.) and material used for construction or discharged must be free from toxic pollutants in toxic amounts (see section 307 of the CWA).

19. Mitigation. The District Engineer will consider the factors discussed below when determining the acceptability of appropriate and practicable mitigation necessary to offset adverse effects on the aquatic environment that are more than minimal.

(a) The project must be designed and constructed to avoid and minimize adverse effects to waters of the US to the maximum extent practicable at the project site (i.e., on site).

(b) Mitigation in all its forms (avoiding, minimizing, rectifying, reducing or compensating) will be required to the extent necessary to ensure that the adverse effects to the aquatic environment are minimal.

(c) Compensatory mitigation at a minimum one-for-one ratio will be required for all wetland impacts requiring a PCN, unless the District Engineer determines in writing that some other form of mitigation would be more environmentally appropriate and provides a project-specific waiver of this requirement. Consistent with National policy, the District Engineer will establish a preference for restoration of wetlands as compensatory mitigation, with preservation used only in exceptional circumstances.

(d) Compensatory mitigation (i.e., replacement or substitution of aquatic resources for those impacted) will not be used to increase the acreage losses allowed by the acreage limits of some of the NWPs. For example, 1/4-acre of wetlands cannot be created to change a 3/4-acre loss of wetlands to a 1/2-acre loss associated with NWP 39 verification. However, 1/2-acre of created wetlands can be used to reduce the impacts of a 1/2-acre loss of wetlands to the minimum impact level in order to meet the minimal impact requirement associated with NWPs.

(e) To be practicable, the mitigation must be available and capable of being done considering costs, existing technology, and logistics in light of the overall project purposes. Examples of mitigation that may be appropriate and practicable include, but are not limited to: reducing the size of the project; establishing and maintaining wetland or upland vegetated buffers to protect open waters such as streams; and replacing losses of aquatic resource functions and values by creating, restoring, enhancing, or preserving similar functions and values, preferably in the same watershed.

(f) Compensatory mitigation plans for projects in or near streams or other open waters will normally include a requirement for the establishment, maintenance, and legal protection (e.g., easements, deed restrictions) of vegetated buffers to open waters. In many cases, vegetated buffers will be the only compensatory mitigation required. Vegetated buffers should consist of native species. The width of the vegetated buffers required will address documented water quality or aquatic habitat loss concerns. Normally, the vegetated buffer will be 25 to 50 feet wide on each side of the stream, but the District Engineers may require slightly wider vegetated buffers to address documented water quality or habitat loss concerns. Where both wetlands and open waters exist on the project site, the Corps will determine the appropriate compensatory mitigation (e.g., stream buffers or wetlands compensation) based on what is best for the aquatic environment on a watershed basis. In cases where vegetated buffers are determined to be the most appropriate form of compensatory mitigation, the District Engineer may waive or reduce the requirement to provide wetland compensatory mitigation for wetland impacts.

(g) Compensatory mitigation proposals submitted with the "notification" may be either conceptual or detailed. If conceptual plans are approved under the verification, then the Corps will condition the verification to require detailed plans be submitted and approved by the Corps prior to construction of the authorized activity in waters of the US.

(h) Permittees may propose the use of mitigation banks, in-lieu fee arrangements or separate activity specific compensatory mitigation. In all cases that require compensatory mitigation, the mitigation provisions will specify the party responsible for accomplishing and/or complying with the mitigation plan.

20. Spawning Areas. Activities, including structures and work in navigable waters of the US or discharges of dredged or fill material, in spawning areas during spawning seasons must be avoided to the maximum extent practicable. Activities that result in the physical destruction (e.g., excavate, fill, or smother downstream by substantial turbidity) of an important spawning area are not authorized.

21. Management of Water Flows. To the maximum extent practicable, the activity must be designed to maintain preconstruction downstream flow conditions (e.g., location, capacity, and flow rates). Furthermore, the activity must not permanently restrict or impede the passage of normal or expected high flows (unless the primary purpose of the fill is to impound waters) and the structure or discharge of dredged or fill material must withstand expected high flows. The activity must, to the maximum extent practicable, provide for retaining excess flows from the site, provide for maintaining surface flow rates from the site similar to preconstruction conditions, and provide for not increasing water flows from the project site, relocating water, or redirecting water flow beyond preconstruction conditions. Stream channelizing will be reduced to the minimal amount necessary, and the activity must, to the maximum extent practicable, reduce adverse effects such as flooding or erosion downstream and upstream of the project site, unless the activity is part of a larger system designed to manage water flows. In most cases, it will not be a requirement to conduct detailed studies and monitoring of water flow.

This condition is only applicable to projects that have the potential to affect waterflows. While appropriate measures must be taken, it is not necessary to conduct detailed studies to identify such measures or require monitoring to ensure their effectiveness. Normally, the Corps will defer to state and local authorities regarding management of water flow.

22. Adverse Effects from Impoundments. If the activity creates an impoundment of water, adverse effects to the aquatic system due to the acceleration of the passage of water, and/or the restricting its flow shall be minimized to the maximum extent practicable. This includes structures and work in navigable waters of the US, or discharges of dredged or fill material.

23. Waterfowl Breeding Areas. Activities, including structures and work in navigable waters of the US or discharges of dredged or fill material, into breeding areas for migratory waterfowl must be avoided to the maximum extent practicable.

24. Removal of Temporary Fills. Any temporary fills must be removed in their entirety and the affected areas returned to their preexisting elevation.

25. Designated Critical Resource Waters. Critical resource waters include, NOAA-designated marine sanctuaries, National Estuarine Research Reserves, National Wild and Scenic Rivers, critical habitat for Federally listed threatened and endangered species, coral reefs, state natural heritage sites, and outstanding national resource waters or other waters officially designated by a state as having particular environmental or ecological significance and identified by the District Engineer after notice and opportunity for public comment. The District Engineer may also designate additional critical resource waters after notice and opportunity for comment.

(a) Except as noted below, discharges of dredged or fill material into waters of the US are not authorized by NWPs 7, 12, 14, 16, 17, 21, 29, 31, 35, 39, 40, 42, 43, and 44 for any activity within, or directly affecting, critical resource waters, including wetlands adjacent to such waters. Discharges of dredged or fill materials into waters of the US may be authorized by the above NWPs in National Wild and Scenic Rivers if the activity complies with General Condition 7. Further, such discharges may be authorized in designated critical habitat for Federally listed threatened or endangered species if the activity complies with General Condition 11 and the USFWS or the NMFS has concurred in a determination of compliance with this condition.

(b) For NWPs 3, 8, 10, 13, 15, 18, 19, 22, 23, 25, 27, 28, 30, 33, 34, 36, 37, and 38, notification is required in accordance with General Condition 13, for any activity proposed in the designated critical resource waters including wetlands adjacent to those waters. The District Engineer may authorize activities under these NWPs only after it is determined that the impacts to the critical resource waters will be no more than minimal.

26. Fills Within 100-Year Floodplains. For purposes of this General Condition, 100-year floodplains will be identified through the existing Federal Emergency Management Agency's (FEMA) Flood Insurance Rate Maps or FEMA-approved local floodplain maps.

(a) **Discharges in Floodplain; Below Headwaters.** Discharges of dredged or fill material into waters of the US within the mapped 100-year floodplain, below headwaters (i.e. five cfs), resulting in permanent above-grade fills, are not authorized by NWPs 39, 40, 42, 43, and 44.

(b) **Discharges in Floodway; Above Headwaters.** Discharges of dredged or fill material into waters of the US within the FEMA or locally mapped floodway, resulting in permanent above-grade fills, are not authorized by NWPs 39, 40, 42, and 44.

(c) The permittee must comply with any applicable FEMA-approved state or local floodplain management requirements.

27. Construction Period. For activities that have not been verified by the Corps and the project was commenced or under contract to commence by the expiration date of the NWP (or modification or revocation date), the work must be completed within 12 months after such date (including any modification that affects the project).

For activities that have been verified and the project was commenced or under contract to commence within the verification period, the work must be completed by the date determined by the Corps.

For projects that have been verified by the Corps, an extension of a Corps approved completion date maybe requested. This request must be submitted at least one month before the previously approved completion date.

D. Further Information

1. District Engineers have authority to determine if an activity complies with the terms and conditions of an NWP.

2. NWPs do not obviate the need to obtain other Federal, state, or local permits, approvals, or authorizations required by law.

3. NWPs do not grant any property rights or exclusive privileges.

4. NWPs do not authorize any injury to the property or rights of others.

5. NWPs do not authorize interference with any existing or proposed Federal project.

E. Definitions

Best Management Practices (BMPs): BMPs are policies, practices, procedures, or structures implemented to mitigate the adverse environmental effects on surface water quality resulting from development. BMPs are categorized as structural or nonstructural. A BMP policy may affect the limits on a development.

Compensatory Mitigation: For purposes of Section 10/404, compensatory mitigation is the restoration, creation, enhancement, or in exceptional circumstances, preservation of wetlands and/or other aquatic resources for the purpose of compensating for unavoidable adverse impacts which remain after all appropriate and practicable avoidance and minimization has been achieved.

Creation: The establishment of a wetland or other aquatic resource where one did not formerly exist.

Enhancement: Activities conducted in existing wetlands or other aquatic resources that increase one or more aquatic functions.

Ephemeral Stream: An ephemeral stream has flowing water only during and for a short duration after, precipitation events in a typical year. Ephemeral stream beds are located above the water table year-round. Groundwater is not a source of water for the stream. Runoff from rainfall is the primary source of water for stream flow.

Farm Tract: A unit of contiguous land under one ownership that is operated as a farm or part of a farm.

Flood Fringe: That portion of the 100-year floodplain outside of the floodway (often referred to as "floodway fringe").

Floodway: The area regulated by Federal, state, or local requirements to provide for the discharge of the base flood so the cumulative increase in water surface elevation is no more than a designated amount (not to exceed one foot as set by the National Flood Insurance Program) within the 100-year floodplain.

Independent Utility: A test to determine what constitutes a single and complete project in the Corps regulatory program. A project is considered to have independent utility if it would be constructed absent the construction of other projects in the project area. Portions of a multi-phase project that depend upon other phases of the project do not have independent utility. Phases of a project that would be constructed even if the other phases were not built can be considered as separate single and complete projects with independent utility.

Intermittent Stream: An intermittent stream has flowing water during certain times of the year, when groundwater provides water for stream flow. During dry periods, intermittent streams may not have flowing water. Runoff from rainfall is a supplemental source of water for stream flow.

Loss of Waters of the US: Waters of the US that include the filled area and other waters that are permanently adversely affected by flooding, excavation, or drainage because of the regulated activity. Permanent adverse effects include permanent above-grade, at-grade, or below-grade fills that change an aquatic area to dry land, increase the bottom elevation of a waterbody, or change the use of a waterbody. The acreage of loss of waters of the US is the threshold measurement of the impact to existing waters for determining whether a project may qualify for an NWP; it is not a net threshold that is calculated after considering compensatory mitigation that may be used to offset losses of aquatic functions and values. The loss of stream bed includes the linear feet of stream bed that is filled or excavated. [Insert: "Impacts to ephemeral streams are not included in the linear foot measurement of loss of stream bed for the purpose of determining compliance with the linear foot limits of NWPs 39, 40, 42, and 43."] Waters of the US temporarily filled, flooded, excavated, or drained, but restored to preconstruction contours and elevations after construction, are not included in the measurement of loss of waters of the US. Impacts to ephemeral waters are only not included in the acreage or linear foot measurements of loss of waters of the US or loss of stream bed, for the purpose of determining compliance with the threshold limits of the NWPs. [Delete previous sentence]

Non-tidal Wetland: A non-tidal wetland is a wetland (i.e., a water of the US) that is not subject to the ebb and flow of tidal waters. The definition of a wetland can be found at 33 CFR 328.3(b). Non-tidal wetlands contiguous to tidal waters are located landward of the high tide line (i.e., spring high tide line).

Open Water: An area that, during a year with normal patterns of precipitation, has standing or flowing water for sufficient duration to establish an ordinary high water mark. Aquatic vegetation within the area of standing or flowing water is either non-emergent, sparse, or absent. Vegetated shallows are considered to be open waters. The term "open water" includes rivers, streams, lakes, and ponds. For the purposes of the NWPs, this term does not include ephemeral waters.

Perennial Stream: A perennial stream has flowing water year-round during a typical year. The water table is located above the stream bed for most of the year. Groundwater is the primary source of water for stream flow. Runoff from rainfall is a supplemental source of water for stream flow.

Permanent Above-grade Fill: A discharge of dredged or fill material into waters of the US, including wetlands, that results in a substantial increase in ground elevation and permanently converts part or all of the waterbody to dry land. Structural fills authorized by NWPs 3, 25, 36, etc. are not included.

Preservation: The protection of ecologically important wetlands or other aquatic resources in perpetuity through the implementation of appropriate legal and physical mechanisms. Preservation may include protection of upland areas adjacent to wetlands as necessary to ensure protection and/or enhancement of the overall aquatic ecosystem.

Restoration: Re-establishment of wetland and/ or other aquatic resource characteristics and function(s) at a site where they have ceased to exist, or exist in a substantially degraded state.

Riffle and Pool Complex: Riffle and pool complexes are special aquatic sites under the 404(b)(1) Guidelines. Riffle and pool complexes sometimes characterize steep gradient sections of streams. Such stream sections are recognizable by their hydraulic characteristics. The rapid movement of water over a course substrate in riffles results in a rough flow, a turbulent surface, and high dissolved oxygen levels in the water. Pools are deeper areas associated with riffles. A slower stream velocity, a streaming flow, a smooth surface, and a finer substrate characterize pools.

Single and Complete Project: The term "single and complete project" is defined at 33 CFR 330.2(i) as the total project proposed or accomplished by one owner/developer or partnership or other association of owners/developers (see definition of independent utility). For linear projects, the "single and complete project" (i.e., a single and complete crossing) will apply to each crossing of a separate water of the US (i.e., a single waterbody) at that location. An exception is for linear projects crossing a single waterbody several times at separate and distant locations: each crossing is considered a single and complete project. However, individual channels in a braided stream or river, or individual arms of a large, irregularly shaped wetland or lake, etc., are not separate waterbodies.

Stormwater Management: Stormwater management is the mechanism for controlling stormwater runoff for the purposes of reducing downstream erosion, water quality degradation, and flooding and mitigating the adverse effects of changes in land use on the aquatic environment.

Stormwater Management Facilities: Stormwater management facilities are those facilities, including but not limited to, stormwater retention and detention ponds and BMPs, which retain water for a period of time to control runoff and/or improve the quality (i.e., by reducing the concentration of nutrients, sediments, hazardous substances and other pollutants) of stormwater runoff.

Stream Bed: The substrate of the stream channel between the ordinary high water marks. The substrate may be bedrock or inorganic particles that range in size from clay to boulders. Wetlands contiguous to the stream bed, but outside of the ordinary high water marks, are not considered part of the stream bed.

Stream Channelization: The manipulation of a stream channel to increase the rate of water flow through the stream channel. Manipulation may include deepening, widening, straightening, armoring, or other activities that change the stream cross-section or other aspects of stream channel geometry to increase the rate of water flow through the stream channel. A channelized stream remains a water of the US, despite the modifications to increase the rate of water flow.

Tidal Wetland: A tidal wetland is a wetland (i.e., water of the US) that is inundated by tidal waters. The definitions of a wetland and tidal waters can be found at 33 CFR 328.3(b) and 33 CFR 328.3(f), respectively. Tidal waters rise and fall in a predictable and measurable rhythm or cycle due to the gravitational pulls of the moon and sun. Tidal waters end where the rise and fall of the water surface can no longer be practically measured in a predictable rhythm due to masking by other waters, wind, or other effects. Tidal wetlands are located channelward of the high tide line (i.e., spring high tide line) and are inundated by tidal waters two times per lunar month, during spring high tides.

Vegetated Buffer: A vegetated upland or wetland area next to rivers, streams, lakes, or other open waters which separates the open water from developed areas, including agricultural land. Vegetated buffers provide a variety of aquatic habitat functions and values (e.g., aquatic habitat for fish and other aquatic organisms, moderation of water temperature changes, and detritus for aquatic food webs) and help improve or maintain local water quality. A vegetated buffer can be established by maintaining an existing vegetated area or planting native trees, shrubs, and herbaceous plants on land next to open waters. Mowed lawns are not considered vegetated buffers because they provide little or no aquatic habitat functions and values. The establishment and maintenance of vegetated buffers is a method of compensatory mitigation that can be used in conjunction with the restoration, creation, enhancement, or preservation of aquatic habitats to ensure that activities authorized by NWPs result in minimal adverse effects to the aquatic environment. (See General Condition 19.)

Vegetated Shallows: Vegetated shallows are special aquatic sites under the 404(b)(1) Guidelines. They are areas that are permanently inundated and under normal circumstances have rooted aquatic vegetation, such as seagrasses in marine and estuarine systems and a variety of vascular rooted plants in freshwater systems.

Waterbody: A waterbody is any area that in a normal year has water flowing or standing above ground to the extent that evidence of an ordinary high water mark is established. Wetlands contiguous to the waterbody are considered part of the waterbody. [FR Doc. 02–539 Filed 1–14–02; 8:45 am] BILLING CODE 3710–92–P

APPENDIX F

USACE and USEPA Memoranda of Agreement

TABLE OF CONTENTS

U.S. ENVIRONMENTAL PROTECTION AGENCY

Memorandum of Agreement Between the Assistant Administrator for External Affairs and Water U.S. Environmental Protection Agency and the Assistant Secretary of the Army for Civil Works Concerning Regulation of Discharges of Solid Waste Under the Clean Water Act

January 23, 1986

Basis of Agreement

Whereas the Clean Water Act has as its principal objective the requirement "to restore and maintain the chemical, physical, and biological integrity of the Nation's waters; and,

Whereas Section 301 of the Clean Water Act prohibits the discharge of any pollutant into waters of the United States except in compliance with Sections 301, 302, 303, 306, 307, 318, 402, and 404 of the Act; and

Whereas EPA, and States approved by EPA, have been vested with authority to permit discharges of pollutants, other than dredged or fill material, into the waters of the United States pursuant to Section 402 of the Clean Water Act that satisfy the requirements of the Act and regulations developed to administer this program promulgated in 40 CFR 122–125; and

Whereas the Army, and States approved by EPA, have been vested with authority to permit discharges of dredged or fill material into waters of the United States that satisfy the requirements of the Act and regulations developed to administer this program promulgated in 33 CFR 320 et seq. and 40 CFR 230 et seq.; and

Whereas the definitions of the term "fill material" contained in the aforementioned regulations have created uncertainty as to whether Section 402 of the Act or Section 404 is intended to regulate discharges of solid waste materials into waters of the United States for the purpose of disposal of waste; and

Whereas the Resource Conservation and Recovery Act Amendments of 1984 (RCRA) require that certain steps be taken to improve the control of solid waste; and

Whereas interim control of such discharges is necessary to ensure sound management of the Nation's waters and to avoid complications in enforcement actions taken against persons discharging pollutants into waters of the United States without a permit;

The undersigned agencies do hereby agree to use their respective abilities cooperatively in an interim program to control the discharges of solid waste material into waters of the United States.

Procedures

When either agency is aware of a proposed or an unpermitted discharge of solid waste into waters of the United States, the agency will notify the discharger of the prohibition against such discharges as provided in Section 301 of the Clean Water Act. Such notice is not a prerequisite for an enforcement action by either agency.

Normally, if an activity in B.1 above warrants action, EPA will issue an administrative order or file a complaint under Section 309 to control the discharge.

In issuing a notice of violation or administrative order or in filing a complaint, it is not necessary in order to demonstrate a violation of Section 301(a) of the Clean Water Act to identify which permit a permitless discharge should have had. However, after an enforcement action has commenced, a question may be raised by the court, discharger, or other party as to whether a particular discharge having the effect of replacing an aquatic area with dry land or of changing the bottom elevation of a water body meets the primary purpose test for "fill material" in the Corps definition (33 CFR 323.2(k)). For example, such question may be raised in connection with a defense or it may be relevant to the relief to be granted or the terms of a settlement.

To avoid any impediment to prompt resolution of the enforcement action, if such a question arises, a discharge normally will be considered to meet the definition of "fill material" in 33 CFR 323.2(k) for each specific case by consideration of the following factors:

The discharge has as its primary purpose or has one principle purpose of multi-purposes to replace a portion of the waters of the United States with dry land or to raise the bottom elevation.

The discharge results from activities such as road construction or other activities where the material to be discharged is generally identified with construction-type activities.

A principal effect of the discharge is physical loss or physical modification of waters of the United Sates, including smothering of aquatic life or habitat.

The discharge is heterogeneous in nature and of the type normally associated with sanitary landfill discharges.

On the other hand, in the situation in paragraph B.3., a pollutant (other than dredged material) will normally be considered by EPA and the Corps to be subject to Section 402 if it is a discharge in liquid, semi-liquid, or suspended form or if it is a discharge of solid material of homogeneous nature normally associated with single industry wastes, and form a fixed conveyance, or if trucked, from a single site and set of known processes. These materials include placer mining wastes, phosphate mining wastes, titanium mining wastes, sand and gravel wastes, fly ash, and drilling muds. As appropriate, EPA and the Corps will identify additional such materials.

While this document addresses enforcement cases, prospective dischargers who apply for a permit will be encouraged to use the above criteria for purposes of project planning. If a prospective discharger applies for a Section 404 permit based on the considerations in paragraph B.4, or for a Section 402 permit based on the considerations in paragraph B.5, the application will normally be accepted for processing. If a prospective discharger applies for a 404 permit for discharge of materials that might be hazardous, he shall be advised that the discharges of wastes to waters of the United States that are hazardous under RCRA are unlikely to comply with the Section 404(b)(1) Guidelines. To facilitate processing of applications for permits under Sections 402 or 404 for discharges covered by this agreement, an application for such discharge shall not be accepted for processing until the applicant has provided a determination signed by the State or appropriate interstate agency that the proposed discharge will comply with applicable provisions of State law including applicable water quality standards, or evidence of waiver by the State or interstate agency. As mandated under the Clean Water Act, neither a 402 nor a 404 permit will be issued for a discharge of toxic pollutants in toxic amounts. Prospective applicants for a 402 permits shall be advised that the proposed discharge will be evaluated for compliance with the Act, in particular with Sections 101(a), 301, 303, 304, 307, 402, and 405 of the Act.

Determination of Permit

In enforcement cases, where a question arises under paragraph B.3 as to which permit would be required for a permitless discharge, the enforcing agency will determine whether the criteria in paragraph B.4 or B.5, if either, have been satisfied, with concurrence from the other agency. If the enforcing agency concludes that neither set of the criteria has been met and additional analysis is required to determine which Section applies, or if the necessary concurrence is not forthcoming promptly, the Division Engineer and the Regional Administrator (or designees) will consult and determine which permit program is applicable.

In non-enforcement situations, the agency receiving an application shall determine whether it meets the criteria in paragraphs 4 or 5, as the case may be. If the agency determines that the criteria applicable to its permit program have not been meet, it will ask the other agency to determine whether the criteria for the latter's permit have been met. If neither agency determines that the criteria for its permit program have been met, the Division Engineer and the RA (or their designee) shall consult and determine which agency shall process the application in question.

Publication in the Federal Register

Since this Memorandum of Agreement clarifies the definition of fill material with respect to discharges of solid waste into waters of the United States, the parties in this agreement shall jointly publish it in the Federal Register within 45 days after it has been signed.

Effective Date

This agreement shall take effect 90 days after the date of the last signature below and will continue in effect until modified or revoked by agreement of both parties, or revoked by either party alone upon six months written notice.

This agreement automatically expires at such time as EPA has submitted its Report to Congress on the Results of Study of the Adequacy of the Existing Subtitle D Criteria and has published a Notice of Proposed Revisions to the Subtitle D Criteria in the Federal Register, unless the agencies mutually agree that extension of this agreement is needed.

Jennifer Joy Manson /s/
Assistant Administrator for External Affairs,
U.S. Environmental Protection Agency
January 22, 1986

Robert K. Dawson /s/
Assistant Secretary of the Army
(Civil Works)
January 17, 1986

Larry Jensen /s/
Assistant Administrator for Water
U.S. Environmental Protection Agency
January 23, 1986

Memorandum of Agreement Between the Department of the Army and the Environmental Protection Agency Concerning Federal Enforcement for the Section 404 Program of the Clean Water Act

I. PURPOSE AND SCOPE

The United States Department of the Army (Army) and the United States Environmental Protection Agency (EPA) hereby establish policy and procedures pursuant to which they will undertake federal enforcement of the dredged and fill material permit requirements ("Section 404 program") of the Clean Water Act (CWA). The U.S Army Corps of Engineers (Corps) and EPA have enforcement authorities for the Section 404 program, as specified in Sections 301(a), 308, 309, 404(n), and 404(s) of the CWA. In addition, the 1987 Amendments to the CWA (the Water Quality Act of 1987) provide new administrative penalty authority under Section 309(g) for violations of the Section 404 program. For purposes of effective administration of these statutory authorities, this Memorandum of Agreement (MOA) sets forth an appropriate allocation of enforcement responsibilities between EPA and the Corps. The prime goal of the MOA is to strengthen the Section 404 enforcement program by using the expertise, resources and initiative of both agencies in a manner which is effective and efficient in achieving the goals of the CWA.

II. POLICY

A. General. It shall be the policy of the Army and EPA to maintain the integrity of the program through federal enforcement of Section 404 requirements. The basic premise of this effort is to establish a framework for effective Section 404 enforcement with very little overlap. EPA will conduct initial on-site investigations when it is efficient with respect to available time, resources and/or expenditures, and use its authorities as provided in this agreement. In the majority of enforcement cases the Corps, because it has more field resources, will conduct initial investigations and use its authorities as provided in this agreement. This will allow each agency to play a role in enforcement which concentrates its resources in those areas for which its authorities and expertise are best suited. The Corps and EPA are encouraged to consult with each other on cases involving novel or important legal issues and/or technical situations. Assistance from the U.S. Fish and Wildlife Service (FWS), the National Marine Fisheries Service (NMFS) and other federal, state, tribal and local agencies will be sought and accepted when appropriate.

B. Geographic Jurisdictional Determinations. Geographic jurisdictional determinations for a specific case will be made by the investigating agency. If asked for an oral decision, the investigator will caution that oral statements regarding jurisdiction are not an official agency determination. Each agency will advise the other of any problem trends that they become aware of through case by case determinations and initiate interagency discussions or other action to address the issue. (Note: Geographic jurisdictional determinations for "special case" situations and interpretation of Section 404(f) exemptions for "special Section 404(f) matters" will be handled in accordance with the MOA on Geographical Jurisdiction and Section 404(f) of the Section 404 Program.)

C. Violation Determinations. The investigating agency shall be responsible for violation determinations, for example, the need for a permit. Each agency will advise the other of any problem trends that they become aware of through case by case determinations and initiate interagency discussions or other action to address the issue.

D. Lead Enforcement Agency. The Corps will act as the lead enforcement agency for all violations of Corps-issued permits. The Corps will also act as the lead enforcement agency for unpermitted discharge violations which do not meet the criteria for forwarding to EPA, as listed in Section III.D. of this MOA. EPA will act as the lead enforcement agency on all unpermitted discharge violations which meet those criteria. The lead enforcement agency will complete the enforcement action once an investigation has established that a violation exists. A lead enforcement agency decision with regard to any issue in a particular case, including a decision that no enforcement action be taken, is final for that case. This provision does not preclude the lead enforcement agency from referring the matter to the other agency under Sections III.D.2 and III.D.4 of this MOA.

E. Environmental Protection Measures. It is the policy of both agencies to avoid permanent environmental harm caused by the violator's activities by requiring remedial actions or ordering removal and restoration. In those cases where a complete remedy/removal is not appropriate, the violator may be required, in addition to other legal remedies which are appropriate (e.g., payment of administrative penalties) to provide compensatory mitigation to compensate for the harm caused by such illegal actions. Such compensatory mitigation activities shall be placed as an enforceable requirement upon a violator as authorized by law.

III. PROCEDURES

A. Flow chart. The attached flow chart provides an outline of the procedures EPA and the Corps will follow in enforcement cases involving unpermitted discharges. The procedures in (B.), (C.), (D.), (E.) and (F.) below are in a sequence in which they could occur. However, these procedures may be combined in an effort to expedite the enforcement process.

B. Investigation. EPA, if it so requests and upon prior notification to the Corps, will be the investigating agency for unpermitted activities occurring in specially defined geographic areas (e.g., a particular wetland type, areas declared a "special case" within the meaning of the MOA on Geographical Jurisdiction and Section 404(f) of the Section 404 Program). Timing of investigations will be commensurate with agency resources and potential environmental damage. To reduce the potential for duplicative federal effort, each agency should verify prior to initiating an investigation that the other agency does not intend or has not already begun an investigation of the same reported violation. If a violation exists, a field investigation report will be prepared which at a minimum provides a detailed description of the illegal activity, the existing environmental setting, initial view on potential impacts and a recommendation on the need for initial corrective measures. Both agencies agree that investigations must be conducted in a professional, legal manner that will not prejudice future enforcement action on the case. Investigation reports will be provided to the agency selected as the lead on the case.

C. Immediate Enforcement Action. The investigating or lead enforcement agency should inform the responsible parties of the violation and inform them that all illegal activity should cease pending further federal action. A notification letter or administrative order to that effect will be sent in the most expeditious manner. If time allows, an order for initial corrective measures may be included with the notification letter or administrative order. Also, if time allows, input from other federal, state, tribal and local agencies will be considered when determining the need for such initial corrective measures. In all cases the Corps will provide EPA a copy of its violation letters and EPA will provide the Corps copies of its §308 letters and/or §309 administrative orders. These communications will include language requesting the other agency's views and recommendations on the case. The violator will also be notified that the other agency has been contacted.

D. Lead Enforcement Agency Selection. Using the following criteria, the investigating agency will determine which agency will complete action on the enforcement case:

1. EPA will act as the lead enforcement agency when an unpermitted activity involves the following:

a. Repeat Violator(s);

b. Flagrant Violation(s);

c. Where EPA requests a class of cases or a particular case; or

d. The Corps recommends that an EPA administrative penalty action may be warranted.

2. The Corps will act as the lead enforcement agency in all other unpermitted cases not identified in Part III D.1. above. Where EPA notifies the Corps that, because of limited staff resources or other reasons, it will not take action on a specific case, the Corps may take action commensurate with resource availability.

3. The Corps will act as the lead enforcement agency for Corps-issued permit condition violations.

4. Where EPA requests the Corps to take action on a permit condition violation, this MOA establishes a "right of first refusal" for the Corps. Where the Corps notifies EPA that, because of limited staff resources or other reasons, it will not take an action on a permit condition violation case, the EPA may take action commensurate with resource availability. However, a determination by the Corps that the activity is in compliance with the permit will represent a final enforcement decision for that case.

E. Enforcement Response. The lead enforcement agency shall determine, based on its authority, the appropriate enforcement response taking into consideration any views provided by the other agency. An appropriate enforcement response may include an administrative order, administrative penalty complaint, a civil or criminal judicial referral or other appropriate formal enforcement response.

F. Resolution. The lead enforcement agency shall make a final determination that a violation is resolved and notify interested parties so that concurrent enforcement files within another agency can be closed. In addition, the lead enforcement agency shall make arrangements for proper monitoring when required for any remedy/removal, compensatory mitigation or other corrective measures.

G. After-the-Fact Permits. No after-the-fact (ATF) permit application shall be accepted until resolution has been reached through an appropriate enforcement response as determined by the lead enforcement agency (e.g., until all administrative, legal and/or corrective action has been completed, or a decision has been made that no enforcement action is to be taken).

IV. RELATED MATTERS

A. Interagency Agreements. The Army and EPA are encouraged to enter into interagency agreements with other federal, state, tribal and local agencies which will provide assistance to the Corps and EPA in pursuit of Section 404 enforcement activities. For example, the preliminary enforcement site investigations or post-case monitoring activities required to ensure compliance with any enforcement order can be delegated to third parties (e.g., FWS) who agree to assist Corps/ EPA in compliance efforts. However, only the Corps or EPA may make a violation determination and/or pursue an appropriate enforcement response based upon information received from a third party.

B. Corps/EPA Field Agreements. Corps Division or District offices and their respective EPA Regional offices are encouraged to enter into field level agreements to more specifically implement the provisions of this MOA.

C. Data Information Exchange. Data which would enhance either agency's enforcement efforts should be exchanged between the Corps and EPA where available. At a minimum, each agency shall begin to develop a computerized data list of persons receiving ATF permits or that have been subject to a Section 404 enforcement action subsequent to February 4, 1987 (enactment date of the 1987 Clean Water Act Amendments) in order to provide historical compliance data on persons found to have illegally discharged. Such information will help in an administrative penalty action to evaluate the statutory factor concerning history of a violator and will help to determine whether pursuit of a criminal action is appropriate.

V. GENERAL

A. The procedures and responsibilities of each agency specified in this MOA may be delegated to subordinates consistent with established agency procedures.

B. The policy and procedures contained within this MOA do not create any rights, either substantive or procedural, enforceable by any party regarding an enforcement action brought by either agency or by the U.S. Deviation or variance from these MOA procedures will not constitute a defense for violators or others concerned with any Section 404 enforcement action.

C. Nothing in this document is intended to diminish, modify or otherwise affect the statutory or regulatory authorities of either agency. All formal guidance interpreting this MOA shall be issued jointly.

D. This agreement shall take effect 60 days after the date of the last signature below and will continue in effect for five years unless extended, modified or revoked by agreement of both parties, or revoked by either party alone upon six months written notice, prior to that time.

Robert W. Page
Assistant Secretary of the Army
(Civil Works)

Rebecca W. Hanmer
Acting Assistant Administrator for Water
U.S. Environmental Protection Agency

Memorandum of Agreement Between the Department of the Army and the Environmental Protection Agency Concerning the Determination of Mitigation Under the Clean Water Act Section 404(b)(1) Guidelines

February 6, 1990

I. PURPOSE

The United States Environmental Protection Agency (EPA) and the United States Department of the Army (Army) hereby articulate the policy and procedures to be used in the determination of the type and level of mitigation necessary to demonstrate compliance with the Clean Water Act (CWA) Section 404(b)(1) Guidelines ("Guidelines"). This Memorandum of Agreement (MOA) expresses the explicit intent of the Army and EPA to implement the objective of the CWA to restore and maintain the chemical, physical and biological integrity of the Nation's waters, including wetlands. This MOA is specifically limited to the Section 404 Regulatory Program and is written to provide guidance for agency field personnel on the type and level of mitigation which demonstrates compliance with requirements in the Guidelines. The policies and procedures discussed herein are consistent with current Section 404 regulatory practices and are provided in response to questions that have been raised about how the Guidelines are implemented. The MOA does not change the substantive requirements of the Guidelines. It is intended to provide guidance regarding the exercise of discretion under the Guidelines.

Although the Guidelines are clearly applicable to all discharges of dredged or fill material, including general permits and Corps of Engineers (Corps) civil works projects, this MOA focuses on standard permits (33 CFR 325(b)(1)).[1] This focus is intended solely to reflect the unique procedural aspects associated with the review of standard permits, and does not obviate the need for other regulated activities to comply fully with the Guidelines. EPA and Army will seek to develop supplemental guidance for other regulated activities consistent with the policies and principles established in this document.

This MOA provides guidance to Corps and EPA personnel for implementing the Guidelines and must be adhered to when

1. Standard permits are those individual permits which have been processed through application of the Corps public interest review procedures (33 CFR 325) and EPA's Section 404(b)(1) Guidelines, including public notice and receipt of comments. Standard permits do not include letters of permission, regional permits, nationwide permits, or programmatic permits.

considering mitigation requirements for standard permit applications. The Corps will use this MOA when making its determinations of compliance with the Guidelines with respect to mitigation for standard permit applications. EPA will use this MOA in developing its position on compliance with the Guidelines for proposed discharges and will reflect this MOA when commenting on standard permit applications.

II. POLICY

A. The Council on Environmental Quality (CEQ) has defined mitigation in its regulations at 40 CFR 1508.20 to include: avoiding impacts, minimizing impacts, rectifying impacts, reducing impacts over time, and compensating for impacts. The Guidelines establish environmental criteria which must be met for activities to be permitted under Section 404.[2] The type of mitigation enumerated by CEQ are compatible with the requirements of the Guidelines; however, as a practical matter, they can be combined to form three general types: avoidance, minimization and compensatory mitigation. The remainder of this MOA will speak in terms of these general types of mitigation.

B. The Clean Water Act and the Guidelines set forth a goal of restoring and maintaining existing aquatic resources. The Corps will strive to avoid adverse impacts and offset unavoidable adverse impacts to existing aquatic resources, and for wetlands, will strive to achieve a goal of no overall net loss of values and functions. In focusing the goal on no overall net loss to wetlands only, EPA and Army have explicitly recognized the special significance of the nation's wetlands resources. This special recognition of wetlands resources does not in any manner diminish the value of other waters of the United States, which are often of high value. All waters of the United States, such as streams, rivers, lakes, etc., will be accorded the full measure of protection under the Guidelines, including the requirements for appropriate and practicable mitigation. The determination of what level of mitigation constitutes "appropriate" mitigation is based solely on the values and functions of the aquatic resource that will be impacted. "Practicable" is defined at Section 230.3(q) of the Guidelines.[3] However, the level of mitigation determined to be appropriate and practicable under Section 230.10(d) may lead to individual permit decisions which do not fully meet this goal because the mitigation measures necessary to meet this goal are not feasible, not practicable, or would accomplish only inconsequential reductions in impacts. Consequently, it is recognized that no net loss of wetlands functions and values may not be achieved in each and every permit action. However, it remains a goal of the Section 404 regulatory program to contribute to the national goal of no overall net loss of the nation's remaining wetlands base. EPA and Army are committed to working with others through the Administration's interagency task force and other avenues to help achieve this national goal.

C. In evaluating standard Section 404 permit applications, as a practical matter, information on all facets of a project, including potential mitigation, is typically gathered and reviewed at the same time. The Corps, except as indicated below, first makes a determination that potential impact have been avoided to the maximum extent practicable; remaining unavoidable impacts will then be mitigated to the extent appropriate and practicable by requiring steps to minimize impacts, and, finally, compensate for aquatic resource values. This sequence is considered satisfied where the proposed mitigation is in accordance with specific provisions of a Corps and EPA approved comprehensive plan that ensures compliance with the compensation requirements of the Section 404(b)(1) Guidelines (examples of such comprehensive plans may include Special Area Management Plans, Advanced Identification areas (Section 230.80) and State Coastal Zone Management Plans). It may be appropriate to deviate from the sequence when EPA and the Corps agree the proposed discharge is necessary to avoid environmental harm (e.g. to protect a natural aquatic community from saltwater intrusion, chemical contamination, or other deleterious physical or chemical impacts), or EPA and the Corps agree that the proposed discharge can reasonably be expected to result in environmental gain or insignificant environmental losses.

In determining "appropriate and practicable" measures to offset unavoidable impact, such measures should be appropriate to the scope and degree of those impacts and practicable in terms of cost, existing technology, and logistics in light of overall project purposes. The Corps will give full consideration to the views of the resource agencies when making this determination.

1. *Avoidance.*[4] Section 230.10(a) allows permit issuance for only the least environmentally damaging practicable alternative.[5] The thrust of this section on alternatives is avoidance of impacts. Section 230.10(a) requires that no discharge shall be permitted if there is a practicable alternative to the proposed discharge which would have less adverse impact to the aquatic ecosystem, so long as the alternative does not have other significant adverse environmental consequences. In addition, Section 230.10(a)(3) sets forth rebuttable presumptions that 1) alternatives for non-water dependent activities that do not involve special aquatic sites[6] are available and 2) alternatives that do not involve special aquatic sites have less adverse impact on the aquatic environment. Compensatory mitigation may not be used as a method to reduce environmental impacts in the evaluation of the least environmentally damaging practicable alternatives for the purposes of requirements under Section 230.10(a).

2. *Minimization.* Section 230.10(d) states that appropriate and practicable steps to minimize the adverse impacts will be required through project modifications and permit conditions. Subpart H of the Guidelines describes several (but not all) means of minimizing impacts of an activity.

3. *Compensatory Mitigation.* Appropriate and practicable compensatory mitigation is required for unavoidable adverse impacts which remain after all appropriate and practicable minimization has been required. Compensatory actions (e.g., restoration of existing degraded wetlands or creation of man-made wetlands) should be undertaken when practicable, in areas adjacent or continuous to the discharge site (on-site compensatory mitigation). If on-site compensatory mitigation is not practicable, off-site compensatory mitigation should be undertaken in the same geographic area if practicable (i.e., in close proximity and, to the extent possible, the same watershed). In determining compensatory mitigation, the functional values lost by the resource to be impacted must be considered. Generally, in-kind compensatory mitigation is preferable to out-of-kind. There is continued uncertainty regarding the success of wetland creation or other habitat development. Therefore, in determining the nature and extent of habitat development of this type, careful consideration should be given to

2. (except where Section 404(b)(2) applies).

3. Section 230.3(q) of the Guidelines reads as follows: "The term practicable means available and capable of being done after taking into consideration cost, existing technology, and logistics in light of overall project purposes." (Emphasis supplied.)

4. Avoidance as used in Section 404(b)(1) Guidelines and this MOA does not include compensatory mitigation.

5. It is important to recognize that there are circumstances where the impacts of the project are so significant that even if alternatives are not available, the discharge may not be permitted regardless of the compensatory mitigation proposed (40 CFR 230.10(c)).

6. Special aquatic sites include sanctuaries and refuges, wetlands, mud flats, vegetated shallows, coral reefs and riffle pool complexes.

its likelihood of success. Because the likelihood of success is greater and the impacts to potentially valuable uplands are reduced, restoration should be the first option considered.

In the situation where the Corps is evaluating a project where a permit issued by another agency requires compensatory mitigation, the Corps may consider that mitigation as part of the overall application for purposes of public notice, but avoidance and minimization shall still be sought.

Mitigation banking may be an acceptable form of compensatory mitigation under specific criteria designed to ensure an environmentally successful bank. Where a mitigation bank has been approved by EPA and the Corps for purposes of providing compensatory mitigation for specific identified projects, use of that mitigation bank for those particular projects is considered as meeting the objective of Section II.C.3 of this MOA, regardless of the practicability of other forms of compensatory mitigation. Additional guidance on mitigation banking will be provided. Simple purchase or "preservation" of existing wetlands resources may in only exceptional circumstances be accepted as compensatory mitigation. EPA and Army will develop specific guidance for preservation in the context of compensatory mitigation at a later date.

III. OTHER PROCEDURES

A. Potential applicants for major projects should be encouraged to arrange preapplication meetings with the Corps and appropriate federal, state, or Indian tribal, and local authorities to determine requirements and documentation required for proposed permit evaluations. As a result of such meetings, the applicant often revises a proposal to avoid or minimize adverse impacts after developing an understanding of the Guidelines requirements by which a future Section 404 permit decision will be made, in addition to gaining understanding of other state or tribal, or local requirements. Compliance with other statutes, requirements and reviews, such as NEPA and the Corps public interest review, may not in and of themselves satisfy the requirements prescribed in the Guidelines.

B. In achieving the goals of the CWA, the Corps will strive to avoid adverse impacts and offset unavoidable adverse impacts to existing aquatic resources. Measures which can accomplish this can be identified only through resource assessments tailored to the site performed by qualified professionals because ecological characteristics of each aquatic site are unique. Functional values should be assessed by applying aquatic site assessment techniques generally recognized by experts in the field and/or the best professional judgement of federal and state agency representatives, provided such assessments fully consider ecological functions included in the Guidelines. The objective of mitigation for unavoidable impacts is to offset environmental losses. Additionally for wetlands, such mitigation should provide, at a minimum, one for one functional replacement (i.e., no net loss of values), with an adequate margin of safety to reflect the expected degree of success associated with the mitigation plan, recognizing that this minimum requirement may not be appropriate and practicable and thus may not be relevant in all cases, as discussed in Section II.B of this MOA.[7] In the absence of more definitive information on the functions and values of specific wetland sites, a minimum of 1 to 1 acreage replacement may be used as a reasonable surrogate for no net loss of functions and values. However, this ratio may be greater where the functional values of the area being impacted are demonstrably high and the replacement wetlands are of lower functional value or the likelihood of success of the mitigation project is low. Conversely, the ration may be less than 1 to 1 for areas where the functional values associated with the area being impacted are demonstrably low and the likelihood of success associated with the mitigation proposal is high.

C. The Guidelines are the environmental standards for Section 404 permit issuance under the CWA. Aspects of a proposed project may be affected through a determination of requirements needed to comply with the Guidelines to achieve these CWA environmental goals.

D. Monitoring is an important aspect of mitigation, especially in areas of scientific uncertainty. Monitoring should be directed toward determining whether permit conditions are complied with and whether the purpose intended to be served by the conditions are actually achieved. Any time it is determined that a permittee is in non-compliance with the mitigation requirements of the permit, the Corps will take action in accordance with 33 CFR Part 326. Monitoring should not be required for purposes other than these, although information for other uses may accrue from the monitoring requirements. For projects to be permitted involving mitigation with higher levels of scientific uncertainty, such as some forms of compensatory mitigation, long term monitoring, reporting and potential remedial action should be required. This can be required of the applicant through permit conditions.

E. Mitigation requirements shall be conditions of standard Section 404 permits. Army regulations authorize mitigation requirements to be added as special conditions to an Army permit to satisfy legal requirements (e.g. conditions necessary to satisfy the Guidelines) [33 CFR 325.4(a)]. This ensures legal enforceability of the mitigation conditions and enhances the level of compliance. If the mitigation plan necessary to ensure compliance with the Guidelines is not reasonable implementable or enforceable, the permit shall be denied.

F. Nothing in this document, is intended to diminish, modify or otherwise affect the statutory or regulatory authorities of the agencies involved. Furthermore, formal policy guidance on or interpretation of this document shall be issued jointly.

G. This MOA shall take affect on February 8, 1990, and will apply to those completed standard permit applications which are received on or after that date. This MOA may be modified or revoked by agreement of both parties, or revoked by either party alone upon six (6) months written notice.

Robert W. Page /s/
Assistant Secretary of
the Army, Civil Works
February 6, 1990

LaJuna S. Wilcher /s/
Assistant Administrator
for Water, U.S. Environmental
Protection Agency
February 6, 1990

7. For example, there are certain areas where, due to hydrological conditions, the technology for restoration or creation of wetlands may not be available at present, or may otherwise be impracticable. In addition, avoidance, minimization, and compensatory mitigation may not be practicable where there is a high proportion of land which is wetlands. EPA and Army, at present, are discussing with representatives of the oil industry, the potential for a program of accelerated rehabilitation of abandoned oil facilities on the North Slope to serve as a vehicle for satisfying necessary compensation requirements.

Clean Water Act Section 404(q) Memorandum of Agreement Between the Environmental Protection Agency and the Department of the Army

August 11, 1992

Authority: Section 404(q) of the Clean Water Act, 33 U.S. C. 1344(q)

Purpose: Establish policies and procedures to implement Section 404(q) of the Clean Water Act to "minimize, to the maximum extent practicable, duplication, needless paperwork and delays in the issuance of permits."

Applicability: This agreement shall apply to Regulatory authorities under: a) Section 10 of the Rivers and Harbors Act of 1989; b) Section 404 of the Clean Water Act; and c) Section 103 of the Marine Protection, Research and Sanctuaries Act.

General Rules: Policy and procedures for the Department of the Army Regulatory Program are established in 33 CFR Parts 320 through 330, and 40 CFR Part 230.

Organization: This Memorandum of Agreement (MOA) is subdivided into four distinct parts. The procedure for each part are specific to that part and do not necessarily relate to other parts. For example, different signature levels are established for Parts II, III, and IV.

PART I—BACKGROUND

1. The Army Corps of Engineers is solely responsible for making final permit decisions pursuant to Section 10, Section 404(a), and Section 103, including final determinations of compliance with the Corps permit regulations, the Section 404(b)(1) Guidelines, and Section 7(a)(2) of the Endangered Species Act. As such, the Corps will act as the project manager for the evaluation of all permit applications. As the project manager, the Corps is responsible for requesting and evaluating information concerning all permit applications. The Corps will obtain and utilize this information in a manner that moves, as rapidly as practical, the regulatory process towards a final permit decision. The Corps will not evaluate applications as a project opponent or advocate–but instead will maintain an objective evaluation, fully considering all relevant factors. The Corps will fully consider EPA's comments when determining with the National Environmental Policy Act, and other relevant statutes, regulations, and policies. The Corps will also fully consider the EPA's views when determining whether to issue the permit, to issue the permit with conditions and/or mitigation, or to deny the permit.

2. It is recognized that the EPA has an important role in the Department of the Army Regulatory Program under the Clean Water Act, National Environmental Policy Act, and other relevant statutes. When providing comments, only substantive, project-related information (within EPA's area of expertise and authority) on the impacts of activities being evaluated by the Corps and appropriate and practicable measures to mitigate adverse impacts will be submitted. Pursuant to its authority under Section 404(b)(1) of the Clean Water Act, the EPA may provide comments to the Corps identifying its views regarding compliance with the Section 404(b)(1) Guidelines. The comments will be submitted within the time frames established in this agreement and applicable regulation.

3. National or regional issues relating to resources, policy, procedures, and regulation interpretation, can be elevated by either agency to their respective Washington Headquarters for resolution as prescribed in Part III–EVALUATION OF POLICY ISSUES. Individual permit decisions will not be delayed during the policy issue elevation process. Elevation of issues related to specific individual permit cases will be limited to those cases that involve aquatic resources of national importance. Procedures for elevation of such specific cases are provided in PART IV–ELEVATION OF INDIVIDUAL PERMIT DECISIONS.

4. For projects of other Federal agencies and Federally assisted projects for which a Federal agency takes responsibility for environmental analysis and documentation, Army will accept, where appropriate and legally permissible, the environmental documentation and decisions of those agencies.

5. This agreement does not diminish either Agency's authority to decide whether a particular individual permit should be granted, including determining whether the project is in compliance with the Section 404(b)(1) Guidelines, or the Administrator's authority under Section 404(c) of the Clean Water Act.

6. The officials identified in this MOA cannot delegate their responsibilities unless specifically provided for in this MOA.

7. Days referred to in this MOA are calendar days. If the end of the specified time period falls on a weekend or a holiday, the last calendar day will be the first business day following the weekend or holiday. The end of the specific time period shall mean the close of the business day on the last day of the specified time period.

8. This agreement is effective immediately upon the date of the last signature and will continue in effect until modified or revoked by agreement of both parties, or revoked by either party alone upon six months written notice.

9. The Memorandum of Agreement between the Administrator of the Environmental Protection Agency and the Secretary of the Army of the Section 404(q) of the Clean Water Act dated November 12, 1985, is terminated. Those permit applications which have been elevated to the Assistant Secretary of the Army for Civil Works (ASA(CW)) under the November 12, 1985, MOA shall be processed according to its terms. Those permit applications for which Notices of Intent to Issue have been sent by the District Engineer in accordance with paragraph 7.b of the November 12, 1985, MOA shall be governed by that MOA. All other permit applications shall be governed by this agreement. For permit applications where the basic or extended comment period has closed before the signature date of this MOA, the Regional Administrator has 15 calendar days from the date of the last signature below to indicate which individual permit cases will be governed under Part IV by sending the District Engineer the letter required in Part IV, paragraph 3(b).

PART II—COORDINATION PROCEDURES

1. Purpose: The purpose of Part II is to provide and encourage communication and full consideration of each agencies' views concerning proposed projects within the resource limits of each agency and the time constraints of the regulatory process.

2. District Engineers and the Regional Administrators are encouraged to develop, within six months of the date of this MOA, written procedures to ensure effective interagency coordination and to discuss issues, expedite comments, foster strong professional partnerships and cooperative working relationships. These professional partnerships will be based on EPA providing substantive, project specific comments and the Corps giving full consideration to EPA's recommendations as the Corps makes its determination of compliance with the Section 404(b)(1) Guidelines and the decision on the permit application. The procedures will encourage, to the extent possible:

   a. interagency pre-application consultation with prospective applicants;

   b. interagency site visits;

   c. interagency meeting(s) with applicants;

   d. cooperation in acquiring and conveying site specific information needed by either agency to fulfill its responsibilities;

   e. consistent with the time frames set forth in this MOA, an informal process for the timely resolution of issues at the field level to ensure that the permit evaluation proceeds as rapidly as practical.

3. The Regional Administrator will inform the District Engineer, in writing, of the EPA officials who are authorized to provide official EPA comments, including, where appropriate, by category of activity or geographic area. All official EPA comments will be signed by either the Regional Administrator or the designated official or an individual acting for the Regional Administrator or acting for the designated EPA official. Two officials will be designated in EPA Region X to provide for special circumstances in Alaska. Comments signed by any of the above mentioned officials will be considered EPA's response in accordance with Part II of this MOA. Notwithstanding the above, certain actions described in Part IV require the actual signature of the Regional Administrator or Acting Regional Administrator.

4. The Corps will ensure the timely receipt (within 2–3 days from the date of issuance) of public notices by EPA. EPA comments will be submitted in writing during the basic comment period specified in the public notice. To the maximum extent practical, EPA will immediately provide the Corps project manager with a faxed copy of its signed comments. Where the basic comment period is less than 30 calendar days and the situation is an emergency, the District Engineer (or designee) shall, upon written or electronically transmitted request of an official authorized to provide official EPA comments, extend the comment period to 30 calendar days. An extension beyond 30 calendar days from the date of the public notice, must be requested in writing by the Regional Administrator or designee. The written request must be received three calendar days prior to the end of the basic comment period and must demonstrate the reason for the extension (e.g., a joint coordination meeting occurs near the end of the comment period and EPA needs additional time to prepare substantive comments). The District Engineer or his designee will respond, in writing, within three calendar days of receipt of the request letter. If the District Engineer or his designee denies the request for extension within three calendar days prior to the end of the basic comment period, the EPA will have five calendar days from the receipt of the denial letter to submit final EPA comments. The maximum comment period, including extension, will not exceed 60 calendar days, unless sought by the applicant.

5. Consistent with the procedures in Part IV, at the conclusion of the comment period, the Corps will proceed to final action on the permit application. The Corps will consider all comments submitted by EPA pursuant to Part IV, paragraph 3(a) and 3(b).

6. The Corps may, in certain cases, request additional comments from or discuss issues relevant to the project with EPA after the close of the comment period to either clarify matters or obtain information relevant to the permit decision.

7. Consistent with Part IV, if the District Engineer's decision is to issue the permit over the objections of the EPA Regional Administrator or to issue the permit without conditions recommended by the EPA Regional Administrator, the District Engineer will send a copy of the decision to the EPA commenting official.

PART III—ELEVATION OF POLICY ISSUES

1. Purpose: The purpose of Part III is to provide procedures for policy issue coordination and resolution.

2. If either agency considers that the nature of an action or series of action raises concerns regarding the application of existing policy or procedure, or procedural failures in agency coordination, the District or Division Engineer, or Regional Administrator (or designee) may initiate policy implementation review between the District and/or Division Engineer (or designee) and the EPA Regional Administrator (or designee) through written notification. The written notification will describe the issue in sufficient detail and provide recommendations for resolving the issue. The District Engineer or Division Engineer (or designee), depending on the level of the issue, or the Regional Administrator (or designee) will resolve the issue within 60 calendar days of receipt of written notification to initiate policy implementation review.

3. In the context of Part III of this MOA, "resolve" means to review the issue, obtain the views of the requesting party, discuss those views as appropriate, fully consider those views, and then make the final determination, in writing, regarding the particular resource, policy, procedure, or regulation interpretation.

4. If during consultation, the Regional Administrator (or designee) or the Corps (District Engineer or Division Engineer, or designee) determine the issue cannot should not be resolved at the field level, or that an issue has broader implications beyond the Division, the RA and Division Engineer will so notify the Assistant Administrator, Office of Water (AAOW) and the ASA(CW), through the District of Civil Works, respectively, in writing. Such notification will describe the nature of the issue and the reasons why the issue cannot, or should not, be resolved at the District or Division level or Regional level. (e.g., national policy issue)

5. Either the AAOW or the ASA(CW) may initiate informal or formal consultation concerning unresolved regional or national issues by meeting within 30 calendar days of receipt of notification under paragraph 4. above, or within 30 calendar days of receipt of notification of a policy or procedural issue or issues raised directly at Headquarters level. Within 60 calendar days of that meeting, the agencies will agree to provide direction, guidance or joint guidance (e.g., general guidance on the Section 404(b)(1) Guidelines), where appropriate in response to the issues raised in 4., above.

6. At no time should individual permit decisions be delayed pending resolution of policy issues pursuant to Part III of this MOA. Similarly, *changes*, in policy (i.e., new policies) that occur as a result of Part III should not affect applicants who have submitted a complete permit application prior to implementation of such policy change.

7. Upon resolving a particular policy or procedure, the Corps will determine if the policy is of sufficient importance to warrant public comment. All decisions will be implemented pursuant to the requirement of the Administrative Procedures Act, including public notice and comment rulemaking as necessary.

PART IV—ELEVATION OF INDIVIDUAL PERMIT DECISIONS

1. Purpose: The purpose of PART IV is to provide the exclusive procedures for the elevation of specific individual permit cases. *The elevation of specific individual permit cases will be limited to those cases that involve aquatic resources of national importance.* For example, cases that do not meet this resource value threshold cannot be elevated under this Part over a dispute concerning practicable alternatives. More specifically, the elevation of individual permit cases should be limited to those cases where the net loss (i.e., after considering mitigation) from the project (i.e., within the scope of impacts being evaluated by the Corps), will result in unacceptable adverse effects to aquatic resources of national importance. As a basis for comparison, these cases will cause resource damages similar in magnitude to cases evaluated under Section 404(c) of the Clean Water Act. The final decision on the need to elevate a specific individual permit case and any subsequent case specific policy guidance rest solely with the ASA(CW).

2. Because delays associated with the process described within this Part IV can be costly to the regulated public, every effort will be taken to ensure that the process under paragraph 3(b) of this Part will be initiated only when absolutely necessary. Generic issues concerning the use of this Part IV may be elevated by either party using the procedures in Part III.

3. The following procedures will be utilized for the elevation of specific individual permit cases:

FIELD LEVEL PROCEDURES

(a) Within the basic or extended comment period the Regional Administrator (or designee) must notify the District Engineer by letter that in the opinion of EPA the project *may* result in substantial and unacceptable impacts to aquatic resources of national importance as defined in paragraph 1 of this Part.

(b) For those individual permit cases identified in paragraph 3(a), within 25 calendar days after the end of the basic or extended comment period the Regional Administrator must notify the District Engineer by letter (signed by the Regional Administrator) that in EPA's opinion the discharge will have a substantial and unacceptable impact on aquatic resources of national importance. The opinion will clearly state in detail:

(1) why there *will* be substantial and unacceptable impacts to aquatic resource of national importance as defined in paragraph 1 of this Part and;

(2) why the specific permit must be modified, conditioned, or denied to protect the aquatic resource of natural importance. The opinion, which should explain how the agency determination was made, should be based on site specific information and relate directly to matters with EPA's authority and expertise. A signed copy of the EPA letter should be immediately faxed to the Corps regulatory project manager.

(c) Notice of Intent to Proceed:

(1) If, following the receipt of the notification in Part IV paragraph 3(b), the District Engineer's proposed permit decision is contrary to the stated EPA written recommendation in paragraph 3(b), the District Engineer wil, within five calendar days of his proposed decision, forward a copy of the draft permit and decision document by overnight mail to the Wetlands Division Director.

(2) If, following the receipt of the notification in Part IV paragraph 3(b), the District Engineer believes that his proposed decision resolves the written concerns raised by EPA pursuant to paragraph 3(b), the District Engineer will, within five calendar days of his proposed decision, forward a copy of the draft permit and decision document by overnight mail to the Wetlands Division Director.

(3) Alternatively, if the District Engineer, prior to reaching a decision on the permit (e.g., the final decision is pending resolution of issues not related to the concerns raised by EPA), determines that the project has been modified or conditioned sufficiently so there are no longer substantial adverse impacts on aquatic resources of national importance, the District Engineer will notify the Wetlands Division Director, by letter including such project modifications and/or conditions that resolve EPA's concerns raised in paragraph 3(b).

(d) Within 15 calendar days from receipt of the draft permit under paragraphs 3(c)(1) or 3(c)(2) or notification under paragraph 3(c)(3), the Regional Administrator will notify the District Engineer by faxed letter (signed by the Regional Administrator or the Acting Regional Administrator) that:

(1) the Regional Administrator will not request higher level review; or

(2) the Regional Administrator has forwarded the issue to the AAOW with a recommendation to request review by the ASA(CW).

(e) When the Regional Administrator requests elevation pursuant to paragraph 3(d)(2) of this Part the District Engineer will hold in abeyance the issuance of a permit pending completion of the Headquarters level review outlined below. Further, the District Engineer will provide CECW-OR and ASA(CW) a copy of the Regional Administrator's letter notifying the District Engineer of the intent to request higher level review.

AGENCY HEADQUARTERS REVIEW (AS NECESSARY)

(f) Within 20 calendar days from the Regional Administrator's letter notifying the District Engineer of the intent to request higher level review (paragraph 3(d)(2)), the AAOW will either:

(1) notify the ASA(CW) that the AAOW will not request further review (the ASA(CW) will immediately notify CECW-OR of the AAOW's decision, CECW-OR will immediately notify the district regulatory chief); or

(2) request the ASA(CW) to review the permit decision document.

(g) Within 30 calendar days from the AAOW's request for review, the ASA(CW), through the Director of Civil Works, will review the permit decision document and either:

(1) inform the District Engineer to proceed with final action on the permit decision; or

(2) inform the District Engineer to proceed with final action in accordance with case specific policy guidance; or

(3) make the final permit decision in accordance with 33 CFR 325.8

(h) The ASA(CW) will immediately notify the AAOW in writing of its decision in paragraph 3(g) above. The EPA reserves the right to proceed with Section 404(c). To assist the EPA in reaching a decision whether to exercise its Section 404(c) authority, the District Engineer will provide EPA a copy of the Statement of Findings/Record of Decision prepared in support of a permit decision after the ASA(CW) review. The permit shall not be issued during a period of 10 calendar days after such notice unless it contains a condition that no activity may take place pursuant to the permit until such 10th day, or if the EPA has initiated a Section 404(c) proceeding during such 10 day period, until the Section 404(c) proceedings is concluded and subject to the final determination in such proceeding.

MarthaG. Protho /s/
Acting Assistant Administrator for Water
Environmental Protection Agency
August 11, 1992

Nancy P. Dorn /s/
Assistant Secretary of the
Army of Civil Works
Department of the Army
August 11, 1992

Amendment to the January 19, 1989, Department of the Army/Environmental Protection Agency Memorandum of Agreement Concerning the Determination of the Geographic Jurisdiction of the Section 404 Program and the Application of the Exemptions Under Section 404(f) of the Clean Water Act

January 4, 1993

In order to assure consistency and predictability in wetland determinations made by the two agencies, the following amendment to the January 19, 1989, Department of the Army/Environmental Protection Agency Memorandum of Agreement concerning the determination of the geographic jurisdiction of the Section 404 program and the application of the exemptions under Section 404(f) of the Clean Water Act is hereby adopted:

Effective on the date of the last signature below, the second sentence of paragraph 2 of the "Policy" section is amended to read as follows (new language is *italicized*; deletions are lined-out): "In making ~~its~~ their determinations, the Corps and EPA will adhere to the "~~Federal Manual for Identifying and Delineating Jurisdictional Wetlands~~ *"Corps of Engineers Wetlands Delineation Manual" (Waterways Experiment Station Technical Report Y-87-1, January 1987)* and EPA guidance on isolated waters, and other guidance, interpretations, and regulations issued by EPA to clarify EPA positions on geographic jurisdiction and exemptions."

/Signed/ 4 Jan 93
Nancy P. Dorn
Assistant Secretary of the Army
(Civil Works)

/Signed/ 4 Jan 93
LaJuana S. Wilcher
Assistant Administrator for Water
Environmental Protection Agency

Memorandum of Agreement Between the Department of the Army and the Environmental Protection Agency Concerning the Determination of the Geographic Jurisdiction of the Section 404 Program and the Application of the Exemptions Under Section 404(f) of the Clean Water Act

January 19, 1989

I. PURPOSE AND SCOPE. The United States Department of the Army (Army) and the United States Environmental Protection Agency (EPA) hereby establish the policy and procedures pursuant to which they determine the geographic jurisdictional scope of waters of the United States for purposes of section 404 and the application of the exemptions under section 404(f) of the Clean Water Act (CWA).

The Attorney General of the United States issued an opinion on September 5, 1979, that the Administrator of EPA (Administrator) has the ultimate authority under the CWA to determine the geographic jurisdictional scope of section 404 waters of the United States and the application of the section 404(f) exemptions. Pursuant to this authority and for purposes and effective administration of the 404 program, this Memorandum of Agreement (MOA) sets forth an appropriate allocation of responsibilities between the EPA and the U.S. Army Corps of Engineers (Corps) to determine geographic jurisdiction of section 404 program and the applicability of the exemptions under section 404(f) of the CWA.

II. POLICY. It shall be the policy of the Army and EPA for the Corps to continue to perform the majority of the geographic jurisdictional determinations and determinations of the applicability of the exemptions under section 404(f) as part of the Corps role in administering the section 404 regulatory program. It shall also be the policy of the Army and EPA that the Corps shall fully implement EPA guidance on determining the geographic extent of section 404 jurisdiction and applicability of the 404(f) exemptions.

Case-specific determinations made pursuant to the terms of this MOA will be binding on the Government and represent the Government's position in any subsequent Federal action or litigation regarding the case. In making its determinations, the Corps will implement and adhere to the "Federal manual for Identifying and Delineating Jurisdictional Wetlands," EPA guidance on isolated waters, and other guidance, interpretations, and regulations issued by EPA to clarify EPA positions on geographic jurisdiction and exemptions. All future programmatic guidance, interpretations, and regulations on geographic jurisdiction, and exemptions shall be developed by EPA with input from the Corps; however, EPA will be considered the lead agency and will make the final decision if the agencies disagree.

III. DEFINITIONS

A. Special Case. A special case is a circumstance where EPA makes the final determination of the geographic jurisdictional scope of waters of the United States for purposes of section 404.

Special cases may be designated in generic or project-specific situations where significant issues or technical difficulties are anticipated or exist, concerning the determination of the geographic jurisdictional scope of waters of the United States for purposes of section 404 and, where clarifying guidance is or is likely to be needed. Generic special cases will be designated by easily identifiable political or geographic subdivisions such as township, county, parish, state, EPA region, or Corps division or district. EPA will ensure that generic special cases are marked on maps or some other clear format and provided to the appropriate District Engineer (DE).

B. Special 404 (f) Matters. A special 404(f) matter is a circumstance where EPA makes the final determination of the applicability of exemptions under section 404(f) of the CWA.

A special 404(f) matter may be designated in generic or project-specific situations where significant issues or technical difficulties are anticipated or exist, concerning the applicability of exemptions under section 404(f), and where clarifying guidance is, or is likely to be needed. Generic special 404(f) matters will be designated by easily identifiable political or geographic subdivisions such as township, county, parish, state, EPA region, or Corps division or district and by specific 404(f) exemption (e.g., 404(f)(1)(A)).

IV. PROCEDURES

A. Regional Lists. Each regional administrator (RA) shall maintain a regional list of current designated special cases and special 404(f) matters within each region, including documentation, if appropriate, that there are no current designated special cases or special 404(f) matters in the region.

The RA shall create an initial regional list and transmit it to the appropriate DE within 30 days of the date of the last signature on this MOA. In order to be eligible for a regional list, the designated special cases and special 404(f) matter must be approved by the Administrator. (NOTE: Those geographic areas designated as current special cases pursuant to the 1980 Memorandum of Understanding on Geographic Jurisdiction of the Section 404 Program, may be incorporated into the initial regional lists without additional approval by the Administrator based on township, county, parish, state or other appropriate designation, as described in paragraph III.A. of this MOA will no longer be designated by forest cover type.)

B. Change to the Regional Lists. Changes to the regional lists shall be proposed by the RA and approved by the Administrator and may include additions to, amendments to, or deletions from the regional lists. When the RA proposes an addition, amendment, or deletion to the regional list, the RA shall forward the proposal to EPA Headquarters for review and approval. When the RA proposes an addition or amendment in writing or by phone to the appropriate Corps DE, the Corps will not make a final geographic jurisdictional determination within the proposed special case area for a period of ten working days from the date of the RA's notification. The Corps may proceed to make determinations in the proposed special case area after the ten day period if it has not been provided final notification of EPA Headquarters approval of the RA's proposed changes. Deletions to the regional list do not become effective until a revised regional list, approved by EPA Headquarters, is provided to the appropriate DE.

C. *Project Reviews.* The DE shall review section 404 preapplication inquiries, permit applications, and other matters brought to his attention, which involve the discharge of dredged or fill material into waters of the United States to determine if a current designated special case or special 404(f) matter is involved.

(1) *Special Cases/Special 404(f) Matters.* For those projects involving a current designated special case or special 404(f) matter, the DE shall request that the RA make the final determination of the geographic jurisdictional scope of waters of the United States for purposes of section 404 or applicability of the exemptions under section 404(f). The RA shall make the final determination, subject to discretionary review by EPA Headquarters, and transmit it to the DE, and to the applicant/inquirer.

(2) *Non-Special Cases/Non-Special 404(f) Matters.* For those projects not involving a current designated special case or special 404(f) matter, the DE shall make final determinations and communicate those determinations without a requirement for prior consultation with EPA.

D. *Determination of Special Cases or Special 404(f) Matters.* When the special case or special 404(f) matter has been designated on a project specific basis, issuance of the Final determination by the RA will serve as guidance relevant to the specific facts of each particular situation, and will terminate the special case or special 404(f) matter designation. When the special case or special 404(f) matter has been designated on a generic basis, EPA Headquarters will develop, in consultation with Army, relevant programmatic guidance for determining geographic jurisdictional scope of waters of the United States for the purpose of section 404 or the applicability of exemptions under section 404(f). Special cases and special 404(f) matters designated on a generic basis remain in effect until (1) a deletion from the regional list is proposed and processed according to paragraph IV-B of this MOA, or (2) EPA Headquarters issues programmatic guidance that addresses the relevant issues and specifically deletes the special case or special 404(f) matter from the regional list(s), whichever occurs first.

E. *Uncertainties Regarding Special Cases/Special 404(f) Matters.* Should any uncertainties arise in determining whether a particular action involves a current designated special case or special 404(f) matter, the DE shall consult with the RA. Upon completion of the consultation, the RA will make the final determination as to whether the action involves a current designated special case or special 404(f) matter.

F. *Compliance Tracking.* In order to track the DE's compliance with EPA guidance, the DE shall make his files available for inspection by the RA at the district office, including field notes and data sheets utilized in making final determinations as well any photographs of the site that may be available. Copies of final geographic jurisdictional determinations will be provided to the RA upon request at no cost to EPA unless the sample size exceeds 10 percent of the number of determinations for the sample period. Copies in excess of a 10 percent sample will be provided at EPA expense. To ensure that EPA is aware of determinations being made for which notification is not forwarded through the public notice process, the Corps will provide copies to EPA of all final determinations of no geographic jurisdiction and all final determinations that an exemption under Section 404(f) is applicable. Should EPA become aware of any problem trends with the DE's implementation of guidance, EPA shall initiate interagency discussions to address the issue.

V. RELATED ACTIONS.

A. *Enforcement Situations.* For those investigations made pursuant to the 1989 Enforcement MOA between Army and EPA concerning Federal enforcement of section 404 of the CWA, which involve areas that are current designated special cases, the RA shall make the final determination of the geographic jurisdictional scope of waters of the United States for purposes of section 404. The RA's determination is subject to discretionary review by EPA Headquarters, and will be binding regardless of which agency is subsequently designated lead enforcement agency pursuant to the 1989 Enforcement MOA. For those investigations not involving special cases, the agencies will proceed in accordance with the provisions of the 1989 Enforcement MOA.

For those investigations made pursuant to the 1999 Enforcement MOA between Army and EPA concerning Federal enforcement of section 404 of the CWA, which involve current designated special 404(f) matters, the RA shall make the final determination of the applicability of the exemptions under section 404(f). The RA determination is subject to discretionary review by EPA Headquarters, and is binding regardless of which agency is subsequently designated lead enforcement agency pursuant to the 1989 Enforcement MOA. For those investigations not involving special 404(f) matters, the agencies will proceed in accordance with the provisions of the 1989 Enforcement MOA.

B. *Advanced Identification.* EPA may elect to make the final determination of the geographic jurisdictional scope of waters of the United States for purposes of section 404, as part of the advanced identification of disposal sites under 40 CFR 230.80, subject to discretionary review by EPA Headquarters, and regardless of whether the areas involved are current designated special cases, unless the DE has already made a final geographic jurisdictional determination. Any determinations under this section shall be completed in accordance with paragraph IV of this MOA.

C. *404(c) Actions.* EPA may elect to make the final determination of the geographic jurisdictional scope of waters of the United States for purposes of section 404(c) of the CWA.

VI. GENERAL PROVISIONS

A. All final determinations must be in writing and signed by either the DE or RA. Final determination of the DE or RA made pursuant to this MOA or the 1980 Memorandum of Understanding on Geographic Jurisdiction of the Section 404 Program, will be binding on the Government and represent the Government's position in any subsequent Federal action or litigation concerning that final determination.

B. The procedures and responsibilities of each agency specified in this MOA may be delegated to appropriate subordinates consistent with established agency procedure. Headquarters procedures and responsibilities specified in the MOA may only be delegated within headquarters.

C. Nothing in this document is intended to diminish, modify, or otherwise affect the statutory or regulatory authorities of either agency.

D. This agreement shall take effect and supercede the April 23, 1980, Memorandum of Understanding on Geographic Jurisdiction of the Section 404 Program on the 60th day after the date of the last signature below and will continue in effect for five years, unless extended, modified or revoked by agreement of both parties, or revoked by either party alone upon six months written notice, prior to that time.

//Signed//John S. Doyle, Jr., Deputy, January 19, 1989 for

Robert W. Page
Assistant Secretary of the Army
(Civil Works)

//Signed//January 19, 1989
Rebecca W. Hanmer
Acting Assistant Administrator for Water
U. S. Environmental Protection Agency

Memorandum of Agreement Among the Department of Agriculture, the Environmental Protection Agency, the Department of the Interior, and the Department of the Army Concerning the Delineation of Wetlands for Purposes of Section 404 of the Clean Water Act and Subtitle B of the Food Security Act

January 6, 1994

I. BACKGROUND

The Departments of the Army, Agriculture, and the Interior, and the Environmental Protection Agency (EPA) recognize fully that the protection of the Nation's remaining wetlands is an important objective that will be supported through the implementation of the Wetland Conservation (Swampbuster) provision of the Food Security Act (FSA) and Section 404 of the Clean Water Act (CWA). The agencies further recognize and value the important contribution of agricultural producers to our society, our economy, and our environment. We are committed to insuring that Federal wetlands programs are administered in a manner that minimizes the impacts on affected landowners to the fullest possible extent consistent with the important goal of protecting wetlands. We are also committed to minimizing duplication and inconsistencies between Swampbuster and the CWA Section 404 program. On August 24, 1993, the Administration announced a comprehensive package of reforms that will improve both the protection of wetlands and make wetlands programs more fair and flexible for landowners, including the Nation's agricultural producers. This Memorandum of Agreement (MOA) implements one of over 40 components of the Administration's Wetlands Plan.

II. PURPOSE AND APPLICABILITY

A. PURPOSE. The purpose of this MOA is to specify the manner in which wetland delineations and certain other determinations of waters of the United States made by the U.S. Department of Agriculture (USDA) under the FSA will be relied upon for purposes of CWA Section 404. While this MOA will promote consistency between CWA and FSA wetlands programs, it is not intended in any way to diminish the protection of these important aquatic resources In this regard, all signatory agencies to this MOA will ensure that wetlands programs are administered in a manner consistent with the objectives and requirements of applicable laws, implementing regulations, and guidance.

B. APPLICABILITY

1. The Administrator of EPA has the ultimate authority to determine the geographic scope of waters of the United States subject to jurisdiction under the CWA, including the Section 404 regulatory program. Consistent with a current MOA between EPA and the Department of the Army, the Army Corps of Engineers (Corps) conducts jurisdictional delineations associated with the day-to-day administration of the Section 404 program.

2. The Secretary of the USDA, acting through the Chief of the Soil Conservation Service (SCS), has the ultimate authority to determine the geographic scope of wetlands for FSA purposes and to make delineations relative to the FSA, in consultation with the Department of the Interior, Fish and Wildlife Service (FWS).

III. DEFINITION OF AGRICULTURAL LANDS

For the purposes of this MOA, the term "agricultural lands" means those lands intensively used and managed for the production of food or fiber to the extent that the natural vegetation has been removed and cannot be used to determine whether the area meets applicable hydrophytic vegetation criteria in making a wetland delineation.

A. Areas that meet the above definition may include intensely used and managed cropland, hayland, pasture land, orchards, vineyards, and areas which support wetland crops (e.g., cranberries, taro, watercress, rice). For example, lands intensively used and managed for pasture or hayland where the natural vegetation has been removed and replaced with planted grasses or legumes such as ryegrass, bluegrass, or alfalfa, are considered agricultural lands for the purposes of this MOA.

B. Agricultural lands do not include range lands, forest lands, wood lots, or tree farms. Further, lands where the natural vegetation has not been removed, even though that vegetation may be regularly grazed or mowed and collected as forage or fodder (e.g., uncultivated meadows and prairies, salt hay), are not considered agricultural lands for the purposes of this MOA.

Other definitions for the purposes of this MOA are listed below in Section VI.

IV. ALLOCATION OF RESPONSIBILITY

A. In accordance with the terms and procedures of this MOA, wetland delineations made by SCS on agricultural lands, in consultation with FWS, will be accepted by EPA and the Corps for the purposes of determining Section 404 wetland jurisdiction. In addition, EPA and the Corps will accept SCS wetland delineations on non-agricultural lands that are either narrow bands immediately adjacent to, or small pockets interspersed among, agricultural lands. SCS is responsible for making wetland delineations for agricultural lands whether or not the person who owns, manages, or operates the land is a participant in USDA programs.

B. Lands owned or operated by a USDA program participant that are not agricultural lands and for which a USDA program participant requests a wetland delineation, will be delineated by SCS in coordination with the Corps, or EPA as appropriate, and in consultation with FWS. Final wetland delineations conducted by SCS pursuant to the requirements of this paragraph shall not be revised by SCS except where an opportunity for coordination and consultation is provided to the other signatory agencies.

C. SCS may conduct delineations of other waters for the purposes of Section 404 of the CWA, such as lakes, ponds, and streams, in coordination with the Corps, or EPA as appropriate, on lands OF which SCS is otherwise engaged in wetland delineations pursuant to paragraphs IV.A or IV.B of this MOA. Delineations of "other waters" will not be made until the interagency oversight team convened pursuant to Section V.B.2 has agreed on appropriate local procedures and guidance for making such delineations.

D. For agricultural lands, the signatory agencies will use the procedures for delineating wetlands as described in the National Food Security Act Manual, Third Edition (NFSAM). For areas that are not agricultural lands, SCS will use the 1987 Corps Wetland Delineation Manual, with current national Corps guidance, to make wetland delineations applicable to Section 404.

E. Delineations on "agricultural lands" must be performed by personnel who are trained in the use of the NFSAM. Delineations on other lands and waters must be performed by personnel who are trained in the use of the 1987 Corps Wetland Delineation Manual. This MOA includes provisions for the appropriate interagency delineation training below in Section V.E.

F. In the spins of the agencies' commitment to develop agreed upon methods for use m making wetland delineations, subsequent .revisions or amendments to the Corps 1987 manual or portions of the NPSAM affecting the wetland delineation procedures upon which this agreement is based will require the concurrence of the four signatory agencies.

G. A final written wetland delineation made by SCS pursuant to the terms of this MOA will be adhered to by all the signatory agencies and will be effective for a period of five years from the date the delineation is made final, unless new information warrants revision of the delineation before the expiration date. Such new information may include, for example, data on landscape changes caused by a major flood, or a landowner's notification of intent to abandon agricultural use and the return of wetland conditions on a prior converted cropland in accordance with Section 1222 of the FSA, SCS will update wetland delineations on this five-year cycle. Circumstances under which SCS wetland delineations made prior to the effective date of this agreement will be considered as final for Section 404 purposes are addressed in Paragraph V.C.

H. Within the course of administering their Swampbuster responsibilities, SCS and FWS will provide landowners/operators general written information (i.e., EPA/Corps fact sheets) regarding the CWA Section 404 program permit requirements, general permits, and exemptions. The SCS and FWS will not, however, provide opinions regarding the applicability of CWA Section 404 permit requirements or exemptions.

I. USDA will maintain documentation of all final written SCS wetland delineations and record the appropriate label and boundary information on an official wetland delineation map. USDA will make this information available to the signatory agencies upon request.

J. In pursuing enforcement activities, the signatory agencies will rely upon delineations made by the lead agency, as clarified below, providing a single Federal delineation for potential violations of Section 404 or Swampbuster. Nothing in this MOA will diminish, modify, or otherwise affect existing EPA and Corps enforcement authorities under the CWA and clarified in the 1989 "EPA/Army MOA Concerning Federal Enforcement for the Section 404 Program of the Clean Water Act." EPA, the Corps, and SCS may gather information based on site visits or other means to provide additional evidentiary support for a wetland delineation which is the subject of a potential or ongoing CWA Section 404 or Swampbuster enforcement action.

K. For those lands where SCS has not made a final written wetland delineation, and where the Corps or EPA is pursuing a potential CWA violation, the lead agency for the CWA enforcement action will conduct a jurisdictional delineation for the purposes of Section 404 and such delineations will be used by SCS for determining Swampbuster jurisdiction and potential Swampbuster violations. For those lands where the Corps has not made a final written wetland delineation, and where SCS is pursuing a potential Swampbuster violation, SCS will make a final written wetland delineation consistent with Sections IV.A, IV.B, and IV.C of this MOA and provide copies to the Corps and EPA. Such delineations will be used by the Corps and EPA for the purism of determining potential violations of the CWA. In circumstances in which either the Corps or EPA is pursuing a potential CWA violation on land that is subject to an ongoing SCS appeal, a wetland delineation will be conducted by the Corps or EPA in consultation with SCS and FWS.

L. In making wetland delineations, the agencies recognize that discharges of dredged or fill material that are not authorized under Section 404 cannot eliminate Section 404 jurisdiction, and that wetlands that were converted as a result of unauthorized discharges remain subject to Section 404 regulation.

V. PROCEDURES

Accurate and consistent wetland delineations are critical to the success of this MOA. For this reason, the signatory agencies will work cooperatively at the field level to: 1) achieve interagency concurrence on mapping conventions used by SCS for wetland delineations on agricultural lands, 2) provide EPA and Corps programmatic review of SCS delineations, and 3) certify wetland delineations in accordance with Section 1222(a)(2) of the FSA, as amended. The following sections describe the procedures that will be followed to accomplish these objectives.

A. MAPPING CONVENTIONS

1. Each SCS State Conservationist will take the lead in convening representatives of the Corps, EPA, FWS, and SCS to obtain the written concurrence of each of the signatory agencies, within 120 calendar days of the effective date of this MOA, on a set of mapping conventions for use in making wetland delineations. Only mapping conventions concurred upon by all signatory agencies will be used by SCS for wetland delineations.

2. If interagency consensus on mapping conventions is not reached within 120 days -of the date of this MOA, the State Conservationist will refer documentation of the unresolved issues to the Chief of SCS. The Chief of SCS will immediately forward copies of the State Conservationist's documentation of unresolved issues to the Corps Director of Civil Works; the EPA Director of the Office of Wetlands, Oceans, and Watersheds; and the FWS Director. Immediately thereafter, the Chief of SCS or an appropriate designee will lead necessary discussions to achieve interagency concurrence on resolution of outstanding issues, and will forward documentation of

the resolution to the State Conservationist and the appropriate Headquarters offices of the signatory agencies.

3. Once interagency concurrence on mapping conventions is obtained, such mapping conventions will be used immediately in place of the earlier mapping conventions.

4. Agreed-upon mapping conventions developed at the state level will be documented and submitted, for each state, through the Chief of SCS to the Headquarters of each of the signatory agencies. State-level agreements will be reviewed by the Headquarters of the signatory agencies for the purpose of ensuring national consistency.

B. DELINEATION PROCESS REVIEW AND OVERSIGHT

1. This MOA emphasizes the need to ensure consistency in the manner in which wetlands are identified for CWA and FSA purposes, and provides a number of mechanisms to increase meaningful interagency coordination and consultation in order for the agencies to world toward meeting this goal. In this regard, the agencies believe it is critical that efforts for achieving consistency be carefully monitored and evaluated. Consequently, this MOA establishes a monitoring and review process that will be used to provide for continuous improvement in the wetland delineation process specified in this MOA.

2. EPA will lead the signatory agencies in establishing interagency oversight teams at the state level to conduct periodic review of wetland delineations conducted under the provisions of this MOA. These reviews will include delineations done by SCS pursuant to Sections IV.A, IV.B, and [V.C of this MOA and delineations done by EPA or the Corps pursuant to Section IV.K of this MOA. These reviews also will include changes to wetland delineations resulting from the SCS appeals process, as well as disagreements regarding allocation of responsibility. These reviews will occur, at a minimum, on a quarterly basis for the first year, on a semi annual basis for the second year, and annually thereafter. In addition, a review will be initiated whenever one or more of the signatory agencies believes a significant issue needs to be addressed. The purpose of each review will be to evaluate the accuracy of an appropriate sample of wetland delineations. When feasible, this will include actual field verifications of wetland delineations. Should the interagency oversight team identify issues regarding the implementation of this MOA or wetland delineations conducted under the provisions of this MOA, the team will work to resolve those issues and reach agreement on any necessary corrective actions. Each review, and any necessary corrective action, will be documented in a report to be distributed to the signatory agencies' appropriate field and Headquarters offices.

3. In situations in which the interagency oversight team identifies and reports unresolved issues concerning wetland delineations conducted under the provisions of this MOA, including changes to wetland delineations resulting from the SCS appeals process, the Headquarters offices of the signatory agencies will informally review the issue and work to reach agreement on any necessary corrective actions. This informal process notwithstanding, the EPA Regional Administrator or the Corps District Engineer may, at any time, propose to designate a geographic area as a Special cases.

4. Similar to the terms of the current Memorandum of Agreement between the Department of the Army and the EPA Concerning the Determination of the Geographic Jurisdiction of the Section 404 Program and the Application of the Exemptions under Section 404(f) of the CWA, the EPA Regional Administrator or the Corps District Engineer may propose to designate a geographic area, or a particular wetland type within a designated geographic area, as a special case. A special case may be designated only after the interagency oversight team (EPA, Corps, SCS, and FWS) has reviewed the relevant issues and been unable to reach a consensus on an appropriate resolution. Special cases will be designated by an easily identifiable political or geographic subdivision, such as a township, county, parish, state, EPA Region, or Corps division or district, and will be marked on maps or using some other clear format and provided to the appropriate EPA' Corps, FWS, and SCS field offices. Proposed designations of special cases will not be effective until approved by EPA or Corps Headquarters, as appropriate.

5. Upon proposing a special case, the EPA Regional Administrator or Corps District Engineer, as appropriate, will notify the appropriate SCS State Conservationist in writing. Following notification of the proposed designation, SCS will not make wetland delineations for the purposes of CWA jurisdiction within the proposed special case for a period of 20 working days from the date of the notification. SCS may proceed to make wetland delineations for CWA purposes in the proposed special case after the 20-day period if the SCS State Conservationist has not been notified by the EPA Regional Administrator or Corps District Engineer of approval of the proposed special case designation by EPA Headquarters or the Corps Director of Civil Works, as appropriate.

6. Following approval of the proposed special case, the Corps, or EPA as appropriate, will make final CWA wetland delineations in the special case area, rather than SCS. In addition, the referring field office (i.e., either the EPA Regional Administrator or Corps District Engineer) will develop draft guidance relevant to the specific issues raised by the special case and forward the draft guidance to its Headquarters office. The Headquarters office of the agency which designated the special case will develop final guidance after consulting with the signatory agencies' Headquarters offices. EPA concurrence will be required for final guidance for any special case designated by the Corps. Special cases remain in effect until final guidance is issued by the Headquarters office of the agency which designated the special case or the designation is withdrawn by the EPA Regional Administrator or Corps District Engineer, as appropriate.

C. RELIANCE ON PREVIOUS SCS WETLAND DELINEATIONS FOR CWA PURPOSES

1. Section 1222 of the FSA, as amended by the Food Agriculture Conservation and Trade Act, provides that SCS will certify SCS wetland delineations made prior to November 28, 1990. The intent of this process is to ensure the accuracy of wetland delineations conducted prior to November 28, 1990, for the purposes of the FSA. This certification process also will provide a useful basis for establishing reliance on wetland delineations for CWA purposes. All certifications done after the effective date of this MOA that are done using mapping conventions will use the agreed-upon mapping conventions pursuant to Section V.A of this MOA.

2. Written SCS wetland delineations for lands identified in Section IV.A of this MOA conducted prior to the effective date of this MOA will be used for purposes of establishing CWA jurisdiction, subject to the provisions of Section V.C.3 below. If such SCS wetland delineations are subsequently modified or revised through updated certification, these modifications or revisions will supersede the previous delineations for purposes of establishing CWA jurisdiction. Written SCS wetland delineations for lands identified in Sections IV.B and IV.C of this MOA conducted prior to the effective date of this MOA will require coordination with the Corps, or EPA as appropriate, before being used for purposes of determining CWA jurisdiction.

3. As part of the certification effort, SCS will establish priorities to certify SCS wetland delineations. In addition to responding to requests from individual landowners

who feel their original wetland determinations were made in error, SCS will give priority to certifying those wetland delineations where at least two of the four signatory agencies represented on the interagency oversight team convened pursuant to Section V.B.2 of this MOA agree that SCS wetland delineations in a particular area, or a generic class of SCS wetland delineations in a particular area, raise issues regarding their accuracy based on current guidance. These priority areas will be identified only after mapping conventions are agreed upon pursuant to Section V.A of this MOA. Identification of these high priority certification needs shall be made at the level of the SCS State Conservationist, FWS Regional Director, EPA Regional Administrator, and the Corps District Engineer. Following identification of these high priority certification needs, the SCS State Conservationist will immediately notify the affected landowner(s), by letter, that the relevant SCS wetland delineations have been identified as a high priority for being certified under Section 1222 of the FSA. In addition, the notification will inform the landowner that while previous wetland delineations remain valid for purposes of the FSA until certification or certification update is completed, the landowner will need to contact the Corps before proceeding with discharges of dredged or fill material. This communication by the landowner will enable the Corps to review the wetland delineation to establish whether it can be used for purposes of CWA jurisdiction. The SCS State Conservationist will initiate, within 30 calendar days of landowner notification, corrective measures to resolve the wetland delineation accuracy problem.

D. APPEALS. Landowners for whom SCS makes wetland delineations for either Swampbuster or Section 404 will be afforded the opportunity to appeal such wetland delineations through the SCS appeals process. In circumstances where an appeal is made and the State Conservationist is considering a change in the original delineation, the State Conservationist will notify the Corps District Engineer and the EPA Regional Administrator to provide the opportunity for their participation and input on the appeal. FWS also will be consulted consistent with the requirements of current regulations. The Corps and EPA reserve the right, on a case-by-case basis, to determine that a revised delineation resulting from an appeal is not valid for purposes of Section 404 jurisdiction.

E. TRAINING

1. SCS, in addition to FWS and EPA, wild continue to participate in the interagency wetland delineation training sponsored by the Corps, which is based on the most current manual used to delineate wetlands for purposes of Section 404. Completion of this training will be a prerequisite for field staff of all signatory agencies who delineate wetlands on non-agricultural lands using the 1987 Corps Wetland Delineation Manual.

2. The interagency wetland delineation training will address agency wetland delineation responsibilities as defined by this MOA, including SCS NFSAM wetland delineation procedures.

3. Field offices of the signatory agencies are encouraged to provide supplemental interagency wetland delineation training (i.e., in addition to that required in paragraph IV.E), as necessary, to prepare SCS field staff for making Section 404 wetland delineations. For training on the use of the 1987 Corps Wetland Delineation Manual, such supplemental training will rely on the training materials used for the Corps delineation training program and will provide an equivalent level of instruction.

VI. DEFINITIONS

A. "Coordination" means that SCS will contact the Corps, or EPA as appropriate, and provide an opportunity for review, comment, and approval of the findings of SCS prior to making a final delineation. The Corps, or EPA as appropriate, will review the proposed delineation and respond to SCS regarding its acceptability for CWA Section 404 purposes within 45 days of receipt of all necessary information. SCS will not issue a final delineation until agreement is reached between SCS and the Corps or EPA, as appropriate.

B. "Consultation" means that SCS, consistent with current provisions of the FSA, will provide FWS opportunity for full participation in the action being taken and for timely review and comment on the findings of SCS prior to a final wetland delineation pursuant to the requirements of the FSA.

C. A "wetland delineation" is any determination of the presence of wetlands and their boundaries.

D. A "special case" for the purposes of this MOA refers to those geographic areas or wetland types where the Corps or EPA will mace final CWA wetland delineations.

E. "Signatory agencies" means the EPA and the Departments of Army (acting through the Corps), Agriculture (acting through SCS), and Interior (acting through FWS).

F. "USDA program participant" means individual landowners/operators eligible to receive USDA program benefits covered under Title XII of the Food Security Act of 1985, as amended by the Food, Agriculture, Conservation and Trade Act of 1990.

VII. GENERAL

A. The policy and procedures contained within this MOA do not create any rights, either substantive or procedural, enforceable by any party regarding an enforcement action brought by the United States. Deviation or variance from the administrative procedures included in this MOA will not constitute a defense for violators or others concerned with any Section 404 enforcement action.

B. Nothing in this MOA is intended to diminish, modify, or otherwise affect statutory or regulatory authorities of any of the signatory agencies. All formal guidance interpreting this MOA and background materials upon which this MOA is based will be issued jointly by the agencies.

C. Nothing in this MOA will be construed as indicating a financial commitment by SCS, the Corps, EPA, or FWS for the expenditure of funds except as authorized in specific appropriations.

D. This MOA will take effect on the date of the last signature below and will continue in effect until modified or revoked by agreement of all signatory agencies, or revoked by any of the signatory agencies alone upon go days written notice. Modifications to this MOA may be made by mutual agreement and Headquarters level approval by all the signatory agencies. Such modifications will take effect upon signature of the modified document by all the signatory agencies.

E. The signatory agencies will refer delineation requests to the appropriate agency pursuant to this MOA.

APPENDIX G

USACE Regulatory Guidance Letters and Memoranda to the Field

Table of Contents

COMPENSATORY MITIGATION PROJECTS

REGULATORY GUIDANCE LETTER 02-2

SUBJECT: Guidance on Compensatory Mitigation Projects for Aquatic Resource Impacts Under the Corps Regulatory Program Pursuant to Section 404 of the Clean Water Act and Section 10 of the Rivers and Harbors Act of 1899

DATE: 24 December 2002

1. PURPOSE AND APPLICABILITY

a. PURPOSE. Under existing law the Corps requires compensatory mitigation to replace aquatic resource functions unavoidably lost or adversely affected by authorized activities. This Regulatory Guidance Letter (RGL) clarifies and supports the national policy for "no overall net loss" of wetlands and reinforces the Corps commitment to protect waters of the United States, including wetlands. Permittees must provide appropriate and practicable mitigation for authorized impacts to aquatic resources in accordance with the laws and regulations. Relevant laws, regulations, and guidance are listed in Appendix A. This guidance does not modify existing mitigation policies, regulations, or guidance. However, it does supercede RGL 01-1 that was issued October 31, 2001. Districts will consider the requirements of other Federal programs when implementing this guidance.

b. APPLICABILITY. This guidance applies to all compensatory mitigation proposals associated with permit applications submitted for approval after this date.

2. GENERAL CONSIDERATIONS

Districts will use watershed and ecosystem approaches when determining compensatory mitigation requirements, consider the resource needs of the watersheds where impacts will occur, and also consider the resource needs of neighboring watersheds. When evaluating compensatory mitigation plans, Districts should consider the operational guidelines developed by the National Research Council (2001) for creating or restoring ecologically self-sustaining wetlands. These operational guidelines, which are in Appendix B, will be provided to applicants who must implement compensatory mitigation projects.

a. WATERSHED APPROACH. A watershed-based approach to aquatic resource protection considers entire systems and their constituent parts. Districts will recognize the authorities of, and rely on the expertise of, tribal, state, local, and other Federal resource management programs. During the permit evaluation process, Districts will coordinate with these entities and take into account zoning regulations, regional council and metropolitan planning organization initiatives, special area management planning initiatives, and other factors of local public interest. Watersheds will be identified, for accounting purposes, using the U.S. Geologic Survey's Hydrologic Unit Codes. Finally, applicants will be encouraged to provide compensatory mitigation projects that include a mix of habitats such as open water, wetlands, and adjacent uplands. When viewed from a watershed perspective, such projects often provide a greater variety of functions.

b. CONSISTENCY AND COMPATIBILITY. Districts will coordinate proposed mitigation plans with tribes, states, local governments, and other Federal agencies consistent with existing laws, regulation, and policy guidance to ensure that applicants' mitigation plans are consistent with watershed needs and compatible with adjacent land uses. Districts will evaluate applicants' mitigation proposals giving full consideration to comments and recommendations from tribes, states, local governments, and other Federal agencies. Districts may coordinate on a case-by-case basis during the application evaluation process, or on programmatic basis to promote consistent and timely decision making.

c. IMPACTS AND COMPENSATION. Army regulations require appropriate and practicable compensatory mitigation to replace functional losses to aquatic resources, including wetlands. Districts will determine what level of mitigation is "appropriate" based upon the functions lost or adversely affected as a result of impacts to aquatic resources. When determining "practicability," Districts will consider the availability of suitable locations, constructibility, overall costs, technical requirements, and logistics. There may be instances where permit decisions do not meet the "no overall net loss of wetlands" goal because compensatory mitigation would be impracticable, or would only achieve inconsequential reductions in impacts. Consequently, the "no overall net loss of wetlands goal" may not be achieved for each and every permit action, although all Districts will strive to achieve this goal on a cumulative basis, and the Corps will achieve the goal programmatically.

d. MEASURING IMPACTS AND COMPENSATORY MITIGATION. The Corps has traditionally used acres as the standard measure for determining impacts and required mitigation for wetlands and other aquatic resources, primarily because useful functional assessment methods were not available. However, Districts are encouraged to increase their reliance on functional assessment methods. Districts will determine, on a case-by-case basis, whether to use a functional assessment or acreage surrogates for determining mitigation and for describing authorized impacts. Districts will use the same approach to determine losses (debits) and gains (credits) in terms of amounts, types, and location(s) for describing both impacts and compensatory mitigation.

1. FUNCTIONAL ASSESSMENT. The objective is to offset environmental losses resulting from authorized activities. The ecological characteristics of aquatic sites are unique. Therefore, when possible, Districts should use a functional assessment by qualified professionals to determine impacts and compensatory mitigation requirements. Districts should determine functional scores using aquatic site assessment techniques generally accepted by experts in the field or the best professional judgment of Federal, tribal, and state agency representatives, fully considering ecological functions included in the 404 (b)(1) Guidelines. When a District uses a functional assessment method, e.g., a Hydrogeomorphic Assessment or Wetland Rapid Assessment Procedure, the District will make the method available to applicants for planning mitigation.

2. FUNCTIONAL REPLACEMENT. For wetlands, the objective is to provide, at a minimum, one-to-one functional replacement, i.e., no net loss of functions, with an adequate margin of safety to reflect anticipated success. Focusing on the replacement of the functions provided by a wetland, rather than only calculation of acreage impacted or restored, will in most cases provide a more accurate and effective way to achieve the environmental performance objectives of the no net loss policy. In some cases, replacing the functions provided by one wetland area can be achieved by another, smaller wetland; in other cases, a larger replacement wetland may be needed to replace the functions of the wetland impacted by development. Thus, for example, on an acreage basis, the ratio should be greater than one-to-one where the impacted functions are demonstrably high and the replacement wetlands are of lower function. Conversely, the ratio may be less than one-to-one where the functions associated with the area being impacted are demonstrably low and the replacement wetlands are of higher function.

3. FUNCTIONAL CHANGES. Districts may account for functional changes by

recording them as site-specific debits and credits as defined below.

a. CREDIT. A unit of measure, e.g., a functional capacity unit in the Hydrogeomorphic Assessment Method, representing the gain of aquatic function at a compensatory mitigation site; the measure of function is typically indexed to the number of acres of resource restored, established, enhanced, or protected as compensatory mitigation.

b. DEBIT. A unit of measure, e.g., a functional capacity unit in the Hydrogeomorphic Assessment Method, representing the loss of aquatic function at a project site; the measure of function is typically indexed to the number of acres impacted by issuance of the permit.

4. ACREAGE SURROGATE. In the absence of more definitive information on the functions of a specific wetland site, a minimum one-to-one acreage replacement may be used as a reasonable surrogate for no net loss of functions. For example, information on functions might be lacking for enforcement actions that generate after-the-fact permits or when there is no appropriate method to evaluate functions. When Districts require one-to-one acreage replacement, they will inform applicants of specific amounts and types of required mitigation. Districts will provide rationales for acreage replacement and identify the factors considered when the required mitigation differs from the one-to-one acreage surrogate.

5. STREAMS. Districts should require compensatory mitigation projects for streams to replace stream functions where sufficient functional assessment is feasible. However, where functional assessment is not practical, mitigation projects for streams should generally replace linear feet of stream on a one-to-one basis. Districts will evaluate such surrogate proposals carefully because experience has shown that stream compensation measures are not always practicable, constructible, or ecologically desirable.

e. WETLAND PROJECT TYPES. Although the following definitions were developed to characterize wetland projects, the principles they reflect may also be useful for decisions on other aquatic resource projects.

1. ESTABLISHMENT (CREATION). The manipulation of the physical, chemical, or biological characteristics present to develop a wetland on an upland or deepwater site, where a wetland did not previously exist. Establishment results in a gain in wetland acres.

2. RESTORATION. The manipulation of the physical, chemical, or biological characteristics of a site with the goal of returning natural or historic functions to a former or degraded wetland. For the purpose of tracking net gains in wetland acres, restoration is divided into:

A. RE-ESTABLISHMENT. The manipulation of the physical, chemical, or biological characteristics of a site with the goal of returning natural or historic functions to a former wetland. Re-establishment results in rebuilding a former wetland and results in a gain in wetland acres.

B. REHABILITATION. The manipulation of the physical, chemical, or biological characteristics of a site with the goal of repairing natural or historic functions of a degraded wetland. Rehabilitation results in a gain in wetland function but does not result in a gain in wetland acres.

3. ENHANCEMENT. The manipulation of the physical, chemical, or biological characteristics of a wetland (undisturbed or degraded) site to heighten, intensify, or improve specific function(s) or to change the growth stage or composition of the vegetation present. Enhancement is undertaken for specified purposes such as water quality improvement, flood water retention, or wildlife habitat. Enhancement results in a change in wetland function(s) and can lead to a decline in other wetland functions, but does not result in a gain in wetland acres. This term includes activities commonly associated with enhancement, management, manipulation, and directed alteration.

4. PROTECTION/MAINTENANCE (PRESERVATION). The removal of a threat to, or preventing the decline of, wetland conditions by an action in or near a wetland. This term includes the purchase of land or easements, repairing water control structures or fences, or structural protection such as repairing a barrier island. This term also includes activities commonly associated with the term preservation. Preservation does not result in a gain of wetland acres and will be used only in exceptional circumstances.

f. PRESERVATION CREDIT. Districts may give compensatory mitigation credit when existing wetlands, or other aquatic resources are preserved in conjunction with establishment, restoration, and enhancement activities. However, Districts should only consider credit when the preserved resources will augment the functions of newly established, restored, or enhanced aquatic resources. Such augmentation may be reflected in the amount of credit attributed to the entire mitigation project. In exceptional circumstances, the preservation of existing wetlands or other aquatic resources may be authorized as the sole basis for generating credits as mitigation projects. Natural wetlands provide numerous ecological benefits that restored wetlands cannot provide immediately and may provide more practicable long-term ecological benefits. If preservation alone is proposed as mitigation, Districts will consider whether the wetlands or other aquatic resources: 1) perform important physical, chemical or biological functions, the protection and maintenance of which is important to the region where those aquatic resources are located; and, 2) are under demonstrable threat of loss or substantial degradation from human activities that might not otherwise be avoided. The existence of a demonstrable threat will be based on clear evidence of destructive land use changes that are consistent with local and regional (i.e., watershed) land use trends, and that are not the consequence of actions under the permit applicant's control.

g. ON-SITE AND OFF-SITE MITIGATION. Districts may require on-site, off-site, or a combination of on-site and off-site mitigation to maintain wetland functional levels within watersheds. Mitigation should be required, when practicable, in areas adjacent or contiguous to the discharge site (on-site compensatory mitigation). On-site mitigation generally compensates for locally important functions, e.g., local flood control functions or unusual wildlife habitat. However, off-site mitigation may be used when there is no practicable opportunity for on-site mitigation, or when off-site mitigation provides more watershed benefit than on-site mitigation, e.g., is of greater ecological importance to the region of impact. Off-site mitigation will be in the same geographic area, i.e., in close proximity to the authorized impacts and, to the extent practicable, in the same watershed. In choosing between on-site or off-site compensatory mitigation, Districts will consider: 1) likelihood for success; 2) ecological sustainability; 3) practicability of long-term monitoring and maintenance or operation and maintenance; and, 4) relative costs of mitigation alternatives.

h. IN-KIND AND OUT-OF-KIND MITIGATION. Districts may require in-kind, out-of-kind, or a combination of in-kind and out-of-kind, compensatory mitigation to achieve functional replacement within surrounding watersheds. In-kind compensation for a wetland loss involves replacement of a wetland area by establishing, restoring, enhancing, or protecting and maintaining a wetland area of the same physical and functional type. In-kind replacement generally is required when the impacted resource is locally important. Out-of-kind compensation for a wetland loss involves replacement of a wetland area by establishing, restoring, enhancing, or protecting and maintaining an aquatic resource of different physical and functional type. Out-of-kind mitigation is appropriate when it is practicable and provides more environmental or watershed benefit than in-kind compensation (e.g., of greater ecological importance to the region of impact).

i. BUFFERS. Districts may require that compensatory mitigation for projects in wetlands or other aquatic resources include the establishment and maintenance of buffers to ensure that the overall mitigation project performs as expected. Buffers are upland or riparian areas that separate wetlands or other aquatic resources from developed areas and agricultural lands. Buffers typically consist of native plant communities (i.e., indigenous species) that reflect the local landscape and ecology. Buffers enhance or provide a variety of aquatic habitat functions including habitat for wildlife and other organisms, runoff filtration, moderation of water temperature changes, and detritus for aquatic food webs. Additional guidance regarding the appropriate use of buffers as a component of compensatory mitigation is forthcoming.

1. UPLAND AREAS. Under limited circumstances, Districts may give credit for inclusion of upland areas within a compensatory mitigation project to the degree that the protection and management of such areas is an enhancement of aquatic functions and increases the overall ecological functioning of the mitigation site, or of other aquatic resources within the watershed (see Federal Mitigation Banking Guidance and Nationwide Permit General Condition 19). Such enhancement may be reflected in the amount of credit attributed to the mitigation project. Districts will evaluate and document the manner and extent to which upland areas augment the functions of wetland or other aquatic resources. The establishment of buffers in upland areas may only be authorized as mitigation if the District determines that this is best for the aquatic environment on a watershed basis. In making this determination, Districts will consider whether the wetlands or other aquatic resources being buffered: 1) perform important physical, chemical, or biological functions, the protection and maintenance of which is important to the region where those aquatic resources are located; and 2) are under demonstrable threat of loss or substantial degradation from human activities that might not otherwise be avoided.

2. RIPARIAN AREAS. Districts may give credit for inclusion of riparian areas within a compensatory mitigation project to the degree that the protection and management of such areas is an enhancement of aquatic functions and increases the overall ecological functioning of the mitigation site, or of other aquatic resources within the watershed. Such enhancement may be reflected in the amount of credit attributed to the mitigation project. Districts will evaluate and document the manner and extent to which riparian areas augment the functions of streams or other aquatic resources. The establishment of buffers in riparian areas may only be authorized as mitigation if the District determines that this is best for the aquatic environment on a watershed basis. In making this determination, Districts will consider whether the streams or other aquatic resources being buffered: 1) perform important physical, chemical, or biological functions, the protection and maintenance of which is important to the region where those aquatic resources are located; and 2) are under demonstrable threat of loss or substantial degradation from human activities that might not otherwise be avoided.

j. COMPENSATORY MITIGATION ALTERNATIVES. Permit applicants may propose the use of mitigation banks, in-lieu fee arrangements, or separate activity-specific projects.

k. PUBLIC REVIEW AND COMMENT

1. INDIVIDUAL PERMITS. Proposed compensatory mitigation will be made available for public review and comment, consistent with the form (mitigation bank, in-lieu fee arrangement, or separate activity-specific compensatory mitigation project) of proposed compensation. Although, as a matter of regulation at 33 CFR 325.1 (d)(9), compensatory mitigation plans are not required before the Corps can issue a public notice, Districts should encourage applicants, during pre-application consultation, to provide mitigation plans with applications to facilitate timely and effective review. Public Notices should indicate the form of proposed compensatory mitigation and include information on components of the compensatory mitigation plan. If mitigation plans are available, synopses may be included in Public Notices and the complete plans made available for inspection at District offices. If mitigation plans are available and reproducible, Districts will forward copies to Federal, tribal, and state resource agencies. Districts should not delay issuing Public Notices when mitigation plans are not submitted with otherwise complete applications proposing impacts to aquatic resources.

2. GENERAL PERMITS. Requests for nationwide and regional general permit verifications are not subject to public notice and comment. However, general permit compensatory mitigation provisions or requirements are published for public comment at the time general permits are proposed for issuance or reissuance. Additional review of case-specific mitigation plans should be consistent with the conditions of the Nationwide or Regional Permit. Public review and comment should be provided for proposed mitigation banks and in-lieu-fee arrangements consistent with the Banking Guidance and In-lieu-fee Guidance provisions.

l. PERMIT SPECIAL CONDITIONS. Districts will include in individual permits, and general permit verifications that contain a wetland compensatory mitigation requirement, special conditions that identify: 1) the party(s) responsible for meeting any or all components of compensatory mitigation requirements; 2) performance standards for determining compliance; and, 3) other requirements such as financial assurances, real estate assurances, monitoring programs, and the provisions for short and long-term maintenance of the mitigation site. Special conditions may include, by reference, the compensatory mitigation plan, monitoring requirements and a contingency mitigation plan. Permittees are responsible for assuring that activity-specific compensatory mitigation projects are implemented successfully and protected over the long-term. If mitigation banks or in-lieu fee arrangements are used to provide the mitigation, the party(s) identified as responsible for administering those facets of the bank or the in-lieu fee arrangement become liable for implementation and performance.

m. TIMING OF MITIGATION CONSTRUCTION. Construction should be concurrent with authorized impacts to the extent practicable. Advance or concurrent mitigation can reduce temporal losses of aquatic functions and facilitate compliance. In some circumstances it may be acceptable to allow impacts to aquatic resources to occur before accomplishing compensatory mitigation, for example, in cases where construction of the authorized activity would disturb or harm on site compensatory mitigation work or where a simple restoration project is required. Some Federal-aid highway projects have legal and contractual requirements regarding the timing of mitigation that conflict with the policy to accomplish advance or concurrent mitigation. For compensatory mitigation involving in-lieu-fee arrangements or mitigation banks, the guidance applicable to those forms of mitigation should be followed with respect to timing of mitigation site development. After-the-fact mitigation may also be required for permits issued in emergencies or from an enforcement action.

n. COMPENSATORY MITIGATION ACCOMPLISHED AFTER OVERALL PROJECT CONSTRUCTION. In general, when impacts to aquatic resources are authorized before mitigation is initiated, Districts will require: 1) a Corps-approved mitigation plan; 2) a secured mitigation project site; 3) appropriate financial assurances in place; and, 4) legally protected, adequate water rights where necessary. Initial physical and biological improvements in the mitigation plan generally should be completed no later than the first full growing season following the impacts from authorized activities. If beginning the initial improvements within that time frame is not practicable, then other measures that mitigate for the consequences of temporal losses should be included in the mitigation plan.

o. GENERAL PERMITS. For activities authorized by general permits, Districts may recommend consolidated compensatory mitigation projects such as mitigation banks and in-lieu fee programs where such sources of compensatory mitigation are available. Consolidated mitigation facilitates a watershed approach to mitigating impacts to waters of the United States. For regional general permits associated with Special Area Management Plans or other types of watershed plans, the District may also recommend the use of mitigation banks or in-lieu-fee arrangements, consistent with the guidance for those forms of compensation.

3. COMPENSATORY MITIGATION PLANS

Districts will strive to discuss compensatory mitigation proposals with applicants during pre-application consultation. If this does not occur, the scope and specificity of proposed compensatory mitigation plans merely represent the applicant's view of what is necessary, a view that may not be acceptable to the Corps or other governmental authorities. At the earliest opportunity, Districts will advise applicants of the mitigation sequencing requirements of the Section 404(b)(1) Guidelines, or what is required for general permits. Compensation is the last step in the sequencing requirements of the Section 404 (b)(1) Guidelines. Thus, for standard permit applications, Districts should not require detailed compensatory mitigation plans until they have established the unavoidable impact. In all circumstances, the level of information provided regarding mitigation should be commensurate with the potential impact to aquatic resources, consistent with the guidance from Regulatory Guidance Letter 93-2 on the appropriate level of analysis for compliance with the Section 404 (b)(1) Guidelines. Districts will identify for applicants the pertinent factors for this determination (e.g., watershed considerations, local or state requirements, uncertainty, out-of-kind compensation, protection and maintenance requirements, etc.). Districts also will identify for applicants the rationale to be used (e.g., best professional judgment, Hydrogeomorphic Assessment Method, Wetland Rapid Assessment Procedure, etc.) for determining allowable impact and required compensatory mitigation. Applicants will be encouraged to submit appropriate compensatory mitigation proposals with individual permit applications or general permit pre-construction notices. The components listed below form the basis for development of compensatory mitigation plans.

a. BASELINE INFORMATION. As part of the permit decision Districts will include approved, written compensatory mitigation plans describing the location, size, type, functions and amount of impact to aquatic and other resources, as well as the resources in the mitigation project. In addition, they should describe the size, e.g., acreage of wetlands, length and width of streams, elevations of existing ground at the mitigation site, historic and existing hydrology, stream substrate and soil conditions, and timing of the mitigation. Baseline information may include quantitative sampling data on the physical, chemical, and biological characteristics of the aquatic resources at both the proposed mitigation site and the impact site. This documentation will support the compensatory mitigation requirement.

b. GOALS AND OBJECTIVES. Compensatory mitigation plans should discuss environmental goals and objectives, the aquatic resource type(s), e.g., hydrogeomorphic (HGM) regional wetland subclass, Rosgen stream type, Cowardin classification, and functions that will be impacted by the authorized work, and the aquatic resource type(s) and functions proposed at the compensatory mitigation site(s). For example, for impacts to tidal fringe wetlands the mitigation goal may be to replace lost finfish and shellfish habitat, lost estuarine habitat, or lost water quality functions associated with tidal backwater flooding. The objective statement should describe the amount, i.e., acres, linear feet, or functional changes, of aquatic habitat that the authorized work will impact and the amount of compensatory mitigation needed to offset those impacts, by aquatic resource type.

c. SITE SELECTION. Compensatory mitigation plans should describe the factors considered during the site selection process and plan formulation including, but not limited to:

1. WATERSHED CONSIDERATIONS. Mitigation plans should describe how the site chosen for a mitigation project contributes to the specific aquatic resource needs of the impacted watershed. A compensatory mitigation project generally should be in the same watershed. The further removed geographically that the mitigation is, the greater is the need to demonstrate that the proposed mitigation will reasonably offset authorized impacts.

2. PRACTICABILITY. The mitigation plan should describe site selection in terms of cost, existing technology, and logistics.

3. AIR TRAFFIC. Compensatory mitigation projects that have the potential to attract waterfowl and other bird species that might pose a threat to aircraft will be sited consistent with the Federal Aviation Administration Advisory Circular on Hazardous Wildlife Attractants on or near Airports (AC No: 150/5200-33, 5/1/97).

d. MITIGATION WORK PLAN. Compensatory mitigation work plans should contain written specifications and work descriptions, including, but not limited to: 1) boundaries of proposed restoration, establishment, enhancement, or preserved areas (e.g., maps and drawings); 2) construction methods, timing and sequence; 3) source of water supply and connections to existing waters and proximity to uplands; 4) native vegetation proposed for planting; 5) allowances for natural regeneration from an existing seed bank or planting; 6) plans for control of exotic invasive vegetation; 7) elevation(s) and slope(s) of the proposed mitigation area to ensure they conform with required elevation and hydrologic requirements, if practicable, for target plant species; 8) erosion control measures; 9) stream or other open water geomorphology and features such as riffles and pools, bends, deflectors, etc.; and 10) a plan outlining site management and maintenance.

e. PERFORMANCE STANDARDS. Compensatory mitigation plans will contain written performance standards for assessing whether mitigation is achieving planned goals. Performance standards will become part of individual permits as special conditions and be used for performance monitoring. Project performance evaluations will be performed by the Corps, as specified in the permits or special conditions, based upon monitoring reports. Adaptive management activities may be required to adjust to unforeseen or changing circumstances, and responsible parties may be required to adjust mitigation projects or rectify deficiencies. The project performance evaluations will be used to determine whether the environmental benefits or "credit(s)" for the entire project equal or exceed the environmental impact(s) or "debit(s)" of authorized activities. Performance standards for compensatory mitigation sites will be based on quantitative or qualitative characteristics that can be practicably measured. The performance standards will be indicators that demonstrate that the mitigation is developing or has developed into the desired habitat. Performance standards will vary by geographic region and aquatic habitat type, and may be developed through interagency coordination at the regional level. Performance standards for wetlands can be derived from the criteria in the 1987 Corps of Engineers Wetlands Delineation Manual, such as the duration of soil saturation required to meet the wetland hydrology criterion, or variables and associated functional capacity indices in hydrogeomorphic assessment method regional guidebooks. Performance standards may also be based on reference wetlands.

f. PROJECT SUCCESS. Compensatory mitigation plans will identify all parties responsible for compliance with the mitigation plan and their role in the mitigation project. The special conditions for the permit will identify these responsibilities as required above. Restoration projects provide the greatest potential for success in terms of functional compensation; however, each type has utility and may be used for compensatory mitigation.

g. SITE PROTECTION. Compensatory mitigation plans should include a written description of the legal means for protecting mitigation area(s), and permits will be conditioned accordingly. The wetlands, uplands, riparian areas, or other aquatic resources in a mitigation project should be permanently protected, in most cases, with appropriate real estate instruments, e.g., conservation easements, deed restrictions, transfer of title to Federal or state resource agencies or nonprofit conservation organizations. Generally, conservation easements held by tribal, state or local governments, other Federal agencies, or non-governmental groups, such as land trusts, are preferable to deed restrictions. Homeowners' associations should be used for these purposes only in exceptional circumstances, such as when the association is responsible for community open spaces with restrictive covenants. Districts may require third party monitoring if necessary to ensure permanent protection. In no case will the real estate instrument require a Corps official's signature. Also, Districts will not approve a requirement that results in the Federal government holding deed restrictions on properties, or that contains real estate provisions committing Corps Districts to any interest in the property in question, unless proper statutory authority is identified that authorizes such an arrangement.

h. CONTINGENCY PLAN. Compensatory mitigation plans should include contingency plans for unanticipated site conditions or changes. For example, contingency plans may identify financial assurance mechanisms that could be used to implement remedial measures to correct unexpected problems. Additionally, contingency plans will allow for modifications to performance standards if mitigation projects are meeting compensatory mitigation goals, but in unanticipated ways. Finally, contingency plans could address the circumstances that might result in no enforcement or remedial action if forces beyond the control of responsible parties adversely impact mitigation sites. In any case, Districts will determine the course of action to be taken in the event of unexpected conditions based on the goals and objectives for the mitigation project, the performance standards, and the provisions of the contingency plan.

i. MONITORING AND LONG-TERM MANAGEMENT. Compensatory mitigation plans will identify the party(s) responsible for accomplishing, maintaining, and monitoring the mitigation. Districts will require monitoring plans with a reporting frequency sufficient for an inspector to determine compliance with performance standards and to identify remedial action. Monitoring will be required for an adequate period of time, normally 5 to 10 years, to ensure the project meets performance standards. Corps permits will require permanent compensatory mitigation unless otherwise noted in the special conditions of the permit. Districts may take enforcement action even after the identified monitoring period, if there has been a violation.

j. FINANCIAL ASSURANCES. Compensatory mitigation plans will identify the party responsible for providing and managing any financial assurances and contingency funds set aside for remedial measures to ensure mitigation success. This includes identifying the party that will provide for long-term management and protection of the mitigation project. Financial assurances should be commensurate with the level of impact and the level of compensatory mitigation required. Permit conditions for minimal and low impact projects are generally sufficient for enforcing performance standards and requiring compliance, without the requirement of additional financial assurances. Financial assurances should be sufficient to cover contingency actions such as a default by the responsible party, or a failure to meet performance standards. District Engineers will generally emphasize financial assurances when the authorized impacts occur prior to successful completion of the mitigation, to include the monitoring period. Financial assurances may be in the form of performance bonds, irrevocable trusts, escrow accounts, casualty insurance, letters of credit, legislatively enacted dedicated funds for government operated banks or other approved instruments. Such assurances may be phased-out or reduced, once the project has been demonstrated functionally mature and self-sustaining in accordance with performance standards.

Financial assurances for third party mitigation should be consistent with existing guidance (e.g., Federal Guidance for the Establishment, Use and Operation of Mitigation Banks, and the Federal Guidance on the Use of In-Lieu-Fee Arrangements for Compensatory Mitigation under Section 404 of the Clean Water Act and Section 10 of the Rivers and Harbors Act). The District will determine project success, and the need to use financial assurances to carry out remedial measures, in accordance with the project performance standards.

4. DURATION

This guidance remains effective unless revised or rescinded.

FOR THE COMMANDER:

/signed/

ROBERT H. GRIFFIN
Major General, U.S. Army
Director of Civil Works

Encl

Appendix A
Authorities

This RGL is issued in accordance with the following statutes, regulations, and policies. It is intended to clarify provisions within these existing authorities and does not establish new requirements.

a. Clean Water Act Section 404 [33 USC 1344].

b. Rivers and Harbors Act of 1899 Section 10 [33 USC 403 et seq.].

c. Environmental Protection Agency, Section 404(b)(1) Guidelines [40 CFR Part 230]. Guidelines for Specification of Disposal Sites for Dredged or Fill Material.

d. Department of the Army, Section 404 Permit Regulations [33 CFR Parts 320–331]. Policies for evaluating permit applications to discharge dredged or fill material.

e. Memorandum of Agreement between the Environmental Protection Agency and the Department of the Army Concerning the Determination of Mitigation under the Clean Water Act Section 404(b)(1) Guidelines [February 6, 1990].

f. Federal Guidance for the Establishment, Use, and Operation of Mitigation Banks [November 28, 1995].

g. Federal Guidance on the Use of In-Lieu-Fee Arrangements for Compensatory Mitigation under Section 404 of the Clean Water Act and Section 10 of the Rivers and Harbors Act [November 7, 2000]

h. Title XII of the Food Security Act of 1985 as amended by the Farm Security and Rural Investment Act of 2002 [16 USC 3801 et seq.].

i. National Environmental Policy Act [42 USC 4321 et seq.], including the Council on Environmental Quality's implementing regulations [40 CFR Parts 1500–1508].

j. Fish and Wildlife Coordination Act [16 USC 661 et seq.].

k. Fish and Wildlife Service Mitigation Policy [46 FR pages 7644–7663, 1981].

l. Magnuson Fishery Conservation and Management Act [16 USC 1801 et seq.].

m. National Marine Fisheries Service Habitat Conservation Policy [48 FR pages 53142–53147, 1983].

n. The Transportation Equity Act for the 21st Century (TEA-21)

o. Federal Aviation Administration Advisory Circular on Hazardous Wildlife Attracts on or near Airports (AC No: 150/5200-33, 5/1/97)

p. Endangered Species Act of 1973, as amended [16 U.S.C. 1531 et seq.]

q. Migratory Bird Treaty Act [16 U.S.C. 703 et seq.]

r. Issuance of Nationwide Permits [67 FR 2020–2095, January 15, 2002]

Appendix B

Taken from *Operational Guidelines for Creating or Restoring Self-Sustaining Wetlands,* National Research Council 'Compensating for Wetland Losses Under The Clean Water Act,' June 2001 (Chapter 7, pp. 123–128).

1. CONSIDER THE HYDROGEOMORPHIC AND ECOLOGICAL LANDSCAPE AND CLIMATE.

Whenever possible locate the mitigation site in a setting of comparable landscape position and hydrogeomorphic class. Do not generate atypical "hydrogeomorphic hybrids"; instead, duplicate the features of reference wetlands or enhance connectivity with natural upland landscape elements (Gwin et al. 1999).

Regulatory agency personnel should provide a landscape setting characterization of both the wetland to be developed and, using comparable descriptors, the proposed mitigation site. Consider conducting a cumulative impact analysis at the landscape level based on templates for wetland development (Bedford 1999). Landscapes have natural patterns that maximize the value and function of individual habitats. For example, isolated wetlands function in ways that are quite different from wetlands adjacent to rivers. A forested wetland island, created in an otherwise grassy or agricultural landscape, will support species that are different from those in a forested wetland in a large forest tract. For wildlife and fisheries enhancement, determine if the wetland site is along ecological corridors such as migratory flyways or spawning runs. Constraints also include landscape factors. Shoreline and coastal wetlands adjacent to heavy wave action have historically high erosion rates or highly erodible soils, and often heavy boat wakes. Placement of wetlands in these locations may require shoreline armoring and other protective engineered structures that are contrary to the mitigation goals and at cross-purposes to the desired functions.

Even though catastrophic events cannot be prevented, a fundamental factor in mitigation plan design should be how well the site will respond to natural disturbances that are likely to occur. Floods, droughts, muskrats, geese, and storms are expected natural disturbances and should be accommodated in mitigation designs rather than feared. Natural ecosystems generally recover rapidly from natural disturbances to which they are adapted. The design should aim to restore a series of natural processes at the mitigation sites to ensure that resilience will have been achieved.

2. ADOPT A DYNAMIC LANDSCAPE PERSPECTIVE. Consider both current and future watershed hydrology and wetland location. Take into account surrounding land use and future plans for the land. Select sites that are, and will continue to be, resistant to disturbance from the surrounding landscape, such as preserving large buffers and connectivity to other wetlands. Build on existing wetland and upland systems. If possible, locate the mitigation site to take advantage of refuges, buffers, green spaces, and other preserved elements of the landscape. Design a system that utilizes natural processes and energies, such as the potential energy of streams as natural subsidies to the system. Flooding rivers and tides transport great quantities of water, nutrients, and organic matter in relatively short time periods, subsidizing the wetlands open to these flows as well as the adjacent rivers, lakes, and estuaries.

3. RESTORE OR DEVELOP NATURALLY VARIABLE HYDROLOGICAL CONDITIONS. Promote naturally variable hydrology, with emphasis on enabling fluctuations in water flow and level, and duration and frequency of change, representative of other comparable wetlands in the same landscape setting. Preferably, natural hydrology should be allowed to become reestablished rather than finessed through active engineering devices to mimic a natural hydroperiod. When restoration is not an option, favor the use of passive devices that have a higher likelihood to sustain the desired hydroperiod over long term. Try to avoid designing a system dependent on water-control structures or other artificial infrastructure that must be maintained in perpetuity in order for wetland hydrology to meet the specified design. In situations where direct (in-kind) replacement is desired, candidate mitigation sites should have the same basic hydrological attributes as the impacted site.

Hydrology should be inspected during flood seasons and heavy rains, and the annual and extremeevent flooding histories of the site should be reviewed as closely as possible. A detailed hydrological study of the site should be undertaken, including a determination of the potential interaction of groundwater with the proposed wetland. Without flooding or saturated soils, for at least part of the growing season, a wetland will not develop. Similarly, a site that is too wet will not support the desired biodiversity. The tidal cycle and stages are important to the hydrology of coastal wetlands.

4. WHENEVER POSSIBLE, CHOOSE WETLAND RESTORATION OVER CREATION. Select sites where wetlands previously existed or where nearby wetlands still exist. Restoration of wetlands has been observed to be more feasible and sustainable than creation of wetlands. In restored sites the proper substrate may be present, seed sources may be on-site or nearby, and the appropriate hydrological conditions may exist or may be more easily restored.

The U.S. Army Corps of Engineers (Corps) and Environmental Protection Agency (EPA) Mitigation Memorandum of Agreement states that, "because the likelihood of success is greater and the impacts to potentially valuable uplands are reduced, restoration should be the first option considered" (Fed. Regist. 60(Nov. 28):58605). The Florida Department of Environmental Regulation (FDER 1991a) recommends an emphasis on restoration first, then enhancement, and, finally, creation as a last resort. Morgan and Roberts (1999) recommend encouraging the use of more restoration and less creation.

5. AVOID OVER-ENGINEERED STRUCTURES IN THE WETLAND'S DESIGN. Design the system for minimal maintenance. Set initial conditions and let the system develop. Natural systems should be planned to accommodate biological systems. The system of plants, animals, microbes, substrate, and water flows should be developed for self-maintenance and self-design. Whenever possible, avoid manipulating wetland processes using approaches that require continual maintenance. Avoid hydraulic control structures and other engineered structures that are vulnerable to chronic failure and require maintenance and replacement. If necessary to design in structures,

such as to prevent erosion until the wetland has developed soil stability, do so using natural features, such as large woody debris. Be aware that more specific habitat designs and planting will be required where rare and endangered species are among the specific restoration targets.

Whenever feasible, use natural recruitment sources for more resilient vegetation establishment. Some systems, especially estuarine wetlands, are rapidly colonized, and natural recruitment is often equivalent or superior to plantings (Dawe et al. 2000). Try to take advantage of native seed banks, and use soil and plant material salvage whenever possible. Consider planting mature plants as supplemental rather than required, with the decision depending on early results from natural recruitment and invasive species occurrence. Evaluate on-site and nearby seed banks to ascertain their viability and response to hydrological conditions. When plant introduction is necessary to promote soil stability and prevent invasive species, the vegetation selected must be appropriate to the site rather than forced to fit external pressures for an ancillary purpose (e.g., preferred wildlife food source or habitat).

6. PAY PARTICULAR ATTENTION TO APPROPRIATE PLANTING ELEVATION, DEPTH, SOIL TYPE, AND SEASONAL TIMING. When the introduction of species is necessary, select appropriate genotypes. Genetic differences within species can affect wetland restoration outcomes, as found by Seliskar (1995), who planted cordgrass (*Spartina alterniflora*) from Georgia, Delaware, and Massachusetts into a tidal wetland restoration site in Delaware. Different genotypes displayed differences in stem density, stem height, below-ground biomass, rooting depth, decomposition rate, and carbohydrate allocation. Beneath the plantings, there were differences in edaphic chlorophyll and invertebrates.

Many sites are deemed compliant once the vegetation community becomes established. If a site is still being irrigated or recently stopped being irrigated, the vegetation might not survive. In other cases, plants that are dependent on surface-water input might not have developed deep root systems. When the surface-water input is stopped, the plants decline and eventually die, leaving the mitigation site in poor condition after the Corps has certified the project as compliant.

7. PROVIDE APPROPRIATELY HETEROGENEOUS TOPOGRAPHY. The need to promote specific hydroperiods to support specific wetland plants and animals means that appropriate elevations and topographic variations must be present in restoration and creation sites. Slight differences in topography (e.g., micro- and meso-scale variations and presence and absence of drainage connections) can alter the timing, frequency, amplitude, and duration of inundation. In the case of some less-studied, restored wetland types, there is little scientific or technical information on natural microtopography (e.g., what causes strings and flarks in patterned fens or how hummocks in fens control local nutrient dynamics and species assemblages and subsurface hydrology are poorly known). In all cases, but especially those with minimal scientific and technical background, the proposed development wetland or appropriate example(s) of the target wetland type should provide a model template for incorporating microtopography.

Plan for elevations that are appropriate to plant and animal communities that are reflected in adjacent or close-by natural systems. In tidal systems, be aware of local variations in tidal flooding regime (e.g., due to freshwater flow and local controls on circulation) that might affect flooding duration and frequency.

8. PAY ATTENTION TO SUBSURFACE CONDITIONS, INCLUDING SOIL AND SEDIMENT GEOCHEMISTRY AND PHYSICS, GROUNDWATER QUANTITY AND QUALITY, AND INFAUNAL COMMUNITIES. Inspect and characterize the soils in some detail to determine their permeability, texture, and stratigraphy. Highly permeable soils are not likely to support a wetland unless water inflow rates or water tables are high. Characterize the general chemical structure and variability of soils, surface water, groundwater, and tides. Even if the wetland is being created or restored primarily for wildlife enhancement, chemicals in the soil and water may be significant, either for wetland productivity or bioaccumulation of toxic materials. At a minimum, these should included chemical attributes that control critical geochemical or biological processes, such as pH, redox, nutrients (nitrogen and phosphorus species), organic content and suspended matter.

9. CONSIDER COMPLICATIONS ASSOCIATED WITH CREATION OR RESTORATION IN SERIOUSLY DEGRADED OR DISTURBED SITES. A seriously degraded wetland, surrounded by an extensively developed landscape, may achieve its maximal function only as an impaired system that requires active management to support natural processes and native species (NRC 1992). It should be recognized, however, that the functional performance of some degraded sites may be optimized by mitigation, and these considerations should be included if the goal of the mitigation is water- or sediment-quality improvement, promotion of rare or endangered species, or other objectives best served by locating a wetland in a disturbed landscape position. Disturbance that is intense, unnatural, or rare can promote extensive invasion by exotic species or at least delay the natural rates of redevelopment. Reintroducing natural hydrology with minimal excavation of soils often promotes alternative pathways of wetland development. It is often advantageous to preserve the integrity of native soils and to avoid deep grading of substrates that may destroy natural below-ground processes and facilitate exotic species colonization (Zedler 1996).

10. CONDUCT EARLY MONITORING AS PART OF ADAPTIVE MANAGEMENT. Develop a thorough monitoring plan as part of an adaptive management program that provides early indication of potential problems and direction for correction actions. The monitoring of wetland structure, processes, and function from the onset of wetland restoration or creation can indicate potential problems. Process monitoring (e.g., water-level fluctuations, sediment accretion and erosion, plant flowering, and bird nesting) is particularly important because it will likely identify the source of a problem and how it can be remedied. Monitoring and control of nonindigenous species should be a part of any effective adaptive management program. Assessment of wetland performance must be integrated with adaptive management. Both require understanding the processes that drive the structure and characteristics of a developing wetland. Simply documenting the structure (vegetation, sediments, fauna, and nutrients) will not provide the knowledge and guidance required to make adaptive "corrections" when adverse conditions are discovered. Although wetland development may take years to decades, process-based monitoring might provide more sensitive early indicators of whether a mitigation site is proceeding along an appropriate trajectory.

FEDERAL GUIDANCE ON THE USE OF IN-LIEU-FEE ARRANGEMENTS FOR COMPENSATORY MITIGATION UNDER SECTION 404 OF THE CLEAN WATER ACT AND SECTION 10 OF THE RIVERS AND HARBORS ACT

FEDERAL REGISTER VOL. 65, NO. 216

Pages 66914-66917
Tuesday, November 7, 2000

AGENCIES: Corps of Engineers, Department of the Army, DOD; Environmental Protection Agency; Fish and Wildlife Service, Interior; and National Marine Fisheries Service, National Oceanic and Atmospheric Administration, Commerce.

ACTION: Notice

SUMMARY: The Army Corps of Engineers (Corps), Environmental Protection Agency (EPA), Fish and Wildlife Service (FWS) and National Marine Fisheries Service (NMFS) are issuing final policy guidance regarding the use of in-lieu-fee arrangements for the purpose of providing compensation for adverse impacts to wetlands and other aquatic resources. Compensatory mitigation projects are designed to replace aquatic resource functions and values that are adversely impacted under the Clean Water Act Section 404 and Rivers and Harbors Act Section 10 regulatory programs. These mitigation objectives are stated in regulation, the 1990 Memorandum of Agreement on mitigation between Environmental Protection Agency (EPA) and the Department of the Army, the November 28, 1995, Federal Guidance on the Establishment, Use and Operation of Mitigation Banks ("Banking Guidance"), and other relevant policy. The advent of in-lieu-fee approaches to mitigation has highlighted the importance of several fundamental objectives that the agencies established for determining what constitutes appropriate compensatory mitigation. The purpose of this memorandum is to clarify the manner in which in-lieu-fee mitigation may serve as an effective and useful approach to satisfy compensatory mitigation requirements and meet the Administration's goal of no overall net loss of wetlands. This in-lieu-fee guidance elaborates on the discussion of in-lieu-fee mitigation arrangements in the Banking Guidance by outlining the circumstances where in-lieu-fee mitigation may be used, consistent with existing regulations and policy.

EFFECTIVE DATE: The effective date is October 31, 2000.

FOR FURTHER INFORMATION CONTACT: Mr. Jack Chowning (Corps) at (202) 761-4614; Ms. Lisa Morales (EPA) at (202) 260-6013; Mr. Mark Matusiak (FWS) at (703) 358-2183; Ms. Susan-Marie Stedman (NMFS) at (301) 713-2325.

SUPPLEMENTARY INFORMATION. This notice publishes interagency guidance regarding the use of in-lieu-fee arrangements for the purpose of providing compensation for adverse impacts to wetlands and other aquatic resources. Any comments or questions on the document may be directed to the persons listed above in the section entitled: **FOR FURTHER INFORMATION CONTACT.**

Dated: October 20, 2000.
Michael L. Davis,
Deputy Assistant Secretary (Civil Works),
Department of the Army.

Dated: October 20, 2000.
Robert H. Wayland III,
Director, Office of Wetlands,
Oceans, and Watersheds,
Environmental Protection Agency.

Dated: October 31, 2000.
Jamie Clark,
Director, Fish and Wildlife Service,
Department of the Interior.

Dated: October 25, 2000.
Scott B. Gudes,
Deputy Under Secretary for
Oceans and Atmosphere,
National Oceanic and
Atmospheric Administration,
Department of Commerce.

Memorandum to the Field

SUBJECT: Federal Guidance on the Use of In-Lieu-Fee Arrangements for Compensatory Mitigation Under Section 404 of the Clean Water Act and Section 10 of the Rivers and Harbors Act

I. PURPOSE

Compensatory mitigation projects are designed to replace aquatic resource functions and values that are adversely impacted under the Clean Water Act Section 404 and Rivers and Harbors Act Section 10 regulatory programs. These mitigation objectives are stated in regulation, the 1990 Memorandum of Agreement on mitigation between Environmental Protection Agency (EPA) and the Department of the Army, the November 28, 1995, Federal Guidance on the Establishment, Use and Operation of Mitigation Banks ("Banking Guidance"), and other relevant policy. The advent of in-lieu fee approaches to mitigation has highlighted the importance of several fundamental objectives that the agencies established for determining what constitutes appropriate compensatory mitigation. The purpose of this memorandum is to clarify the manner in which in-lieu-fee mitigation may serve as an effective and useful approach to satisfy compensatory mitigation requirements and meet the Administration's goal of no overall net loss of wetlands. This in-lieu-fee guidance elaborates on the discussion of in-lieu-fee mitigation arrangements in the Banking Guidance by outlining the circumstances where in-lieu-fee mitigation may be used, consistent with existing regulations and policy.

II. BACKGROUND

A. "In-lieu-fee" mitigation occurs in circumstances where a permittee provides funds to an in-lieu-fee sponsor instead of either completing project specific mitigation or purchasing credits from a mitigation bank approved under the Banking Guidance.

B. A fundamental precept of the Section 404(b)(1) Guidelines is that no discharge of dredged or fill material in waters of the U.S. may be permitted unless appropriate and practicable steps have been taken to minimize all adverse impacts associated with the discharge. (40 CFR 230.10(d)) Specifically, the Section 404(b)(1) Guidelines establish a mitigation sequence, under which compensatory mitigation is required to offset wetland losses after all appropriate and practicable steps have been taken to first avoid and then minimize wetland impacts. Compliance with these mitigation sequencing requirements is an essential environmental safeguard to ensure that CWA objectives for the protection of wetlands are achieved. The Section 404 permit program relies on the use of compensatory mitigation to offset unavoidable wetlands impacts by replacing lost wetland functions and values.

C. The agencies further clarified their mitigation policies in a Memorandum of Agreement (MOA) between the EPA and the Department of the Army Concerning the Determination of Mitigation under the Clean Water Act Section 404(b)(1) Guidelines (February 6, 1990). That document reiterates that "the Clean Water Act and the Guidelines set forth a goal of restoring and maintaining existing aquatic resources. The Corps will strive to avoid adverse impacts and offset unavoidable adverse impacts to existing aquatic resources, and for wetlands, will strive to achieve a goal of no overall net loss of values and functions." Moreover, the MOA clarifies that mitigation "should be undertaken, when practicable, in areas adjacent or contiguous to the discharge site," and that "if on-site compensatory mitigation is not practicable, off-site compensatory mitigation should be undertaken in the same geographic area if practicable (i.e., in close proximity and, to the extent possible, the same watershed)." As outlined in the MOA, the agencies have also agreed that "generally, in-kind compensatory mitigation is preferable to out-of-kind." The MOA further states that mitigation banking may be an acceptable form of compensatory mitigation. The agencies recognize the general preference for restoration over other forms of mitigation, given the increased chance for ecological success.

D. Pursuant to these standards, project-specific mitigation for authorized impacts has been used by permittees to offset unavoidable impacts. Project-specific mitigation generally consists of restoration, creation, or enhancement of aquatic resources that are similar to the aquatic resources of the impacted area, and is often located on the project site or adjacent to the impact area. Permittees providing project specific mitigation have a U.S. Army Corps of Engineers (Corps) approved mitigation plan detailing the site, source of hydrology, types of aquatic resource to be restored, success criteria, contingency measures, and an annual reporting requirement. The mitigation and monitoring plan becomes part of the Section 404 authorization in the form of a special condition. The permittee is responsible for complying with all terms and conditions of the authorization and would be in violation of their authorization if the mitigation did not comply with the approved plan.

E. In 1995, the agencies issued the Banking Guidance. Consistent with that guidance, permittees may purchase mitigation credits from an approved bank. Mitigation banks will generally be functioning in advance of project impacts and thereby reduce the temporal losses of aquatic functions and values and reduce uncertainty over the ecological success of the mitigation. Mitigation banking instruments are reviewed and approved by an interagency Mitigation Banking Review Team (MBRT). The MBRT ensures that the banking instrument appropriately addresses the physical and legal characteristics of the bank and how the bank will be established and operated (e.g., classes of wetlands and/or other aquatic resources proposed for inclusion in the bank, geographic service area where credits may be sold, wetland classes or other aquatic resource impacts suitable for compensation, methods for determining credits and debits). The bank sponsor is responsible for the operation and maintenance of the bank during its operational life, as well as the long-term management and ecological success of the wetlands and/or other aquatic resources, and must provide financial assurances.

F. The Banking Guidance describes in-lieu-fee mitigation as follows: "... in-lieu-fee, fee mitigation, or other similar arrangements, wherein funds are paid to a natural resource management entity for implementation of either specific or general wetland or other aquatic resource development project, are not considered to meet the definition of mitigation banking because they do not typically provide compensatory mitigation in advance of project impacts. Moreover, such arrangements do not typically provide a clear timetable for the initiation of mitigation efforts. The Corps, in consultation with the other agencies, may find circumstances where such arrangements are appropriate so long as they meet the requirements that would otherwise apply to an offsite, prospective mitigation effort and provides adequate assurances of success and timely implementation. In such cases, a formal agreement between the sponsor and the agencies, similar to a banking instrument, is necessary to define the conditions under which its use is considered appropriate."

III. USE OF IN-LIEU-FEE MITIGATION IN THE REGULATORY PROGRAM

In light of the above considerations and in order to ensure that decisions regarding the use of in-lieu-fee mitigation are made more consistently with existing provisions of agency regulations and permit policies, the following clarification is provided. It is organized in a tiered manner to reflect and incorporate the agencies' broader mitigation policies, and is based on relative assurances of ecological success.

A. IMPACTS AUTHORIZED UNDER INDIVIDUAL PERMIT: In-lieu-fee agreements may be used to compensate for impacts authorized by individual permit if the in-lieu-fee arrangement is developed (or revised, if an existing agreement), reviewed, and approved using the process established for mitigation banks in the Banking Guidance. MBRTs should review applications from such in-lieu-fee sponsors to ensure that such agreements are consistent with the Banking Guidance.

B. IMPACTS AUTHORIZED UNDER GENERAL PERMIT: As a general matter, in-lieu-fee mitigation should only be used to compensate for impacts to waters of the U.S. authorized by a Section 404 general permit, as described below:

1. WHERE "ON-SITE" MITIGATION IS AVAILABLE AND PRACTICABLE: As a general matter, compensatory mitigation that is completed on or adjacent to the site of the impacts it is designed to offset (i.e., project-specific mitigation done by permittees consistent with Corps approved mitigation plans) is preferable to mitigation conducted off-site (i.e., mitigation bank or in-lieu-fee mitigation). The agencies' preference for on-site mitigation, indicated in the 1990 Memorandum of Agreement on mitigation between the EPA and the Department of the Army, should not preclude the use of a mitigation bank or in-lieu-fee mitigation when there is no practicable opportunity for on-site compensation, or when use of a bank or in-lieu-fee mitigation is environmentally preferable to on-site compensation, consistent with the provisions in paragraph 2 below.

2. WHERE "ON-SITE" MITIGATION IS NOT AVAILABLE OR PRACTICABLE: Except as noted below in a. or b., where on-site mitigation is not available, practicable, or determined to be less environmentally desirable, use of a mitigation bank is preferable to in-lieu-fee mitigation where permitted impacts are within the service area of a mitigation bank approved to sell mitigation credits, and those credits are available. Use of a mitigation bank is also preferable over in-lieu-fee mitigation where both the available in-lieu-fee arrangement and the service area of an approved mitigation bank are outside of the watershed of the permitted project impacts, unless the mitigation bank is determined on a case by case basis to not be practicable and environmentally desirable.

a. WHERE MITIGATION BANK DOES NOT PROVIDE "IN-KIND" MITIGATION: In those circumstances where wetlands impacts proposed for general permit authorization are within the service area of an approved mitigation bank with available credits, but the impacted wetland type is not identified by the Mitigation Banking Instrument for compensation within such bank, then the authorized impact may be compensated through an in-lieu-fee arrangement, subject to the considerations described in Section IV below, if the in-lieu fee arrangement would provide in-kind restoration as mitigation.

b. WHERE MITIGATION BANK DOES NOT PROVIDE RESTORATION, CREATION, OR ENHANCEMENT MITIGATION: In those circumstances where wetlands impacts proposed for general permit authorization are within the service area of an approved mitigation bank, but the only available credits are through preservation, then the authorized impact may be compensated through an in-lieu-fee arrangement subject to the considerations described in Section IV below, if the in-lieu fee arrangement would provide in kind restoration as mitigation.

IV. PLANNING, ESTABLISHMENT, AND USE OF IN-LIEU-FEE MITIGATION ARRANGEMENTS

This section describes the basic considerations that should be addressed for any proposed use of in-lieu-fee mitigation to offset unavoidable impacts associated with a discharge authorized under a general permit described in Section III above.

A. PLANNING CONSIDERATIONS

1. QUALIFIED ORGANIZATIONS: Given the goal to ensure long-term mitigation success, the Corps, in consultation with the other

Federal agencies, should carefully evaluate the demonstrated performance of natural resource management organizations (*e.g.*, governmental organizations, land trusts) prior to approving them to manage in-lieu fee arrangements. In fact, given the unique strengths and specialties of such organizations, it may be useful for the Corps, in consultation with other Federal resource agencies, to establish formal arrangements with several natural resource management organizations to ensure there are sufficient options to effectively replace lost functions and values. In any event, in-lieu-fee arrangements and subsequent modifications should be made in consultation with the other Federal agencies and only after an opportunity for public notice and comment has been afforded.

2. OPERATIONAL INFORMATION: Those organizations considered qualified to implement formal in-lieu-fee arrangements should work in advance with the Corps to ensure that authorized impacts will be offset fully on a project-by-project basis consistent with Section 10/404 permit requirements. As detailed in the paragraphs that follow, organizations should supply the Corps with information in advance on (1) potential sites where specific restoration projects or types of restoration projects are planned, (2) the schedule for implementation, (3) the type of mitigation that is most ecologically appropriate on a particular parcel, and (4) the financial, technical, and legal mechanisms to ensure long-term mitigation success. The Corps should ensure that the formal in-lieu-fee arrangements and project authorizations contain distinct provisions that clearly state that the legal responsibility for ensuring mitigation terms are satisfied fully rests with the organization accepting the in-lieu-fee. In-lieu-fee sponsors should be able to demonstrate approval of all necessary State and local permits and authorizations. In-lieu-fee sponsors (*e.g.*, State) should notify the Corps and MBRT if the service area of any mitigation bank overlaps the jurisdiction in which their in-lieu-fees may be spent.

3. WATERSHED PLANNING: Local watershed planning efforts, as a general matter, identify wetlands and other aquatic resources that have been degraded and usually have established a prioritization list of restoration needs. In-lieu-fee mitigation projects should be planned and developed to address the specific resource needs of a particular watershed.

4. SITE SELECTION: The Federal agencies and in-lieu-fee sponsor should give careful consideration to the ecological suitability of a site for achieving the goal and objectives of compensatory mitigation (*e.g.*, possess the physical, chemical and biological characteristics to support the desired aquatic resources and functions, preferably in-kind restoration or creation of impacted aquatic resources). The location of the site relative to other ecological features, hydrologic sources, and compatibility with adjacent land uses and watershed management plans shall be considered by the Federal agencies during the evaluation process.

5. TECHNICAL FEASIBILITY: In-lieu-fee mitigation should be planned and designed to be self-sustaining over time to the extent possible. The techniques for establishing aquatic resources must be carefully selected. The restoration of historic or substantially degraded aquatic resources (*e.g.*, prior-converted cropland, farmed wetlands) utilizing proven techniques increases the likelihood of success and typically does not result in the loss of other valuable resources. Thus, restoration should be the first option considered for siting in-lieu fee mitigation. This guidance recognizes that in some circumstances aquatic resources must be actively managed to ensure their sustainability. Furthermore, long-term maintenance requirements may be necessary and appropriate in some cases (*e.g.*, to maintain fire dependent habitat communities in the absence of natural fire, to control invasive exotic plant species). Proposed mitigation techniques should be well-understood and reliable. When uncertainties surrounding the technical feasibility of a proposed mitigation technique exist, appropriate arrangements may be phased-out or reduced once the attainment of prescribed performance standards is demonstrated. In any event, a plan detailing specific performance standards should be submitted to ensure the technical success of the project can be evaluated.

6. ROLE OF PRESERVATION: As described in the Banking Guidance, simple purchase or "preservation" of existing wetlands may be accepted as compensatory mitigation only in exceptional circumstances. Mitigation credit may be given when existing wetlands and/or other aquatic resources are preserved in conjunction with restoration, creation or enhancement activities, and when it is demonstrated that the preservation will augment the functions of the restored, created or enhanced aquatic resource.

7. COLLECTION OF FUNDS: Funds collected under any in-lieu-fee arrangement should be used for replacing wetlands functions and values and not to finance non-mitigation programs and priorities (*e.g.*, education projects, research). Funds collected should be based upon a reasonable cost estimate of all funds needed to compensate for the impacts to wetlands or other waters that each permit is authorized to offset. Funds collected should ensure a minimum of one-for-one acreage replacement, consistent with existing regulation and permit conditions. Land acquisition and initial physical and biological improvements should be completed by the first full growing season following collection of the initial funds. However, because site improvements associated with in-lieu fee mitigation may take longer to initiate, initial physical and biological improvements may be completed no later than the second full growing season where (1) initiation by the first full growing season is not practicable, (2) mitigation ratios are raised to account for increased temporal losses of aquatic resource functions and values, and (3) the delay is approved in advance by the Corps.

8. MONITORING AND MANAGEMENT: The in-lieu-fee sponsor is responsible for securing adequate funds for the operation and maintenance of the mitigation sites. The wetlands and/or other aquatic resources in the mitigation site should be protected in perpetuity with appropriate real estate arrangements (*e.g.*, conservation easements, transfer of title to Federal or State resource agency or non-profit conservation agency). Such arrangements should effectively restrict harmful activities (*e.g.*, incompatible uses) that might otherwise jeopardize the purpose of the compensatory mitigation. In addition, there should be appropriate schedules for regular (*e.g.*, annual) monitoring reports to document funds received, impacts permitted, how funds were disbursed, types of projects funded, and the success of projects conducted under the in-lieu-fee arrangement. The Corps, in conjunction with other Federal and State agencies, should evaluate the reports and conduct regular reviews to ensure that the arrangement is operating effectively and consistent with agency policy and the specific agreement. The Corps will track all uses of in-lieu-fee arrangements and report those figures by public notice on an annual basis.

B. ESTABLISHMENT OF IN-LIEU-FEE AGREEMENTS

A formal in-lieu-fee agreement, consistent with the planning provisions above, should be established by the sponsor with the Corps, in consultation with the other agencies. It may be appropriate to establish an "umbrella" arrangement for the establishment and operation of multiple sites. In such circumstances, the need for supplemental information (*e.g.*, site specific plans) should be addressed in specific in-lieu-fee agreements. The inlieu- fee agreement should contain:

1. a description of the sponsor's experience and qualifications with respect to providing compensatory mitigation;

2. potential site locations, baseline conditions at the sites, and general plans that indicate what kind of wetland compensation can be provided (*e.g.*, wetland type,

restoration or other activity, proposed time line, *etc.*);

3. geographic service area;

4. accounting procedures;

5. methods for determining fees and credits;

6. a schedule for conducting the activities that will provide compensatory mitigation or a requirement that projects will be started within a specified time after impacts occur;

7. performance standards for determining ecological success of mitigation sites;

8. reporting protocols and monitoring plans;

9. financial, technical and legal provisions for remedial actions and responsibilities (*e.g.*, contingency fund);

10. financial, technical and legal provisions for long-term management and maintenance (*e.g.*, trust); and

11. provision that clearly states that the legal responsibility for ensuring mitigation terms are fully satisfied rests with the organization accepting the fee.

In cases where initial establishment of in-lieu-fee compensatory mitigation involves a discharge into waters of the United States requiring Section 10/404 authorization, submittal of a Section 10/404 application should be accompanied by the in-lieu-fee agreement.

V. GENERAL

A. EFFECT OF GUIDANCE. This guidance does not change the substantive requirements of the Section 10/404 regulatory program. Rather, it interprets and provides guidance and procedures for the use of in-lieu fee mitigation consistent with existing regulations. The policies set out in this document are not final agency action, but are intended solely as guidance. The guidance is not intended, nor can it be relied upon, to create any rights enforceable by any party in litigation with the United States. This guidance does not establish or affect legal rights or obligations, establish a binding norm on any party and it is not finally determinative of the issues addressed. Any regulatory decisions made by the agencies in any particular matter addressed by this guidance will be made by applying the governing law and regulations to the relevant facts.

B. DEFINITIONS. Unless otherwise noted, the terms used in this guidance have the same definitions as those terms in the Banking Guidance. Note that as part of the Administration's Clean Water Action Plan, the Federal agencies have proposed a tracking system to more accurately account for wetland losses and gains that includes definitions of terms such as restoration used in wetland programs. Future notice will be given when these definitions will be applied to Section 10/404 regulatory program.

C. EFFECTIVE DATE. This guidance is effective immediately on the date of the last signature below. Therefore, existing in-lieu-fee arrangements or agreements should be reviewed and modified as necessary in light of the above.

D. CONVERSION TO BANKS: If requested by the in-lieu-fee sponsor, the Corps, in conjunction with the other Federal agencies, will provide assistance and recommendations on the steps necessary to convert individual in-lieu fee arrangements to mitigation banks, consistent with the Banking Guidance.

E. FUTURE REVISIONS. The agencies are supporting a comprehensive, independent evaluation of the effectiveness of compensatory mitigation by the National Academy of Sciences. The technical results of this evaluation are expected to be used by the public to improve the quality of wetlands and aquatic resource restoration, creation, and enhancement. The agencies will take note of the results of this evaluation and other relevant information to make any necessary revisions to guidance on compensatory mitigation, to ensure the greatest opportunity for ecological success of restored, created, and enhanced wetlands and other aquatic resources. At a minimum, a review of the use of this guidance will be initiated no later than 12 months after the effective date.

Michael L. Davis,
Deputy Assistant Secretary
(Civil Works),
Department of the Army.

Robert H. Wayland III,
Director, Office of Wetlands,
Oceans, and Watersheds
Environmental Protection Agency.

Jamie Clark,
Director, Fish and Wildlife Service,
Department of the Interior.

Scott B. Gudes,
Deputy Under Secretary
for Oceans and Atmosphere,
National Oceanic and
Atmospheric Administration,
Department of Commerce.

"DEEP-RIPPING" IN WETLANDS

REGULATORY GUIDANCE LETTER 96-02

SUBJECT: Applicability of Exemptions under Section 404(f) to "Deep-Ripping" Activities in Wetlands

DATE: 12 December 1996

EXPIRES: 31 December 2001

Memorandum to the Field

Department of the Army, U.S. Army Corps of Engineers

United States Environmental Protection Agency

SUBJECT: Applicability of Exemptions under Section 404(f) to "Deep-Ripping" Activities in Wetlands

PURPOSE: The purpose of this memorandum is to clarify the applicability of exemptions provided under Section 404(f) of the Clean Water Act (CWA) to discharges associated with "deep-ripping" and related activities in wetlands.[1]

BACKGROUND

1. Section 404(f)(1) of the CWA exempts from the permit requirement certain discharges associated with normal farming, forestry, and ranching practices in waters of the United States, including wetlands. Discharges into waters subject to the Act associated with farming, forestry, and ranching practices identified under Section 404(f)(1) do not require a permit except as provided under Section 40.4(f)(2).

2. Section 404(f)(1) does not provide a total automatic exemption for all activities related to agricultural silvicultural or ranching practices. Rather, Section 404(f)(1) exempts only those activities specifically identified in paragraphs (A) through (F), and "other activities of essentially the same character as named" [44 FR 34264]. For example, Section 404(f)(1)(A) lists discharges of dredged or fill material from "normal farming, silviculture and ranching activities, such as plowing, seeding, cultivating,

1. As this guidance addresses primarily agricultural-related activities, characterizations of such practices have been developed in consultation with experts at the U.S. Department of Agriculture (USDA), Natural Resources Conservation Service.

minor drainage, harvesting for the production of food, fiber, and forest products, or upland soil and water conservation practices."

3. Section 404(f)(1)(A) is limited to activities that are part of an "established (i.e., ongoing) farming, silviculture, or ranching operation." This "established" requirement is intended to reconcile the dual intent reflected in the legislative history that although Section 40.4 should not unnecessarily restrict farming, forestry, or ranching from continuing at a particular site, discharge activities which could destroy wetlands or other waters should be subject to regulation.

4. EPA and Corps regulations [40 CFR 230 and 33 CFR 320] and preamble define in some detail the specific "normal" activities fisted in Section 404(f)(1)(A). Three points may be useful in the current context:

a. As explained in the preamble to the 1979 proposed regulations, the words "such as" have been consistently interpreted as restricting the section "to the activities *named* in the statute and other activities of essentially the same character as named," and "preclude the extension of the exemption... to activities that are unlike those named." [44 FR 34264].

b. Plowing is specifically defined in the regulations not to include the redistribution of surface material in a manner which converts wetlands areas to uplands [See 40 CFR 233.35(a)(1)(iii)(D)].

c. Discharges associated with activities that establish an agricultural operation in wetlands where previously ranching had been conducted, represents a "change in use" within the meaning of Section 404(f)(2). Similarly, discharges that establish forestry practices in wetlands historically subject to agriculture also represent a change in use of the site (See 40 CFR 233.35(c)].

5. The statute includes a provision at Section 404(f)(2) that "recaptures" or reestablishes the permit requirement for those otherwise exempt discharges which:

a. convert an area of the waters of the U.S. to a new use, *and*

b. impair the flow or circulation of waters of the U.S. *or* reduce the reach of waters of the U.S.

Conversion of an area of waters of the U.S. to uplands triggers both provisions (a) and (b) above. Thus, at a minimum any otherwise exempt discharge that results in the conversion of waters of the U.S. to upland is recaptured under Section 404(f)(2) and requires a permit. It should be noted that in order to trigger the recapture provisions of Section 404(f)(2), the discharges themselves need not be the sole cause of the destruction of the wetland or other change in use or sole cause of the reduction or impairment of reach, flow, or circulation of waters of the U.S. Rather, the discharges need only be "incidental to" or "part of" an activity which is intended to or will forseeably bring about that result. Thus, in applying Section 404(f)(2), one must consider discharges in context, rather than isolation.

ISSUE

1. Questions have been raised involving "deep-ripping" and related activities in wetlands and whether discharges associated with these actions fall within the exemptions at Section 404(f)(1)(A). In addition, the issue has been raised whether, if such activities fall within the exemption, they would be recaptured under Section 404(f)(2).

2. "Deep-ripping" is defined as the mechanical manipulation of the soil to break up or pierce highly compacted, impermeable or slowly permeable subsurface soil layers, or other similar kinds of restrictive soil layers. These practices are typically used to break up these subsoil layers (e.g., impermeable soil layer, hardpan) as part of the initial preparation of the soil to establish an agricultural or silvicultural operation. Deep-ripping and related activities are also used in established farming operations to break up highly compacted soil. Although deep-ripping and related activities may be required more than once, the activity is typically not an annual practice. Deep-ripping and related activities are undertaken to improve site drainage and facilitate deep root growth, and often occur to depths greater than 16 inches and, in some cases, exceeding 4 feet below the surface. As such it requires the use of heavy equipment, including bulldozers, equipped with ripper-blades, shanks, or chisels often several feet in length. Deep-ripping and related activities involve extending the blades to appropriate depths and dragging them through the soil to break up the restrictive layer.

3. Conversely, plowing is defined in EPA and Corps regulations [40 CFR 230 and 33 CFR 320] as "all forms of primary tillage... used... for the breaking up, cutting, turning over, or stirring of soil to prepare it for the planting of crops" [40 CFR 232.3(d)(4)]. As a general matter, normal plowing activities involve the annual or at least regular, preparation of soil prior to seeding or other planting activities. According to USDA, plowing generally involves the use of a blade, chisel or series of blades, chisels, or discs, usually 8–10 inches in length pulled behind a farm vehicle to prepare the soil for the planting of annual crops or to support an ongoing farming practice. Plowing is commonly used to break up the surface of the soil to maintain soil tilth and to facilitate infiltration throughout the upper root zone.

DISCUSSION

1. Plowing in wetlands is exempt from regulation consistent with the following circumstances:

a. it is conducted as part of an ongoing, established agricultural, silvicultural or ranching operation; and

b. the plowing is not incidental to an activity that results in the immediate or gradual conversion of wetlands to non-waters.

2. Deep-ripping and related activities are distinguishable from plowing and similar practices (e.g., discing, harrowing) with regard to the purposes and circumstances under which it is conducted, the nature of the equipment that is used, and its effect, including in particular the impacts to the hydrology of the site.

a. Deep-ripping and related activities are commonly conducted to depths exceeding 16 inches, and as deep as 6–8 feet below the soil surface to break restrictive soil layers and improve water drainage at sites that have not supported deeper rooting crops. Plowing depths, according to USDA, rarely exceed one foot into the soil and not deeper than 16 inches without the use of special equipment involving special circumstances. As such, deep- ripping and related activities typically involve the use of special equipment, including heavy mechanized equipment and bulldozers, equipped with elongated ripping blades, shanks, or chisels often several feet in length. Moreover, while plowing is generally associated with ongoing operations, deep-ripping and related activities are typically conducted to prepare a site for establishing crops not previously planted at the site. Although deep-ripping may have to be redone at regular intervals in some circumstances to maintain proper soil drainage, the activity is typically not an annual or routine practice.

b. Frequently, deep-ripping and related activities are conducted as a preliminary step for converting a "natural" system or for preparing rangeland for a new use such as farming or silviculture. In those instances, deep ripping and related activities are often required to break up naturally-occurring impermeable or slowly permeable subsurface soil layers to facilitate proper root growth. For example, for certain depressional wetlands types such as vernal pools, the silica-cemented hardpan (durapan) or other restrictive layer traps precipitation and seasonal runoff creating ponding and saturation conditions at the soil surface. The presence of these impermeable or slowly permeable subsoil layers is essential to support the hydrology of the system. Once these layers are disturbed by activities such as deep-ripping, the hydrology of the system is disturbed and the wetland is often destroyed.

c. In contrast, there are other circumstances where activities such as deep-ripping and related activities are a standard practice of an established on-going farming operation. For example, in parts of the Southeast, where there are deep soils having a high clay content, mechanized farming practices can lead to the compaction of the soil below the sod surface. It may be necessary to break up, on a regular although not annual basis, these restrictive layers in order to allow for normal root development and infiltration. Such activities may require special equipment and can sometimes occur to depths greater than 16 inches. However, because of particular physical conditions, including the presence of a water table at or near the surface for part of the growing season, the activity typically does not have the effect of impairing the hydrology of the system or otherwise altering the wetland characteristics of the site.

CONCLUSION

1. When deep-ripping and related activities are undertaken as part of an *established on-going* agricultural silvicultural or ranching operation, to break up compacted soil layers *and* where the hydrology of the site will not be altered such that it would result in conversion of waters of the U.S. to upland, such activities are exempt under Section 404(f)(1)(A).

2. Deep-ripping and related activities in wetlands are not part of a normal ongoing activity, and therefore not exempt, when such practices are conducted in association with efforts to establish for the first time (or when a previously established operation was abandoned) an agricultural silvicultural or ranching operation. In addition, deep-ripping and related activities are not exempt in circumstances where such practices would trigger the "recapture" provision of Section 404(f)(2):

a. Deep-ripping to establish a farming operation at a site where a ranching or forestry operation was in place is a change in use of such a site. Deep-ripping and related activities that also have the effect of altering or removing the wetland hydrology of the site would trigger Section 404(f)(2) and such ripping would require a permit.

b. Deep-ripping a site that has the effect of converting wetlands to non-waters would also trigger Section 404(f)(2) and such ripping would require a permit.

3. It is the agencies' experience that certain wetland types are particularly vulnerable to hydrological alteration as a result of deep-ripping and related activities. Depressional wetland systems such as prairie potholes, vernal pools and playas whose hydrology is critically dependent upon the presence of an impermeable or slowly permeable subsoil layer are particularly sensitive to disturbance or alteration of this subsoil layer. Based upon this experience, the agencies have concluded that, as a general matter, deep-ripping and similar practices, consistent with the descriptions above, conducted in prairie potholes, vernal pools, playas, and similar depressions wetlands destroy the hydrological integrity of these wetlands. In these circumstances, deep-ripping in prairie potholes, vernal pools, and playas is recaptured under Section 404(f)(2) and requires a permit under the Clean Water Act.

Robert H Wayland III
Director
Office of Wetlands,
Oceans and Watersheds
U.S. Environmental
Protection Agency

Daniel R Burns, P.E.
Chief, Operations,
Construction and Readiness Division
Directorate of Civil Works
U.S. Army Corps of Engineers

FEDERAL GUIDANCE FOR THE ESTABLISHMENT, USE AND OPERATION OF MITIGATION BANKS

FEDERAL REGISTER VOLUME 60, NUMBER 228

Page 58605-58614
November 28, 1995

AGENCIES: Corps of Engineers, Department of the Army, DOD; Environmental Protection Agency; Natural Resources Conservation Service, Agriculture; Fish and Wildlife Service, Interior; and National Marine Fisheries Service, National Oceanic and Atmospheric Administration, Commerce.

ACTION: Notice

SUMMARY: The Army Corps of Engineers (Corps), Environmental Protection Agency (EPA), National Resources Conservation Service (NRCS), Fish and Wildlife Service (FWS) and National Marine Fisheries Service (NMFS) are issuing final policy guidance regarding the establishment, use and operation of mitigation banks for the purpose of providing compensation for adverse impacts to wetlands and other aquatic resources. The purpose of this guidance is to clarify the manner in which mitigation banks may be used to satisfy mitigation requirements of the Clean Water Act (CWA) Section 404 permit program and the wetland conservation provisions of the Food Security Act (FSA) (i.e., "Swampbuster" provisions). Recognizing the potential benefits mitigation banking offers for streamlining the permit evaluation process and providing more effective mitigation for authorized impacts to wetlands, the agencies encourage the establishment and appropriate use of mitigation banks in the Section 404 and "Swampbuster" programs.

DATES: The effective date of this Memorandum to the Field is December 28, 1995.

FOR FURTHER INFORMATION CONTACT: Mr. Jack Chowning (Corps) at (202) 761-1781; Mr. Thomas Kelsch (EPA) at (202) 260-8795; Ms. Sandra Byrd (NRCS) at (202) 690-3501; Mr. Mark Miller (FWS) at (703) 358-2183; Ms. Susan-Marie Stedman (NMFS) at (301) 713-2325.

SUPPLEMENTARY INFORMATION: Mitigating the environmental impacts of necessary development actions on the Nation's wetlands and other aquatic resources is a central premise of Federal wetlands programs. The CWA Section 404 permit program relies on the use of compensatory mitigation to offset unavoidable damage to wetlands and other aquatic resources through, for example, the restoration or creation of wetlands. Under the "Swampbuster" provisions of the FSA, farmers are required to provide mitigation to offset certain conversions of wetlands for agricultural purposes in order to maintain their program eligibility.

Mitigation banking has been defined as wetland restoration, creation, enhancement, and in exceptional circumstances, preservation undertaken expressly for the purpose of compensating for unavoidable wetland losses in advance of development actions, when such compensation cannot be achieved at the development site or would not be as environmentally beneficial. It typically involves the consolidation of small, fragmented wetland mitigation projects into one large contiguous site. Units of restored, created, enhanced or preserved wetlands are expressed as "credits" which may subsequently be withdrawn to offset "debits" incurred at a project development site.

Ideally, mitigation banks are constructed and functioning in advance of development impacts, and are seen as a way of reducing uncertainty in the CWA Section 404 permit program or the FSA "Swampbuster" program by having established compensatory mitigation credit available to an applicant. By consolidating compensation requirements, banks can more effectively replace lost wetland functions within a watershed, as well as provide economies of scale relating to the planning, implementation, monitoring and management of mitigation projects.

On August 23, 1993, the Clinton Administration released a comprehensive package of improvements to Federal wetlands programs which included support for the use of mitigation banks. At that same time, EPA and the Department of the Army issued interim guidance clarifying the role of mitigation banks in the Section 404 permit program and providing general guidelines for their establishment and use. In that document it was acknowledged that additional guidance would be developed, as necessary, following completion of the first phase of the Corps Institute for Water Resources national study on mitigation banking.

The Corps, EPA, NRCS, FWS and NMFS provided notice [60 FR 12286; March 6, 1995] of a proposed guidance on the policy of the Federal government regarding the establishment, use and operation of mitigation banks. The proposed guidance was based, in part, on the experiences to date with mitigation banking, as well as other environmental, economic and institutional issues identified through the Corps national study. Over 130 comments were received on the proposed guidance. The final guidance is based on full and thorough consideration of the public comments received.

A majority of the letters received supported the proposed guidance in general, but suggested modifications to one or more parts of the proposal. In response to these comments, several changes have been made to further clarify the provisions and make other modifications, as necessary, to ensure effective establishment and use of mitigation banks. One key issue on which the agencies received numerous comments focused on the timing of credit withdrawal. In order to provide additional clarification of the changes made to the final guidance in response to comments, the agencies wish to emphasize that it is our intent to ensure that decisions to allow credits to be withdrawn from a mitigation bank in advance of bank maturity be make on a case-by-case basis to best reflect the particular ecological and economic circumstances of each bank. The percentage of advance credits permitted for a particular bank may be higher or lower than the 15 percent example included in the proposed guidance. The final guidance is being revised to eliminate the reference to a specific percentage in order to provide needed flexibility. Copies of the comments and the agencies' response to significant comments are available for public review. Interested parties should contact the agency representatives for additional information.

This guidance does not change the substantive requirements of the Section 404 permit program or the FSA "Swampbuster" program. Rather, it interprets and provides internal guidance and procedures to the agency field personnel for the establishment, use and operation of mitigation banks consistent with existing regulations and policies of each program. The policies set out in this document are not final agency action, but are intended solely as guidance. The guidance is not intended, not can it be relied upon, to create any rights enforceable by any party in litigation with the United States. The guidance does not establish or affect legal rights or obligations, establish a binding norm on any party and it is not finally determinative of the issues addressed. Any regulatory decisions made by the agencies in any particular matter addressed by this guidance will be made by applying the governing law and regulations to the relevant facts. The purpose of the document is to provide policy and technical guidance to encourage the effective use of mitigation banks as a means of compensating for the authorized loss of wetlands and other aquatic resources.

John H. Zirschky,
Acting Assistant Secretary (Civil Works),
Department of the Army.

Robert Perciasepe,
Assistant Administrator for Water,
Environmental Protection Agency.

James R. Lyons,
Assistant Secretary,
Natural Resources and Environment,
Department of Agriculture.

George T. Frampton, Jr.,
Assistant Secretary for
Fish and Wildlife and Parks,
Department of the Interior.

Douglas K. Hall,
Assistant Secretary for
Oceans and Atmosphere,
Department of Commerce.

Memorandum to the Field

SUBJECT: Federal Guidance for the Establishment, Use and Operation of Mitigation Banks

I. INTRODUCTION

A. PURPOSE AND SCOPE OF GUIDANCE

This document provides policy guidance for the establishment, use and operation of mitigation banks for the purpose of providing compensatory mitigation for authorized adverse impacts to wetlands and other aquatic resources. This guidance is provided expressly to assist Federal personnel, bank sponsors, and others in meeting the requirements of Section 404 of the Clean Water Act (CWA), Section 10 of the Rivers and Harbors Act, the wetland conservation provisions of the Food Security Act (FS) (i.e., "Swampbuster"), and other applicable Federal statutes and regulations. The policies and procedures discussed herein are consistent with current requirements of the Section 10/404 regulatory program and "Swampbuster" provisions and are intended only to clarify the applicability of existing requirements to mitigation banking.

The policies and procedures discussed herein are applicable to the establishment, use and operation of public mitigation banks, as well as privately-sponsored mitigation banks, including third party banks (e.g. entrepreneurial banks).

B. BACKGROUND

For purposes of this guidance, mitigation banking means the restoration, creation, enhancement and, in exceptional circumstances, preservation of wetlands and/or other aquatic resources expressly for the purpose of providing compensatory mitigation in advance of authorized impacts to similar resources.

The objective of a mitigation bank is to provide for the replacement of the chemical, physical and biological functions of wetlands and other aquatic resources which are lost as a result of authorized impacts. Using appropriate methods, the newly established functions are quantified as mitigation "credits" which are available for use by the bank sponsor or by other parties to compensate for adverse impacts (i.e., "debits"). Consistent with mitigation policies established under the Council on Environmental Quality Implementing Regulations (CEQ regulations) (40 CFR Part 1508.20), and the Section 404(b)(1) Guidelines (Guidelines) (40 CFR Part 230), the use of credits may only be authorized for purposes of complying with Section 10/404 when adverse impacts are unavoidable. In addition, for both the Section 10/404 and "Swampbuster" programs, credits may only be authorized when on-site compensation is either not practicable or use of a mitigation bank is environmentally preferable to on-site compensation. Prospective bank sponsors should not construe or anticipate participation in the establishment of a mitigation bank as ultimate authorization for specific projects, as excepting such projects from any applicable requirements, or as preauthorizing the use of credits from that bank for any particular project.

Mitigation banks provide greater flexibility to applicants needing to comply with mitigation requirements and can have several advantages over individual mitigation projects, some of which are listed below:

1. It may be more advantageous for maintaining the integrity of the aquatic ecosystem to consolidate compensatory mitigation into a single large parcel or contiguous parcels when ecologically appropriate;

2. Establishment of a mitigation bank can bring together financial resources, planning and scientific expertise not practicable to many project-specific compensatory mitigation proposals. This consolidation of resources can increase the potential for the establishment and long- term management of successful mitigation that maximizes opportunities for contributing to biodiversity and/or watershed function;

3. Use of mitigation banks may reduce permit processing times and provide more cost-effective compensatory mitigation opportunities for projects that qualify;

4. Compensatory mitigation is typically implemented and functioning in advance of project impacts, thereby reducing temporal losses of aquatic functions and uncertainty over whether the mitigation will be successful in offsetting project impacts;

5. Consolidation of compensatory mitigation within a mitigation bank increases the efficiency of limited agency resources in the review and compliance monitoring of mitigation projects, and thus improves the reliability of efforts to restore, create or enhance wetlands for mitigation purposes.

6. The existence of mitigation banks can contribute towards attainment of the goal for no overall net loss of the Nation's wetlands by providing opportunities to compensate for authorized impacts when mitigation might not otherwise be appropriate or practicable.

II. POLICY CONSIDERATIONS

The following policy considerations provide general guidance for the establishment, use and operation of mitigation banks. It is the agencies' intent that this guidance be applied to mitigation bank proposals submitted for approval on or after the effective date of this guidance and to those in early stages of planning or development. It is not intended that this policy be retroactive for mitigation banks that have already received agency approval. While it is recognized that individual mitigation banking proposals may vary, it is the intent of this guidance that the fundamental precepts be applicable to future mitigation banks.

For the purposes of Section 10/104, and consistent with the CEQ regulations, the Guidelines, and the Memorandum of Agreement Between the Environmental Protection Agency (EPA) and the Department of the Army Concerning the Determination of Mitigation under the Clean Water Act Section 404(b)(1) Guidelines, mitigation means sequentially avoiding impacts, minimizing impacts, and compensating for remaining unavoidable impacts. Compensatory mitigation, under Section 10/404, is the restoration, creation, enhancement, or in exceptional circumstances, preservation of wetlands and/or other aquatic resources for the purpose of compensating for unavoidable adverse impacts. A site where wetlands and/or other aquatic resources are restored, created, enhanced, or in exceptional circumstances, preserved expressly for the purpose of providing compensatory mitigation in advance of authorized impacts to similar resources is a mitigation bank.

A. AUTHORITIES

This guidance is established in accordance with the following statutes, regulations, and policies. It is intended to clarify provisions within these existing authorities and does to establish any new requirements.

1. Clean Water Act Section 404 (33 U.S.C. 1344).

2. Rivers and Harbors Act of 1899 Section 10 (33 U.S.C. 403 et seq.)

3. Environmental Protection Agency, Section 404(b)(1) Guidelines (40 CFR Part 230). Guidelines for Specification of Disposal Sites for Dredged or Fill Material.

4. Department of the Army, Section 404 Permit Regulations (33 CFR Parts 320-330). Policies for evaluating permit applications to discharge dredged or fill material.

5. Memorandum of Agreement between the Environmental Protection Agency and the Department of the Army Concerning the Determination of Mitigation under the Clean Water Act Section 404(b)(1) Guidelines (February 6, 1990).

6. Title XII Food Security Act of 1985 as amended by the Food, Agriculture, Conservation and Trade Act of 1990 (16 U.S.C. 3801 et seq.).

7. National Environmental Policy Act (42 U.S.C. 4321 et seq.), including the Council on Environmental Quality's implementing regulations (40 CFR Parts 1500–1508).

8. Fish and Wildlife Coordination Act (16 U.S.C. 661 et seq.).

9. Fish and Wildlife Service Mitigation Policy (46 FR pages 7644–7663, 1981).

10. Magnuson Fishery Conservation and Management Act (16 U.S.C. 1801 et seq.).

11. National Marine Fisheries Service Habitat Conservation Policy (48 FR pages 53142–53147, 1983).

The policies set out in this document are not final agency action, but are intended solely as guidance. The guidance is not intended, nor can it be relied upon, to create any rights enforceable by any party in litigation with the United States. This guidance does not establish or affect legal rights or obligations, establish a binding norm on any party and it is not finally determinative of the issues addressed. Any regulatory decisions made by the agencies in any particular matter addressed by this guidance will be made by applying the governing law and regulations to the relevant facts.

B. PLANNING CONSIDERATIONS

1. GOAL SETTING

The overall goal of a mitigation bank is to provide economically efficient and flexible mitigation opportunities, while fully compensating for wetland and other aquatic resource losses in a manner that contributes to the long-term ecological functioning of the watershed within which the bank is to be located. The goal will include the need to replace essential aquatic functions which are anticipated to be lost through authorized activities within the bank's service area. In some cases, banks may also be used to address other resource objectives that have been identified in a watershed management plan or other resource assessment. It is desirable to set the particular objectives for a mitigation bank (i.e., the type and character of wetlands and/ or aquatic resources to be established) in advance of site selection. The goal and objectives should be driven by the anticipated mitigation need; the site selected should support achieving the goal and objectives.

2. SITE SELECTION

The agencies will give careful consideration to the ecological suitability of a site for achieving the goal and objectives of a bank, i.e., that it posses the physical, chemical and biological characteristics to support establishment of the desired aquatic resources and functions. Size and location of the site relative to other ecological features, hydrologic sources (including the availability of water rights), and compatibility with adjacent land uses and watershed management plans are important factors for consideration. It also is important that ecologically significant aquatic or upland resources (e.g., shallow subtidal habitat, mature forests), cultural sites, or habitat for Federally or State-listed threatened and endangered species are not compromised in the process of establishing a bank. Other significant factors for consideration include, but are not limited to, development trends (i.e., anticipated land use changes), habitat status and trends, local or regional goals for the restoration or protection of particular habitat types or functions (e.g., re-establishment of habitat corridors or habitat for species of concern), water quality and floodplain management goals, and the relative potential for chemical contamination of the wetlands and/ or other aquatic resources.

Banks may be sited on public or private lands. Cooperative arrangements between public and private entities to use public lands for mitigation banks may be acceptable. In some circumstances, it may be appropriate to site banks on Federal, state, tribal or locally-owned resource management areas (e.g., wildlife management areas, national or state forests, public parks, recreation areas). The siting of banks on such lands may be acceptable if the internal policies of the public agency allow use of its land for such purposes, and the public agency grants approval. Mitigation credits generated by banks of this nature should be based solely on those values in the bank that are supplemental to the public program(s) already planned or in place, that is, baseline values represented by existing or already planned public programs, including preservation value, should not be counted toward bank credits.

Similarly, Federally-funded wetland conservation projects undertaken via separate authority and for other purposes, such as the Wetlands Reserve Program, Farmer's Home Administration fee title transfers or conservation easements, and Partners for Wildlife Program, cannot be used for the purpose of generating credits within a mitigation bank. However, mitigation credit may be given for activities undertaken in conjunction with, but supplemental to, such programs in order to maximize the overall ecological benefit of the conservation project.

3. TECHNICAL FEASIBILITY

Mitigation banks should be planned and designed to be self-sustaining over time to the extent possible. The techniques for establishing wetlands and/or other aquatic resources must be carefully selected, since this science is constantly evolving. The restoration of historic or substantially-degraded wetlands and/or other aquatic resources (e.g., prior-converted cropland, farmed wetlands) utilizing proven techniques increases the likelihood of success and typically does not result in the loss of other valuable resources. Thus, restoration should be the first option considered when siting a bank. Because of the difficulty in establishing the correct hydrologic conditions associated with many creation projects and the tradeoff in wetland functions involved with certain enhancement activities, these methods should only be considered where there are adequate assurances to ensure success and that the project will result in an overall environmental benefit.

In general, banks which involve complex hydraulic engineering features and/ or questionable water sources (e.g., pumped) are most costly to develop, operate and maintain, and have a higher risk of failure than banks designed to function with little or no human intervention. The former situations should only be considered where there are adequate assurances to ensure success. This guidance recognizes that in some circumstances wetlands must be actively managed to ensure their viability and sustainability. Furthermore, long-term maintenance requirements may be necessary and appropriate in some cases (e.g., to maintain fire-dependent plant communities in the absence of natural fire; to control invasive exotic plant species).

Proposed mitigation techniques should be well-understood and reliable. When uncertainties surrounding the technical feasibility of a proposed mitigation technique exist, appropriate arrangements (e.g., financial assurances, contingency plans, additional monitoring requirements) should be in place to increase the likelihood of success. Such arrangements may be phased-out or reduced once the attainment of prescribed performance standards is demonstrated.

4. ROLE OF PRESERVATION

Credit may be given when existing wetlands and/or other aquatic resources are preserved in conjunction with restoration, creation or enhancement activities, and when it is demonstrated that the preservation will augment the functions of the restored, created or enhanced aquatic resource. Such augmentation may be reflected in the total number of credits available from the bank.

In addition, the preservation of existing wetlands and/or other aquatic resources in perpetuity may be authorized as the sole basis for generating credits in mitigation banks only in exceptional circumstances, consistent with existing regulations, policies and guidance. Under such circumstances, preservation may be accomplished through the implementation of appropriate legal mechanisms (e.g., transfer of deed, deed restrictions, conservation easement) to protect wetlands and/or other aquatic resources, accompanied by implementation of appropriate changes in land use or other physical changes as necessary (e.g., installation of restrictive fencing).

Determining whether preservation is appropriate as the sole basis for generating credits at a mitigation bank requires careful judgment regarding a number of factors. Consideration must be given to whether wetlands and/or other aquatic resources proposed for preservation (1) perform physical or biological functions, the preservation of which is important to the region in which the aquatic resources are located, and (2) are under demonstrable threat of loss or substantial degradation due to human activities that might not otherwise be expected to be restricted. The existence of a demonstrable threat will be based on clear evidence of destructive land use changes which are consistent with local and regional land use trends and are not the consequence of actions under the control of the bank sponsor. Wetlands and other aquatic resources restored under the Conservation Reserve Program or similar programs requiring only temporary conservation easements may be eligible for banking credit upon termination of the original easement if the wetlands are provided permanent protection and it would otherwise be expected that the resources would be converted upon termination of the easement. The number of mitigation credits available from a bank that is based solely on preservation should be based on the functions that would otherwise be lost or degraded if the aquatic resources were not preserved, and the timing of such loss or degradation. As such, compensation for aquatic resource impacts will typically require a greater number of acres from a preservation bank than from a bank which is based on restoration, creation or enhancement.

5. INCLUSION OF UPLAND AREAS

Credit may be given for the inclusion of upland areas occurring within a bank only to the degree that such features increase the overall ecological functioning of the bank. If such features are included as part of a bank, it is important that they receive the same protected status as the rest of the bank and be subject to the same operational procedures and requirements. The presence of upland areas may increase the per-unit value of the aquatic habitat in the bank. Alternatively, limited credit may be given to upland areas protected within the bank to reflect the functions inherently provided by such areas (e.g., nutrient and sediment filtration of stormwater runoff, wildlife habitat diversity) which directly enhance or maintain the integrity of the aquatic ecosystem and that might otherwise be subject to threat of loss or degradation. An appropriate functional assessment methodology should be used to determine the manner and extent to which such features augment the functions of restored, created or enhanced wetlands and/or other aquatic resources.

6. MITIGATION BANKING AND WATERSHED PLANNING

Mitigation banks should be planned and developed to address the specific resource needs of a particular watershed. Furthermore, decisions regarding the location, type of wetlands and/or other aquatic resources to be established, and proposed uses of a mitigation bank are most appropriately made within the context of a comprehensive watershed plan. Such watershed planning efforts often identify categories of activities having minimal adverse effects on the aquatic ecosystem and that, therefore, could be authorized under a general permit. In order to reduce the potential cumulative effects of such activities, it may be appropriate to offset these types of impacts through the use of a mitigation bank established in conjunction with a watershed plan.

C. ESTABLISHMENT OF MITIGATION BANKS

1. PROSPECTUS

Prospective bank sponsors should first submit a prospectus to the Army Corps of Engineers (Corps) or Natural Resources Conservation Service (NRCS) [1] to initiate the planning and review process by the appropriate agencies. Prior to submitting a prospectus, bank sponsors are encouraged to discuss their proposal with the appropriate agencies (e.g., pre-application coordination).

It is the intent of the agencies to provide practical comments to the bank sponsors regarding the general need for and technical feasibility of proposed banks. Therefore, bank sponsors are encouraged to include in the prospectus sufficient information concerning the objectives for the bank and how it will be established and operated to allow the agencies to provide such feedback. Formal agency involvement and review is initiated with submittal of a prospectus.

2. MITIGATION BANKING INSTRUMENTS

Information provided in the prospectus will serve as the basis for establishing the mitigation banking instrument. All mitigation banks need to have a banking instrument as documentation of agency concurrence on the objectives and administration of the bank. The banking instrument should describe in detail the physical and legal characteristics of the bank, and how the bank will be established and operated. For regional banking programs sponsored by a single entity (e.g., a state transportation agency), it may be appropriate to establish an "umbrella" instrument for the establishment and operation of multiple bank sites. In such circumstances, the need for supplemental site-specific information (e.g., individual site plans) should be addressed in the banking instrument. The banking instrument will be signed by the bank sponsor and the concurring regulatory and resource agencies represented on the Mitigation Bank Review Team (section II.C.2). The following information should be addressed, as appropriate, within the banking instrument:

a. Bank goals and objectives;

b. Ownership of bank lands;

c. Bank size and classes of wetlands and/or other aquatic resources proposed for inclusion in the bank, including a site plan and specifications;

d. Description of baseline conditions at the bank site;

e. Geographic service area;

f. Wetland classes or other aquatic resource impacts suitable for compensation;

g. Methods for determining credits and debits;

h. accounting procedures;

i. Performance standards for determining credit availability and bank success;

j. Reporting protocols and monitoring plan;

k. Contingency and remedial actions and responsibilities;

l. Financial assurances;

m. Compensation ratios;

n. Provisions for long-term management and maintenance.

The terms and conditions of the banking instrument may be amended, in accordance with the procedures used to establish the instrument and subject to agreement by the signatories.

In cases where initial establishment of the mitigation bank involves a discharge into waters of the United States requiring Section 10/404 authorization, the banking instrument will be made part of a Department of the Army permit for that discharge. Submittal of an individual permit application should be accompanied by a sufficiently- detailed prospectus to allow for concurrent processing of each. Preparation of a banking instrument, however, should not alter the normal permit evaluation process timeframes. A bank sponsor may proceed with activities for the construction of a bank subsequent to receiving the Department of the Army authorization. It should be noted, however, that a bank sponsor who proceeds in the absence of a banking instrument does so at his/her own risk.

In cases where the mitigation bank is established pursuant to the FSA, the banking instrument will be included in the plan developed or approved by NRCS and the Fish and Wildlife Service (FWS).

3. AGENCY ROLES AND COORDINATION

Collectively, the signatory agencies to the banking instrument will comprise the Mitigation Bank Review Team (MBRT). Representatives from the Corps, EPA, FWS, National Marine Fisheries Service (NMFS) and NRCS, as appropriate given the projected use for the bank, should typically comprise the MBRT. In addition, it is appropriate for representatives from state, tribal and local regulatory and resource agencies to participate

1. The Corps will typically serve as the lead agency for the establishment of mitigation banks. Bank sponsors proposing establishment of mitigation banks solely for the purpose of complying with the "Swampbuster" provisions of FSA should submit their prospectus to the NRCS.

where an agency has authorities and/or mandates directly affecting or affected by the establishment, use or operation of a bank. No agency is required to sign a banking instrument; however, in signing a banking instrument, an agency agrees to the terms of that instrument.

The Corps will serve as Chair of the MBRT, except in cases where the bank is proposed solely for the purpose of complying with the FSA, in which case NRCS will be the MBRT Chair. In addition, where a bank is proposed to satisfy the requirements of another Federal, state, tribal or local program, it may be appropriate for the administering agency to serve as co-Chair of the MBRT.

The primary role of the MBRT is to facilitate the establishment of mitigation banks through the development of mitigation banking instruments. Because of the different authorities and responsibilities of each agency represented on the MBRT, there is a benefit in achieving agreement on the banking instrument. For this reason, the MBRT will strive to obtain consensus on its actions. The Chair making final decisions regarding the terms and conditions of the banking instrument where consensus cannot otherwise be reached within a reasonable timeframe (e.g., 90 days from the date of submittal of a complete prospectus). The MBRT will review and seek consensus on the banking instrument and final plans for the restoration, creation, enhancement, and/or preservation of wetlands and other aquatic resources.

Consistent with its authorities under Section 10/404, the Corps is responsible for authorizing use of a particular mitigation bank on a project-specific basis and determining the number and availability of credits required to compensate for proposed impacts in accordance with the terms of the banking instrument. Decisions rendered by the Corps must fully consider review agency comments submitted as part of the permit evaluation process. Similarly, the NRCS, in consultation with the FWS, will make the final decision pertaining to the withdrawal of credits from banks as appropriate mitigation pursuant to FSA.

4. ROLE OF THE BANK SPONSOR

The bank sponsor is responsible for the preparation of the banking instrument in consultation with the MBRT. The bank sponsor should, therefore, have sufficient opportunity to discuss the content of the banking instrument with the MBRT. The bank sponsor is also responsible for the overall operation and management of the bank in accordance with the terms of the banking instrument, including the preparation and distribution of monitoring reports and accounting statements/ledger, as necessary.

5. PUBLIC REVIEW AND COMMENT

The public should be notified of and have an opportunity to comment on all bank proposals. For banks which require authorization under an individual Section 10/404 permit or a state, tribal or local program that involves a similar public notice and comment process, this condition will typically be satisfied through such standard procedures. For other proposals, the Corps or NRCS, upon receipt of a complete banking prospectus, should provide notification of the availability of the prospectus for a minimum 21-day public comment period. Notification procedures will be similar to those used by the Corps in the standard permit review process. Copies of all public comments received will be distributed to the other members of the MBRT and the bank sponsor for full consideration in the development of the final banking instrument.

6. DISPUTE RESOLUTION PROCEDURE

The MBRT will work to reach consensus on its actions in accordance with this guidance. It is anticipated that all issues will be resolved by the MBRT in this manner.

a. DEVELOPMENT OF THE BANKING INSTRUMENT

During the development of the banking instrument, if any agency representative considers that a particular decision raises concern regarding the application of existing policy or procedures, an agency may request, through written notification, that the issue be reviewed by the Corps District Engineer, or NRCS State Conservationist, as appropriate. Said notification will describe the issue in sufficient detail and provide recommendations for resolution. Within 20 days, the District Engineer or State Conservationist (as appropriate) will consult with the notifying agency(ies) and will resolve the issue. The resolution will be forwarded to the other MBRT member agencies. The bank sponsor may also request the District Engineer or State Conservationist review actions taken to develop the banking instrument if the sponsor believes that inadequate progress has been made on the instrument by the MBRT.

b. APPLICATION OF THE BANKING INSTRUMENT

As previously stated, the Corps and NRCS are responsible for making final decisions on a project-specific basis regarding the use of a mitigation bank for purposes of Section 10/404 and FSA, respectively. In the event an agency on the MBRT is concerned that a proposed use may be inconsistent with the terms of the banking instrument, that agency may raise the issue to the attention of the Corps or NRCS through the permit evaluation process. In order to facilitate timely and effective consideration of agency comments, the Corps or NRCS, as appropriate, will advise the MBRT agencies of a proposed use of a bank. The Corps will fully consider comments provided by the review agencies regarding mitigation as part of the permit evaluation process. The NCRS will consult with FWA is making its decisions pertaining to mitigation.

If, in the view of an agency on the MBRT, an issued permit or series of permits reflects a pattern of concern regarding the application of the terms of the banking instrument, that agency may initiate review of the concern by the full MBRT through written notification to the MBRT Chair. The MBRT Chair will convene a meeting of the MBRT, or initiate another appropriate forum for communication, typically within 20 days of receipt of notification, to resolve concerns. Any such effort to address concerns regarding the application of a banking instrument will not delay any decision pending before the authorizing agency (e.g., Corps or NRCS).

D. CRITERIA FOR USE OF A MITIGATION BANK

1. PROJECT APPLICABILITY

All activities regulated under Section 10/404 may be eligible to use a mitigation bank as compensation for unavoidable impacts to wetlands and/or other aquatic resources. Mitigation banks established for FSA purposes may be debited only in accordance with the mitigation and replacement provisions of 7 CFR Part 12.

Credits from mitigation banks may also be used to compensate for environmental impacts authorized under other programs (e.g., state or local **wetland** regulatory programs, NPDES program, Corps civil works projects, Superfund removal and remedial actions). In no case may the same credits be used to compensate for more than one activity; however, the same credits may be used to compensate for an activity which requires authorization under more than one program.

2. RELATIONSHIP TO MITIGATION REQUIREMENTS

Under the existing requirements of Section 10/404, all appropriate and practicable steps must be undertaken by the applicant to first avoid and then minimize adverse impacts to aquatic resources, prior to authorization to use a particular mitigation bank. Remaining unavoidable impacts must be compensated to the extent appropriate and practicable. For both the Section 10/404 and "Swampbuster" programs, requirements for compensatory mitigation may be satisfied through the use of mitigation banks when either on-site compensation is not practicable or use of the mitigation bank is environmentally preferable to on-site compensation.

It is important to emphasize that applicants should not expect that establishment of, or purchasing credits from, a mitigation bank will necessarily lead to a determination of compliance with applicable mitigation requirements (i.e., Section 404(b)(1) Guidelines or FSA Manual), or as excepting projects from any applicable requirements.

3. GEOGRAPHIC LIMITS OF APPLICABILITY

The service area of a mitigation bank is the area (e.g., watershed, county) wherein a bank can reasonably be expected to provide appropriate compensation for impacts to wetlands and/or other aquatic resources. This area should be designated in the banking instrument. Designation of the service area should be based on consideration of hydrologic and biotic criteria, and be stipulated in the banking instrument. Use of a mitigation bank to compensate for impacts beyond the designated service area may be authorized, on a case-by-case basis, where it is determined to be practicable and environmentally desirable.

The geographic extent of a service area should, to the extent environmentally desirable, be guided by the cataloging unit of the "Hydrologic Unit map of the United States" (USGS, 1980) and the ecoregion of the "Ecoregions of the United States" (James M. Omernik, EPA, 1986) or section of the "Descriptions of the Ecoregions of the United States" (Robert G. Bailey, USDA, 1980). It may be appropriate to use other classification systems developed at the state or regional level for the purpose of specifying bank service areas, when such systems compare favorably in their objectives and level of detail. In the interest of the integrating banks with other resource management objectives, bank service areas may encompass larger watershed areas if the designation of such areas is supported by local or regional management plans (e.g., Special Area Management Plans, Advance Identification), State Wetland Conservation Plans or other Federally sponsored or recognized resource management plans. Furthermore, designation of a more inclusive service area may be appropriate for mitigation banks whose primary purpose is to compensate for linear projects that typically involve numerous small impacts in several different watersheds.

4. USE OF A MITIGATION BANK VS. ON-SITE MITIGATION

The agencies' preference for on-site mitigation, indicated in the 1990 Memorandum of Agreement on mitigation between the EPA and the Department of the Army, should not preclude the use of a mitigation bank when there is no practicable opportunity for on-site compensation, or when use of a bank is environmentally preferable to on-site compensation. On-site mitigation may be preferable where there is a practicable opportunity to compensate for important local functions including local flood control functions, habitat for a species or population with a very limited geographic range or narrow environmental requirements, or where local water quality concerns dominate.

In choosing between on-site mitigation and use of a mitigation bank, careful consideration should be given to the likelihood for successfully establishing the desired habitat type, the compatibility of the mitigation project with adjacent land uses, and the practicability of long-term monitoring and maintenance to determine whether the effort will be ecologically sustainable, as well as the relative cost of mitigation alternatives. In general, use of a mitigation bank to compensate for minor aquatic resource impacts (e.g., numerous, small impacts associated with linear projects; impacts authorized under nationwide permits) is preferable to on-site mitigation. With respect to larger aquatic resource impacts, use of a bank may be appropriate if it is capable of replacing essential physical and/or biological functions of the aquatic resources which are expected to be lost or degraded. Finally, there may be circumstances warranting a combination of on-site and off-site mitigation to compensate for losses.

5. IN-KIND VS. OUT-OF-KIND MITIGATION DETERMINATIONS

In the interest of achieving functional replacement, in-kind compensation of aquatic resource impacts should generally be required. Out-of-kind compensation may be acceptable if it is determined to be practicable and environmentally preferable to in-kind compensation (e.g., of greater ecological value to a particular region). However, non-tidal wetlands should typically not be used to compensate for the loss or degradation of tidal wetlands. Decisions regarding out-of-kind mitigation are typically made on a case-by-case basis during the permit evaluation process. The banking instrument may identify circumstances in which it is environmentally desirable to allow out-of-kind compensation within the context of a particular mitigation bank (e.g., for banks restoring a complex of associated wetland types). Mitigation banks developed as part of an area-wide management plan to address a specific resource objective (e.g., restoration of a particularly vulnerable or valuable wetland habitat type) may be such an example.

6. TIMING OF CREDIT WITHDRAWAL

The number of credits available for withdrawal (i.e., debiting) should generally be commensurate with the level of aquatic functions attained at a bank at the time of debiting. The level of function may be determined through the application of performance standards tailored to the specific restoration, creation or enhancement activity at the bank site or through the use of an appropriate functional assessment methodology.

The success of a mitigation bank with regard to its capacity to establish a healthy and fully functional aquatic system relates directly to both the ecological and financial stability of the bank. Since financial considerations are particularly critical in early stages of bank development, it is generally appropriate, in cases where there is adequate financial assurance and where the likelihood of the success of the bank is high, to allow limited debiting of a percentage of the total credits projected for the bank at maturity. Such determinations should take into consideration the initial capital costs needed to establish the bank, and the likelihood of its success. However, it is the intent of this policy to ensure that those actions necessary for the long-term viability of a mitigation bank be accomplished prior to any debiting of the bank. In this regard, the following minimum requirements should be satisfied prior to debiting: (1) banking instrument and mitigation plans have been approved; (2) bank site has been secured; and (3) appropriate financial assurances have been established. In addition, initial physical and biological improvements should be completed no later than the first full growing season following initial debiting of a bank. The temporal loss of functions associated with the debiting of projected credits may justify the need for requiring higher compensation ratios in such cases. For mitigation banks which propose multiple-phased construction, similar conditions should be established for each phase.

Credits attributed to the preservation of existing aquatic resources may become available for debiting immediately upon implementation of appropriate legal protection accompanied by appropriate changes in land use or other physical changes, as necessary.

7. CREDITING/DEBITING/ ACCOUNTING PROCEDURES

Credits and debits are the terms used to designate the units of trade (i.e., currency) in mitigation banking. Credits represent the accrual or attainment of aquatic functions at a bank; debits represent the loss of aquatic functions at an impact or project site. Credits are debited from a bank when they are used to offset aquatic resource impacts (e.g. for the purpose of satisfying Section 10/404 permit or FSA requirements).

An appropriate functional assessment methodology (e.g., Habitat Evaluation Procedures, hydrogeomorphic approach to wetlands functional assessment, other regional assessment methodology) acceptable to all signatories should be used to assess wetland and/or other aquatic resource restoration, creation and enhancement activities within a mitigation bank, and to quantify the amount of available credits. The range of functions to be assessed will depend upon the assessment methodology identified in the banking instrument. The same methodology should be used to assess both credits and debits. If an appropriate functional assessment methodology is impractical to employ, acreage may be used as a surrogate for measuring function. Regardless of the method employed, the number of credits should reflect the difference between site conditions under the with-and without-bank scenarios.

The bank sponsor should be responsible for assessing the development of the bank and submitting appropriate documentation of such assessments to the authorizing agency(ies), who will distribute the documents to the other members of the MBRT for review. Members of the MBRT are encouraged to conduct regular (e.g., annual) on-site inspections, as appropriate, to monitor bank performance. Alternatively, functional assessments may be conducted by a team representing involved resources and regularly agencies and other appropriate parties. The number of available credits in a mitigation bank may need to be adjusted to reflect actual conditions.

The banking instrument should require that bank sponsors establish and maintain an accounting system (i.e., ledger) which documents the activity of all mitigation bank accounts. Each time an approved debit/ credit transaction occurs at a given bank, the bank sponsor should submit a statement to the authorizing agency(ies). The bank sponsor should also generate an annual ledger report for all mitigation bank accounts to be submitted to the MBRT Chair for distribution to each member of the MBRT.

Credits may be sold to third parties. The cost of mitigation credits to a third party is determined by the bank sponsor.

8. PARTY RESPONSIBLE FOR BANK SUCCESS

The bank sponsor is responsible for assuring the success of the debited restoration, creation, enhancement and preservation activities at the mitigation bank, and it is therefore extremely important that an enforceable mechanism be adopted establishing the responsibility of the bank sponsor to develop and operate the bank properly. Where authorization under Section 10/404 and/or FSA is necessary to establish the bank, the Department of the Army permit or NRCS plan should be conditioned to ensure that provisions of the banking instrument are enforceable by the appropriate agency(ies). In circumstances where establishment of a bank does not require such authorization, the details of the bank sponsor's responsibilities should be delineated by the relevant authorizing agency (e.g., the Corps in the case of Section 10/404 permits) in any permit in which the permittee's mitigation obligations are met through use of the bank. In addition, the bank sponsor should sign such permits for the limited purpose of meeting those mitigation responsibilities, thus confirming that those responsibilities are enforceable against the bank sponsor if necessary.

E. LONG-TERM MANAGEMENT, MONITORING AND REMEDIATION

1. BANK OPERATIONAL LIFE

The operational life of a bank refers to the period during which the terms and conditions of the banking instrument are in effect. With the exception of arrangements for the long-term management and protection in perpetuity of the wetlands and/or other aquatic resources, the operational life of a mitigation bank terminates at the point when (1) Compensatory mitigation credits have been exhausted or banking activity is voluntarily terminated with written notice by the bank sponsor provided to the Corps or NRCS and other members of the MBRT, and (2) it has been determined that the debited bank is functionally mature and/or self-sustaining to the degree specified in the banking instrument.

2. LONG-TERM MANAGEMENT AND PROTECTION

The wetlands and/or other aquatic resources in a mitigation bank should be protected in perpetuity with appropriate real estate arrangements (e.g., conservation easements, transfer of title to Federal or State resource agency or non-profit conservation organization). Such arrangements should effectively restrict harmful activities (i.e., incompatible uses[2]) that might otherwise jeopardize the purpose of the bank. In exceptional circumstances, real estate arrangements may be approved which dictate finite protection for a bank (e.g., for coastal protection projects which prolong the ecological viability of the aquatic system). However, in no case should finite protection extend for a lesser time than the duration of project impacts for which the bank is being used to provide compensation.

The bank sponsor is responsible for securing adequate funds for the operation and maintenance of the bank during its operational life, as well as for the long-term management of the wetlands and/or other aquatic resources, as necessary. The banking instrument should identify the entity responsible for the ownership and long-term management of the wetlands and/or other aquatic resources. Where needed, the acquisition and protection of water rights should be secured by the bank sponsor and documented in the banking instrument.

3. MONITORING REQUIREMENTS

The bank sponsor is responsible for monitoring the mitigation bank in accordance with monitoring provisions identified in the banking instrument to determine the level of success and identify problems requiring remedial action. Monitoring provisions should be set forth in the banking instrument and based on scientifically sound performance standards prescribed for the bank. monitoring should be conducted at time intervals appropriate for the particular project type and until such time that the authorizing agency(ies), in consultation with the MBRT, are confident that success is being achieved (i.e., performance standards are attained). The period for monitoring will typically be five years; however, it may be necessary to extend this period for projects requiring more time to reach a stable condition (e.g., forested wetlands) or where remedial activities were undertaken. Annual monitoring reports should be submitted to the authorizing agency(ies), who is responsible for distribution to the other members of the MBRT, in accordance with the terms specified in the banking instrument.

2. For example, certain silvicultural practices (e.g. clear cutting and/or harvests on short-term rotations) may be incompatible with the objectives of a mitigation bank. In contrast, silvicultural practices such as long-term rotations, selective cutting, maintenance of vegetation diversity, and undisturbed buffers are more likely to be considered a compatible use.

4. REMEDIAL ACTION

The banking instrument should stipulate the general procedures for identifying and implementing remedial measures at a bank, or any portion thereof. Remedial measures should be based on information contained in the monitoring reports (i.e., the attainment of prescribed performance standards), as well as agency site inspections. The need for remediation will be determined by the authorizing agency(ies) in consultation with the MBRT and bank sponsor.

5. FINANCIAL ASSURANCES

The bank sponsor is responsible for securing sufficient funds or other financial assurances to cover contingency actions in the event of bank default or failure. Accordingly, banks posing a greater risk of failure and where credits have been debited, should have comparatively higher financial sureties in place, than those where the likelihood of success is more certain. In addition, the bank sponsor is responsible for securing adequate funding to monitor and maintain the bank throughout its operational life, as well as beyond the operational life if not self-sustaining. Total funding requirements should reflect realistic cost estimates for monitoring, long-term maintenance, contingency and remedial actions.

Financial assurances may be in the form of performance bonds, irrevocable trusts, escrow accounts, casualty insurance, letters of credit, legislatively-enacted dedicated funds for government operate banks or other approved instruments. Such assurances may be phased-out or reduced, once it has been demonstrated that the bank is functionally mature and/or self-sustaining (in accordance with performance standards).

F. OTHER CONSIDERATIONS

1. IN-LIEU-FEE MITIGATION ARRANGEMENTS

For purposes of this guidance, in-lieu-fee, fee mitigation, or other similar arrangements, wherein funds are paid to a natural resource management entity for implementation of either specific or general wetland or other aquatic resource development projects, are not considered to meet the definition of mitigation banking because they do not typically provide compensatory mitigation in advance of project impacts. Moreover, such arrangements do not typically provide a clear timetable for the initiation of mitigation efforts. The Corps, in consultation with the other agencies, may find there are circumstances where such arrangements are appropriate so long as they meet the requirements that would otherwise apply to an offsite, prospective mitigation effort and provides adequate assurances of success and timely implementation. In such cases, a formal agreement between the sponsor and the agencies, similar to a banking instrument, is necessary to define the conditions under which its use is considered appropriate.

2. SPECIAL CONSIDERATIONS FOR "SWAMPBUSTER"

Current FSA legislation limits the extent to which mitigation banking can be used for FSA purposes. Therefore, if a mitigation bank is to be used for FSA purposes, it must meet the requirements of FSA.

III. DEFINITIONS

For the purposes of this guidance document the following terms are defined:

A. AUTHORIZING AGENCY. Any Federal, state, tribal or local agency that has authorized a particular use of a mitigation bank as compensation for an authorized activity; the authorizing agency will typically have the enforcement authority to ensure that the terms and conditions of the banking instrument are satisfied.

B. BANK SPONSOR. Any public or private entity responsible for establishing and, in most circumstances, operating a mitigation bank.

C. COMPENSATORY MITIGATION. For purposes of Section 10/404, compensatory mitigation is the restoration, creation, enhancement, or in exceptional circumstances, preservation of wetlands and/or other aquatic resources for the purpose of compensating for unavoidable adverse impacts which remain after all appropriate and practicable avoidance and minimization has been achieved.

D. CONSENSUS. The term consensus, as defined herein, is a process by which a group synthesizes its concerns and ideas to form a common collaborative agreement acceptable to all members. While the primary goal of consensus is to reach agreement on an issue by all parties, unanimity may not always be possible.

E. CREATION. The establishment of a **wetland** or other aquatic resource where one did not formerly exist.

F. CREDIT. A unit of measure representing the accrual or attainment of aquatic functions at a mitigation bank; the measure of function is typically indexed to the number of wetland acres restored, created, enhanced or preserved.

G. DEBIT. A unit of measure representing the loss of aquatic functions at an impact or project site.

H. ENHANCEMENT. Activities conducted in existing wetlands or other aquatic resources which increase one or more aquatic functions.

I. MITIGATION. For purposes of Section 10/404 and consistent with the Council on Environmental Quality regulations, the Section 404(b)(1) Guidelines and the Memorandum of Agreement Between the Environmental Protection Agency and the Department of the Army Concerning the Determination of Mitigation under the Clean Water Act Section 404(b)(1) Guidelines, mitigation means sequentially avoiding impacts, minimizing impacts, and compensating for remaining unavoidable impacts.

J. MITIGATION BANK. A mitigation bank is a site where wetlands and/ or other aquatic resources are restored, created, enhanced, or in exceptional circumstances, preserved expressly for the purpose of providing compensatory mitigation in advance of authorized impacts to similar resources. For purposes of Section 10/404, use of a mitigation bank may only be authorized when impacts are unavoidable.

K. MITIGATION BANK REVIEW TEAM (MBRT). An interagency group of Federal, state, tribal and/or local regulatory and resource agency representatives which are signatory to a banking instrument and oversee the establishment, use and operation of a mitigation bank.

L. PRACTICABLE. Available and capable of being done after taking into consideration cost, existing technology, and logistics in light of overall project purposes.

M. PRESERVATION. The protection of ecologically important wetlands or other aquatic resources in perpetuity through the implementation of appropriate legal and physical mechanisms. Preservation may include protection of upland areas adjacent to wetlands as necessary to ensure protection and/or enhancement of the aquatic ecosystem.

N. RESTORATION. Re-establishment of **wetland** and/or other aquatic resource characteristics and function(s) at a site where they have ceased to exist, or exist in a substantially degraded state.

O. SERVICE AREA. The service area of a mitigation bank is the designated area (e.g., watershed, county) wherein a bank can reasonably be expected to provide appropriate compensation for impacts to wetlands and/or other aquatic resources.

John H. Zirschky,
Acting Assistant Secretary (Civil Works),
Department of the Army.

Robert Perciasepe,
Assistant Administrator for Water,
Environmental Protection Agency.

Thomas R. Hebert,
Acting Undersecretary for Natural
Resources and Environment,
Department of Agriculture.

Robert P. Davison,
Acting Assistant Secretary for
Fish and Wildlife and Parks,
Department of the Interior.

Douglas K. Hall,
Assistant Secretary for
Oceans and Atmosphere,
Department of Commerce.

APPLICATION OF BEST MANAGEMENT PRACTICES TO MECHANICAL SILVICULTURAL SITE PREPARATION ACTIVITIES FOR THE ESTABLISHMENT OF PINE PLANTATIONS IN THE SOUTHEAST

MEMORANDUM TO THE FIELD

United States
Environmental Protection Agency
Office of Wetlands,
Oceans and Watersheds
Washington, D.C. 20460

United States
Department of the Army
U.S. Amy Corps of Engineers
Washington, D.C. 20314

November 28, 1995

SUBJECT: Application of Best Management Practices to Mechanical Silvicultural Site Preparation Activities for the Establishment of Pine Plantations in the Southeast

This memorandum [1] clarifies the applicability of forested wetlands best management practices to mechanical silvicultural site preparation activities for the establishment of pine plantations in the southeast Mechanical silvicultural site preparation activities [2] conducted in accordance with the best management practices discussed below, which are designed to minimize impacts to the aquatic ecosystem, will not require a Clean Water Act Section 404 permit. These best management practices further recognize that certain wetlands should not be subject to unpermitted mechanical silvicultural site preparation activities because of the adverse nature of potential impacts associated with these activities on these sites.

This memorandum recognizes State expertise that is reflected in the development and implementation of regionally specific best management practices (BMPs) associated with forestry activities in wetlands. Such BMPs encourage sound silvicultural operations while providing protection of certain wetlands functions and values. The U.S. Army Corps of Engineers (Corps) and the U.S. Environmental Protection Agency (EPA) believe that it is appropriate to apply the Clean Water Act Section 404 program in a manner that builds from, and, is consistent with, this State experience. The Agencies will support and assist State efforts to build upon these BMPs at the State level to ensure that mechanical silvicultural site preparation is conducted in a manner that best reflects the specific wetlands resource protection and management goals of each State.

INTRODUCTION

Forested wetlands exhibit a wide variety of water regimes, soils, and vegetation types that in turn provide a myriad of functions and values. The States in the Southeast contain forested wetlands systems that in many cases are also subject to ongoing timber operations. In developing silvicultural BMPs, States have identified those specific forestry practices that will protect water quality. This guidance was developed to respond to questions regarding the applicability of Section 404 to mechanical silvicultural site preparation activities. EPA and the Corps relied extensively on existing State knowledge to protect aquatic ecosystems with BMPs, including the types of wetlands, types of activities, and BMPs described below.

This memorandum reflects information gathered from the southeastern United States, where mechanical silvicultural site preparation activities are associated with the establishment of pine plantations in wetlands.[3] As such, this memorandum, and particularly the descriptions of wetlands, activities, and BMPs, necessarily focus on this area of the country. However, the guidance presented is generally applicable when addressing mechanical silvicultural site preparation activities in wetlands elsewhere in the country.

CIRCUMSTANCES WHERE MECHANICAL SILVICULTURAL SITE PREPARATION ACTIVITIES REQUIRE A PERMIT

The States, in coordination with the forestry community and the public, have recognized that mechanical silvicultural site preparation activities may have measurable and significant impacts on aquatic ecosystems when conducted in wetlands that are permanently

1. This guidance is written to provide interpretation and clarification of existing EPA and Corps regulations and does not change any substantive requirements of these regulations. This memorandum is further intended to provide clarification regarding the exercise of discretion under agent agency regulations.

2. Mechanical silvicultural site preparation a include shearing, raking, ripping chopping, windrowing, piling, and other similar physical methods used to cut, break apart, or move logging debris following harvest for the establishment of pine plantations.

3. Information was considered from the following States in the Southeast: Virginia, North Carolina, South Carolina, Georgia, Florida, Tennessee, Alabama, Mississippi, Louisiana, and Arkansas.

flooded, intermittently exposed, and semi-permanently flooded, and in certain additional wetland communities that exhibit aquatic functions and values that are more susceptible to impacts from these activities. For the wetland types identified in this section, it is most effective to evaluate proposals for site preparation and potential associated environmental effects on a case-by-case basis as part of the individual permit process. Therefore, mechanical silvicultural site preparation activities in the areas listed below require a permit.[4]

A permit will be required in the following areas unless they have been so altered through past practices (including the installation and continuous maintenance of water management structures) as to no longer exhibit the distinguishing characteristics described below (see "Circumstances Where Mechanical Silvicultural Site Preparation Activities Do Not Require a Permit" below). Of course, discharges incidental to activities in any wetlands that convert waters of the United States to non-waters always require authorization under Clean Water Act Section 404.

1) PERMANENTLY FLOODED, INTERMITTENTLY EXPOSED, AND SEMI-PERMANENTLY FLOODED WETLANDS. The hydrology of permanently flooded wetland systems is characterized by water that covers the land surface throughout the year in all years. The hydrology of intermittently exposed wetlands is characterized by surface water that is present throughout the year except in years of extreme drought. The hydrology of semi-permanently flooded wetlands is characterized by surface water that persists throughout the growing season in most years and, when it is absent, the water table is usually at or very near the land surface.[5] Examples typical of these wetlands include Cypress-Gum Swamps, Muck and Peat Swamps, and Cypress Strands/Domes.

2) RIVERINE BOTTOMLAND HARDWOOD WETLANDS: seasonally flooded (or wetter) bottomland hardwood wetlands within the first or second bottoms of the floodplains of river systems. Site-specific characteristics of hydrology, soils, vegetation, and the presence of alluvial features elaborated in paragraphs a, b, and c below will be determinative of the boundary of riverine bottomland hardwood wetlands. National Wetlands Inventory maps can pride a useful reference for the general location of thes wetlands on the landscape.

a) the hydrologic characteristics included in this definition refer to seasonally flooded or wetter river floodplain sites where overbank flooding has resulted in alluvial features such as well-defined floodplains, bottoms/terraces, natural levees, and backswamps. For the purposes of this guidance definition, "seasonally flooded" bottomland hardwood wetlands are characterized by surface water that is present for extended periods, especially early in the growing season[6] (usually greater than 14 consecutive days), but is absent by the end of the season in most years. When surface water is absent, the water table is often near the land surface. Field indicators of the presence of surface water include water-stained leaves, drift lines, and water marks on trees.

b) the vegetative characteristics included in this definition refer to forested wetlands where hardwoods dominate the canopy. For the purposes of this guidance definition, riverine bottomland hardwoods do not include sites in which greater than 25% of the canopy is pine.

c) the soil characteristics included in this definition refer to listed hydric soils that are poorly drained or very poorly drained. For the purposes of this guidance definition, riverine bottomland hardwoods do not include sites with hydric soils that are somewhat poorly drained or that, at a particular site, do not demonstrate chroma, concretions, and other field characteristics verifying it as a hydric soil.

3) WHITE CEDAR SWAMPS: wetlands, greater than one acre in headwaters and greater than five acres elsewhere, underlain by peat of greater than one meter, and vegetated by natural white cedar representing more than 50% of the basal area, where the total basal area for all tree species is 60 square feet or greater.

4) CAROLINA BAY WETLANDS: oriented, elliptical depressions with a sand rim, either a) underlain by clay-based soils and vegetated by cypress; or, b) underlain by peat of greater than one-half meter and typically vegetated with an overstory of Red, Sweet, and Loblolly Bays.

5) NON-RIVERINE FOREST WETLANDS: wetlands in this group are rare, high quality wet forests, with mature vegetation, located on the Southeastern coastal plain, whose hydrology is dominated by high water tables. Two forest community types fall into this group.[7]

a) NON-RIVERINE WET HARDWOOD FORESTS—poorly drained mineral soil interstream flats (comprising 10 or more contiguous acres), typically on the margins of large peatland areas, seasonally flooded or saturated by high water tables, with vegetation dominated (greater than 50% of basal area per acre) by swamp chestnut oak, cherrybark oak, or laurel oak alone or in combination.

b) NON-RIVERINE SWAMP FORESTS—very poorly drained flats (comprising 5 or more contiguous acres), with organic soils or mineral soils with high organic content, seasonally to frequently flooded or saturated by high water tables, with vegetation dominated by bald cypress, pond cypress, swamp tupelo, water tupelo, or Atlantic white cedar alone or in combination.

The term "high quality" used in this characterization refers to generally undisturbed forest stands, whose character is not significantly affected by human activities (e.g., forest management). Non-riverine Forest wetlands dominated by red maple, sweetgum, or loblolly pine alone or in combination are not considered to be of high quality, and therefore do not require a permit.

6) LOW POCOSIN WETLANDS: central, deepest parts of domed peatlands on poorly drained instream flats, underlain by peat soils greater than one meter, typically vegetated by a dense layer of short shrubs.

7) WET MARL FORESTS: hardwood forest wetlands underlain with poorly drained marl-derived, high pH softs.

8) TIDAL FRESHWATER MARSHES: wetlands regularly or irregularly flooded by freshwater with dense herbaceous vegetation, on the margins of estuaries or drowned rivers or creeks.

9) MARITIME GRASSLANDS, SHRUB SWAMPS, AND SWAMP FORESTS: barrier island wetlands in dune swales and flats, underlain by wet mucky or sandy soils, vegetated by wetland herbs, shrubs, and trees.

CIRCUMSTANCES WHERE MECHANICAL SILVICULTURAL SITE PREPARATION ACTIVITIES DO NOT REQUIRE A PERMIT

Mechanical silvicultural site preparation activities in wetlands that are seasonally flooded, intermittently flooded, temporarily flooded, or saturated, or in existing pine plantations and other silvicultural sites (except as listed above), minimize impacts to the aquatic ecosystem and do not require a permit if conducted

4. The community descriptions draw extensively from: Schafale, M.P., and A.S. Weakley. 1990. Classification of the Natural Communities of North Carolina. North Carolina Natural Heritage Program Raleigh, NC. 325pp.

5. Cowardin, L.M., et al. 1979. Classification of wetlands and deepwater habitats of the United States. U.S. Fish and Wildlife Service, Washington, DC. 131pp.

6. Consistent with the 1987 Corps of Engineers Wetlands Delineation Manual, growing season starting and ending dates are determined by the 28 degrees F or lower temperature threshold.

7. These forest types are a subset of those described in Schafale and Weakley, 1990.

according to the BMPs listed below. Of course, silvicultural practices conducted in uplands never require a Clean Water Act Section 404 permit.

The hydrology of seasonally flooded wetlands is characterized by surface water that is present for extended periods, especially early in the growing season, but is absent by the end of the season in most years (when surface water is absent, the water table is often near the surface). The hydrology of intermittently flooded wetland systems is characterized by substrate that is usually exposed, but where surface water is present for variable periods without detectable seasonable periodicity. The hydrology of temporarily flooded wetlands is characterized by surface water that is present for brief periods during the growing season, but also by a water table that usually lies well below the soil surface for most of the season. The hydrology of saturated wetlands is characterized by substrate that is saturated to the surface for extended periods during the growing season, but also by surface water that is seldom present.[8] Examples typical of these wetlands include Pine Flatwoods, Pond Pine Woodlands, and Wet Flats (e.g., certain pine/hardwood forests).

BEST MANAGEMENT PRACTICES

Every State in the Southeast has developed BMPs for forestry to protect water quality and all but two have also developed specific BMPs for forested wetlands. These BMPs have been developed because silvicultural practices have the potential to result in impacts to the aquatic ecosystem. Mechanical silvicultural site preparation activities include shearing, raking, ripping, chopping, windrowing, piling, and other similar physical methods used to cut, break apart, or move logging debris following harvest. Impacts such as soil compaction, turbidity, erosion, and hydrologic modifications can result if not effectively controlled by BMPs. States have developed BMPs that address not only types of wetlands and types of activties, but also detail specific measures to protect water quality through establishing special management zones, practices for stream crossings, and practices for forest road construction.

In developing forested wetlands BMPs, States in the Southeast have recognized that certain silvicultural site preparation techniques are more effective when conducted in areas that have drier water regimes. The BMPs stated below represent a composite of State expertise to protect water quality from silvicultural impacts. These BMPs also address the location, as well as the nature, of activities. The Corps and EPA believe that these forested wetlands BMPs are effective in protecting water quality and therefore are adopting them to protect these functions and values considered under Section 404.

The following forested wetlands BMPs are designed to minimize the impacts associated with mechanical silvicultural site preparation activities in circumstances where these activities do not require a permit (authorization from the Corps is necessary for discharges associated with silvicultural site preparation in wetlands described above as requiring a permit[9]). The BMPs include, at a minimum, the following:

1) position shear blades or rakes at or near the soil surface and windrow, pile, and otherwise move logs and logging debris by methods that minimize dragging or pushing through the soil to minimize soil disturbance associated with shearing, raking, and moving trees, stumps, brush, and other unwanted vegetation;

2) conduct activities in such a manner as to avoid excessive soil compaction and maintain soil tilth;

3) arrange windrows in such a manner as to limit erosion, overland flow, and runoff;

4) prevent disposal or storage of lop or logging debris in streamside management zones–defined areas adjacent to streams, lakes, and other waterbodies–to protect water quality;

5) maintain the natural contour of the site and ensure that activities do not immediately or gradually convert the wetland to a non-wetland; and

6) conduct activities with appropriate water management mechanisms to minimize off-site water quality impacts.

IMPLEMENTATION

EPA and the Corps will continue to work closely with State forestry agencies to promote the implementation of consistent and effective BMPs that facilitate sound silvicultural practices. In those States where no BMPs specific to mechanical silvicultural site preparation activities in forested wetlands are currently in place, EPA and the Corps will coordinate with those States to develop BMPs. In the interim, mechanical silvicultural site preparation activities conducted in accordance with this guidance will not require a Section 404 permit.

In order to ensure consistency in the application of this guidance over time, changes to the vegetation of forested wetlands associated with human activities conducted after the issuance of this guidance will not alter its applicabiety. For example, this guidance is not intended to establish the requirement for a permit for mechanical silvicultural site preparation where tree harvesting results in the establishment of site characteristics for which a permit would otherwise be required (e.g., where the selective cutting of naturally occurring pine in a Riverine Bottomland Hardwood wetland site with originally greater than 25% pine in the canopy results in a site "where hardwoods dominate the canopy"). In a similar manner, while harvesting of timber consistent with the requirements of Section 404(f) is exempt from regulation and natural changes (e.g., wildfire, succession) may change site characteristics, human manipulation of the vegetative characteristics of a site does not alter its status for the purposes of this guidance (e.g., removal of all the Atlantic White Cedar in an Atlantic White Cedar Swamp does not eliminate the need for a permit for mechanical silvicultural site preparation if the area would have required a permit before the removal of the trees).

Finally, the Agencies will encourage efforts at the State level to identify additional wetlands which may be of special concern and could be incorporated into State BMPs and cooperative programs, initiatives, and partnerships to protect these wetlands. To facilitate this effort, stakeholders are encouraged to develop a process after the issuance of this guidance to identify and protect unique and rare wetland sites on lands of the participating stakeholders. EPA and the Corps will monitor the application of this guidance, progress with conserving special wetland sites through cooperative programs and initiatives, and consider any new information, such as advances in silvicultural practices, improvements to State BMPs, or data relevant to potential impacts to wetlands, to determine whether the list of wetlands subject to the permit requirement should be modified or other revisions to this guidance are appropriate.

FURTHER INFORMATION

The Corps and EPA will work closely with the States, forestry community, and public to answer any questions that may arise with regard to this guidance. For further information on this memorandum, please contact Mr. John Goodin of EPA's. Wetlands Division at (202) 260-9910 or Mr. Sam Colinson of the Corps of Engineer's Regulatory Branch at (202) 761-0199. The public may also contact:

8. Cowardin et al, 1979.

9. Contact the nearest Corps District listed at the end of this document for further information.

EPA

Region IV
Tom Welburn
(404)347-3871 ext. 6507

Region VI
Bill Cox
(214)665-6680

Region III
Barbara D'Angelo
(215)597-9301

CORPS

Corps Wilmington District
Wayne Wright
(910)251-4630

Corps Charleston District
Bob Riggs
(803)727-4330

Corps Savannah District
Nick Ogden
(912)652-5768

Corps Jacksonville District
John Hall
(904)232-1666

Corps Norfolk District
Woody Poore
(804)441-7068

Corps Mobile District
Ron Krizman
(334)690-2658

Corps Little Rock District
Lou Cockman
(501)324-5296

Corps Memphis District
Larry Watson
(901)544-3471

Corps Nashville District
Randy Castleman
(615)736-5181

Corps New Orleans District
Ron Ventola
(504)862-2255

Corps Vicksburg District
Beth Guynes
(601)631-5276

/S/

Robert H. Wayland, III
Director, Office of Wetlands,
Oceans, and Watersheds
U.S. Environmental
Protection Agency

/S/

Michael L. Davis
Chief Regulatory Branch
U.S. Army Corps of Engineers

FLEXIBILITY AFFORDED TO SMALL LANDOWNERS

REGULATORY GUIDANCE LETTER 95-01

SUBJECT: Guidance on Individual Permit Flexibility for Small Landowners

DATE: 31 March 1995

EXPIRES: 31 December 2000

1. Enclosed is a memorandum for the field signed by the Acting Assistant Secretary of the Army (Civil Works) and the Environmental Protection Agency dated 6 March 1995. This memorandum provides guidance on flexibility that the U.S. Army Corps of Engineers should apply when making determinations of compliance with the Section 404(b)(1) Guidelines with regard to the alternatives analysis.

2. This memorandum should be implemented immediately. It constitutes an important aspect of the President's Plan for protecting the Nation's wetlands, "Protecting America's Wetlands: A Fair, Flexible, and Effective Approach" (published 24 August 1993.).

3. This guidance expires on 31 December 2000 unless sooner revised or rescinded.

FOR THE DIRECTOR OF CIVIL WORKS:

DANIEL R. BURNS, P.E.
Chief, Operations, Construction,
and Readiness Division
Directorate of Civil Works

Memorandum for the Field

In order to clearly affirm the flexibility afforded to small landowners under Section 404 of the Clean Water Act, this policy clarifies that for discharges of dredged or fill material affecting up to two acres of non-tidal wetlands for the construction or expansion of a home or farm building, or expansion of a small business, it is presumed that alternatives located on property not currently owned by the applicant are not practicable under the Section 404(b)(1) Guidelines.

Specifically, for those activities involving discharges of dredged or fill material affecting up to two acres into jurisdictional wetlands for:

1. the construction or expansion of a single family home and attendant features, such as a driveway, garage, storage shed, or septic field;

2. the construction or expansion of a barn or other farm building; or

3. the expansion of a small business facility;

which are not otherwise covered by a general permit, it is presumed that alternatives located on property not currently owned by the applicant are not practicable under the Section 404(b)(1) Guidelines. The Guidelines' requirements to appropriately and practicably minimize and compensate for any adverse environmental impacts of such activities remain.

DISCUSSION

The Clean Water Act Section 404 regulatory program provides that the Army Corps of Engineers evaluate permit applications of the discharge of dredged or fill material into waters of the U.S., including wetlands, in accordance with regulatory requirements of the Section 404(b)(1) Guidelines ("Guidelines"). The Guidelines are the substantive environmental criteria used in evaluating discharges of dredged or fill material.

The Section 404(b)(1) Guidelines establish a mitigation sequence that provides a sound framework to ensure that the environmental impacts of permitted actions are acceptable. Under this framework, there is a three-step sequence for mitigating potential adverse impacts to the aquatic environment associated with a proposed discharge–first avoidance, then minimization, and lastly compensation for unavoidable impacts to aquatic resources.

The Guidelines' mitigation sequence is designed to establish a consistent approach to be used in ensuring that all practicable measures have been taken to reduce potential adverse impacts associated with proposed projects in wetlands and other aquatic systems. The Guidelines define the term "practicable" as "available and capable of being done [by the applicant] after taking into consideration cost, existing technology, and logistics in light of overall project purposes." (40 CFR 230.3(q)). The first step in the sequence requires the evaluation of potential alternative sites under § 230.10(a) of the Guidelines, to locate the proposed project so that aquatic impacts are avoided to the extend practicable.

This policy statement clarifies that, for the purposes of the alternative analysis, it is presumed that practicable alternatives are limited to property owned by the permit applicant in circumstances involving certain small projects affecting less than two acres of non-tidal wetlands. This presumption is consistent with the practicability considerations required under the Guidelines and reflects the nature of the projects to which the presumptions applies–specifically, the construction or expansion of a single family home and attendant features, the construction or expansion of a barn or other farm building, or the expansion of a business. For such small projects that would solely expand an existing structure, the basic project purpose is so tied to the existing structures owned by the applicant, that it would be highly unusual that the project could be practicably located on other sites not owned by the applicant. In these cases, such as construction of driveways, garages, or storage sheds, or with home and barn additions, proximity to the existing structure is typically a fundamental aspect of the project purpose.

In the evaluation of potential practicable alternatives, the Guidelines do not exclude the consideration of sites that, while not currently owned by the permit applicant, could reasonably be obtained to satisfy the project purpose. However, it is the experience of the Army Corps of Engineers and EPA that areas not currently owned by the applicant have, in the great majority of circumstances, not been determined to the practicable alternatives in cases involving the small landowner activities describe above. Cost, availability, and logistical and capability considerations inherent in the determination of practicability under the Guidelines have been the basis for this conclusion by the agencies.

The agencies recognize that the presumption characterized in this policy statement may be rebutted in certain circumstances. For example, a more thorough review of practicable alternatives would be warranted for individual sites comprising a subdivision of homes, if following issuance of this policy statement, a real estate developer subdivided a large, contiguous wetlands parcel into numerous parcels. In addition, the presumption is applicable to the expansion of existing small business facilities. Small businesses are typically confined to only one location and with economic and logistical limitations that generally preclude the availability of practicable alternative locations to meet their expansion needs. Conversely, larger businesses with multiple locations and greater resources are expected to consider opportunities to practicably avoid adverse aquatic impacts by evaluating off-site alternatives.

Finally, it is important to note that this presumption of practicable alternatives is intended to apply to the individual permit process. Alternatives are not evaluated for activities covered by general permits. Many activities related to the construction or expansion of a home, farm, or business, are already covered by a general permit. In addition, in conjunction with the issuance of this policy statement, a nationwide general permit authorizing discharges related to single family residential development is being proposed and will be available for public comment.

ROBERT PERCIASEPE
Assistant Administrator for Water
U.S. Environmental Protection Agency

JOHN ZIRSCHKY
Acting Assistant Secretary of the Army
(Civil Works)

GEOGRAPHIC JURISDICTIONAL DETERMINATIONS

REGULATORY GUIDANCE LETTER 94-01

SUBJECT: Expiration of Geographic Jurisdictional Determinations

DATE: 23 May 1994

EXPIRES: 31 December 1999

1. Regulatory Guidance Letter (RGL) 90-6, Subject: Expiration Dates for Wetlands Jurisdictional Delineations" is extended until 31 December 1999, subject to the following revisions.

2. This guidance should be applied to all jurisdictional determinations for all waters of the United States made pursuant to Section 10 of the Rivers and Harbors Act of 1899, Section 404 of the Clean Water Act, and Section 103 of the Marine Protection Research and Sanctuaries Act of 1972.

3. To be consistent with paragraph IV.A. of the 6 January 1994, interagency Memorandum of Agreement Concerning the Delineation of Wetlands for Purposes of Section 404 of the Clean Water Act and Subtitle B of the Food Security Act, all U.S. Army Corps of Engineers geographic jurisdictional determinations shall be in writing and normally remain valid for a period of five years. The Corps letter (see paragraph 4. (d) of RGL 90-6) should include a statement that the jurisdictional determination is valid for a period of five years from the date of the letter unless new information warrants revision of the determination before the expiration date.

4. For wetland jurisdictional delineations the "effective date of this RGL" referred to in paragraphs 4 and 5 of RGL 90-6 was and remains 14 August 1990. For jurisdictional determinations, other than wetlands jurisdictional delineations, the "effective date of this RGL" referred to in paragraphs 4 and 5 of RGL 90-6 will be the date of this RGL.

5. Previous Corps written jurisdictional determinations, including wetland jurisdictional delineations, with a validity period of three years remain valid for the stated period of three years. The district engineer is not required to issue new letters to extend such period from three years to a total of five years. However, if requested to do so, the district engineer will normally extend the three year period to a total of five years unless new information warrants a new jurisdictional determination.

6. Districts are not required to issue a public notice on this guidance but may do so at their discretion.

7. This guidance expires on 31 December 1999 unless sooner revised or rescinded.

FOR THE DIRECTOR
OF CIVIL WORKS:

JOHN P. ELMORE, P.E.
Chief, Operations,
Construction and Readiness Division
Directorate of Civil Works

404(B)(1) FLEXIBILITY AND MITIGATION BANKING

REGULATORY GUIDANCE LETTER 93-02

SUBJECT: Guidance on Flexibility of the 404(b)(1) Guidelines and Mitigation Banking

DATE: 23 August 1993

EXPIRES: 31 December 1998

1. Enclosed are two guidance documents signed by the Office of the Assistant Secretary of the Army (Civil Works) and the Environmental Protection Agency. The first document provides guidance on the flexibility that the U.S. Army Corps of Engineers should be utilizing when making determinations of compliance with the Section 404(b)(1) Guidelines, particularly with regard to the alternatives analysis. The second document provides guidance on the use of mitigation banks as a means of providing compensatory mitigation for Corps regulatory decisions.

2. Both enclosed guidance documents should be implemented immediately. These guidance documents constitute an important aspect of the President's plan for protecting the Nation's wetlands, "Protecting America's Wetlands: A Fair, Flexible and Effective Approach" (published on 24 August 1993).

3. This guidance expires 31 December 1998 unless sooner revised or rescinded.

FOR THE DIRECTOR
OF CIVIL WORKS:

JOHN P. ELMORE, P.E.
Chief, Operations,
Construction and Readiness Division
Directorate of Civil Works

Memorandum to the Field

United States
Environmental Protection Agency
Office of Wetlands,
Oceans and Watersheds
Washington, D.C. 20460

United States
Department of the Army
U.S. Army Corps of Engineers
Washington, D.C. 20314

SUBJECT: Appropriate Level of Analysis Required For Evaluating Compliance With the Section 404)b)(1) Guidelines Alternatives Requirements

1. Purpose: The purpose of this memorandum is to clarify the appropriate level of analysis required for evaluating compliance with

the Clean Water Act Section 404(b)(1) Guidelines requirements for consideration of alternatives. 40 CFR 230.10(a). Specifically, this memorandum describes the flexibility afforded by the Guidelines to make regulatory decisions based on the relative severity of the environmental impact of proposed discharges of dredged or fill material into waters of the United States.

2. Background: The Guidelines are the substantive environmental standards by which all Section 404 permit applications are evaluated. The Guidelines, which are binding regulations, were published by the Environmental Protection Agency at 40 CFR Part 230 on December 24, 1980. The fundamental precept of the Guidelines is that discharges of dredged or fill material into waters of the United States, including wetlands, should not occur unless it can be demonstrated that such discharges, either individually or cumulatively, will not result in unacceptable adverse effects on the aquatic ecosystem. The Guidelines specifically require that "no discharge of dredged or fill material shall be permitted if there is a practicable alternative to the proposed discharge which would have less adverse impact on the aquatic ecosystem, so long as the alternative does not have other significant adverse environmental consequences." 40 CFR 230.10(a). Based on this provision, the applicant is required in every case (irrespective of whether the discharge site is a special aquatic site or whether the activity associated with the discharge is water dependent) to evaluate opportunities for use of non-aquatic areas and other aquatic sites that would result in less adverse impact on the aquatic ecosystem. A permit cannot be issued, therefore, in circumstances where a less environmentally damaging practicable alternative for the proposed discharge exists (except as provided for under Section 404(b)(2)).

3. Discussion: The Guidelines are, as noted above, binding regulations. It is important to recognize, however, that this regulatory status does not limit the inherent flexibility provided in the Guidelines for implementing these provisions. The preamble to the Guidelines is very clear in this regard:

> Of course, as the regulation itself makes clear, a certain amount of flexibility is still intended. For example, while the ultimate conditions of compliance are "regulatory," the Guidelines allow some room for judgement in determining what must be done to arrive at a conclusion that those conditions have or have not been met.
>
> Guidelines Preamble, "Regulations versus Guidelines," *45 Federal Register 85336* (December 24, 1980)

Notwithstanding this flexibility, the record must contain sufficient information to demonstrate that the proposed discharge compiles with the requirements of Section 230.10(a) of the Guidelines. The amount of information needed to make such a determination and the level of scrutiny required by the Guidelines is commensurate with the severity of the environmental impact (as determined by the functions of the aquatic resource and the nature of the proposed activity) and the scope/cost of the project.

a. Analysis Associated with Minor Impacts:

The Guidelines do not contemplate that the same intensity of analysis will be required for all types of projects but instead envision a correlation between the scope of the evaluation and the potential extent of adverse impacts on the aquatic environment. The introduction to Section 230.10(a) recognizes that the level of analysis required may vary with the nature and complexity of each individual case:

> Although all requirements in Section 230.10 must be met, the compliance evaluation procedures will vary to reflect the seriousness of the potential for adverse impacts on the aquatic ecosystems posed by specific dredged or fill material discharge activities.
>
> 40 CFR 230.10

Similarly, Section 230.6 ("Adaptability") makes clear that the Guidelines:

> allow evaluation and documentation for a variety of activities, ranging from those with large, complex impacts on the aquatic environment to those for which the impact is likely to be innocuous. It is unlikely that the Guidelines will apply in their entirely to any one activity, no matter how complex. It is anticipated that substantial numbers of permit applications will be for minor, routine activities that have little, if any, potential for significant degradation of the aquatic environment. *It generally is not intended or expected that extensive testing, evaluation or analysis will be needed to make findings of compliance in such routine cases.*
>
> 40 CFR 230.6 (9) (emphasis added)

Section 230.6 also emphasizes that when making determinations of compliance with the Guidelines, users:

> must recognize the different levels of effort that should be associated with varying degrees of impact and require or prepare commensurate documentation. *The level of documentation should reflect the significance and complexity of the discharge activity.*
>
> 40 CFR 230.6 (b) (emphasis added)

Consequently, the Guidelines clearly afforded flexibility to adjust the stringency of the alternatives review for projects that would have only minor impacts. Minor impacts are associated with activities that generally would have little potential to degrade the aquatic environment and include one, and frequently more, of the following characteristics: are located in aquatic resources of limited natural function; are small in size and cause little direct impact; have little potential for secondary or cumulative impacts; or cause only temporary impacts. It is important to recognize, however, that in some circumstances even small or temporary fills result in substantial impacts, and that in such cases a more detailed evaluation is necessary. The Corps Districts and EPA Regions will, through the standard permit evaluation process, coordinate with the U.S. Fish and Wildlife Service, National Marine Fisheries Service and other appropriate state and Federal agencies in evaluating the likelihood that adverse impacts would result from a particular proposal. It is not appropriate to consider compensatory mitigation in determining whether a proposed discharge will cause only minor impacts for purposes of the alternatives analysis required by Section 230.10(a).

In reviewing projects that have the potential only for minor impacts on the aquatic environment, Corps and EPA field offices are directed to consider, in coordination with state and Federal resource agencies, the following factors:

i. Such projects by their nature should not cause or contribute to significant degradation individually or cumulatively. Therefore, it generally should not be necessary to conduct or require detailed analyses to determine compliance with Section 230.10(c).

ii. Although sufficient information must be developed to determine whether the proposed activity is in the fact the least damaging practicable alternative, the Guidelines do not require an elaborate search for practicable alternatives if it is reasonably anticipated that there are only minor differences between the environmental impacts of the proposed activity and potentially practicable alternatives. This decision will be made after consideration of resource agency comments on the proposed project. It often makes sense to examine first whether potential alternatives would result in no identifiable or discernible difference in impact on the aquatic ecosystem. Those alternatives that do not may be eliminated from the analysis since Section 230.10(a) of the Guidelines only prohibits discharges when a practicable alternative exists which would have *less adverse impact on the aquatic ecosystem.* Because evaluating practicability is generally the more difficult aspect of the alternatives analysis, this approach should save time and effort for both the applicant and the regulatory agencies.* By initially focusing the alternatives analysis on the question of impacts on the aquatic ecosystem, it may be impossible to limit (or in some instances eliminate altogether) the number of alternatives that have to be evaluated for practicability.

iii. When it is determined that there is no identifiable or discernible difference in adverse impact on the environment between the applicant's proposed alternative and all other practicable alternatives, then the applicant's alternative is considered as satisfying the requirements of Section 230.10(a).

iv. Even where a practicable alternative exists that would have less adverse impact on the aquatic ecosystem, the Guidelines allow it to be rejected if it would have "other significant adverse environment consequences." 40 CFR 230.10(A). As explained in the preamble, this allows for consideration of "evidence of damages to other ecosystems in deciding whether there is a 'better' alternative." Hence, in applying the alternatives analysis required by the Guidelines, it is not appropriate to select an alternative where minor impacts on the aquatic environment are avoided at the cost of substantial impacts to other natural environmental values.

v. In cases of negligible or trivial impacts (e.g., small discharges to construct individual driveways), it may be possible to conclude that no alternative location could result in less adverse impact on the aquatic environment within the meaning of the Guidelines. In such cases, it may not be necessary to conduct an offsite alternatives analysis but instead require only any practicable onsite minimization.

This guidance concerns application of the Section 404(b)(1) Guidelines to projects with minor impacts. Projects which may cause more than minor impacts on the aquatic environment, either individually or cumulatively, should be subjected to a proportionately more detailed level of analysis to determine compliance or noncompliance with the Guidelines. Projects which cause substantial impacts, in particular, must be thoroughly evaluated through the standard permit evaluation process to determine compliance with all provisions of the Guidelines.

b. Relationship between the Scope of Analysis and the Scope/Cost of the Proposed Project:

The Guidelines provide the Corps and EPA with discretion for determining the necessary level of analysis to support a conclusion as to whether or not an alternative is practicable. Practicable alternatives are those alternatives that are "available and capable of being done after taking into consideration cost, existing technology, and logistics in light of overall project purposes." 40 CFR 230.10(a)(2). The preamble to the Guidelines provides clarification on how cost is to be considered in the determination of practicability:

> Our intent is to consider those alternatives which are *reasonable in terms of the overall scope/cost of the proposed project.* The term economic [for which the term "cost" was substituted in the final rule] might be construed to include consideration of the applicant's financial standing, or investment, or market share, a cumbersome inquiry which is not necessarily material to the objectives of the Guidelines.
>
> Guidelines Preamble, "Alternatives," *Federal Register 85339* (December 24, 1980) (emphasis added).

Therefore, the level of analysis required for determining which alternatives are practicable will vary depending on the type of project proposed. The determination of what constitutes an unreasonable expense should generally consider whether the projected cost is substantially greater that the costs normally associated with the particular type of project. Generally, as the scope/cost of the project increases, the level of analysis should also increase. To the extent the Corps obtains information on the costs associated with the project, such information may be considered when making a determination of what constitutes an unreasonable expense.

The preamble to the Guidelines also states that "[i]f an alleged alternative is unreasonably expensive to the applicant, the alternative is not, 'practicable.'" Guidelines Preamble, "Economic Factors," *45 Federal Register 85343* (December 24, 1980). Therefore, to the extent that the individual homeowners and small businesses may typically be relevant consideration in determining what constitutes a practicable alternative. It is important to emphasize, however, that it is not a particular applicant's financial standing that is the primary consideration for determining practicability, but rather characteristics of the project and what constitutes a reasonable expense for these projects that are most relevant to practicability determinations.

4. The burden of proof to demonstrate compliance with the Guidelines rests with the applicant; where insufficient information is provided to determine compliance, the Guidelines require that no permit be issued. 40 CFR 230.12(a)(3)(iv).

5. A reasonable, common sense approach in applying the requirements of the Guidelines' alternatives analysis is fully consistent with sound environmental protection. The Guidelines clearly contemplate that reasonable direction should be applied based on the nature of the aquatic resource and potential impacts of a proposed activity in determining compliance with the alternatives test. Such an approach encourages effective decision making and fosters a better understanding and enhanced confidence in the Section 404 program.

6. This guidance is consistent with the February 6, 1990 "Memorandum of Agreement Between the Environmental Protection Agency and the Department of the Army Concerning The Determination of Mitigation under the Clean Water Act Section 404(b)(1) Guidelines."

ROBERT H. WAYLAND, III
Director, Office of Wetlands,
Oceans, and Watersheds
U.S. Environmental
Protection Agency

MICHAEL L. DAVIS
Office of the Assistant
Secretary of the Army
(Civil Works)
Department of the Army

Memorandum to the Field

United States
Environmental Protection Agency
Office of Wetlands,
Oceans, and Watersheds
Washington, D.C. 20460

United States
Department of the Army
U.S. Army Corps of Engineers
Washington, D.C. 20314

SUBJECT: Establishment and Use of Wetland Mitigation Banks In the Clean Water Act Section 404 Regulatory Program

1. This memorandum provides guidelines for the establishment and use of wetland mitigation banks in the Clean Water Act Section 404

* In certain instances, however, it may be easier to examine practicability first. Some projects may be so site-specific (e.g. erosion control, bridge replacement) that no offsite alternative could be practicable. In such cases the alternatives analysis may appropriately be limited to onsite options only.

regulatory program. This memorandum serves as interim guidance pending completion of Phase I by the Corps of Engineers' Institute for Water Resources study on wetland mitigation banking* at which time this guidance will be reviewed and any appropriate revisions will be incorporated into final guidelines.

2. For purposes of this guidance, wetland mitigation banking refers to the restoration, creation, enhancement, and, in exceptional circumstances, preservation of wetlands or other aquatic habitats expressly for the purpose of providing compensatory mitigation in advance of discharges into wetlands permitted under the Section 404 regulatory program. Wetland mitigation banks can have several advantages over individual mitigation projects, some of which are listed below:

a. Compensatory mitigation can be implemented and functioning in advance of project impacts, thereby reducing temporal losses of wetland functions and uncertainty over whether the mitigation will be successful in offsetting wetland losses.

b. It may be more ecologically advantageous for maintaining the integrity of the aquatic ecosystem to consolidate compensatory mitigation for impacts to many smaller, isolated or fragmented habitats into a single large parcel or contiguous parcels.

c. Development of a wetland mitigation bank can bring together financial resources and planning and scientific expertise not practicable to many individual mitigation proposals. This consolidation of resources can increase the potential for the establishment and long-term management of successful mitigation.

d. Wetland mitigation banking proposals may reduce regulatory uncertainty and provide more cost-effective compensatory mitigation opportunities.

3. The Section 404(b)(1) Guidelines (Guidelines), as clarified by the "Memorandum of Agreement Concerning the Determination of Mitigation under the Section 404(b)(1) Guidelines" (Mitigation MOA) signed February 6, 1990, by the Environmental Protection Agency and the Department of the Army, establish a mitigation sequence that is used in the evaluation of individual permit applications. Under this sequence, all appropriate and practicable steps must be undertaken by the applicant to first avoid and then minimize adverse impacts to the aquatic ecosystem. Remaining unavoidable impacts must then be offset through compensatory mitigation to the extent appropriate and practicable. Requirements for compensatory mitigation may be satisfied through the use of wetland mitigation banks, so long as their use is consistent with standard practices for evaluating compensatory mitigation proposals outlined in the Mitigation MOA. It is important to emphasize that, given the mitigation sequence requirements described above, permit applicants should not anticipate that the establishment of, or participation in, a wetland mitigation bank will ultimately lead to a determination of compliance with the Section 404(b)(1) Guidelines without adequate demonstration that impacts associated with the proposed discharge have been avoided and minimized to the extent practicable.

4. The agencies' preference for on-site, in-kind compensatory mitigation does not preclude the use of wetland mitigation banks where it has been determined by the Corps, or other appropriate permitting agency, in coordination with the Federal resource agencies through the standard permit evaluation process, that the use of a particular mitigation bank as compensation for proposed wetland impacts would be appropriate for offsetting impacts to the aquatic ecosystem. In making such a determination, careful consideration must be given to wetland functions, landscape position, and affected species populations at both the impact and mitigation bank sites. In addition, compensation for wetland impacts should occur, where appropriate and practicable, within the same watershed as the impact site. Where a mitigation bank is being developed in conjunction with a wetland resource planning initiative (e.g., Special Area Management Plan, State Wetland Conservation Plan) to satisfy particular wetland restoration objectives, the permitting agency will determine, in coordination with the Federal resource agencies, whether use of the bank should be considered an appropriate form of compensatory mitigation for impacts occurring within the same watershed.

5. Wetland mitigation banks should generally be in place and functional before credits may be used to offset permitted wetland losses. However, it may be appropriate to allow incremental distribution of credits corresponding to the appropriate stage of successful establishment of wetland functions. Moreover, variable mitigation ratios (credit acreage to impacted wetland acreage) may be used in such circumstances to reflect the wetland functions attained at a bank site at a particular point in time. For example, higher ratios would be required when a bank is not yet fully functional at the time credits are to be withdrawn.

6. Establishment of each mitigation bank should be accompanied by the development of a formal written agreement (e.g., memorandum of agreement) among the Corps, EPA, other relevant resource agencies, and those parties who will own, develop, operate or otherwise participate in the bank. The purpose of the agreement is to establish clear guidelines for establishment and use of the mitigation bank. A wetlands mitigation bank may also be established through issuance of a Section 404 permit where establishing the proposed bank involves a discharge of dredged or fill material into waters of the United States. The banking agreement or, where applicable, special conditions of the permit establishing the bank should address the following considerations, where appropriate:

a. location of the mitigation bank;

b. goals and objectives for the mitigation project;

c. identification of bank sponsors and participants;

d. development and maintenance plan;

e. evaluation methodology acceptable to all signatories to establish bank credits and assess bank success in meeting the project goals and objectives;

f. specific accounting procedures for tracking crediting and debiting;

g. geographic area of applicability;

h. monitoring requirements and responsibilities;

i. remedial action responsibilities including funding; and

j. provisions for protecting the mitigation bank in perpetuity.

Agency participation in a wetlands mitigation banking agreement may not, in any way, restrict or limit the authorities and responsibilities of the agencies.

7. An appropriate methodology, acceptable to all signatories, should be identified and used to evaluate the success of wetland restoration and creation efforts within the mitigation bank and to identify the appropriate stage of development for issuing mitigation credits. A full range of wetland functions should be assessed. Functional evaluations of the mitigation bank should generally be conducted by a multi-disciplinary team representing involved resource and regulatory agencies and other appropriate parties. The

* The Corps of Engineers Institute for Water Resources, under the authority of Section 307(d) of the Water Resources Development Act of 1990, is undertaking a comprehensive two-year review and evaluation of wetland mitigation banking to assist in the development of a national policy on this issue. The interim summary report documenting the results of the first phase of the study is scheduled for completion in the fall of 1993.

same methodology should be used to determine the functions and values of both credits and debits. As an alternative, credits and debits can be based on acres of various types of wetlands (e.g., National Wetland Inventory classes). Final determinations regarding debits and credits will be made by the Corps, or other appropriate permitting agency, in consultation with Federal resource agencies.

8. Permit applications may draw upon the available credits of a third party mitigation bank (i.e., a bank developed and operated by an entity other than the permit applicant). The Section 404 permit, however, must state explicitly that the permittee remains responsible for ensuring that the mitigation requirements are satisfied.

9. To ensure legal enforceability of the mitigation conditions, use of mitigation bank credits must be conditioned in the Section 404 permit by referencing the banking agreement or Section 404 permit establishing the bank; however, such a provision should not limit the responsibility of the Section 404 permittee for satisfying all legal requirements of the permit.

ROBERT H. WAYLAND, III
Office of Wetlands,
Oceans, and Watersheds
U.S. Environmental
Protection Agency

MICHAEL L. DAVIS
Office of the Assistant
Secretary of the Army (Civil Works)
Department of the Army

SPECIAL AREA MANAGEMENT PLANS (SAMPS)

REGULATORY GUIDANCE LETTER 92-03

SUBJECT: Extension of RGL 86-10, Special Area Management Plans (SAMPS)

DATE: 19 August 1992

EXPIRES: 31 December 1997

RGL 86-10, subject: "Special Area Management Plans (SAMPS)" is extended until 31 December 1997 unless sooner revised or rescinded.

FOR THE DIRECTOR
OF CIVIL WORKS

JOHN P. ELMORE
Chief, Operations,
Construction and Readiness Division
Directorate of Civil Works

"NORMAL CIRCUMSTANCES" FOR CROPPED WETLANDS

REGULATORY GUIDANCE LETTER 90-07

SUBJECT: Clarification of the Phrase "Normal Circumstances" as it Pertains to Cropped Wetlands

DATE: 26 September 1990

EXPIRES: 31 December 1993

1. The purpose of this regulatory guidance letter (RGL) is to clarify the concept of "normal circumstances" as currently used in the Army Corps of Engineers definition of wetlands (33 CFR 328.3(b)), with respect to cropped wetlands.

2. Since 1977, the Corps and the Environmental Protection Agency (EPA) have defined wetlands as:

> "areas that are inundated or saturated by surface or groundwater at a frequency and duration sufficient to support, and that under *normal circumstances* do support, a prevalence of vegetation typically adapted for life in saturated soil conditions..." (33 CFR 328.3(b)) (emphasis added).

While "normal circumstances" has not been defined by regulation, the Corps previously provided guidance on this subject in two expired "normal circumstances" RGLs (RGLs 82-2 and 86-9). These RGLs did not specifically deal with the issue of wetland conversion for purpose of crop production.

3. When the Corps adopted the Federal Manual for Identifying and Delineating Jurisdictional Wetlands (Manual) on 10 January 1989, the Corps chose to define "normal circumstances" in a manner consistent with the definition used by the Soil Conservation Service (SCS) in its administration of the Swampbuster provisions of the Food Security Act of 1985 (FSA). Both the SCS and the Manual interpret "normal circumstances" as the soil and hydrologic conditions that are normally present, without regard to whether the vegetation has been removed [7 CFR 12.31(b)(2)(i)] [Manual page 71].

4. The primary consideration in determining whether a disturbed area qualifies as a section 404 wetland under "normal circumstances" involves an evaluation of the extent and relative permanence of the physical alteration of wetlands hydrology and hydrophytic vegetation. In addition, consideration is given to the purpose and cause of the physical alterations to hydrology and vegetation. For example, we have always maintained that areas where individuals have destroyed

hydrophytic vegetation in an attempt to eliminate the regulatory requirements of section 404 remain part of the overall aquatic system, and are subject to regulation under section 404. In such a case, where the Corps can determine or reasonably infer that the purpose of the physical disturbance to hydrophytic vegetation was to avoid regulation, the Corps will continue to assert section 404 jurisdiction.

5. The following guidance is provided regarding how the concept of "normal circumstances" applies to areas that are in agricultural crop production:

a. "Prior converted cropland" is defined by the SCS (Section 512.15 of the National Food Security Act Manual, August 1988) as wetlands which were both manipulated (drained or otherwise physically altered to remove excess water from the land) and cropped before 23 December 1985, to the extent that they no longer exhibit important wetland values. Specifically, prior converted cropland is inundated for no more than 14 consecutive days during the growing season. Prior converted cropland generally does not include pothole or playa wetlands. In addition, wetlands that are seasonally flooded or ponded for 15 or more consecutive days during the growing season are not considered prior converted cropland.

b. "Farmed wetlands" are wetlands which were both manipulated and cropped before 23 December 1985, but which continue to exhibit important wetland values. Specifically, farmed wetlands include cropped potholes, playas, and areas with 15 or more consecutive days (or 10 percent of the growing season, whichever is less) of inundation during the growing season.

c. The definition of "normal circumstances" found at page 71 of the Manual is based upon the premise that for certain altered wetlands, even though the vegetation has been removed by cropping, the basic soil and hydrological characteristics remain to the extent that hydrophytic vegetation would return if the cropping ceased. This assumption is valid for "farmed wetlands" and as such these areas are subject to regulation under section 404.

d. In contrast to "farmed wetlands," "prior converted croplands" generally have been subject to such extensive and relatively permanent physical hydrological modifications and alteration of hydrophytic vegetation that the resultant cropland constitutes the "normal circumstances" for purposes of section 404 jurisdiction. Consequently, the "normal circumstances" of prior converted croplands generally do not support a "prevalence of hydrophytic vegetation" and as such are not subject to regulation under section 404. In addition, our experience and professional judgment lead us to conclude that because of the magnitude of hydrological alterations that have most often occurred on prior converted cropland, such cropland meets, minimally if at all, the Manual's hydrology criteria.

e. If prior converted cropland is abandoned (512.17 National Food Security Act Manual as amended, June 1990) and wetland conditions return, then the area will be subject to regulation under section 404. An area will be considered abandoned if for five consecutive years there has been no cropping, management or maintenance activities related to agricultural production. In this case, positive indicators of all mandatory wetlands criteria, including hydrophytic vegetation, must be observed.

f. For the purposes of section 404, the final determination of whether an area is a wetland under normal circumstances will be made pursuant to the 19 January 1989 Army/EPA Memorandum of Agreement on geographic jurisdiction. For those cropped areas that have previously been designated as "prior converted cropland" or "farmed wetland" by the SCS, the Corps will rely upon such a designation to the extent possible. For those cropped areas that have not been designated "prior converted cropland" or "farmed wetland" by the SCS, the Corps will consult with SCS staff and make appropriate use of SCS data in making a determination of "normal circumstances" for section 404 purposes. Although every effort should be made at the field level to resolve Corps/SCS differences in opinion on the proper designation of cropped wetlands, the Corps will make the final determination of section 404 jurisdiction. However, in order to monitor implementation of this RGL, cases where the Corps and SCS fail to agree on designation of prior converted cropland or farmed wetlands should be documented and a copy of the documentation forwarded to CECW-OR.

6. This policy is applicable to section 404 of the Clean Water Act only.

7. This guidance expires 31 December 1993 unless sooner revised or rescinded.

FOR THE COMMANDER:

PATRICK J. KELLY
Major General, USA
Director of Civil Works

WETLAND JURISDICTIONAL DELINEATIONS

REGULATORY GUIDANCE LETTER 90-06

SUBJECT: Expiration Dates for Wetlands Jurisdictional Delineations

DATE: 14 August 1990

EXPIRES: 31 December 1993

1. Recently, questions have been raised regarding the length of time that wetlands jurisdictional delineations remain valid. In light of the need for national consistency in this area, the guidance in paragraph 4(a)–(d) below is provided. This guidance is subject to the provisions in paragraphs 5, 6, and 7.

2. Since wetlands are affected over time by both natural and man-made activities, we can expect local changes in wetland boundaries. As such, wetlands jurisdictional delineations will not remain valid for an indefinite period of time.

3. The purpose of this guidance is to provide a consistent national approach to reevaluating wetlands delineations. This provides greater certainty to the regulated public and ensures their ability to rely upon wetlands jurisdictional delineations for a definite period of time.

4. a. Written wetlands jurisdictional delineations made before the effective date of this guidance, without a specific time limit imposed in the Corps written delineation, will remain valid for a period of two years from the effective date of this Regulatory Guidance Letter (RGL).

b. Written wetlands jurisdictional delineations made before the effective date of this guidance, with a specified time imposed in the Corps written delineation, will be valid until the date specified.

c. Oral delineations (i.e., not verified in writing by the Corps) are no longer valid as of the effective date of this RGL.

d. As specified in the 20 March 1989, Memorandum of Agreement Between the Department of the Army and the Environmental Protection Agency Concerning the Determination of the Geographic Jurisdiction of the Section 404 Program and the Application of the Exemptions Under Section 404(f) of the Clean Water Act (MOA), all wetlands jurisdictional delineations (including those prepared by the project proponent or consultant and verified by the Corps) shall be put in writing. Generally this should be in the form of a letter to the project proponent. The Corps letter shall include a statement that the wetlands jurisdictional delineation is valid for a period of three years from the date of the letter unless new information warrants revision of the delineation before the expiration date. Longer periods, not to exceed five years, may be provided where the nature and duration of a proposed project so warrant. The delineation should be supported by proper documentation. Generally the project proponent should be given the opportunity to complete the delineation and provide the supporting documentation subject to the Corps verification. However, the Corps will complete the delineation and documentation at the project proponent's request, consistent with other work priorities.

5. The guidance in paragraph 4(a)–(b) above does not apply to completed permit applications [33 CFR 325.1(d)(9)] received before the effective date of this RGL, or where the applicant can fully demonstrate that substantial resources have been expended or committed based on a previous Corps jurisdictional delineation (e.g., final engineering design work, contractual commitments for construction, or purchase or long term leasing of property will, in most cases, be considered a substantial commitment of resources). However, district engineers cannot rely upon the expenditure or commitment of substantial resources to validate an otherwise expired delineation for more than five years from the expiration dates noted in paragraph 4(a)–(b). At the end of the five year period a new delineation would be required. In certain rare cases, it may be appropriate to honor a previous oral wetlands delineation when the applicant can fully demonstrate a substantial expenditure or commitment of resources. However, the presumption is that oral delineations are not valid and acceptance of such must be based on clear evidence and equities of the particular case. This determination is left to the discretion of the district engineer.

6. When making wetlands jurisdictional delineations it is very important to have complete and accurate documentation which substantiates the Corps decision (e.g., data sheets, etc). Documentation must allow a reasonably accurate replication of the delineation at a future date. In this regard, documentation will normally include information such as data sheets, maps, sketches, and in some cases surveys.

7. This guidance does not alter or supercede any provisions of law, regulations, or any interagency agreement between Army and EPA. Further, this guidance does not impair the Corps discretion to revise wetlands jurisdictional delineations where new information so warrants.

8. Each district shall issue a public notice on this guidance no later than 1 September 1990. The public notice shall contain the full text of this RGL.

9. This guidance expires on 31 December 1993 unless sooner revised or rescinded.

FOR THE DIRECTOR
OF CIVIL WORKS:

JOHN P. ELMORE
Chief, Operations,
Construction and Readiness Division
Directorate of Civil Works

LANDCLEARING ACTIVITIES

REGULATORY GUIDANCE LETTER 90-05

SUBJECT: Landclearing Activities Subject to Section 404 Jurisdiction

DATE: 18 July 1990

EXPIRES: 31 December 1992

1. The purpose of this guidance is to interpret the statutory and regulatory definitions of "discharge of a pollutant" (CWA section 502(12) and 33 CFR 327.2(f)) to the effect that land-clearing activities using mechanized equipment such as backhoes or bulldozers with sheer blades, rakes, or discs constitute point source discharges and are subject to section 404 jurisdiction when they take place in wetlands which are waters of the United States.

2. In *Avoyelles Sportsmen's League, Inc. v. Marsh,* 715 F. 2d 897, 923–24 (5th Cir. 1983) the court stated that the term "discharge" may reasonably be understood to include "redeposit" and concluded that the term "discharge" covers the redepositing of soil taken from wetlands such as occurs during mechanized landclearing activities. Although the court in Avoyelles did not decide whether all landclearing activities constitute a discharge, it is our position that mechanized landclearing activities in jurisdictional wetlands result in a redeposition of soil that is subject to regulation under section 404. Some limited exceptions may occur, such as

cutting trees above the soil's surface with a chain saw, but as a general rule, mechanized landclearing is a regulated activity.

3. As with any discharge subject to section 404, each case must be reviewed to determine if the discharge qualifies for a regional or nationwide permit, or for an exemption under section 404(f). This guidance is not intended to alter the exemptions for normal farming or silviculture activities under section 404(f).

4. This interpretation alters in some respects the guidance provided by previous Regulatory Guidance Letters (RGLs) on Landclearing (in particular RGL 85-4) and FOAs should exercise appropriate enforcement discretion with regard to properties whose owners have previously been informed that no permit is required for such landclearing based on the prior RGLs. The guidance in this RGL should apply to property which has not been cleared, unless the owner can demonstrate that he has committed substantial resources towards the clearing, in reliance on earlier Corps guidance, to the extent that it would be inequitable to apply this guidance.

5. This guidance expires on 31 December 1992 unless sooner modified or rescinded.

FOR THE DIRECTOR
OF CIVIL WORKS:

JOHN P. ELMORE
Chief, Operations,
Construction and
Readiness Division
Directorate of Civil Works

CLEAN WATER ACT SECTION 404 REGULATORY PROGRAM AND AGRICULTURAL ACTIVITIES

MEMORANDUM FOR THE FIELD

United States
Environmental Protection Agency
United States
Department of the Army

May 3, 1990

A number of questions have recently been raised about the applicability of the Clean Water Act Section 404 Regulatory Program to agriculture. This memorandum is intended to assist Section 404 field personnel in responding to those questions and to assure that the program is implemented in a consistent manner. At the outset, we should emphasize that we respect and support the underlying purposes of the Clean Water Act regarding the exemption from Section 404 permitting requirements for "normal farming" activities. The exemptions (at Section 404(f) of the Act) recognize that American agriculture fulfills the vitally important public need for supplying abundant and affordable food and fiber and it is our intent to assure that the exemptions are appropriately implemented.

What are normal farming activities? Who makes that determination? Can agricultural producers plant crops in wetland areas that have been farmed for many years? These are questions that have generated significant confusion and concern in the agricultural community. This memorandum will explain the extent of the Section 404 program and clarify some misunderstandings that may exist in the field. Therefore we encourage you to widely distribute this memorandum.

WHAT IS SECTION 404?

The Federal Water Pollution Control Act Amendments of 1972 established the Section 404 Regulatory Program. Under this Act, it is unlawful to discharge dredged or fill material into waters of the United States without first receiving authorization (usually a permit) from the Corps, unless the discharge is covered under an exemption. The term "waters of the United States" defines the extent of geographic jurisdiction of the Section 404 program. The term includes such waters as rivers, lakes, streams, tidal waters, and most wetlands. A discharge of dredged or fill material involves the physical placement of soil, sand, gravel, dredged material or other such materials into the waters of the United States. Section 404(f) exemptions, which were added in 1977, provide that discharges that are part of normal farming, ranching, and forestry activities associated with an active and continuous ("ongoing") farming or forestry operation generally do not require a Section 404 permit.

With this background in mind, we can now turn to the issues that are the focus of concern. As previously noted, Section 404(f) exempts discharges of dredged or fill material into waters of the United States associated with certain normal agricultural activities. Of course, activities that do not involve a discharge of dredged or fill material into waters of the United States never require a Section 404 permit. Further as provided in the Interagency Federal Manual for Identifying and Delineating Jurisdictional Wetlands, while a site is effectively and legally drained to the extent that it no longer meets the regulatory wetlands hydrology criteria (as interpreted by the Interagency Manual), it is not a wetland subject to jurisdiction under Section 404 of the Clean Water Act.

WHAT IS THE "NORMAL FARMING" ACTIVITIES EXEMPTION?

The Clean Water Act exempts from the Section 404 program discharges associated with normal farming, ranching, and forestry activities such as plowing, cultivating, minor drainage, and harvesting for the production of food, fiber, and forest products, or upland soil and water conservation practices (Section 404(f)(1)(A)). To be exempt, these activities must be part of an established, ongoing operation. For example, if a farmer has been plowing, planting and harvesting in wetlands, he can continue to do so without the need for a Section 404 permit, so long as he does not convert the wetlands to dry land. Activities which convert a wetland which has not been used for farming or forestry into such uses are not considered part of an established operation, and are not exempt. For example, the conversion of a bottomland hardwood wetland to crop production is not exempt.

In determining whether an activity is part of an established operation, several points need to be considered. First, the specific farming activity need not itself have been ongoing as long as it is introduced as part of an ongoing farming operation. For example, if crops have been grown and harvested on a regular basis, the mere addition or change of a cultivation technique (e.g., discing between crop rows to control weeds rather than using herbicides) is considered to be part of the established farming operation. Second, the planting of different agricultural crops as part of an established rotation (e.g., soybeans to rice) is exempt. Similarly, the rotation of rice and crawfish production is also exempt (construction of fish ponds is not an exempt activity and is addressed below). Third, the resumption of agricultural production in areas laying fallow as part of a normal rotational cycle are considered to be part of an established operation and would be exempted under Section 404(f). However, if a wetland area has not been used for farming for so long that it would require hydrological modifications (modifications to the surface or groundwater flow) that would result in a discharge of dredged or fill material, the farming operation would no longer be established or ongoing.

As explained earlier, normal farming operations include cultivating, harvesting, minor drainage, plowing, and seeding. While these terms all have common, everyday definitions, it is important to recognize that these terms have specific, regulatory meanings in relation to the Section 404(f) exemptions. For example, plowing that is exempt under Section 404(f) means all mechanical means of manipulating soil, including land leveling, to prepare it for the planting of crops. However, grading activities that would change any area of waters of the United States, including wetlands, into dry land are not exempt. Minor drainage that is exempt under Section 404(f) is limited to discharges associated with the continuation of established wetland crop production (e.g., building rice levees) or the connection of upland crop drainage facilities to waters of the United States. In addition, minor drainage also refers to the emergency removal of blockages that close or constrict existing drainageways used as part of an established crop production. Minor drainage is defined such that it does not include discharges associated with the construction of ditches which drain or significantly modify any wetlands or aquatic areas considered as waters of the United States. Seeding that is exempt under Section 404(f) includes not only the placement of seeds themselves, but also the placement of soil beds for seeds or seedlings on established farm or forest lands. Cultivating under Section 404(f) includes physical methods of soil treatment to aid and improve the growth, quality, or yield of established crops. Except as provided under Section 404(f)(2) as explained below, construction or maintenance of irrigation ditches or maintenance of drainage ditches is also exempt.

Recognizing area and regional differences in normal farming practices, EPA and the Corps agree to develop additional definitions of normal farming practices in consultation with the designated Land Grant Colleges and the Cooperative Extension Services. We also further encourage our field staffs to utilize the expertise in these colleges and agricultural services in the ongoing implementation of the Section 404 program.

WHEN DOES THE NORMAL FARMING ACTIVITY EXEMPTION NOT APPLY?

Sections 404(f)(2) provides that discharges related to activities that change the use of the waters of the United States, including wetlands, and reduce the reach, or impair the flow or circulation of waters of the United States are not exempted. This "recapture" provision involves a two-part test that results in an activity being considered not exempt when both parties are met: 1) does the activity represent a "new use" of the wetland, and 2) would the activity result in a "reduction in reach/impairment of flow or circulation" of waters of the United States? Consequently, any discharge of dredged or fill material that results in the destruction of the wetlands character of an area (e.g., it conversion to uplands due to new or expanded drainage) is considered a change in the waters of the United States, and by definition, a reduction of their reach and is not exempt under Section 404(f). In addition, Section 404(f)(1) of the Act provides that discharges that contain toxic pollutants listed under Section 307 are not exempted and must be permitted.

However, discharges that are not exempt are not necessarily prohibited. Non-exempted discharges must first be authorized either through a general or individual Section 404 permit before they are initiated.

WHAT ARE GENERAL PERMITS?

Even if a farming activity is one that does not fall under an exemption and a permit is required, some farming activities are eligible for General Permits. Section 404(e) of the Act authorizes the Corps, after notice and opportunity for public hearing, to issue General Permits on a State, regional or nationwide basis for certain categories of activities involving a discharge of dredged or fill material in waters of the United States. Such activities must be similar in nature and cause only minimal adverse environmental effects. *Discharges authorized under a General Permit may proceed without applying to the Corps for an individual permit. However, in some circumstances, conditions associated with a General Permit may require that persons wishing to discharge under that permit must notify the Corps or other designated State or local agency before the discharge takes place.* A list of current General Permits is available from each Corps District Office, as well as information regarding notification requirements or other relevant conditions.

RICE FARMING

Questions have arisen regarding the relationship of the Section 404 program to rice farming. We understand these concerns, and recently have initiated actions that will allow farmers to understand better the regulatory program and provide more efficient and equitable mechanisms for implementing provisions of the Section 404 program.

In an April 19, 1990 letter responding to a request from Senator Patrick J. Leahy, Chairman and 11 members of the Senate Committee on Agriculture, Nutrition and Forestry, we stated our position that discharges of dredged or fill material associated with the construction of rice levees for rice farming in wetlands which are in established agricultural crop production are "normal farming activities" within the meaning of Section 404(f)(1)(A) and are therefore exempt from Section 404 regulation under the following conditions:

1. the purpose of these levees is limited to the maintenance and manipulation of shallow water levels for the production of rice crops; and

2. consistent with current agricultural practices associated with rice cultivation,

   – the height of the rice levees should *generally* not exceed 24 inches above their base; and

   – the material to be discharged for levee construction should generally be derived exclusively from the distribution of soil immediately adjacent to the constructed levee.

Land leveling for rice farming in wetlands which are in established crop production also is a "normal farming activity" within the meaning of Section 404(f)(1)(A) and is therefore exempt from Section 404 regulation.

FISH PONDS

We are developing a General Permit authorizing discharges of dredged or fill material associated with the construction of levees and ditches for the construction of fish ponds in wetlands that were in agricultural crop production prior to December 23, 1985. A draft General Permit has been developed by the Vicksburg District, Army Corps of Engineers and should be issued by June 1, 1990. This General Permit should serve as a model permit for other areas of the country and this activity will be considered for a nationwide General Permit.

It should be made clear, however, that the Section 404(f) exemption for "normal farming activities" and the General Permit being developed for fish ponds apply only to the use of wetlands which are already in use for agricultural crop production. These provisions do not apply to 1) wetlands that were once in use for agricultural crop production but have lain idle so long that modifications to the hydrologic regime are necessary to resume crop production or, 2) the conversion of naturally vegetated wetlands to agriculture, such as the conversion of bottomland hardwood wetlands to agriculture.

LIMITATIONS OF THE SECTION 404(f) EXEMPTIONS

It should be emphasized that the use of Section 404(f) exemptions does not affect Section 404 jurisdiction. For example, the fact that an activity in wetlands is exempted as normal farming practices does not authorize the filling of the wetland for the construction of buildings without a Section 404 permit. Similarly, a Section 404 permit would be required for the discharge of dredged or fill material associated with draining a wetland and converting it to dry land.

ENFORCEMENT

Given that the normal farming practices as described above are exempt from regulation under Section 404, neither EPA nor the Corps will initiate enforcement actions against farmers or other persons for engaging in such normal farming activities. Further, there will be no enforcement against actions that meet the description of activities covered by, and any conditions contained in, general permits issued by the Corps.

CONCLUSION

Proper implementation of the Section 404 program is an issue of extreme importance to the nation. We encourage you to distribute this memorandum not only to your staffs but to the public at large so that there will be a better general understanding of the program and how it operates.

LaJuana S. Wilcher /s/
Assistant Administrator for Water
U.S. Environmental
Protection Agency
May 3, 1990

Roland W. Page
Assistant Secretary of
the Army, Civil Works
May 3, 1990

SPECIAL AREA MANAGEMENT PLANS (SAMPS)

REGULATORY GUIDANCE LETTER 86-10

SUBJECT: Special Area Management Plans (SAMPS)

DATE: 2 October 1986

EXPIRES: 31 December 1988

1. The 1980 Amendments to the Coastal Zone Management Act define the SAMP process as "a comprehensive plan providing for natural resource protection and reasonable coastal-dependent economic growth containing a detailed and comprehensive statement of policies, standards and criteria to guide public and private uses of lands and waters; and mechanisms for timely implementation in specific geographic areas within the coastal zone." This process of collaborative interagency planning within a geographic area of special sensitivity is just as applicable in non-coastal areas.

2. A good SAMP reduces the problems associated with the traditional case-by-case review. Developmental interests can plan with predictability and environmental interests are assured that individual and cumulative impacts are analyzed in the context of broad ecosystem needs.

3. Because SAMPs are very labor intensive, the following ingredients should usually exist before a district engineer becomes involved in a SAMP:

a. The area should be environmentally sensitive and under strong developmental pressure.

b. There should be a sponsoring local agency to ensure that the plan fully reflects local needs and interests.

c. Ideally there should be full public involvement in the planning and development process.

d. All parties must express a willingness at the outset to conclude the SAMP process with a definitive regulatory product (see next paragraph).

4. An ideal SAMP would conclude with two products: 1) appropriate local/state approvals and a Corps general permit (GP) or abbreviated processing procedure (APP) for activities in specifically defined situations; and 2) a local/state restriction and/or an Environmental Protection Agency (EPA) 404(c) restriction (preferably both) for undesirable activities. An individual permit review may be conducted for activities that do not fall into either category above. However, it should represent a small number of the total cases addressed by the SAMP. We recognize that an ideal SAMP is difficult to achieve, and, therefore, it is intended to represent an upper limit rather than an absolute requirement.

5. Do not assume that an environmental impact statement is automatically required to develop a SAMP.

6. EPA's program for advance identification of disposal areas found at 40 CFR 230.80 can be integrated into a SAMP process.

7. In accordance with this guidance, district engineers are encouraged to participate in development of SAMPs. However, since development of a SAMP can require a considerable investment of time, resources, and money, the SAMP process should be entered only if it is likely to result in a definitive regulatory product as defined in paragraph 4 above.

8. This guidance expires 31 December 1988 unless sooner revised or rescinded.

FOR THE CHIEF
OF ENGINEERS:

"NORMAL CIRCUMSTANCES" IN WETLAND DEFINITION

REGULATORY GUIDANCE LETTER 86-09

SUBJECT: Clarification of "Normal Circumstances" in the Wetland Definition (33 CFR 323.2 (c))

DATE: August 27, 1986

EXPIRES: December 31, 1988

1. This letter will serve to continue the guidance originally issued as RGL 82-2, regarding Corps policy on land-use conversion as it concerns regulatory jurisdiction. Specifically, the guidance addresses situations involving changes in the physical characteristics of a wetland which cause the area to lose or gain characteristics which would alter its status of "waters of the United States" for purposes of the Section 404 regulatory program.

2. The current definition of "waters of the United States" delineates wetlands as follows, at 33 CFR 323.2(c) The term wetlands means those areas that are inundated or saturated by surface or ground water at a frequency and duration sufficient to support, and that under normal circumstances do support, a prevalence of vegetation typically adapted for life in saturated soil conditions. Wetlands generally include swamps, marshes, bogs, and similar areas.

The regulations now in force cover the actual discharge of dredged or fill material into "wetlands," as they are a part of the "waters of the United States." However, these regulations do not discuss what effect the conversion of a wetland to other uses (e.g., agricultural) has upon regulatory jurisdiction, once the land-use conversion has been accomplished.

3. As was stated in RGL 82-2, it is our intent under Section 404 to regulate discharges of dredged or fill material into the aquatic system as it exists and not as it may have existed over a record period of time. The wetland definition is designed to achieve this intent. It pertains to an existing wetland and requires that the area be inundated or saturated by water at a frequency and duration sufficient to support aquatic vegetation. We do not intend to assert jurisdiction over those areas that once were wetlands and part of an aquatic system, but which, in the past, have been transformed into dry land for various purposes. Neither do we intend the definition of "wetlands" to be interpreted as extending to abnormal situations including non-aquatic areas that have aquatic vegetation. Thus, we have listed swamps, bogs, and marshes at the end of the definition at 323.2(c) to further clarify our intent to include only truly aquatic areas.

4. The use of the phrase "under normal circumstances" is meant to respond to those situations in which an individual would attempt to eliminate the permit review requirements of Section 404 by destroying the aquatic vegetation, and to those areas that are not aquatic but experience an abnormal presence of aquatic vegetation. Several instances of destruction of aquatic vegetation to eliminate Section 404 jurisdiction have actually occurred. Because those areas would still support aquatic vegetation "under normal circumstances," they remain a part of the overall aquatic system intended to be protected by the Section 404 program; therefore, jurisdiction still exists. On the other hand, the abnormal presence of aquatic vegetation in a non-aquatic area would not be sufficient to include that area within the Section 404 program.

5. Many areas of wetlands converted in the past to other uses would, if left unattended for a sufficient period of time, revert to wetlands solely through the devices of nature. However, such natural circumstances are not what is meant by "normal circumstances" in the definition quoted above. "Normal circumstances" are determined on the basis of an area's characteristics and use, at present and in the recent past. Thus, if a former wetland has been converted to another use (other than by recent unpermitted action not subject to 404(f) or 404(r) exemptions) and that use alters its wetland characteristics to such an extent that it is no longer a "water of the United States," that area will no longer come under the Corps regulatory jurisdiction for purposes of Section 404. However, if the area is abandoned and over time regains wetland characteristics such that it meets the definition of "wetlands," then the Corps 404 jurisdiction has been restored.

6. This policy is applicable to Section 404 authority only, not to Section 10.

7. This guidance expires 31 December 1988 unless sooner revised or rescinded.

FOR THE CHIEF
OF ENGINEERS:

APPENDIX H

USACE District Regulatory Offices—Contact Information

Provided by USACE as of August 2003

Contact the relevant district office to determine the specific requirements associated with USACE regulatory evaluation of your proposed project. For changes made after publication of this book, please check the USACE web site (www.usace.army.mil/inet/functions/cw/cecwo/reg/district.htm).

PLEASE NOTE: Most districts boundaries are based in whole or in part upon state boundaries; however, because Corps districts cover wide geographic areas, the district office that will evaluate your activity may not be located in the state in which your project is proposed to be sited. District field units are numerous, however, and you may be further directed to a field office geographically closer to your project site.

ALASKA

U.S. Army Corps of Engineers
Alaska District
Attention: CEPOA-CO-R
P.O. Box 898
Anchorage, AK 99506-0898
907-753-2712 tel
907-753-5567 fax
www.poa.usace.army.mil/reg/default.htm

ALBUQUERQUE

U.S. Army Corps of Engineers
Albuquerque District
Attention: CESPA-OD-R
4101 Jefferson Plaza NE
Albuquerque, NM 87109-3435
505-342-3283 tel
505-342-3498 fax
www.spa.usace.army.mil/reg

BALTIMORE

U.S. Army Corps of Engineers
Baltimore District
Attention: CENAB-OP-R
P.O. Box 1715
Baltimore, MD 21203-1715
410-962-3670 tel
410-962-8024 fax
www.nab.usace.army.mil/Regulatory

BUFFALO

U.S. Army Corps of Engineers
Buffalo District
Attention: CELRB-TD-R
1776 Niagara Street
Buffalo, NY 14207-3199
716-879-4313 tel
716-879-4310 fax
www.lrb.usace.army.mil/orgs/reg/index.htm

CHARLESTON

U.S. Army Corps of Engineers
Charleston District
Attention: CESAC-RD
P.O. Box 919
Charleston, SC 29402-0919
843-329-8044 tel
843-329-2332 fax
www.sac.usace.army.mil/permits/index.html

CHICAGO

U.S. Army Corps of Engineers
Chicago District
Attention: CELRC-CO-RE
111 North Canal Street, Suite 600
Chicago, IL 60606-7206
312-353-6400 tel
312-353-4110 fax
www.lrc.usace.army.mil/co-r

DETROIT

U.S. Army Corps of Engineers
Detroit District
Attention: CELRE-CO-L
P.O. Box 1027
Detroit, MI 48231-1027
313-226-2432 tel
313-226-6763 fax
www.lre.usace.army.mil/index.cfm?chn_id=1928

FT. WORTH

U.S. Army Corps of Engineers
Ft. Worth District
Attention: CESWF-PER-R
P.O. Box 17300
Ft. Worth, TX 76102-0300
817-886-1731 tel
817-886-6493 fax
www.swf.usace.army.mil/regulatory/index.html

GALVESTON

U.S. Army Corps of Engineers
Galveston District
Attention: CESWG-PE-R
P.O. Box 1229
Galveston, TX 77553-1229
409-766-3930 tel
409-766-3931 fax
www.swg.usace.army.mil/reg

HUNTINGTON

U.S. Army Corps of Engineers
Huntington District
Attention: CELRH-OR-F
502 8th Street
Huntington, WV 25701-2070
304-529-5487 tel
304-529-5085 fax
www.lrh.usace.army.mil/or/permits

HONOLULU

U.S. Army Corps of Engineers
Honolulu District
Attention: CEPOH-EC-R
Building 230, Fort Safter
Honolulu, HI 96858-5440
808-438-9258 tel
808-438-4060 fax
www.poh.usace.army.mil/regulatory.asp

JACKSONVILLE

U.S. Army Corps of Engineers
Jacksonville District
Attention: CESAJ-RD
P.O. Box 4970
701 San Marco Boulevard
Jacksonville, FL 32207
904-232-1666 tel
904-232-1684 fax
www.saj.usace.army.mil/permit/index.html

KANSAS CITY

U.S. Army Corps of Engineers
Kansas City District
Attention: CENWK-OD-R
700 Federal Building
601 East 12th Street
Kansas City, MO 64106-2896
816-983-3990 tel
816-426-2321 fax
www.nwk.usace.army.mil/regulatory/regulatory.htm

LITTLE ROCK

U.S. Army Corps of Engineers
Little Rock District
Attention: CESWL-PR-R
P.O. Box 867
Little Rock, AR 72203-0867
501-324-5296 tel
501-324-6013 fax
www.swl.usace.army.mil/regulatory

LOS ANGELES

U.S. Army Corps of Engineers
Los Angeles District
Attention: CESPL-CO-R
911 Wilshire Boulevard
P.O. Box 2711
Los Angeles, CA 90053-2325
213-452-3406 tel
213-452-4196 fax
www.spl.usace.army.mil/regulatory

LOUISVILLE

U.S. Army Corps of Engineers
Louisville District
Attention: CELRL-OP-F
P.O. Box 59
Louisville, KY 40401-0059
502-315-6733 tel
502-315-6677 fax
www.lrl.usace.army.mil/orf/default.htm

MEMPHIS

U.S. Army Corps of Engineers
Memphis District
Attention: CEMVM-CO-R
Clifford Davis Federal Building
Room B-202
Memphis, TN 38103-1894
901-544-3471 tel
901-544-0211 fax
www.mvm.usace.army.mil/environmental.htm

MOBILE

U.S. Army Corps of Engineers
Mobile District
Attention: CESAM-OP-S
P.O. Box 2288
Mobile, AL 36628-00001
334-690-2658 tel
334-690-2660 fax
www.sam.usace.army.mil/op/reg

NASHVILLE

U.S. Army Corps of Engineers
Nashville District
Attention: CELRN-OP-F
3701 Bell Road
Nashville, TN 37214-2660
615-369-7500 tel
615-369-7501 fax
www.orn.usace.army.mil/cof/default.htm

NEW ENGLAND

U.S. Army Corps of Engineers
New England District
Attention: CENAE-R-PT
696 Virginia Road
Concord, MA 01742-2751
978-318-8338 tel
978-318-8303 fax
www.nae.usace.army.mil/reg/index.htm

NEW ORLEANS

U.S. Army Corps of Engineers
New Orleans District
Attention: CEMVN-OD-S
P.O. Box 60267
New Orleans, LA 70160-0267
504-862-2255 tel
504-862-2289 fax
www.mvn.usace.army.mil/ops/regulatory

NEW YORK

U.S. Army Corps of Engineers
New York District
Attention: CENAN-OP-R
26 Federal Plaza
New York, NY 10278-0090
212-264-3996 tel
212-264-4260 fax
www.nan.usace.army.mil/business/buslinks/regulat/index.htm

NORFOLK

U.S. Army Corps of Engineers
Norfolk District
Attention: CENAO-TS-G
803 Front Street
Norfolk, VA 23510-1096
757-441-7068 tel
757-441-7678 fax
www.nao.usace.army.mil/regulatory/regulatory.html

OMAHA

U.S. Army Corps of Engineers
Omaha District
Attention: CENWO-OP-NR
P.O. Box 5
Omaha, NE 68101-0005
402-221-4211 tel
402-221-4939 fax
www.nwo.usace.army.mil/html/op-r/regwebpg.htm

PHILADELPHIA

U.S. Army Corps of Engineers
Philadelphia District
Attention: CENAP-OP-R
100 Penn Square East
2nd and Chestnut Street
Philadelphia, PA 19107-3390
215-656-6725 tel
215-656-6724 fax
www.nap.usace.army.mil/cenap-op/ops.htm

PITTSBURGH

U.S. Army Corps of Engineers
Pittsburgh District
Attention: CELRP-OP-F
Federal Building
1000 Liberty Avenue
Pittsburgh, PA 15222-4186
412-395-7155 tel
412-644-4211 fax
www.lrp.usace.army.mil/or/or-f/permits.htm

PORTLAND

U.S. Army Corps of Engineers
Portland District
Attention: CENWP-OP-G
P.O. Box 2946
Portland, OR 97208-2946
503-808-4371 tel
503-808-4375 fax
www.nwp.usace.army.mil/op/g

ROCK ISLAND

U.S. Army Corps of Engineers
Rock Island District
Attention: CEMVR-OD-P
Clock Tower Building
Rock Island, IL 61201-2004
309-794-5370 tel
309-794-5191 fax
www.mvr.usace.army.mil/regulatory/regulatorydivisionhomepage.htm

SACRAMENTO

U.S. Army Corps of Engineers
Sacramento District
Attention: CESPK-CO-R
1325 J Street
Sacramento, CA 95814-2922
916-557-5250 tel
916-557-6877 fax
www.spk.usace.army.mil

ST. LOUIS

U.S. Army Corps of Engineers
St. Louis District
Attention: CEMVS-CO-F
1222 Spruce Street
St. Louis, MO 63103-2833
314-331-8575 tel
314-331-8741 fax
www.mvs.usace.army.mil/permits

ST. PAUL

U.S. Army Corps of Engineers
St. Paul District
Attention: CEMVP-CO-R
Army Corps of Engineers Centre
190 Fifth Street East
St. Paul, MN 55101-1638
651-290-5354 tel
651-290-5330 fax
www.mvp.usace.army.mil/regulatory

SAN FRANCISCO

U.S. Army Corps of Engineers
San Francisco District
Attention: CESPN-OR-R
333 Market Street, 8th floor
San Francisco, CA 94105-2197
415-977-8461 tel
415-977-8343 fax
www.spn.usace.army.mil/regulatory

SAVANNAH

U.S. Army Corps of Engineers
Savannah District
Attention: CESAS-OP-F
P.O. Box 889
Savannah, GA 31402-0889
912-652-5768 tel
912-652-5995 fax
http://144.3.144.48/permit.htm

SEATTLE

U.S. Army Corps of Engineers
Seattle District
Attention: CENWS-OD-RG
P.O. Box 3755
(4735 East Marginal Way South)
Seattle, WA 98124-3755
206-764-3495 tel
206-764-6602 fax
www.nws.usace.army.mil/

TULSA

U.S. Army Corps of Engineers
Tulsa District
Attention: CESWT-PE-R
1645 S. 101st East Avenue
Tulsa, OK 74128-4609
918-669-7401 tel
918-669-4306 fax
www.swt.usace.army.mil/permits/permits.cfm

VICKSBURG

U.S. Army Corps of Engineers
Vicksburg District
Attention: CEMVK-OD-F
4155 Clay Street
Vicksburg, MS 39183-3435
601-631-5276 tel
601-631-5459 fax
www.mvk.usace.army.mil/offices/od/odf

WALLA WALLA

U.S. Army Corps of Engineers
Walla Walla District
Attention: CENWW-OD-RF
201 North 3rd Street
Walla Walla, WA 99362
509-527-7151 tel
509-527-7820 fax
www.nww.usace.army.mil/html/offices/op/rf/rfhome.htm

WILMINGTON

U.S. Army Corps of Engineers
Wilmington District
Attention: CESAW-RG
P.O. Box 1890
Wilmington, NC 28402-1890
910-251-4944 tel
910-251-4025 fax
www.saw.usace.army.mil/wetlands/regtour.htm

APPENDIX I

USACE Wetland Delineation Form

DATA FORM
ROUTINE WETLAND DETERMINATION
(1987 COE Wetlands Delineation Manual)

Project/Site: ______________________ Date: ____________

Applicant/Owner: ______________________ County: ____________

Investigator: ______________________ State: ____________

Do Normal Circumstances exist on the site? Yes No Community ID: ________

Is the site significantly disturbed (Atypical Situation)? Yes No Transect ID: ________

Is the area a potential Problem Area? Yes No Plot ID: ________

(If needed, explain on reverse.)

VEGETATION

Dominant Plant Species	Stratum	Indicator	Dominant Plant Species	Stratum	Indicator
1. ________	____	____	9. ________	____	____
2. ________	____	____	10. ________	____	____
3. ________	____	____	11. ________	____	____
4. ________	____	____	12. ________	____	____
5. ________	____	____	13. ________	____	____
6. ________	____	____	14. ________	____	____
7. ________	____	____	15. ________	____	____
8. ________	____	____	16. ________	____	____

Percent of Dominant Species that are OBL, FACW or FAC (excluding FAC–)

HYDROLOGY

_ Recorded Data (Describe in Remarks):
- ___ Stream, Lake, or Tide Gauge
- ___ Aerial Photographs
- ___ Other

_ No Recorded Data Available

Field Observations:

Depth of Surface Water: ______ (in.)

Depth to Free Water in Pit: ______ (in.)

Depth to Saturated Soil: ______ (in.)

Wetland Hydrology Indicators

Primary Indicators:
- ___ Inundated
- ___ Saturated in Upper 12 Inches
- ___ Water Marks
- ___ Drift Lines
- ___ Sediment Deposits
- ___ Drainage Patterns in Wetlands

Secondary Indicators (2 or more required):
- ___ Oxidized Root Channels in Upper 12 Inches
- ___ Water-Stained Leaves
- ___ Local Soil Survey Data
- ___ FAC-Neutral Test
- ___ Other (Explain in Remarks)

Remarks:

SOILS

Map Unit Name
(Series and Phase): ______________________________

Drainage Class: ____________

Field Observations
Confirm Mapped Type? Yes No

Taxonomy (Subgroup): ______________________________

Profile Description

Depth (inches)	Horizon	Matrix Color (Munsell Moist)	Mottle Colors (Munsell Moist)	Mottle Abundance/Contrast	Texture, Concretions Structure etc.
____	____	________	________	________	________
____	____	________	________	________	________
____	____	________	________	________	________
____	____	________	________	________	________
____	____	________	________	________	________
____	____	________	________	________	________

Hydric Soil Indicators:

___ Histosol
___ Histic Epipedon
___ Sulfidic Odor
___ Aquic Moisture Regime
___ Reducing Conditions
___ Gleyed or Low-Chroma Colors

___ Concretions
___ High Organic Content in Surface Layer in Sandy Soils
___ Organic Streaking in Sandy Soils
___ Listed on Local Hydric Soils List
___ Listed on National Hydric Soils List
___ Other (Explain in Remarks)

Remarks:

WETLAND DETERMINATION

Hydrophytic Vegetation Present? Yes No (Circle)
Wetland Hydrology Present? Yes No
Hydric Soils Present? Yes No

(Circle)
Is this Sampling Point Within a Wetland? Yes No

Remarks:

Approved by HQUSACE 3/92

APPENDIX J

Section 404/ Section 10 Permit Application Form

APPLICATION FOR DEPARTMENT OF THE ARMY PERMIT (33 CFR 325)	OMB APPROVAL NO. 0710-0003 Expires

The Public burden for this collection of information is estimated to average 10 hours per response, although the majority of applications should require 5 hours or less. This includes the time for reviewing instructions, searching existing data sources, gathering and maintaining the data needed and completing and reviewing the collection of information. Send comments regarding this burden estimate or any other aspect of this collection of information, including suggestions for reducing this burden, to Department of Defense, Washington Headquarters Service Directorate of Information Operations and Reports, 1215 Jefferson Davis Highway, Suite 1204, Arlington, VA 22202-4302; and to the Office of Management and Budget, Paperwork Reduction Project (0710-0003), Washington, DC 20503. Respondents should be aware that notwithstanding any other provision of law no person shall be subject to any penalty for failing to comply with a collection of information if it does not display a currently valid OMB control number. Please DO NOT RETURN your form to either of those addresses. Completed applications must be submitted to the District Engineer having jurisdiction over the location of the proposed activity.

PRIVACY ACT STATEMENT

Authorities: Rivers and Harbors Act, Section 10, 33 USC 403; Clean Water Act, Section 404, 33 USC 1344; Marine Protection, Research and Sanctuaries Act, Section 103, 33 USC 1413. Principal Purpose: Information provided on this form will be used in evaluating the application for a permit. Routine Uses: This information may be shared with the Department of Justice and other federal, state, and local government agencies. Submission of requested information is voluntary, however, if information is not provided the permit application cannot be evaluated nor can a permit be issued.

One set of original drawings or good reproducible copies which show the location and character of the proposed activity must be attached to this application (see sample drawings and instructions) and be submitted to the District Engineer having jurisdiction over the location of the proposed activity. An application that is not completed in full will be returned.

(ITEMS 1 THRU 4 TO BE FILLED BY THE CORPS)

1. APPLICATION NO.	2. FIELD OFFICE CODE	3. DATE RECEIVED	4. DATE APPLICATION COMPLETED

(ITEMS BELOW TO BE FILLED BY APPLICANT)

5. APPLICANT'S NAME	8. AUTHORIZED AGENT'S NAME AND TITLE *(an agent is not required)*
6. APPLICANT'S ADDRESS	9. AGENT'S ADDRESS
7. APPLICANT'S PHONE NUMBERS WITH AREA CODE a. Residence b. Business	10. AGENT'S PHONE NUMBERS WITH AREA CODE a. Residence b. Business

11. STATEMENT OF AUTHORIZATION

I hereby authorize ____________________ to act in my behalf as my agent in the processing of this application and to furnish, upon request, supplemental information in support of this permit application.

APPLICANT'S SIGNATURE DATE

NAME, LOCATION AND DESCRIPTION OF PROJECT OR ACTIVITY

12. PROJECT NAME OR TITLE *(see instructions)*

13. NAME OF WATERBODY, IF KNOWN *(if applicable)*	14. PROJECT STREET ADDRESS *(if applicable)*
15. LOCATION OF PROJECT COUNTY STATE	

16. OTHER LOCATION DESCRIPTIONS, IF KNOWN (see instructions)

17. DIRECTIONS TO THE SITE

ENG FORM 4345, Jul 97 EDITION OF SEP 94 IS OBSOLETE

18. Nature of Activity *(Description of project, include all features)*

19. Project Purpose *(Describe the reason or purpose of the project, see instructions)*

USE BLOCKS 20-22 IF DREDGED AND/OR FILL MATERIAL IS TO BE DISCHARGED

20. Reason(s) for Discharge

21. Type(s) of Material Being Discharged and the Amount of Each Type in Cubic Yards

22. Surface Area in Acres of Wetlands or Other Waters Filled *(see instructions)*

23. Is Any Portion of the Work Already Complete? Yes ________ No ________ IF YES, DESCRIBE THE COMPLETED WORK

24. Addresses of Adjoining Property Owners, Lessees, etc., Whose Property Adjoins the Waterbody (if more than can be entered here, please attach a supplemental list).

25. List of Other Certifications or Approvals/Denials Received from other Federal, State, or Local Agencies for Work Described in This Application

AGENCY	TYPE APPROVAL*	IDENTIFICATION NUMBER	DATE APPLIED	DATE APPROVED	DATE DENIED

*Would include but is not restricted to zoning, building and flood plain permits

26. Application is hereby made for a permit or permits to authorize the work described in this application. I certify that the information in this application is complete and accurate. I further certify that I possess the authority to undertake the work described herein or am acting as the duly authorized agent of the applicant.

SIGNATURE OF APPLICANT ________ DATE ________ SIGNATURE OF AGENT ________ DATE ________

The application must be signed by the person who desires to undertake the proposed activity (applicant) or it may be signed by a duly authorized agent if the statement in block 11 has been filled out and signed.

18 U.S.C. Section 1001 provides that: Whoever, in any manner within the jurisdiction of any department or agency of the United States knowingly and willfully falsifies, conceals, or covers up any trick, scheme, or disguises a material fact or makes any false, fictitious or fraudulent statements or representations or makes or uses any false writing or document knowing same to contain any false, fictitious or fraudulent statements or entry, shall be fined not more than $10,000 or imprisoned not more than five years or both.

Instructions For Preparing A Department of the Army Permit Application

Blocks 1 thru 4. To be completed by Corps of Engineers.

Block 5. APPLICANT'S NAME. Enter the name of the responsible party or parties. If the responsible party is an agency, company, corporation or other organization, indicate the responsible officer and title. If more than one party is associated with the application, please attach a sheet with the necessary information marked Block 5.

Block 6. ADDRESS OF APPLICANT. Please provide the full address of the party or parties responsible for the application. If more space is needed, attach an extra sheet of paper marked Block 6.

Block 7. APPLICANT PHONE NUMBERS. Please provide the number where you can usually be reached during normal business hours.

Blocks 8. AUTHORIZED AGENT'S NAME AND TITLE. Indicate name of individual or agency, designated by you, to represent you in this process. An agent can be an attorney, builder, contractor, engineer or any other person or organization. Note: An agent is not required.

Blocks 9 AND 10. AGENT'S ADDRESS AND TELEPHONE NUMBER. Please provide the complete mailing address of the agent, along with the telephone number where he/she can be reached during normal business hours.

Block 11. STATEMENT OF AUTHORIZATION. To be completed by applicant if an agent is to be employed.

Block 12. PROPOSED PROJECT NAME OR TITLE. Please provide name identifying the proposed project (i.e., Landmark Plaza, Burned Hills Subdivision or Edsall Commercial Center).

Block 13. NAME OF WATERBODY. Please provide the name of any stream, lake, marsh or other waterway to be directly impacted by the activity. If it is a minor (no name) stream, identify the waterbody the minor stream enters.

Block 14. PROPOSED PROJECT STREET ADDRESS. If the proposed project is located at a site having a street address (not a box number), please enter it here.

Block 15. LOCATION OF PROPOSED PROJECT. Enter the county and state where the proposed project is located. If more space is required, please attach a sheet with the necessary information marked Block 15.

Block 16. OTHER LOCATION DESCRIPTIONS. If available, provide the Section, Township and Range of the site and/or the latitude and longitude. You may also provide description of the proposed project location, such as lot numbers, tract numbers or you may choose to locate the proposed project site from a known point (such as the right descending bank of Smith Creek, one mile down from the Highway 14 bridge). If a large river or stream, include the river mile of the proposed project site, if known.

Block 17. DIRECTIONS TO THE SITE. Provide directions to the site from a known location or landmark. Include highway and street numbers as well as names. Also provide distances from known locations and any other information that would assist in locating the site.

Block 18. NATURE OF ACTIVITY. Describe the overall activity or project. Give appropriate dimensions of structures such as wingwalls, dikes (identify the materials to be used in construction, as well as the methods by which the work is to be done), or excavations (length, width, and height). Indicate whether discharge of dredged or fill material is involved. Also, identify any structure to be constructed on a fill, piles or float supported platforms.

The written descriptions and illustrations are an important part of the application. Please describe, in detail, what you wish to do. If more space is needed, attach an extra sheet of paper marked Block 18.

Block 19. PROPOSED PROJECT PURPOSE. Describe the purpose and need for the proposed project. What will it be used for and why? Also include a brief description of any related activities to be developed as the result of the proposed project. Give the approximate dates you plan to both begin and complete all work.

Block 20. REASON(S)FOR DISCHARGE. If the activity involves the discharge of dredged and/or fill material into a wetland or other waterbody, including the temporary placement of material,explain the specific purpose of the placement of the material (such as erosion control).

Instructions For Preparing A
Department of the Army Permit Application

Block 21. TYPES OF MATERIAL BEING DISCHARGED AND THE AMOUNT OF EACH TYPE IN CUBIC YARDS. Describe the material to be discharged and amount of each material to be discharged within Corps jurisdiction.Please be sure this description will agree with your illustrations. Discharge material includes: rock, sand, clay, concrete, etc.

Block 22. SURFACE AREAS OF WETLANDS OR OTHER WATERS FILLED. Describe the area to be filled at each location. Specifically identify the surface areas, or part thereof, to be filled. Also include the means by which the discharge is to be done (backhoe, dragline, etc.). If dredged material is to be discharged on an upland site, identify the site and the steps to be taken (if necessary) to prevent runoff from the dredged material back into a waterbody. If more space is needed, attach an extra sheet of paper marked Block 22.

Block 23. IS ANY PORTION OF THE WORK ALREADY COMPLETE? Provide any background on any part of the proposed project already completed. Describe the area already developed, structures completed, any dredged or fill material already discharged, the type of material, volume in cubic yards, acres filled, if a wetland or other waterbody (in acres or square feet). If the work was done under an existing Corps permit, identify the authorization if possible.

Block 24. NAMES AND ADDRESSES OF ADJOINING PROPERTY OWNERS, LESSEES, ETC., WHOSE PROPERTY ADJOINS THE PROJECT SITE. List complete names and full mailing addresses of the adjacent property owners (public and private)lessees, etc., whose property adjoins the waterbody or aquatic site where the work is being proposed so that they may be notified of the proposed activity (usually by public notice). If more space is needed, attach an extra sheet of paper marked "Block 24".

Block 25. INFORMATION ABOUT APPROVALS OR DENIALS BY OTHER AGENCIES. You may need the approval of other Federal, state or local agencies for your project. Identify any applications you have submitted and the status, if any (approved or denied)of each application. You need not have obtained all other permits before applying for a Corps permit.

Block 26. SIGNATURE OF APPLICANT OR AGENT. The application must be signed by the owner or other authorized party (agent). This signature shall be an affirmation that the party applying for the permit possesses the requisite property rights to undertake the activity applied for (including compliance with special conditions, mitigation, etc.).

DRAWINGS AND ILLUSTRATIONS - DRAWINGS AND ILLUSTRATIONS

Three types of illustrations are needed to properly depict the work to be undertaken. These illustrations or drawings are identified as a Vicinity Map, a Plan View or a Typical Cross-Section Map. Identify each illustration with a figure or attachment number.

Please submit one original, or good quality copy, of all drawings on 8 1/2 x 11 inch plain white paper (tracing paper or film may be substituted). Use the fewest number of sheets necessary for your drawings or illustrations.

Each illustration should identify the project, the applicant, and the type of illustration (vicinity map, plan view or cross-section). While illustrations need not be professional (many small, private project illustrations are prepared by hand), they should be clear, accurate and contain all necessary information.

APPENDIX K

Section 404 Clean Water Act Case Law Summaries

Zabel v. Tabb

430 F. 2d 199 (5th Cir. 1970)

Zabel and Russell (Plaintiffs) instituted a law suit to compel the U.S. Army Corps of Engineers (USACE) to issue a Section 10 Rivers and Harbors Act (Section 10) permit to dredge and fill in the navigable waters of Boca Ciega Bay, in Pinellas County near St. Petersburg, Florida. The Plaintiffs desired to dredge and fill on their property in the Bay for a trailer park, with a bridge or culvert to their adjoining upland. The Plaintiffs applied to USACE for a permit which was denied on factually substantial ecological reasons even though the project would not interfere with navigation, flood control, or the production of power. The trial court held that USACE had no authority to consider anything except interference with navigation. Since there was no interference with navigation, the trial court ordered USACE to issue the permit regardless of evidence related to environmental harm. On appeal the Court of Appeals (Court) reversed, holding that nothing in the statutory structure compels USACE to close its eyes to all that others see or think they see. USACE was entitled, if not required, to consider ecological factors and, being persuaded by them, to deny a permit "in protection of the public interest that might have been granted routinely five, ten, or fifteen years ago before man's explosive increase made all, including Congress, aware of civilization's potential destruction from breathing its own polluted air and drinking its own infected water and the immeasurable loss from silent-spring-like disturbance of nature's economy."

The Court also dismissed the Plaintiffs taking claim on the basis of navigable servitude. The rationale was that due to the fact of the waters and underlying land being subject to the paramount servitude (navigable servitude) in the Federal government, there was no taking.

United States v. Holland

373 F. Supp. 665 (M.D. Fla. 1974)

Defendant (Holland) was ordered by USACE to stop the illegal discharge of sand, dirt, dredged spoil, and biological materials into man-made canals and mangrove wetlands, as the fill would result in the elimination of the normal ebb and flow of tides over the property. While USACE claimed that its authority to regulate water pollution under the Federal Water Pollution Control Act (FWPCA) of 1972 encompassed these intertidal wetlands, Holland argued that his fill activities did not trigger USACE jurisdiction, because he was not placing fill in a navigable waterway. The court agreed with USACE, finding that the mangrove wetland's hydrologic connection to important fisheries warranted protection under the FWPCA.

The court reasoned that, under the FWPCA, USACE was given the authority to regulate "any discharge of any pollutant by any person" to navigable waters and that this included all "waters of the United States." The court relied on amendments to the FWPCA that expanded the definition of navigability to include both navigable waters and wetlands adjacent to navigable waters and wetlands that have an effect on interstate commerce. The court concluded that Congress is not limited by the "navigable waters" test in its authority to control pollution under the Commerce Clause. Instead, the court stated that the FWPCA is "intended to reach water-bodies such as these canals," as the crucial information is not that the canals are man-made, but that they were used to "convey pollutants without a permit."

The court also concluded that the mean high-water line is no limit to federal authority under the FWPCA, reasoning that "while the line remains a valid demarcation for other purposes, it has no rational connection to the aquatic ecosystems which the FWPCA is intended to protect.... Congress has wisely determined that federal authority over water pollution properly rests on the Commerce Clause and... the Commerce Clause gives Congress ample authority to reach activities above the mean high-water line that pollute the waters of the United States."

In requiring Holland to obtain a permit for the placement of fill, the court believed it had not restricted him from developing the land, but simply had required him to obtain the correct permit under the law. In the final opinion, the court suggested that any costs incurred by Holland to obtain a permit would be inconsequential compared to the cost of neglecting the area's ecological importance.

Natural Resources Defense Council v. Callaway

392 F. Supp. 685 (D.D.C. 1975)

The secretary of the Army (Callaway) and the chief of USACE were found to have exceeded their authority by limiting the statutory definition of "navigable waters" under the Clean Water Act (CWA) by adopting regulations that defined navigable waters in its literal sense. In challenging this self-imposed limitation on USACE's regulatory jurisdiction, the Natural Resources Defense Council (NRDC) brought action against the defendants, claiming that Congress had intended for USACE's jurisdiction under "navigable waters" to be exercised to the "maximum extent permissible" under the Commerce Clause of the Constitution. This broad definition would include territorial seas and waters that were non-navigable but shared a hydrologic connection to navigable waters. The court agreed and ordered USACE to revise its definition in accordance with the mandate of the act as interpreted by the court. The court stated that the term should not be limited to the "traditional tests of navigability," and that by doing so the defendants has "acted unlawfully and in derogation of their responsibilities under the Clean Water Act."

Kaiser Aetna v. United States

444 U.S. 164 (1979)

In this takings case involving navigable servitude, Kaiser Aetna (Petitioner) owned Kaupa Pond on the island of Oahu, Hawaii, contiguous to a navigable bay and the Pacific Ocean. The Petitioner developed the pond into a marina. In doing so, the pond had been connected to the bay by dredging. Prior to dredging the Petitioner contacted USACE who informed it that no permit was required.

Subsequently, a dispute arose between the Petitioner and USACE over whether a Section 10 of the Rivers and Harbors Act (Section 10) permit was required for additional work and the issue of allowing public access to the pond. This dispute resulted in the government bringing a suit against the Petitioner attempting to require public access to the pond based on the pond being subject to navigable servitude as a result of it being connected with navigable waters via the dredging that had occurred. The trial court held that the pond was a "navigable water of the United States" and therefore, subject to USACE's Section 10 jurisdiction. The trial court also held that despite navigable servitude, the government lacked the power to require public access to the pond without payment of just compensation.

The Ninth Circuit agreed with the trial court's decision that the pond was subject to USACE's Section 10 jurisdiction but reversed the trial court's holding that the navigable servitude did not require petitioners to grant public access to the pond. The Ninth Circuit reasoned that the "federal regulatory authority over navigable waters... and the right of public use cannot consistently be separated. It is the public right of navigational use that renders regulatory control necessary in the public interest."

The Supreme Court (Court) framed the issue as "whether the [Ninth Circuit] erred in holding that petitioners' improvements to Kuapa Pond caused its original character to be so altered that it became subject to an overriding federal navigational servitude, thus converting into a public aquatic park that which petitioners had invested millions of dollars in improving on the assumption that it was a privately owned pond.... " The Court reversed the Ninth Circuit. In doing so it stated that if the government wishes to make what was formerly Kuapa Pond into a public aquatic park after Petitioners have proceeded as far as they have here, it may not, without invoking its eminent domain power and paying just compensation. The Court also stated that although the pond falls within the definition of "navigable waters," the government's attempt to create a public right of access to the improved pond goes so far beyond ordinary regulation or improvement for navigation.

United States v. Board of Trustees of Florida Keys Community College

531 F. Supp. 267 (S.D. Fla. 1981)

This strict liability case involves a suit brought by the United States (U.S.) to restore allegedly filled wetlands. The wetlands were filled by a construction company with the consent of the Board of Trustees of Florida Keys Community College (Defendants), the owner of the wetlands. The U.S sought restoration of an open water slough on the edge of the Florida Bay, which it contended was filled in violation of the Rivers and Harbors Act and the Clean Water Act. The Defendants conceded their failure to obtain a Clean Water Act permit but contended that the violation was not intentional and that the remedy was unwarranted. The Court begins it opinion stating that "'[b]eauty is in the eye of the beholder.' Here one party's unsightly swamp was another party's endangered wildlife refuge."

In finding the Defendants liable, the Court stated that civil liability (strict liability) under the Clean Water Act and the Rivers and Harbors Act is predicated on either (1) performance, or (2) responsibility for or control over performance of the work, in the absence of the necessary federal permit. This was based on the rationale that civil liability under the CWA is not limited to intentional violations. "The regulatory provisions of the [Clean Water Act] were written without regard to intentionality,... making the person responsible for the discharge of any pollutant strictly liable.... Willful or negligent violations of the [Clean Water Act] are separately punishable by criminal penalties." The Court did however, in this instance find that the remedy of restoring the wetland was inappropriate.

Deltona Corp. v. Alexander

682 F. 2d 888 (11 Cir. 1982)

Defendant (Deltona) appealed USACE's denial of Section 404 permits for dredge and fill activities associated with their development project. In 1964, the Deltona Corporation had purchased 10,300 acres of island land in Florida, intending to convert the entire area into a planned community development with "finger canals" providing aquatic access to waterfront homes. When development began, USACE authority was limited to the Rivers and Harbors Act, but in the latter stages of development in the early 1970s, USACE had additional authority and responsibilities under the newly passed CWA and NEPA. Section 404 of the CWA required USACE to consider environmental factors when issuing permits authorizing dredge and fill activities such as those associated with the defendant's project.

In 1973, after obtaining local and state permits, Deltona applied to USACE for a CWA Section 10 Streambed Alteration Agreement (now administered by state agencies) and Section 404 permits for the last three construction areas as required by the CWA. USACE denied two of the three Section 404 permits, stating that the proposed project's destruction of 2,152 acres of mangrove swamps and 735 acres of bay bottom was "contrary to the Corps' wetland conservation policies." One of the permits was granted because construction had proceeded beyond the point where significant natural resources could be conserved.

Deltona argued that USACE did not have the authority to deny the remaining fill permits, claiming that the agency had known about the development activities and had "unofficially endorsed them" in earlier years by granting Section 10 permits under the Rivers and Harbors Act. The court disagreed, holding that USACE had acted within its authority when it denied the fill permit. The court noted that the government cannot be stopped "when it exercises its sovereign powers for the benefit of the public." The court further noted that "[t]he act of granting a Section 404 permit is unquestionably an exercise of the government's sovereign power to protect the public interest. In fact, the entire rationale behind the Section 404 permit is to ensure that the public interest in environmental safety and quality is preserved." The defendant also argued for a trial to establish the extent of USACE's jurisdiction over upland versus wetlands in the project area, but the court simply stated that a wetland determination is a matter best left to the expertise of USACE. Instead of a trial, the court recommended that Deltona enter into a dialogue with USACE to resolve such issues.

Hough v. Marsh

557 F. Supp. 74 (D.Mass. 1982)

In 1976, a .25 acre wetland was identified by developers as a section of land that must be filled as an essential part of the construction of two residences and a tennis court on a larger three-acre plot of land on Martha's Vineyard, Massachusetts. After receiving local authorization for the development, the landowner deposited fill without applying for a Section 404 permit, claiming that he had never received a letter from USACE. USACE issued a cease-and-desist order and ordered the defendant to apply for a fill permit as required by the CWA. In May 1981, USACE announced a decision to issue the permit despite 259 written objections to the project during the public notice period. The plaintiffs, a group of 10 residents, claimed that in issuing the permit, USACE had violated three environmental regulations: NEPA, the CWA, and the National Historic Preservation Act (NHPA).

The district court agreed and held in favor of the plaintiffs on five separate issues. First, the court held that USACE had failed to adequately address an issue of public concern through a hearing, specifically, the fact that the fill and eventual construction of the residences would have an impact on several residents' current view of a coastal lighthouse. Second, the court held that USACE failed to require the applicant to supply viable alternatives to the project as required by the CWA. The agency had only submitted one piece of evidence supporting its finding of no alternatives. The evidence consisted of a written letter from a realtor that did not identify any other potential lots for development in or around the desired development area or why the buildings needed to be constructed side by side. Third, the court found that USACE failed to require the permit applicants to obtain two necessary local permits prior to issuing the Section 404 permit. The planning board had established a zoning rule to protect tidal marsh that required the issuance of a special permit to fill such areas, and an additional local law established a ban against filling beach areas for the purposes of constructing a private residence.

In its fourth holding, the court found that USACE did not adequately address cumulative and economic impacts that would result from granting the fill permit, because the agency's analysis did not discuss that a local sightseeing bus would have to drastically amend its itinerary to account for the loss of the view of the historic lighthouse. This was found by the court to be a significant impact warranting discussion. Lastly, the court held that USACE had violated the NHPA by placing only one phone call to the Massachusetts Historical Commission to discuss the lighthouse's status in regard to the National Register of Historic Places. In addition, USACE did not address impacts on the lighthouse in its permit approval and did not list the lighthouse in its discussion of the area's historic values. The court stated that, although the lighthouse was not on the official list, the NHPA was still triggered and USACE was therefore required to fully discuss the proposed project's effects on the historic feature. In light of its five holdings, the court ordered USACE to reevaluate issuing the Section 404 permit after holding a public hearing and addressing impacts in accordance with all applicable regulations.

Buttrey v. United States

690 F. 2d 1186 (5th Cir. 1982)

John Buttrey and John Buttrey Development Inc. (Plaintiffs) were developers of a subdivision in Slidell, Louisiana, known as Magnolia Forest. On May 5, 1980, USACE issued a cease-and-desist order advising Plaintiffs that his placement of fill in a wetland area was regulated by USACE and that initiating such work without a permit violated Section 404 of the Clean Water Act. In November of the same year, USACE issued another cease-and-desist order related to the construction of a levee and fill of a wetland that triggered both Section 404 of the Clean Water Act and Section 10 of the Rivers and Harbors Act. Plaintiffs filed a suit for declaratory and injunctive relief challenging the jurisdiction of USACE among other claims. The trial court dismissed the Plaintiffs' suit and the Plaintiff appealed. The only issue on brought to the Court of Appeals (Court) was the constitutionality of Congress' delegation of the authority embodied in Section 404 of the Clean Water Act to USACE. The Court held that USACE has authority to issue a cease-and-desist order to developer for construction and dredging in wetlands without permit, because Congress properly delegated authority to USACE in the Clean Water Act, and that delegation does not violate the Commerce Clause of the U.S. Constitution.

1902 Atlantic Ltd. v. Hudson,

574 F. Supp. 1381 (E.D. Va. 1983)

In this regulatory takings case, landowner, 1902 Atlantic Ltd. (Atlantic) requested a Section 404 permit from USACE to fill approximately 11 acres of a tidally influenced borrow pit. The man-made, non-navigable borrow pit had been created during excavation activities associated with the creation of a nearby highway overpass. Subsequent to the excavation of the borrow pit, unknown persons created a channel that connected the borrow pit to a tributary of Chesapeake Bay, allowing tidal waters to enter the borrow pit. The tidal influence and saltwater provided habitat and soil conditions that supported several species of wetland vegetation. The wetland, as it was defined by USACE, covered approximately 32,000 square feet (slightly less than .75 of an acre), and was considered jurisdictional under both Section 404 of the CWA and Section 10 of the Rivers and Harbors Act. Atlantic proposed to fill the borrow pit to create an upland area for industrial development in accordance with local zoning regulations, to raise the tax base of the city, and to provide jobs.

USACE denied the permit request numerous times despite Atlantic's modification of the proposed development to lessen impacts on wetlands from 32,000 square feet to approximately 15,000 square feet, with mitigation composed of replanting all impacted wetlands at a one-to-one ratio. USACE's stated basis for denying the project was that it was contrary to the Environmental Protection Agency's (EPA's) Section 404(b) guidelines and USACE's Wetlands Policy. It was argued that the 15,000 square feet of wetlands would be destroyed, the creation of an industrial site was not a water-dependent activity, and other alternative upland sites were available for the project. Atlantic filed suit, claiming that USACE had acted in an arbitrary and capricious manner by denying its permit. In addition, Atlantic argued that the denial of the permit by USACE was a "taking" in violation of the Fifth Amendment.

The court phrased the issue presented to it as "whether [USACE] may prevent the owner of this property from making a non-water dependent use of the borrow pit where no economically feasible water-dependent purpose for the property exists." The plaintiff conceded USACE's jurisdiction over the borrow pit pursuant to Section 404 but contested its jurisdiction under Section 10 of the Rivers and Harbors Act of 1899. The court first recognized that the definition of navigable waters varies depending on which context (statute) is used. The court then determined that USACE did not have jurisdiction pursuant to Section 10 of the Rivers and Harbors Act, based on the act's dominant theme of promoting and protecting navigation. The court found that the borrow pit was not navigable. In fact, the public has not and does not use the borrow pit for navigable purposes, it has no value or benefit to commerce, and the filling of the pit in accordance with Atlantic's revised plan will not have any adverse effect on commerce, on navigation, or on the navigable capacity of any nearby waters. Based on these facts, the court rejected USACE's argument for a strict application of the ebb and flow of the tide test for determining "navigable waters." The court went on to hold that USACE acted arbitrarily and capriciously in denying Atlantic's fill permit. The court reasoned that USACE had placed undue importance on the water dependency regulation and failed to properly consider all factors, and therefore failed to conduct a proper balancing test. The court also found the denial of the permit to amount to a regulatory taking.

The Avoyelles Sportsmen's League, Inc. v. Marsh

715 F. 2d 897 (5th Cir. 1983)

In this case, which held that backhoes and bulldozers can be "point sources," a landowner began clearing a 20,000-acre tract of forested land for conversion to agriculture use. USACE, upon being informed, ordered a halt to the clearing pending a wetland delineation. After USACE examined the vegetation, soil conditions, and hydrology of the 20,000-acre tract, USACE determined that approximately 35 percent of the tract was a wetland. Believing that USACE had greatly underestimated the wetland acreage, Avoyelles Sportsmen's League (Avoyelles) brought an action to enjoin the clearing of the land. Avoyelles claimed that the clearing would result in the discharge of dredged and fill material into waters of the United States, violating Sections 301(a), 404, and 402 of the CWA. Avoyelles also requested that the court declare the entire tract a wetland and require the defendants to obtain the appropriate permits prior to continuing its activities. The district court issued both a temporary restraining order and subsequently, a preliminary injunction pending trial. As part of the order granting the preliminary injunction, the district court ordered the U.S. Environmental Protection Agency (EPA) (oversight agency of USACE) to prepare a final wetlands determination requiring that all interested parties have the opportunity to participate in the administrative proceedings. The EPA's final wetland determination concluded that approximately 80 percent of the land was a wetland.

At trial, the district court conducted a *de novo* review, as opposed to relying on the EPA's final wetland determination's administrative record, holding that a Section 404 permit was required for the land-clearing activities and that over 90 percent of the land was a wetland. The Court of Appeals (Court) determined that the trial court had improperly conducted a *de novo* review of the EPA's final wetland determination. The Court held that the proper standard of review was the arbitrary and capricious standard. Instead of remanding the case back to the trial court to apply the appropriate standard, the Court conducted its own review of the administrative record and determined that the EPA's final wetland determination was not arbitrary and capricious. In doing so the Court of Appeals also determined that the use of backhoes and bulldozers in the clearing process constituted point sources within the meaning of the CWA, and therefore upheld the trial court's determination that a Section 404 permit was required prior to resuming clearing activities.

Buttrey v. United States

690 F. 2d 1186 (5th Cir. 1982),
cert. denied, 461 U.S. 927 (1983)

In 1980, USACE issued two cease-and-desist orders to the defendant (Buttrey) for illegal dredge and fill activities that occurred as part of Buttrey's subdivision development in wetland areas adjacent to the Morgan River in Louisiana. Despite the cease-and-desist orders, surveillance, and unannounced inspections by USACE, the defendant did not comply with orders to apply for a Section 404 permit under the CWA and a Section 10 permit of the Rivers and Harbors Act. Instead, Buttrey brought suit against USACE, claiming that Congress' delegation of authority under Section 404 of the CWA was unconstitutional as he believed dredge-and-fill regulations were "not essential or necessary to the purpose of an Army."

The court disagreed, holding that it was constitutional for Congress to grant authority under the Commerce Clause to USACE to regulate the discharge of dredged or fill material into navigable waters of the United States. The court reasoned that from the earliest days of the U.S. government improvement of navigable waterways and flood control management, which includes the regulation of adjacent wetlands, have historically been performed by USACE under its civil functions. Therefore, the court did not find USACE's exercise of regulatory powers under Section 404 to be "the kind of thing which would offend" the constitutionally required separation of military and civil power. Instead, the court held that, because the constitutional power relied upon for the CWA was the commerce clause and not the war powers clause and because administration of Section 404 permits did not "infringe upon any other provisions of the Constitution," the plaintiff's claim was dismissed.

Shoreline Associates v. Marsh

555 F. Supp. 169 (D.C. Md. 1983)

A developer, Shoreline Associates (Shoreline) asked the court to reverse and remand a determination of USACE denying a Section 10 permit under the Rivers and Harbors Act and a Section 301 permit under the CWA to dredge and fill 8.2 acres of tidal wetland. Shoreline argued that USACE's denial of the permit was arbitrary and capricious and not supported by substantial evidence, and constituted an abuse of discretion.

The EPA, U.S. Fish and Wildlife Service, and National Marine Fisheries Service, along with the public and conservation groups, independently filed written comments with USACE calling for a denial of Shoreline's permit application. Shoreline was provided the opportunity to review these written comments, provide written responses, submit written expert testimony in support of its application, and meet with USACE to discuss the application in person. USACE denied the permit, finding the activities proposed by Shoreline to be contradictory to USACE and EPA guidelines and not in the public interest.

The court held that in denying the permit USACE complied with procedural due process requirements and followed the appropriate regulatory criteria by evaluating "probable impacts which the proposed activity may have on the public interest." The applicable regulations required that "[t]he benefits which reasonably may be expected to accrue from the proposal must be balanced against its reasonably forseeable detriments." Factors for making such a determination include, "conservation, economics, aesthetics, general environmental concerns, historic values, fish and wildlife values, flood damage prevention, land use, navigation, recreation, water supply, water quality, energy needs, safety, food production, and, in general, the needs and welfare of the people." In reviewing the record, the court was presented with site-specific evidence that each of these criteria had been evaluated with respect to Shoreline's project. The court also determined that Shoreline had been given every opportunity afforded to it under procedural due process to communicate with USACE while it was making a permit approval determination. The court noted Shoreline had been given several opportunities to make written submissions and was also offered several opportunities for face-to-face informal meetings. Further, Shoreline never requested an adjudicatory hearing on the matter. The court felt that USACE provided adequate opportunities for "written communications" and informal meetings and therefore found no error in either the procedural due process or the factual basis of USACE's decision to deny Shoreline's permit application.

United States v. Tull

769 F. 2d 182 (4th Cir. 1985)

The U.S. brought suit against a Virginia developer for unpermitted discharge of fill material into navigable waters and wetlands adjacent to navigable waters. Between 1975 and 1982, the defendant (Tull) developed numerous properties for resale as mobile home lots on Chincoteague Island in Virginia by filling wetlands with soil and grading the lots. The properties all had wetland characteristics and were adjacent to a navigable waterway, the Fowling Gut. Despite USACE's issuance of a cease-and-desist order and injunction on certain properties, the defendant continued fill activities without applying for a Section 404 permit. The defendant's activities included the fill of an extension of the navigable waterway referred to as Fowling Gut. The Court of Appeals (Court) in affirming the trail court's decision held that: (1) regulation of the wetlands at issue did not violate the Commerce Clause; (2) the definition of wetlands is not unconstitutionally vague and is sufficiently definitive to give a person of ordinary intelligence fair notice; (3) the Government is not equitably estopped based on the assertion that USACE misled the property owner in regard to whether a permit was required; and (4) sufficient evidence was present to the trial court to support its conclusion that the Fowling Gut was subject to the ebb and flow of the tide and therefore was a navigable water. The Court also held that the property owner was not entitled to a jury trial; however, this last issue was reversed by the Supreme Court (*Tull v. United States,* 481 U.S. 412 (1987)).

Quivira Mining Co. v. U.S. Environmental Protection Agency

765 F. 2d 126 (10th Cir. 1985), cert. denied, 474 U.S. 1055 (1986)

In this jurisdictional case, Quivira Mining Company (Quivira) appealed two determinations by the EPA denying review of the issuance of discharge permits under the CWA. Quivira challenged EPA's authority to regulate discharge of pollutants from uranium mining and milling activities into gullies or arroyos. Quivira argued that the arroyos at issue were not "waters of the United States," and thus outside EPA's permit jurisdiction under the CWA. Both intermittent waterbodies had a surface connection with navigable waters only occasionally during times of heavy rainfall. Absent heavy rains, the arroyos contain surface flows for only a short distance from Quivira's points of discharge.

In support of its position, Quivira argued that the court was required to conduct a *de novo* review of EPA's factual determinations because of their jurisdictional nature. The court rejected this argument and applied the more deferential "substantial evidence" standard, emphasizing that Quivira's position would give the EPA's determinations no special weight in any appeal challenging the issuance of a discharge permit. Further, the court stated that adopting such an argument "would contradict the clear meaning of the Administrative Procedures Act and would deprive the appellate courts of the expertise of the agency in determining such matters."

Next, the court applied the substantial evidence standard to the issue of whether the discharges were subject to regulation under the CWA. After reviewing the administrative record, the court held that substantial evidence supported that the arroyos, despite not being navigable-in-fact, were waters of the United States and thus subject to EPA regulation. In reaching its decision, the court relied on case law interpreting the CWA's reference to navigable waters as extending as far as possible under the Commerce Clause. Factually, the court relied on the existence of surface flows that occasionally provided surface connection with navigable waters. The court also made note of the fact that waters from the arroyos flow into underground aquifers that in turn flow into navigable-in-fact streams. Based on these facts, the court concluded that sufficient impacts on interstate commerce existed to satisfy the Commerce Clause.

Riverside Irrigation District v. Andrews

758 F. 2d 508 (10th Cir. 1985)

This case concerns the denial of use of a nationwide permit, based on indirect impacts of the discharge related to decreases in water quantity and their impact on endangered species. USACE denied a proposal for a nationwide permit for the deposit of dredged material required for the construction of Wildcat Dam and Reservoir on Wildcat Creek, a tributary of the South Platte River. The developers had hoped to obtain a nationwide permit to discharge non-toxic dredged material during construction of a dam, but the permit request was denied and developers were instructed by USACE to apply for an individual permit. USACE's basis for denying the nationwide permit was that the proposed deposit did not meet the nationwide permit's requirement that the discharge not affect USFWS-designated critical habitat. In this case, USACE found that the "increased use of water that the resulting reservoir would facilitate would deplete the stream flow and endanger a critical habitat of the whooping crane, an endangered species."

The developer filed a lawsuit seeking declaratory and injunctive relief and review of USACE's action. The developer argued that USACE had exceeded its authority when, in determining the applicability of a nationwide permit, it considered the effect of decreases in water quantity caused by the storage of water in the reservoir. The developer's position was that USACE was required to look only at the direct, onsite effects of the discharge, particularly effects on water quality and not indirect, downstream effects on water quantity. The developer also argued that to the extent that USACE could look to effects on water quantity, those effects had to be direct effects of the discharge as opposed to the indirect effects resulting from the overall construction project once complete (less downstream water). The developer also asserted that USACE could not deny the permit because such a denial contradicts the state's rights to allocate water in its jurisdiction as established by the Wallop Amendment (The Wallop Amendment was created to protect the water allocation right afforded to states by Congress.)

The court held that USACE had not violated its authority in denying the nationwide permit based on indirect effects. The court upheld USACE's finding that the proposed project would have an effect, "albeit indirect," on critical habitat of an endangered species, and thus was ineligible for a nationwide permit. In explaining its decision, the court pointed to the expressed language of the controlling statutes and regulations, authorizing USACE "to consider downstream effects of changes in water quantity as well as on-site changes in water quality in determining whether a proposed discharge qualifies for a nationwide permit." The court held that USACE had not violated the Wallop Amendment. The decision was based on express language in the CWA stating that USACE is required to consider changes in water quantity during permit review and the fact that the Wallop Amendment cannot nullify the specific grant of jurisdiction found in the CWA.

United States v. Riverside Bayview Homes

474 U.S. 121 (1985)

In this case, the U.S. Supreme Court (Supreme Court) was presented with the issue of whether USACE had exceeded its authority under the CWA. USACE had required landowners to obtain permits before discharging fill material into wetlands adjacent to a navigable water bodies and their tributaries. In 1975, USACE issued regulations redefining "waters of the United States" and extending its jurisdiction to include freshwater wetlands adjacent to other jurisdictional waters. Subsequently, Riverside Bayview Homes, Inc. (Bayview) began placing fill materials in an 80-acre parcel of low-lying marshy land near the shores of Lake St. Clair in Michigan. Because Bayview had failed to obtain a CWA Section 404 permit, USACE filed suit demanding injunctive relief enjoining the filling of the property without a USACE permit.

The district court granted USACE's injunction; however, Bayview appealed and the Court of Appeals remanded the case back to the district court for consideration of USACE's 1977 amendment to the definition of a wetland. The 1977 amendment eliminated a reference to periodic inundation and made other minor changes. The district court once again held the property subject to USACE's permit authority, and Bayview again appealed. The Court of Appeals interpreted the regulation to encompass only those wetlands that were subject to flooding by adjacent navigable waters based partially on concerns over a broader definition possibly resulting in a taking of private property. The Supreme Court reversed the lower court, upholding USACE's jurisdiction over wetlands adjacent to navigable water bodies and their tributaries irrespective of whether inundated or saturated by adjacent navigable waters or other water sources, e.g., groundwater.

In reaching its decision, the Supreme Court quickly dismissed the lower court's concern regarding possible takings that may have resulted from USACE's interpretation of its regulation. The Supreme Court reasoned that the existence of a possible takings claim provided no legal basis for limiting USACE's jurisdiction. The Supreme Court reviewed the plain language of the regulation and found it to clearly state that saturation or inundation by either surface or groundwater is sufficient to bring an area within the definition of a wetland, if wetland vegetation was present. The Supreme Court then reviewed the validity of the regulation by looking at the intended purpose of the CWA via both the statutory language and the legislative record. The Supreme Court found the intent of Congress to be one of broad application for the protection of the aquatic ecosystems. In addition, the Supreme Court noted that the proposed 1977 legislative amendments limiting USACE's jurisdiction over wetlands were hotly debated and eventually abandoned. Further, in 1977, Congress ended up adopting legislation supporting regulation of discharges to wetlands that evidenced Congress' intent to afford protection to these areas.

Bersani v. U.S. Environmental Protection Agency

850 F. 2d 36 (2d Cir. 1988),
cert. denied, 489 U.S. 1089 (1989)

In this case, it was held that an alternatives analysis pursuant to Section 404(b)(1) of the CWA can require an examination of alternatives available at the time the applicant entered the market. A developer (Bersani) challenged the EPA's final determination (vetoing USACE's approval) that prohibited the development of a shopping mall on a wetland known as Sweeden's Swamp. The proposed development included extensive mitigation, both on- and off-site, to offset the destruction of 32.2 acres of wetland habitat. EPA, after considering the project's practical alternatives under Section 404(b) of the CWA, found that the project would have "unacceptable adverse impacts" on wildlife under Section 404(c) of the CWA.

At the district court, Bersani argued that neither statute nor regulation authorized the EPA to consider Section 404(b) practical alternatives in its determination of whether a project would result in unacceptable adverse impacts under Section 404(c). The court rejected this argument, finding that regulations allow the use of relevant portions of Section 404(b) and that EPA had used the practical alternatives issue to find unacceptable adverse impacts on wildlife, a category listed in Section 404(c). Next, Bersani argued that if in fact the EPA was allowed to consider practical alternatives in its Section 404(c) analysis, then EPA was erroneous in its conclusion that a feasible alternative existed.

The district court upheld EPA's determination that a feasible alternative existed at a location previously considered by Bersani. Bersani asserted that the site had been considered and passed over by six other shopping mall developers over the last 15 years and that EPA had impermissibly substituted its judgment for that of the marketplace in an area in which it lacked expertise. The district court found that the EPA decision was not erroneous based on evidence in the record demonstrating that the site was suitable for a shopping mall virtually identical to that proposed by Bersani. It also found that a competitor had reached a similar conclusion that the location was suitable for development based on market analysis. The record also established that EPA had reviewed each of the features that Bersani found objectionable about the site in reaching its decision.

The district court discussed the "availability" of the site. Bersani argued that "available" meant presently available at the time of agency decision on the permit application. EPA's position was that the determination should be made at the time the applicant "entered that market, began its investigation, or comprehensively evaluated the area." The EPA reasoned that "it is reasonable and consistent to expect applicants planning non-water dependent projects to demonstrate that construction on uplands was considered before the use of wetlands was contemplated." This position was contrary to USACE's interpretation that alternatives were to be considered contemporaneously with the application. Regardless, the court held that EPA's interpretation of the regulations was reasonable and found that the record supported EPA's finding that the alternative site was available at the time Bersani "entered the market."

The Court of Appeals affirmed the district court's decision holding that: (1) the market theory is consistent with both the regulatory language and past practice; (2) EPA's interpretation, while not necessarily entitled to deference, is reasonable; and (3) EPA's application of the regulation is supported by the administrative record.

Leslie Salt Co. v. United States

896 F. 2d 354 (9th Cir. 1990),
cert. denied, 489 U.S. 1126 (1991)

This case concerns the effect man-made changes to property, done without the owner's consent, have on determining USACE's jurisdiction. Leslie Salt Co. (Leslie) challenged USACE's assertion of jurisdiction over 153 acres of private property in the San Francisco Bay area. In 1919, the property was converted into its current uses of pasture and livestock grazing (one-third) and a salt crystallization plant (two-thirds), that processed salt from water that was pumped to the site from the San Francisco Bay. The property did have standing water for short periods after winter rainstorms, but did not support wetland vegetation. Between 1980 and 1985, community complaints about dust caused Leslie to till the soil to reduce dust and to meet air quality standards. The tilling resulted in the creation of furrows that were more hospitable to plants. Around the same time, a California Department of Transportation (Caltrans) highway project that had several components on the property forced Leslie to allow the construction of various ditches, culverts, and drainages on the property, which resulted in tidal waters entering the property. In addition, Caltrans and the U.S. Fish and Wildlife Service breached a levee on the wildlife refuge adjacent to the property that allowed water to flow up the culverts. In 1985, USACE issued a cease-and-desist order to Leslie upon notification that it was digging a feeder ditch and siltation pond on a portion of a crystallizer. Leslie complied, but one year later, when USACE issued its second cease-and-desist order to stop Leslie's construction of a plug to prevent tide water from entering the facility through a culvert installed by Caltrans, Leslie brought an action against USACE.

The trial court held that changes to the property caused by the government do not create jurisdiction. The trial court reasoned that a contrary holding would allow USACE to expand its own jurisdiction by creating wetlands where none existed before. The Court of Appeals (Court) disagreed with the trial court and reversed its holding. The Court pointed out that USACE was not directly responsible for flooding Leslie's property. Accordingly, USACE did not create the wetland conditions and thereby attempt to expand its own jurisdiction. The Court then stated that USACE's jurisdiction does not depend on how the property at issue became a water of the U.S. If USACE's regulations under CWA jurisdiction harm a landowner, his or her appropriate response is to seek damages through inverse condemnation proceedings, not restrict USACE's jurisdiction. The Court also reversed the trial courts alternative rationales (normalcy and adjacency) for denying USACE jurisdiction due to the fact that these alternative rationales were based on USACE inappropriately expanding its own jurisdiction.

The trial courts decision regarding jurisdiction over the crystallizers was based on the regulations creating a distinction between artificial and natural waters. The Court rejected this distinction. The Court pointed out that USACE regulations created an exemption only for those artificially created waters that are currently being used for commercial purposes, and that even then they are subject to jurisdiction based on a case-by-case review. Here, the crystallizers had not been used for commercial purposes for decades. The Court also rejected the trial court's holding that crystallizers and calcium chloride pits were not jurisdictional based on their being dry for most of the year. The Court referencing the regulations enumeration of two seasonal water features (intermittent streams and playa lakes) shows that the seasonal nature of the ponding does not impede USACE's jurisdiction.

Sylvester v. U.S. Army Corps of Engineers

871 F. 2d 817 (9th Cir. 1989)

This case addresses USACE and its discretionary authority to limit the scope of NEPA on projects involving both federal and non-federal actions. The development project at issue involved the construction of a resort facility, ski lifts, and golf course. The golf course required the fill of 11 acres of wetland habitat that would be used by guests and visitors of the Squaw Valley Resort. The golf course was being constructed concurrently with the other resort facilities. The developers had obtained all local permits, prepared an environmental impact report (EIR) required by the state of California, and created plans to mitigate impacts on wetland areas. In addition, USACE's issuance of the Section 404 permit required significant wetland mitigation and a requirement that certain state agency conditions were complied with. In issuing the Section 404 permit, USACE chose not to prepare an environmental impact statement (EIS) because it believed the golf course (the only aspect of the project within its jurisdiction) was not directly linked to or essential to the entire resort project. Sylvester (a citizen) brought action against USACE because he felt that USACE had violated NEPA by approving the fill of wetland without analyzing all impacts associated with development of the entire resort. Sylvester believed that the fill of wetlands and the entire resort project were interdependent and that USACE was obligated to prepare an EIS pursuant to NEPA because the project as a whole would have significant environmental impacts. The district court found in favor of Sylvester and imposed a temporary restraining order and preliminary injunction against the developers until USACE could produce an EIS analyzing environmental impacts of the entire resort project.

Upon review, the Court of Appeals (Court) reversed and remanded the lower court's decision. The Court held that USACE's interpretation of the statute was permissible and the fill of the wetland (golf course) and resort project were not interdependent. Accordingly, USACE was not required to prepare an EIS to evaluate the impacts of the entire project but was allowed to limit its review to the golf course. In support of this holding, the Court cited USACE's NEPA regulations. The regulations direct the USACE district engineer to determine the appropriateness of preparing an environmental assessment (EA) or EIS for a project in situations where only one component of a larger project requires a federal permit. The regulations contain factors to assist in determining when USACE has control and responsibility for portions of the project outside its jurisdiction for NEPA purposes.

Tabb Lakes, Ltd. v. United States

885 F. 2d 866 (4th Cir. 1989)

The district court held that when USACE adopted an interpretation of its jurisdiction including "waters which are used or could be used as habitat by other migratory birds which cross state lines" (Migratory Bird Rule), they were required to comply with the Administrative Procedures Act (APA). The court held that USACE must follow the APA because the rule represented a new material addition to the coverage of the CWA. In reaching this decision, the court determined that the new rule did not fall under the exception for an interpretive rule on existing regulation. For a rule to meet the interpretive rule exception to the APA, the rule must be "explanatory and 'simply state what the administrative agency thinks the statute means'.... Conversely, rules which implement a statute and create new legal obligations are substantive and must be promulgated pursuant to the notice-and-comment procedures of the APA."

Mulberry Hills Development Corp. v. United States

772 F. Supp. 1553 (D. Md. 1990)

In this ripeness case, Mulberry (Plaintiff) was the owner and developer of a 62-acre tract of land located in Talbot County, Maryland. USACE performed a field review of the property and utilizing the three parameters of soil conditions, hydrology and types of vegetation, determined that a portion of the property was a wetland. Based on this determination USACE issued a cease-and-desist order to the Plaintiff requiring that it seek a Clean Water Act permit prior to filling any additional wetland. The Plaintiff filed suit asking for a preliminary injunction to prevent USACE from enforcing its cease-and-desist order and enforcing Section 404 of the Clean Water Act. The Federal district court denied the motion because: (1) even though the parties agreed that wetlands existed on the property and that fill had occurred, a delineation of the wetlands had yet to take place; (2) USACE had yet to apply the wetland delineation guidance that the Plaintiff was challenging; (3) the Clean Water Act prohibits pre-enforcement judicial review; and (4) potential economic injury to Plaintiff is inadequate to warrant court review.

Southern Pine Associates v. U.S.

912 F. 2d 713 (4th Cir. 1990)

In this case dealing with subject matter jurisdiction of the court (ripeness) to hear a case, Southern Pine Associates and VIOC Construction Inc. (Plaintiffs) appealed the trial court's order dismissing their complaint and petition for a temporary restraining order for lack of jurisdiction. Southern Pines owned 293 acres of land located in Chesapeake, Virginia. VIOC had a contract with Southern Pines and was involved in the clearing and building of 40 acres of the property. The EPA issued a "Findings of Violation and Order for Compliance" to Plaintiffs stating that they had violated the Clean Water Act by discharging fill material into wetlands without a permit. In a cover letter accompanying the order, EPA asked Plaintiffs to provide information about the site for it to review in order to make a "final determination of the boundaries of the wetlands that fall under the jurisdiction of the Clean Water Act." The Plaintiffs halted all work, denied EPA access to the site, and filed a complaint and petition for a temporary restraining order. They alleged that EPA's assertion of jurisdiction over the property created an actual controversy and argued that EPA lacks jurisdiction over the site based on the wetlands not being adjacent to any body of water. The trial court dismissed case for lack of subject matter jurisdiction.

The Court of Appeals (Court) in reviewing this case distinguished it from other cases allowing pre-enforcement judicial review (prior to final determination by agency) by pointing out that the "statutory structure and history of the Clean Water Act provides clear and convincing evidence that Congress intended to exclude this type of action [pre-enforcement]. The Court also stated that "[t]he structure of these environmental statutes indicates that Congress intended to allow EPA to act to address environmental problems quickly and without becoming immediately entangled in litigation. Finally, in affirming the trial courts dismissal of the Plaintiff's action, the Court emphasized that the Plaintiff's fifth amendment rights had not been violated since they are not subject to an injunction or penalties until EPA pursues an enforcement proceeding.

Hoffman Group Inc. v. United States Environmental Protection Agency

902 F. 2d 567 (7th Cir. 1990)

This case addresses the subject matter jurisdiction of a court in reviewing an order issued by an agency that is not a "final determination." Hoffman Group Inc. (Plaintiff) brought this action challenging the Findings of Violation and Compliance Order issued by the United States Environmental Protection Agency (EPA) for alleged violation of the Clean Water Act. Plaintiff owned property known as the Victoria Crossing subdivision in the Village of Hoffman Estates, Illinios. USACE had determined that the Plaintiff had placed unauthorized fill into 6.2 acres of wetland. USACE and Plaintiff negotiated a mitigation plan in conjunction with an after-the-fact permit that was rejected by EPA. The EPA issued a compliance order directing Plaintiff to cease further discharges and to submit a plan to restore the wetlands to their original condition.

After receiving EPA's compliance order, the Plaintiff filed suit seeking declaratory and injunctive relief preventing EPA from enforcing its order. Plaintiff's suit was dismissed on the grounds that it was seeking impermissible pre-enforcement review of an agency action taken under the Clean Water Act. In affirming the trial court's ruling the Court of Appeals (Court) focused on the intent of Congress in drafting and adopting the Clean Water Act. "In drafting the Clean Water Act, Congress chose to make assessed administrative penalties subject to review while at the same time it chose not to make a compliance order judicially reviewable unless the EPA decides to bring a civil suit to enforce it."

Avella v. United States Army

916 F. 2d 721 (11th Cir. 1990)

In this case the Court of Appeals (Court) affirmed the trial court's ruling dismissing the Dr. Avella's (Plaintiff) case for lack of subject matter jurisdiction. The case involved USACE's negative response to the Plaintiff's request for a determination that a Clean Water Act nationwide permit was applicable to his property. The trial court held that it lacked jurisdiction to hear the case because: (1) the response from USACE was not an order, license, or other agency action susceptible to review; (2) even if USACE's response was the type of action, it was not a final action with binding legal affect under the Administrative Procedures Act; and (3) review was unavailable under the Declaratory Judgment Act. The Court of Appeals affirmed based on the same rationale and emphasizing that the Declaratory Judgment Act is not a source of jurisdiction of subject matter jurisdiction for the court and that subject matter jurisdiction must be established on its own.

Hobbs v. United States

947 F. 2d 941 (4th Cir. 1991)

The facts are as follows, Phillip and Dorothy Hobbs owned about 169 acres in Ware Neck, Gloucester County, Virginia. Their son and daughter-in-law owned an additional 37 acres in the county, collectively they owned approximately 206 acres. The Hobbses cleared, drained, and constructed roads on approximately 50 acres of wetlands to create hay fields. The Hobbses stated that they did not apply for a permit because they were under the impression that they were not required to do so. This was based on overhearing a conversation between the United States Fish and Wildlife Service and USACE representative who visited their property and a statement to Phillip by the USACE representative that "it was a grey area of enforcement for the Corps of Engineers... and he didn't see where the [Hobbses] had... done anything wrong." Subsequently, EPA issued the Hobbses a Finding of Violation and Compliance Order.

The Hobbses responded by filing a suit challenging the EPA order and the trial court issued a preliminary injunction allowing them to continue to farm. However, EPA filed a counterclaim claiming violation of the Clean Water Act and the Hobbses' claim was finally dismissed. The trial court granted summary motion to the U.S. on its claim against the Hobbses for violation of the Clean Water Act. The Court of Appeals (Court) affirmed the trial court's ruling finding the Hobbses in violation of the Clean Water Act.

On appeal the Hobbses raised a number of arguments only a few of which will be discussed here. One argument that the court rejected was that the EPA's wetland delineation manual could not be used to determine the existence of wetlands on their property since it was not properly adopted as required by the Administrative Procedures Act (APA). The Court held that the APA did not apply in this case since the manual merely clarified and explained existing law. The last issue to be discussed here involves the Hobbses claim that their due process rights had been violated. They asserted that they were precluded from presenting evidence of their good faith belief that the administrative orders were invalid, that they had no means to contest the administrative orders, and they were not provided with prior notice or an opportunity to be heard before issuance of the administrative order. The Court found this argument to be without merit based on the Clean Water Act imposing strict liability making "good faith" irrelevant. Further, administrative orders do not in themselves impose penalties and therefore, such notice and opportunity to be heard was not required prior to or upon issuance of such orders. The penalties being appealed from resulted from a jury trial that determined that the Hobbses had violated the Clean Water Act.

Golden Gate Audubon Soc., Inc. v. U.S. Army Corps of Engineers
796 F. Supp. 1306 (N.D. Cal. 1992)

This case addresses the jurisdiction of USACE as it relates to areas that began being converted to dry land prior to 1975, the year in which USACE defined its jurisdiction over adjacent wetlands. The Golden Gate Audubon Society and other environmental groups filed this case against USACE, the EPA, the Port of Oakland (Port), and others. The suit challenged USACE's finding that it did not have jurisdiction over an area that had begun being filled before the passage of the CWA. USACE had been previously instructed by the court on remand to make a finding as to whether the Port had transformed the site into dry land by 1975. If the site had been transformed to dry land by 1975, USACE could find that dry land was its normal circumstances, since the regulatory definition of wetlands does not extend USACE's jurisdiction retroactively over areas that have been transformed into dry land. This case concerns the finding that USACE made as a result of the court's opinion remanding this matter.

USACE's findings concluded that the site was converted to uplands by 1975 based on the site being completely filled by 1972. This fill had destroyed the previously existing wetlands prior to the enactment of the CWA. USACE also found that biological wetlands were present at the site in 1975, but concluded that, since the site had previously been converted to dry land, these wetlands were non-jurisdictional. USACE also found that wetlands had reemerged on the site from 1972 to 1986. These wetlands reemerged in both abandoned and unabandoned areas of the site. USACE determined that the wetlands located in the abandoned areas were jurisdictional since this was the "normal circumstances" of those areas. However, the wetlands located on the areas that were not abandoned were considered non-jurisdictional based on wetlands not being the normal circumstances of the site. USACE also found other areas of the site to fall under its jurisdiction based on tidal influence.

In reviewing the findings of USACE, the court concluded that USACE had interpreted the CWA too narrowly. The regulation defined wetlands as "those areas that are inundated or saturated by surface or ground water at a frequency and duration sufficient to support, and that under normal circumstances do support, a prevalence of vegetation typically adapted for life in saturated soil conditions." The 1975 expansion of USACE's jurisdiction brought all non-exempt wetlands, as defined above, under USACE's jurisdiction. Accordingly, the question was not whether the wetlands were the normal circumstances of the site but simply whether the wetlands existed as of 1975. USACE had inappropriately decided that the normal circumstances of the unabandoned areas of the site were fill areas with ongoing fill activities and other construction and thus non-jurisdictional, despite the existence of wetlands. The court stated that "the [USACE's] 'under normal circumstances' analysis is legally flawed in several respects, and thus not entitled to this Court's deference." The court went on to say that USACE's two-step analysis, which first looks to whether wetlands are present and then determines whether those wetlands are normal, has the result of imposing an additional jurisdictional requirement "which is circular, redundant and simply does not exist." The court concluded that "it is impossible to state that the 'normal circumstances' of an area which contains wetlands is anything other than 'wetlands'."

James City County, Va. v. U.S. Environmental Protection Agency
955 F. 2d 254 (4th Cir. 1992)

James City County (County) challenged the EPA's veto of a USACE decision to approve a Section 404 permit allowing the construction of a dam and water reservoir. In 1984, the County applied to USACE for a permit to place fill for the construction of a dam and reservoir for providing supplemental water for the projected population growth of the County. The project involved the creation of a lake that would flood 425 acres of wetlands. USACE, U.S. Fish and Wildlife Service, National Marine Fisheries Service, and EPA jointly completed an environmental impact statement (EIS). Upon completion of the EIS, USACE issued the notice of intent to issue its permit. EPA then vetoed USACE's decision.

USACE had determined that there were no practicable, environmentally superior alternatives to the project. Upon reviewing the USACE decision under Section 404(c), the EPA concluded in its final determination that the project would result in severe direct and cumulative loss of wildlife habitat and would result in serious impacts to and/or losses of wildlife species. However, EPA also stated that there were viable alternatives to the project including: the construction of three smaller dams as opposed to one; increased use of groundwater; desalination; and increased conservation. The County filed suit and the district court overturned EPA's veto, ordered USACE to issue the permit, and denied the EPA's request for a remand.

EPA appealed the district court's decision to the Court of Appeals (Court). The Court affirmed the district court's decision in part and remanded to the district court for further remand to the EPA to reconsider its veto in light of the Court's decision. The Court's rationale for deciding in favor of the County was based on the lack of substantial evidence establishing practicable alternative water sources (where water-dependent activity exists with no presumption to the contrary). None of the alternatives considered by EPA in its veto were determined to be practicable based on the evidence in the record. The alternatives either involved increased costs, were prohibited or not supported by the applicable jurisdictional authority, were experimental, or would not provide a sufficient quantity of water. The Court remanded the case to the EPA, giving the agency 60 days to review the permit to determine whether it should be vetoed based solely on environmental impacts as opposed to the existence of practicable alternatives.

Save Our Community v. U.S. Environmental Protection Agency
971 F. 2d 1155 (5th Cir. 1992)

This case held that the drainage of a wetland without evidence of a discharge of a pollutant was not within USACE's Section 404 jurisdiction. A group of local citizens (SOC) filed suit against the EPA because EPA did not require developers Trinity Valley Reclamation Inc. (Trinity) to obtain a permit for the draining of several ponds in Ferris, Texas. Prior to draining the ponds, Trinity consulted with the U.S. Fish and Wildlife Service, the EPA, and USACE (agencies). The agencies advised Trinity that the ponds in question were under USACE jurisdiction, but that no permit was required as long as Trinity did not discharge pollutants into the ponds. Nevertheless, Trinity mitigated the loss of wetland habitat by creating other wetland areas in another location on the property. Irrespective of the understanding between Trinity and the agencies, the district court enjoined Trinity and required it to obtain a Section 404 permit from USACE before continuing its drainage activities. In finding that Section 404 jurisdiction *existed,* the district court refused to base its decision on the existence of a discharge of pollutants and instead held that drainage alone amounted to an end run around the CWA and required a permit irrespective of whether a discharge was involved.

Trinity and the EPA appealed the decision claiming that SOC lacked standing and that without a discharge of pollutants no violation exists, and no jurisdiction is created. The Court of Appeals (Court) held that SOC had standing to bring the case but that "without the existence of an effluent discharge of some kind, there is no coverage under Section 404."

On the issue of standing, Trinity asserted that SOC's individual members lacked standing to bring the suit; consequently, SOC lacked standing to represent its members' interests. The Court rejected this argument finding that affidavits submitted to the Court established that the individual members had suffered sufficient injury based on owning property near the ponds and enjoyment of the wildlife and open space. The Court also found that the injury suffered by SOC's members was "fairly traceable" to Trinity's failure to obtain a permit, based on the direct relationship between the drainage of the ponds and the reduction in value for wildlife. It was also determined that SOC's members alleged injury would be redressed by a favorable decision enjoining the draining of the ponds.

The rationale for the Court requiring the existence of a discharge of some kind of effluent to establish jurisdiction under Section 404 was based on the language of the statute, regulations, and case law. The Court noted that "[i]n precise language, the CWA prohibits 'the discharge of any pollutant by any person' except as authorized by specific sections of the Act." SOC argued that, if in fact a discharge of pollutant is required, the district court did in fact make a finding of fact that there were some "minor" discharges resulting from erosion caused by the draining activity. The Court concluded that the record was incomplete on the issue of "discharge" and that material issues of fact remained. Accordingly, the Court dissolved the injunction, reversed the judgment, and remanded the case back to the lower court.

North Carolina Wildlife Federation v. Tulloch

E.D. N.C. 1992, Civil No. C90-713-CIV-5-BO

No case summary available since parties settled prior to court reaching a decision. *See* 58 Federal Register 45008, August 25, 1993.

Hoffman Homes, Inc. v. Administrator, U.S. Environmental Protection Agency

999 F. 2d 256 (7th Cir. 1993)

This case addressed two questions. The first issue involved whether or not USACE and EPA could reasonably interpret CWA's Section 404 regulations to allow migratory birds to trigger jurisdiction under the Commerce Clause. The second issue concerned whether substantial evidence supported EPA's determination that the wetland at issue was suitable or potential habitat for migratory birds prior to being filled.

During the construction of a new subdivision, Hoffman Homes, Inc. (Hoffman), filled two wetland areas. The area at issue was one acre in size and contained impermeable clay that collected water during wet weather. This wetland was not hydrologically connected or adjacent to any navigable waters. Hoffman agreed with USACE that the wetland in question had been filled, but asserted that the wetland did not meet the definition of waters of the United States. EPA argued that, based on the potential for the wetlands to be used as habitat for migratory birds, that jurisdiction existed.

Upon appeal, the Court of Appeals (Court) held that EPA's interpretation of the CWA allowing migratory birds to constitute a connection between a wetland and interstate commerce was reasonable. In doing so, the Court acknowledged that "millions of people annually spend more than a billion dollars on hunting, trapping, and observing migratory birds" and that "protecting endangered birds preserves interstate commerce in species and interstate movement of persons wishing to study species." The Court next turned its attention to whether EPA's conclusion that the wetland was suitable or potential habitat for migratory birds was supported by substantial evidence. The Court determined that insufficient evidence existed to support EPA's determination. The Court concluded that evidence of migratory birds being located in nearby jurisdictional wetlands did not support a finding that the wetland at issue was suitable habitat for these birds. The Court pointed to the fact that the record contained no evidence of migratory birds ever being present on the wetland and that the evidence of its potential value as habitat amounted to "mere speculation." The Court concluded that the migratory birds "are better judges of what is suitable for their welfare than are we.... Having avoided the [wetland] the migratory birds have spoken and submitted their own evidence." Hoffman was not required to pay the fine issued by EPA.

Roanoke River Basin Ass'n v. Hudson

991 F. 2d 132 (4th Cir. 1993),
cert. denied, 510 U.S. 864 (1993)

The Roanoke River Basin Association (RRBA) challenged USACE's issuance of a Section 404 permit to the City of Virginia Beach (City). The permit was issued for construction of an 85-mile pipeline that would transport 60 million gallons of potable water a day from a lake in the Roanoke River basin to Virginia Beach. The RRBA claimed that, among other deficiencies, USACE's analysis failed to consider "relevant information" related to the biological impacts of the pipeline, and that an environmental impact statement (EIS) was necessary for compliance with NEPA. Upon review of the administrative record, the district court found that USACE's analysis was adequate, with the exception of the analysis of impacts on striped bass and the necessary factual support of the City's projected water needs. These two issues were remanded to USACE, a supplemental independent analysis was undertaken, and in compliance with NEPA, a supplemental environmental assessment and revised Finding of No Significant Impact was issued. The revised documents supported former findings, but amended the Section 404 permit to include a mitigation measure requiring the City to "allow use of some of its stored water during the bass' spawning period in the unlikely event that it might be needed to maintain the flow of the river." The district court found that this reduced all adverse effects to a less-than-significant level, and upheld the permit issuance. On appeal by RRBA, the Court of Appeals (Court) affirmed this finding (*RRBA v. Hudson,* 940 F. 2d 58, 66 (4th Cir. 1991)) and, in this related case, denied the RRBA attorney's fees and expenses. The Court held that while the RRBA had prevailed on two issues, this was not enough to result in an award of fees under the Equal Access to Justice Act. In support of their holding, the Court stated that "in light of the magnitude of the project, [USACE's] significant tasks under relevant environmental statutes and regulations, and [USACE's] overall 'careful analysis of the environmental and non-environmental facts', were not unreasonable in fact or law."

Mills v. United States

36 F. 3d 1052 (11th Cir. 1994),
cert. denied, 514 U.S. 1112 (1995)

This case addresses whether Congress unconstitutionally delegated its legislative authority to USACE in allowing USACE to define waters of the U.S. to include wetlands, in light of the fact that this definition can be used as the basis for criminal charges. The defendants, a father and son (Mills), purchased two lots of land in 1986 for the purpose of development. The property was adjacent to a waterfront, but contained mostly mature pine, oak, gum, bay, and magnolia trees, as well as shrubs, on apparently dry land. Although there was evidence of a depressional feature on the land that carried rainwater runoff, the land itself had "no standing water on it, and did not appear to be a marsh, swamp, or bog." After dumping sand to level the property in preparation for building, the Mills were found guilty of violating the CWA for placing fill material into a regulated wetland without a permit. In 1989, they were both sentenced to 21 months in jail for the violation. The Mills filed an appeal claiming that their convictions should be voided. The Mills argued among other things, that Congress has unconstitutionally delegated its legislative authority to USACE.

The trial court (court) held that it was "constrained by Supreme Court precedent to conclude that the Clean Water Act does not impermissibly delegate legislative power to [USACE]." The court began by stating that because the critical terms for this case were "utterly non-definitive" under the CWA, USACE's decision to define the term "waters of the U.S." to include wetlands was permissible. In addition, the court held that Congress had, in fact, appropriately delegated its criminal lawmaking authority to USACE because it had provided an "intelligible principal" when it wrote the CWA. Therefore, despite the defendants argument that there was "no statute that makes it a federal crime to place clean, unpolluted sand on dry appearing land," the court indicated that in light of the Supreme Court opinion, it was left with no choice but to uphold USACE's authority to "flesh out the statute" and "define the elements of a felony offense." The court did add that despite the Supreme Court's support of USACE's regulatory authority over wetlands, that "it is doubtful that [the Supreme Court] realized that [USACE's] definition extends to land that appears to be dry, but which may have some saturated-soil vegetation, as is the situation here, or that it would define the elements of a felony offense." The Court of Appeals affirmed.

United States v. Pozsgai

999 F. 2d 719 (3d Cir. 1993),
cert. denied, 510 U.S. 1110 (1994)

In this case, the court upheld USACE's authority to regulate isolated wetlands via the Commerce Clause, as well as the definition of "waters of the U.S." The defendant (Pozsgai) had deposited fill material consisting of concrete rubble, earth, and building scraps into a 14-acre site in Morrisville, Pennsylvania. Upon investigation of the incident, USACE determined that the entire site was a wetland, based on the vegetation, soils, and presence of standing water, and that Pozsgai must obtain a CWA Section 404 permit. The defendant ignored both a cease-and-desist order and subsequent court order to restore the property to its original state. The district court jury found the defendant in violation of 40 counts of unpermitted discharges and sentenced him to three years' prison, five years' probation, as well as a monetary fine of $200,000. Pozsgai argued that the material he had deposited was not a "pollutant" and that the material added was not added into water but was placed in a wetland, which should not be considered a water of the U.S. The defendant argued in the alternative that the wetland was not "adjacent" within the meaning of the CWA regulations, and that the government had failed to establish that the fill of the wetland affected interstate commerce.

The Court of Appeals (Court) affirmed the decision of the lower court. The Court held that "dredged spoil, rock, and sand" is specifically identified in the CWA as a pollutant, and upheld USACE's definition of waters of the U.S. that included wetlands. In addition, the Court found that the wetlands in question were "adjacent" because they were adjacent to a small tributary of the Pennsylvania Canal, which was a tributary of the Delaware Canal, a water that supported interstate commerce of coal. For this reason, the Court held that it was not necessary for the government to prove that fill of wetlands would affect interstate commerce, as it satisfied the adjacency test.

Loveladies Harbor Inc., v. United States

28 F. 3d 1171 (Fed. Cir. 1994)

In this regulatory takings case, Loveladies Harbor Inc. (Loveladies) argued that by denying a CWA Section 404 permit, USACE had destroyed the economic value of land proposed for development. Of the original 250 acres Loveladies purchased in 1958, 51 acres had not been developed before the CWA was passed in 1972. Of the 51 remaining acres, 50 were wetland habitat that Loveladies identified as a potential development area. After several years of negotiations and ultimately in accordance with a New Jersey Superior Court order, the New Jersey Department of Environmental Protection granted Loveladies permission to fill 12.5 acres of the remaining wetlands. Having obtained state-level approvals, Loveladies applied to USACE for a Section 404 permit. USACE denied the permit. Loveladies filed a claim in the court of federal claims for just compensation under the Fifth Amendment. That court found in favor of Loveladies and valued the land prior to permit denial to be $2,658,000 and after permit denial to be worth $12,500. The land's substantial value decrease and relatively low value to the public were cited as factors supporting that "the plaintiffs have shown that their private interest in developing and utilizing their property outweighs the public value in preserving these wetlands."

The Court of Appeals (Court) affirmed the lower court's decision. The Court also briefly addressed the nuisance issue USACE used in its defense. USACE stated that, because it believed the public would consider the fill to be a nuisance, the case should fall under the Mugler Rule. The Mugler Rule was created to prevent a private party from collecting just compensation under the Fifth Amendment if its actions might be considered a noxious public nuisance. The Court found that USACE did not clearly show that the actions of Loveladies could be "seen as so offensive to the public sensibility as to warrant no constitutional protection" as required under the Mugler Rule.

United States v. Banks

115 F. 3d 916 (11th Cir. 1997),
cert. denied, 522 U.S. 1075 (1998)

The primary issue addressed in this case is the statute of limitations for enforcement of violations of the CWA. In 1980, Park B. Banks (Banks) purchased three lots of property in Florida for the purpose of creating a coconut farm and building a new home. In 1983, a USACE biologist informed Banks that his development activities were in violation of CWA Section 404 regulations and that he must obtain a permit. A cease-and-desist order that threatened enforcement actions was issued shortly thereafter. Banks applied for an "after-the-fact permit" but it was denied by USACE. Despite the fact that he was denied a fill permit, Banks continued to discharge fill into the wetland and in 1988 purchased and developed additional land adjacent to his original property. USACE issued another four cease-and-desist orders in 1990, notifying Banks that his actions were illegal. In December 1991, the government filed suit against Banks requesting an injunction and required Banks to fully restore all affected wetlands. Banks argued that pursuant to the concurrent remedy rule, which bars claims for equitable relief for violations of law when the statue of limitations has expired, the court could not grant equitable relief to USACE.

The court held that "because Congress did not expressly indicate otherwise in the statutory language of [28 U.S.C. section] 2462 (Banks' basis for the statute of limitations argument), its provisions apply only to civil penalties; the government's equitable claims against Banks are not barred." Short of a direct order from Congress, the statute of limitation does not apply to claims brought by the federal government in its sovereign capacity as opposed to vindicating a private interest. In this case, the court held that the federal government was acting in the public's interest and that Banks' argument that the statute of limitations had passed was not relevant in light of the federal government's sovereign capacity to protect the public interest. The Court of Appeals (Court) upheld the decision of the lower court and, in accordance with USACE's request, ordered Banks to restore the wetlands to their undisturbed condition by removing the fill and otherwise implementing a restoration plan. Banks was also required to pay an "appropriate" civil penalty.

Banks raised issues related to whether his property qualified as a wetland, whether his lands were adjacent wetlands, and the applicability of Nationwide Permit 26. The Court rejected each of these arguments.

United States v. James J. Wilson

133 F. 3d 251 (4th Cir. 1997)

This case primarily concerned a challenge to USACE authority to assert jurisdiction over isolated wetlands lacking any direct or indirect surface connection to interstate waters, navigable waters, or interstate commerce. The decision also discusses "sidecasting" and the proper *mens rea* (criminal intent) requirement for violations of the CWA.

The case involved the felony convictions of a James J. Wilson, Interstate General Co., L.P., and St. Charles Associates, L.P. (defendants), charged with knowingly discharging fill material and excavated dirt into wetlands on four separate parcels without a permit, in violation of the CWA. Evidence was introduced that the defendants had attempted to drain at least three of four parcels containing wetlands by digging ditches. The dirt from these ditches was then deposited next to the ditch (sidecasted). Evidence was also presented that the defendants deposited fill dirt and gravel on three of the parcels. This was all done without obtaining permits from USACE. The district court sentenced Wilson to 21 months' imprisonment and one year supervised release and fined him $1 million. It fined the other two defendants $3 million and placed them on five years' probation. The defendants were also required to implement a wetlands restoration and mitigation plan. The Court of Appeals (Court), in reviewing the district court's decision, held that USACE's definition of waters of the U.S., which includes those waters whose degradation "could affect" interstate commerce, exceeded its authority under the CWA, as limited by the Commerce Clause. Additionally, the Court found that the district court had failed to require *mens rea* with respect to each element of an offense defined by the CWA. Based on these holdings, the Court reversed the district court's convictions and remanded the case for a new trial.

The defendants argued that by "allowing the jury to find a nexus with interstate commerce based on whether activities 'could affect' interstate commerce, the court authorized a 'limitless view of federal jurisdiction'," greatly exceeded that allowed under the Commerce Clause. The Court distinguished the facts at issue here from the Supreme Court's decision in *United States v. Riverside Bayview Homes, Inc.*, 474 U.S. 121 (1995), upholding USACE's jurisdiction over adjacent wetlands. The Court noted that in *Bayview Homes* the Supreme Court upheld USACE's regulation defining water of the U.S. "in the context of a wetland 'that actually abuts on a navigable waterway'." Accordingly, the Court held that the lower court's instructions to "the jury that waters of the U.S. included adjacent wetlands 'even without a direct or indirect surface connection to other waters of the [U.S.]'... intolerably stretches the ordinary meaning of the word 'adjacent' and the phrase 'waters of the [U.S.]' to include wetlands remote from any interstate navigable waters."

In addition, the Court held that "sidecasting" resulting solely from the digging of ditch in a wetland without the introduction of additional fill material does not violate the CWA and that a jury instruction stating that the CWA prohibited it was improper. The Court held that if a wetland is successfully dried out by the digging of a ditch, the addition of fill material does not violate the CWA "because adding fill to dry land cannot be construed to be polluting the waters of the [U.S.]." The Court also held that to find a person guilty of a criminal offense under the CWA, *mens rea* "requires the government to prove the defendant's knowledge of facts meeting each essential element of the substantive offense, but need not prove that the defendant know his conduct to be illegal."

National Mining Congress v. U.S. Army Corps of Engineers

145 F. 3d 1399 (D.C. Cir. 1998)

In 1993, USACE and EPA created the so-called "Tulloch rule" in response to litigation. The result of the Tulloch rule was to remove the *de minimis* exemption from the definition of "discharge of dredged material," as used in Section 404 of the CWA. This rule expanded the definition of "discharge of dredged material" to include "any addition of dredged material into, including any redeposit of dredged material within, the waters of the United States." This new definition included "incidental fallback" that spills from a bucket or shovel during dredging activities, resulting in a regulated discharge to the site from which it was removed. Enacting the Tulloch rule, USACE assumed jurisdictional authority over nearly all excavation, mining, land clearing, ditching, and channelization activities occurring in waters of the United States.

The Tulloch rule was challenged in the district court by the American Mining Congress (AMC). The district court invalidated the Tulloch rule holding that the rule exceeded the scope of the agencies' statutory authority. *American Mining Congress v. U.S. Army Corps of Engineers,* 951 F. Supp. 267 (D.D.C. 1997). USACE and EPA appealed the lower court's decision and the Court of Appeals (Court) affirmed. The Court based its decision on the plain language of the CWA defining "discharge," as "any addition of any pollutant to navigable waters from any point source." The court emphasized that the statutory term "addition" cannot reasonably be interpreted to encompass the removal of material from the water of the United States where a small portion of it happens to fall back. "Because incidental fallback represents a net withdrawal, not an addition, of material, it cannot be a discharge." The Court also relied on Section 10 of the Rivers and Harbors Act of 1899, which gave USACE jurisdiction over dredging. The court reasoned that the legislature had intended Section 404 of the CWA to give USACE jurisdiction over "discharges" while Section 10 of the Rivers and Harbors Act was intended to cover dredging activities.

Resource Investments Inc. v. U.S. Army Corps of Engineers

151 F. 3d 1162 (9th Cir. 1998)

The case presents the issue of whether USACE is authorized under Section 404 of the CWA to require a Section 404 permit for the construction of a municipal solid waste landfill (landfill) affecting wetlands.

Resource Investment, Inc. (Resource), a private company, sought to construct and operate a landfill on a 320-acre site in Pierce County, Washington. The landfill would occupy 168 acres of the 320-acre site and require clearing, excavating, filling, and grading of approximately 21.6 acres of the site's 70 acres of wetlands. Resource's plan included mitigation of the wetlands loss by creating, preserving, restoring, and enhancing wetlands on a dedicated 85-acre onsite wetland mitigation area. The landfill complied with the Tacoma-Pierce County Solid Waste Management Plan (County Waste Plan), which was consistent with the State of Washington's solid waste permit program developed pursuant to the Resource Conservation and Recovery Act (RCRA) and approved by the EPA. Pursuant to this program, Resource had successfully demonstrated that a number of criteria were met ensuring the protection of wetlands, drinking water, and wildlife, including a practicable-alternatives analysis and mitigation of impacts on wetlands.

In addition to complying with the County Waste Plan, Resource also filed an application with USACE for a Section 404 permit. The permit was denied on the grounds that it had failed to demonstrate the unavailability of practicable alternatives for waste disposal that were less environmentally damaging and that it was not in the public's interest because it would cause significant degradation of wetlands and cause unacceptable risk to groundwater. Resource sought review of USACE's decision by the district court. The district court affirmed the denial of the permit, finding the decision not to be arbitrary, capricious, contrary to law, or an abuse of discretion.

On appeal, Resource argued that USACE lacked authority under Section 404 of the CWA to require a permit because under RCRA the regulation of landfills lies solely with the EPA or states with solid waste permit programs approved by EPA. As an issue of statutory interpretation, the Court of Appeals (Court) reviewed that matter *de novo*. The Court held that the proposed project was a landfill and that EPA, rather than USACE, had authority under RCRA. The Court based its decision on the fact that refuse in a landfill does not fall within the definition of either "dredged material" or "fill material" and in fact falls within an exception to the definition of fill material under the CWA. The Court also pointed to a letter and memorandum exchanged between USACE and EPA establishing the agencies' understanding that landfills were to be regulated by EPA under RCRA. The Court also concluded that USACE's interpretation was unreasonable because it would create regulatory overlap since both agencies would be applying the same criteria with potentially inconsistent results.

Florida Rock Industries, Inc. v. United States

18 F. 3d 1560 (Fed. Cir. 1994),
cert. denied, 513 U.S. 1109 (1995)

In this regulatory takings case, Florida Rock Industries, Inc. (Florida Rock) purchased a 1,560-acre parcel of land composed mainly of wetlands for the purpose of mining limestone. The property was purchased by Florida Rock in 1972, shortly before the enactment of the CWA. In 1978, Florida Rock began mining without a permit. USACE directed Florida Rock to cease and desist all mining and related activities until a CWA Section 404 permit was obtained. In compliance with the order and at USACE's request for a permit prescribing a three-year mining plan, Florida Rock's application indicated that 98 acres would be required to meet its needs. USACE denied the application. Florida Rock filed suit against USACE in the court of federal claims (trial court) under the Fifth Amendment claiming an uncompensated regulatory taking and arguing all economic value of the land had been destroyed.

The trial court concluded that the only viable economic use of the property was limestone mining and held that the value of the property after the taking was negligible. The trial court's decision was vacated by the Court of Appeals (Court) and remanded for a determination of "fair market value." On remand, the trial court discarded evidence of the land being worth $6,100 per acre and actual offers to purchase the land for approximately $4,000 per acre. The trial court discarded the evidence based on an erroneous argument made by Florida Rock that the Court in its remand had required the trial court to only consider buyers with full knowledge of the regulatory scheme. Accordingly, the issue currently on appeal is what impact the regulatory imposition had on the economic use and value of the property.

The Court rejected the trial court's analysis that led to the conclusion that all economically beneficial use of the land was taken by USACE's permit denial. The Court stated that the trial court's reliance on Florida Rock's argument discounting all of the "comparable sales values on the principle that none of the purchasers were sufficiently sophisticated and knowledgeable... was in error–contrary to our instructions [on remand], contrary to generally accepted understandings of market valuations, and finally, contrary to the working assumptions of a free market." The Court went on to state that "[w]hen the market provides a well-substantiated value for a property, a court may not substitute its own judgment as to what is a wise investment." The Court vacated the trial court's decision and remanded for further proceedings consistent with its opinion.

In *dicta,* the Court also discussed the dilemma created in regulatory takings cases regarding when reduction in value becomes a partial taking that merits compensation.

Wetlands Action Network v. U.S. Army Corps of Engineers

222 F. 3d 1105 (9th Cir. 2000), cert. denied, 122 S. Ct. 41 (2001)

In this NEPA case, Maguire Thomas Partners-Playa Vista (MTP-PV) was planning to build a large-scale mixed development at the Playa Vista property in Los Angeles, California. The property contained 186 acres of wetlands and MTP-PV proposed to fill 21.4 acres of these. Because of the project size, MTP-PV consulted with USACE and decided to divide the permit process into three separate applications corresponding with the phases of the overall project. The first permit application proposed to fill 16 acres of scattered wetlands along with the creation of 52 acres of freshwater wetlands. The second permit application would be for the restoration and creation of approximately 230 acres of salt marsh and the third permit for a marina development and flood control channel filling 10 acres of wetlands.

MTP-PV submitted its application to USACE for the first phase of the project and subsequently responded to numerous comments from federal agencies. After revising its alternatives analysis and conducting additional studies, the permit was issued along with the associated environmental assessment (EA) and Finding of No Significant Impact (FONSI). The EA stated that the division of the project into three applications was logical and did not amount to piecemealing. The EA also concluded that it did not need to evaluate the upland portions of the overall project because they were outside of its jurisdiction and could exist independently. The EA concluded that, because of the net increase in wetlands, the project would not have a significant impact on the quality of the human environment and that an environmental impact statement (EIS) was not required.

Wetlands Action Network and California Public Interest Research Group (collectively WAN) filed suit and succeeded in having the district court hold that: (1) USACE had violated NEPA by improperly limiting its analysis to the impacts of the activities covered by the permit as opposed to impacts of the whole project; and (2) that USACE should have prepared an EIS as opposed to an EA based on the untested nature of the freshwater wetlands, lack of a fully developed mitigation plan, and controversy over the effects. The Court of Appeals (Court) reversed the lower court decision, upholding USACE's decision not to consider the impacts of the whole project. This decision was based on USACE regulations and support in the record for the conclusion that the three phases of the project were not connected actions because each had independent utility. The Court also upheld USACE's decision not to prepare an EIS. In support of its holding, the Court found that the FONSI was based on relevant and substantial data and that any negative impacts would be mitigated, and that the mitigation plan had been reviewed by federal agencies at the time the permit was issued.

Borden Ranch Partnership v. U.S. Army Corps of Engineers

261 F. 3d 810 (9th Cir. 2001), cert. granted, 122 S. Ct. 2355 (2002)

This case held that a tractor performing deep ripping is a point source that results in the discharge of pollutants under the CWA when performing this activity in wetlands. A real estate developer, Angelo Tsakopoulos, purchased Borden Ranch, an 8,400-acre ranch located in the Central Valley of California. The intent was to convert the ranch from its past use as rangeland for cattle to vineyards, orchards, and subdivided property. The ranch contained significant hydrological features, including vernal pools, swales, and intermittent drainages. These hydrologic features depend upon a dense layer of soil, called a "restrictive layer" or "clay pan," that prevents surface water from penetrating deeply into the soil. In order to allow the roots of the vines and trees to penetrate this restrictive layer, Tsakopoulos utilized a farming method know as "deep ripping." Deep ripping involves a tractor or bulldozer dragging large 4- to-7-foot-long metal prongs through the soil penetrating the restrictive layer. Tsakopoulos and USACE disagreed regarding USACE's authority to regulate deep ripping in wetlands. This disagreement eventually led to the EPA issuing an Administrative Order to Tsakopoulos, who in turned filed suit challenging the authority of USACE and EPA to regulate deep ripping.

The district court determined that Tsakopoulos was in violation of the CWA and imposed a $1,500,000 penalty to be reduced to $500,000 if restoration measures were performed. Tsakopoulos appealed to the Ninth Circuit Court of Appeals (Court), which affirmed the district court's holding that deep ripping of protected wetlands is subject to the jurisdiction of USACE and EPA. The Court determined that the deep ripper or plow was a point source resulting in the discharge of a pollutant. The Court analogized to case law to support its reasoning that deep ripping removed soil from the wetlands, thereby transforming it into "dredged spoil," a statutory pollutant. The Court concluded that "Congress determined that plain dirt, once excavated from waters of the [U.S.], could not be redeposited into those waters without causing harm to the environment." The Court also held that Tsakopoulos could not qualify under the farming exemption to the CWA based on an exclusion to the exemption called the "recapture provision." This provision requires a permit for farming activities that convert the land to a new use involving substantial hydrological alterations. The Court concluded that deep ripping "radically altered the hydrological regime of the protected wetlands" and therefore, was not exempt.

Solid Waste Agency of Northern Cook County v. U.S. Army Corps of Engineers

531 U.S. 159 (2001)

In a 5-4 decision, the U.S. Supreme Court (Supreme Court) held USACE's "Migratory Bird Rule" to be inconsistent with the CWA. USACE had informally adopted the Migratory Bird Rule in 1986 in an effort to clarify the reach of its jurisdiction. USACE stated that jurisdiction under Section 404 of the CWA extended to intrastate waters that are or would be used as habitat for birds protected by migratory bird treaties, used as habitat by other migratory birds that cross state lines, or used to irrigate crops sold in interstate commerce.

This case involved the attempt by the Solid Waste Agency of Northern Cook County (SWANCC) to utilize an abandoned gravel pit with permanent and seasonal isolated ponds as a potential disposal site for nonhazardous waste. Pursuant to the Migratory Bird Rule, USACE claimed jurisdictional authority over the ponds and required that a CWA Section 404 permit be obtained. The ponds were determined not to be wetlands but, based on the presence of migratory birds, were determined to be waters of the U.S. After its permit was denied, SWANCC filed suit. SWANCC argued that USACE had exceeded its statutory authority in interpreting the CWA to cover nonnavigable, isolated, intrastate waters based on the presence of migratory birds. In the alternative, SWANCC argued that Congress lacked the power under the Commerce Clause to grant such regulatory jurisdiction; however, the Supreme Court did not find it necessary to consider this argument.

In reaching its decision, the Supreme Court concluded that USACE's interpretation of the statute read the term "navigable waters" out of the CWA. Further, USACE did not sufficiently establish Congress' intent for the Supreme Court to uphold an interpretation invoking "the outer limits of Congress' power." The Supreme Court expressed concern that the Migratory Bird Rule, by allowing regulation of isolated waters, would alter "the federal-state framework by permitting federal encroachment upon a traditional state power that Congress in the CWA expressly chose to preserve.

The dissenting opinion noted that the court had drawn a new jurisdictional line that invalidates the Migratory Bird Rule as well as USACE's jurisdiction of all waters except for "actually navigable waters, their tributaries, and wetlands adjacent to each." Citing the 1972 amendments to the CWA as its primary example, the dissenting opinion argued that the goals of federal water regulation began an "ambitious and comprehensive" shift towards environmental protection rather than focusing on protecting navigability.

Headwaters, Inc. v. Talent Irrigation District

243 F. 3d 526 (9th Cir. 2001)

This case extended the jurisdiction of the CWA to irrigation canals. In May 1996, the Talent Irrigation District (TID) applied Magnicide H, an aquatic herbicide, to the Talent Canal. Shortly after the application of the herbicide, the Oregon Department of Fish and Wildlife found that, due to a faulty water containment system, the herbicide had leaked out of the canal and killed 92,000 juvenile steelhead trout in a connected river named Bear Creek. Headwaters, Inc. and Oregon Natural Resources Council Action (collectively Headwaters) filed suit under the citizen's suit provision of the CWA. Headwaters alleged that TID violated the CWA when it discharged herbicides into the irrigation canals, and through the canals into Bear Creek without a National Pollution Discharge Elimination System (NPDES) permit. TID argued that the herbicide was regulated under the Federal Insecticide Fungicide and Rodenticide Act (FIFRA) and therefore not subject to the CWA. TID based this argument on the fact that the herbicide had an EPA-approved label and that the label did not mention a permit requirement. The district court found that Headwaters had standing to bring a suit under the citizen's suit provision of the CWA, that the irrigation canals were waters of the U.S., and that the herbicide was a pollutant under the CWA. However, the district court held that no NPDES permit was required, based on adequate regulation under FIFRA. The Court of Appeals (Court) reversed and remanded the lower court's decision.

In reversing the lower court, the Court agreed with the district court's conclusion that the irrigation canals were waters of the U.S. The Court reasoned that the irrigation canals were actually tributaries and therefore fell under the jurisdiction of the CWA. The Court refuted TID's argument that the canals were "closed systems" during the application of the herbicide and therefore not waters of the U.S., by pointing to the fish kills as evidence to the contrary. Regardless, "[e]ven tributaries that flow intermittently are 'waters of the [U.S.]'" However, the Court disagreed with the lower court on whether an NPDES permit was required. Noting their different, yet complementary purposes, the Court harmonized the statutes stating "that the registration and labeling of Magnicide H under FIFRA does not preclude the need for a permit under the CWA." The Court pointed out that the EPA, when it grants NPDES permits, is able to consider local and environmental conditions that cannot be considered in the FIFRA labeling approval process. The Court concluded that the EPA-approved label under FIFRA did not eliminate TID's responsibility to obtain an NPDES permit.

United States v. Deaton

332 F.3d 698 (4th Cir. 2003)

In this jurisdictional case, James and Rebecca Deaton (Defendants) were sued by the government for violating the Clean Water Act by failing to obtain a permit from USACE before digging a 1,100-foot ditch across their 12-acre parcel of land depositing excavated dirt in wetlands on their property. The contractor who dug the ditch piled the excavated dirt on either side of the ditch, a practice known as sidecasting.

USACE asserted jurisdiction based on the Defendants' wetlands being adjacent to, and draining into, a roadside ditch whose waters eventually flow in the navigable Wicomico River and Chesapeake Bay. Water flowing into the roadside ditch takes a winding, 32-mile path to the Chesapeake Bay. USACE issued a stop-work order but after a lengthy period of unsuccessful negotiations the government filed a civil complaint alleging that the Defendants had violated the Clean Water Act. The trial court held that the sidecasting did not constitute the discharge of a pollutant under the Clean Water Act; however, the Court of Appeals (Court) reversed the lower court decision. The Court in remanding the case back to the trial court for further proceedings held that "the Clean Water Act's definition of discharge as 'any addition of any pollutant to navigable water' encompasses sidecasting in a wetland."

Subsequent to the Court's remand order, the Supreme Court decided *Solid Waste Agency of Northern Cook County v. United States Army Corps of Engineers,* 531 U.S. 159 (2001) (*SWANCC*). Based on the new guidance for determining USACE's jurisdiction provided by *SWANCC,* the Defendants filed a motion with the trial court to reconsider. The trial court denied the motion holding: (1) that the wetlands are adjacent to the roadside ditch that is tributary to a navigable water; (2) that there is a hydrological connection between the wetlands and the navigable waters and therefore, *SWANCC* does not bar jurisdiction; and (3) that protecting the Defendant's wetlands is reasonably related to Congress's authority under the Commerce Clause to protect navigable waters as channels of commerce. The Defendants appealed.

The Defendants' main argument was that USACE had no jurisdiction over the roadside ditch, and as a result no jurisdiction over their wetlands. The Court held that Congress's power under the Commerce Clause to protect navigable waters allows it to regulate the discharge of pollutants that flow into the roadside ditch. Congress delegated part of this authority to USACE in the Clean Water Act. USACE, in turn, has promulgated a regulation, that extends its jurisdiction to tributaries of navigable waters. This regulation represents a reasonable interpretation of the Clean Water Act entitled to deference.

APPENDIX L

Wetlands Information Web Pages

As of August 2003

UNITED STATES

Government Agencies

Avisory Council on Historic Preservation
www.achp.gov

California Wetlands Information System
http://ceres.ca.gov/wetlands

EPA Compliance and Enforcement Page
www.epa.gov/compliance

EPA Water Page
www.epa.gov/OW

EPA Office of Wetlands, Oceans, and Watersheds
www.epa.gov/owow

NOAA Fisheries, NMFS Home Page
www.nmfs.noaa.gov

NOAA Fisheries, NMFS Endangered Species
www.nmfs.noaa.gov/endangered.htm

NOAA Ocean and Coastal Resource Management
www.ocrm.nos.noaa.gov

NRCS Hydric Soils of the U.S.
http://soils.usda.gov/use/hydric

NRCS Wetlands Conservation Compliance
www.nrcs.usda.gov/programs/wetlands

NRCS Wetlands Reserve Program
www.nrcs.usda.gov/programs/wrp

NRCS Wetland Science Institute
www.pwrc.usgs.gov/WLI

State Historic Preservation Officers by State
http://grants.cr.nps.gov/shpos/Get_All_SHPOs.cfm

Texas Wetland Information Network
www.glo.state.tx.us/wetnet

USACE Environmental Laboratory Wetlands
www.wes.army.mil/el/wetlands/wetlands.html

USACE Institute for Water Resources
www.iwr.usace.army.mil

USACE Home Page
www.usace.army.mil

USACE Regulatory Program Headquarters Home Page
www.usace.army.mil/inet/functions/cw/cecwo/reg/index.htm

USGS National Wetlands Research Center
www.nwrc.usgs.gov

USFWS Branch of Habitat Assessment
www.nwi.fws.gov/bha

USFWS Endangered Species Program
http://endangered.fws.gov

USFWS Home Page
www.fws.gov

USFWS National Wetlands Inventory
www.nwi.fws.gov

Wild & Scenic Rivers
www.nps.gov/rivers

Non-Governmental Organizations

The Association of State Wetland Managers, Inc.
http://aswm.org

Ducks Unlimited
www.ducks.org

Endangered Species and Wetlands Report
www.eswr.com

The National Wetlands Coalition
http://thenwc.org/home.htm

National Wetlands Conservation Alliance
http://users.erols.com/wetlandg

National Wetlands Newsletter
www2.eli.org/nwn/nwnmain.cfm

National Wildlife Federation–Wetlands
www.nwf.org/wetlands

Sierra Club-Wetlands
www.sierraclub.org/wetlands

Society for Ecological Restoration
www.ser.org

Society of Wetland Scientists
www.sws.org

University of Southern Alabama Wetland Treatment Systems Research Page
www.eng.usouthal.edu/civil/usace/wetland/wetland-index.htm

Water Shedss: Information on Wetlands
www.water.ncsu.edu/watershedss/info/wetlands/manage.html

Wetlands Regulation Center
www.wetlands.com

INTERNATIONAL

BirdLife International
www.birdlife.org.uk

Environment Australia–
Inland Waters, Wetlands
www.ea.gov.au/water/
wetlands/index.html

Kushiro International
Wetland Centre
www.kiwc.net/english/main.html

Partners for Wetlands
www.partnersforwetlands.org

The Ramsar Convention
on Wetlands
www.ramsar.org

Wetlands International
www.wetlands.org

Wildfowl and Wetlands Trust
www.wwt.org.uk

World Conservation Union
http://iucn.org

World Environmental
Organization
www.world.org

Glossary

adaptive management

A systematic process for continually improving management practices by learning from the outcome of a monitoring program.

adjacent wetlands

Wetlands that are bordering, contiguous, or neighboring to other waters of the United States. Wetlands separated from other waters of the United States by man-made dikes or barriers, natural river berms, beach dunes and the like are "adjacent wetlands."

advanced identification

Method for identifying the suitability of wetland sites for future disposal of dredged or fill material.

anaerobic condition

Lack of oxygen in either a gaseous or dissolved form.

base flow

The stream flow that is caused by groundwater inputs to a stream channel, as opposed to the stream flow that occurs in response to stormwater runoff.

buffer (area, zone, or habitat)

An intervening upland area or other form of barrier that separates aquatic resources from developed or disturbed areas and reduces impacts on the aquatic resources that may result from human activities.

compensatory mitigation

Mitigation implemented to offset impacts on aquatic areas that cannot be offset through impact avoidance and minimization. Types of compensatory mitigation include habitat establishment (creation), restoration, enhancement, and protection/maintenance.

credit

A unit of measure (e.g., functional capacity units in HGM) representing the gain of aquatic functions at a compensatory mitigation site; the measure of function is typically indexed to the number of acres of resources restored, established, enhanced, rehabilitated, or protected/maintained as compensatory mitigation.

debit

Unit of measure (e.g., functional capacity units in HGM) representing the loss of aquatic functions at an impact site or project site; the measure of function is typically indexed to the number of acres [of resources] lost or impacted by issuance of the permit.

delineation

The identification and mapping of the boundaries of wetlands and other waters of the United States.

Department of the Army (DA) permit

Same as a Section 404 permit. Also, the permit issued by USACE under Section 10 of the Rivers and Harbors Act.

discharge

The placement of dredged or fill material into waters of the United States, including wetlands, that results in more than a minimal effect on the aquatic system, including redeposition of material during excavation, mechanized land clearing, and ditching.

dredged material

Material removed from waters of the United States.

EPA Section 404(b)(1) Guidelines

Regulations put forth by EPA that provide the standards for unacceptable adverse impacts on and preferred mitigation procedures for waters of the United States, including wetlands, used to determine whether a Section 404 permit should be issued.

ephemeral stream

Stream that has flowing water only during and for a short duration after precipitation events in a typical year. Ephemeral streambeds are always located above the water table. Groundwater is not a source of water for the stream. Runoff from rainfall is the primary source of water for stream flow.

facultative hydrophyte

Plant that can tolerate wetland conditions but can also survive in nonwetland habitats.

fill material

Any material used for the primary purpose of replacing an aquatic area with dry land or of changing the bottom elevation of a waterbody.

floodplain

That portion of a river valley, adjacent to the channel, that is built of sediments deposited during the present flow regime of the river and is covered with water when the river overflows its banks at flood stages. Sometimes referred to in terms of its recurrence interval (e.g., 2-, 10-, or 100-year floodplain).

functional assessment

Qualitative or quantitative evaluation conducted to determine the types and levels of functions a wetland or other water provides.

function

Condition or natural process that occurs in wetlands, streams, and other waters.

General permit

A permit for a specific class of activities within a specified area issued by USACE, authorizing the discharge of dredged or fill material into waters of the United States.

Habitat Conservation Plan

Plan, prepared in accordance with Section 10 of the federal Endangered Species Act, that describes the anticipated effect of a proposed taking of a federally-listed threatened or endangered species and how that take will be minimized and mitigated, thereby allowing development to proceed while promoting listed species conservation.

habitat enhancement

The manipulation of the physical, chemical, or biological characteristics of a wetland (undisturbed or degraded) site to heighten, intensify, or improve specific function(s) or to change the growth stage or composition of the vegetation present. Enhancement is undertaken for a specified purpose(s) such as water quality improvement, flood water retention, or wildlife habitat.

habitat establishment

The manipulation of the physical, chemical, or biological characteristics present to develop a wetland on a nonwetland or deepwater site, where a wetland did not previously exist.

habitat evaluation procedure (HEP)

A method used by USFWS to assess and score the quality of habitat for wildlife.

habitat protection/ maintenance

The removal of a threat to, or preventing the decline of, wetland conditions by an action in or near a wetland.

habitat restoration

The manipulation of the physical, chemical, or biological characteristics of a site with the goal of returning natural/ historic functions to a former or degraded wetland.

headwaters

Non-tidal rivers, streams, and their lakes and impoundments, including adjacent wetlands, that are part of a surface tributary system to an interstate or navigable water of the United States upstream of the point on the river or stream at which the average annual flow is less than five cubic feet per second.

hydric soil

A soil that formed under conditions of saturation, flooding, or ponding long enough during the growing season to develop anaerobic conditions in the upper part.

hydrogeomorphic (HGM) classification

Approach to classifying wetlands that places emphasis on those hydrologic and geomorphic aspects of wetlands (as opposed to biotic) that are responsible for maintaining many of wetland functions.

hydrology

The distribution and movement of surface water and groundwater in an ecosystem.

hydrophyte

Plant that can grow in wetland conditions (hydric soils and wetland hydrology); literally "water plant."

impact avoidance

Elimination of all project impacts on aquatic areas through project design.

impact minimization

Process of limiting project impacts on aquatic areas by designing or redesigning a project.

independent utility

Test to determine what constitutes a single and complete project in the USACE regulatory program. A project is considered to have independent utility if it could be constructed absent the construction of other projects in the project area.

Individual permit

The permit issued by USACE under Section 404 of the Clean Water Act for authorizing the discharge of dredged or fill material into waters of the United States for an individual project for which a specific review was conducted.

in-kind mitigation

Mitigation that results in aquatic habitat that is the same habitat type and provides similar functions and values as the aquatic habitat removed.

in-lieu fee

Funds provided by a project proponent to another party (such as a sponsor), who implements compensatory mitigation at a site other than the proponent's development site.

intermittent stream

Stream that has flowing water during certain times of the year, when groundwater provides water for stream flow. During dry periods, intermittent streams may not have flowing water. Runoff from rainfall is an additional source of water for stream flow.

isolated waters

Those non-tidal waters of the United States that are: 1) not part of a surface tributary system to interstate or navigable waters of the United States; and 2) not adjacent to such tributary waterbodies.

jurisdictional determination

The final USACE decision of its jurisdiction of a site under Section 404 of the CWA.

jurisdictional waters

See waters of the United States.

letter of credit

Written confirmation from a project proponent's financial institution that verifies that the proponent has adequate funds to implement a mitigation program and to ensure that the mitigation is successful.

letter of permission

Type of standard Section 404/10 permit issued through an abbreviated processing procedure that includes coordination with federal and state fish and wildlife agencies and a public interest evaluation, but without the publishing of a public notice.

mitigation

Actions or project design features that reduce impacts on wetlands and other waters by avoiding, minimizing, or compensating for adverse effects.

mitigation banking

Use of a single site, suitable for enhancement, restoration, and/or creation of wetlands and other waters, for the mitigation of impacts on wetlands and other waters that result from more than one project at other sites.

mitigation banking instrument

A document that describes the physical and legal characteristics of a mitigation bank and how it will be operated.

monitoring

Collecting data over time to document the success or failure of a mitigation site.

Nationwide permit (NWP)

A general permit issued by USACE for the entire United States.

navigable waters

Waterways that are, could be, or were used to transport interstate or foreign commerce.

nonwetland

Area that does not meet the criteria as a wetland under Section 404 of the Clean Water Act.

obligate hydrophyte

Plant dependent on wetland conditions for growth and reproduction. Under the USFWS indicator status definition, a plant species that occurs almost always (estimated probability 99 percent) in wetlands under natural conditions, but that may also occur rarely (estimated probability one percent) in nonwetlands.

offsite mitigation

Mitigation that occurs on a site distant from the site of impact.

onsite mitigation

Mitigation that occurs contiguous with or near the site of impact, typically on the same project site.

ordinary high water mark (OHWM)

That line on the shore established by the fluctuations of water and indicated by physical characteristics such as a clear, natural line impressed on the bank, shelving, changes in the character of soil, destruction of terrestrial vegetation, the presence of litter and debris, or other appropriate means that consider the characteristics of the surrounding areas.

other waters of the United States

A water of the United States that is not a special aquatic site.

out-of-kind mitigation

Mitigation that results in wetlands or other waters that are not the same habitat type or do not replace the functions and values of the wetlands or other waters removed.

peak flow

The maximum rate (expressed in cubic feet per second) of stream flow occurring from a storm event of a specified magnitude (as in a 10-year storm).

perennial stream

Stream that has flowing water year-round during a typical year. The water table is located above the streambed for most of the year. Groundwater is a primary source of water for stream flow. Runoff from rainfall is an additional source of water for stream flow.

perennial wetland

Wetland in which soil saturation or ponded water is present year-round. Also referred to as permanent wetland.

performance standard

Specified goal of a mitigation plan that must be met for mitigation to be determined successful.

permanent wetland

See perennial wetland.

preapplication meeting

A meeting between the project proponent and USACE conducted prior to application for a Section 404 permit and often attended by EPA, USFWS, and relevant state or local agencies.

pre-construction notification

Notification made to USACE by a project proponent of their intent to perform fill activities that would be authorized under one or more nationwide permits.

programmatic general permit

Permit issued by a USACE district or division engineer where a local, state, or other federal program provides protections for waters and wetlands that achieve the objectives for the Section 404 permit program.

public interest review

USACE evaluation under the Clean Water Act of the probable impacts of a proposed project and its intended use in the public interest.

reference wetland

An existing wetland, typically of relatively high function and value or representing a relatively undisturbed habitat, that serves as a point of comparison to a mitigation wetland.

regional general permit

Section 404/10 Permit issued by a USACE district or division engineer that authorizes a class of activities within a geographic region that are similar in nature and have minimal individual or cumulative environmental effects.

riparian habitat

Vegetation associated with river, stream, or lake banks and floodplains.

saturated soil

Soil that contains as much water as it can physically hold; virtually all pore spaces are filled.

Section 10 Permit

Permit issued by USACE under Section 10 of the Rivers and Harbors Act for authorizing activities that involve obstruction of, excavation in, or filling of navigable waters.

Section 404(b)(1) Alternatives Analysis

Qualitative evaluation conducted to determine whether a practicable alternative to a proposed project exists such that there would be a less adverse impact on the aquatic ecosystem. *See* EPA Section 404(b)(1) Guidelines.

Section 404(b)(1) Guidelines

See EPA Section 404(b)(1) Guidelines.

Section 404 permit

The permit issued by USACE under Section 404 of the Clean Water Act for authorizing the discharge of dredged or fill material into waters of the United States, including wetlands; also known as USACE permit, fill permit, Department of the Army permit, DA permit, individual permit, 404 permit.

special aquatic site

Specific types of waters of the United States for which mitigation requirements are more stringent under Section 404(b)(1) Guidelines. *See* chapter 3, Jurisdictional Limits for a list of special aquatic sites and their definitions.

Special Area Management Plan

Comprehensive plan providing for natural resource protection and reasonable coastal-dependent economic growth containing a detailed and comprehensive statement of policies, standards, and criteria to guide public and private uses of lands and waters; and mechanisms for timely implementation in specific geographic areas within the coastal zone. This definition from the Coastal Zone Management Act has been adopted by USACE for any area of the United States and not just the coastal zone.

standard permit

A permit for a specific activity (project) that may be issued only after an individual application is submitted and the formal review process is completed by USACE. Standard permits include individual permits and, where applicable, letters of permission.

takings

The appropriation, including excessive regulation, that amounts to an appropriation of private property by the federal government.

upland

Type of landform typically associated with an elevated area having well-drained soils. Not to be confused with nonwetland (*see* definition). Upland is often used to mean nonwetland, as in "upland soils" or "upland plants."

value

Recognized benefit that wetlands and other waters provides to people.

water balance (water budget)

A mathematical model that takes into account gains of water (e.g., precipitation, surface inflow, groundwater inflow) and losses of water (e.g., deep percolation, evapotranspiration) into and from a planned wetland to determine if it would have the appropriate depth and duration of wetland hydrology for the habitat being considered.

Waters of the United States

Water bodies that are regulated under Section 404 of the Clean Water Act.

wetland

Those areas that are inundated or saturated by surface or groundwater at a frequency and duration sufficient to support, and that under normal circumstances do support, a prevalence of vegetation typically adapted for life in saturated soil conditions.

List of Acronyms

ACHP = Advisory Council on Historic Preservation
ADID = Advanced identification
APA = Administrative Procedures Act
APE = Area of potential effect
APP = Abbreviated permit process
ARNI = Aquatic resource of national importance
BA = Biological assessment
BO = Biological opinion
CAFRA = Coastal Area Facility Review Act
CFR = Code of Federal Regulations
CWA = Clean Water Act
CZMA = Coastal Zone Management Act
CZMP = Coastal zone management program
DA = Department of the Army
DFG = California Department of Fish and Game
DOI = Department of Interior
EA = Environmental assessment
EIS = Environmental impact statement
EPA = United States Environmental Protection Agency
ERP = Environmental Resource Permit
ESA = Endangered Species Act
FERC = Federal Energy Regulatory Commission
FOIA = Freedom of Information Act
FONSI = Finding of no significant impact
FWCA = Fish and Wildlife Coordination Act
GIS = Geographic information system
GMA = Growth Management Act
HCP = Habitat conservation plan
HGM = Hydrogeomorphic
HPA = Hydraulic Project Approval
IP = Individual permit
JARPA = Joint aquatic resource permits application
LEDPA = Least environmentally damaging practicable alternative
LOP = Letter of permission
MGL = Massachusetts General Law
MOA = Memorandum of Agreement
MOU = Memorandum of understanding
MSSW = Management and Storage of Surface Waters Program
NEPA = National Environmental Policy Act
NHPA = National Historic Preservation Act
NJSA = New Jersey Statutes Annotated
NOAA = National Oceanic and Atmospheric Administration
NPDES = National Pollutant Discharge Elimination System
NRCS = Natural Resources Conservation Service
NRDC = National Resources Defense Council
NWI = National Wetlands Inventory
NWP = Nationwide permit
OHWM = Ordinary high water mark
PCN = Preconstruction notification
PGP = Programmatic general permit
RCRA = Resource Conservation and Recovery Act
RCW = Revised Code of Washington
RGL = Regulatory guidance letter
RGP = Regional general permit
ROD = Record of decision
RPA = Reasonable and prudent alternative
RPM = Reasonable and prudent measure
SAMP = Special area management plan
SEPA = State Environmental Policy Act
SHPO = State Historic Preservation Officer
SMA = Shoreline Management Act
SPGP = State programmatic general permit
SWANCC = Solid Waste Agency of Northern Cook County
THPO = Tribal Historic Preservation Officer
TMDL = Total maximum daily load
USACE = United States Army Corps of Engineers
USDA = United States Department of Agriculture
USFS = United States Forest Service
USFWS = United States Fish and Wildlife Service
USCG = United States Coast Guard
WDFW = Washington Department of Fish and Wildlife
WPCA = Water Pollution Control Act
WRPP = Wetland Resource Permitting Program
WSRA = National Wild and Scenic Rivers Act

Printed References

Brinson, M.M. 1993. A hydrogeomorphic classification for wetlands. August 1992–Final Report. (Technical Report WRP-DE-4.) U.S. Army Corps of Engineers, Waterways Experiment Station.

Commission of Geosciences, Environment and Resources. 1992. *Restoration of aquatic ecosystems: science, technology, and public policy, national research council.* Washington, D.C.: National Academy Press.

Committee on Mitigating Wetland Losses. 2001. *Compensating for wetland losses under the Clean Water Act.* Board on Environmental Studies and Toxicology, Water Science and Technology Board, Division on Earth and Life Sciences, National Research Council. Washington, D.C.: National Academy Press.

Cowardin, L.M., V. Carter, F.C. Golet, and E.T. LaRoe. 1979. *Classification of wetlands and deepwater habitats of the United States.* (FWS/OBS-79/31.) Washington, D.C.: U.S. Fish and Wildlife Service.

Dahl, T.E. 1990. *Wetland losses in the United States: 1780s to 1980s.* Washington, D.C.: U.S. Fish and Wildlife Service.

Department of the Army and Environmental Protection Agency. 1990. Memorandum of agreement concerning the determination of mitigation under the Clean Water Act Section 404(b)1 guidelines. Washington, D.C.

Department of the Army, Environmental Protection Agency, Natural Resources Conservation Service, U.S. Fish and Wildlife Service, and National Oceanographic and Atmospheric Administration. 1995. Federal guidance on the establishment, use, and operation of mitigation banks. Federal Register 60(228). Washington, D.C.

Department of the Army, Environmental Protection Agency, U.S. Fish and Wildlife Service, and National Oceanographic and Atmospheric Administration. 2000. Federal guidance on the use of in-lieu fee arrangements for compensatory mitigation under Section 404 of the Clean Water Act and Section 10 of the Rivers and Harbors Act 65(216). Federal Register: November 7, 2000. Washington, D.C.

Environmental Law Institute. 1993. *Wetland mitigation banking: an Environmental Law Institute report.* Washington, D.C.

Hollings, C.S. (ed.). 1978. *Adaptive environmental management and assessment.* Chichester: Wiley.

Kershner, J.L. 1997. Monitoring and adaptive management. In J.E. Williams, C.A. Wood, and Michael P. Dombeck (eds.). *Watershed restoration: principles and practices.* Bethesda, Maryland: American Fisheries Society.

Mitsch, W., and J.G. Gosselink. 1993. *Wetlands.* Second edition. New York: Van Nostrand Reinhold.

Odum, Eugene P., 1971. Fundamentals of Ecology, Third Edition. Philadelphia: W.B. Saunders Company.

Reed, P.B., Jr. 1988. *National list of plant species that occur in wetlands: California (Region 0).* Biological Report 88 (26.10). Washington, D.C: U.S. Fish and Wildlife Service.

Rockwell, David, 1998. The Nature of North America. New York: The Berkeley Publishing Company.

Salvesen, D. 1990. *Wetlands: mitigating and regulating development impacts.* Washington, D.C: The Urban Land Institute.

Shabman, L., D. King, and P. Scodari. 1993. Wetland mitigation success through credit market systems. *Environmental Concern Wetland Journal* 5(2):9–12.

Short, K. 1988. *Mitigation banking.* USFWS Biological Report 88(41). Washington, D.C: Research and Development, National Ecology Research Center, U.S. Fish and Wildlife Service.

Smith, D.R., A. Amman, C. Bartoldus, and M.M. Brinson. 1995. An approach for assessing wetland functions using hydrogeomorphic classification, reference wetlands, and functional indices. October 1995, Final Report. (Technical Report WRP-DE-9.) Waterways Experiment Station, U.S. Army Corps of Engineers.

Smith, R.D. 2000. Assessment of Riparian Ecosystem Integrity in the San Diego

Creek Watershed, Orange County, California. Prepared for U.S. Army Corps of Engineers Los Angeles District. U.S. Army Corps of Engineers, Waterways Experiment Station, Vicksburg, Mississippi.

Sullivan, M.E., and M.E. Richardson. 1993. *Functions and values of the Verde River riparian ecosystem and an assessment of adverse impacts to resources: a supporting document for the initiation of the Verde River advanced identification.* Prepared for U.S. Environmental Protection Agency, Region 9, San Francisco, California. Phoenix, Arizona: U.S. Fish and Wildlife Service.

U.S. Army Corps of Engineers. 1987. *Corps of Engineers wetlands delineation manual.* (Technical Report Y-87-1.) Prepared for the Department of the Army, Washington, D.C. Vicksburg, Michigan.

________. 1991. *Habitat mitigation and monitoring proposal guidelines.* San Francisco District. San Francisco, California.

________. 1993. *Habitat mitigation and monitoring proposal guidelines.* Los Angeles District Regulatory Branch. Los Angeles, California.

________. 2001. Regulatory guidance letter No. 01-1, guidance for the establishment and maintenance of compensatory mitigation projects under the USACE regulatory program pursuant to Section 404(a) of the Clean Water Act and Section 10 of the Rivers and Harbors Act on 1899. Washington, D.C.

U.S. Department of Interior. 1994. *The impact of federal programs on wetlands.* Volume II. A report to Congress by the Secretary of the Interior. March. Washington, D.C.

U.S. Environmental Protection Agency. 1991. *Mitigation banking guidance: U.S. Environmental Protection Agency Region 9, San Francisco, California.* Region 9. San Francisco, California.

U.S. Fish and Wildlife Service. 2000. *Status and trends of wetlands in the conterminous United States 1986–1997.* Washington, D.C.

Vepraskas, M.J. 1992. *Redoximorphic features for identifying aquic conditions (Technical Bulletin 301.)* Raleigh, NC: North Carolina State University.

Walters, C. 1986. *Adaptive management of renewable resources.* New York: Macmillan.

Whittaker, Robert H. 1975. Communities and Ecosystems, Second Edition. New York: MacMillan.

Table of Authorities

Index

D

E

L

M

N

O

P

R

S

T

U

V

W

Notes

Notes

Eminent Domain
A Step-by-Step Guide to the Acquisition of Real Property

Explains the processes California public agencies must follow to acquire private property for public purposes through eminent domain. Includes case law, legal references, tips, a table of authorities, sample letters and forms, a glossary, and an index.

Richard G. Rypinski • 2002 (second) edition

Exactions and Impact Fees in California
A Comprehensive Guide to Policy, Practice, and the Law

Designed to help public officials, citizens, attorneys, planners, and developers understand exactions. With practice tips, case studies, drawings, and photos to illustrate key considerations and legal principles.

William W. Abbott, Peter M. Detwiler, M. Thomas Jacobson, Margaret Sohagi, and Harriet A. Steiner
2001 (second) edition w/ 2002 Supplement

Guide to California Planning

Describes how planning really works in California, how cities, counties, developers, and citizen groups all interact with each other to shape California communities and the California landscape, for better and for worse. Recipient of the California Chapter APA Award for Planning Education.

William Fulton
1999 (second) edition

Guide to Hazardous Materials and Waste Management

Valuable reference for students and professionals in the field, adapted from a popular course that has trained hundreds of environmental managers. Provides the information necessary to understand and manage hazardous materials and wastes.

Jon W. Kindschy, Marilyn Kraft, and Molly Carpenter • 1997 edition

Subdivision Map Act Manual

All-new reference with the latest information and practice tips needed to understand Subdivision Map Act legal provisions, recent court-made law, and the review and approval processes. With the full text of the Map Act, practice tips, a comprehensive table of authorities, and an index.

Daniel J. Curtin, Jr. and Robert E. Merritt • 2003 edition

Redevelopment in California

Definitive guide to both the law and practice of redevelopment in California cities and counties, together with codes, case law, and commentary. Contains short articles, notes, photographs, charts, graphs, and illustrative time schedules.

David F. Beatty et al. • 2004 (third) edition

Telecommunications

Detailed summary and analysis of federal and state laws governing the location and regulation of physical facilities including cable, telephone, and wireless systems (cellular, paging, and Internet), satellite dishes, and antennas. With practice tips, photos, a glossary, table of authorities, and an index.

Paul Valle-Riestra • 2002